Basic College Mathematics with Early Integers

Third Edition

Elayn Martin-Gay

University of New Orleans

Boston Columbus Hoboken Indianapolis New York San Francisco
Amsterdam Cape Town Dubai London Madrid Milan Munich Paris Montréal Toronto
Delhi Mexico City São Paulo Sydney Hong Kong Seoul Singapore Taipei Tokyo

Editorial Director, Mathematics: *Christine Hoag*
Editor-in-Chief: *Michael Hirsch*
Acquisitions Editor: *Mary Beckwith*
Project Manager: *Lauren Morse*
Project Team Lead: *Christina Lepre*
Assistant Editor: *Matthew Summers*
Editorial Assistant: *Megan Tripp*
Executive Development Editor: *Dawn Nuttall*
Program Team Lead: *Karen Wernholm*
Program Manager: *Patty Bergin*
Cover and Illustration Design: *Tamara Newnam*
Program Design Lead: *Heather Scott*
Director, Course Production: *Ruth Berry*
Executive Content Manager, MathXL: *Rebecca Williams*
Senior Content Developer, TestGen: *John Flanagan*
Media Producer: *Audra Walsh*
Director of Marketing, Mathematics: *Roxanne McCarley*
Senior Marketing Manager: *Rachel Ross*
Marketing Assistant: *Kelly Cross*
Senior Author Support/Technology Specialist: *Joe Vetere*
Senior Procurement Specialist: *Carol Melville*
Interior Design, Production Management, Answer Art, and Composition:
  Integra Software Services Pvt. Ltd.
Text Art: *Scientific Illustrators*

Acknowledgments of third party content appear on page xvi, which constitutes an extension of this copyright page.

PEARSON, ALWAYS LEARNING, and MYMATHLAB are exclusive trademarks in the U.S. and/or other countries owned by Pearson Education, Inc. or its affiliates.

Library of Congress Cataloging-in-Publication Data
Martin-Gay, K. Elayn
    Basic college mathematics with early integers/ Elayn Martin-Gay,
University of New Orleans. – 3rd edition.
    pages cm
  ISBN-13: 978-0-13-386471-7 (alk. paper)
  ISBN-10: 0-13-386471-5 (alk. paper)
1. Mathematics–Textbooks.   2. Numbers, Natural—Textbooks.   I. Title.
  QA39.3.M374 2016
  510–dc23   2014030418

1 2 3 4 5 6 7 8 9 10—CRK—19 18 17 16 15

www.pearsonhighered.com

ISBN-10: 0-13-386471-5 (Student Edition)
ISBN-13: 978-0-13-386471-7

This book is dedicated to students everywhere—
and we should all be students. After all, is there anyone among
us who really knows too much? Take that hint and continue
to learn something new every day of your life.

Best of wishes from a fellow student:
Elayn Martin-Gay

Contents

Statistics and Probability   511

Introduction to Algebra   562

Geometry   624

Appendices

Preface

Basic College Mathematics with Early Integers, **Third Edition** was written to provide a solid foundation in the basics of college mathematics, including the topics of whole numbers, integers, fractions, decimals, ratio and proportion, percent, and measurement as well as introductions to geometry, statistics and probability, and algebra topics. Integers are introduced in Chapter 2 and integrated throughout the text. This allows students to gain confidence and mastery by working with integers throughout the course. Specific care was taken to make sure students have the most up-to-date relevant text preparation for their next mathematics course or for nonmathematical courses that require an understanding of basic mathematical concepts. I have tried to achieve this by writing a user-friendly text that is keyed to objectives and contains many worked-out examples. As suggested by AMATYC and the NCTM Standards (plus Addenda), real-life and real-data applications, data interpretation, conceptual understanding, problem solving, writing, cooperative learning, appropriate use of technology, number sense, estimation, critical thinking, and geometric concepts are emphasized and integrated throughout the book.

The many factors that contributed to the success of the previous editions have been retained. In preparing the Third Edition, I considered comments and suggestions of colleagues, students, and many users of the prior edition throughout the country.

What's New in the Third Edition?

- **The Martin-Gay Program** has been revised and enhanced with a new design in the text and MyMathLab® to actively encourage students to use the text, video program, and Video Organizer as an integrated learning system.

- **The New Video Organizer** is designed to help students take notes and work practice exercises while watching the Interactive Lecture Series videos (available in MyMathLab and on DVD). All content in the Video Organizer is presented in the same order as it is presented in the videos, making it easy for students to create a course notebook and build good study habits.

  — Covers all of the video examples in order.

  — Provides ample space for students to write down key definitions and properties.

  — Includes "Play" and "Pause" button icons to prompt students to follow along with the author for some exercises while they try others on their own.

  The Video Organizer is available in a loose-leaf, notebook-ready format. It is also available for download in MyMathLab.

- **Vocabulary, Readiness & Video Check** questions have been added prior to every section exercise set. These exercises quickly check a student's understanding of new vocabulary words. The **readiness** exercises center on a student's understanding of a concept that is necessary in order to continue to the exercise set. **New Video Check questions for the Martin-Gay Interactive Lecture videos** are now included in every section for each learning objective. **These exercises are all available for assignment in MyMathLab** and are a great way to assess whether students have viewed and understood the key concepts presented in the videos.

- **New Student Success Tips Videos** are 3- to 5-minute video segments designed to be daily reminders to students to continue practicing and maintaining good organizational and study habits. They are organized in three categories and are available in MyMathLab and the Interactive Lecture Series. The categories are:

1. Success Tips that apply to any course in college in general, such as Time Management.

2. Success Tips that apply to any mathematics course. One example is based on understanding that mathematics is a course that requires homework to be completed in a timely fashion.

3. Section- or Content-specific Success Tips to help students avoid common mistakes or to better understand concepts that often prove challenging. One example of this type of tip is how to apply the order of operations to simplify an expression.

- **Interactive DVD Lecture Series**, featuring your text author (Elayn Martin-Gay), provides students with active learning at their own pace. The videos offer the following resources and more:

    A complete lecture for each section of the text highlights key examples and exercises from the text. "Pop-ups" reinforce key terms, definitions, and concepts.

    An interface with menu navigation features allows students to quickly find and focus on the examples and exercises they need to review.

    Interactive Concept Check exercises measure students' understanding of key concepts and common trouble spots.

    New Student Success Tips Videos.

- **The Interactive DVD Lecture Series** also includes the following resources for test prep:

    The Chapter Test Prep Videos help students during their most teachable moment—when they are preparing for a test. This innovation provides step-by-step solutions for the exercises found in each Chapter Test. For the Third Edition, the chapter test prep videos are also available on YouTube™. The videos are captioned in English and Spanish.

    The Practice Final Exam Videos help students prepare for an end-of-course final. Students can watch full video solutions to each exercise in the Practice Final Exam at the end of this text.

- **The Martin-Gay MyMathLab** course has been updated and revised to provide more exercise coverage, including assignable video check questions and an expanded video program. There are section lecture videos for every section, which students can also access at the specific objective level; Student Success Tips videos; and an increased number of watch clips at the exercise level to help students while doing homework in MathXL. Suggested homework assignments have been premade for assignment at the instructor's discretion.

- **New MyMathLab Ready to Go Courses** (access code required) provide students with all the same great MyMathLab features that you're used to, but make it easier for instructors to get started. Each course includes preassigned homework and quizzes to make creating your course even simpler. Ask your Pearson representative about the details for this particular course or to see a copy of this course.

Key Pedagogical Features

The following key features have been retained and/or updated for the Third Edition of the text:

Problem-Solving Process This is formally introduced in Chapter 1 with a four-step process that is integrated throughout the text. The four steps are **Understand, Translate, Solve,** and **Interpret.** The repeated use of these steps in a variety of examples shows their wide applicability. Reinforcing the steps can increase students' comfort level and confidence in tackling problems.

Exercise Sets Revised and Updated   The exercise sets have been carefully examined and extensively revised. Special focus was placed on making sure that even- and odd-numbered exercises are paired and that real-life applications were updated.

Examples   Detailed, step-by-step examples were added, deleted, replaced, or updated as needed. Many examples reflect real life. Additional instructional support is provided in the annotated examples.

Practice Exercises   Throughout the text, each worked-out example has a parallel Practice exercise. These invite students to be actively involved in the learning process. Students should try each Practice exercise after finishing the corresponding example. Learning by doing will help students grasp ideas before moving on to other concepts. Answers to the Practice exercises are provided at the bottom of each page.

Helpful Hints   Helpful Hints contain practical advice on applying mathematical concepts. Strategically placed where students are most likely to need immediate reinforcement, Helpful Hints help students avoid common trouble areas and mistakes.

Concept Checks   This feature allows students to gauge their grasp of an idea as it is being presented in the text. Concept Checks stress conceptual understanding at the point-of-use and help suppress misconceived notions before they start. Answers appear at the bottom of the page. Exercises related to Concept Checks are included in the exercise sets.

Mixed Practice Exercises   In the section exercise sets, these exercises require students to determine the problem type and strategy needed to solve it just as they would need to do on a test.

Integrated Reviews   This unique mid-chapter exercise set (and notes where appropriate) helps students assimilate new skills and concepts that they have learned separately over several sections. These reviews provide yet another opportunity for students to work with "mixed" exercises as they master the topics.

Vocabulary Check   This feature provides an opportunity for students to become more familiar with the use of mathematical terms as they strengthen their verbal skills. These appear at the end of each chapter before the Chapter Highlights. Vocabulary, Readiness & Video exercises provide practice at the section level.

Chapter Highlights   Found at the end of every chapter, these contain key definitions and concepts with examples to help students understand and retain what they have learned and help them organize their notes and study for tests.

Chapter Review   The end of every chapter contains a comprehensive review of topics introduced in the chapter. The Chapter Review offers exercises keyed to every section in the chapter, as well as Mixed Review exercises that are not keyed to sections.

Chapter Test and Chapter Test Prep Videos   The Chapter Test is structured to include those exercises that involve common student errors. The **Chapter Test Prep Videos** gives students instant access to a step-by-step video solution of each exercise in the Chapter Test.

Cumulative Review   This review follows every chapter in the text (except Chapter 1). Each odd-numbered exercise contained in the Cumulative Review is an earlier worked example in the text that is referenced in the back of the book along with the answer.

Writing Exercises ↖ These exercises occur in almost every exercise set and require students to provide a written response to explain concepts or justify their thinking.

Applications Real-world and real-data applications have been thoroughly updated, and many new applications are included. These exercises occur in almost every exercise set and show the relevance of mathematics and help students gradually and continuously develop their problem-solving skills.

Review Exercises These exercises occur in each exercise set (except in Chapter 1) and are keyed to earlier sections. They review concepts learned earlier in the text that will be needed in the next section or chapter.

Exercise Set Resource Icons Located at the opening of each exercise set, these icons remind students of the resources available for extra practice and support:

See Student Resources descriptions on page xv for details on the individual resources available.

Exercise Icons These icons facilitate the assignment of specialized exercises and let students know what resources can support them.

- ▶ DVD Video icon: exercise worked on the Interactive DVD Lecture Series.
- △ Triangle icon: identifies exercises involving geometric concepts.
- ↖ Pencil icon: indicates a written response is needed.
- ▦ Calculator icon: optional exercises intended to be solved using a scientific or graphing calculator.

Group Activities Found at the end of each chapter, these activities are for individual or group completion, and are usually hands-on or data-based activities that extend the concepts found in the chapter, allowing students to make decisions and interpretations and to think and write about algebra.

Optional: Calculator Exploration Boxes and Calculator Exercises The optional Calculator Explorations provide keystrokes and exercises at appropriate points to give students an opportunity to become familiar with these tools. Section exercises that are best completed by using a calculator are identified by ▦ for ease of assignment.

Student and Instructor Resources

Video Organizer

Designed to help students take notes and work practice exercises while watching the Interactive Lecture Series videos. The Video Organizer:

- Covers all of the video examples in order
- Provides ample space for students to write down key definitions and rules
- Includes "Play" and "Pause" button icons to prompt students to follow along with the author for some exercises while they try others on their own
- Includes Student Success Tips Outline and Questions

Available in loose-leaf, notebook-ready format and in MyMathLab. Answers to exercises available to instructors in MyMathLab.

Interactive DVD Lecture Series Videos

Provides students with active learning at their pace.
The videos offer:

- A complete lecture for each text section. The interface allows easy navigation to examples and exercises students need to review.
- Interactive Concept Check exercises
- Student Success Tips Videos
- Practice Final Exam
- Chapter Test Prep Videos

Student Solutions Manual

Provides completely worked-out solutions to the odd-numbered section exercises; all exercises in the Integrated Reviews, Chapter Reviews, Chapter Tests, and Cumulative Reviews

Annotated Instructor's Edition

Contains all the content found in the student edition, plus the following:

- Answers to exercises on the same text page
- Teaching Tips throughout the text placed at key points

Instructor's Resource Manual with Tests and Mini-Lectures

- Mini-lectures for each text section
- Additional practice worksheets for each section
- Several forms of test per chapter—free response and multiple choice
- Answers to all items

Instructor's Solutions Manual
TestGen® (Available for download from the IRC)

Instructor-to-Instructor Videos—available in the Instructor Resources section of the MyMathLab course.

Online Resources
MyMathLab® (access code required)

MathXL® (access code required)

Acknowledgments

There are many people who helped me develop this text, and I will attempt to thank some of them here. Emily Keaton and Cindy Trimble were *invaluable* for contributing to the overall accuracy of the text. Dawn Nuttall was *invaluable* for her many suggestions and contributions during the development and writing of this Third Edition. Allison Campbell and Lauren Morse provided guidance throughout the production process.

A very special thank you goes to my editor, Mary Beckwith, for being there 24/7/365, as my students say. And my thanks to the staff at Pearson for all their support: Heather Scott, Patty Bergin, Matt Summers, Michelle Renda, Roxanne McCarley, Rachel Ross, Michael Hirsch, Chris Hoag, and Paul Corey.

I would like to thank the following reviewers for their input and suggestions:

Anita Aikman, *Collin County Community College*
Sheila Anderson, *Housatonic Community College*
Adrianne Arata, *College of the Siskiyous*
Cedric Atkins, *Mott Community College*
Laurel Berry, *Bryant & Stratton College*
Connie Buller, *Metropolitan Community College*
Lisa Feintech, *Cabrillo College*
Chris Ford, *Shasta College*
Cindy Fowler, *Central Piedmont Community College*
Pam Gerszewski, *College of the Albemarle*
Doug Harley, *Del Mar College*

Sonya Johnson, *Central Piedmont Community College*
Deborah Jones, *High Tech College*
Nancy Lange, *Inver Hills Community College*
Jean McArthur, *Joliet Junior College*
Carole Shapero, *Oakton Community College*
Jennifer Strehler, *Oakton Community College*
Tanomo Taguchi, *Fullerton College*
Leigh Ann Wheeler, *Greenville Technical Community College*
Valerie Wright, *Central Piedmont Community College*

I would also like to thank the following dedicated group of instructors who participated in our focus groups, Martin-Gay Summits, and our design review for the series. Their feedback and insights have helped to strengthen this edition of the text. These instructors include:

Billie Anderson, *Tyler Junior College*
Cedric Atkins, *Mott Community College*
Lois Beardon, *Schoolcraft College*
Laurel Berry, *Bryant & Stratton College*
John Beyers, *University of Maryland*
Bob Brown, *Community College of Baltimore County–Essex*
Lisa Brown, *Community College of Baltimore County–Essex*
NeKeith Brown, *Richland College*
Gail Burkett, *Palm Beach State College*
Cheryl Cantwell, *Seminole State College*
Ivette Chuca, *El Paso Community College*
Jackie Cohen, *Augusta State College*
Julie Dewan, *Mohawk Valley Community College*
Monette Elizalde, *Palo Alto College*
Kiel Ellis, *Delgado Community College*
Janice Ervin, *Central Piedmont Community College*
Richard Fielding, *Southwestern College*

Dena Frickey, *Delgado Community College*
Cindy Gaddis, *Tyler Junior College*
Gary Garland, *Tarrant County Community College*
Kim Ghiselin, *State College of Florida*
Nita Graham, *St. Louis Community College*
Kim Granger, *St. Louis Community College*
Pauline Hall, *Iowa State University*
Pat Hussey, *Triton College*
Dorothy Johnson, *Lorain County Community College*
Sonya Johnson, *Central Piedmont Community College*
Ann Jones, *Spartanburg Community College*
Irene Jones, *Fullerton College*
Paul Jones, *University of Cincinnati*
Mike Kirby, *Tidewater Community College*
Kathy Kopelousous, *Lewis and Clark Community College*

Tara LaFrance, *Delgado Community College*
John LaMaster, *Indiana Purdue University Fort Wayne*
Nancy Lange, *Inver Hills Community College*
Judy Langer, *Westchester Community College*
Lisa Lindloff, *McLennan Community College*
Sandy Lofstock, *St. Petersburg College*
Kathy Lovelle, *Westchester Community College*
Nicole Mabine, *North Lake College*
Jean McArthur, *Joliet Junior College*
Kevin McCandless, *Evergreen Valley College*
Ena Michael, *State College of Florida*
Daniel Miller, *Niagara County Community College*
Marcia Molle, *Metropolitan Community College*
Carol Murphy, *San Diego Miramar College*
Charlotte Newsom, *Tidewater Community College*
Cao Nguyen, *Central Piedmont Community College*
Greg Nguyen, *Fullerton College*
Eric Oilila, *Jackson Community College*

Linda Padilla, *Joliet Junior College*
Armando Perez, *Laredo Community College*
Davidson Pierre, *State College of Florida*
Marilyn Platt, *Gaston College*
Chris Riola, *Moraine Valley Community College*
Carole Shapero, *Oakton Community College*
Janet Sibol, *Hillsborough Community College*
Anne Smallen, *Mohawk Valley Community College*
Barbara Stoner, *Reading Area Community College*
Jennifer Strehler, *Oakton Community College*
Ellen Stutes, *Louisiana State University Eunice*
Tanomo Taguchi, *Fullerton College*
Robyn Toman, *Anne Arundel Community College*
MaryAnn Tuerk, *Elgin Community College*
Walter Wang, *Baruch College*
Leigh Ann Wheeler, *Greenville Technical Community College*
Darlene Williams, *Delgado Community College*
Valerie Wright, *Central Piedmont Community College*

A special thank you to those students who participated in our design review: Katherine Browne, Mike Bulfin, Nancy Canipe, Ashley Carpenter, Jeff Chojnachi, Roxanne Davis, Mike Dieter, Amy Dombrowski, Kay Herring, Todd Jaycox, Kaleena Levan, Matt Montgomery, Tony Plese, Abigail Polkinghorn, Harley Price, Eli Robinson, Avery Rosen, Robyn Schott, Cynthia Thomas, and Sherry Ward.

Elayn Martin-Gay

About the Author

Elayn Martin-Gay has taught mathematics at the University of New Orleans for more than 25 years. Her numerous teaching awards include the local University Alumni Association's Award for Excellence in Teaching, and Outstanding Developmental Educator at University of New Orleans, presented by the Louisiana Association of Developmental Educators.

Prior to writing textbooks, Elayn Martin-Gay developed an acclaimed series of lecture videos to support developmental mathematics students in their quest for success. These highly successful videos originally served as the foundation material for her texts. Today, the videos are specific to each book in the Martin-Gay series. The author has also created Chapter Test Prep Videos to help students during their most "teachable moment"—as they prepare for a test—along with Instructor-to-Instructor videos that provide teaching tips, hints, and suggestions for each developmental mathematics course, including basic mathematics, prealgebra, beginning algebra, and intermediate algebra.

Elayn is the author of 12 published textbooks as well as multimedia, interactive mathematics, all specializing in developmental mathematics courses. She has also published series in Algebra 1, Algebra 2, and Geometry. She has participated as an author across the broadest range of educational materials: textbooks, videos, tutorial software, and courseware. This provides an opportunity of various combinations for an integrated teaching and learning package offering great consistency for the student.

Applications Index

The Whole Numbers

1

A Selection of Resources for Success in This Mathematics Course

Textbook

Instructor

MyMathLab and MathXL

Video Organizer

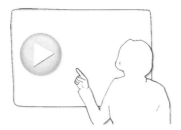

Interactive Lecture Series

Whole numbers are the basic building blocks of mathematics. The whole numbers answer the question "How many?"

This chapter covers basic operations on whole numbers. Knowledge of these operations provides a good foundation on which to build further mathematical skills.

For more information about the resources illustrated above, read Section 1.1.

1.1 Study Skill Tips for Success in Mathematics

Objectives

A Get Ready for This Course.

B Understand Some General Tips for Success.

C Know How to Use This Text.

D Know How to Use Video and Notebook Organizer Resources.

E Get Help as Soon as You Need It.

F Learn How to Prepare for and Take an Exam.

G Develop Good Time Management.

Before reading Section 1.1, you might want to ask yourself a few questions.

1. When you took your last math course, were you organized? Were your notes and materials from that course easy to find, or were they disorganized and hard to find—if you saved them at all?

2. Were you satisfied—really satisfied—with your performance in that course? In other words, do you feel that your outcome represented your best effort?

If the answer is "no" to these questions, then it is time to make a change. Changing to or resuming good study skill habits is not a process you can start and stop as you please. It is something that you must remember and practice each and every day. To begin, continue reading this section.

Objective A Getting Ready for This Course

Now that you have decided to take this course, remember that a *positive attitude* will make all the difference in the world. Your belief that you can succeed is just as important as your commitment to this course. Make sure you are ready for this course by having the time and positive attitude that it takes to succeed.

Make sure that you are familiar with the way that this course is being taught. Is it a traditional course, in which you have a printed textbook and meet with an instructor? Is it taught totally online, and your textbook is electronic and you e-mail your instructor? Or is your course structured somewhere in between these two methods? (Not all of the tips that follow will apply to all forms of instruction.)

Also make sure that you have scheduled your math course for a time that will give you the best chance for success. For example, if you are also working, you may want to check with your employer to make sure that your work hours will not conflict with your course schedule.

On the day of your first class period, double-check your schedule and allow yourself extra time to arrive on time in case of traffic problems or difficulty locating your classroom. Make sure that you are aware of and bring all necessary class materials.

Objective B General Tips for Success

Below are some general tips that will increase your chance for success in a mathematics class. Many of these tips will also help you in other courses you may be taking.

Most important! Organize your class materials. In the next couple pages, many ideas will be presented to help you organize your class materials—notes, any handouts, completed homework, previous tests, etc. In general, you MUST have these materials organized. All of them will be valuable references throughout your course and when studying for upcoming tests and the final exam. One way to make sure you can locate these materials when you need them is to use a three-ring binder. This binder should be used solely for your mathematics class and should be brought to each and every class and/or lab. This way, any material can be immediately inserted in a section of this binder and will be there when you need it.

Form study groups and/or exchange names and e-mail addresses. Depending on how your course is taught, you may want to keep in contact with your fellow students. Some ways of doing this are to form a study group—whether in person or through the Internet. Also, you may want to ask if anyone is interested in exchanging e-mail addresses or any other form of contact.

Helpful Hint

MyMathLab® and MathXL®
When assignments are turned in online, keep a hard copy of your complete written work. You will need to refer to your written work to be able to ask questions and to study for tests later.

2

Choose to attend all class periods. If possible, sit near the front of the classroom. This way, you will see and hear the presentation better. It may also be easier for you to participate in classroom activities.

Do your homework. You've probably heard the phrase "practice makes perfect" in relation to music and sports. It also applies to mathematics. You will find that the more time you spend solving mathematics exercises, the easier the process becomes. Be sure to schedule enough time to complete your assignments before the due date assigned by your instructor.

Check your work. Review the steps you took while working a problem. Learn to check your answers in the original exercises. You may also compare your answers with the "Answers to Selected Exercises" section in the back of the book. If you have made a mistake, try to figure out what went wrong. Then correct your mistake. If you can't find what went wrong, **don't** erase your work or throw it away. Show your work to your instructor, a tutor in a math lab, or a classmate. It is easier for someone to find where you had trouble if he or she looks at your original work.

Learn from your mistakes and be patient with yourself. Everyone, even your instructor, makes mistakes. (That definitely includes me—Elayn Martin-Gay.) Use your errors to learn and to become a better math student. The key is finding and understanding your errors.

Was your mistake a careless one, or did you make it because you can't read your own math writing? If so, try to work more slowly or write more neatly and make a conscious effort to carefully check your work.

Did you make a mistake because you don't understand a concept? Take the time to review the concept or ask questions to better understand it.

Did you skip too many steps? Skipping steps or trying to do too many steps mentally may lead to preventable mistakes.

Know how to get help if you need it. It's all right to ask for help. In fact, it's a good idea to ask for help whenever there is something that you don't understand. Make sure you know when your instructor has office hours and how to find his or her office. Find out whether math tutoring services are available on your campus. Check on the hours, location, and requirements of the tutoring service.

Don't be afraid to ask questions. You are not the only person in class with questions. Other students are normally grateful that someone has spoken up.

Turn in assignments on time. This way, you can be sure that you will not lose points for being late. Show every step of a problem and be neat and organized. Also be sure that you understand which problems are assigned for homework. If allowed, you can always double-check the assignment with another student in your class.

Objective C Knowing and Using Your Text

Flip through the pages of this text or view the e-text pages on a computer screen. Start noticing examples, exercise sets, end-of-chapter material, and so on. Every text is organized in some manner. Learn the way this text is organized by reading about and then finding an example in your text of each type of resource listed below. Finding and using these resources throughout your course will increase your chance of success.

- *Practice Exercises.* Each example in every section has a parallel Practice exercise. As you read a section, try each Practice exercise after you've finished the corresponding example. Answers are at the bottom of the page. This "learn-by-doing" approach will help you grasp ideas before you move on to other concepts.

- *Symbols at the Beginning of an Exercise Set.* If you need help with a particular section, the symbols listed at the beginning of each exercise set will remind you of the resources available.

Helpful Hint

MyMathLab® and MathXL®
If you are doing your homework online, you can work and re-work those exercises that you struggle with until you master them. Try working through all the assigned exercises twice before the due date.

Helpful Hint

MyMathLab® and MathXL®
If you are completing your homework online, it's important to work each exercise on paper before submitting the answer. That way, you can check your work and follow your steps to find and correct any mistakes.

Helpful Hint

MyMathLab® and MathXL®
Be aware of assignments and due dates set by your instructor. Don't wait until the last minute to submit work online.

- *Objectives.* The main section of exercises in each exercise set is referenced by an objective, such as **A** or **B**, and also an example(s). There is also often a section of exercises entitled "Mixed Practice," which is referenced by two or more objectives or sections. These are mixed exercises written to prepare you for your next exam. Use all of this referencing if you have trouble completing an assignment from the exercise set.

- *Icons (Symbols).* Make sure that you understand the meaning of the icons that are beside many exercises. ⊙ tells you that the corresponding exercise may be viewed on the video Lecture Series that corresponds to that section. ＼ tells you that this exercise is a writing exercise in which you should answer in complete sentences. △ tells you that the exercise involves geometry.

- *Integrated Reviews.* Found in the middle of each chapter, these reviews offer you a chance to practice—in one place—the many concepts that you have learned separately over several sections.

- *End-of-Chapter Opportunities.* There are many opportunities at the end of each chapter to help you understand the concepts of the chapter.

  Vocabulary Checks contain key vocabulary terms introduced in the chapter.

  Chapter Highlights contain chapter summaries and examples.

  Chapter Reviews contain review problems. The first part is organized section by section and the second part contains a set of mixed exercises.

  Chapter Tests are sample tests to help you prepare for an exam. The Chapter Test Prep Videos found in the Interactive Lecture Series, MyMathLab, and YouTube provide the video solution to each question on each Chapter Test.

  Cumulative Reviews start at Chapter 2 and are reviews consisting of material from the beginning of the book to the end of that particular chapter.

- *Student Resources in Your Textbook.* You will find a **Student Resources** section at the back of this textbook. It contains the following to help you study and prepare for tests:

  Study Skill Builders contain study skills advice. To increase your chance for success in the course, read these study tips and answer the questions.

  Bigger Picture—Study Guide Outline provides you with a study guide outline of the course, with examples.

  Practice Final provides you with a Practice Final Exam to help you prepare for a final.

- *Resources to Check Your Work.* The **Answers to Selected Exercises** section provides answers to all odd-numbered section exercises and to all integrated review, chapter review, chapter test, and cumulative review exercises. Use the **Solutions to Selected Exercises** to see the worked-out solution to every other odd-numbered exercise.

Helpful Hint

MyMathLab®

In MyMathLab, you have access to the following video resources:

- Lecture Videos for each section
- Chapter Test Prep Videos

Use these videos provided by the author to prepare for class, review, and study for tests.

Objective D Knowing and Using Video and Notebook Organizer Resources ▶

Video Resources

Below is a list of video resources that are all made by me—the author of your text, Elayn Martin-Gay. By making these videos, I can be sure that the methods presented are consistent with those in the text.

- *Interactive DVD Lecture Series.* Exercises marked with a ▶ are fully worked out by the author on the DVDs and within MyMathLab. The lecture series provides approximately 20 minutes of instruction per section and is organized by Objective.

- *Chapter Test Prep Videos.* These videos provide solutions to all of the Chapter Test exercises worked out by the author. They can be found in MyMathLab, the Interactive Lecture series, and You Tube. This supplement is very helpful before a test or exam.
- *Student Success Tips.* These video segments are about 3 minutes long and are daily reminders to help you continue practicing and maintaining good organizational and study habits.
- *Final Exam Videos.* These video segments provide solutions to each question. These videos can be found within MyMathLab and the Interactive Lecture Series.

Notebook Organizer Resources

The resources below are in three-ring notebook ready form. They are to be inserted in a three-ring binder and completed. Both resources are numbered according to the sections in your text to which they refer.

- *Video Organizer.* This organizer is closely tied to the Interactive Lecture (Video) Series. Each section should be completed while watching the lecture video on the same section. Once completed, you will have a set of notes to accompany the Lecture (Video) Series section by section.
- *Student Organizer.* This organizer helps you study effectively through note-taking hints, practice, and homework while referencing examples in the text and examples in the Lecture Series.

Objective E   Getting Help

If you have trouble completing assignments or understanding the mathematics, get help as soon as you need it! This tip is presented as an objective on its own because it is so important. In mathematics, usually the material presented in one section builds on your understanding of the previous section. This means that if you don't understand the concepts covered during a class period, there is a good chance that you will not understand the concepts covered during the next class period. If this happens to you, get help as soon as you can.

Where can you get help? Many suggestions have been made in this section on where to get help, and now it is up to you to get it. Try your instructor, a tutoring center, or a math lab, or you may want to form a study group with fellow classmates. If you do decide to see your instructor or go to a tutoring center, make sure that you have a neat notebook and are ready with your questions.

> **Helpful Hint**   **MyMathLab® and MathXL®**
>
> - Use the **Help Me Solve This** button to get step-by-step help for the exercise you are working. You will need to work an additional exercise of the same type before you can get credit for having worked it correctly.
> - Use the **Video** button to view a video clip of the author working a similar exercise.

Objective F   Preparing for and Taking an Exam

Make sure that you allow yourself plenty of time to prepare for a test. If you think that you are a little "math anxious," it may be that you are not preparing for a test in a way that will ensure success. The way that you prepare for a test in mathematics is important. To prepare for a test:

1. Review your previous homework assignments.
2. Review any notes from class and section-level quizzes you have taken. (If this is a final exam, also review chapter tests you have taken.)
3. Review concepts and definitions by reading the Chapter Highlights at the end of each chapter.
4. Practice working out exercises by completing the Chapter Review found at the end of each chapter. (If this is a final exam, go through a Cumulative Review. There is one found at the end of each chapter except Chapter 1. Choose the review found at the end of the latest chapter that you have covered in your course.) *Don't stop here!*

> **Helpful Hint**   **MyMathLab® and MathXL®**
>
> Review your written work for previous assignments. Then, go back and re-work previous assignments. Open a previous assignment, and click **Similar Exercise** to generate new exercises. Re-work the exercises until you fully understand them and can work them without help features.

5. It is important that you place yourself in conditions similar to test conditions to find out how you will perform. In other words, as soon as you feel that you know the material, get a few blank sheets of paper and take a sample test. There is a Chapter Test available at the end of each chapter, or you can work selected problems from the Chapter Review. Your instructor may also provide you with a review sheet. During this sample test, do not use your notes or your textbook. Then check your sample test. If your sample test is the Chapter Test in the text, don't forget that the video solutions are in MyMathLab, the Interactive Lecture Series, and YouTube. If you are not satisfied with the results, study the areas that you are weak in and try again.

6. On the day of the test, allow yourself plenty of time to arrive at where you will be taking your exam.

When taking your test:

1. Read the directions on the test carefully.

2. Read each problem carefully as you take the test. Make sure that you answer the question asked.

3. Watch your time and pace yourself so that you can attempt each problem on your test.

4. If you have time, check your work and answers.

5. Do not turn your test in early. If you have extra time, spend it double-checking your work.

Objective G Managing Your Time

As a college student, you know the demands that classes, homework, work, and family place on your time. Some days you probably wonder how you'll ever get everything done. One key to managing your time is developing a schedule. Here are some hints for making a schedule:

1. Make a list of all of your weekly commitments for the term. Include classes, work, regular meetings, extracurricular activities, etc. You may also find it helpful to list such things as laundry, regular workouts, grocery shopping, etc.

2. Next, estimate the time needed for each item on the list. Also make a note of how often you will need to do each item. Don't forget to include time estimates for the reading, studying, and homework you do outside of your classes. You may want to ask your instructor for help estimating the time needed.

3. In the exercise set that follows, you are asked to block out a typical week on the schedule grid given. Start with items with fixed time slots like classes and work.

4. Next, include the items on your list with flexible time slots. Think carefully about how best to schedule items such as study time.

5. Don't fill up every time slot on the schedule. Remember that you need to allow time for eating, sleeping, and relaxing! You should also allow a little extra time in case some items take longer than planned.

6. If you find that your weekly schedule is too full for you to handle, you may need to make some changes in your workload, classload, or other areas of your life. You may want to talk to your advisor, manager or supervisor at work, or someone in your college's academic counseling center for help with such decisions.

1.1 Exercise Set MyMathLab®

1. What is your instructor's name?

2. What are your instructor's office location and office hours?

3. What is the best way to contact your instructor?

4. Do you have the name and contact information of at least one other student in class?

5. Will your instructor allow you to use a calculator in this class?

6. Why is it important that you write step-by-step solutions to homework exercises and keep a hard copy of all work submitted online?

7. Is there a tutoring service available on campus? If so, what are its hours? What services are available?

8. Have you attempted this course before? If so, write down ways that you might improve your chances of success during this attempt.

9. List some steps that you can take if you begin having trouble understanding the material or completing an assignment. If you are completing your homework in MyMathLab® and MathXL®, list the resources you can use for help.

10. How many hours of studying does your instructor advise for each hour of instruction?

11. What does the ＼ icon in this text mean?

12. What does the △ icon in this text mean?

13. What does the ● icon in this text mean?

14. Search the minor columns in your text. What are Practice exercises?

15. When might be the best time to work a Practice exercise?

16. Where are the answers to Practice exercises?

17. What answers are contained in this text and where are they?

18. What are Study Skill Tips of the Day and where are they?

19. What and where are Integrated Reviews?

20. How many times is it suggested that you work through the homework exercises in MathXL® before the submission deadline?

21. How far in advance of the assigned due date is it suggested that homework be submitted online? Why?

22. Chapter Highlights are found at the end of each chapter. Find the Chapter 1 Highlights and explain how you might use it and how it might be helpful.

23. Chapter Reviews are found at the end of each chapter. Find the Chapter 1 Review and explain how you might use it and how it might be helpful.

24. Chapter Tests are found at the end of each chapter. Find the Chapter 1 Test and explain how you might use it and how it might be helpful when preparing for an exam on Chapter 1. Include how the Chapter Test Prep Videos may help. If you are working in MyMathLab® and MathXL®, how can you use previous homework assignments to study?

25. What is the Video Organizer? Explain the contents and how it might be used.

26. What is the Student Organizer? Explain the contents and how it might be used.

27. Read or reread objective **G** and fill out the schedule grid on the next page.

	Monday	Tuesday	Wednesday	Thursday	Friday	Saturday	Sunday
4:00 a.m.							
5:00 a.m.							
6:00 a.m.							
7:00 a.m.							
8:00 a.m.							
9:00 a.m.							
10:00 a.m.							
11:00 a.m.							
12:00 p.m.							
1:00 p.m.							
2:00 p.m.							
3:00 p.m.							
4:00 p.m.							
5:00 p.m.							
6:00 p.m.							
7:00 p.m.							
8:00 p.m.							
9:00 p.m.							
10:00 p.m.							
11:00 p.m.							
Midnight							
1:00 a.m.							
2:00 a.m.							
3:00 a.m.							

1.2   Place Value, Names for Numbers, and Reading Tables

Objectives

A Find the Place Value of a Digit in a Whole Number.

B Write a Whole Number in Words and in Standard Form.

C Write a Whole Number in Expanded Form.

D Read Tables.

The **digits** 0, 1, 2, 3, 4, 5, 6, 7, 8, and 9 can be used to write numbers. For example, the **whole numbers** are

    0, 1, 2, 3, 4, 5, 6, 7, 8, 9, 10, 11, . . .

and the **natural numbers** are 1, 2, 3, 4, 5, 6, 7, 8, 9, 10, 11, . . .

    The three dots (. . .) after each 11 means that these lists continue indefinitely. That is, there is no largest whole number. The smallest whole number is 0. Also, there is no largest natural number. The smallest natural number is 1.

Objective A   Finding the Place Value of a Digit in a Whole Number

The position of each digit in a number determines its **place value.** For example, the distance (in miles) between the planet Mercury and the planet Earth can be represented by the whole number 48,337,000. Next is a place-value chart for this whole number.

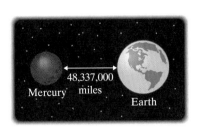

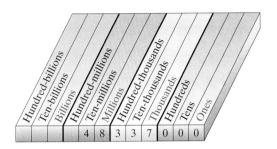

The two 3s in 48,337,000 represent different amounts because of their different placements. The place value of the 3 on the left is hundred-thousands. The place value of the 3 on the right is ten-thousands.

| Examples | Find the place value of the digit 3 in each whole number. |

1. 396,418
hundred-thousands

2. 93,192
thousands

3. 534,275,866
ten-millions

■ Work Practice 1–3

Practice 1–3

Find the place value of the digit 8 in each whole number.
1. 38,760,005
2. 67,890
3. 481,922

Objective B Writing a Whole Number in Words and in Standard Form

A whole number such as 1,083,664,500 is written in **standard form.** Notice that commas separate the digits into groups of three, starting from the right. Each group of three digits is called a **period.** The names of the first four periods are shown in red.

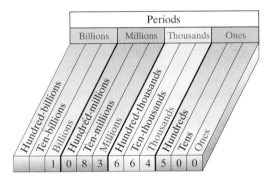

Writing a Whole Number in Words

To write a whole number in words, write the number in each period followed by the name of the period. (The ones period is usually not written.) This same procedure can be used to read a whole number.

For example, we write 1,083,664,500 as

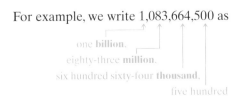

one **billion,**
eighty-three **million,**
six hundred sixty-four **thousand,**
five hundred

Helpful Hint Notice the commas after the name of each period.

Answers
1. millions **2.** hundreds
3. ten-thousands

Helpful Hint

The name of the ones period is not used when reading and writing whole numbers. For example,

9,265

is read as

"nine **thousand,** two hundred sixty-five."

Practice 4–6

Write each number in words.

4. 67

5. 395

6. 12,804

Examples Write each number in words.

4. 85   eighty-five

5. 126   one hundred twenty-six

6. 27,034 twenty-seven thousand, thirty-four

■ Work Practice 4–6

Helpful Hint

The word "and" is *not* used when reading and writing whole numbers. It is used when reading and writing mixed numbers and some decimal values, as shown later in this text.

Practice 7

Write 321,670,200 in words.

Example 7 Write 106,052,447 in words.

Solution: 106,052,447 is written as

one hundred six **million,** fifty-two **thousand,** four hundred forty-seven

■ Work Practice 7

✓**Concept Check** True or false? When writing a check for $2600, the word name we write for the dollar amount of the check is "two thousand sixty." Explain your answer.

Practice 8–11

Write each number in standard form.

8. twenty-nine

9. seven hundred ten

10. twenty-six thousand, seventy-one

11. six million, five hundred seven

Writing a Whole Number in Standard Form

To write a whole number in standard form, write the number in each period, followed by a comma.

Examples Write each number in standard form.

8. sixty-one   61   **9.** eight hundred five   805

10. nine thousand, three hundred eighty-six

9,386   or   9386

11. two million, five hundred sixty-four thousand, three hundred fifty

2,564,350

■ Work Practice 8–11

Answers

4. sixty-seven   **5.** three hundred ninety-five   **6.** twelve thousand, eight hundred four   **7.** three hundred twenty-one million, six hundred seventy thousand, two hundred   **8.** 29   **9.** 710   **10.** 26,071   **11.** 6,000,507

✓**Concept Check Answer**

false

Helpful Hint

A comma may or may not be inserted in a four-digit number. For example, both

9,386   and   9386

are acceptable ways of writing nine thousand, three hundred eighty-six.

Objective C Writing a Whole Number in Expanded Form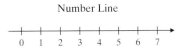

The place value of a digit can be used to write a number in expanded form. The **expanded form** of a number shows each digit of the number with its place value. For example, 5672 is written in expanded form as

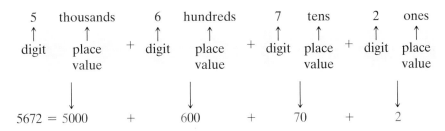

| Example 12 | Write 2,706,449 in expanded form. |

Solution: $2,000,000 + 700,000 + 6000 + 400 + 40 + 9$

Work Practice 12

Practice 12

Write 1,047,608 in expanded form.

We can visualize whole numbers by points on a line. The line below is called a **number line.** This number line has equally spaced marks for each whole number. The arrow to the right simply means that the whole numbers continue indefinitely. In other words, there is no largest whole number.

Number Line

We will study number lines further in Section 1.5.

Objective D Reading Tables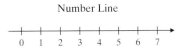

Now that we know about place value and names for whole numbers, we introduce one way that whole numbers may be presented. **Tables** are often used to organize and display facts that involve numbers. The table on the next page shows the ten countries with the most Nobel Prize winners since the inception of the Nobel Prize in 1901, and the categories of the prizes. The numbers for the Economics prize reflect the winners since 1969, when this category was established. (The numbers may seem large for two reasons: first, the annual Nobel Prize is often awarded to more than one individual, and second, several award winners hold dual citizenship, so they are counted in two countries.)

Answer
12. $1,000,000 + 40,000 + 7000 + 600 + 8$

Countries with Most Nobel Prize Winners, 1901–2013							
Country	Chemistry	Economics	Literature	Peace	Physics	Physiology & Medicine	Total
United States	68	54	11	22	90	97	342
United Kingdom	27	8	8	9	23	31	106
Germany	27	1	8	4	22	17	79
France	7	1	15	8	13	11	55
Sweden	4	2	9	5	4	8	32
Switzerland	6	0	2	3	4	6	21
Russia (USSR)	1	1	4	2	10	2	20
Japan	6	0	2	1	6	2	17
Netherlands	3	1	0	1	9	2	16
Italy	1	0	6	1	4	3	15

Source: Based on data from Encyclopaedia Britannica, Inc.

For example, by reading from left to right along the row marked "United States," we find that the United States has 68 Chemistry, 54 Economics, 11 Literature, 22 Peace, 90 Physics, and 97 Physiology and Medicine Nobel Prize winners.

Example 13   Use the Nobel Prize Winner table to answer each question.

a. How many total Nobel Prize winners are from Sweden?

b. Which countries shown have fewer Nobel Prize winners than Russia?

Solution:

a. Find "Sweden" in the left column. Then read from left to right until the "Total" column is reached. We find that Sweden has 32 Nobel Prize winners.

b. Russia has 20 Nobel Prize winners. Japan has 17, Netherlands has 16, and Italy has 15, so they have fewer Nobel Prize winners than Russia.

▣ Work Practice 13

Practice 13

Use the Nobel Prize Winner table to answer the following questions:

a. How many Nobel Prize winners in Literature come from France?

b. Which countries shown have more than 60 Nobel Prize winners?

Answers

13. a. 15   **b.** United States, United Kingdom, and Germany

Vocabulary, Readiness & Video Check

Use the choices below to fill in each blank.

standard form   period   whole

expanded form   place value   words

1. The numbers 0, 1, 2, 3, 4, 5, 6, 7, 8, 9, 10, 11, 12, … are called _____ numbers.

2. The number 1,286 is written in _____.

3. The number "twenty-one" is written in _____.

4. The number $900 + 60 + 5$ is written in _____.

5. In a whole number, each group of three digits is called a(n) _____.

6. The _____ of the digit 4 in the whole number 264 is ones.

Martin-Gay Interactive Videos   *Watch the section lecture video and answer the following questions.*

See Video 1.2

Objective A   **7.** In ⊞ Example 1, what is the place value of the digit 6?

Objective B   **8.** Complete this statement based on ⊞ Example 3: To read (or write) a number, read from _____ to _____.

Objective C   **9.** In ⊞ Example 5, what is the expanded-form value of the digit 8?

Objective D   **10.** Use the table given in ⊞ Example 6 to determine which mountain in the table is the shortest.

1.2 Exercise Set MyMathLab®

Objective A *Determine the place value of the digit 5 in each whole number. See Examples 1 through 3.*

1. 657

2. 905

3. 5423

4. 6527

5. 43,526,000

6. 79,050,000

7. 5,408,092

8. 51,682,700

Objective B *Write each whole number in words. See Examples 4 through 7.*

9. 354

10. 316

11. 8279

12. 5445

13. 26,990

14. 42,009

15. 2,388,000

16. 3,204,000

17. 24,350,185

18. 47,033,107

Write each number in the sentence in words. See Examples 4 through 7.

19. As of July 2013, the population of Iceland was 315,281. (*Source:* CIA World Factbook)

20. The land area of Belize is 22,806 square kilometers. (*Source:* CIA World Factbook)

21. The Burj Khalifa, in Dubai, United Arab Emirates, a hotel and office building, is the world's tallest building at a height of 2717 feet. (*Source:* Council on Tall Buildings and Urban Habitat)

22. As of October 2013, there were 119,948 patients in the United States waiting for an organ transplant. (*Source:* Organ Procurement and Transplantation Network)

23. In 2012, UPS received an average of 32,100,000 online tracking requests each day. (*Source:* UPS)

24. Each year, an estimated 500,000,000 Americans visit carnivals, fairs, and festivals. (*Source:* Outdoor Amusement Business Association)

25. The highest point in Colorado is Mount Elbert, at an elevation of 14,433 feet. (*Source:* U.S. Geological Survey)

26. The highest point in Oregon is Mount Hood, at an elevation of 11,239 feet. (*Source:* U.S. Geological Survey)

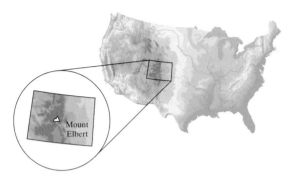

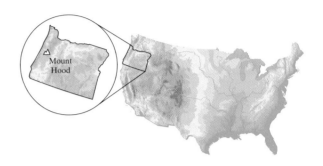

27. In 2013, the Great Internet Mersenne Prime Search, a cooperative computing project, helped find a prime number that has over 17,000,000 digits. (*Source:* Mersenne Research, Inc.)

28. The Goodyear blimp *Eagle* holds 202,700 cubic feet of helium. (*Source:* The Goodyear Tire & Rubber Company)

Write each whole number in standard form. See Examples 8 through 11.

29. Six thousand, five hundred eighty-seven

30. Four thousand, four hundred sixty-eight

31. Fifty-nine thousand, eight hundred

32. Seventy-three thousand, two

33. Thirteen million, six hundred one thousand, eleven

34. Sixteen million, four hundred five thousand, sixteen

35. Seven million, seventeen

36. Two million, twelve

37. Two hundred sixty thousand, nine hundred ninety-seven

38. Six hundred forty thousand, eight hundred eighty-one

Write the whole number in each sentence in standard form. See Examples 8 through 11.

39. After an orbit correction in October 2013, the International Space Station orbited Earth at an average altitude of about four hundred eighteen kilometers. (*Source:* Heavens Above)

40. The average distance between the surfaces of Earth and the Moon is about two hundred thirty-four thousand miles.

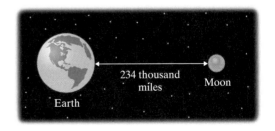

41. La Rinconada, Peru, is the highest town in the world. It is located sixteen thousand, seven hundred thirty-two feet above sea level. (*Source:* Russell Ash: *Top 10 of Everything*)

42. The world's tallest freestanding tower is the Tokyo Sky Tree in Japan. Its height is two thousand eighty feet tall. (*Source:* Council on Tall Buildings and Urban Habitat)

43. The Warner Bros. film *Harry Potter and the Deathly Hallows Part 2* holds the record for U.S./Canada opening day box office gross when it took in approximately ninety-one million, seventy-one thousand dollars on its opening day in 2011. (*Source:* Box Office Mojo)

44. The Buena Vista film *Marvel's The Avengers* set the U.S./Canada record for second-highest opening day box office gross when it took in approximately eighty million, eight hundred fourteen thousand dollars on its opening day in 2012. (*Source:* Box Office Mojo)

45. In 2012, the UPS delivery fleet consisted of one hundred one thousand vehicles. (*Source:* UPS)

46. Morten Andersen, who played football for New Orleans, Atlanta, N.Y. Giants, Kansas City, and Minnesota between 1982 and 2007, holds the record for the most points scored in a career. Over his 25-year career he scored two thousand, five hundred forty-four points. (*Source:* NFL.com)

Objective C *Write each whole number in expanded form. See Example 12.*

47. 406

48. 789

49. 3470

50. 6040

51. 80,774

52. 20,215

53. 66,049

54. 99,032

55. 39,680,000

56. 47,703,029

Objectives B C D Mixed Practice *The table shows the six tallest mountains in New England and their elevations. Use this table to answer Exercises 57 through 62. See Example 13.*

Mountain (State)	Elevation (in feet)
Boott Spur (NH)	5492
Mt. Adams (NH)	5774
Mt. Clay (NH)	5532
Mt. Jefferson (NH)	5712
Mt. Sam Adams (NH)	5584
Mt. Washington (NH)	6288
Source: U.S. Geological Survey	

Elevation in feet

57. Write the elevation of Mt. Clay in standard form and then in words.

58. Write the elevation of Mt. Washington in standard form and then in words.

59. Write the height of Boott Spur in expanded form.

60. Write the height of Mt. Jefferson in expanded form.

61. Which mountain is the tallest in New England?

62. Which mountain is the second tallest in New England?

The table shows the top ten museums in the world in 2012. Use this table to answer Exercises 63 through 68. See Example 13.

Top 10 Museums Worldwide in 2012

Museum	Location	Visitors
Louvre	Paris, France	9,720,000
National Museum of Natural History	Washington, DC, United States	7,600,000
National Air and Space Museum	Washington, DC, United States	6,800,000
The Metropolitan Museum of Art	New York, NY, United States	6,116,000
British Museum	London, United Kingdom	5,576,000
Tate Modern	London, United Kingdom	5,319,000
National Gallery	London, United Kingdom	5,164,000
Vatican Museums	Vatican City	5,065,000
American Museum of Natural History	New York, NY, United States	5,000,000
Natural History Museum	London, United Kingdom	4,936,000

(*Source:* Themed Entertainment Association)

63. Which museum had fewer visitors, the National Gallery in London or the National Air and Space Museum in Washington, DC?

64. Which museum had more visitors, the British Museum in London or the Louvre in Paris?

65. How many people visited the Vatican Museums? Write the number of visitors in words.

66. How many people visited The Metropolitan Museum of Art? Write the number of visitors in words.

67. How many of 2012's top ten museums in the world were located in the United States?

68. How many of 2012's top ten museums in the world were visited by fewer than 6,000,000 people?

Concept Extensions

69. Write the largest four-digit number that can be made from the digits 1, 9, 8, and 6 if each digit must be used once. ____ ____ ____ ____

70. Write the largest five-digit number that can be made using the digits 5, 3, and 7 if each digit must be used at least once. ____ ____ ____ ____ ____

Check to see whether each number written in standard form matches the number written in words. If not, correct the number in words. See the Concept Check in this section.

71.

72.

73. If a number is given in words, describe the process used to write this number in standard form.

74. If a number is written in standard form, describe the process used to write this number in expanded form.

75. In June 2013, the MilkyWay-2, a high-speed computer built by China's National University of Defense Technology, was ranked as the world's fastest computer. Its speed was clocked at nearly 34 petaflops, or more than 34 quadrillion arithmetic operations per second. Look up "quadrillion" (in the American system) and use the definition to write this number in standard form. (*Source:* top500.org)

76. As of December 2012, the national debt of France was approximately $5 trillion. Look up "trillion" (in the American system) and use the definition to write 5 trillion in standard form. (*Source:* CIA World Factbook)

77. The Pro Football Hall of Fame was established on September 7, 1963, in this town. Use the information and the diagram to the right to find the name of the town.
- Alliance is east of Massillon.
- Dover is between Canton and New Philadelphia.
- Massillon is not next to Alliance.
- Canton is north of Dover.

Pro Football
Hall of Fame

OHIO

Adding Whole Numbers and Perimeter

Objective A Adding Whole Numbers

According to Gizmodo, the iPod nano (currently in its seventh generation) is still the best overall MP3 player.

Suppose that an electronics store received a shipment of two boxes of iPod nanos one day and an additional four boxes of iPod nanos the next day. The **total** shipment in the two days can be found by adding 2 and 4.

2 boxes of iPod nanos + 4 boxes of iPod nanos = 6 boxes of iPod nanos

The **sum** (or total) is 6 boxes of iPod nanos. Each of the numbers 2 and 4 is called an **addend,** and the process of finding the sum is called **addition.**

To add whole numbers, we add the digits in the ones place, then the tens place, then the hundreds place, and so on. For example, let's add $2236 + 160$.

$$\begin{array}{r} 2236 \\ +\ 160 \\ \hline 2396 \end{array}$$

Line up numbers vertically so that the place values correspond. Then add digits in corresponding place values, starting with the ones place.

sum of ones
sum of tens
sum of hundreds
sum of thousands

Objectives

A Add Whole Numbers.

B Find the Perimeter of a Polygon.

C Solve Problems by Adding Whole Numbers.

Example 1 Add: $23 + 136$

Solution:
$$\begin{array}{r} 23 \\ +\ 136 \\ \hline 159 \end{array}$$

■ Work Practice 1

Practice 1

Add: $7235 + 542$

When the sum of digits in corresponding place values is more than 9, **carrying** is necessary. For example, to add $365 + 89$, add the ones-place digits first.

Carrying
$$\begin{array}{r} \overset{1}{3}65 \\ +\ 89 \\ \hline 4 \end{array}$$
5 ones + 9 ones = **14 ones** or **1 ten + 4 ones**
Write the 4 ones in the ones place and carry the 1 ten to the tens place.

Next, add the tens-place digits.
$$\begin{array}{r} \overset{1\ 1}{3}65 \\ +\ 89 \\ \hline 54 \end{array}$$
1 ten + 6 tens + 8 tens = **15 tens** or **1 hundred + 5 tens**
Write the 5 tens in the tens place and carry the 1 hundred to the hundreds place.

Next, add the hundreds-place digits.
$$\begin{array}{r} \overset{1\ 1}{3}65 \\ +\ 89 \\ \hline 454 \end{array}$$
1 hundred + 3 hundreds = **4 hundreds**
Write the 4 hundreds in the hundreds place.

Answer
1. 7777

17

Practice 2

Add: 27,364 + 92,977

Example 2 Add: 34,285 + 149,761

Solution:
$$
\begin{array}{r}
\overset{1\ 1\ 1}{34{,}285} \\
+\ 149{,}761 \\
\hline
184{,}046
\end{array}
$$

■ Work Practice 2

✓**Concept Check** What is wrong with the following computation?

$$
\begin{array}{r}
394 \\
+\ 283 \\
\hline
577
\end{array}
$$

Before we continue adding whole numbers, let's review some properties of addition that you may have already discovered. The first property that we will review is the **addition property of 0.** This property reminds us that the sum of 0 and any number is that same number.

Addition Property of 0

The sum of 0 and any number is that number. For example,

$$7 + 0 = 7$$
$$0 + 7 = 7$$

Next, notice that we can add any two whole numbers in any order and the sum is the same. For example,

$$4 + 5 = 9 \quad \text{and} \quad 5 + 4 = 9$$

We call this special property of addition the **commutative property of addition.**

Commutative Property of Addition

Changing the **order** of two addends does not change their sum. For example,

$$2 + 3 = 5 \quad \text{and} \quad 3 + 2 = 5$$

Another property that can help us when adding numbers is the **associative property of addition.** This property states that when adding numbers, the grouping of the numbers can be changed without changing the sum. We use parentheses to group numbers. They indicate which numbers to add first. For example, let's use two different groupings to find the sum of $2 + 1 + 5$.

$$(2 + 1) + 5 = 3 + 5 = 8$$

Also,

$$2 + (1 + 5) = 2 + 6 = 8$$

Both groupings give a sum of 8.

Answer

2. 120,341

✓**Concept Check Answer**

forgot to carry 1 hundred to the hundreds place

Associative Property of Addition

Changing the **grouping** of addends does not change their sum. For example,

$$3 + (5 + 7) = 3 + 12 = 15 \quad \text{and} \quad (3 + 5) + 7 = 8 + 7 = 15$$

The commutative and associative properties tell us that we can add whole numbers using any order and grouping that we want.

When adding several numbers, it is often helpful to look for two or three numbers whose sum is 10, 20, and so on. Why? Adding multiples of 10 such as 10 and 20 is easier.

Example 3   Add: $13 + 2 + 7 + 8 + 9$

Solution:

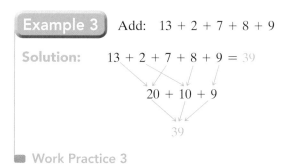

$$13 + 2 + 7 + 8 + 9 = 39$$
$$20 + 10 + 9$$
$$39$$

■ Work Practice 3

Practice 3

Add: $11 + 7 + 8 + 9 + 13$

Feel free to use the process of Example 3 anytime when adding.

Example 4   Add: $1647 + 246 + 32 + 85$

Solution:

$$
\begin{array}{r}
{\scriptstyle 1\,2\,2} \\
1647 \\
246 \\
32 \\
+ \quad 85 \\
\hline
2010
\end{array}
$$

■ Work Practice 4

Practice 4

Add: $19 + 5042 + 638 + 526$

Objective B   Finding the Perimeter of a Polygon

In geometry addition is used to find the perimeter of a polygon. A **polygon** can be described as a flat figure formed by line segments connected at their ends. (For more review, see Appendix A.3.) Geometric figures such as triangles, squares, and rectangles are called polygons.

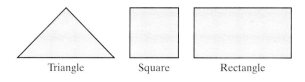

Triangle   Square   Rectangle

The **perimeter** of a polygon is the *distance around* the polygon. This means that the perimeter of a polygon is the sum of the lengths of its sides.

Answers

3. 48   **4.** 6225

Practice 5

Find the perimeter of the polygon shown. (A centimeter is a unit of length in the metric system.)

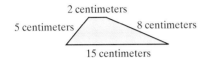

2 centimeters
5 centimeters   8 centimeters
15 centimeters

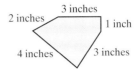 **Example 5** Find the perimeter of the polygon shown.

3 inches
2 inches
1 inch
4 inches   3 inches

Solution: To find the perimeter (distance around), we add the lengths of the sides.

2 in. + 3 in. + 1 in. + 3 in. + 4 in. = 13 in.

The perimeter is 13 inches.

■ Work Practice 5

To make the addition appear simpler, we will often not include units with the addends. If you do this, make sure units are included in the final answer.

Practice 6

A park is in the shape of a triangle. Each of the park's three sides is 647 feet. Find the perimeter of the park.

Example 6 Calculating the Perimeter of a Building

The world's largest commercial building under one roof is the flower auction building of the cooperative VBA in Aalsmeer, Netherlands. The floor plan is a rectangle that measures 776 meters by 639 meters. Find the perimeter of this building. (A meter is a unit of length in the metric system.) (*Source: The Handy Science Answer Book*, Visible Ink Press)

Solution: Recall that opposite sides of a rectangle have the same length. To find the perimeter of this building, we add the lengths of the sides. The sum of the lengths of its sides is

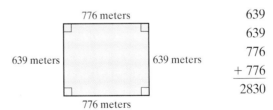

776 meters
639 meters   639 meters
776 meters

639
639
776
+ 776
2830

The perimeter of the building is 2830 meters.

■ Work Practice 6

Objective C Solving Problems by Adding

Often, real-life problems occur that can be solved by adding. The first step in solving any word problem is to *understand* the problem by reading it carefully.

Descriptions of problems solved through addition *may* include any of these key words or phrases:

Addition		
Key Words or Phrases	**Examples**	**Symbols**
added to	5 added to 7	7 + 5
plus	0 plus 78	0 + 78
increased by	12 increased by 6	12 + 6
more than	11 more than 25	25 + 11
total	the total of 8 and 1	8 + 1
sum	the sum of 4 and 133	4 + 133

Answers
5. 30 cm **6.** 1941 ft

To solve a word problem that involves addition, we first use the facts given to write an addition statement. Then we write the corresponding solution of the real-life problem. It is sometimes helpful to write the statement in words (brief phrases) and then translate to numbers.

Example 7 Finding the Number of Vehicles Sold in the United States

In 2011, a total of 12,734,424 passenger vehicles were sold in the United States. In 2012, total passenger vehicle sales in the United States had increased by 1,705,636 vehicles. Find the total number of passenger vehicles sold in the United States in 2012. (*Source:* Alliance of Automobile Manufacturers)

Solution: The key phrase here is "had increased by," which suggests that we add. To find the number of vehicles sold in 2012, we add the increase, 1,705,636, to the number of vehicles sold in 2011.

In Words		Translate to Numbers
Number sold in 2011	⟶	12,734,424
+ increase	⟶	+ 1,705,636
Number sold in 2012	⟶	14,440,060

The number of passenger vehicles sold in the United States in 2012 was 14,440,060.

■ Work Practice 7

Graphs can be used to visualize data. The graph shown next is called a **bar graph.** For this bar graph, the height of each bar is labeled above the bar. To check this height, follow the top of each bar to the vertical line to the left. For example, the first bar is labeled 185. Follow the top of that bar to the left until the vertical line is reached, between 180 and 200, but closer to 180, or 185.

Example 8 Reading a Bar Graph

In the following graph, each bar represents a country and the height of each bar represents the number of threatened mammal species identified in that country.

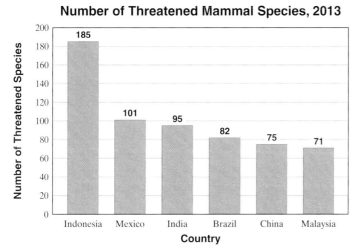

Number of Threatened Mammal Species, 2013

Source: International Union for Conservation of Nature

Practice 7

Georgia produces 70 million pounds of freestone peaches per year. The second largest U.S. producer of peaches, South Carolina, produces 50 million more freestone peaches than Georgia. How much does South Carolina produce? (*Source:* farms.com)

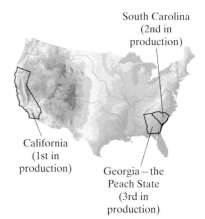

South Carolina (2nd in production)

California (1st in production)

Georgia—the Peach State (3rd in production)

Practice 8

Use the graph in Example 8 to answer the following:

a. Which country shown has the fewest threatened mammal species?

b. Find the total number of threatened mammal species for Brazil, India, and Mexico.

Answers

7. 120 million lb
8. a. Malaysia   **b.** 278

(Continued on next page)

a. Which country shown has the greatest number of threatened mammal species?

b. Find the total number of threatened mammal species for Malaysia, China, and Indonesia.

Solution:

a. The country with the greatest number of threatened mammal species corresponds to the tallest bar, which is Indonesia.

b. The key word here is "total." To find the total number of threatened mammal species for Malaysia, China, and Indonesia, we add.

In Words		Translate to Numbers
Malaysia	$\longrightarrow$	71
China	$\longrightarrow$	75
Indonesia	$\longrightarrow$	+ 185
	Total	331

The total number of threatened mammal species for Malaysia, China, and Indonesia is 331.

■ Work Practice 8

Calculator Explorations Adding Numbers

To add numbers on a calculator, find the keys marked $+$ and $=$ or ENTER.

For example, to add 5 and 7 on a calculator, press the keys 5 $+$ 7 then $=$ or ENTER.

The display will read ☐ 12 .

Thus, 5 + 7 = 12.

To add 687, 981, and 49 on a calculator, press the keys 687 $+$ 981 $+$ 49 then $=$ or ENTER.

The display will read ☐ 1717 .

Thus, 687 + 981 + 49 = 1717. (Although entering 687, for example, requires pressing more than one key, here numbers are grouped together for easier reading.)

Use a calculator to add.

1. 89 + 45

2. 76 + 97

3. 285 + 55

4. 8773 + 652

5.
```
    985
   1210
    562
 +   77
```

6.
```
    465
   9888
    620
 + 1550
```

Vocabulary, Readiness & Video Check

Use the choices below to fill in each blank. Some choices may be used more than once.

sum	order	addend	associative
perimeter	number	grouping	commutative

1. The sum of 0 and any number is the same _____.

2. The sum of any number and 0 is the same _____.

3. In 35 + 20 = 55, the number 55 is called the _____ and 35 and 20 are each called a(n) _____.

4. The distance around a polygon is called its _____.

5. Since $(3 + 1) + 20 = 3 + (1 + 20)$, we say that changing the _____ in addition does not change the sum. This property is called the _____ property of addition.

6. Since $7 + 10 = 10 + 7$, we say that changing the _____ in addition does not change the sum. This property is called the _____ property of addition.

Martin-Gay Interactive Videos *Watch the section lecture video and answer the following questions.*

See Video 1.3

Objective A **7.** Complete this statement based on the lecture before ▥ Example 1: To add whole numbers, we line up _____ values and add from _____ to _____.

Objective B **8.** In ▥ Example 4, the perimeter of what type of polygon is found? How many addends are in the resulting addition problem?

Objective C **9.** In ▥ Example 6, what key word or phrase indicates addition?

1.3 **Exercise Set** MyMathLab®

Objective A *Add. See Examples 1 through 4.*

1. 14
  + 22

2. 27
  + 31

3. 62
  + 230

4. 37
  + 542

5. 12
  13
  + 24

6. 23
  45
  + 30

▷ **7.** 5267
  + 132

8. 236
  + 6243

9. $53 + 64$

10. $41 + 74$

11. $22 + 490$

12. $35 + 470$

13. $22,781 + 186,297$

14. $17,427 + 821,059$

▷ **15.** 8
  9
  2
  5
  + 1

16. 3
  5
  8
  5
  + 7

17. 6
  21
  14
  9
  + 12

18. 12
  4
  8
  26
  + 10

19. 81
  17
  23
  79
  + 12

20. 64
  28
  56
  25
  + 32

21. $62 + 18 + 14$

22. $23 + 49 + 18$

23. $40 + 800 + 70$

24. $30 + 900 + 20$

25. $7542 + 49 + 682$

26. $1624 + 32 + 976$

27. 24 + 9006 + 489 + 2407

28. 16 + 1056 + 748 + 7770

29.
```
  627
  628
+ 629
```

30.
```
  427
  383
+ 229
```

31.
```
  6820
  4271
+ 5626
```

32.
```
  6789
  4321
+ 5555
```

33.
```
  507
  593
+  10
```

34.
```
  864
   33
+ 356
```

35.
```
  4200
  2107
+ 2692
```

36.
```
  5000
  1400
+ 3021
```

37.
```
     49
    628
   5 762
+ 29,462
```

38.
```
     26
    582
   4 763
+ 62,511
```

39.
```
  121,742
   57,279
   26,586
+ 426,782
```

40.
```
  504,218
  321,920
   38,507
+ 594,687
```

Objective B *Find the perimeter of each figure. See Examples 5 and 6.*

△ **41.**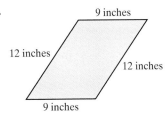
9 inches
12 inches
12 inches
9 inches

△ **42.**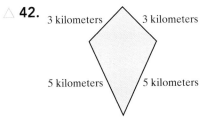
3 kilometers   3 kilometers
5 kilometers   5 kilometers

▷△ **43.**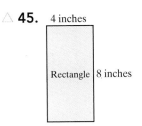
7 feet   8 feet
10 feet

△ **44.**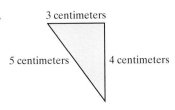
3 centimeters
5 centimeters   4 centimeters

△ **45.**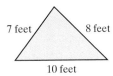
4 inches
Rectangle | 8 inches

△ **46.**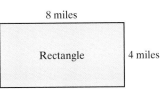
8 miles
Rectangle   4 miles

△ **47.**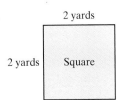
2 yards
2 yards   Square

△ **48.**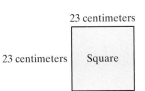
23 centimeters
23 centimeters   Square

△ **49.**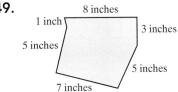
8 inches
1 inch
3 inches
5 inches
5 inches
7 inches

△ **50.**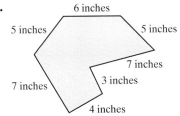
6 inches
5 inches   5 inches
7 inches
7 inches   3 inches
4 inches

△ **51.**

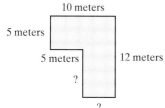

△ **52.**

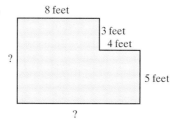

Objectives A B C Mixed Practice–Translating *Solve. See Examples 1 through 8.*

⊙ **53.** Find the sum of 297 and 1796.

54. Find the sum of 802 and 6487.

55. Find the total of 76, 39, 8, 17, and 126.

56. Find the total of 89, 45, 2, 19, and 341.

57. What is 452 increased by 92?

58. What is 712 increased by 38?

59. What is 2686 plus 686 plus 80?

60. What is 3565 plus 565 plus 70?

61. The estimated population of Florida was 19,318 thousand in 2012. If is projected to increase by 1823 thousand by 2020. What is Florida's projected population in 2020? (*Source:* U.S. Census Bureau, Florida Office of Economic & Demographic Research)

62. The estimated population of California was 38,041 thousand in 2012. It is projected to increase by 2603 thousand by 2020. What is California's projected population in 2020? (*Source:* U.S. Census Bureau, California Department of Finance)

⊙ **63.** The highest point in South Carolina is Sassafras Mountain at 3560 feet above sea level. The highest point in North Carolina is Mt. Mitchell, whose peak is 3124 feet increased by the height of Sassafras Mountain. Find the height of Mt. Mitchell. (*Source:* U.S. Geological Survey)

64. The distance from Kansas City, Kansas, to Hays, Kansas, is 285 miles. Colby, Kansas, is 98 miles farther from Kansas City than Hays. Find the total distance from Kansas City to Colby.

△ **65.** Leo Callier is installing an invisible fence in his backyard. How many feet of wiring are needed to enclose the yard below?

△ **66.** A homeowner is considering installing gutters around her home. Find the perimeter of her rectangular home.

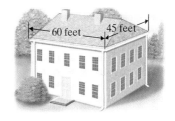

67. The tallest waterfall in the United States is Yosemite Falls in Yosemite National Park in California. Yosemite Falls is made up of three sections, as shown in the graph. What is the total height of Yosemite Falls? (*Source:* U.S. Department of the Interior)

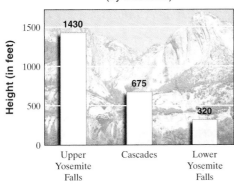

Tallest U.S. Waterfall
(by Sections)

68. Jordan White, a nurse at Mercy Hospital, is recording fluid intake on a patient's medical chart. During his shift, the patient had the following types and amounts of intake measured in cubic centimeters (cc). What amount should Jordan record as the total fluid intake for this patient?

Oral	Intravenous	Blood
240	500	500
100	200	
355		

69. In 2012, Harley-Davidson sold 172,251 of its motorcycles domestically. In addition, 77,598 Harley-Davidson motorcycles were sold internationally. What was the total number of Harley-Davidson motorcycles sold in 2012? (*Source:* Harley-Davidson, Inc.)

70. Hank Aaron holds Major League Baseball's record for the most runs batted in over his career. He batted in 1305 runs from 1954 to 1965. He batted in another 992 runs from 1966 until he retired in 1976. How many total runs did Hank Aaron bat in during his career in professional baseball?

71. During August 2013, a total of 999,040 vehicles were produced in the United States. During the same period, a total of 475,671 vehicles were produced in Canada and Mexico. What was the total number of vehicles produced in North America in August 2013? (*Source:* WardsAuto InfoBank)

72. In 2012, the country of New Zealand had 26,767,000 more sheep than people. If the human population of New Zealand in 2012 was 4,433,000, what was the sheep population? (*Source:* Statistics New Zealand)

73. The largest permanent Monopoly board is made of granite and located in San Jose, California. Find the perimeter of the square playing board.

31 ft

31 ft

74. The smallest commercially available jigsaw puzzle (with a minimum of 1000 pieces) is manufactured in Hong Kong, China. (*Source: Guinness World Records*) Find the exact perimeter of this rectangular-shaped puzzle in millimeters.

182 millimeters
(about 7 in.)

257 millimeters
(about 10 in.)

75. In 2013, there were 2657 Gap Inc. (Gap, Banana Republic, Old Navy) stores located in the United States and 438 located outside the United States. How many Gap Inc. stores were located worldwide? (*Source:* Gap Inc.)

76. Wilma Rudolph, who won three gold medals in track and field events in the 1960 Summer Olympics, was born in 1940. Allyson Felix, who also won three gold medals in track and field events but in the 2012 Summer Olympics, was born 45 years later. In what year was Allyson Felix born?

The table shows the number of Target stores in ten states. Use this table to answer Exercises 77 through 82.

| The Top States for Target Stores in 2012 ||
State	Number of Stores
California	257
Florida	123
Illinois	89
Michigan	59
Minnesota	75
New York	67
Ohio	64
Pennsylvania	63
Texas	149
Virginia	57
(*Source:* Target Corporation)	

77. Which state has the most Target stores?

78. Which of the states listed in the table has the fewest Target stores?

79. What is the total number of Target stores located in the three states with the most Target stores?

80. How many Target stores are located in the ten states listed in the table?

81. Which pair of neighboring states has more Target stores combined, New York and Pennsylvania or Michigan and Ohio?

82. Target operates stores in 49 states. There are 775 Target stores located in the states not listed in the table. How many Target stores are in the United States?

83. The state of Delaware has 2997 miles of urban highways and 3361 miles of rural highways. Find the total highway mileage in Delaware. (*Source:* U.S. Federal Highway Administration)

84. The state of Rhode Island has 5260 miles of urban highways and 1225 miles of rural highways. Find the total highway mileage in Rhode Island. (*Source:* U.S. Federal Highway Administration)

Concept Extensions

85. In your own words, explain the commutative property of addition.

86. In your own words, explain the associative property of addition.

87. Give any three whole numbers whose sum is 100.

88. Give any four whole numbers whose sum is 25.

89. Add: 56,468,980 + 1,236,785 + 986,768,000

90. Add: 78,962 + 129,968,350 + 36,462,880

Check each addition below. If it is incorrect, find the correct answer. See the Concept Check in this section.

91.	**92.**	**93.**	**94.**
566	773	14	19
932	659	173	214
+871	+481	86	49
2369	1913	+257	+651
		520	923

1.4 Subtracting Whole Numbers

Objectives

A Subtract Whole Numbers.

B Solve Problems by Subtracting Whole Numbers.

Objective A Subtracting Whole Numbers

If you have $5 and someone gives you $3, you have a total of $8, since 5 + 3 = 8. Similarly, if you have $8 and then someone borrows $3, you have $5 left. **Subtraction** is finding the **difference** of two numbers.

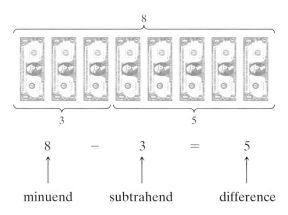

In this example, 8 is the **minuend,** and 3 is the **subtrahend.** The **difference** between these two numbers, 8 and 3, is 5.

Notice that addition and subtraction are very closely related. In fact, subtraction is defined in terms of addition.

$$8 - 3 = 5 \text{ because } 5 + 3 = 8$$

This means that subtraction can be *checked* by addition, and we say that addition and subtraction are reverse operations.

Example 1   Subtract. Check each answer by adding.

a. $12 - 9$ **b.** $22 - 7$ **c.** $35 - 35$ **d.** $70 - 0$

Solution:

a. $12 - 9 = 3$ because $3 + 9 = 12$
b. $22 - 7 = 15$ because $15 + 7 = 22$
c. $35 - 35 = 0$ because $0 + 35 = 35$
d. $70 - 0 = 70$ because $70 + 0 = 70$

■ **Work Practice 1**

Practice 1

Subtract. Check each answer by adding.
a. $14 - 6$
b. $20 - 8$
c. $93 - 93$
d. $42 - 0$

Look again at Examples 1(c) and 1(d).

1(c) $35 - 35 = 0$
         same   difference
         number   is 0

1(d) $70 - 0 = 70$
         a number   difference is the
         minus 0   same number

These two examples illustrate the subtraction properties of 0.

Subtraction Properties of 0

The difference of any number and that same number is 0. For example,

$$11 - 11 = 0$$

The difference of any number and 0 is that same number. For example,

$$45 - 0 = 45$$

To subtract whole numbers we subtract the digits in the ones place, then the tens place, then the hundreds place, and so on. When subtraction involves numbers of two or more digits, it is more convenient to subtract vertically. For example, to subtract 893 − 52,

$$
\begin{array}{r}
8\,9\,3 \\
-\ \ 5\,2 \\
\hline
8\,4\,1
\end{array}
$$
← minuend
← subtrahend
← difference

3 − 2
9 − 5
8 − 0

Line up the numbers vertically so that the minuend is on top and the place values correspond. Subtract in corresponding place values, starting with the ones place.

To check, add.

difference
+ subtrahend
minuend

or

$$
\begin{array}{r}
841 \\
+\ \ 52 \\
\hline
893
\end{array}
$$
← Since this is the original minuend, the problem checks.

Practice 2

Subtract. Check by adding.
a. 9143 − 122
b. 978 − 851

Example 2 Subtract: 7826 − 505. Check by adding.

Solution:
$$
\begin{array}{r}
7826 \\
-\ 505 \\
\hline
7321
\end{array}
$$

Check:
$$
\begin{array}{r}
7321 \\
+\ 505 \\
\hline
7826
\end{array}
$$

Work Practice 2

Subtracting by Borrowing

When subtracting vertically, if a digit in the second number (subtrahend) is larger than the corresponding digit in the first number (minuend), **borrowing** is necessary. For example, consider

$$
\begin{array}{r}
8\,|1 \\
-\ 6\,|3
\end{array}
$$

Since the 3 in the ones place of 63 is larger than the 1 in the ones place of 81, borrowing is necessary. We borrow 1 ten from the tens place and add it to the ones place.

Borrowing

$$
8 - 1 = 7 \rightarrow
\begin{array}{r}
\overset{7\ 11}{8\,\rlap{/}1} \\
-\ 6\,3
\end{array}
$$
← 1 ten + 1 one = 11 ones

tens   ten   tens

Now we subtract the ones-place digits and then the tens-place digits.

Practice 3

Subtract. Check by adding.
a.
$$
\begin{array}{r}
697 \\
-\ 49
\end{array}
$$
b.
$$
\begin{array}{r}
326 \\
-245
\end{array}
$$
c.
$$
\begin{array}{r}
1234 \\
-\ 822
\end{array}
$$

$$
\begin{array}{r}
\overset{7\ 11}{8\,\rlap{/}1} \\
-6\,3 \\
\hline
1\,8
\end{array}
$$
← 11 − 3 = 8
7 − 6 = 1

Check:
$$
\begin{array}{r}
18 \\
+63 \\
\hline
81
\end{array}
$$
The original minuend.

Example 3 Subtract: 543 − 29. Check by adding.

Solution:
$$
\begin{array}{r}
\overset{3\ 13}{5\,4\,\rlap{/}3} \\
-\ \ 2\,9 \\
\hline
5\,1\,4
\end{array}
$$

Check:
$$
\begin{array}{r}
514 \\
+\ 29 \\
\hline
543
\end{array}
$$

Work Practice 3

Answers
2. a. 9021 **b.** 127
3. a. 648 **b.** 81 **c.** 412

Sometimes we may have to borrow from more than one place. For example, to subtract $7631 - 152$, we first borrow from the tens place.

$$
\begin{array}{r}
763\cancel{1} \\
-\ 152 \\
\hline
9
\end{array}
$$
$\leftarrow 11 - 2 = 9$

In the tens place, 5 is greater than 2, so we borrow again. This time we borrow from the hundreds place.

$$
\begin{array}{r}
76\cancel{3}\cancel{1} \\
-\ 152 \\
\hline
7479
\end{array}
$$

Check:
$$
\begin{array}{r}
7479 \\
+\ 152 \\
\hline
7631
\end{array}
$$ The original minuend.

Example 4 Subtract: $900 - 174$. Check by adding.

Solution: In the ones place, 4 is larger than 0, so we borrow from the tens place. But the tens place of 900 is 0, so to borrow from the tens place we must first borrow from the hundreds place.

$$
\begin{array}{r}
\overset{8\ \ \ 10}{\cancel{9}\ \cancel{0}\ 0} \\
-\ 1\ 7\ 4
\end{array}
$$

Now borrow from the tens place.

$$
\begin{array}{r}
\overset{8\ \ 9}{\cancel{9}\ \overset{10}{\cancel{0}}\ \overset{10}{\cancel{0}}} \\
-\ 1\ 7\ 4 \\
\hline
7\ 2\ 6
\end{array}
$$

Check:
$$
\begin{array}{r}
726 \\
+174 \\
\hline
900
\end{array}
$$

Work Practice 4

Practice 4

Subtract. Check by adding.

a.
$$
\begin{array}{r}
400 \\
-164
\end{array}
$$

b.
$$
\begin{array}{r}
1000 \\
-\ 762
\end{array}
$$

Objective B Solving Problems by Subtracting

Often, real-life problems occur that can be solved by subtracting. The first step in solving any word problem is to *understand* the problem by reading it carefully.

Descriptions of problems solved through subtraction *may* include any of these key words or phrases:

Subtraction		
Key Words or Phrases	**Examples**	**Symbols**
subtract	subtract 5 from 8	$8 - 5$
difference	the difference of 10 and 2	$10 - 2$
less	17 less 3	$17 - 3$
less than	2 less than 20	$20 - 2$
take away	14 take away 9	$14 - 9$
decreased by	7 decreased by 5	$7 - 5$
subtracted from	9 subtracted from 12	$12 - 9$

Helpful Hint Be careful when solving applications that suggest subtraction. Although order *does not* matter when adding, order *does* matter when subtracting. For example, $20 - 15$ and $15 - 20$ do not simplify to the same number.

Answers
4. a. 236 **b.** 238

✓**Concept Check** In each of the following problems, identify which number is the minuend and which number is the subtrahend.

a. What is the result when 6 is subtracted from 40?

b. What is the difference of 15 and 8?

c. Find a number that is 15 fewer than 23.

To solve a word problem that involves subtraction, we first use the facts given to write a subtraction statement. Then we write the corresponding solution of the real-life problem. It is sometimes helpful to write the statement in words (brief phrases) and then translate to numbers.

Practice 5

The radius of Uranus is 15,759 miles. The radius of Neptune is 458 miles less than the radius of Uranus. What is the radius of Neptune? (*Source:* National Space Science Data Center)

Example 5   Finding the Radius of a Planet

The radius of Jupiter is 43,441 miles. The radius of Saturn is 7257 miles less than the radius of Jupiter. Find the radius of Saturn. (*Source:* National Space Science Data Center)

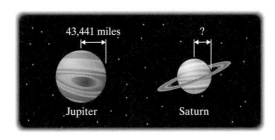

43,441 miles   ?

Jupiter   Saturn

Solution:

In Words		Translate to Numbers
radius of Jupiter	⟶	$\overset{13}{\cancel{4}\,\overset{13}{\cancel{3}},\overset{3}{\cancel{4}}\,\overset{3}{\cancel{4}}\,\overset{11}{\cancel{1}}}$ 43,441
− 7257	⟶	− 7 257
radius of Saturn	⟶	36,184

The radius of Saturn is 36,184 miles.

■ Work Practice 5

Helpful Hint   Since subtraction and addition are reverse operations, don't forget that a subtraction problem can be checked by adding.

Practice 6

During a sale, the price of a new suit is decreased by $47. If the original price was $92, find the sale price of the suit.

Example 6   Calculating Miles per Gallon

A subcompact car gets 42 miles per gallon of gas. A full-size car gets 17 miles per gallon of gas. Find the difference between the subcompact car miles per gallon and the full-size car miles per gallon.

Solution:

In Words		Translate to Numbers
subcompact miles per gallon	⟶	$\overset{3}{\cancel{4}}\overset{12}{\cancel{2}}$
− full-size miles per gallon	⟶	−1 7
difference in miles per gallon		2 5

The difference in the subcompact car miles per gallon and the full-size car miles per gallon is 25 miles per gallon.

■ Work Practice 6

Answers

5. 15,301 miles   **6.** $45

✓**Concept Check Answers**

a. minuend: 40; subtrahend: 6

b. minuend: 15; subtrahend: 8

c. minuend: 23; subtrahend: 15

Helpful Hint

Once again, because subtraction and addition are reverse operations, don't forget that a subtraction problem can be checked by adding.

Calculator Explorations Subtracting Numbers

To subtract numbers on a calculator, find the keys marked $\boxed{-}$ and $\boxed{=}$ or $\boxed{\text{ENTER}}$.

For example, to find $83 - 49$ on a calculator, press the keys $\boxed{83}\ \boxed{-}\ \boxed{49}$ then $\boxed{=}$ or $\boxed{\text{ENTER}}$.

The display will read $\boxed{\qquad 34\ }$.

Thus, $83 - 49 = 34$.

Use a calculator to subtract.

1. $865 - 95$ **2.** $76 - 27$

3. $147 - 38$ **4.** $366 - 87$

5. $9625 - 647$ **6.** $10,711 - 8925$

Vocabulary, Readiness & Video Check

Use the choices below to fill in each blank.

0	minuend	difference
number	subtrahend	

1. The difference of any number and that same number is _____.

2. The difference of any number and 0 is the same _____.

3. In $37 - 19 = 18$, the number 37 is the _____, and the number 19 is the _____.

4. In $37 - 19 = 18$, the number 18 is called the _____.

Find each difference.

5. $6 - 6$ **6.** $93 - 93$ **7.** $600 - 0$ **8.** $5 - 0$

Martin-Gay Interactive Videos

See Video 1.4

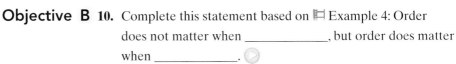

Watch the section lecture video and answer the following questions.

Objective A **9.** In ⊞ Example 2, explain how we end up subtracting 7 from 12 in the ones place. ◯

Objective B **10.** Complete this statement based on ⊞ Example 4: Order does not matter when _____, but order does matter when _____. ◯

1.4 Exercise Set MyMathLab®

Objective A *Subtract. Check by adding. See Examples 1 and 2.*

1.   67
      − 23

2.   72
      − 41

3.   389
      − 124

4.   572
      − 321

5.   167
      − 32

6.   286
      − 45

7. 2677 − 423

8. 5766 − 324

9. 6998 − 1453

10. 4912 − 2610

11.   749
       − 149

12.   257
       − 257

Subtract. Check by adding. See Examples 1 through 4.

13.   62
       − 37

14.   55
       − 29

15.   70
       − 25

16.   80
       − 37

17.   938
       − 792

18.   436
       − 275

19.   922
       − 634

20.   674
       − 299

21.   600
       − 432

22.   300
       − 149

23.   142
       − 36

24.   773
       − 29

25.   923
       − 476

26.   813
       − 227

27.   6283
       − 560

28.   5349
       − 720

29.   533
       − 29

30.   724
       − 16

31.   200
       − 111

32.   300
       − 211

33.   1983
       − 1904

34.   1983
       − 1914

35.   56,422
       − 16,508

36.   76,652
       − 29,498

37. 50,000 − 17,289

38. 40,000 − 23,582

39. 7020 − 1979

40. 6050 − 1878

41. 51,111 − 19,898

42. 62,222 − 39,898

Objective B *Solve. See Examples 5 and 6.*

43. Subtract 5 from 9.

44. Subtract 9 from 21.

45. Find the difference of 41 and 21.

46. Find the difference of 16 and 5.

47. Subtract 56 from 63.

48. Subtract 41 from 59.

49. Find 108 less 36.

50. Find 25 less 12.

51. Find 12 subtracted from 100.

52. Find 86 subtracted from 90.

53. Professor Graham is reading a 503-page book. If she has just finished reading page 239, how many more pages must she read to finish the book?

54. When a couple began a trip, the odometer read 55,492. When the trip was over, the odometer read 59,320. How many miles did they drive on their trip?

55. In 2008, the hole in the Earth's ozone layer over Antarctica was about 25 million square kilometers in size. By 2012, the hole had shrunk to about 18 million square kilometers. By how much did the hole shrink from 2008 to 2012? (*Source:* NASA Ozone Watch)

56. Bamboo can grow to 98 feet while Pacific giant kelp (a type of seaweed) can grow to 197 feet. How much taller is the kelp than the bamboo?

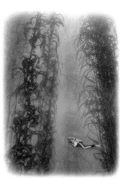

Bamboo   Kelp

A river basin is the geographic area drained by a river and its tributaries. The Mississippi River Basin is the third largest in the world and is divided into six sub-basins, whose areas are shown in the following bar graph. Use this graph for Exercises 57 through 60.

57. Find the total U.S. land area drained by the Upper Mississippi and Lower Mississippi sub-basins.

58. Find the total U.S. land area drained by the Ohio and Tennessee sub-basins.

59. How much more land is drained by the Missouri sub-basin than the Arkansas Red-White sub-basin?

60. How much more land is drained by the Upper Mississippi sub-basin than the Lower Mississippi sub-basin?

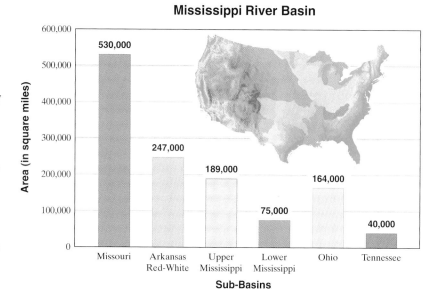

Mississippi River Basin

Area (in square miles)

- Missouri: 530,000
- Arkansas Red-White: 247,000
- Upper Mississippi: 189,000
- Lower Mississippi: 75,000
- Ohio: 164,000
- Tennessee: 40,000

Sub-Basins

61. The peak of Mt. McKinley in Alaska is 20,320 feet above sea level. The peak of Long's Peak in Colorado is 14,255 feet above sea level. How much higher is the peak of Mt. McKinley than Long's Peak? (*Source:* U.S. Geological Survey)

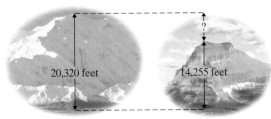

Mt. McKinley, Alaska   Long's Peak, Colorado

62. On January 12, 1916, the city of Indianapolis, Indiana, had the greatest temperature change in a day. It dropped 58 degrees. If the high temperature was 68° Fahrenheit, what was the low temperature?

63. The Oroville Dam, on the Feather River, is the tallest dam in the United States at 754 feet. The Hoover Dam, on the Colorado River, is 726 feet high. How much taller is the Oroville Dam than the Hoover Dam? (*Source:* U.S. Bureau of Reclamation)

64. A new iPhone with 32 GB costs $299. Jocelyn Robinson has $713 in her savings account. How much will she have left in her savings account after she buys the iPhone? (*Source:* Apple, Inc.)

65. The distance from Kansas City to Denver is 645 miles. Hays, Kansas, lies on the road between the two and is 287 miles from Kansas City. What is the distance between Hays and Denver?

66. Pat Salanki's blood cholesterol level is 243. The doctor tells him it should be decreased to 185. How much of a decrease is this?

67. A new DVD player with remote control costs $295. A college student has $914 in her savings account. How much will she have left in her savings account after she buys the DVD player?

68. A stereo that regularly sells for $547 is discounted by $99 in a sale. What is the sale price?

69. The population of Arizona is projected to grow from 6499 thousand in 2012 to 9129 thousand in 2032. What is Arizona's projected population increase over this time? (*Source:* Arizona Department of Administration)

70. In 1996, the centennial of the Boston Marathon, the official number of participants was 38,708. In 2013, there were 11,869 fewer participants. How many official participants were there for the 2013 Boston Marathon? (*Source:* Boston Athletic Association)

The decibel (dB) is a unit of measurement for sound. Every increase of 10 dB is a tenfold increase in sound intensity. The bar graph below shows the decibel levels for some common sounds. Use this graph for Exercises 71 through 74.

71. What is the dB rating for live rock music?

72. Which is the quietest of all the sounds shown in the graph?

73. How much louder is the sound of snoring than normal conversation?

74. What is the difference in sound intensity between live rock music and loud television?

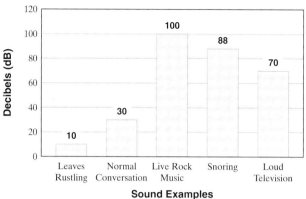

Decibel Levels for Common Sounds

75. The 113th Congress has 535 senators and representatives. Of these, 189 were registered Boy Scouts at some time in their lives. How many members of the 113th Congress were never Boy Scouts? (*Source: Boy Scouts of America*)

76. In the United States, there were 28,799 tornadoes from 1990 through 2012. In all, 13,205 of these tornadoes occurred from 1990 through 2000. How many tornadoes occurred during the period after 2000? (*Source:* Storm Prediction Center, National Weather Service)

77. Until recently, the world's largest permanent maze was located in Ruurlo, Netherlands. This maze of beech hedges covers 94,080 square feet. A new hedge maze using hibiscus bushes at the Dole Plantation in Wahiawa, Hawaii, covers 100,000 square feet. How much larger is the Dole Plantation maze than the Ruurlo maze? (*Source: The Guinness Book of Records*)

78. There were only 27 California condors in the entire world in 1987. To date, the number has increased to an estimated 223 living in the wild. How much of an increase is this? (*Source:* California Department of Fish and Wildlife)

The bar graph shows the top five U.S. airports according to number of passengers arriving and departing in 2013. Use this graph to answer Exercises 79 through 82.

79. Which airport is the busiest?

80. Which airports have 60 million passengers or fewer per year?

81. How many more passengers per year does the Chicago O'Hare International Airport have than the Denver International Airport?

82. How many more passengers per year does the Hartsfield-Jackson Atlanta International Airport have than the Dallas/Ft. Worth International Airport?

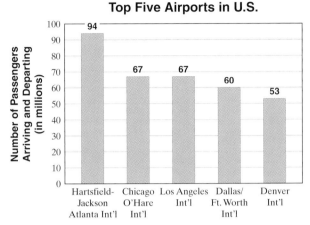

Top Five Airports in U.S.

Source: Airports Council International

Solve.

83. Two seniors, Jo Keen and Trudy Waterbury, were candidates for student government president. Who won the election if the votes were cast as follows? By how many votes did the winner win?

Class	Candidate	
	Jo	**Trudy**
Freshman	276	295
Sophomore	362	122
Junior	201	312
Senior	179	18

84. Two students submitted advertising budgets for a student government fund-raiser.

	Student A	**Student B**
Radio ads	$600	$300
Newspaper ads	$200	$400
Posters	$150	$240
Handbills	$120	$170

If $1200 is available for advertising, how much excess would each budget have?

Mixed Practice (*Sections 1.3 and 1.4*) *Add or subtract as indicated.*

85.
```
  986
+  48
```

86.
```
  986
−  48
```

87. 76 − 67

88. 80 + 93 + 17 + 9 + 2

89.
```
  9000
−  482
```

90.
```
  10,000
−  1786
```

91.
```
   10,962
    4851
+   7063
```

92.
```
   12,468
    3211
+   1988
```

Concept Extensions

For each exercise, identify which number is the minuend and which number is the subtrahend. See the Concept Check in this section.

93.
```
  48
−  1
```

94.
```
  2863
− 1904
```

95. Subtract 7 from 70.

96. Find 86 decreased by 25.

Identify each answer as correct or incorrect. Use addition to check. If the answer is incorrect, then write the correct answer.

97.
$$\begin{array}{r} 741 \\ -\ 56 \\ \hline 675 \end{array}$$

98.
$$\begin{array}{r} 478 \\ -\ 89 \\ \hline 389 \end{array}$$

99.
$$\begin{array}{r} 1029 \\ -\ 888 \\ \hline 141 \end{array}$$

100.
$$\begin{array}{r} 7615 \\ -\ 547 \\ \hline 7168 \end{array}$$

Fill in the missing digits in each problem.

101.
$$\begin{array}{r} 526\_ \\ -\ 2\_85 \\ \hline 28\_4 \end{array}$$

102.
$$\begin{array}{r} 10,\_4\_ \\ -\ 8\ 5\ \_4 \\ \hline \_710 \end{array}$$

103. Is there a commutative property of subtraction? In other words, does order matter when subtracting? Why or why not?

104. Explain why the phrase "Subtract 7 from 10" translates to "10 − 7."

105. The local college library is having a Million Pages of Reading promotion. The freshmen have read a total of 289,462 pages; the sophomores have read a total of 369,477 pages; the juniors have read a total of 218,287 pages; and the seniors have read a total of 121,685 pages. Have they reached a goal of one million pages? If not, how many more pages need to be read?

1.5 Rounding and Estimating

Objective A Rounding Whole Numbers

Rounding a whole number means approximating it. A rounded whole number is often easier to use, understand, and remember than the precise whole number. For example, instead of trying to remember the Minnesota state population as 5,197,621, it is much easier to remember it rounded to the nearest million: 5,000,000, or 5 million people.(*Source: World Almanac*)

Recall from Section 1.2 that the line below is called a number line. To **graph** a whole number on this number line, we darken the point representing the location of the whole number. For example, the number 4 is graphed below.

$$\begin{array}{ccccccccc} 0 & 1 & 2 & 3 & 4 & 5 & 6 & 7 \end{array}$$

On a number line, the whole number 36 is closer to 40 than 30, so 36 rounded to the nearest ten is 40.

The whole number 52 is closer to 50 than 60, so 52 rounded to the nearest ten is 50.

Objectives

A Round Whole Numbers.

B Use Rounding to Estimate Sums and Differences.

C Solve Problems by Estimating.

Minnesota Population: 5,197,621 or about 5 million

In trying to round 25 to the nearest ten, we see that 25 is halfway between 20 and 30. It is not closer to either number. In such a case, we round to the larger ten, that is, to 30.

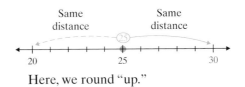

Here, we round "up."

To round a whole number without using a number line, follow these steps:

Rounding Whole Numbers to a Given Place Value

Step 1: Locate the digit to the right of the given place value.

Step 2: If this digit is 5 or greater, add 1 to the digit in the given place value and replace each digit to its right by 0.

Step 3: If this digit is less than 5, replace it and each digit to its right by 0.

Practice 1

Round to the nearest ten.

a. 57

b. 641

c. 325

Example 1   Round 568 to the nearest ten.

Solution: 5 6 ⑧   The digit to the right of the tens place is the ones
             ↑   place, which is circled.
      tens place

        5 6 ⑧   Since the circled digit is 5 or greater, add 1 to the 6 in
      ↑   ↖   the tens place and replace the digit to the right by 0.
   Add 1.   Replace
          with 0.

We find that 568 rounded to the nearest ten is 570.

▨ Work Practice 1

Practice 2

Round to the nearest thousand.

a. 72,304

b. 9222

c. 671,800

Example 2   Round 278,362 to the nearest thousand.

Solution:

The number 278,362 rounded to the nearest thousand is 278,000.

▨ Work Practice 2

Answers

1. a. 60   **b.** 640   **c.** 330

2. a. 72,000   **b.** 9000   **c.** 672,000

Example 3 Round 248,982 to the nearest hundred.

Solution:

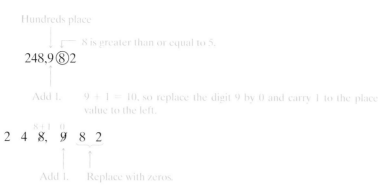

Hundreds place

8 is greater than or equal to 5.

248,9⑧2

Add 1.   9 + 1 = 10, so replace the digit 9 by 0 and carry 1 to the place value to the left.

8+1 0
2 4 8, 9 8 2

Add 1.   Replace with zeros.

The number 248,982 rounded to the nearest hundred is 249,000.

■ Work Practice 3

✓**Concept Check** Round each of the following numbers to the nearest *hundred*. Explain your reasoning.

a. 59   **b.** 29

Objective B Estimating Sums and Differences

By rounding addends, minuends, and subtrahends, we can estimate sums and differences. An estimated sum or difference is appropriate when the exact number is not necessary. Also, an estimated sum or difference can help us determine if we made a mistake in calculating an exact amount. To estimate the sum below, round each number to the nearest hundred and then add.

768	rounds to	800
1952	rounds to	2000
225	rounds to	200
+ 149	rounds to	+ 100
		3100

The estimated sum is 3100, which is close to the **exact** sum of 3094.

Example 4 Round each number to the nearest hundred to find an estimated sum.

```
   294
   625
  1071
 + 349
```

Solution:

Exact:		**Estimate:**
294	rounds to	300
625	rounds to	600
1071	rounds to	1100
+ 349	rounds to	+ 300
		2300

The estimated sum is 2300. (The exact sum is 2339.)

■ Work Practice 4

Practice 5

Round each number to the nearest thousand to find an estimated difference.

$$3785$$
$$-2479$$

Example 5 Round each number to the nearest hundred to find an estimated difference.

$$4725$$
$$-2879$$

Solution:

Exact:		**Estimate:**
4725	rounds to	4700
− 2879	rounds to	− 2900
		1800

The estimated difference is 1800. (The exact difference is 1846.)

▪ Work Practice 5

Objective C Solving Problems by Estimating

Making estimates is often the quickest way to solve real-life problems when solutions do not need to be exact.

Practice 6

Tasha Kilbey is trying to estimate how far it is from Gove, Kansas, to Hays, Kansas. Round each given distance on the map to the nearest ten to estimate the total distance.

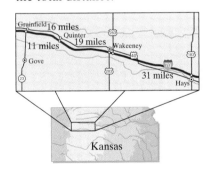

Example 6 Estimating Distances

A driver is trying to quickly estimate the distance from Temple, Texas, to Brenham, Texas. Round each distance given on the map to the nearest ten to estimate the total distance.

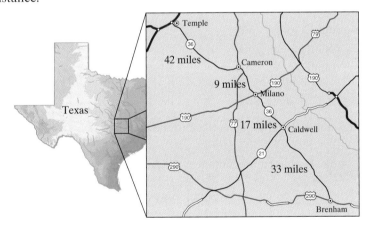

Solution:

Exact Distance:		**Estimate:**
42	rounds to	40
9	rounds to	10
17	rounds to	20
+ 33	rounds to	+ 30
		100

It is approximately 100 miles from Temple to Brenham. (The exact distance is 101 miles.)

▪ Work Practice 6

Answers

5. 2000   **6.** 80 mi

Example 7 Estimating Data

In three recent months, the numbers of tons of mail that went through Hartsfield-Jackson Atlanta International Airport were 635, 687, and 567. Round each number to the nearest hundred to estimate the tons of mail that passed through this airport.

Solution:

Exact Tons of Mail:		Estimate:
635	rounds to	600
687	rounds to	700
+567	rounds to	+600
		1900

The approximate tonnage of mail that moved through Atlanta's airport over this period was 1900 tons. (The exact tonnage was 1889 tons.)

▧ Work Practice 7

Practice 7

In 2012, Ecuador topped the International Union for Conservation of Nature's Red List of Threatened Species. At that time, Ecuador was home to 139 threatened bird and mammal species, 316 threatened other animal species, and 1842 threatened plant species. Round each number to the nearest hundred to estimate the total number of threatened species in Ecuador. (*Source: International Union for Conservation of Nature*)

Answer

7. 2200 total threatened species

Vocabulary, Readiness & Video Check

Use the choices below to fill in each blank.

60	rounding	exact
70	estimate	graph

1. To _____ a number on a number line, darken the point representing the location of the number.

2. Another word for approximating a whole number is _____.

3. The number 65 rounded to the nearest ten is _____, but the number 61 rounded to the nearest ten is _____.

4. A(n) _____ number of products is 1265, but a(n) _____ is 1000.

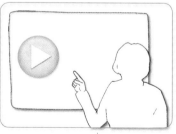

Martin-Gay Interactive Videos

See Video 1.5 🍎

Watch the section lecture video and answer the following questions.

Objective A 5. In ▤ Example 1, when rounding the number to the nearest ten, why do we replace the digit 3 with a 4? ◯

Objective B 6. As discussed in ▤ Example 3, explain how a number line can help us understand how to round 22 to the nearest ten. ◯

Objective C 7. What is the significance of the circled digit in each height value in ▤ Example 5? ◯

1.5 Exercise Set MyMathLab®

Objective A *Round each whole number to the given place. See Examples 1 through 3.*

1. 423 to the nearest ten

2. 273 to the nearest ten

3. 635 to the nearest ten

4. 846 to the nearest ten

5. 2791 to the nearest hundred

6. 8494 to the nearest hundred

7. 495 to the nearest ten

8. 898 to the nearest ten

9. 21,094 to the nearest thousand

10. 82,198 to the nearest thousand

11. 33,762 to the nearest thousand

12. 42,682 to the nearest ten-thousand

13. 328,495 to the nearest hundred

14. 179,406 to the nearest hundred

15. 36,499 to the nearest thousand

16. 96,501 to the nearest thousand

17. 39,994 to the nearest ten

18. 99,995 to the nearest ten

19. 29,834,235 to the nearest ten-million

20. 39,523,698 to the nearest million

Complete the table by estimating the given number to the given place value.

		Tens	Hundreds	Thousands
21.	5281			
22.	7619			
23.	9444			
24.	7777			
25.	14,876			
26.	85,049			

Round each number to the indicated place.

27. The University of California, Los Angeles, had a total undergraduate enrollment of 27,941 students in fall 2012. Round this number to the nearest thousand. (*Source:* UCLA)

28. In 2012, there were 12,997 Burger King restaurants worldwide. Round this number to the nearest thousand. (*Source:* Burger King Worldwide, Inc.)

29. Kareem Abdul-Jabbar holds the NBA record for points scored, a total of 38,387 over his NBA career. Round this number to the nearest thousand. (*Source:* National Basketball Association)

30. It takes 60,149 days for Neptune to make a complete orbit around the Sun. Round this number to the nearest hundred. (*Source:* National Space Science Data Center)

31. In 2013, the most valuable brand in the world was Apple, having just overtaken the longtime leader, Coca-Cola. The estimated brand value of Apple was $98,316,000,000. Round this to the nearest billion. (*Source:* Interbrand)

32. According to the 2013 Population Clock, the population of the United States was 316,797,189 in October 2013. Round this population figure to the nearest million. (*Source:* U.S. Census population clock)

33. The average salary for a Boston Red Sox baseball player during the 2013 season was $5,021,850. Round this average salary to the nearest thousand. (*Source:* CBSSports.com)

34. In FY 2013, the Procter & Gamble Company had $84,167,000,000 in sales. Round this sales figure to the nearest billion. (*Source:* Procter & Gamble)

35. In the United States in 2012, the travel industry generated $128,800,000,000 in tax revenue for local, state, and federal governments. Round this travel-related tax revenue figure to the nearest billion. (*Source:* U.S. Travel Association)

36. U.S. farms produced 3,149,166,000 bushels of soybeans in 2013. Round the soybean production figure to the nearest ten-million. (*Source:* U.S. Department of Agriculture)

Objective B *Estimate the sum or difference by rounding each number to the nearest ten. See Examples 4 and 5.*

37.
```
   39
   45
   22
 + 17
```

38.
```
   52
   33
   15
 + 29
```

39.
```
   449
 - 373
```

40.
```
   555
 - 235
```

Estimate the sum or difference by rounding each number to the nearest hundred. See Examples 4 and 5.

41.
```
   1913
   1886
 + 1925
```

42.
```
   4050
   3133
 + 1220
```

43.
```
   1774
 - 1492
```

44.
```
   1989
 - 1870
```

45.
```
   3995
   2549
 + 4944
```

46.
```
   799
   1655
 + 271
```

Three of the given calculator answers below are incorrect. Find them by estimating each sum.

47. 463 + 219   602

48. 522 + 785   1307

49. 229 + 443 + 606   1278

50. 542 + 789 + 198   2139

51. 7806 + 5150   12,956

52. 5233 + 4988   9011

Helpful Hint   Estimation is useful to check for incorrect answers when using a calculator. For example, pressing a key too hard may result in a double digit, while pressing a key too softly may result in the digit not appearing in the display.

Objective C *Solve each problem by estimating. See Examples 6 and 7.*

53. An appliance store advertises three refrigerators on sale at $899, $1499, and $999. Round each cost to the nearest hundred to estimate the total cost.

54. Suppose you scored 89, 97, 100, 79, 75, and 82 on your biology tests. Round each score to the nearest ten to estimate your total score.

55. The distance from Kansas City to Boston is 1429 miles and from Kansas City to Chicago is 530 miles. Round each distance to the nearest hundred to estimate how much farther Boston is from Kansas City than Chicago is.

56. The Gonzales family took a trip and traveled 588, 689, 277, 143, 59, and 802 miles on six consecutive days. Round each distance to the nearest hundred to estimate the distance they traveled.

57. The peak of Mt. McKinley, in Alaska, is 20,320 feet above sea level. The top of Mt. Rainier, in Washington, is 14,410 feet above sea level. Round each height to the nearest thousand to estimate the difference in elevation of these two peaks.(*Source:* U.S. Geological Survey)

58. A student is pricing new car stereo systems. One system sells for $1895 and another system sells for $1524. Round each price to the nearest hundred dollars to estimate the difference in price of these systems.

59. In 2012, the United States Postal Service delivered 159,859,000,000 pieces of mail. In 2011, it delivered 168,297,000,000 pieces of mail. Round each number to the nearest billion to estimate how much the mail volume decreased from 2011 to 2012. (*Source:* United States Postal Service)

60. Round each distance given on the map to the nearest ten to estimate the total distance from North Platte, Nebraska, to Lincoln, Nebraska.

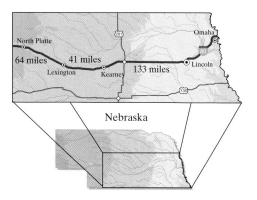

61. Head Start is a national program that provides developmental and social services for America's low-income preschool children ages three to five. Enrollment figures in Head Start programs showed an increase from 1,125,209 in 2011 to 1,128,030 in 2012. Round each number of children to the nearest thousand to estimate this increase. (*Source:* U.S. Department of Health and Human Services)

62. Enrollment figures at a local community college showed an increase from 49,713 credit hours in 2012 to 51,746 credit hours in 2013. Round each number to the nearest thousand to estimate the increase.

Mixed Practice (Sections 1.2 and 1.5) *The following table shows the top five leading U.S. advertisers in 2012 and the amount of money spent that year on advertising. Complete this table. The first line is completed for you. (Source: Ad Age DataCenter)*

	Advertiser	Amount Spent on Advertising in 2012 (in millions of dollars)	Amount Written in Standard Form	Standard Form Rounded to Nearest Ten-Million	Standard Form Rounded to Nearest Hundred-Million
	Procter & Gamble Co.	$4829	$4,829,000,000	$4,830,000,000	$4,800,000,000
63.	General Motors Co.	$3067			
64.	Comcast Corp.	$2989			
65.	AT&T	$2910			
66.	Verizon Communications	$2381			

Concept Extensions

67. Find one number that when rounded to the nearest hundred is 5700.

68. Find one number that when rounded to the nearest ten is 5700.

69. A number rounded to the nearest hundred is 8600.
   a. Determine the smallest possible number.
   b. Determine the largest possible number.

70. On August 23, 1989, it was estimated that 1,500,000 people joined hands in a human chain stretching 370 miles to protest the fiftieth anniversary of the pact that allowed what was then the Soviet Union to annex the Baltic nations in 1939. If the estimate of the number of people is to the nearest hundred-thousand, determine the largest possible number of people in the chain.

71. In your own words, explain how to round a number to the nearest thousand.

72. In your own words, explain how to round 9660 to the nearest thousand.

73. Estimate the perimeter of the rectangle by first rounding the length of each side to the nearest ten.

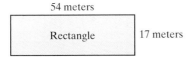

54 meters

Rectangle   17 meters

74. Estimate the perimeter of the triangle by first rounding the length of each side to the nearest hundred.

5950 miles   7693 miles

8203 miles

Objectives

A Use the Properties of Multiplication.

B Multiply Whole Numbers.

C Multiply by Whole Numbers Ending in Zero(s).

D Find the Area of a Rectangle.

E Solve Problems by Multiplying Whole Numbers.

Multiplication Shown as Repeated Addition Suppose that we wish to count the number of laptops provided in a computer class. The laptops are arranged in 5 rows, and each row has 6 laptops.

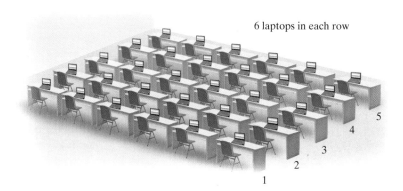

6 laptops in each row

Adding 5 sixes gives the total number of laptops. We can write this as $6 + 6 + 6 + 6 + 6 = 30$ laptops. When each addend is the same, we refer to this as **repeated addition.**

Multiplication is repeated addition but with different notation.

$$
\underbrace{6 + 6 + 6 + 6 + 6}_{\substack{5 \text{ addends; each} \\ \text{addend is } 6}} = \underbrace{5}_{\substack{(\text{number of} \\ \text{addends}) \text{ factor}}} \times \underbrace{6}_{\substack{(\text{each addend}) \\ \text{factor}}} = \underbrace{30}_{\text{product}}
$$

The $\times$ is called a **multiplication sign.** The numbers 5 and 6 are called **factors.** The number 30 is called the **product.** The notation 5×6 is read as "five times six." The symbols $\cdot$ and ( ) can also be used to indicate multiplication.

$$5 \times 6 = 30, \quad 5 \cdot 6 = 30, \quad (5)(6) = 30, \quad \text{and} \quad 5(6) = 30$$

✓ Concept Check

a. Rewrite $5 + 5 + 5 + 5 + 5 + 5 + 5$ using multiplication.

b. Rewrite 3×16 as repeated addition. Is there more than one way to do this? If so, show all ways.

Objective A Using the Properties of Multiplication

As with addition, we memorize products of one-digit whole numbers and then use certain properties of multiplication to multiply larger numbers. (If necessary, review the multiplication of one-digit numbers in Appendix A.2)

Notice that when any number is multiplied by 0, the result is always 0. This is called the **multiplication property of 0.**

✓ Concept Check Answers
a. $7 \times 5 = 35$
b. $16 + 16 + 16 = 48$; yes,
$3 + 3 + 3 + 3 + 3 + 3 + 3 + 3 + 3 + 3 + 3 + 3 + 3 + 3 + 3 + 3 = 48$

Multiplication Property of 0

The product of 0 and any number is 0. For example,

$5 \cdot 0 = 0$ and $0 \cdot 8 = 0$

Also notice in Appendix A.2 that when any number is multiplied by 1, the result is always the original number. We call this result the **multiplication property of 1.**

Multiplication Property of 1

The product of 1 and any number is that same number. For example,

$1 \cdot 9 = 9$ and $6 \cdot 1 = 6$

Example 1   Multiply.

a. 6×1 **b.** $0(18)$ **c.** $1 \cdot 45$ **d.** $(75)(0)$

Solution:

a. $6 \times 1 = 6$ **b.** $0(18) = 0$
c. $1 \cdot 45 = 45$ **d.** $(75)(0) = 0$

■ Work Practice 1

Practice 1

Multiply.
a. 3×0
b. $4(1)$
c. $(0)(34)$
d. $1 \cdot 76$

Like addition, multiplication is commutative and associative. Notice that when multiplying two numbers, the order of these numbers can be changed without changing the product. For example,

$3 \cdot 5 = 15$ and $5 \cdot 3 = 15$

This property is the **commutative property of multiplication.**

Commutative Property of Multiplication

Changing the **order** of two factors does not change their product. For example,

$9 \cdot 2 = 18$ and $2 \cdot 9 = 18$

Another property that can help us when multiplying is the **associative property of multiplication.** This property states that when multiplying numbers, the grouping of the numbers can be changed without changing the product. For example,

$(2 \cdot 3) \cdot 4 = 6 \cdot 4 = 24$

Also,

$2 \cdot (3 \cdot 4) = 2 \cdot 12 = 24$

Both groupings give a product of 24.

Answers
1. a. 0   **b.** 4   **c.** 0   **d.** 76

Associative Property of Multiplication

Changing the **grouping** of factors does not change their product. From the previous page, we know that for example,

$$(2 \cdot 3) \cdot 4 = 2 \cdot (3 \cdot 4)$$

With these properties, along with the **distributive property,** we can find the product of any whole numbers. The distributive property says that multiplication **distributes** over addition. For example, notice that $3(2 + 5)$ simplifies to the same number as $3 \cdot 2 + 3 \cdot 5$.

$$3(2 + 5) = 3(7) = 21$$

$$3 \cdot 2 + 3 \cdot 5 = 6 + 15 = 21$$

Since $3(2 + 5)$ and $3 \cdot 2 + 3 \cdot 5$ both simplify to 21, then

$$3(2 + 5) = 3 \cdot 2 + 3 \cdot 5$$

Notice in $3(2 + 5) = 3 \cdot 2 + 3 \cdot 5$ that each number inside the parentheses is multiplied by 3.

Distributive Property

Multiplication distributes over addition. For example,

$$2(3 + 4) = 2 \cdot 3 + 2 \cdot 4$$

Practice 2

Rewrite each using the distributive property.
a. $5(2 + 3)$
b. $9(8 + 7)$
c. $3(6 + 1)$

Example 2 Rewrite each using the distributive property.

a. $3(4 + 5)$ **b.** $10(6 + 8)$ **c.** $2(7 + 3)$

Solution: Using the distributive property, we have

a. $3(4 + 5) = 3 \cdot 4 + 3 \cdot 5$
b. $10(6 + 8) = 10 \cdot 6 + 10 \cdot 8$
c. $2(7 + 3) = 2 \cdot 7 + 2 \cdot 3$

■ Work Practice 2

Objective B Multiplying Whole Numbers

Let's use the distributive property to multiply $7(48)$. To do so, we begin by writing the expanded form of 48 (see Section 1.2) and then applying the distributive property.

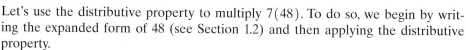

$$
\begin{aligned}
7(48) &= 7(40 + 8) && \text{Write 48 in expanded form.} \\
&= 7 \cdot 40 + 7 \cdot 8 && \text{Apply the distributive property.} \\
&= 280 + 56 && \text{Multiply.} \\
&= 336 && \text{Add.}
\end{aligned}
$$

Answers

2. a. $5(2 + 3) = 5 \cdot 2 + 5 \cdot 3$
b. $9(8 + 7) = 9 \cdot 8 + 9 \cdot 7$
c. $3(6 + 1) = 3 \cdot 6 + 3 \cdot 1$

This is how we multiply whole numbers. When multiplying whole numbers, we will use the following notation.

First:

$$\begin{array}{r} \overset{5}{48} \\ \times\ 7 \\ \hline 336 \end{array}$$ ← $7 \cdot 8 = 56$ Write 6 in the ones place and carry 5 to the tens place.

Next:

$$\begin{array}{r} \overset{5}{48} \\ \times\ 7 \\ \hline 336 \end{array}$$ $7 \cdot 4 + 5 = 28 + 5 = 33$

The product of 48 and 7 is 336.

Example 3 Multiply:

a. $\begin{array}{r} 25 \\ \times\ 8 \end{array}$

b. $\begin{array}{r} 246 \\ \times\ 5 \end{array}$

Solution:

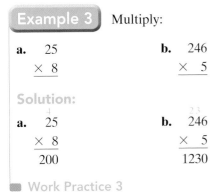

a. $\begin{array}{r} \overset{4}{25} \\ \times\ 8 \\ \hline 200 \end{array}$

b. $\begin{array}{r} \overset{2\,3}{246} \\ \times\ 5 \\ \hline 1230 \end{array}$

■ Work Practice 3

Practice 3

Multiply.

a. $\begin{array}{r} 36 \\ \times\ 4 \end{array}$

b. $\begin{array}{r} 132 \\ \times\ 9 \end{array}$

To multiply larger whole numbers, use the following similar notation. Multiply 89×52.

Step 1

$$\begin{array}{r} \overset{1}{89} \\ \times\ 52 \\ \hline 178 \end{array}$$ ← Multiply 89×2.

Step 2

$$\begin{array}{r} \overset{4}{89} \\ \times\ 52 \\ \hline 178 \\ 4450 \end{array}$$ ← Multiply 89×50.

Step 3

$$\begin{array}{r} 89 \\ \times\ 52 \\ \hline 178 \\ 4450 \\ \hline 4628 \end{array}$$ ← Add.

The numbers 178 and 4450 are called **partial products.** The sum of the partial products, 4628, is the product of 89 and 52.

Example 4 Multiply: 236×86

Solution: $\begin{array}{r} 236 \\ \times\ 86 \\ \hline 1416 \\ 18880 \\ \hline 20,296 \end{array}$
  ← 6(236)
  ← 80(236)
  Add.

■ Work Practice 4

Practice 4

Multiply.

a. $\begin{array}{r} 594 \\ \times\ 72 \end{array}$

b. $\begin{array}{r} 306 \\ \times\ 81 \end{array}$

Example 5 Multiply: 631×125

Solution: $\begin{array}{r} 631 \\ \times\ 125 \\ \hline 3155 \\ 12620 \\ 63100 \\ \hline 78,875 \end{array}$
  ← 5(631)
  ← 20(631)
  ← 100(631)
  Add.

■ Work Practice 5

Practice 5

Multiply.

a. $\begin{array}{r} 726 \\ \times\ 142 \end{array}$

b. $\begin{array}{r} 288 \\ \times\ 4 \end{array}$

Answers
3. a. 144 **b.** 1188
4. a. 42,768 **b.** 24,786
5. a. 103,092 **b.** 1152

✔**Concept Check** Find and explain the error in the following multiplication problem.

$$\begin{array}{r} 102 \\ \times\ 33 \\ \hline 306 \\ \underline{306} \\ 612 \end{array}$$

Objective C Multiplying by Whole Numbers Ending in Zero(s) ▶

Interesting patterns occur when we multiply by a number that ends in zeros. To see these patterns, let's multiply a number, say 34, by 10, then 100, then 1000.

1 zero
↓
$34 \cdot 10 = 340$ 1 zero attached to 34.

2 zeros
$34 \cdot 100 = 3400$ 2 zeros attached to 34.

3 zeros
$34 \cdot 1000 = 34,000$ 3 zeros attached to 34.

These patterns help us develop a shortcut for multiplying by whole numbers ending in zeros.

To multiply by 10, 100, 1000, and so on,
    Form the product by attaching the number of zeros in that number to the other factor.
        For example, $41 \cdot 100 = 4100$.
        2 zeros ⎯⎯⎯↑

Examples Multiply.

6. $176 \cdot 1000 = 176,000$ Attach 3 zeros.

7. $2041 \cdot 100 = 204,100$ Attach 2 zeros.

■ Work Practice 6–7

We can use a similar format to multiply by any whole number ending in zeros. For example, since

$$15 \cdot 500 = 15 \cdot 5 \cdot 100,$$

we find the product by multiplying 15 and 5, then attaching two zeros to the product.

$$\begin{array}{r} \overset{2}{1}5 \\ \times\ 5 \\ \hline 75 \end{array} \quad 15 \cdot 500 = 7500$$

Practice 6–7

Multiply.
6. $75 \cdot 100$
7. $808 \cdot 1000$

Answers
6. 7500 **7.** 808,000

✔**Concept Check Answer**
$$\begin{array}{r} 102 \\ \times\ 33 \\ \hline 306 \\ \underline{3060} \\ 3366 \end{array}$$

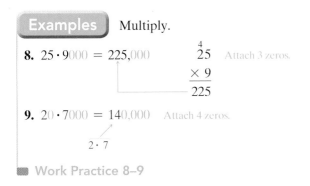

Examples Multiply.

8. $25 \cdot 9000 = 225,000$ Attach 3 zeros.

$$\begin{array}{r} \overset{4}{25} \\ \times\ 9 \\ \hline 225 \end{array}$$

9. $20 \cdot 7000 = 140,000$ Attach 4 zeros.

$2 \cdot 7$

■ Work Practice 8–9

Practice 8–9

Multiply.
8. $35 \cdot 3000$
9. $600 \cdot 600$

Objective D Finding the Area of a Rectangle

A special application of multiplication is finding the **area** of a region. Area measures the amount of surface of a region. For example, we measure a plot of land or the living space of a home by its area. The figures below show two examples of units of area measure. (A centimeter is a unit of length in the metric system.)

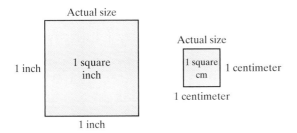

For example, to measure the area of a geometric figure such as the rectangle below, count the number of square units that cover the region.

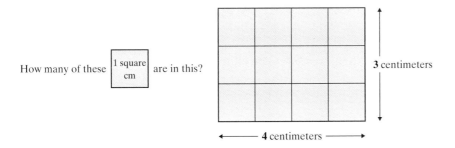

This rectangular region contains 12 square units, each 1 square centimeter. Thus, the area is 12 square centimeters. This total number of squares can be found by counting or by multiplying **4 · 3** (length · width).

$$\begin{aligned} \text{Area of a rectangle} &= \text{length} \cdot \text{width} \\ &= (4 \text{ centimeters})(3 \text{ centimeters}) \\ &= 12 \text{ square centimeters} \end{aligned}$$

In this section, we find the areas of rectangles only. In later sections, we will find the areas of other geometric regions.

Helpful Hint

Notice that area is measured in **square** units while perimeter is measured in units.

Answers
8. 105,000   **9.** 360,000

Practice 10

The state of Wyoming is in the shape of a rectangle whose length is 360 miles and whose width is 280 miles. Find its area.

Example 10 Finding the Area of a State

The state of Colorado is in the shape of a rectangle whose length is 380 miles and whose width is 280 miles. Find its area.

Solution: The area of a rectangle is the product of its length and its width.

Area = length · width

= (380 miles)(280 miles)

= 106,400 square miles

The area of Colorado is 106,400 square miles.

🔲 Work Practice 10

Objective E Solving Problems by Multiplying

There are several words or phrases that indicate the operation of multiplication. Some of these are as follows:

Multiplication		
Key Words or Phrases	**Examples**	**Symbols**
multiply	multiply 5 by 7	$5 \cdot 7$
product	the product of 3 and 2	$3 \cdot 2$
times	10 times 13	$10 \cdot 13$

Many key words or phrases describing real-life problems that suggest addition might be better solved by multiplication instead. For example, to find the **total** cost of 8 shirts, each selling for $27, we can either add

$$27 + 27 + 27 + 27 + 27 + 27 + 27 + 27$$

or we can multiply 8(27).

Practice 11

A particular computer printer can print 16 pages per minute in color. How many pages can it print in 45 minutes?

Example 11 Finding DVD Space

A digital video disc (DVD) can hold about 4800 megabytes (MB) of information. How many megabytes can 12 DVDs hold?

Solution: Twelve DVDs will hold 12 × 4800 megabytes.

In Words		Translate to Numbers
megabytes per disc	→	4800
× DVDs	→	× 12
		9600
		48000
total megabytes		57,600

Twelve DVDs will hold 57,600 megabytes.

🔲 Work Practice 11

Answers

10. 100,800 sq mi   **11.** 720 pages

Example 12 Budgeting Money

Suzanne Scarpulla and a friend plan to take their children to the Georgia Aquarium in Atlanta, the world's largest aquarium. The peak hour ticket price for each child is $29 and for each adult, $35. If five children and two adults plan to go, how much money is needed for admission? (*Source:* GeorgiaAquarium.org)

Solution: If the price of one child's ticket is $29, the cost for 5 children is $5 \times 29 = \$145$. The price of one adult ticket is $35, so the cost for two adults is $2 \times 35 = \$70$. The total cost is:

In Words		Translate to Numbers
cost for 5 children	$\longrightarrow$	145
+ cost for 2 adults	$\longrightarrow$	+ 70
total cost		215

The total cost is $215.

■ Work Practice 12

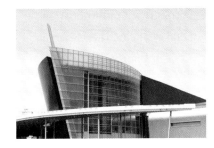

Practice 12

Ken Shimura purchased DVDs and CDs through a club. Each DVD was priced at $11, and each CD cost $9. Ken bought eight DVDs and five CDs. Find the total cost of the order.

Example 13 Estimating Word Count

The average page of a book contains 259 words. Estimate, rounding each number to the nearest hundred, the total number of words contained on 212 pages.

Solution: The exact number of words is 259×212. Estimate this product by rounding each factor to the nearest hundred.

259 rounds to 300
$\times 212$ rounds to $\times 200,$

$300 \times 200 = 60,000$

$3 \cdot 2 = 6$

There are approximately 60,000 words contained on 212 pages.

■ Work Practice 13

Practice 13

If an average page in a book contains 163 words, estimate, rounding each number to the nearest hundred, the total number of words contained on 391 pages.

Calculator Explorations Multiplying Numbers

To multiply numbers on a calculator, find the keys marked $\boxed{\times}$ and $\boxed{=}$ or $\boxed{\text{ENTER}}$. For example, to find $31 \cdot 66$ on a calculator, press the keys $\boxed{31}$ $\boxed{\times}$ $\boxed{66}$ then $\boxed{=}$ or $\boxed{\text{ENTER}}$. The display will read $\boxed{2046}$. Thus, $31 \cdot 66 = 2046$.

Use a calculator to multiply.

1. 72×48 **2.** 81×92

3. $163 \cdot 94$ **4.** $285 \cdot 144$

5. $983(277)$ **6.** $1562(843)$

Answers
12. $133 **13.** 80,000 words

Vocabulary, Readiness & Video Check

Use the choices below to fill in each blank.

area	grouping	commutative	1	product	length
factor	order	associative	0	distributive	number

1. The product of 0 and any number is _____.

2. The product of 1 and any number is the _____.

3. In $8 \cdot 12 = 96$, the 96 is called the _____ and 8 and 12 are each called a(n) _____.

4. Since $9 \cdot 10 = 10 \cdot 9$, we say that changing the _____ in multiplication does not change the product. This property is called the _____ property of multiplication.

5. Since $(3 \cdot 4) \cdot 6 = 3 \cdot (4 \cdot 6)$, we say that changing the _____ in multiplication does not change the product. This property is called the _____ property of multiplication.

6. _____ measures the amount of surface of a region.

7. Area of a rectangle = _____ $\cdot$ width.

8. We know $9(10 + 8) = 9 \cdot 10 + 9 \cdot 8$ by the _____ property.

Martin-Gay Interactive Videos   Watch the section lecture video and answer the following questions.

See Video 1.6

Objective A 9. The expression in ▣ Example 3 is rewritten using what property?

Objective B 10. During the multiplication process for ▣ Example 5, why is a single zero placed at the end of the second partial product?

Objective C 11. Explain two different approaches to solving the multiplication problem $50 \cdot 900$ in ▣ Example 7.

Objective D 12. Why are the units to the answer to ▣ Example 8 not just meters? What are the correct units?

Objective E 13. In ▣ Example 9, why can "total" imply multiplication as well as addition?

1.6 Exercise Set MyMathLab

Objective A *Multiply. See Example 1.*

1. $1 \cdot 24$ **2.** $55 \cdot 1$ **3.** $0 \cdot 19$ **4.** $27 \cdot 0$

5. $8 \cdot 0 \cdot 9$ **6.** $7 \cdot 6 \cdot 0$ **7.** $87 \cdot 1$ **8.** $1 \cdot 41$

Use the distributive property to rewrite each expression. See Example 2.

9. $6(3 + 8)$ **10.** $5(8 + 2)$ **11.** $4(3 + 9)$

12. $6(1 + 4)$ **13.** $20(14 + 6)$ **14.** $12(12 + 3)$

Objective B *Multiply. See Example 3.*

15. 64
× 8

16. 79
× 3

17. 613
× 6

18. 638
× 5

19. 277 × 6

20. 882 × 2

21. 1074 × 6

22. 9021 × 3

Objectives A B Mixed Practice *Multiply. See Examples 1 through 5.*

23. 89
× 13

24. 91
× 72

25. 421
× 58

26. 526
× 23

27. 306
× 81

28. 708
× 21

29. (780)(20)

30. (720)(80)

31. (495)(13)(0)

32. (593)(47)(0)

33. (640)(1)(10)

34. (240)(1)(20)

35. 1234 × 39

36. 1357 × 79

37. 609 × 234

38. 807 × 127

39. 8649
× 274

40. 1234
× 567

41. 589
× 110

42. 426
× 110

43. 1941
× 2035

44. 1876
× 1407

Objective C *Multiply. See Examples 6 through 9.*

45. 8 × 100

46. 6 × 100

47. 11 × 1000

48. 26 × 1000

49. 7406 · 10

50. 9054 · 10

51. 6 · 4000

52. 3 · 9000

53. 50 · 900

54. 70 · 300

55. 41 · 80,000

56. 27 · 50,000

Objective D Mixed Practice (*Section 1.3*) *Find the area and the perimeter of each rectangle. See Example 10.*

57.
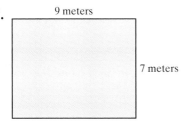
9 meters

7 meters

58. 3 inches
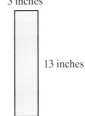
13 inches

59. 17 feet

40 feet

60. 25 centimeters
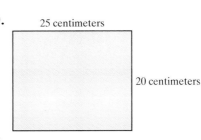
20 centimeters

Objective E Mixed Practice *(Section 1.5)* *Estimate the products by rounding each factor to the nearest hundred. See Example 13.*

61. 576×354 **62.** 982×650 **63.** 604×451 **64.** 111×999

Without actually calculating, mentally round, multiply, and choose the best estimate.

65. $38 \times 42 =$
  a. 16
  b. 160
  c. 1600
  d. 16,000

66. $2872 \times 12 =$
  a. 2872
  b. 28,720
  c. 287,200
  d. 2,872,000

67. $612 \times 29 =$
  a. 180
  b. 1800
  c. 18,000
  d. 180,000

68. $706 \times 409 =$
  a. 280
  b. 2800
  c. 28,000
  d. 280,000

Objectives D E **Mixed Practice–Translating** *Solve. See Examples 10 through 13.*

69. Multiply 80 by 11.

70. Multiply 70 by 12.

71. Find the product of 6 and 700.

72. Find the product of 9 and 900.

73. Find 2 times 2240.

74. Find 3 times 3310.

75. One tablespoon of olive oil contains 125 calories. How many calories are in 3 tablespoons of olive oil? (*Source: Home and Garden Bulletin No. 72,* U.S. Department of Agriculture)

76. One ounce of hulled sunflower seeds contains 14 grams of fat. How many grams of fat are in 8 ounces of hulled sunflower seeds? (*Source: Home and Garden Bulletin No. 72,* U.S. Department of Agriculture)

77. The textbook for a course in biology costs $94. There are 35 students in the class. Find the total cost of the biology books for the class.

78. The seats in a lecture hall are arranged in 14 rows with 34 seats in each row. Find how many seats are in this room.

79. Cabot Creamery is packing a pallet of 20-lb boxes of cheddar cheese to send to a local restaurant. There are five layers of boxes on the pallet, and each layer is four boxes wide by five boxes deep.
  a. How many boxes are in one layer?
  b. How many boxes are on the pallet?
  c. What is the weight of the cheese on the pallet?

80. An apartment building has *three floors*. Each floor has five rows of apartments with four apartments in each row.
  a. How many apartments are on 1 floor?
  b. How many apartments are in the building?

△ **81.** A plot of land measures 80 feet by 110 feet. Find its area.

△ **82.** A house measures 45 feet by 60 feet. Find the floor area of the house.

△ **83.** The largest hotel lobby can be found at the Hyatt Regency in San Francisco, CA. It is in the shape of a rectangle that measures 350 feet by 160 feet. Find its area.

△ **84.** Recall from an earlier section that the world's largest commercial building under one roof is the flower auction building of the cooperative VBA in Aalsmeer, Netherlands. The floor plan is a rectangle that measures 776 meters by 639 meters. Find the area of this building. (*Source: The Handy Science Answer Book,* Visible Ink Press)

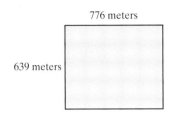

776 meters

639 meters

85. A pixel is a rectangular dot on a graphing calculator screen. If a graphing calculator screen contains 62 pixels in a row and 94 pixels in a column, find the total number of pixels on a screen.

86. A certain compact disc (CD) can hold 700 megabytes (MB) of information. How many MB can 17 discs hold?

87. A line of print on a computer contains 60 characters (letters, spaces, punctuation marks). Find how many characters there are in 35 lines.

88. An average cow eats 3 pounds of grain per day. Find how much grain a cow eats in a year. (Assume 365 days in 1 year.)

89. One ounce of Planters® Dry Roasted Peanuts has 170 calories. How many calories are in 8 ounces? (*Source:* Kraft Foods)

90. One ounce of Planters® Dry Roasted Peanuts has 14 grams of fat. How many grams of fat are in 16 ounces? (*Source:* Kraft Foods)

91. The Thespian club at a local community college is ordering T-shirts. T-shirts size S, M, or L cost $10 each and T-shirts size XL or XXL cost $12 each. Use the table below to find the total cost. (The first row is filled in for you.)

T-Shirt Size	Number of Shirts Ordered	Cost per Shirt	Cost per Size Ordered
S	4	$10	$40
M	6		
L	20		
XL	3		
XXL	3		

92. The student activities group at North Shore Community College is planning a trip to see the local minor league baseball team. Tickets cost $5 for students, $7 for nonstudents, and $2 for children under 12. Use the following table to find the total cost.

Person	Number of Persons	Cost per Person	Cost per Category
Student	24	$5	$120
Nonstudent	4		
Children under 12	5		

93. Celestial Seasonings of Boulder, Colorado, is a tea company that specializes in herbal teas, accounting for over $100,000,000 in herbal tea blend sales in the United States annually. Their plant in Boulder has bagging machines capable of bagging over 1000 bags of tea per minute. If the plant runs 24 hours a day, how many tea bags are produced in one day? (*Source:* Celestial Seasonings)

94. The number of "older" Americans (ages 65 and older) has increased fourteenfold since 1900. If there were 3 million "older" Americans in 1900, how many were there in 2012? (*Source:* U.S. Census Bureau)

Mixed Practice (*Sections 1.3, 1.4, 1.6*) *Perform each indicated operation.*

95.
$$\begin{array}{r} 128 \\ + 7 \\ \hline \end{array}$$

96.
$$\begin{array}{r} 126 \\ - 8 \\ \hline \end{array}$$

97.
$$\begin{array}{r} 134 \\ \times 16 \\ \hline \end{array}$$

98. $47 + 26 + 10 + 231 + 50$

99. Find the sum of 19 and 4.

100. Find the product of 19 and 4.

101. Find the difference of 19 and 4.

102. Find the total of 19 and 4.

Concept Extensions

Solve. See the first Concept Check in this section.

103. Rewrite $7 + 7 + 7 + 7$ using multiplication.

104. Rewrite $11 + 11 + 11 + 11 + 11 + 11$ using multiplication.

105. a. Rewrite $3 \cdot 5$ as repeated addition.
b. Explain why there is more than one way to do this.

106. a. Rewrite $4 \cdot 5$ as repeated addition.
b. Explain why there is more than one way to do this.

Find and explain the error in each multiplication problem. See the second Concept Check in this section.

107.
$$\begin{array}{r} 203 \\ \times 14 \\ \hline 812 \\ 203 \\ \hline 1015 \end{array}$$

108.
$$\begin{array}{r} 31 \\ \times 50 \\ \hline 155 \end{array}$$

Fill in the missing digits in each problem.

109.
$$\begin{array}{r} 4\_ \\ \times \_3 \\ \hline 126 \\ 3780 \\ \hline 3906 \end{array}$$

110.
$$\begin{array}{r} \_7 \\ \times 6\_ \\ \hline 171 \\ 3420 \\ \hline 3591 \end{array}$$

111. Explain how to multiply two 2-digit numbers using partial products.

112. In your own words, explain the meaning of the area of a rectangle and how this area is measured.

113. A window washer in New York City is bidding for a contract to wash the windows of a 23-story building. To write a bid, the number of windows in the building is needed. If there are 7 windows in each row of windows on 2 sides of the building and 4 windows per row on the other 2 sides of the building, find the total number of windows.

114. During the 2012–2013 regular season, Kevin Durant of the Oklahoma City Thunder led the NBA in total points scored. He scored 139 three-point field goals, 592 two-point field goals, and 679 free throws (worth one point each). How many points did Kevin Durant score during the 2012–2013 regular season? (*Source:* NBA)

1.7 Dividing Whole Numbers

Suppose three people pooled their money and bought a raffle ticket at a local fund-raiser. Their ticket was the winner, and they won a $75 cash prize. They then divided the prize into three equal parts so that each person received $25.

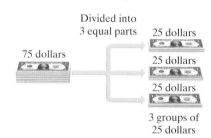

Objectives

A Divide Whole Numbers.

B Perform Long Division.

C Solve Problems That Require Dividing by Whole Numbers.

D Find the Average of a List of Numbers.

Objective A Dividing Whole Numbers

The process of separating a quantity into equal parts is called **division.** The division above can be symbolized by several notations.

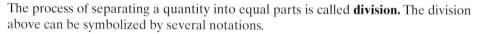

$$\begin{array}{c} \text{quotient} \\ \downarrow \\ 25 \\ 3\overline{)75} \end{array} \leftarrow \text{dividend}$$

divisor

$$\frac{75}{3} = 25 \leftarrow \text{quotient}$$

dividend

divisor

$$75 \div 3 = 25$$

dividend   divisor

quotient

$$75/3 = 25$$

dividend   quotient

divisor

(In the notation $\frac{75}{3}$, the bar separating 75 and 3 is called a **fraction bar.**) Just as subtraction is the reverse of addition, division is the reverse of multiplication. This means that division can be checked by multiplication.

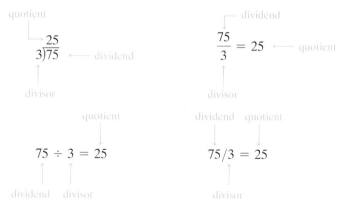

$$\begin{array}{c} 25 \\ 3\overline{)75} \end{array} \quad \text{because} \quad 25 \cdot 3 = 75$$

$$\boxed{\text{Quotient}} \cdot \boxed{\text{Divisor}} = \boxed{\text{Dividend}}$$

Since multiplication and division are related in this way, you can use your knowledge of multiplication facts (or study Appendix A.2) to review quotients of one-digit divisors if necessary.

Practice 1

Find each quotient. Check by multiplying.

a. $9\overline{)72}$

b. $40 \div 5$

c. $\dfrac{24}{6}$

Example 1 Find each quotient. Check by multiplying.

a. $42 \div 7$ **b.** $\dfrac{64}{8}$ **c.** $3\overline{)21}$

Solution:

a. $42 \div 7 = 6$ because $6 \cdot 7 = 42$

b. $\dfrac{64}{8} = 8$ because $8 \cdot 8 = 64$

c. $3\overline{)21}^{\,7}$ because $7 \cdot 3 = 21$

Work Practice 1

Practice 2

Find each quotient. Check by multiplying.

a. $\dfrac{7}{7}$ **b.** $5 \div 1$

c. $1\overline{)11}$ **d.** $4 \div 1$

e. $\dfrac{10}{1}$ **f.** $21 \div 21$

Example 2 Find each quotient. Check by multiplying.

a. $1\overline{)7}$ **b.** $12 \div 1$ **c.** $\dfrac{6}{6}$ **d.** $9 \div 9$ **e.** $\dfrac{20}{1}$ **f.** $18\overline{)18}$

Solution:

a. $1\overline{)7}^{\,7}$ because $7 \cdot 1 = 7$ **b.** $12 \div 1 = 12$ because $12 \cdot 1 = 12$

c. $\dfrac{6}{6} = 1$ because $1 \cdot 6 = 6$ **d.** $9 \div 9 = 1$ because $1 \cdot 9 = 9$

e. $\dfrac{20}{1} = 20$ because $20 \cdot 1 = 20$ **f.** $18\overline{)18}^{\,1}$ because $1 \cdot 18 = 18$

Work Practice 2

Example 2 illustrates the important properties of division described next:

Division Properties of 1

The quotient of any number (except 0) and that same number is 1. For example,

$$8 \div 8 = 1 \qquad \dfrac{5}{5} = 1 \qquad 4\overline{)4}^{\,1}$$

The quotient of any number and 1 is that same number. For example,

$$9 \div 1 = 9 \qquad \dfrac{6}{1} = 6 \qquad 1\overline{)3}^{\,3} \qquad \dfrac{0}{1} = 0$$

Practice 3

Find each quotient. Check by multiplying.

a. $\dfrac{0}{7}$ **b.** $8\overline{)0}$

c. $7 \div 0$ **d.** $0 \div 14$

Example 3 Find each quotient. Check by multiplying.

a. $9\overline{)0}$ **b.** $0 \div 12$ **c.** $\dfrac{0}{5}$ **d.** $\dfrac{3}{0}$

Solution:

a. $9\overline{)0}^{\,0}$ because $0 \cdot 9 = 0$ **b.** $0 \div 12 = 0$ because $0 \cdot 12 = 0$

c. $\dfrac{0}{5} = 0$ because $0 \cdot 5 = 0$

Answers

1. a. 8 **b.** 8 **c.** 4 **2. a.** 1 **b.** 5
c. 11 **d.** 4 **e.** 10 **f.** 1 **3. a.** 0
b. 0 **c.** undefined **d.** 0

d. If $\dfrac{3}{0}$ = a *number,* then the *number* times 0 = 3. Recall from Section 1.6 that any number multiplied by 0 is 0 and not 3. We say, then, that $\dfrac{3}{0}$ is **undefined.**

■ Work Practice 3

Example 3 illustrates important division properties of 0.

Division Properties of 0

The quotient of 0 and any number (except 0) is 0. For example,

$$0 \div 9 = 0 \qquad \frac{0}{5} = 0 \qquad 14\overline{)0}^{\,0}$$

The quotient of any number and 0 is not a number. We say that

$$\frac{3}{0}, \quad 0\overline{)3}, \quad \text{and} \quad 3 \div 0$$

are **undefined.**

Objective B Performing Long Division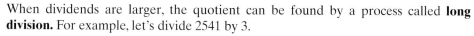

When dividends are larger, the quotient can be found by a process called **long division.** For example, let's divide 2541 by 3.

$$\text{divisor} \longrightarrow 3\overline{)2541}$$
$$\uparrow$$
$$\text{dividend}$$

We can't divide 3 into 2, so we try dividing 3 into the first two digits.

$$\begin{array}{r} 8 \\ 3\overline{)2541} \end{array} \qquad 25 \div 3 = 8 \text{ with 1 left, so our best estimate is 8. We place 8 over the 5 in 25.}$$

Next, multiply 8 and 3 and subtract this product from 25. Make sure that this difference is less than the divisor.

$$\begin{array}{r} 8 \\ 3\overline{)2541} \\ -24 \\ \hline 1 \end{array} \qquad \begin{array}{l} 8(3) = 24 \\ 25 - 24 = 1, \text{ and 1 is less than the divisor 3.} \end{array}$$

Bring down the next digit and go through the process again.

$$\begin{array}{r} 84 \\ 3\overline{)2541} \\ -24\downarrow \\ \hline 14 \\ -12 \\ \hline 2 \end{array} \qquad \begin{array}{l} 14 \div 3 = 4 \text{ with 2 left} \\ \\ \\ 4(3) = 12 \\ 14 - 12 = 2 \end{array}$$

Once more, bring down the next digit and go through the process.

$$\begin{array}{r} 847 \\ 3\overline{)2541} \\ -24 \\ \hline 14 \\ -12\downarrow \\ \hline 21 \\ -21 \\ \hline 0 \end{array} \qquad \begin{array}{l} 21 \div 3 = 7 \\ \\ \\ \\ 7(3) = 21 \\ 21 - 21 = 0 \end{array}$$

The quotient is 847. To check, see that 847 × 3 = 2541.

Practice 4

Divide. Check by multiplying.

a. $4908 \div 6$

b. $2212 \div 4$

c. $753 \div 3$

Example 4 Divide: $3705 \div 5$. Check by multiplying.

Solution:

$$\begin{array}{r} 7 \\ 5)\overline{3705} \\ -35\!\downarrow \\ \hline 20 \end{array}$$
$37 \div 5 = 7$ with 2 left. Place this estimate, 7, over the 7 in 37.
$7(5) = 35$
$37 - 35 = 2$, and 2 is less than the divisor 5.
Bring down the 0.

$$\begin{array}{r} 74 \\ 5)\overline{3705} \\ -35 \\ \hline 20 \\ -20\!\downarrow \\ \hline 05 \end{array}$$
$20 \div 5 = 4$
$4(5) = 20$
$20 - 20 = 0$, and 0 is less than the divisor 5.
Bring down the 5.

$$\begin{array}{r} 741 \\ 5)\overline{3705} \\ -35 \\ \hline 20 \\ -20\!\downarrow \\ \hline 5 \\ -5 \\ \hline 0 \end{array}$$
$5 \div 5 = 1$
$1(5) = 5$
$5 - 5 = 0$

Check:

$$\begin{array}{r} 741 \\ \times \quad 5 \\ \hline 3705 \end{array}$$

■ Work Practice 4

Practice 5

Divide and check by multiplying.

a. $7)\overline{2128}$

b. $9)\overline{45,900}$

Example 5 Divide and check: $1872 \div 9$

Solution:

$$\begin{array}{r} 208 \\ 9)\overline{1872} \\ -18\!\downarrow\downarrow \\ \hline 07 \\ -0\!\downarrow \\ \hline 72 \\ -72 \\ \hline 0 \end{array}$$
$2(9) = 18$
$18 - 18 = 0$; bring down the 7.
$0(9) = 0$
$7 - 0 = 7$; bring down the 2.
$8(9) = 72$
$72 - 72 = 0$

Check: $208 \cdot 9 = 1872$

■ Work Practice 5

Answers

4. a. 818 **b.** 553 **c.** 251

5. a. 304 **b.** 5100

Naturally, quotients don't always "come out even." Making 4 rows out of 26 chairs, for example, isn't possible if each row is supposed to have exactly the same number of chairs. Each of 4 rows can have 6 chairs, but 2 chairs are still left over.

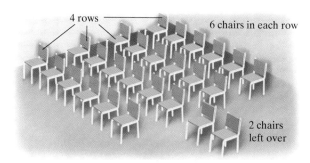

4 rows — 6 chairs in each row

2 chairs left over

We signify "leftovers" or **remainders** in this way:

$$\begin{array}{r} 6 \ \text{R } 2 \\ 4\overline{)26} \end{array}$$

The **whole number part of the quotient** is 6; the **remainder part of the quotient** is 2. Checking by multiplying,

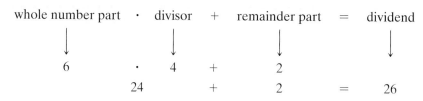

whole number part · divisor + remainder part = dividend

    6   ·   4   +   2
          24   +   2   =   26

Example 6   Divide and check: $2557 \div 7$

Solution:

$$\begin{array}{r} 365 \ \text{R } 2 \\ 7\overline{)2557} \\ -21 \\ \hline 45 \\ -42 \\ \hline 37 \\ -35 \\ \hline 2 \end{array}$$

$3(7) = 21$
$25 - 21 = 4$; bring down the 5.
$6(7) = 42$
$45 - 42 = 3$; bring down the 7.
$5(7) = 35$
$37 - 35 = 2$; the remainder is 2.

Check: 365 · 7 + 2 = 2557

whole number part · divisor + remainder part = dividend

■ Work Practice 6

Practice 6

Divide and check.

a. $4\overline{)939}$

b. $5\overline{)3287}$

Answers

6. a. 234 R 3   **b.** 657 R 2

Practice 7

Divide and check.

a. $9\overline{)81,605}$

b. $4\overline{)23,310}$

Example 7 Divide and check: $56,717 \div 8$

Solution:

$$
\begin{array}{r}
7089 \ \text{R } 5 \\
8\overline{)56717} \\
\underline{-56}\downarrow \\
07 \\
\underline{-0}\downarrow \\
71 \\
\underline{-64}\downarrow \\
77 \\
\underline{-72} \\
5
\end{array}
$$

$7(8) = 56$

Subtract and bring down the 7.

$0(8) = 0$

Subtract and bring down the 1.

$8(8) = 64$

Subtract and bring down the 7.

$9(8) = 72$

Subtract. The remainder is 5.

Check: $7089 \quad \cdot \quad 8 \quad + \quad 5 \quad = \quad 56,717$

whole number part $\cdot$ divisor $+$ remainder part $=$ dividend

■ Work Practice 7

When the divisor has more than one digit, the same pattern applies. For example, let's find $1358 \div 23$.

$$
\begin{array}{r}
5 \\
23\overline{)1358} \\
\underline{-115}\downarrow \\
208
\end{array}
$$

$135 \div 23 = 5$ with 20 left over. Our estimate is 5.

$5(23) = 115$

$135 - 115 = 20$. Bring down the 8.

Now we continue estimating.

$$
\begin{array}{r}
59 \ \text{R } 1 \\
23\overline{)1358} \\
\underline{-115} \\
208 \\
\underline{-207} \\
1
\end{array}
$$

$208 \div 23 = 9$ with 1 left over.

$9(23) = 207$

$208 - 207 = 1$. The remainder is 1.

To check, see that $59 \cdot 23 + 1 = 1358$.

Practice 8

Divide: $8920 \div 17$

Example 8 Divide: $6819 \div 17$

Solution:

$$
\begin{array}{r}
401 \ \text{R } 2 \\
17\overline{)6819} \\
\underline{-68}\downarrow \\
01 \\
\underline{-0}\downarrow \\
19 \\
\underline{-17} \\
2
\end{array}
$$

$4(17) = 68$

Subtract and bring down the 1.

$0(17) = 0$

Subtract and bring down the 9.

$1(17) = 17$

Subtract. The remainder is 2.

To check, see that $401 \cdot 17 + 2 = 6819$.

■ Work Practice 8

Answers

7. a. 9067 R 2 **b.** 5827 R 2

8. 524 R 12

Example 9 Divide: 51,600 ÷ 403

Solution:

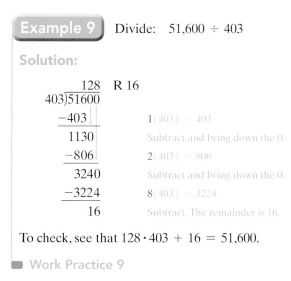

```
        128   R 16
403)51600
   −403↓
    1130
    −806↓
     3240
    −3224
       16
```

1(403) = 403
Subtract and bring down the 0.
2(403) = 806
Subtract and bring down the 0.
8(403) = 3224
Subtract. The remainder is 16.

To check, see that $128 \cdot 403 + 16 = 51{,}600$.

■ Work Practice 9

Practice 9

Divide: 33,282 ÷ 678

Division Shown as Repeated Subtraction To further understand division, recall from Section 1.6 that addition and multiplication are related in the following manner:

$$\underbrace{3 + 3 + 3 + 3}_{\text{4 addends; each addend is 3}} = 4 \times 3 = 12$$

In other words, multiplication is repeated addition. Likewise, division is repeated subtraction.

For example, let's find

$$35 \div 8$$

by repeated subtraction. Keep track of the number of times 8 is subtracted from 35. We are through when we can subtract no more because the difference is less than 8.

35 ÷ 8:   Repeated Subtraction

```
 35 }
 −8 }  1 time

 27 }
 −8 }  2 times

 19 }
 −8 }  3 times

 11 }
 −8 }  4 times
  3  ⟵ Remainder
    (We cannot subtract 8 again.)
```

35 dollars

8 dollars — 1 time
8 dollars — 2 times
8 dollars — 3 times
8 dollars — 4 times
3 dollars left over

Thus, $35 \div 8 = 4\text{ R }3$.

To check, perform the same multiplication as usual and finish by adding in the remainder.

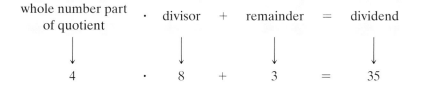

whole number part of quotient · divisor + remainder = dividend

4 · 8 + 3 = 35

Objective C Solving Problems by Dividing

Below are some key words and phrases that may indicate the operation of division:

Division		
Key Words or Phrases	**Examples**	**Symbols**
divide	divide 10 by 5	$10 \div 5$ or $\dfrac{10}{5}$
quotient	the quotient of 64 and 4	$64 \div 4$ or $\dfrac{64}{4}$
divided by	9 divided by 3	$9 \div 3$ or $\dfrac{9}{3}$
divided or shared equally among	\$100 divided equally among five people	$100 \div 5$ or $\dfrac{100}{5}$
per	100 miles per 2 hours	$\dfrac{100 \text{ miles}}{2 \text{ hours}}$

✔**Concept Check** Determine whether each of the following is the correct way to represent "the quotient of 60 and 12." Explain your answer.

a. $12 \div 60$

b. $60 \div 12$

Practice 10

Three students bought 171 blank CDs to share equally. How many CDs did each person get?

Example 10 Finding Shared Earnings

Three college students share a paper route to earn money for expenses. The total in their fund after expenses was \$2895. How much is each person's equal share?

Solution:

In words: Each person's share = total money ÷ number of persons

Translate: Each person's share = 2895 ÷ 3

Then

$$
\begin{array}{r}
965 \\
3\overline{)2895} \\
\underline{-27} \\
19 \\
\underline{-18} \\
15 \\
\underline{-15} \\
0
\end{array}
$$

Each person's share is \$965.

■ Work Practice 10

Answer

10. 57 CDs

✔**Concept Check Answers**

a. incorrect   **b.** correct

Example 11　Dividing Number of Downloads

As part of a promotion, an executive receives 238 cards, each good for one free song download. If she wants to share them evenly with 19 friends, how many download cards will each friend receive? How many will be left over?

Solution:

In words:	Number of cards for each person	=	number of cards	÷	number of friends

Translate:	Number of cards for each person	=	238	÷	19

$$
\begin{array}{r}
12 \text{ R } 10 \\
19\overline{)238} \\
-19 \\
\hline
48 \\
-38 \\
\hline
10
\end{array}
$$

Each friend will receive 12 download cards. The cards cannot be divided equally among her friends since there is a nonzero remainder. There will be 10 download cards left over.

■ Work Practice 11

Practice 11

Printers can be packed 12 to a box. If 532 printers are to be packed but only full boxes are shipped, how many full boxes will be shipped? How many printers are left over and not shipped?

Objective D Finding Averages

A special application of division (and addition) is finding the average of a list of numbers. The **average** of a list of numbers is the sum of the numbers divided by the *number* of numbers.

$$
\text{average} = \frac{\text{sum of numbers}}{\textit{number} \text{ of numbers}}
$$

Example 12　Averaging Scores

A mathematics instructor is checking a simple program she wrote for averaging the scores of her students. To do so, she averages a student's scores of 75, 96, 81, and 88 by hand. Find this average score.

Solution:　To find the average score, we find the sum of the student's scores and divide by 4, the number of scores.

$$
\begin{array}{r}
75 \\
96 \\
81 \\
+88 \\
\hline
340 \text{ sum}
\end{array}
\qquad
\text{average} = \frac{340}{4} = 85
\qquad
\begin{array}{r}
85 \\
4\overline{)340} \\
-32 \\
\hline
20 \\
-20 \\
\hline
0
\end{array}
$$

The average score is 85.

■ Work Practice 12

Practice 12

To compute a safe time to wait for reactions to occur after allergy shots are administered, a lab technician is given a list of elapsed times between administered shots and reactions. Find the average of the times 4 minutes, 7 minutes, 35 minutes, 16 minutes, 9 minutes, 3 minutes, and 52 minutes.

Answers
11. 44 full boxes; 4 printers left over
12. 18 minutes

Calculator Explorations **Dividing Numbers**

To divide numbers on a calculator, find the keys marked
$\boxed{÷}$ and $\boxed{=}$ or $\boxed{\text{ENTER}}$. For example, to find $435 ÷ 5$
on a calculator, press the keys $\boxed{435}$ $\boxed{÷}$ $\boxed{5}$ then $\boxed{=}$ or
$\boxed{\text{ENTER}}$. The display will read $\boxed{87}$. Thus, $435 ÷ 5 = 87$.

Use a calculator to divide.

1. $848 ÷ 16$ **2.** $564 ÷ 12$

3. $95\overline{)5890}$ **4.** $27\overline{)1053}$

5. $\dfrac{32,886}{126}$ **6.** $\dfrac{143,088}{264}$

7. $0 ÷ 315$ **8.** $315 ÷ 0$

Vocabulary, Readiness & Video Check

Use the choices below to fill in each blank. Some choices may be used more than once.

1	number	divisor	dividend
0	undefined	average	quotient

1. In $90 ÷ 2 = 45$, the answer 45 is called the _____ , 90 is called the _____ , and 2 is called the

_____ .

2. The quotient of any number and 1 is the same _____ .

3. The quotient of any number (except 0) and the same number is _____ .

4. The quotient of 0 and any number (except 0) is _____ .

5. The quotient of any number and 0 is _____ .

6. The _____ of a list of numbers is the sum of the numbers divided by the _____ of numbers.

Martin-Gay Interactive Videos

See Video 1.7

Watch the section lecture video and answer the following questions.

Objective A **7.** Look at ▣ Examples 6–8. What number can never be the divisor in division? ▶

Objective B **8.** In ▣ Example 10, how many 102s are in 21? How does this result affect the quotient? ▶

9. What calculation would you use to check the answer in ▣ Example 10? ▶

Objective C **10.** In ▣ Example 11, what is the importance of knowing that the distance to each hole is the same? ▶

Objective D **11.** As shown in ▣ Example 12, what two operations are used when finding an average? ▶

1.7 **Exercise Set** MyMathLab

Objective A *Find each quotient. See Examples 1 through 3.*

1. $54 \div 9$

2. $72 \div 9$

3. $36 \div 3$

4. $24 \div 3$

5. $0 \div 8$

6. $0 \div 4$

7. $31 \div 1$

8. $38 \div 1$

9. $\dfrac{18}{18}$

10. $\dfrac{49}{49}$

11. $\dfrac{24}{3}$

12. $\dfrac{45}{9}$

13. $26 \div 0$

14. $\dfrac{12}{0}$

15. $26 \div 26$

16. $6 \div 6$

17. $0 \div 14$

18. $7 \div 0$

19. $18 \div 2$

20. $18 \div 3$

Objectives A B **Mixed Practice** *Divide and then check by multiplying. See Examples 1 through 5.*

21. $3\overline{)87}$

22. $5\overline{)85}$

23. $3\overline{)222}$

24. $8\overline{)640}$

25. $3\overline{)1014}$

26. $4\overline{)2104}$

27. $\dfrac{30}{0}$

28. $\dfrac{0}{30}$

29. $63 \div 7$

30. $56 \div 8$

31. $150 \div 6$

32. $121 \div 11$

Divide and then check by multiplying. See Examples 6 and 7.

33. $7\overline{)479}$

34. $7\overline{)426}$

35. $6\overline{)1421}$

36. $3\overline{)1240}$

37. $305 \div 8$

38. $167 \div 3$

39. $2286 \div 7$

40. $3333 \div 4$

Divide and then check by multiplying. See Examples 8 and 9.

41. $55\overline{)715}$

42. $23\overline{)736}$

43. $23\overline{)1127}$

44. $42\overline{)2016}$

45. $97\overline{)9417}$

46. $44\overline{)1938}$

47. $3146 \div 15$

48. $7354 \div 12$

49. $6578 \div 13$

50. $5670 \div 14$

51. $9299 \div 46$

52. $2505 \div 64$

53. $\dfrac{12{,}744}{236}$

54. $\dfrac{5781}{123}$

55. $\dfrac{10{,}297}{103}$

56. $\dfrac{23{,}092}{240}$

57. $20{,}619 \div 102$

58. $40{,}853 \div 203$

59. $244{,}989 \div 423$

60. $164{,}592 \div 543$

Divide. See Examples 1 through 9.

61. $7\overline{)119}$

62. $8\overline{)104}$

63. $7\overline{)3580}$

64. $5\overline{)3017}$

65. $40\overline{)85,312}$

66. $50\overline{)85,747}$

67. $142\overline{)863,360}$

68. $214\overline{)650,560}$

Objective C Translating *Solve. See Examples 10 and 11.*

69. Find the quotient of 117 and 5.

70. Find the quotient of 94 and 7.

71. Find 200 divided by 35.

72. Find 116 divided by 32.

73. Find the quotient of 62 and 3.

74. Find the quotient of 78 and 5.

75. Martin Thieme teaches American Sign Language classes for $65 per student for a 7-week session. He collects $2145 from the group of students. Find how many students are in the group.

76. Kathy Gomez teaches Spanish lessons for $85 per student for a 5-week session. From one group of students, she collects $4930. Find how many students are in the group.

77. The gravity of Jupiter is 318 times as strong as the gravity of Earth, so objects on Jupiter weigh 318 times as much as they weigh on Earth. If a person would weigh 52,470 pounds on Jupiter, find how much the person weighs on Earth.

78. Twenty-one people pooled their money and bought lottery tickets. One ticket won a prize of $5,292,000. Find how many dollars each person received.

79. An 18-hole golf course is 5580 yards long. If the distance to each hole is the same, find the distance between holes.

80. A truck hauls wheat to a storage granary. It carries a total of 5768 bushels of wheat in 14 trips. How much does the truck haul each trip if each trip it hauls the same amount?

81. There is a bridge over highway I-35 every three miles. The first bridge is at the beginning of a 265-mile stretch of highway. Find how many bridges there are over 265 miles of I-35.

82. The white stripes dividing the lanes on a highway are 25 feet long, and the spaces between them are 25 feet long. Let's call a "lane divider" a stripe followed by a space. Find how many whole "lane dividers" there are in 1 mile of highway. (A mile is 5280 feet.)

83. Ari Trainor is in the requisitions department of Central Electric Lighting Company. Light poles along a highway are placed 492 feet apart. The first light pole is at the beginning of a 1-mile strip. Find how many poles he should order for the 1-mile strip of highway. (A mile is 5280 feet.)

84. Professor Lopez has a piece of rope 185 feet long that she wants to cut into pieces for an experiment in her physics class. Each piece of rope is to be 8 feet long. Determine whether she has enough rope for her 22-student class. Determine the amount extra or the amount short.

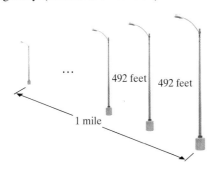

85. Broad Peak in Pakistan is the twelfth-tallest mountain in the world. Its elevation is 26,400 feet. A mile is 5280 feet. How many miles tall is Broad Peak? (*Source:* National Geographic Society)

86. Arian Foster of the Houston Texans led the NFL in touchdowns during the 2012 regular football season, scoring a total of 102 points from touchdowns. If a touchdown is worth 6 points, how many touchdowns did Foster make during 2012? (*Source:* National Football League)

87. Find how many yards are in 1 mile. (A mile is 5280 feet; a yard is 3 feet.)

88. Find how many whole feet are in 1 rod. (A mile is 5280 feet; 1 mile is 320 rods.)

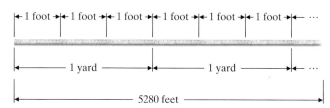

Objective D *Find the average of each list of numbers. See Example 12.*

89. 10, 24, 35, 22, 17, 12

90. 37, 26, 15, 29, 51, 22

91. 205, 972, 210, 161

92. 121, 200, 185, 176, 163

93. 86, 79, 81, 69, 80

94. 92, 96, 90, 85, 92, 79

The normal monthly temperatures in degrees Fahrenheit for Salt Lake City, Utah, is given in the graph. Use this graph to answer Exercises 95 and 96. (Source: National Climatic Data Center)

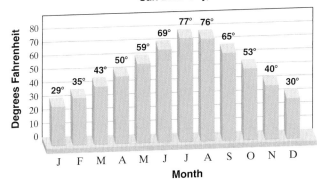

95. Find the average temperature for June, July, and August.

96. Find the average temperature for October, November, and December.

Mixed Practice (Sections 1.3, 1.4, 1.6, 1.7) *Perform each indicated operation. Watch the operation symbol.*

97. 82 + 463 + 29 + 8704

98. 23 + 407 + 92 + 7011

99. 546
　　× 28

100. 712
　　× 54

101. 722
　　− 43

102. 712
　　− 54

103. $\dfrac{45}{0}$

104. $\dfrac{0}{23}$

105. 228 ÷ 24

106. 304 ÷ 31

Concept Extensions

Match each word phrase to the correct translation. (Not all letter choices will be used.) See the Concept Check in this section.

107. The quotient of 40 and 8

108. The quotient of 200 and 20

a. $20 \div 200$ **b.** $200 \div 20$
c. $40 \div 8$ **d.** $8 \div 40$

109. 200 divided by 20

110. 40 divided by 8

The following table shows the top eight countries with the most Nobel Prize winners through 2013. Use this table to answer Exercises 111 and 112. (Source: Based on data from Encyclopaedia Britannica, Inc.)

111. Find the average number of Nobel Prize winners for the countries shown.

112. Find the average number of Nobel Prize winners per category for the United States.

Countries with Most Nobel Prize Winners, 1901–2013							
Country	Chemistry	Economics	Literature	Peace	Physics	Physiology & Medicine	Total
United States	68	54	11	22	90	97	342
United Kingdom	27	8	8	9	23	31	106
Germany	27	1	8	4	22	17	79
France	7	1	15	8	13	11	55
Sweden	4	2	9	5	4	8	32
Switzerland	6	0	2	3	4	6	21
Russia (USSR)	1	1	4	2	10	2	20
Japan	6	0	2	1	6	2	17

In Example 12 in this section, we found that the average of 75, 96, 81, and 88 is 85. Use this information to answer Exercises 113 and 114.

113. If the number 75 is removed from the list of numbers, does the average increase or decrease? Explain why.

114. If the number 96 is removed from the list of numbers, does the average increase or decrease? Explain why.

115. Without computing it, tell whether the average of 126, 135, 198, 113 is 86. Explain why it is possible or why it is not.

116. Without computing it, tell whether the average of 38, 27, 58, and 43 is 17. Explain why it is possible or why it is not.

117. If the area of a rectangle is 60 square feet and its width is 5 feet, what is its length?

118. If the area of a rectangle is 84 square inches and its length is 21 inches, what is its width?

119. Write down any two numbers whose quotient is 25.

120. Write down any two numbers whose quotient is 1.

121. Find $26 \div 5$ using the process of repeated subtraction.

122. Find $86 \div 10$ using the process of repeated subtraction.

Operations on Whole Numbers

1.
$$\begin{array}{r} 23 \\ 46 \\ +79 \\ \hline \end{array}$$

2.
$$\begin{array}{r} 7006 \\ -\ 451 \\ \hline \end{array}$$

3.
$$\begin{array}{r} 36 \\ \times 45 \\ \hline \end{array}$$

4. $8\overline{)4496}$

5. $1 \cdot 79$

6. $\dfrac{36}{0}$

7. $9 \div 1$

8. $9 \div 9$

9. $0 \cdot 13$

10. $7 \cdot 0 \cdot 8$

11. $0 \div 2$

12. $12 \div 4$

13. $4219 - 1786$

14. $1861 + 7965$

15. $5\overline{)1068}$

16.
$$\begin{array}{r} 1259 \\ \times\ \ 63 \\ \hline \end{array}$$

17. $3 \cdot 9$

18. $45 \div 5$

19.
$$\begin{array}{r} 207 \\ -\ 69 \\ \hline \end{array}$$

20.
$$\begin{array}{r} 207 \\ +\ 69 \\ \hline \end{array}$$

21. $7\overline{)7695}$

22. $9\overline{)1000}$

23. $32\overline{)21,222}$

24. $65\overline{)70,000}$

25. $4000 - 2976$

26. $10,000 - 101$

27.
$$\begin{array}{r} 303 \\ \times 101 \\ \hline \end{array}$$

28. $(475)(100)$

29. Find the total of 57 and 8.

30. Find the product of 57 and 8.

Answers

1. _____
2. _____
3. _____
4. _____
5. _____
6. _____
7. _____
8. _____
9. _____
10. _____
11. _____
12. _____
13. _____
14. _____
15. _____
16. _____
17. _____
18. _____
19. _____
20. _____
21. _____
22. _____
23. _____
24. _____
25. _____
26. _____
27. _____
28. _____
29. _____
30. _____

31. _____

32. _____

33. _____

34. _____

35. _____

36. _____

37. _____

38. _____

39. _____

40. _____

41. _____

42. _____

43. _____

44. _____

45. _____

46. _____

31. Find the quotient of 62 and 9.

32. Find the difference of 62 and 9.

33. Subtract 17 from 200.

34. Find the difference of 432 and 201.

Complete the table by rounding the given number to the given place value.

		Tens	Hundreds	Thousands
35.	9735			
36.	1429			
37.	20,801			
38.	432,198			

Find the perimeter and area of each figure.

△ **39.**

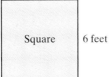

Square   6 feet

△ **40.**

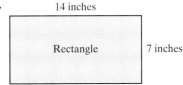

14 inches

Rectangle   7 inches

Find the perimeter of each figure.

△ **41.**

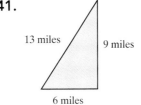

13 miles   9 miles

6 miles

△ **42.**

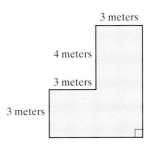

3 meters

4 meters

3 meters

3 meters

Find the average of each list of numbers.

43. 19, 15, 25, 37, 24

44. 108, 131, 98, 159

45. The Mackinac Bridge is a suspension bridge that connects the lower and upper peninsulas of Michigan across the Straits of Mackinac. Its total length is 26,372 feet. The Lake Pontchartrain Bridge is a twin concrete trestle bridge in Slidell, Louisiana. Its total length is 28,547 feet. Which bridge is longer and by how much? (*Sources:* Mackinac Bridge Authority and Federal Highway Administration, Bridge Division)

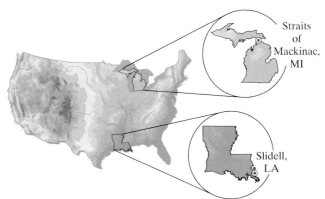

Straits of Mackinac, MI

Slidell, LA

46. The average teenage male American consumes 2 quarts of carbonated soft drinks per day. On average, how many quarts of carbonated soft drinks would be consumed in a year? (Use 365 for the number of days.) (*Source:* American Beverage Association)

1.8 An Introduction to Problem Solving

Objective A Solving Problems Involving Addition, Subtraction, Multiplication, or Division

In this section, we decide which operation to perform in order to solve a problem. Don't forget the key words and phrases that help indicate which operation to use. Some of these are listed below and were introduced earlier in the chapter. Also included are several words and phrases that translate to the symbol " = ":

Addition (+)	Subtraction (−)	Multiplication (·)	Division (÷)	Equality (=)
sum	difference	product	quotient	equals
plus	minus	times	divide	is equal to
added to	subtract	multiply	shared equally	is/was
more than	less than	multiply by	among	yields
increased by	decreased by	of	divided by	
total	less	double/triple	divided into	

The following problem-solving steps may be helpful to you:

Problem-Solving Steps

1. UNDERSTAND the problem. Some ways of doing this are to read and reread the problem, construct a drawing, and look for key words to identify an operation.
2. TRANSLATE the problem. That is, write the problem in short form using words, and then translate to numbers and symbols.
3. SOLVE the problem. It is helpful to estimate the solution by rounding. Then carry out the indicated operation from step 2.
4. INTERPRET the results. *Check* the proposed solution in the stated problem and *state* your conclusions. Write your results with the correct units attached.

Example 1 Calculating the Length of a River

The Hudson River in New York State is 306 miles long. The Snake River in the northwestern United States is 732 miles longer than the Hudson River. How long is the Snake River? (*Source:* U.S. Department of the Interior)

Solution:

1. UNDERSTAND. Read and reread the problem, and then draw a picture. Notice that we are told that Snake River is 732 miles longer than the Hudson River. The phrase "longer than" means that we add.

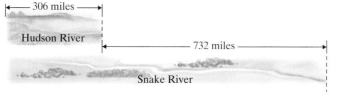

(Continued on next page)

Objectives

A Solve Problems by Adding, Subtracting, Multiplying, or Dividing Whole Numbers.

B Solve Problems That Require More Than One Operation.

Practice 1

The building called 555 California Street is the second-tallest building in San Francisco, California, at 779 feet. The tallest building in San Francisco is the Transamerica Pyramid, which is 74 feet taller than 555 California Street. How tall is the Transamerica Pyramid? (*Source: The World Almanac*)

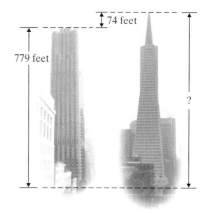

555 California Street   Transamerica Pyramid

Answer
1. 853 ft

2. TRANSLATE.

In words: Snake River   is   732 miles   longer than   the Hudson River

↓   ↓   ↓   ↓   ↓

Translate: Snake River   =   732   +   306

3. SOLVE: Let's see if our answer is reasonable by also estimating. We will estimate each addend to the nearest hundred.

$$\begin{array}{rl} 732 & \text{rounds to} \\ +306 & \text{rounds to} \\ \hline 1038 & \text{exact} \end{array}$$

$$\begin{array}{rl} 700 & \\ +300 & \\ \hline 1000 & \text{estimate} \end{array}$$

4. INTERPRET. *Check* your work. The answer is reasonable since 1038 is close to our estimated answer of 1000. *State* your conclusion: The Snake River is 1038 miles long.

▪ Work Practice 1

Practice 2

Four friends bought a lottery ticket and won $65,000. If each person is to receive the same amount of money, how much does each person receive?

Example 2 Filling a Shipping Order

How many cases can be filled with 9900 cans of jalapeños if each case holds 48 cans? How many cans will be left over? Will there be enough cases to fill an order for 200 cases?

Solution:

1. UNDERSTAND. Read and reread the problem. Draw a picture to help visualize the situation.

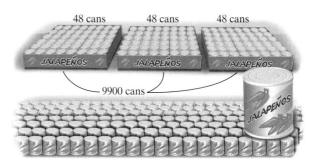

48 cans   48 cans   48 cans

9900 cans

JALAPEÑOS

Since each case holds 48 cans, we want to know how many 48s there are in 9900. We find this by dividing.

2. TRANSLATE.

In words: Number of cases   is   9900   divided by   48

↓   ↓   ↓   ↓   ↓

Translate: Number of cases   =   9900   ÷   48

3. SOLVE: Let's estimate a reasonable solution before we actually divide. Since 9900 rounded to the nearest thousand is 10,000 and 48 rounded to the nearest ten is 50, 10,000 ÷ 50 = 200. Now find the exact quotient.

$$\begin{array}{r} 206 \ \ R12 \\ 48\overline{)9900} \\ -96 \ \ \ \ \\ \hline 300 \\ -288 \\ \hline 12 \end{array}$$

Answer

2. $16,250

4. INTERPRET. *Check* your work. The answer is reasonable since 206 R 12 is close to our estimate of 200. *State* your conclusion: 206 cases will be filled, with 12 cans left over. There will be enough cases to fill an order for 200 cases.

🔲 Work Practice 2

Example 3 Calculating Budget Costs

The director of a learning lab at a local community college is working on next year's budget. Thirty-three new DVD players are needed at a cost of $187 each. What is the total cost of these DVD players?

Solution:

1. UNDERSTAND. Read and reread the problem, and then draw a diagram.

33 DVD Players

$ 187 $ 187   ... $ 187

From the phrase "total cost," we might decide to solve this problem by adding. This would work, but repeated addition, or multiplication, would save time.

2. TRANSLATE.

In words:	Total cost	is	number of DVD players	times	cost of a DVD player
	↓	↓	↓	↓	↓
Translate:	Total cost	=	33	×	$187

3. SOLVE: Once again, let's estimate a reasonable solution.

$$
\begin{array}{r}
187 \\
\times\ 33 \\
\hline
561 \\
5610 \\
\hline
6171
\end{array}
\qquad
\begin{array}{r}
\text{rounds to} \\
\text{rounds to}
\end{array}
\qquad
\begin{array}{r}
200 \\
\times\ 30 \\
\hline
6000
\end{array}
$$

187 rounds to 200
× 33 rounds to × 30
561
6000 estimate
5610
6171 exact

4. INTERPRET. *Check* your work. *State* your conclusion: The total cost of the DVD players is $6171.

🔲 Work Practice 3

Example 4 Calculating a Public School Teacher's Salary

In 2012, the average salary for a public school teacher in California was $69,324. For the same year, the average salary for a public school teacher in Iowa was $17,796 less than this. What was the average public school teacher's salary in Iowa? (*Source:* National Education Association)

Solution:

1. UNDERSTAND. Read and reread the problem. Notice that we are told that the Iowa salary is $17,796 less than the California salary. The phrase "less than" indicates subtraction.

(Continued on next page)

Practice 3

The director of the learning lab also needs to include in the budget a line for 425 blank CDs at a cost of $4 each. What is this total cost for the blank CDs?

Practice 4

In 2012, the average salary for a public school teacher in Alaska was $65,468. For the same year, the average salary for a public school teacher in Hawaii was $11,168 less than this. What was the average public school teacher's salary in Hawaii? (*Source:* National Education Association)

Answers
3. $1700 **4.** $54,300

2. TRANSLATE. Remember that order matters when subtracting, so be careful when translating.

In words: Iowa salary   is   California salary   minus $17,796

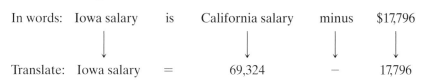

Translate: Iowa salary   =   69,324   −   17,796

3. SOLVE. This time, instead of estimating, let's check by adding.

$$\begin{array}{r} 69{,}324 \\ -17{,}796 \\ \hline 51{,}528 \end{array}$$ **Check:** $$\begin{array}{r} 51{,}528 \\ +17{,}796 \\ \hline 69{,}324 \end{array}$$

4. INTERPRET. *Check* your work. The check is above. *State* your conclusion: The average Iowa teacher's salary in 2012 was $51,528.

▒ Work Practice 4

Objective B Solving Problems That Require More Than One Operation ▶

We must sometimes use more than one operation to solve a problem.

Practice 5

A gardener is trying to decide how much fertilizer to buy for his yard. He knows that his lot is in the shape of a rectangle that measures 90 feet by 120 feet. He also knows that the floor of his house is in the shape of a rectangle that measures 45 feet by 65 feet. How much area of the lot is not covered by the house?

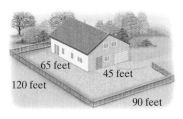

Answer

5. 7875 sq ft

Example 5 Planting a New Garden

A gardener bought enough plants to fill a rectangular garden with length 30 feet and width 20 feet. Because of shading problems from a nearby tree, the gardener changed the width of the garden to 15 feet. If the area is to remain the same, what is the new length of the garden?

Solution:

1. UNDERSTAND. Read and reread the problem. Then draw a picture to help visualize the problem.

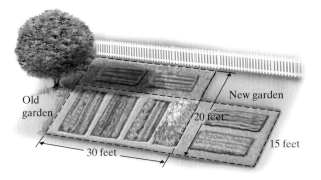

2. TRANSLATE. Since the area of the new garden is to be the same as the area of the old garden, let's find the area of the old garden. Recall that

Area = length × width = 30 feet × 20 feet = 600 square feet

Since the area of the new garden is to be 600 square feet also, we need to see how many 15s there are in 600. This means division. In other words,

In words: New length = Area of garden ÷ New width

Translate: New length = 600 ÷ 15

3. SOLVE.

$$\begin{array}{r} 40 \\ 15\overline{)600} \\ -60 \\ \hline 00 \end{array}$$

4. INTERPRET. *Check* your work. *State* your conclusion: The length of the new garden is 40 feet.

■ Work Practice 5

Vocabulary, Readiness & Video Check

Martin-Gay Interactive Videos   *Watch the section lecture video and answer the following questions.*

Objective A **1.** The answer to the calculations in ▣ Example 3 is 3500. What is the final interpreted solution, written as a sentence? ◉

Objective B **2.** What two operations are used to solve ▣ Example 4? ◉

See Video 1.8 ●

1.8 Exercise Set MyMathLab® ◉

Objective A *Solve. Exercises 1, 2, 11, and 12 have been started for you. See Examples 1 through 4.*

1. 41 increased by 8 is what number?

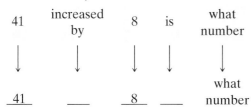

Start the solution:

1. UNDERSTAND the problem. Reread it as many times as needed.
2. TRANSLATE into an equation. (Fill in the blanks below.)

41	increased by	8	is	what number
↓	↓	↓	↓	↓
41	___	8	___	what number

Finish with:

3. SOLVE
4. INTERPRET

2. What is 12 multiplied by 9?

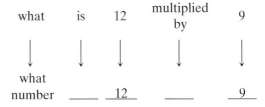

Start the solution:

1. UNDERSTAND the problem. Reread it as many times as needed.
2. TRANSLATE into an equation. (Fill in the blanks below.)

what	is	12	multiplied by	9
↓	↓	↓	↓	↓
what number	___	12	___	9

Finish with:

3. SOLVE
4. INTERPRET

3. What is the quotient of 1185 and 5?

4. 78 decreased by 12 is what number?

5. What is the total of 35 and 7?

6. What is the difference of 48 and 8?

7. 60 times 10 is what number?

8. 60 divided by 10 is what number?

△ **9.** A vacant lot in the shape of a rectangle measures 120 feet by 80 feet.

  a. What is the perimeter of the lot?

  b. What is the area of the lot?

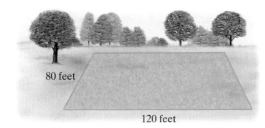

△ **10.** A parking lot in the shape of a rectangle measures 100 feet by 150 feet.

  a. What is the perimeter of the lot?

  b. What is the area of the parking lot?

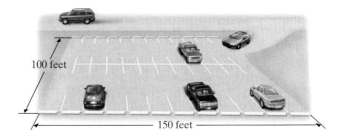

11. A family bought a house for $185,700 and later sold the house for $201,200. How much money did they make by selling the house?

Start the solution:

1. UNDERSTAND the problem. Reread it as many times as needed.

2. TRANSLATE into an equation. (Fill in the blanks below.)

money made	is	selling price	minus	purchase price
↓	↓	↓	↓	↓

money made				
=	___	−	___	

Finish with:

3. SOLVE

4. INTERPRET

12. Three people dream of equally sharing a $147 million lottery. How much would each person receive if they have the winning ticket?

Start the solution:

1. UNDERSTAND the problem. Reread it as many times as needed.

2. TRANSLATE into an equation. (Fill in the blanks below.)

each person's share	is	lottery amount	divided by	number by persons
↓	↓	↓	↓	↓

each person's share				
=	___	÷	___	

Finish with:

3. SOLVE

4. INTERPRET

13. There are 24 hours in a day. How many hours are in a week?

14. There are 60 minutes in an hour. How many minutes are in a day?

15. The Verrazano Narrows Bridge is the longest bridge in New York, measuring 4260 feet. The George Washington Bridge, also in New York, is 760 feet shorter than the Verrazano Narrows Bridge. Find the length of the George Washington Bridge.

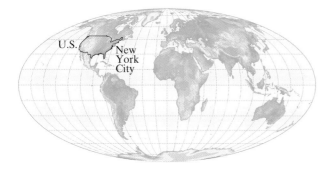

16. In 2013, the Goodyear Tire & Rubber Company began replacing its fleet of nonrigid GZ-20 blimps with new Goodyear NT semi-rigid airships. The new Goodyear NT airship can hold 297,527 cubic feet of helium. Its GZ-20 predecessor held 94,827 fewer cubic feet of helium. How much helium did a GZ-20 blimp hold? (*Source:* Goodyear Tire & Rubber Company)

17. Yellowstone National Park in Wyoming was the first national park in the United States. It was created in 1872. One of the more recent additions to the National Park System is First State National Monument. It was established in 2013. How much older is Yellowstone than First State? (*Source:* National Park Service)

18. Razor scooters were introduced in 2000. Radio Flyer Wagons were first introduced 83 years earlier. In what year were Radio Flyer Wagons introduced? (*Source:* Toy Industry Association, Inc.)

19. Since their introduction, the number of LEGO building bricks that have been sold is equivalent to the world's current population of approximately 6 billion people owning 62 LEGO bricks each. About how many LEGO bricks have been sold since their introduction? (*Source:* LEGO Company)

20. In 2012, the average weekly pay for a home health aide in the United States was about $420. At this rate, how much will a home health aide earn working a 52-week year? (*Source:* Bureau of Labor Statistics)

21. The three most common city names in the United States are Fairview, Midway, and Riverside. There are 287 towns named Fairview, 252 named Midway, and 180 named Riverside. Find the total number of towns named Fairview, Midway, or Riverside.

22. In the game of Monopoly, a player must own all properties in a color group before building houses. The yellow color-group properties are Atlantic Avenue, Ventnor Avenue, and Marvin Gardens. These cost $260, $260, and $280, respectively, when purchased from the bank. What total amount must a player pay to the bank before houses can be built on the yellow properties? (*Source:* Hasbro, Inc.)

23. In 2012, the average weekly pay for a correctional officer in the United States was $840. If such an officer works 40 hours in one week, what is his or her hourly pay? (*Source:* Bureau of Labor Statistics)

24. In 2012, the average weekly pay for a loan officer was $1360. If a loan officer works 40 hours in one week, what is his or her hourly pay? (*Source:* Bureau of Labor Statistics)

25. Three ounces of canned tuna in oil has 165 calories. How many calories does 1 ounce have? (*Source: Home and Garden Bulletin No. 72,* U.S. Department of Agriculture)

26. A whole cheesecake has 3360 calories. If the cheesecake is cut into 12 equal pieces, how many calories will each piece have? (*Source: Home and Garden Bulletin No. 72,* U.S. Department of Agriculture)

27. The average estimated 2012 U.S. population was 313,900,000. Between Memorial Day and Labor Day, 7 billion hot dogs are consumed. Approximately how many hot dogs were consumed per person between Memorial Day and Labor Day in 2012? Divide, but do not give the remainder part of the quotient. (*Source:* U.S. Census Bureau, National Hot Dog and Sausage Council)

28. Adrian Peterson, a running back with the NFL's Minnesota Vikings, scored an average of 5 points per game during the 2012 regular season. He played in a total of 16 games during the season. What was the total number of points he scored during the 2012 football season? (*Source:* National Football League)

29. Macy's, formerly the Federated Department Stores Company, operates a total of 843 Macy's and Bloomingdale's department stores around the country. In 2012, Macy's had sales of approximately $27,685,999,890. What is the average amount of sales made by each of the 843 stores? Divide, but do not give the remainder part of the quotient. (*Source:* Macy's)

30. In 2012, PetSmart employed approximately 52,000 associates and operated roughly 1300 stores. What is the average number of associates employed at each of its stores? (*Source:* PetSmart, Inc.)

31. In 2012, the Museum of Modern Art in New York welcomed 2,805,659 visitors. The J. Paul Getty Museum in Los Angeles received 1,590,608 visitors. How many more people visited the Museum of Modern Art than the Getty Museum? (*Source: The Art Newspaper*)

32. In 2012, Target Corporation operated 1778 stores in the United States. Of these, 149 were in Texas. How many Target Stores were located in states other than Texas? (*Source:* Target Corporation)

33. The length of the southern boundary of the conterminous United States is 1933 miles. The length of the northern boundary of the conterminous United States is 2054 miles longer than this. What is the length of the northern boundary? (*Source:* U.S. Geological Survey)

34. In humans, 14 muscles are required to smile. It takes 29 more muscles to frown. How many muscles does it take to frown?

2054 miles longer

1933 miles

35. An instructor at the University of New Orleans receives a paycheck every four weeks. Find how many paychecks he receives in a year. (A year has 52 weeks.)

36. A loan of $6240 is to be paid in 48 equal payments. How much is each payment?

Objective B *Solve. See Example 5.*

37. Find the total cost of 3 sweaters at $38 each and 5 shirts at $25 each.

38. Find the total cost of 10 computers at $2100 each and 7 boxes of diskettes at $12 each.

39. A college student has $950 in an account. She spends $205 from the account on books and then deposits $300 in the account. How much money is now in the account?

40. The temperature outside was 57°F (degrees Fahrenheit). During the next few hours, it decreased by 18 degrees and then increased by 23 degrees. Find the new temperature.

The table shows the menu from a concession stand at the county fair. Use this menu to answer Exercises 41 and 42.

41. A hungry college student is debating between the following two orders:
 a. a hamburger, an order of onion rings, a candy bar, and a soda.
 b. a hot dog, an apple, an order of french fries, and a soda.
 Which order will be cheaper? By how much?

42. A family of four is debating between the following two orders:
 a. 6 hot dogs, 4 orders of onion rings, and 4 sodas.
 b. 4 hamburgers, 4 orders of french fries, 2 apples, and 4 sodas.
 Will the family save any money by ordering (b) instead of (a)? If so, how much?

Corky's Concession Stand Menu	
Item	**Price**
Hot dog	$3
Hamburger	$4
Soda	$1
Onion rings	$3
French fries	$2
Apple	$1
Candy bar	$2

Objectives A B Mixed Practice *Use the bar graph to answer Exercises 43 through 50. (Source: Internet World Stats)*

43. Which region of the world listed had the greatest number of Internet users in 2012?

44. Which region of the world listed had the least number of Internet users in 2012?

45. How many more Internet users (in millions) did the world region with the most Internet users have than the world region with the fewest Internet users?

46. How many more Internet users did Africa have than the Middle East in 2012?

47. How many more Internet users did North America have than Latin America/Caribbean?

48. Which region of the world had more Internet users, Europe or North America? How many more Internet users did it have?

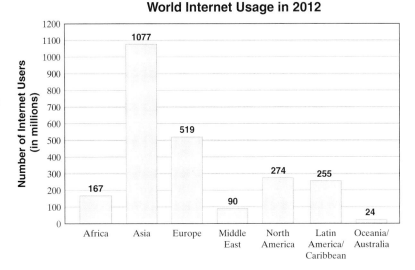

World Internet Usage in 2012

Find the average number of Internet users for the world regions listed in the graph.

49. The two world regions with the greatest number of Internet users.

50. The four world regions with the least number of Internet users.

Solve.

51. The learning lab at a local university is receiving new equipment. Twenty-two computers are purchased for $615 each and three printers for $408 each. Find the total cost for this equipment.

52. The washateria near the local community college is receiving new equipment. Thirty-six washers are purchased for $585 each and ten dryers are purchased for $388 each. Find the total cost for this equipment.

53. The American Heart Association recommends consuming no more than 2400 milligrams of salt per day. (This is about the amount in 1 teaspoon of salt.) How many milligrams of sodium is this in a week?

54. This semester a particular student pays $1750 for room and board, $709 for a meal ticket plan, and $2168 for tuition. What is her total bill?

△ **55.** The Meishs' yard is in the shape of a rectangle and measures 50 feet by 75 feet. In their yard, they have a rectangular swimming pool that measures 15 feet by 25 feet.
   a. Find the area of the entire yard.
   b. Find the area of the swimming pool.
   c. Find the area of the yard that is not part of the swimming pool.

56. The community is planning to construct a rectangular-shaped playground within the local park. The park is in the shape of a square and measures 100 yards on each side. The playground is to measure 15 yards by 25 yards.
   a. Find the area of the entire park.
   b. Find the area of the playground.
   c. Find the area of the park that is not part of the playground.

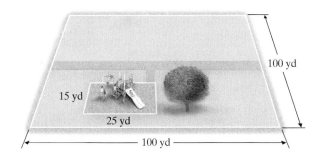

Concept Extensions

57. In 2012, the United States Postal Service handled 170,000,000 articles with return receipt service, which generated revenues of $399,000,000. Round the number of return receipt articles and the return receipt revenues to the nearest hundred-million to estimate the average revenue generated by each return receipt article. (*Source:* United States Postal Service)

58. In 2012, the United States Postal Service handled 40,000,000 pieces of Express Mail, which generated revenues of $802,000,000. Round the Express Mail revenues to the nearest ten-million to estimate the average revenue generated by each piece of Express Mail. (*Source:* United States Postal Service)

59. Write an application of your own that uses the term "bank account" and the numbers 1036 and 524.

1.9 Exponents, Square Roots, and Order of Operations

Objective A Using Exponential Notation

In the product $3 \cdot 3 \cdot 3 \cdot 3 \cdot 3$, notice that 3 is a factor several times. When this happens, we can use a shorthand notation, called an **exponent,** to write the repeated multiplication.

$$3 \cdot 3 \cdot 3 \cdot 3 \cdot 3 \qquad \text{can be written as}$$

3 is a factor 5 times

exponent

3^5 Read as "three to the fifth power."

base

This is called **exponential notation.** The **exponent,** 5, indicates how many times the **base,** 3, is a factor.

The table below shows examples of reading exponential notation in words.

Expression	In Words
5^2	"five to the second power" or "five squared"
5^3	"five to the third power" or "five cubed"
5^4	"five to the fourth power"

Usually, an exponent of 1 is not written, so when no exponent appears, we assume that the exponent is 1. For example, $2 = 2^1$ and $7 = 7^1$.

Examples   Write using exponential notation.

1. $7 \cdot 7 \cdot 7 = 7^3$
2. $3 \cdot 3 = 3^2$
3. $6 \cdot 6 \cdot 6 \cdot 6 \cdot 6 = 6^5$
4. $3 \cdot 3 \cdot 3 \cdot 3 \cdot 17 \cdot 17 \cdot 17 = 3^4 \cdot 17^3$

■ Work Practice 1–4

Objective B Evaluating Exponential Expressions

To **evaluate** an exponential expression, we write the expression as a product and then find the value of the product.

Examples   Evaluate.

5. $9^2 = 9 \cdot 9 = 81$
6. $6^1 = 6$
7. $3^4 = 3 \cdot 3 \cdot 3 \cdot 3 = 81$
8. $5 \cdot 6^2 = 5 \cdot 6 \cdot 6 = 180$

■ Work Practice 5–8

Objectives

A Write Repeated Factors Using Exponential Notation.

B Evaluate Expressions Containing Exponents.

C Evaluate the Square Root of a Perfect Square.

D Use the Order of Operations.

E Find the Area of a Square.

Practice 1–4

Write using exponential notation.

1. $8 \cdot 8 \cdot 8 \cdot 8$
2. $3 \cdot 3 \cdot 3$
3. $10 \cdot 10 \cdot 10 \cdot 10 \cdot 10$
4. $5 \cdot 5 \cdot 4 \cdot 4 \cdot 4 \cdot 4 \cdot 4 \cdot 4$

Practice 5–8

Evaluate.

5. 4^2 6. 7^3
7. 11^1 8. $2 \cdot 3^2$

Answers
1. 8^4 2. 3^3 3. 10^5 4. $5^2 \cdot 4^6$
5. 16   6. 343   7. 11   8. 18

Example 8 illustrates an important property: An exponent applies only to its base. The exponent 2, in $5 \cdot 6^2$, applies only to its base, 6.

Helpful Hint

An exponent applies only to its base. For example, $4 \cdot 2^3$ means $4 \cdot 2 \cdot 2 \cdot 2$.

Helpful Hint

Don't forget that 2^4, for example, is *not* $2 \cdot 4$. The expression 2^4 means repeated multiplication of the same factor.

$$2^4 = 2 \cdot 2 \cdot 2 \cdot 2 = 16, \quad \text{whereas } 2 \cdot 4 = 8$$

✓**Concept Check** Which of the following statements is correct?
a. 3^6 is the same as $6 \cdot 6 \cdot 6$.
b. "Eight to the fourth power" is the same as 8^4.
c. "Ten squared" is the same as 10^3.
d. 11^2 is the same as $11 \cdot 2$.

Objective C Evaluating Square Roots

A **square root** of a number is one of two identical factors of the number. For example,

$$7 \cdot 7 = 49, \text{ so a square root of 49 is 7.}$$

We use this symbol $\sqrt{}$ (called a radical sign) for finding square roots. Since

$$7 \cdot 7 = 49, \text{ then } \sqrt{49} = 7.$$

Practice 9–11

Find each square root.
9. $\sqrt{100}$
10. $\sqrt{4}$
11. $\sqrt{1}$

Examples Find each square root.

9. $\sqrt{25} = 5$ because $5 \cdot 5 = 25$
10. $\sqrt{81} = 9$ because $9 \cdot 9 = 81$
11. $\sqrt{0} = 0$ because $0 \cdot 0 = 0$

■ Work Practice 9–11

Helpful Hint

Make sure you understand the difference between squaring a number and finding the square root of a number.

$$9^2 = 9 \cdot 9 = 81 \quad \sqrt{9} = 3 \text{ because } 3 \cdot 3 = 9$$

Answers
9. 10 **10.** 2 **11.** 1

✓**Concept Check Answer**
b

Not every square root simplifies to a whole number. We will study this more in a later chapter. In this section, we will find square roots of perfect squares only.

Objective D Using the Order of Operations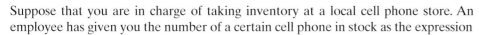

Suppose that you are in charge of taking inventory at a local cell phone store. An employee has given you the number of a certain cell phone in stock as the expression

$$6 + 2 \cdot 30$$

To calculate the value of this expression, do you add first or multiply first? If you add first, the answer is 240. If you multiply first, the answer is 66.

Mathematical symbols wouldn't be very useful if two values were possible for one expression. Thus, mathematicians have agreed that, given a choice, we multiply first.

$$6 + 2 \cdot 30 = 6 + 60 \quad \text{Multiply.}$$
$$= 66 \quad \text{Add.}$$

This agreement is one of several **order of operations** agreements.

Order of Operations

1. Perform all operations within parentheses ( ), brackets [ ], or other grouping symbols such as fraction bars or square roots, starting with the innermost set.
2. Evaluate any expressions with exponents.
3. Multiply or divide in order from left to right.
4. Add or subtract in order from left to right.

Below we practice using order of operations to simplify expressions.

Example 12 Simplify: $2 \cdot 4 - 3 \div 3$

Solution: There are no parentheses and no exponents, so we start by multiplying and dividing, from left to right.

$$2 \cdot 4 - 3 \div 3 = 8 - 3 \div 3 \quad \text{Multiply.}$$
$$= 8 - 1 \quad \text{Divide.}$$
$$= 7 \quad \text{Subtract.}$$

■ Work Practice 12

Practice 12

Simplify: $9 \cdot 3 - 8 \div 4$

Example 13 Simplify: $4^2 \div 2 \cdot 4$

Solution: We start by evaluating 4^2.

$$4^2 \div 2 \cdot 4 = 16 \div 2 \cdot 4 \quad \text{Write } 4^2 \text{ as 16.}$$

Next we multiply or divide *in order* from left to right. Since division appears before multiplication from left to right, we divide first, then multiply.

$$16 \div 2 \cdot 4 = 8 \cdot 4 \quad \text{Divide.}$$
$$= 32 \quad \text{Multiply.}$$

■ Work Practice 13

Practice 13

Simplify: $48 \div 3 \cdot 2^2$

Answers
12. 25 **13.** 64

Practice 14

Simplify: $(10 - 7)^4 + 2 \cdot 3^2$

Example 14 Simplify: $(8 - 6)^2 + 2^3 \cdot 3$

Solution: $(8 - 6)^2 + 2^3 \cdot 3 = 2^2 + 2^3 \cdot 3$ Simplify inside parentheses.

$$= 4 + 8 \cdot 3 \quad \text{Write } 2^2 \text{ as 4 and } 2^3 \text{ as 8.}$$
$$= 4 + 24 \quad \text{Multiply.}$$
$$= 28 \quad \text{Add.}$$

■ Work Practice 14

Practice 15

Simplify:

$36 \div [20 - (4 \cdot 2)] + 4^3 - 6$

Example 15 Simplify: $4^3 + [3^2 - (10 \div 2)] - 7 \cdot 3$

Solution: Here we begin with the innermost set of parentheses.

$$4^3 + [3^2 - (10 \div 2)] - 7 \cdot 3 = 4^3 + [3^2 - 5] - 7 \cdot 3 \quad \text{Simplify inside parentheses.}$$
$$= 4^3 + [9 - 5] - 7 \cdot 3 \quad \text{Write } 3^2 \text{ as 9.}$$
$$= 4^3 + 4 - 7 \cdot 3 \quad \text{Simplify inside brackets.}$$
$$= 64 + 4 - 7 \cdot 3 \quad \text{Write } 4^3 \text{ as 64.}$$
$$= 64 + 4 - 21 \quad \text{Multiply.}$$
$$= 47 \quad \text{Add and subtract from left to right.}$$

■ Work Practice 15

Practice 16

Simplify: $\dfrac{25 + 8 \cdot 2 - 3^3}{2(3 - 2)}$

Example 16 Simplify: $\dfrac{7 - 2 \cdot 3 + 3^2}{5(2 - 1)}$

Solution: Here, the fraction bar is like a grouping symbol. We simplify above and below the fraction bar separately.

$$\frac{7 - 2 \cdot 3 + 3^2}{5(2 - 1)} = \frac{7 - 2 \cdot 3 + 9}{5(1)} \quad \text{Evaluate } 3^2 \text{ and } (2 - 1).$$
$$= \frac{7 - 6 + 9}{5} \quad \text{Multiply } 2 \cdot 3 \text{ in the numerator and multiply 5 and 1 in the denominator.}$$
$$= \frac{10}{5} \quad \text{Add and subtract from left to right.}$$
$$= 2 \quad \text{Divide.}$$

■ Work Practice 16

Practice 17

Simplify: $81 \div \sqrt{81} \cdot 5 + 7$

Example 17 Simplify: $64 \div \sqrt{64} \cdot 2 + 4$

Solution: $64 \div \sqrt{64} \cdot 2 + 4 = 64 \div 8 \cdot 2 + 4$ Find the square root.

$$= 8 \cdot 2 + 4 \quad \text{Divide.}$$
$$= 16 + 4 \quad \text{Multiply.}$$
$$= 20 \quad \text{Add.}$$

■ Work Practice 17

Answers

14. 99 **15.** 61 **16.** 7 **17.** 52

Objective E Finding the Area of a Square

Since a square is a special rectangle, we can find its area by finding the product of its length and its width.

Area of a rectangle = length · width

By recalling that each side of a square has the same measurement, we can use the following procedure to find its area:

$$\begin{aligned} \text{Area of a square} &= \text{length} \cdot \text{width} \\ &= \text{side} \cdot \text{side} \\ &= (\text{side})^2 \end{aligned}$$

Square   Side

Side

Helpful Hint

Recall from Section 1.6 that area is measured in **square** units while perimeter is measured in units.

Example 18   Find the area of a square whose side measures 4 inches.

Solution: Area of a square $= (\text{side})^2$

$= (4 \text{ inches})^2$

$= 16$ square inches

4 inches

The area of the square is 16 square inches.

■ Work Practice 18

Practice 18

Find the area of a square whose side measures 12 centimeters.

Calculator Explorations **Exponents**

To evaluate an exponent such as 4^7 on a calculator, find the keys marked $\boxed{y^x}$ or $\boxed{\wedge}$ and $\boxed{=}$ or $\boxed{\text{ENTER}}$. To evaluate 4^7, press the keys $\boxed{4}$ $\boxed{y^x}$ (or $\boxed{\wedge}$) $\boxed{7}$ then $\boxed{=}$ or $\boxed{\text{ENTER}}$. The display will read $\boxed{16384}$. Thus, $4^7 = 16,384$.

Use a calculator to evaluate.

1. 4^6 **2.** 5^6 **3.** 5^5

4. 7^6 **5.** 2^{11} **6.** 6^8

Order of Operations

To see whether your calculator has the order of operations built in, evaluate $5 + 2 \cdot 3$ by pressing the keys $\boxed{5}$ $\boxed{+}$ $\boxed{2}$ $\boxed{\times}$ $\boxed{3}$ then $\boxed{=}$ or $\boxed{\text{ENTER}}$. If the display reads $\boxed{11}$, your calculator does have the order of operations built in. This means that most of the time,

you can key in a problem exactly as it is written and the calculator will perform operations in the proper order. When evaluating an expression containing parentheses, key in the parentheses. (If an expression contains brackets, key in parentheses.) For example, to evaluate $2[25 - (8 + 4)] - 11$, press the keys $\boxed{2}$ $\boxed{\times}$ $\boxed{(}$ $\boxed{25}$ $\boxed{-}$ $\boxed{(}$ $\boxed{8}$ $\boxed{+}$ $\boxed{4}$ $\boxed{)}$ $\boxed{)}$ $\boxed{-}$ $\boxed{11}$ then $\boxed{=}$ or $\boxed{\text{ENTER}}$.

The display will read $\boxed{15}$.

Use a calculator to evaluate.

7. $7^4 + 5^3$

8. $12^4 - 8^4$

9. $63 \cdot 75 - 43 \cdot 10$

10. $8 \cdot 22 + 7 \cdot 16$

11. $4(15 \div 3 + 2) - 10 \cdot 2$

12. $155 - 2(17 + 3) + 185$

Answer

18. 144 sq cm

Vocabulary, Readiness & Video Check

Use the choices below to fill in each blank.

addition   multiplication   exponent   base
subtraction   division   square root

1. In $2^5 = 32$, the 2 is called the _____ and the 5 is called the _____.
2. To simplify $8 + 2 \cdot 6$, which operation should be performed first? _____
3. To simplify $(8 + 2) \cdot 6$, which operation should be performed first? _____
4. To simplify $9(3 - 2) \div 3 + 6$, which operation should be performed first? _____
5. To simplify $8 \div 2 \cdot 6$, which operation should be performed first? _____
6. The _____ of a whole number is one of two identical factors of the number.

Martin-Gay Interactive Videos   *Watch the section lecture video and answer the following questions.*

See Video 1.9

Objective A 7. In the ⊞ Example 1 expression, what is the 3 called and what is the 12 called? ▷

Objective B 8. As mentioned in ⊞ Example 4, what "understood exponent" does any number we've worked with before have? ▷

Objective C 9. From ⊞ Example 7, how do we know that $\sqrt{64} = 8$? ▷

Objective D 10. List the three operations needed to evaluate ⊞ Example 9 in the order they should be performed. ▷

Objective E 11. As explained in the lecture before ⊞ Example 12, why does the area of a square involve an exponent whereas the area of a rectangle usually does not? ▷

1.9 Exercise Set MyMathLab®

Objective A *Write using exponential notation. See Examples 1 through 4.*

1. $4 \cdot 4 \cdot 4$ 2. $5 \cdot 5 \cdot 5 \cdot 5$ 3. $7 \cdot 7 \cdot 7 \cdot 7 \cdot 7 \cdot 7$ 4. $6 \cdot 6 \cdot 6 \cdot 6 \cdot 6 \cdot 6 \cdot 6$

▷ 5. $12 \cdot 12 \cdot 12$ 6. $10 \cdot 10 \cdot 10$ ▷ 7. $6 \cdot 6 \cdot 5 \cdot 5 \cdot 5$ 8. $4 \cdot 4 \cdot 3 \cdot 3 \cdot 3$

9. $9 \cdot 8 \cdot 8$ 10. $7 \cdot 4 \cdot 4 \cdot 4$ 11. $3 \cdot 2 \cdot 2 \cdot 2 \cdot 2$ 12. $4 \cdot 6 \cdot 6 \cdot 6 \cdot 6$

13. $3 \cdot 2 \cdot 2 \cdot 2 \cdot 2 \cdot 5 \cdot 5 \cdot 5 \cdot 5 \cdot 5$ 14. $6 \cdot 6 \cdot 2 \cdot 9 \cdot 9 \cdot 9 \cdot 9$

Objective B *Evaluate. See Examples 5 through 8.*

15. 8^2 **16.** 6^2 ▶ **17.** 5^3 **18.** 6^3 **19.** 2^5 **20.** 3^5

21. 1^{10} **22.** 1^{12} ▶ **23.** 7^1 **24.** 8^1 **25.** 2^7 **26.** 5^4

27. 2^8 **28.** 3^3 **29.** 4^4 **30.** 4^3 **31.** 9^3 **32.** 8^3

33. 12^2 **34.** 11^2 ▶ **35.** 10^2 **36.** 10^3 **37.** 20^1 **38.** 14^1

39. 3^6 **40.** 4^5 **41.** $3 \cdot 2^6$ **42.** $5 \cdot 3^2$ **43.** $2 \cdot 3^4$ **44.** $2 \cdot 7^2$

Objective C *Find each square root. See Examples 9 through 11.*

▶ **45.** $\sqrt{9}$ **46.** $\sqrt{36}$ ▶ **47.** $\sqrt{64}$ **48.** $\sqrt{121}$

49. $\sqrt{144}$ **50.** $\sqrt{0}$ **51.** $\sqrt{16}$ **52.** $\sqrt{169}$

Objective D *Simplify. See Examples 12 through 16. (This section does not contain square roots.)*

▶ **53.** $15 + 3 \cdot 2$ **54.** $24 + 6 \cdot 3$ ▶ **55.** $14 \div 7 \cdot 2 + 3$ **56.** $100 \div 10 \cdot 5 + 4$

57. $32 \div 4 - 3$ **58.** $42 \div 7 - 6$ **59.** $13 + \dfrac{24}{8}$ **60.** $32 + \dfrac{8}{2}$

61. $6 \cdot 5 + 8 \cdot 2$ **62.** $3 \cdot 4 + 9 \cdot 1$ **63.** $\dfrac{5 + 12 \div 4}{1^7}$ **64.** $\dfrac{6 + 9 \div 3}{3^2}$

65. $(7 + 5^2) \div 4 \cdot 2^3$ **66.** $6^2 \cdot (10 - 8)$ **67.** $5^2 \cdot (10 - 8) + 2^3 + 5^2$

68. $5^3 \div (10 + 15) + 9^2 + 3^3$ **69.** $\dfrac{18 + 6}{2^4 - 2^2}$ **70.** $\dfrac{40 + 8}{5^2 - 3^2}$

71. $(3 + 5) \cdot (9 - 3)$ **72.** $(9 - 7) \cdot (12 + 18)$ ▶ **73.** $\dfrac{7(9 - 6) + 3}{3^2 - 3}$

74. $\dfrac{5(12 - 7) - 4}{5^2 - 18}$

75. $8 \div 0 + 37$

76. $18 - 7 \div 0$

77. $2^4 \cdot 4 - (25 \div 5)$

78. $2^3 \cdot 3 - (100 \div 10)$

79. $3^4 - [35 - (12 - 6)]$

80. $[40 - (8 - 2)] - 2^5$

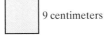 **81.** $(7 \cdot 5) + [9 \div (3 \div 3)]$

82. $(18 \div 6) + [(3 + 5) \cdot 2]$

83. $8 \cdot \left[2^2 + (6 - 1) \cdot 2\right] - 50 \cdot 2$

84. $35 \div \left[3^2 + (9 - 7) - 2^2\right] + 10 \cdot 3$

85. $\dfrac{9^2 + 2^2 - 1^2}{8 \div 2 \cdot 3 \cdot 1 \div 3}$

86. $\dfrac{5^2 - 2^3 + 1^4}{10 \div 5 \cdot 4 \cdot 1 \div 4}$

Simplify. See Examples 12 through 17. (This section does contain square roots.)

87. $6 \cdot \sqrt{9} + 3 \cdot \sqrt{4}$

88. $3 \cdot \sqrt{25} + 2 \cdot \sqrt{81}$

89. $4 \cdot \sqrt{49} - 0 \div \sqrt{100}$

90. $7 \cdot \sqrt{36} - 0 \div \sqrt{64}$

91. $\dfrac{\sqrt{4} + 4^2}{5(20 - 16) - 3^2 - 5}$

92. $\dfrac{\sqrt{9} + 9^2}{3(10 - 6) - 2^2 - 1}$

93. $\sqrt{81} \div \sqrt{9} + 4^2 \cdot 2 - 10$

94. $\sqrt{100} \div \sqrt{4} + 3^3 \cdot 2 - 20$

95. $\left[\sqrt{225} \div (11 - 6) + 2^2\right] + \left(\sqrt{25} - \sqrt{1}\right)^2$

96. $\left[\sqrt{169} \div (20 - 7) + 2^5\right] - \left(\sqrt{4} + \sqrt{9}\right)^2$

97. $7^2 - \left\{18 - \left[40 \div (4 \cdot 2) + \sqrt{4}\right] + 5^2\right\}$

98. $29 - \{5 + 3[8 \cdot (10 - \sqrt{64})] - 50\}$

Objective E Mixed Practice (*Sections 1.3, 1.6*) *Find the area and perimeter of each square. See Example 18.*

 99.

7 meters

△ **100.**
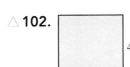
9 centimeters

△ **101.**

23 miles

△ **102.**
41 feet

Concept Extensions

Answer the following true or false. See the Concept Check in this section.

103. "Six to the fifth power" is the same as 6^5.

104. "Seven squared" is the same as 7^2.

105. 2^5 is the same as $5 \cdot 5$.

106. 4^9 is the same as $4 \cdot 9$.

Insert grouping symbols (parentheses) so that each given expression evaluates to the given number.

107. $2 + 3 \cdot 6 - 2$; evaluates to 28

108. $2 + 3 \cdot 6 - 2$; evaluates to 20

109. $24 \div 3 \cdot 2 + 2 \cdot 5$; evaluates to 14

110. $24 \div 3 \cdot 2 + 2 \cdot 5$; evaluates to 15

111. A building contractor is bidding on a contract to install gutters on seven homes in a retirement community, all in the shape shown. To estimate the cost of materials, she needs to know the total perimeter of all seven homes. Find the total perimeter.

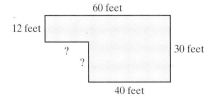

112. The building contractor from Exercise 111 plans to charge $4 per foot for installing vinyl gutters. Find the total charge for the seven homes given the total perimeter answer to Exercise 111.

Simplify.

113. $(7 + 2^4)^5 - (3^5 - 2^4)^2$

114. $25^3 \cdot (45 - 7 \cdot 5) \cdot 5$

115. Write an expression that simplifies to 5. Use multiplication, division, addition, subtraction, and at least one set of parentheses. Explain the process you would use to simplify the expression.

116. Explain why $2 \cdot 3^2$ is not the same as $(2 \cdot 3)^2$.

Chapter 1 Group Activity

Modeling Subtraction of Whole Numbers

A mathematical concept can be represented or modeled in many different ways. For instance, subtraction can be represented by the following symbolic model:

$$11 - 4$$

The following verbal models can also represent subtraction of these same quantities:

> "Four subtracted from eleven" or
> "Eleven take away four"

Physical models can also represent mathematical concepts. In these models, a number is represented by that many objects. For example, the number 5 can be represented by five pennies, squares, paper clips, tiles, or bottle caps.

A physical representation of the number 5

Take-Away Model for Subtraction: 11 – 4

- Start with 11 objects.
- Take 4 objects away.
- How many objects remain?

Start:

Take away 4:

Remain:

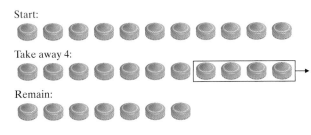

Comparison Model for Subtraction: 11 – 4

- Start with a set of 11 of one type of object and a set of 4 of another type of object.

- Make as many pairs that include one object of each type as possible.

- How many more objects left are in the larger set?

Missing Addend Model for Subtraction: 11 – 4

- Start with 4 objects.
- Continue adding objects until a total of 11 is reached.
- How many more objects were needed to give a total of 11?

Start:

Continue adding objects:

Group Activity

Use an appropriate physical model for subtraction to solve each of the following problems. Explain your reasoning for choosing each model.

1. Sneha has assembled 12 computer components so far this shift. If her quota is 20 components, how many more components must she assemble to reach her quota?

2. Yuko has 14 daffodil bulbs to plant in her yard. She planted 5 bulbs in the front yard. How many bulbs does she have left for planting in the backyard?

3. Todd is 19 years old and his sister Tanya is 13 years old. How much older is Todd than Tanya?

Chapter 1 Vocabulary Check

Fill in each blank with one of the words or phrases listed below.

difference	area	square root
place value	factor	quotient
sum	whole numbers	perimeter

addend	divisor	minuend
subtrahend	exponent	digits
dividend	average	product

1. The _____ are 0, 1, 2, 3, . . .

2. The _____ of a polygon is its distance around or the sum of the lengths of its sides.

3. The position of each digit in a number determines its _____.

4. A(n) _____ is a shorthand notation for repeated multiplication of the same factor.

5. To find the _____ of a rectangle, multiply length times width.

6. A(n) _____ of a number is one of two identical factors of the number.

7. The _____ used to write numbers are 0, 1, 2, 3, 4, 5, 6, 7, 8, and 9.

8. The _____ of a list of numbers is their sum divided by the number of numbers.

Use the facts below for Exercises 9 through 18.

$$2 \cdot 3 = 6 \quad 4 + 17 = 21 \quad 20 - 9 = 11 \quad 5\overline{)35}\,^{7}$$

9. The 5 above is called the _____.

10. The 35 above is called the _____.

11. The 7 above is called the _____.

12. The 3 above is called a(n) _____.

13. The 6 above is called the _____.

14. The 20 above is called the _____.

15. The 9 above is called the _____.

16. The 11 above is called the _____.

17. The 4 above is called a(n) _____.

18. The 21 above is called the _____.

Helpful Hint

▶ Are you preparing for your test? Don't forget to take the Chapter 1 Test on page 108. Then check your answers at the back of the text and use the Chapter Test Prep Videos to see the fully worked-out solutions to any of the exercises you want to review.

1 Chapter Highlights

Definitions and Concepts	Examples
Section 1.2 Place Value, Names for Numbers, and Reading Tables	
The **whole numbers** are 0, 1, 2, 3, 4, 5, ... The position of each digit in a number determines its **place value**. A place-value chart is shown next with the names of the periods given. 	0, 14, 968, 5,268,619

(continued)

Definitions and Concepts	Examples

Section 1.2   Place Value, Names for Numbers, and Reading Tables (*continued*)

To write a whole number in words, write the number in each period followed by the name of the period. (The name of the ones period is not included.)	9,078,651,002 is written as nine billion, seventy-eight million, six hundred fifty-one thousand, two.
To write a whole number in standard form, write the number in each period, followed by a comma.	Four million, seven hundred six thousand, twenty-eight is written as 4,706,028.

Section 1.3   Adding Whole Numbers and Perimeter

To add whole numbers, add the digits in the ones place, then the tens place, then the hundreds place, and so on, carrying when necessary.	Find the sum: $$\begin{array}{r} \overset{211}{} \\ 2689 \leftarrow \text{addend} \\ 1735 \leftarrow \text{addend} \\ +\ \ 662 \leftarrow \text{addend} \\ \hline 5086 \leftarrow \text{sum} \end{array}$$
The **perimeter** of a polygon is its distance around or the sum of the lengths of its sides.	△ Find the perimeter of the polygon shown. 5 feet, 2 feet, 3 feet, 9 feet The perimeter is 5 feet + 3 feet + 9 feet + 2 feet = 19 feet.

Section 1.4   Subtracting Whole Numbers

To subtract whole numbers, subtract the digits in the ones place, then the tens place, then the hundreds place, and so on, borrowing when necessary.	Subtract: $$\begin{array}{r} \overset{8\ 15}{79\cancel{5}4} \leftarrow \text{minuend} \\ -5673 \leftarrow \text{subtrahend} \\ \hline 2281 \leftarrow \text{difference} \end{array}$$

Section 1.5   Rounding and Estimating

Rounding Whole Numbers to a Given Place Value **Step 1:** Locate the digit to the right of the given place value. **Step 2:** If this digit is 5 or greater, add 1 to the digit in the given place value and replace each digit to its right with 0. **Step 3:** If this digit is less than 5, replace it and each digit to its right with 0.	Round 15,721 to the nearest thousand. 15,⑦21 Add 1 — Replace with zeros. Since the circled digit is 5 or greater, add 1 to the given place value and replace digits to its right with zeros. 15,721 rounded to the nearest thousand is 16,000.

Definitions and Concepts	Examples

Section 1.6   Multiplying Whole Numbers and Area

To multiply 73 and 58, for example, multiply 73 and 8, then 73 and 50. The sum of these partial products is the product of 73 and 58. Use the notation to the right.

$$
\begin{array}{r}
73 \quad \leftarrow \text{factor} \\
\times\ 58 \quad \leftarrow \text{factor} \\
\hline
584 \quad \leftarrow 73 \times 8 \\
3650 \quad \leftarrow 73 \times 50 \\
\hline
4234 \quad \leftarrow \text{product}
\end{array}
$$

To find the **area** of a rectangle, multiply length times width.

△ Find the area of the rectangle shown.

11 meters

7 meters

area of rectangle $=$ length $\cdot$ width
$=$ (11 meters)(7 meters)
$=$ 77 square meters

Section 1.7   Dividing Whole Numbers

Division Properties of 0

The quotient of 0 and any number (except 0) is 0.

The quotient of any number and 0 is not a number. We say that this quotient is undefined.

$$\frac{0}{5} = 0$$

$$\frac{7}{0} \text{ is undefined}$$

To divide larger whole numbers, use the process called **long division** as shown to the right.

$$
\begin{array}{r}
507 \ \ \text{R } 2 \quad \leftarrow \text{quotient and remainder} \\
\text{divisor} \rightarrow 14\overline{)7100} \quad \leftarrow \text{dividend} \\
-70\!\downarrow \qquad\qquad 5(14) = 70 \\
\hline
10 \qquad\qquad \text{Subtract and bring down the 0.} \\
-0\!\downarrow \qquad\qquad 0(14) = 0 \\
\hline
100 \qquad\qquad \text{Subtract and bring down the 0.} \\
-98 \qquad\qquad 7(14) = 98 \\
\hline
2 \qquad\qquad \text{Subtract. The remainder is 2.}
\end{array}
$$

To check, see that $507 \cdot 14 + 2 = 7100$.

The **average** of a list of numbers is

$$\text{average} = \frac{\text{sum of numbers}}{\textit{number of numbers}}$$

Find the average of 23, 35, and 38.

$$\text{average} = \frac{23 + 35 + 38}{3} = \frac{96}{3} = 32$$

Definitions and Concepts	Examples

Section 1.8   An Introduction to Problem Solving

Problem-Solving Steps

Suppose that 225 tickets are sold for each performance of a play. How many tickets are sold for 5 performances?

1. UNDERSTAND the problem.

1. UNDERSTAND. Read and reread the problem. Since we want the number of tickets for 5 performances, we multiply.

2. TRANSLATE the problem.

2. TRANSLATE.

number of tickets	is	number of performances	times	tickets per performance
↓		↓	↓	↓
Number of tickets	=	5	·	225

3. SOLVE the problem.

3. SOLVE: See if the answer is reasonable by also estimating.

$$\begin{array}{r} \overset{1\,2}{225} \\ \times\ \ 5 \\ \hline 1125 \end{array} \text{ exact} \qquad \text{rounds to} \qquad \begin{array}{r} 200 \\ \times\ \ 5 \\ \hline 1000 \end{array} \text{ estimate}$$

4. INTERPRET the results.

4. INTERPRET. **Check** your work. The product is reasonable since 1125 is close to our estimated answer of 1000, and **state** your conclusion: There are 1125 tickets sold for 5 performances.

Section 1.9   Exponents, Square Roots, and Order of Operations

An **exponent** is a shorthand notation for repeated multiplication of the same factor.

A **square root** of a number is one of two identical factors of the number.

$$\overset{\text{exponent}}{3^4} = \underbrace{3 \cdot 3 \cdot 3 \cdot 3}_{\text{4 factors of 3}} = 81$$

base

$$\sqrt{36} = 6 \quad \text{because} \quad 6 \cdot 6 = 36$$
$$\sqrt{121} = 11 \quad \text{because} \quad 11 \cdot 11 = 121$$
$$\sqrt{0} = 0 \quad \text{because} \quad 0 \cdot 0 = 0$$

Simplify: $\dfrac{5 + 3^2}{2(7 - 6)}$

Order of Operations

1. Perform all operations within parentheses ( ), brackets [ ], or other grouping symbols such as square roots or fraction bars, starting with the innermost set.

2. Evaluate any expressions with exponents.

3. Multiply or divide in order from left to right.

4. Add or subtract in order from left to right.

The **area of a square** is (side)^2.

Simplify above and below the fraction bar separately.

$$\frac{5 + 3^2}{2(7 - 6)} = \frac{5 + 9}{2(1)} \quad \text{Evaluate } 3^2 \text{ above the fraction bar.}$$
Subtract: $7 - 6$ below the fraction bar.
$$= \frac{14}{2} \quad \begin{array}{l}\text{Add.}\\ \text{Multiply.}\end{array}$$
$$= 7 \quad \text{Divide.}$$

Find the area of a square with side length 9 inches.

$$\begin{aligned} \text{Area of the square} &= (\text{side})^2 \\ &= (9 \text{ inches})^2 \\ &= 81 \text{ square inches} \end{aligned}$$

(1.2) *Determine the place value of the digit 4 in each whole number.*

1. 7640

2. 46,200,120

Write each whole number in words.

3. 7640

4. 46,200,120

Write each whole number in expanded form.

5. 3158

6. 403,225,000

Write each whole number in standard form.

7. Eighty-one thousand, nine hundred

8. Six billion, three hundred four million

The following table shows the Internet and Facebook use of world regions as of June 2012. Use this table to answer Exercises 9 through 12.

World Region	Internet Users	Facebook Users
Africa	167,335,676	51,612,460
Asia	1,076,681,059	254,336,520
Europe	518,512,109	250,934,000
Middle East	90,000,455	23,811,620
North America	273,785,413	184,177,220
Latin America/ Caribbean	254,915,745	188,339,620
Oceania/Australia	24,287,919	14,614,780
(*Source:* Internet World Stats)		

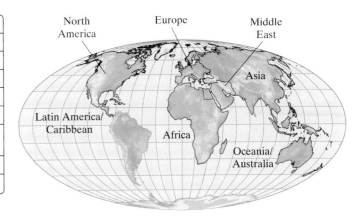

9. Find the number of Internet users in Europe.

10. Find the number of Facebook users in North America.

11. Which world region had the largest number of Facebook users?

12. Which world region had the smallest number of Internet users?

(1.3) *Add.*

13. 17 + 46

14. 28 + 39

15. 25 + 8 + 15

16. 27 + 9 + 41

17. 932 + 24

18. 819 + 21

19. 567 + 7383

20. 463 + 6787

21. 91 + 3623 + 497

22. 82 + 1647 + 238

Solve.

23. Find the sum of 86, 331, and 909.

24. Find the sum of 49, 529, and 308.

25. What is 26,481 increased by 865?

26. What is 38,556 increased by 744?

27. The distance from Chicago to New York City is 714 miles. The distance from New York City to New Delhi, India, is 7318 miles. Find the total distance from Chicago to New Delhi if traveling by air through New York City.

28. Susan Summerline earned salaries of $62,589, $65,340, and $69,770 during the years 2002, 2003, and 2004, respectively. Find her total earnings during those three years.

Find the perimeter of each figure.

29.

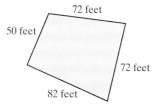

72 feet
50 feet
72 feet
82 feet

30. 11 kilometers   20 kilometers
35 kilometers

(1.4) *Subtract and then check.*

31. 93 − 79   **32.** 61 − 27   **33.** 462 − 397   **34.** 583 − 279   **35.** 4000 − 86   **36.** 8000 − 92

Solve.

37. Subtract 7965 from 25,862.

38. Subtract 4349 from 39,007.

39. Find the increase in population for San Antonio, Texas, from 2000 (population: 1,144,646) to 2012 (population: 1,382,951). (*Source:* U.S. Census Bureau)

40. Find the decrease in population for Detroit, Michigan, from 2000 (population: 951,270) to 2012 (population: 701,475). (*Source:* U.S. Census Bureau)

41. Bob Roma is proofreading the Yellow Pages for his county. If he has finished 315 pages of the total 712 pages, how many pages does he have left to proofread?

42. Shelly Winters bought a new car listed at $28,425. She received a discount of $1599 and a factory rebate of $1200. Find how much she paid for the car.

The following bar graph shows the monthly savings account balance for a freshman attending a local community college. Use this graph to answer Exercises 43 through 46.

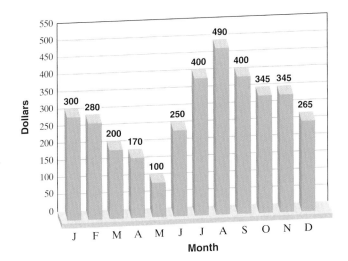

43. During what month was the balance the least?

44. During what month was the balance the greatest?

45. By how much did his balance decrease from February to April?

46. By how much did his balance increase from June to August?

(1.5) *Round to the given place.*

47. 93 to the nearest ten

48. 45 to the nearest ten

49. 467 to the nearest ten

50. 493 to the nearest hundred

51. 4832 to the nearest hundred

52. 57,534 to the nearest thousand

53. 49,683,712 to the nearest million

54. 768,542 to the nearest hundred-thousand

55. In 2012, 126,226,713 Americans cast a ballot in the presidential election. Round this number to the nearest million. (*Source:* CNN)

56. In 2011, there were 98,817 public elementary and secondary schools in the United States. Round this number to the nearest thousand. (*Source:* National Center for Education Statistics)

Estimate the sum or difference by rounding each number to the nearest hundred.

57. 4892 + 647 + 1876

58. 5925 − 1787

59. A group of students took a week-long driving trip and traveled 628, 290, 172, 58, 508, 445, and 383 miles on seven consecutive days. Round each distance to the nearest hundred to estimate the distance they traveled.

60. The estimated 2012 population of Houston, Texas, was 2,160,821, and for San Diego, California, it was 1,338,348. Round each number to the nearest hundred-thousand and estimate how much larger Houston is than San Diego. (*Source:* U.S. Census Bureau)

(1.6) *Multiply.*

61. 273
× 7

62. 349
× 4

63. 47
× 30

64. 69
× 42

65. 20(8)(5)

66. 25(9)(4)

67. 48
× 77

68. 77
× 22

69. $49 \cdot 49 \cdot 0$

70. $62 \cdot 88 \cdot 0$

71. 586
× 29

72. 242
× 37

73. 642
× 177

74. 347
× 129

75. 1026
× 401

76. 2107
× 302

77. $375 \cdot 1000$

78. $108 \cdot 1000$

79. $30 \cdot 400$

80. $50 \cdot 700$

81. $1700 \cdot 3000$

82. $1900 \cdot 4000$

Solve.

83. Find the product of 5 and 230.

84. Find the product of 6 and 820.

85. Multiply 9 and 12.

86. Multiply 8 and 14.

87. One ounce of Swiss cheese contains 8 grams of fat. How many grams of fat are in 3 ounces of Swiss cheese? (*Source: Home and Garden Bulletin No. 72,* U.S. Department of Agriculture)

88. The cost for a South Dakota resident to attend Black Hills State University full-time is $7617 per semester. Determine the cost for 20 students to attend full-time. (*Source:* Black Hills State University)

Find the area of each rectangle.

△**89.**

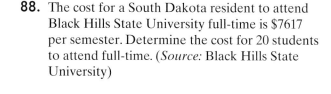

12 miles
5 miles

△**90.** 20 centimeters

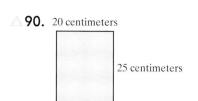

25 centimeters

(1.7) *Divide and then check.*

91. $\dfrac{18}{6}$

92. $\dfrac{36}{9}$

93. $42 \div 7$

94. $35 \div 5$

95. $27 \div 5$

96. $18 \div 4$

97. $16 \div 0$

98. $0 \div 8$

99. $9 \div 9$

100. $10 \div 1$

101. $0 \div 668$

102. $918 \div 0$

103. $5\overline{)167}$

104. $8\overline{)159}$

105. $26\overline{)626}$

106. $19\overline{)680}$

107. $47\overline{)23{,}792}$

108. $53\overline{)48{,}111}$

109. $207\overline{)578{,}291}$

110. $306\overline{)615{,}732}$

Solve.

111. Find the quotient of 92 and 5.

112. Find the quotient of 86 and 4.

113. One foot is 12 inches. Find how many feet there are in 5496 inches.

114. One mile is 1760 yards. Find how many miles there are in 22,880 yards.

115. Find the average of the numbers 76, 49, 32, and 47.

116. Find the average of the numbers 23, 85, 62, and 66.

(1.8) *Solve.*

117. A box can hold 24 cans of corn. How many boxes can be filled with 648 cans of corn?

118. If a ticket to a movie costs $6, how much do 32 tickets cost?

119. In 2012, U.S. companies spent $74 billion on television advertising. By comparison, U.S. companies spent only $69 billion on television advertising in 2011. How much more did U.S. companies spend on television advertising in 2012? (*Source:* Kantar Media)

120. The cost to banks when a person uses an ATM (Automatic Teller Machine) is 27¢. The cost to banks when a person deposits a check with a teller is 48¢ more. How much is this cost?

121. A golf pro orders shirts for the company sponsoring a local charity golfing event. Shirts size large cost $32 while shirts size extra-large cost $38. If 15 large shirts and 11 extra-large shirts are ordered, find the cost.

122. Two rectangular pieces of land are purchased: one that measures 65 feet by 110 feet and one that measures 80 feet by 200 feet. Find the total area of land purchased. (*Hint:* Find the area of each rectangle, then add.)

(1.9) *Simplify.*

123. 7^2 **124.** 5^3 **125.** $5 \cdot 3^2$ **126.** $4 \cdot 10^2$

127. $18 \div 3 + 7$ **128.** $12 - 8 \div 4$ **129.** $\dfrac{5(6^2 - 3)}{3^2 + 2}$ **130.** $\dfrac{7(16 - 8)}{2^3}$

131. $48 \div 8 \cdot 2$ **132.** $27 \div 9 \cdot 3$

133. $2 + 3\left[1^5 + (20 - 17) \cdot 3\right] + 5 \cdot 2$ **134.** $21 - [2^4 - (7 - 5) - 10] + 8 \cdot 2$

Simplify. (These exercises contain square roots.)

135. $\sqrt{81}$ **136.** $\sqrt{4}$ **137.** $\sqrt{1}$ **138.** $\sqrt{0}$

139. $4 \cdot \sqrt{25} - 2 \cdot 7$ **140.** $8 \cdot \sqrt{49} - 3 \cdot 9$

141. $(\sqrt{36} - \sqrt{16})^3 \cdot [10^2 \div (3 + 17)]$ **142.** $(\sqrt{49} - \sqrt{25})^3 \cdot [9^2 \div (2 + 7)]$

143. $\dfrac{5 \cdot 7 - 3 \cdot \sqrt{25}}{2(\sqrt{121} - 3^2)}$ **144.** $\dfrac{4 \cdot 8 - 1 \cdot \sqrt{121}}{3(\sqrt{81} - 2^3)}$

Find the area of each square.

△**145.** A square with side length of 7 meters.   △**146.**

3 inches

Mixed Review

Perform the indicated operations.

147. $375 - 68$ **148.** $729 - 47$ **149.** 723×3 **150.** 629×4

151. $264 + 39 + 598$ **152.** $593 + 52 + 766$ **153.** $13\overline{)5962}$ **154.** $18\overline{)4267}$

155. 1968×36 **156.** 5324×18 **157.** $2000 - 356$ **158.** $9000 - 519$

Round to the given place.

159. 736 to the nearest ten

160. 258,371 to the nearest thousand

161. 1999 to the nearest hundred

162. 44,499 to the nearest ten thousand

Write each whole number in words.

163. 36,911

164. 154,863

Write each whole number in standard form.

165. Seventy thousand, nine hundred forty-three

166. Forty-three thousand, four hundred one

Simplify.

167. 4^3

168. 5^3

169. $\sqrt{144}$

170. $\sqrt{100}$

171. $24 \div 4 \cdot 2$

172. $\sqrt{256} - 3 \cdot 5$

173. $\dfrac{8(7-4)-10}{4^2-3^2}$

174. $\dfrac{(15+\sqrt{9}) \cdot (8-5)}{2^3+1}$

Solve.

175. 36 divided by 9 is what number?

176. What is the product of 2 and 12?

177. 16 increased by 8 is what number?

178. 7 subtracted from 21 is what number?

The following table shows the 2012 and 2013 average Major League Baseball salaries (rounded to the nearest thousand) for the five teams with the largest payrolls for 2013. Use this table to answer Exercises 179 and 180. (Source: CBSSports.com, Associated Press)

Team	2013 Average Salary	2012 Average Salary
Los Angeles Dodgers	$7,469,000	$3,264,000
New York Yankees	$7,151,000	$6,256,000
Philadelphia Phillies	$6,125,000	$5,798,000
Detroit Tigers	$5,708,000	$4,561,000
Boston Red Sox	$5,022,000	$5,094,000

179. How much more was the average salary for a Los Angeles Dodgers player in 2013 than in 2012?

180. How much less was the average Boston Red Sox salary than the average New York Yankee salary in 2013?

181. A manufacturer of drinking glasses ships his delicate stock in special boxes that can hold 32 glasses. If 1714 glasses are manufactured, how many full boxes are filled? Are there any glasses left over?

182. A teacher orders 2 small white boards for $27 each and 8 boxes of dry erase pens for $4 each. What is her total bill before taxes?

Answers

Simplify.

1. Write 82,426 in words.

2. Write "four hundred two thousand, five hundred fifty" in standard form.

1. _____

2. _____

3. $59 + 82$

4. $600 - 487$

5.   496
   $\times\ \ 30$

3. _____

4. _____

5. _____

6. $52,896 \div 69$

7. $2^3 \cdot 5^2$

8. $\sqrt{4} \cdot \sqrt{25}$

6. _____

7. _____

9. $0 \div 49$

10. $62 \div 0$

11. $\left(2^4 - 5\right) \cdot 3$

8. _____

9. _____

10. _____

12. $16 + 9 \div 3 \cdot 4 - 7$

13. $\dfrac{64 \div 8 \cdot 2}{\left(\sqrt{9} - \sqrt{4}\right)^2 + 1}$

11. _____

12. _____

13. _____

14. $2\left[(6 - 4)^2 + (22 - 19)^2\right] + 10$

15. $5698 \cdot 1000$

14. _____

15. _____

16. $8000 \cdot 1400$

17. Round 52,369 to the nearest thousand.

16. _____

17. _____

18. _____

Estimate each sum or difference by rounding each number to the nearest hundred.

18. $6289 + 5403 + 1957$

19. $4267 - 2738$

19. _____

Solve.

20. Subtract 15 from 107.

21. Find the sum of 15 and 107.

22. Find the product of 15 and 107.

23. Find the quotient of 107 and 15.

24. Twenty-nine cans of Sherwin-Williams paint cost $493. How much was each can?

25. Jo McElory is looking at two new refrigerators for her apartment. One costs $599 and the other costs $725. How much more expensive is the higher-priced one?

26. One tablespoon of white granulated sugar contains 45 calories. How many calories are in 8 tablespoons of white granulated sugar? (*Source: Home and Garden Bulletin No. 72, U.S. Department of Agriculture*)

27. A small business owner recently ordered 16 digital cameras that cost $430 each and 5 printers that cost $205 each. Find the total cost for these items.

Find the perimeter and the area of each figure.

△ 28.

| Square | 5 centimeters |

△ 29.

20 yards

| Rectangle | 10 yards |

20. _____

21. _____

22. _____

23. _____

24. _____

25. _____

26. _____

27. _____

28. _____

29. _____

2 Integers and Introduction to Variables

Thus far, we have studied whole numbers, but these numbers are not sufficient for representing many situations in real life. For example, to express 5 degrees below zero or $100 in debt, numbers less than 0 are needed. This chapter is devoted to integers, which include numbers less than 0, and operations on these numbers.

Director James Cameron has made the deepest solo descent so far into the Mariana Trench in the Pacific Ocean. He reached a depth of 35,756 feet in the Deepsea Challenger, shown above. Next, maybe Richard Branson?

The Krubera Cave now holds the title of deepest. In this cave, many new depth records have been set—each one deeper than the last. The latest record is 7188 feet, but who knows how deeply this cave will be explored next?

Where Do We Explore Next?

Throughout this chapter, we present many applications having to do with water depths below sea level and land depths below the surface of Earth by way of mines and caves. Recently, there has been a surge of interest in exploring these depths. Although we have already reached the deepest-known part of our oceans—the Mariana Trench in the Pacific Ocean—cave exploration is a little more tricky. For example, with so many "branches" of a cave, we are never certain that it has been totally explored. New caves are being discovered and explored even as this is written. The deepest-known cave in the world, the Krubera, was not discovered until 2001 by Ukrainian cave explorers. See exercises throughout this chapter.

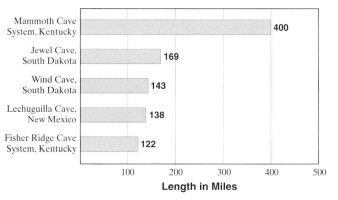

Top 5 Longest Caves in U.S.

Source: National Speleological Society, Geo 2, 2014

2.1 Introduction to Variables and Algebraic Expressions

Objective A Evaluating Algebraic Expressions

Objectives

A Evaluate Algebraic Expressions Given Replacement Values.

B Translate Phrases into Variable Expressions.

Perhaps the most important quality of mathematics is that it is a science of patterns. Communicating about patterns is often made easier by using a letter to represent all the numbers fitting a pattern. We call such a letter a **variable.** For example, in Section 1.3 we presented the addition property of 0, which states that the sum of 0 and any number is that number. We might write

$$0 + 1 = 1$$
$$0 + 2 = 2$$
$$0 + 3 = 3$$
$$0 + 4 = 4$$
$$0 + 5 = 5$$
$$0 + 6 = 6$$
$$\vdots$$

continuing indefinitely. This is a pattern, and all whole numbers fit the pattern. We can communicate this pattern for all whole numbers by letting a letter, such as a, represent all whole numbers. We can then write

$$0 + a = a$$

Using variable notation is a primary goal of learning **algebra.** We now take some important first steps in beginning to use variable notation.

A combination of operations on letters (variables) and numbers is called an **algebraic expression** or simply an **expression.**

Algebraic Expressions

$$3 + x \qquad 5 \cdot y \qquad 2 \cdot z - 1 + x$$

If two variables or a number and a variable are next to each other, with no operation sign between them, the operation is multiplication. For example,

$$2x \quad \text{means} \quad 2 \cdot x$$

and

$$xy \text{ or } x(y) \quad \text{means} \quad x \cdot y$$

Also, the meaning of an exponent remains the same when the base is a variable. For example,

$$x^2 = \underbrace{x \cdot x}_{\text{2 factors of } x} \quad \text{and} \quad y^5 = \underbrace{y \cdot y \cdot y \cdot y \cdot y}_{\text{5 factors of } y}$$

Algebraic expressions such as $3x$ have different values depending on replacement values for the variables—in this case, the variable x. For example, if x is 2, then $3x$ becomes

$$3x = 3 \cdot 2$$
$$= 6$$

If x is 7, then $3x$ becomes

$$3x = 3 \cdot 7$$
$$= 21$$

Replacing a variable in an expression by a number and then finding the value of the expression is called **evaluating the expression** for the variable. When finding the value of an expression, remember to follow the order of operations given in Section 1.9.

Practice 1

Evaluate $x - 2$ if x is 5.

Example 1 Evaluate $x + 7$ if x is 8.

Solution: Replace x with 8 in the expression $x + 7$.

$$x + 7 = 8 + 7 \quad \text{Replace } x \text{ with 8.}$$
$$= 15 \quad \text{Add.}$$

Work Practice 1

When we write a statement such as "x is 5," we can use an equal symbol to represent "is" so that

x is 5   can be written as $x = 5$.

Practice 2

Evaluate $y(x - 3)$ for $x = 3$ and $y = 7$.

Example 2 Evaluate $2(x - y)$ for $x = 8$ and $y = 4$.

Solution: $2(x - y) = 2(8 - 4)$ Replace x with 8 and y with 4.
$$= 2(4) \quad \text{Subtract.}$$
$$= 8 \quad \text{Multiply.}$$

Work Practice 2

Practice 3

Evaluate $\dfrac{y + 6}{x}$ for $x = 2$ and $y = 8$.

Example 3 Evaluate $\dfrac{x - 5y}{y}$ for $x = 21$ and $y = 3$.

Solution: $\dfrac{x - 5y}{y} = \dfrac{21 - 5(3)}{3}$ Replace x with 21 and y with 3.
$$= \dfrac{21 - 15}{3} \quad \text{Multiply.}$$
$$= \dfrac{6}{3} \quad \text{Subtract.}$$
$$= 2 \quad \text{Divide.}$$

Work Practice 3

Practice 4

Evaluate $25 - z^3 + x$ for $z = 2$ and $x = 1$.

Example 4 Evaluate $x^2 + z - 3$ for $x = 5$ and $z = 4$.

Solution: $x^2 + z - 3 = 5^2 + 4 - 3$ Replace x with 5 and z with 4.
$$= 25 + 4 - 3 \quad \text{Evaluate } 5^2.$$
$$= 26 \quad \text{Add and subtract from left to right.}$$

Work Practice 4

Answers

1. 3   **2.** 0   **3.** 7   **4.** 18

Helpful Hint

If you are having difficulty replacing variables with numbers, first replace each variable with a set of parentheses, then insert the replacement number within the parentheses.

Using this method in Example 4, we have:

$$x^2 + z - 3 = ()^2 + () - 3$$
$$= (5)^2 + (4) - 3$$
$$= 25 + 4 - 3$$
$$= 26$$

✓ **Concept Check** What's wrong with the solution to the following problem? Evaluate $3x + 2y$ for $x = 2$ and $y = 3$.

Solution: $3x + 2y = 3(3) + 2(2)$
$$= 9 + 4$$
$$= 13$$

Example 5   The expression $\dfrac{5(F - 32)}{9}$ can be used to write degrees Fahrenheit F as degrees Celsius C. Find the value of this expression for $F = 86$.

Solution: $\dfrac{5(F - 32)}{9} = \dfrac{5(86 - 32)}{9}$

$$= \dfrac{5(54)}{9}$$

$$= \dfrac{270}{9}$$

$$= 30$$

Thus 86°F is the same temperature as 30°C.

■ Work Practice 5

Practice 5

Evaluate $\dfrac{5(F - 32)}{9}$ for $F = 41$.

Objective B Translating Phrases into Variable Expressions ▶

To aid us in solving problems later, we practice translating verbal phrases into algebraic expressions. Recall from Section 1.8 that certain key words and phrases suggest addition, subtraction, multiplication, or division. These are reviewed next.

Addition (+)	Subtraction (−)	Multiplication (·)	Division (÷)
sum	difference	product	quotient
plus	minus	times	divide
added to	subtract	multiply	shared equally among
more than	less than	multiply by	per
increased by	decreased by	of	divided by
total	less	double/triple	divided into

Answer
5. 5

✓ **Concept Check Answer**
The replacement values were switched. If done correctly, we have:

$$3x + 2y = 3(2) + 2(3)$$
$$= 6 + 6$$
$$= 12$$

Practice 6

Write each as an algebraic expression. Use x to represent "a number."

a. Twice a number.

b. 8 increased by a number.

c. 10 minus a number.

d. 10 subtracted from a number.

e. The quotient of a number and 16.

Example 6 Write each as an algebraic expression. Use x to represent "a number."

a. 7 increased by a number.

b. 15 decreased by a number.

c. The product of 2 and a number.

d. The quotient of a number and 5.

e. 2 subtracted from a number.

Solution:

a. In words: 7   increased by   a number
Translate: 7   + x

b. In words: 15   decreased by   a number
Translate: 15   − x

c. In words:
The product of
2   and   a number
Translate: 2   · x or $2x$

d. In words:
The quotient of
a number   and   5
Translate: x ÷   5   or $\dfrac{x}{5}$

e. In words: 2   subtracted from   a number
Translate: x −   2

■ Work Practice 6

Helpful Hint

Remember that order is important when subtracting. Study the order of numbers and variables below.

Phrase	Translation
a number *decreased by* 5	$x - 5$
a number *subtracted from* 5	$5 - x$

Answers

6. a. $2x$ **b.** $8 + x$ **c.** $10 - x$

d. $x - 10$ **e.** $x \div 16$ or $\dfrac{x}{16}$

Vocabulary, Readiness & Video Check

Use the choices below to fill in each blank. Some choices may be used more than once.

evaluating the expression   variable(s)   expression

1. A combination of operations on letters (variables) and numbers is a(n) _____.

2. A letter that represents a number is a(n) _____.

3. $3x - 2y$ is called a(n) _____ and the letters x and y are _____.

4. Replacing a variable in an expression by a number and then finding the value of the expression is called _____.

Martin-Gay Interactive Videos   *Watch the section lecture video and answer the following questions.*

Objective A **5.** Complete this statement based on the lecture before ⊞ Example 1: When a letter and a variable are next to each other, the operation is an understood _____.

Objective B **6.** In ⊞ Example 4, what phrase translates to subtraction?

See Video 2.1

2.1 Exercise Set MyMathLab®

Objective A *Complete the table. The first row has been done for you. See Examples 1 through 5.*

	a	b	$a + b$	$a - b$	$a \cdot b$	$a \div b$
	45	9	54	36	405	5
1.	21	7				
2.	24	6				
3.	152	0				
4.	298	0				
5.	56	1				
6.	82	1				

Evaluate each expression for $x = 2$, $y = 5$, and $z = 3$. See Examples 1 through 5.

7. $3 + 2z$

8. $7 + 3z$

9. $6xz - 5x$

10. $4yz + 2x$

11. $z - x + y$

12. $x + 5y - z$

13. $3x - z$

14. $2y + 5z$

15. $y^3 - 4x$

16. $y^3 - z$

17. $2xy^2 - 6$

18. $3yz^2 + 1$

19. $8 - (y - x)$

20. $5 + (2x - 1)$

21. $y^4 + (z - x)$

22. $x^4 - (y - z)$

23. $\dfrac{6xy}{z}$

24. $\dfrac{8yz}{15}$

25. $\dfrac{2y - 2}{x}$

26. $\dfrac{6 + 3x}{z}$

27. $\dfrac{x + 2y}{z}$

28. $\dfrac{2z + 6}{3}$

29. $\dfrac{5x}{y} - \dfrac{10}{y}$

30. $\dfrac{70}{2y} - \dfrac{15}{z}$

31. $2y^2 - 4y + 3$

32. $3z^2 - z + 10$

33. $(3y - 2x)^2$

34. $(4y + 3z)^2$

35. $(xy + 1)^2$

36. $(xz - 5)^4$

▶ **37.** $2y(4z - x)$

38. $3x(y + z)$

39. $xy(5 + z - x)$

40. $xz(2y + x - z)$

41. $\dfrac{7x + 2y}{3x}$

42. $\dfrac{6z + 2y}{4}$

43. The expression $16t^2$ gives the distance in feet that an object falls after t seconds. Complete the table by evaluating $16t^2$ for each given value of t.

t	1	2	3	4
$16t^2$				

44. The expression $\dfrac{5(F - 32)}{9}$ gives the equivalent degrees Celsius for F degrees Fahrenheit. Complete the table by evaluating this expression for each given value of F.

F	50	59	68	77
$\dfrac{5(F - 32)}{9}$				

Objective B *Write each phrase as a variable expression. Use x to represent "a number." See Example 6.*

45. The sum of a number and five

46. Ten plus a number

47. The total of a number and eight

48. The difference of a number and five hundred

▶ **49.** Twenty decreased by a number

50. A number less thirty

51. The product of 512 and a number

52. A number times twenty

53. A number divided by 2

54. The quotient of seven and a number

55. The sum of seventeen and a number added to the product of five and the number

56. The difference of twice a number, and four

▶ **57.** The product of five and a number

58. The quotient of twenty and a number, decreased by three

59. A number subtracted from 11

60. Twelve subtracted from a number

61. A number less 5

62. The product of a number and 7

63. 6 divided by a number

64. The sum of a number and 7

65. Fifty decreased by eight times a number

66. Twenty decreased by twice a number

Review

Determine the place value of the digit 7 in each whole number. See Section 1.2.

67. 720

68. 2307

69. 67,522

70. 179

Concept Extensions

Solve. See the Concept Check in this section. Determine whether each expression is correctly evaluated for $x = 2$, $y = 0$, and $z = 7$. If incorrect, then correctly evaluate.

71. $2y + 3z \overset{?}{=} 2(2) + 3(7)$
$\overset{?}{=} 4 + 21$
$\overset{?}{=} 25$

72. $2z - 4x \overset{?}{=} 2(7) - 4(2)$
$\overset{?}{=} 14 - 8$
$\overset{?}{=} 6$

73. $\dfrac{4z}{2x} \overset{?}{=} \dfrac{4(7)}{2(2)}$
$\overset{?}{=} \dfrac{28}{4}$
$\overset{?}{=} 7$

74. $\dfrac{2xy}{z} \overset{?}{=} \dfrac{2(2)(0)}{7}$
$\overset{?}{=} \dfrac{0}{7}$
is undefined

Use a calculator to evaluate each expression for $x = 23$ and $y = 72$.

75. $x^4 - y^2$

76. $2(x + y)^2$

77. $x^2 + 5y - 112$

78. $16y - 20x + x^3$

79. If x is a whole number, which expression is the largest: $2x$, $5x$, or $\dfrac{x}{3}$? Explain your answer.

80. If x is a whole number, which expression is the smallest: $2x$, $5x$, or $\dfrac{x}{3}$? Explain your answer.

81. In Exercise **43**, what do you notice about the value of $16t^2$ as t gets larger?

82. In Exercise **44**, what do you notice about the value of $\dfrac{5(F - 32)}{9}$ as F gets larger?

Objectives

A Represent Real-Life Situations with Integers.

B Graph Integers on a Number Line.

C Compare Integers.

D Find the Absolute Value of a Number.

E Find the Opposite of a Number.

F Read Bar Graphs Containing Integers.

Objective A Representing Real-Life Situations

Thus far in this text, all numbers have been 0 or greater than 0. Numbers greater than 0 are called **positive numbers.** However, sometimes situations exist that cannot be represented by a number greater than 0. For example,

5 degrees below 0°

Sea level

20 feet below sea level

To represent these situations, we need numbers less than 0.

Extending the number line to the left of 0 allows us to picture **negative numbers,** which are numbers that are less than 0.

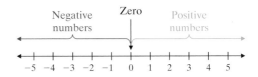

When a single + sign or no sign is in front of a number, the number is a positive number. When a single − sign is in front of a number, the number is a negative number. Together, we call positive numbers, negative numbers, and zero the **signed numbers.**

> −5 indicates "negative five."
>
> 5 and +5 both indicate "positive five."
>
> The number 0 is neither positive nor negative.

Some signed numbers are integers. The **integers** consist of the numbers labeled on the number line above. The integers are

$$\ldots, -3, -2, -1, 0, 1, 2, 3, \ldots$$

Now we have numbers to represent the situations previously mentioned.

5 degrees below 0 $-5°$

20 feet below sea level -20 feet

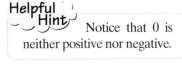

Helpful Hint   Notice that 0 is neither positive nor negative.

Helpful Hint

A − sign, such as the one in −1, tells us that the number is to the left of 0 on a number line. −1 is read "negative one."

A + sign or no sign tells us that a number lies to the right of 0 on a number line. For example, 3 and +3 both mean "positive three."

Example 1 Representing Depth with an Integer

The world's deepest cave is Krubera (or Voronja), in the country of Georgia, located by the Black Sea in Asia. It has been explored to a depth of 7188 feet below the surface of Earth. Represent this position using an integer. (*Source: MessagetoEagle.com and Wikipedia*)

Solution: If 0 represents the surface of Earth, then 7188 feet below the surface can be represented by -7188.

🔳 Work Practice 1

Objective B Graphing Integers

Example 2 Graph 0, −3, 2, and −2 on the number line.

Solution:

🔳 Work Practice 2

Objective C Comparing Integers

We can compare integers by using a number line. For any two numbers graphed on a number line, the number to the **right** is the **greater number** and the number to the **left** is the **smaller number.** Also, the symbols < and > are called **inequality symbols.**

The inequality symbol > means "is greater than" and
the inequality symbol < means "is less than."
For example, both −5 and −7 are graphed on the number line below.

On the graph, −7 is **to the left of** −5, so −7 **is less than** −5, written as

$$-7 < -5$$

We can also write

$$-5 > -7$$

since −5 is **to the right of** −7, so −5 **is greater than** −7.

✓Concept Check Is there a largest positive number? Is there a smallest negative number? Explain.

Practice 1

a. The world's deepest bat colony spends each winter in a New York zinc mine at a depth of 3805 feet. Represent this position with an integer. (*Source: Guinness Book of World Records*)

b. The tamarack tree, a type of conifer, commonly grows at the edge of the arctic tundra and survives winter temperatures of 85 degrees below zero Fahrenheit. Represent this temperature with an integer in degrees Fahrenheit.

Practice 2

Graph −5, −4, 3, and −3 on the number line.

Practice 3

Insert < or > between each pair of numbers to make a true statement.

a. 0   −3

b. −5   5

c. −8   −12

Example 3 Insert < or > between each pair of numbers to make a true statement.

a. −7   7   **b.** 0   −4   **c.** −9   −11

Solution:

a. −7 is to the left of 7 on a number line, so −7 < 7.

b. 0 is to the right of −4 on a number line, so 0 > −4.

c. −9 is to the right of −11 on a number line, so −9 > −11.

▪ Work Practice 3

Helpful Hint If you think of < and > as arrowheads, notice that in a true statement, the arrow always points to the smaller number.

5 > −4   −3 < −1

↑   ↑

smaller   smaller

number   number

Objective D Finding the Absolute Value of a Number

The **absolute value** of a number is the number's distance from 0 on a number line. The symbol for absolute value is | |. For example, |3| is read as "the absolute value of 3."

|3| = 3 because 3 is 3 units from 0.

|−3| = 3 because −3 is 3 units from 0.

Practice 4

Simplify.

a. |−4| **b.** |2|

c. |−8|

Example 4 Simplify.

a. |−2| **b.** |8| **c.** |0|

Solution:

a. |−2| = 2 because −2 is 2 units from 0.

b. |8| = 8 because 8 is 8 units from 0.

c. |0| = 0 because 0 is 0 units from 0.

▪ Work Practice 4

Helpful Hint

Since the absolute value of a number is that number's *distance* from 0, the absolute value of a number is always 0 or positive. It is never negative.

|0| = 0 |−6| = 6

↑   ↑

zero   a positive number

Objective E Finding Opposites

Two numbers that are the same distance from 0 on a number line but are on opposite sides of 0 are called **opposites**.

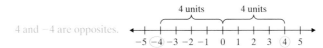

4 and −4 are opposites.

When two numbers are opposites, we say that each is the opposite of the other. Thus **4 is the opposite of −4** and **−4 is the opposite of 4.**

Answers

3. **a.** >   **b.** <   **c.** >

4. **a.** 4   **b.** 2   **c.** 8

The phrase "the opposite of" is written in symbols as "−". For example,

The opposite of   5   is   −5

                    ↓   ↓   ↓   ↓

                    −   (5)   =   −5

The opposite of   −3   is   3

                    ↓   ↓   ↓   ↓

                    −   (−3)   =   3   or

$$-(-3) = 3$$

In general, we have the following:

Opposites

If a is a number, then $-(-a) = a$.

Notice that because "the opposite of" is written as "−", to find the opposite of a number, we place a "−" sign in front of the number.

Example 5 Find the opposite of each number.

a. 11   **b.** −2   **c.** 0

Solution:

a. The opposite of 11 is −11.
b. The opposite of −2 is $-(-2)$ or 2.
c. The opposite of 0 is 0.

Helpful Hint Remember that 0 is neither positive nor negative.

■ Work Practice 5

Practice 5

Find the opposite of each number.

a. 7   **b.** −17

✓**Concept Check** True or false? The number 0 is the only number that is its own opposite.

Example 6 Simplify.

a. $-(-4)$ **b.** $-|-5|$ **c.** $-|6|$

Solution:

a. $-(-4) = 4$ The opposite of negative 4 is 4.

b. $-|-5| = -5$ The opposite of the absolute value of −5 is the opposite of 5, or −5.

c. $-|6| = -6$ The opposite of the absolute value of 6 is the opposite of 6, or −6.

■ Work Practice 6

Practice 6

Simplify.
a. $-|-2|$
b. $-|5|$
c. $-(-11)$

Example 7 Evaluate $-|-x|$ if $x = -2$.

Solution: Carefully replace x with −2; then simplify.

$$-|-x| = -|-(-2)|$$ Replace x with −2.

Then $-|-(-2)| = -|2| = -2$.

■ Work Practice 7

Practice 7

Evaluate $-|x|$ if $x = -9$.

Answers

5. a. −7 b. 17 6. a. −2 b. −5
c. 11 7. −9

✓**Concept Check Answer**

True

Objective F Reading Bar Graphs Containing Integers

The bar graph below shows the average daytime surface temperature (in degrees Fahrenheit) of the eight planets, excluding the newly classified "dwarf planet," Pluto. Notice that a negative temperature is illustrated by a bar below the horizontal line representing 0°F, and a positive temperature is illustrated by a bar above the horizontal line representing 0°F.

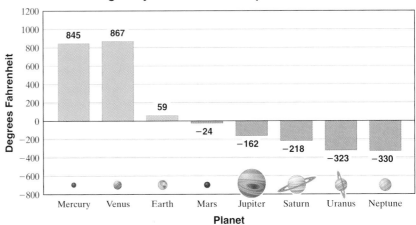

Average Daytime Surface Temperatures of Planets*

Source: The World Almanac
* For some planets, the temperature given is the temperature where the atmospheric pressure equals 1 Earth atmosphere.

Practice 8

Which planet has the highest average daytime surface temperature? What is this temperature?

Answer

8. Venus; 867°F

Example 8 Which planet has the lowest average daytime surface temperature?

Solution: The planet with the lowest average daytime surface temperature is the one that corresponds to the bar that extends the farthest in the negative direction (downward). Neptune has the lowest average daytime surface temperature, −330°F.

■ Work Practice 8

Vocabulary, Readiness & Video Check

Use the choices below to fill in each blank. Not all choices will be used.

opposites	absolute value	right	is less than
inequality symbols	negative	positive	left
signed	integers	is greater than	

1. The numbers . . . , −3, −2, −1, 0, 1, 2, 3, . . . are called _____.
2. Positive numbers, negative numbers, and zero, together, are called _____ numbers.
3. The symbols "<" and ">" are called _____.
4. Numbers greater than 0 are called _____ numbers, while numbers less than 0 are called _____ numbers.
5. The sign "<" means _____ and ">" means _____.
6. On a number line, the greater number is to the _____ of the lesser number.
7. A number's distance from 0 on a number line is the number's _____.
8. The numbers −5 and 5 are called _____.

Martin-Gay Interactive Videos   *Watch the section lecture video and answer the following questions.*

See Video 2.2

Objective A **9.** In ▣ Example 1, what application is used to represent a negative number?

Objective B **10.** In ▣ Example 2, the tick marks are labeled with what kind of numbers on the number line?

Objective C **11.** From ▣ Example 3 and your knowledge of a number line, complete this statement: 0 will always be greater than any of the _____ integers.

Objective D **12.** What is the answer to ▣ Example 5? The absolute value of what other integer has this same answer?

Objective E **13.** Complete this statement based on ▣ Example 10: A negative sign can be translated to the phrase "_____."

Objective F **14.** In ▣ Examples 13 and 14, what other lake has a negative integer elevation?

2.2 Exercise Set MyMathLab®

Objective A *Represent each quantity by an integer. See Example 1.*

1. A worker in a silver mine in Nevada works 1445 feet underground.

2. A scuba diver is swimming 35 feet below the surface of the water in the Gulf of Mexico.

3. The peak of Mount Elbert in Colorado is 14,433 feet above sea level. (*Source:* U.S. Geological Survey)

4. The lowest elevation in the United States is found at Death Valley, California, at an elevation of 282 feet below sea level. (*Source:* U.S. Geological Survey)

5. The record high temperature in Arkansas is 120 degrees above zero Fahrenheit. (*Source:* National Climatic Data Center)

6. The record high temperature in California is 134 degrees above zero Fahrenheit. (*Source:* National Climatic Data Center)

7. The average depth of the Atlantic Ocean is 11,810 feet below its surface. (*Source: The World Almanac,* 2013)

8. The average depth of the Pacific Ocean is 14,040 feet below its surface. (*Source: The World Almanac,* 2013)

9. Sears had a loss of $3140 million for the fiscal year 2011. (*Source:* CNN Money)

10. Rite Aid had a loss of $555 million for the fiscal year 2011. (*Source:* CNN Money)

11. Two divers are exploring the wreck of the *Andrea Doria*, south of Nantucket Island, Massachusetts. Guillermo is 160 feet below the surface of the ocean and Luigi is 147 feet below the surface. Represent each quantity by an integer and determine who is deeper.

12. The temperature on one January day in Minneapolis was 10° below 0° Celsius. Represent this quantity by an integer and tell whether this temperature is cooler or warmer than 5° below 0° Celsius.

13. For the first half of 2013, digital track sales declined 2 percent when compared to the sales in the first half of 2012. (*Source:* Nielsen Sound Scan)

14. In a recent year, the number of CDs shipped to music retailers reflected a 23 percent decrease from the previous year. Write an integer to represent the percent decrease in CDs shipped. (*Source:* Recording Industry Association of America)

Objective B *Graph each integer in the list on the same number line. See Example 2.*

15. 1, 2, 4, 6

16. 3, 5, 2, 0

17. 1, −1, 2, −2, −4

18. 3, −3, 5, −5, 6

19. 0, 2, 5, 7

20. 0, 3, 6, 10

21. 0, −2, −7, −5

22. 0, −7, 3, −6

Objective C *Insert < or > between each pair of integers to make a true statement. See Example 3.*

23. 0   −7

24. −8   0

25. −7   −5

26. −12   −10

27. −30   −35

28. −27   −29

29. −26   26

30. 13   −13

Objective D *Simplify. See Example 4.*

31. |5|

32. |7|

33. |−8|

34. |−19|

35. |0|

36. |100|

37. |−5|

38. |−10|

Objective E *Find the opposite of each integer. See Example 5.*

39. 5   **40.** 8   **41.** −4   **42.** −6

43. 23   **44.** 123   **45.** −85   **46.** −13

Objectives C D E Mixed Practice *Simplify. See Example 6.*

47. $|-7|$ **48.** $|-11|$ **49.** $-|20|$ **50.** $-|43|$

51. $-|-3|$ **52.** $-|-18|$ **53.** $-(-43)$ **54.** $-(-27)$

55. $-|-15|$ **56.** $-|-29|$ **57.** $-(-33)$ **58.** $-(-14)$

Evaluate. See Example 7.

59. $|-x|$ if $x = -6$ **60.** $-|x|$ if $x = -8$ **61.** $-|-x|$ if $x = 2$ **62.** $-|-x|$ if $x = 10$

63. $|x|$ if $x = -32$ **64.** $|x|$ if $x = 32$ **65.** $-|x|$ if $x = 7$ **66.** $|-x|$ if $x = 1$

Insert $<$, $>$, or $=$ between each pair of numbers to make a true statement. See Examples 3 through 6.

67. −12   −6   **68.** −4   −17   **69.** $|-8|$ $|-11|$ **70.** $|-8|$ $|-4|$

71. $|-47|$ $-(-47)$ **72.** $-|17|$ $-(-17)$ **73.** $-|-12|$ $-(-12)$ **74.** $|-24|$ $-(-24)$

75. 0   −9   **76.** −45   0   **77.** $|0|$ $|-9|$ **78.** $|-45|$ $|0|$

79. $-|-2|$ $-|-10|$ **80.** $-|-8|$ $-|-4|$ **81.** $-(-12)$ $-(-18)$ **82.** −22 $-(-38)$

Objectives D E Mixed Practice *Fill in the chart. See Examples 4 through 7.*

	Number	Absolute Value of Number	Opposite of Number
83.	31		
85.			−28

	Number	Absolute Value of Number	Opposite of Number
84.	−13		
86.			90

Objective F *The bar graph shows the elevations of selected lakes. Use this graph For Exercises 87 through 90. (Source: U.S. Geological Survey) See Example 8.*

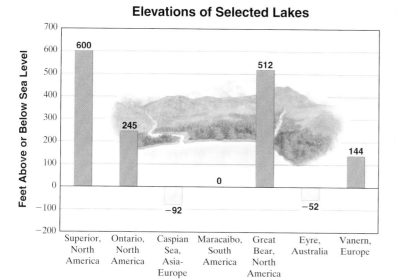

Elevations of Selected Lakes

87. Which lake shown has the lowest elevation?

88. Which lake shown has an elevation at sea level?

89. Which lake shown has the highest elevation?

90. Which lake shown has the second-lowest elevation?

The following bar graph represents the boiling temperature, the temperature at which a substance changes from liquid to gas at standard atmospheric pressure. Use this graph to answer Exercises 91 through 94.

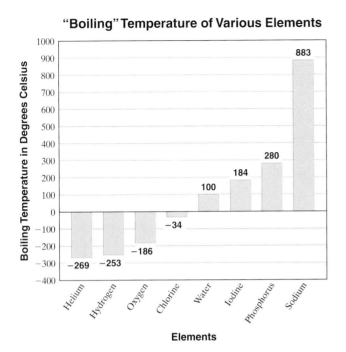

"Boiling" Temperature of Various Elements

91. Which element has a positive boiling temperature closest to that of water?

92. Which element has the lowest boiling temperature?

93. Which element has a boiling temperature closest to $-200°C$?

94. Which element has an average boiling temperature closest to $+300°C$?

Review

Add. See Section 1.3.

95. $0 + 13$

96. $9 + 0$

97. $15 + 20$

98. $20 + 15$

99. $47 + 236 + 77$

100. $362 + 37 + 90$

Concept Extensions

Write the given numbers in order from least to greatest.

101. $2^2, -|3|, -(-5), -|-8|$

102. $|10|, 2^3, -|-5|, -(-4)$

103. $|-1|, -|-6|, -(-6), -|1|$

104. $1^4, -(-3), -|7|, |-20|$

105. $-(-2), 5^2, -10, -|-9|, |-12|$

106. $3^3, -|-11|, -(-10), -4, -|2|$

Choose all numbers for x from each given list that make each statement true.

107. $|x| > 8$
   a. 0   **b.** -5 **c.** 8   **d.** -12

108. $|x| > 4$
   a. 0   **b.** 4   **c.** -1 **d.** -100

109. Evaluate: $-(-|-5|)$

110. Evaluate: $-(-|-(-7)|)$

Answer true or false for Exercises 111 through 115.

111. If $a > b$, then a must be a positive number.

112. The absolute value of a number is *always* a positive number.

113. A positive number is always greater than a negative number.

114. Zero is always less than a positive number.

115. The number $-a$ is always a negative number. (*Hint:* Read "$-$" as "the opposite of.")

116. Given the number line, is it true that $b < a$?

117. Write in your own words how to find the absolute value of a signed number.

118. Explain how to determine which of two signed numbers is larger.

For Exercises 119 and 120, see the Concept Check in this section.

119. Is there a largest negative number? If so, what is it?

120. Is there a smallest positive number? If so, what is it?

2.3 Adding Integers

Objective A Adding Integers

Adding integers can be visualized using a number line. A positive number can be represented on the number line by an arrow of appropriate length pointing to the right, and a negative number by an arrow of appropriate length pointing to the left.

Objectives

A Add Integers.

B Evaluate an Algebraic Expression by Adding.

C Solve Problems by Adding Integers.

Both arrows represent 2 or +2. They both point to the right and they are both 2 units long.

Both arrows represent −3. They both point to the left and they are both 3 units long.

Example 1 Add using a number line: $5 + (-2)$

Solution: To add integers on a number line, such as $5 + (-2)$, we start at 0 on the number line and draw an arrow representing 5. From the tip of this arrow, we draw another arrow representing −2. The tip of the second arrow ends at their sum, 3.

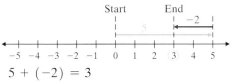

$$5 + (-2) = 3$$

Work Practice 1

Practice 1

Add using a number line:
$5 + (-1)$

Answer

1.

$$5 + (-1) = 4$$

Practice 2

Add using a number line:
$-6 + (-2)$

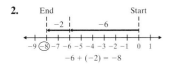

Example 2 Add using a number line: $-1 + (-4)$

Solution: Start at 0 and draw an arrow representing -1. From the tip of this arrow, we draw another arrow representing -4. The tip of the second arrow ends at their sum, -5.

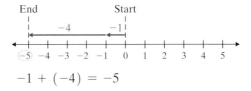

$$-1 + (-4) = -5$$

■ Work Practice 2

Practice 3

Add using a number line:
$-8 + 3$

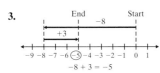

Example 3 Add using a number line: $-7 + 3$

Solution:

■ Work Practice 3

Using a number line each time we add two numbers can be time consuming. Instead, we can notice patterns in the previous examples and write rules for adding signed numbers.

Rules for adding signed numbers depend on whether we are adding numbers with the same sign or different signs. When adding two numbers with the same sign, as in Example 2, notice that the sign of the sum is the same as the sign of the addends.

Adding Two Numbers with the Same Sign

Step 1: Add their absolute values.

Step 2: Use their common sign as the sign of the sum.

Practice 4

Add: $(-3) + (-9)$

Practice 5–6

Add.

5. $-12 + (-3)$

6. $9 + 5$

Example 4 Add: $-2 + (-21)$

Solution:

Step 1: $|-2| = 2, |-21| = 21$, and $2 + 21 = 23$.

Step 2: Their common sign is negative, so the sum is negative:

$$-2 + (-21) = -23$$

■ Work Practice 4

Answers

2.

End ┊   Start
   ┊ -2 ┊   -6
←┼──┼──┼──┼──┼──┼──┼──┼──┼──┼──→
 -9 ⊝ -7 -6 -5 -4 -3 -2 -1 0 1

$-6 + (-2) = -8$

3.

        End   Start
         ┊   -8
    ←──────────────
     ┊ +3
←┼──┼──┼──┼──┼──┼──┼──┼──┼──┼──→
 -9 -8 -7 -6 ⊝ -4 -3 -2 -1 0 1

$-8 + 3 = -5$

4. -12 **5.** -15 **6.** 14

Examples Add.

5. $-5 + (-10) = -15$

6. $2 + 6 = 8$

■ Work Practice 5–6

When adding two numbers with different signs, as in Examples 1 and 3, the sign of the result may be positive or negative, or the result may be 0.

Adding Two Numbers with Different Signs

Step 1: Find the larger absolute value minus the smaller absolute value.

Step 2: Use the sign of the number with the larger absolute value as the sign of the sum.

Example 7   Add: $-14 + 35$

Solution:

Step 1: $|-14| = 14$, $|35| = 35$, and $35 - 14 = 21$.

Step 2: 35 has the larger absolute value and its sign is an understood $+$:

$-14 + 35 = +21$ or 21

■ Work Practice 7

Practice 7

Add: $-3 + 59$

Example 8   Add: $13 + (-17)$

Solution:

Step 1: $|13| = 13$, $|-17| = 17$, and $17 - 13 = 4$.

Step 2: -17 has the larger absolute value and its sign is $-$:

$13 + (-17) = -4$

■ Work Practice 8

Practice 8

Add: $22 + (-28)$

Examples   Add.

9. $-18 + 10 = -8$

10. $12 + (-8) = 4$

11. $0 + (-5) = -5$ The sum of 0 and any number is the number.

■ Work Practice 9–11

Practice 9–11

Add.

9. $-46 + 20$

10. $8 + (-6)$

11. $-2 + 0$

Recall that numbers such as 7 and -7 are called opposites. In general, the sum of a number and its opposite is always 0.

$7 + (-7) = 0$ $-26 + 26 = 0$ $1008 + (-1008) = 0$

opposites   opposites   opposites

If a is a number, then

$-a$ is its opposite. Also,

$\left.\begin{array}{l} a + (-a) = 0 \\ -a + a = 0 \end{array}\right\}$ The sum of a number and its opposite is 0.

Examples   Add.

12. $-21 + 21 = 0$

13. $36 + (-36) = 0$

■ Work Practice 12–13

Practice 12–13

Add.

12. $15 + (-15)$

13. $-80 + 80$

Answers

7. 56   **8.** -6 **9.** -26 **10.** 2

11. -2 **12.** 0   **13.** 0

✔**Concept Check** What is wrong with the following calculation?

$6 + (-22) = 16$

✔**Concept Check Answer**

$6 + (-22) = -16$

In the following examples, we add three or more integers. Remember that by the associative and commutative properties for addition, we may add numbers in any order that we wish. In Examples 14 and 15, let's add the numbers from left to right.

Practice 14

Add: $8 + (-3) + (-13)$

Example 14   Add: $(-3) + 4 + (-11)$

Solution: $(-3) + 4 + (-11) = 1 + (-11)$
$$= -10$$

■ Work Practice 14

Practice 15

Add: $5 + (-3) + 12 + (-14)$

Example 15   Add: $1 + (-10) + (-8) + 9$

Solution: $1 + (-10) + (-8) + 9 = -9 + (-8) + 9$
$$= -17 + 9$$
$$= -8$$

The sum will be the same if we add the numbers in any order. To see this, let's first add the positive numbers together and then the negative numbers together.

$$1 + 9 = 10 \quad \text{Add the positive numbers.}$$
$$(-10) + (-8) = -18 \quad \text{Add the negative numbers.}$$
$$10 + (-18) = -8 \quad \text{Add these results.}$$

The sum is -8.

■ Work Practice 15

Helpful Hint

Don't forget that addition is commutative and associative. In other words, numbers may be added in any order.

Objective B Evaluating Algebraic Expressions

We can continue our work with algebraic expressions by evaluating expressions given integer replacement values.

Practice 16

Evaluate $x + 3y$ for $x = -4$ and $y = 1$.

Example 16   Evaluate $2x + y$ for $x = 3$ and $y = -5$.

Solution: Replace x with 3 and y with -5 in $2x + y$.

$$2x + y = 2 \cdot 3 + (-5)$$
$$= 6 + (-5)$$
$$= 1$$

■ Work Practice 16

Practice 17

Evaluate $x + y$ for $x = -11$ and $y = -6$.

Example 17   Evaluate $x + y$ for $x = -2$ and $y = -10$.

Solution: $x + y = (-2) + (-10)$ Replace x with -2 and y with -10.
$$= -12$$

■ Work Practice 17

Answers

14. -8 **15.** 0 **16.** -1 **17.** -17

Objective C Solving Problems by Adding Integers

Next, we practice solving problems that require adding integers.

Example 18 Calculating Temperature

In Philadelphia, Pennsylvania, the record extreme high temperature is 104°F. Add to this temperature the opposite of 111 degrees, and the result is the record extreme low temperature. Find this temperature. (*Source:* National Climatic Data Center)

Solution:

	extreme low temperature	=	extreme high temperature	add to this	opposite of 111°
In words:					
	↓		↓		↓
Translate:	extreme low temperature	=	104	+	(−111)
		= −7			

The record extreme low temperature in Philadelphia, Pennsylvania, is −7°F.

■ Work Practice 18

Practice 18

If the temperature was −8° Fahrenheit at 6 a.m., and it rose 4 degrees by 7 a.m. and then rose another 7 degrees in the hour from 7 a.m. to 8 a.m., what was the temperature at 8 a.m.?

Answer

18. 3°F

Calculator Explorations Entering Negative Numbers

To enter a negative number on a calculator, find the key marked +/− . (Some calculators have a key marked CHS and some calculators have a special key (−) for entering a negative sign.) To enter the number −2, for example, press the keys 2 +/− . The display will read −2 .

To find −32 + (−131), press the keys
32 +/− + 131 +/− = or
(−) 32 + (−) 131 ENTER
The display will read −163 . Thus
−32 + (−131) = −163.

Use a calculator to perform each indicated operation.

1. −256 + 97

2. 811 + (−1058)

3. 6(15) + (−46)

4. −129 + 10(48)

5. −108,650 + (−786,205)

6. −196,662 + (−129,856)

Vocabulary, Readiness & Video Check

Use the choices below to fill in each blank. Not all choices will be used. (Review Section 1.3 if needed.)

−a   a   0   commutative   associative

1. If n is a number, then $-n + n =$ _____.

2. Since $x + n = n + x$, we say that addition is _____.

3. If a is a number, then $-(-a) =$ _____.

4. Since $n + (x + a) = (n + x) + a$, we say that addition is _____.

Add.

5. $5 + 0$

6. $0 + 3$

7. $0 + (-35)$

8. $(-2) + 0$

9. $-12 + 12$

10. $-9 + 9$

11. $28 + (-28)$

12. $48 + (-48)$

Martin-Gay Interactive Videos

Watch the section lecture video and answer the following questions.

See Video 2.3

Objective A **13.** What is the sign of the sum in ▥ Example 6 and why? ⊙

Objective B **14.** What is the sign of the sum in ▥ Example 8 and why? ⊙

Objective C **15.** What does the answer to ▥ Example 10, −231, mean in the context of the application? ⊙

2.3 Exercise Set MyMathLab® ⊙

Objective A *Add using a number line. See Examples 1 through 3.*

⊙ **1.** −1 + (−6)

2. 9 + (−4)

⊙ **3.** −4 + 7

4. 10 + (−3)

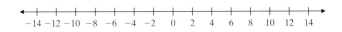

5. −13 + 7

6. −6 + (−5)

Add. See Examples 4 through 13.

7. 23 + 12   **8.** 15 + 42   **9.** −6 + (−2)   **10.** −5 + (−4)

11. −43 + 43   **12.** −62 + 62   ⊙ **13.** 6 + (−2)   **14.** 8 + (−3)

15. −6 + 8   **16.** −8 + 12   **17.** 3 + (−5)   **18.** 5 + (−9)

⊙ **19.** −2 + (−7)   **20.** −6 + (−1)   **21.** −12 + (−12)   **22.** −23 + (−23)

23. −25 + (−32)   **24.** −45 + (−90)   **25.** −123 + (−100)   **26.** −500 + (−230)

27. −7 + 7   **28.** −10 + 10   **29.** 12 + (−5)   **30.** 24 + (−10)

⊙ **31.** −6 + 3   **32.** −8 + 2   **33.** −12 + 3   **34.** −15 + 5

35. 56 + (−26)   **36.** 89 + (−37)   **37.** −37 + 57   **38.** −25 + 65

39. $-42 + 93$　　　　**40.** $-64 + 164$　　　　**41.** $34 + (-67)$　　　　**42.** $42 + (-83)$

43. $124 + (-144)$　　　**44.** $325 + (-375)$　　　▶**45.** $-82 + (-43)$　　　**46.** $-56 + (-33)$

Add. See Examples 14 and 15.

▶**47.** $-4 + 2 + (-5)$　　　**48.** $-1 + 5 + (-8)$　　　**49.** $-52 + (-77) + (-117)$

50. $-103 + (-32) + (-27)$　　**51.** $12 + (-4) + (-4) + 12$　　**52.** $18 + (-9) + 5 + (-2)$

53. $(-10) + 14 + 25 + (-16)$　　**54.** $34 + (-12) + (-11) + 213$

Objective A Mixed Practice *Add. See Examples 1 through 15.*

55. $-8 + (-14) + (-11)$　　　**56.** $-10 + (-6) + (-1)$　　　**57.** $-26 + 5$

58. $-35 + (-12)$　　　　**59.** $5 + (-1) + 17$　　　　**60.** $3 + (-23) + 6$

61. $-14 + (-31)$　　　　**62.** $-100 + 70$　　　　**63.** $13 + 14 + (-18)$

64. $(-45) + 22 + 20$　　　**65.** $-87 + 87$　　　　**66.** $-87 + 0$

67. $-3 + (-8) + 12 + (-1)$　　　　**68.** $-16 + 6 + (-14) + (-20)$

69. $0 + (-103)$　　　　　　　**70.** $94 + (-94)$

Objective B *Evaluate $x + y$ for the given replacement values. See Examples 16 and 17.*

▶**71.** $x = -20$ and $y = -50$　　　　**72.** $x = -1$ and $y = -29$

Evaluate $3x + y$ for the given replacement values. See Examples 16 and 17.

▶**73.** $x = 2$ and $y = -3$　　　　**74.** $x = 7$ and $y = -11$

75. $x = 3$ and $y = -30$　　　　**76.** $x = 13$ and $y = -17$

Objective C Translating *Translate each phrase; then simplify. See Example 18.*

77. Find the sum of -8 and 25.　　　　**78.** Find the sum of -30 and 10.

79. Find the sum of -31, -9, and 30.　　　　**80.** Find the sum of -49, -2, and 40.

Solve. See Example 18.

81. Suppose a deep-sea diver dives from the surface to 215 feet below the surface. He then dives down 16 more feet. Use integers to represent this situation. Then find the diver's present depth.

82. Suppose a diver dives from the surface to 248 meters below the surface and then swims up 6 meters, down 17 meters, down another 24 meters, and then up 23 meters. Use integers to represent this situation. Then find the diver's depth after these movements.

In golf, it is possible to have positive and negative scores. The following table shows the results of the eighteen-hole Round 2 for Jim Furyk and Jason Dufner at the 2013 PGA Championship in Rochester, New York. Use the table to answer Exercises 83 and 84.

Player/Hole	1	2	3	4	5	6	7	8	9	10	11	12	13	14	15	16	17	18
Furyk	−1	0	0	0	0	0	0	0	0	−1	0	0	0	0	0	−1	1	0
Dufner	0	−2	0	−1	−1	0	0	0	0	0	−1	0	−1	0	0	−1	0	0

(*Source:* Professional Golfers' Association)

83. Find the total score for each of the athletes in the round.

84. In golf, the lower score is the winner. Use the result of Exercise 83 to determine who won Round 2.

The following bar graph shows the yearly net income for Apple, Inc. Net income is one indication of a company's health. It measures revenue (money taken in) minus cost (money spent). Use this graph to answer Exercises 85 through 88. (Source: Apple, Inc.)

85. What was the projected net income (in dollars) for Apple, Inc. in 2015?

86. What was the net income (in dollars) for Apple, Inc. in 2001?

87. Find the total net income (in dollars) for the years 2011 and 2013.

88. Find the total net income (in dollars) for the years 2007, 2009, and 2011.

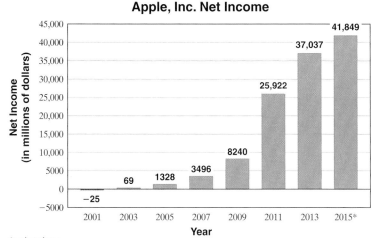

89. The temperature at 4 p.m. on February 2 was −10° Celsius. By 11 p.m. the temperature had risen 12 degrees. Find the temperature at 11 p.m.

90. In some card games, it is possible to have both positive and negative scores. After four rounds of play, Michelle had scores of 14, −5, −8, and 7. What was her total score for the game?

A small company reports the following net incomes. Use this table for Exercises 91 and 92.

91. Find the sum of the net incomes for 2011 and 2012.

92. Find the net income sum for all four years shown.

Year	Net Income (in Dollars)
2010	$75,083
2011	−$10,412
2012	−$1786
2013	$96,398

93. The all-time record low temperature for Texas is −23°F. Florida's all-time record low temperature is 21°F higher than Texas' record low. What is Florida's record low temperature? (*Source:* National Climatic Data Center)

94. The all-time record low temperature for California is −45°F. In Pennsylvania, the lowest temperature ever recorded is 3°F higher than California's all-time low temperature. What is the all-time record low temperature for Pennsylvania? (*Source:* National Climatic Data Center)

95. The deepest spot in the Pacific Ocean is the Mariana Trench, which has an elevation of 10,924 meters below sea level. The bottom of the Pacific's Aleutian Trench has an elevation 3245 meters higher than that of the Mariana Trench. Use a negative number to represent the depth of the Aleutian Trench. (*Source:* Defense Mapping Agency)

96. The deepest spot in the Atlantic Ocean is the Puerto Rico Trench, which has an elevation of 8605 meters below sea level. The bottom of the Atlantic's Cayman Trench has an elevation 1070 meters above the level of the Puerto Rico Trench. Use a negative number to represent the depth of the Cayman Trench. (*Source:* Defense Mapping Agency)

Review

Subtract. See Section 1.4.

97. $44 - 0$ **98.** $91 - 0$ **99.** $200 - 59$ **100.** $400 - 18$

Concept Extensions

101. Name 2 numbers whose sum is −17.

102. Name 2 numbers whose sum is −30.

Each calculation below is incorrect. Find the error and correct it. See the Concept Check in this section.

103. $7 + (-10) \overset{?}{=} 17$

104. $-4 + 14 \overset{?}{=} -18$

105. $-10 + (-12) \overset{?}{=} -120$

106. $-15 + (-17) \overset{?}{=} 32$

For Exercises 107 through 110, determine whether each statement is true or false.

107. The sum of two negative numbers is always a negative number.

108. The sum of two positive numbers is always a positive number.

109. The sum of a positive number and a negative number is always a negative number.

110. The sum of zero and a negative number is always a negative number.

111. In your own words, explain how to add two negative numbers.

112. In your own words, explain how to add a positive number and a negative number.

2.4 Subtracting Integers

Objectives

A Subtract Integers.

B Add and Subtract Integers. ▶

C Evaluate an Algebraic Expression by Subtracting. ▶

D Solve Problems by Subtracting Integers. ▶

In Section 2.2, we discussed the opposite of an integer.

The opposite of 3 is -3.

The opposite of -6 is 6.

In this section, we use opposites to subtract integers.

Objective A Subtracting Integers

To subtract integers, we will write the subtraction problem as an addition problem. To see how to do this, study the examples below.

$$10 - 4 = 6$$
$$10 + (-4) = 6$$

Since both expressions simplify to 6, this means that

$$10 - 4 = 10 + (-4) = 6$$

Also,

$$3 - 2 = 3 + (-2) = 1$$
$$15 - 1 = 15 + (-1) = 14$$

Thus, to subtract two numbers, we add the first number to the opposite of the second number. (The opposite of a number is also known as its **additive inverse**.)

> **Subtracting Two Numbers**
>
> If a and b are numbers, then $a - b = a + (-b)$.

Practice 1–4

Subtract.

1. $12 - 7$ **2.** $-6 - 4$
3. $11 - (-14)$ **4.** $-9 - (-1)$

Examples Subtract.

	Subtraction	=	first number	+	opposite of the second number	
1.	$8 - 5$	=	8	+	(-5)	= 3
2.	$-4 - 10$	=	-4	+	(-10)	= -14
3.	$6 - (-5)$	=	6	+	5	= 11
4.	$-11 - (-7)$	=	-11	+	7	= -4

■ Work Practice 1–4

Practice 5–7

Subtract.

5. $5 - 9$ **6.** $-12 - 4$
7. $-2 - (-7)$

Examples Subtract.

5. $-10 - 5 = -10 + (-5) = -15$

6. $8 - 15 = 8 + (-15) = -7$

7. $-4 - (-5) = -4 + 5 = 1$

■ Work Practice 5–7

Answers

1. 5 **2.** -10 **3.** 25 **4.** -8
5. -4 **6.** -16 **7.** 5

Helpful Hint

To visualize subtraction, try the following:

The difference between $5°F$ and $-2°F$ can be found by subtracting. That is,

$$5 - (-2) = 5 + 2 = 7$$

Can you visually see from the thermometer on the right that there are actually 7 degrees between $5°F$ and $-2°F$?

5° F

0° F

−2° F

7 degrees

✓**Concept Check** What is wrong with the following calculation?

$$-9 - (-6) = -15$$

Example 8 Subtract 7 from -3.

Solution: To subtract 7 *from* -3, we find

$$-3 - 7 = -3 + (-7) = -10$$

■ Work Practice 8

Practice 8

Subtract 5 from -10.

Objective B Adding and Subtracting Integers

If a problem involves adding or subtracting more than two integers, we rewrite differences as sums and add. Recall that by associative and commutative properties, we may add numbers in any order. In Examples 9 and 10, we will add from left to right.

Example 9 Simplify: $7 - 8 - (-5) - 1$

Solution:
$$
\begin{aligned}
7 - 8 - (-5) - 1 &= 7 + (-8) + 5 + (-1) \\
&= -1 + 5 + (-1) \\
&= 4 + (-1) \\
&= 3
\end{aligned}
$$

■ Work Practice 9

Practice 9

Simplify: $-4 - 3 - 7 - (-5)$

Example 10 Simplify: $7 + (-12) - 3 - (-8)$

Solution:
$$
\begin{aligned}
7 + (-12) - 3 - (-8) &= 7 + (-12) + (-3) + 8 \\
&= -5 + (-3) + 8 \\
&= -8 + 8 \\
&= 0
\end{aligned}
$$

■ Work Practice 10

Practice 10

Simplify:
$3 + (-5) - 6 - (-4)$

Objective C Evaluating Expressions

Now let's practice evaluating expressions when the replacement values are integers.

Answers

8. -15 **9.** -9 **10.** -4

✓**Concept Check Answer**

$-9 - (-6) = -9 + 6 = -3$

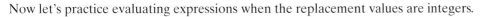

Practice 11

Evaluate $x - y$ for $x = -2$ and $y = 14$.

Example 11 Evaluate $x - y$ for $x = -3$ and $y = 9$.

Solution: Replace x with -3 and y with 9 in $x - y$.

$$
\begin{array}{ccc}
x & - & y \\
\downarrow & \downarrow & \downarrow
\end{array}
$$
$$= (-3) - 9$$
$$= (-3) + (-9)$$
$$= -12$$

■ Work Practice 11

Practice 12

Evaluate $3y - z$ for $y = 9$ and $z = -4$.

Example 12 Evaluate $2a - b$ for $a = 8$ and $b = -6$.

Solution: Watch your signs carefully!

$$
\begin{array}{ccc}
2a & - & b \\
\downarrow & \downarrow & \downarrow
\end{array}
$$
$$= 2 \cdot 8 - (-6) \quad \text{Replace } a \text{ with 8 and } b \text{ with } -6.$$
$$= 16 + 6 \quad \text{Multiply.}$$
$$= 22 \quad \text{Add.}$$

> **Helpful Hint** Watch carefully when replacing variables in the expression $2a - b$. Make sure that all symbols are inserted and accounted for.

■ Work Practice 12

Objective D Solving Problems by Subtracting Integers

Solving problems often requires subtraction of integers.

Practice 13

The highest point in Asia is the top of Mount Everest, at a height of 29,028 feet above sea level. The lowest point is the Dead Sea, which is 1312 feet below sea level. How much higher is Mount Everest than the Dead Sea? (*Source:* National Geographic Society)

Example 13 Finding a Change in Elevation

The highest point in the United States is the top of Mount McKinley, at a height of 20,320 feet above sea level. The lowest point is Death Valley, California, which is 282 feet below sea level. How much higher is Mount McKinley than Death Valley? (*Source:* U.S. Geological Survey)

Solution:

1. **UNDERSTAND.** Read and reread the problem. To find "how much higher," we subtract. Don't forget that since Death Valley is 282 feet *below* sea level, we represent its height by -282. Draw a diagram to help visualize the problem.

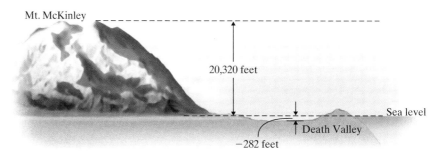

Mt. McKinley

20,320 feet

Sea level

Death Valley

-282 feet

2. **TRANSLATE.**

In words:	how much higher is Mt. McKinley	=	height of Mt. McKinley	minus	height of Death Valley
	↓	↓	↓	↓	↓
Translate:	how much higher is Mt. McKinley	=	20,320	−	(-282)

Copyright 2016 Pearson Education, Inc.

Answers

11. -16 **12.** 31 **13.** 30,340 ft

3. SOLVE:

$$20{,}320 - (-282) = 20{,}320 + 282 = 20{,}602$$

4. INTERPRET. Check and state your conclusion: Mount McKinley is 20,602 feet higher than Death Valley.

■ Work Practice 13

Vocabulary, Readiness & Video Check

Multiple choice: Select the correct lettered response following each exercise.

1. It is true that $a - b =$ _____.

   a. $b - a$ **b.** $a + (-b)$ **c.** $a + b$

2. The opposite of n is _____.

   a. $-n$ **b.** $-(-n)$ **c.** n

3. To evaluate $x - y$ for $x = -10$ and $y = -14$, we replace x with -10 and y with -14 and evaluate _____.

   a. $10 - 14$ **b.** $-10 - 14$ **c.** $-14 - 10$ **d.** $-10 - (-14)$

4. The expression $-5 - 10$ equals _____.

   a. $5 - 10$ **b.** $5 + 10$ **c.** $-5 + (-10)$ **d.** $10 - 5$

Subtract.

5. $5 - 5$ **6.** $7 - 7$ **7.** $8642 - 8642$ **8.** $9012 - 9012$

Martin-Gay Interactive Videos   *Watch the section lecture video and answer the following questions.*

Objective A   **9.** In the lecture before ⊞ Example 1, what can the "opposite" of a number also be called? ○

Objective B   **10.** In ⊞ Example 7, how is the example rewritten in the first step of simplifying and why? ○

Objective C   **11.** In ⊞ Example 8, why do we multiply first? ○

Objective D   **12.** What does the answer to ⊞ Example 9, 263, mean in the context of the application? ○

See Video 2.4 ●

2.4   Exercise Set   MyMathLab®

Objective A   *Subtract. See Examples 1 through 7.*

1. $-5 - (-5)$ **2.** $-6 - (-6)$ **3.** $8 - 3$ **4.** $5 - 2$

▶ **5.** $3 - 8$ **6.** $2 - 5$ **7.** $7 - (-7)$ **8.** $12 - (-12)$

9. $-5 - (-8)$ **10.** $-25 - (-25)$ **11.** $-14 - 4$ **12.** $-2 - 42$

13. $2 - 16$ **14.** $8 - 9$ **15.** $22 - 55$ **16.** $17 - 63$

17. $362 - (-40)$ **18.** $844 - (-20)$ **19.** $-4 - 10$ **20.** $-5 - 8$

21. $-7 - (-3)$ **22.** $-12 - (-5)$ **23.** $16 - 23$ **24.** $16 - 45$

Translating *Translate each phrase; then simplify. See Example 8.*

25. Subtract 17 from -25.   **26.** Subtract 10 from -22.   **27.** Find the difference of -20 and -3.

28. Find the difference of -8 and -13.   **29.** Subtract -11 from 2.   **30.** Subtract -50 from -50.

Mixed Practice (*Sections 2.3, 2.4*) *Add or subtract as indicated.*

31. $-21 + (-17)$ **32.** $-35 + (-11)$ **33.** $9 - 20$ **34.** $7 - 30$

35. $-49 - 78$ **36.** $-105 - 68$ **37.** $48 - 59$ **38.** $86 - 98$

Objective B   *Simplify. See Examples 9 and 10.*

39. $7 - 3 - 2$ **40.** $8 - 4 - 1$ **41.** $12 - 5 - 7$

42. $30 - 7 - 12$ **43.** $-5 - 8 - (-12)$ **44.** $-10 - 6 - (-9)$

45. $-10 + (-5) - 12$ **46.** $-15 + (-8) - 4$ **47.** $12 - (-34) + (-6)$

48. $23 - (-17) + (-9)$ **49.** $-(-6) - 12 + (-16)$ **50.** $-(-9) - 7 + (-23)$

51. $-9 - (-12) + (-7) - 4$ **52.** $-6 - (-8) + (-12) - 7$ **53.** $-3 + 4 - (-23) - 10$

54. $5 + (-18) - (-21) - 2$

Objective C   *Evaluate $x - y$ for the given replacement values. See Examples 11 and 12.*

55. $x = -4$ and $y = 7$ **56.** $x = -7$ and $y = 1$

57. $x = 8$ and $y = -23$ **58.** $x = 9$ and $y = -2$

Evaluate $2x - y$ for the given replacement values. See Examples 11 and 12.

59. $x = 4$ and $y = -4$ **60.** $x = 8$ and $y = -10$

61. $x = 1$ and $y = -18$ **62.** $x = 14$ and $y = -12$

Objective D *Solve. See Example 13.*

The bar graph from Section 2.2 showing the average daytime surface temperature in degrees Fahrenheit of known planets is reprinted below. Notice that a negative temperature is illustrated by a bar below the horizontal line representing 0°F, and a positive temperature is illustrated by a bar above the horizontal line representing 0°F.

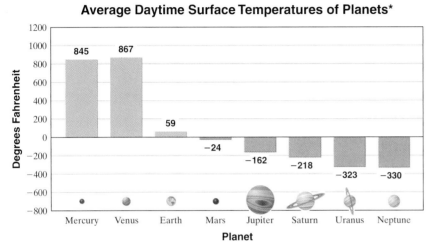

Average Daytime Surface Temperatures of Planets*

*(For some planets, the temperature given is the temperature where the atmospheric pressure equals 1 Earth atmosphere; Source: The World Almanac)

63. Find the difference in temperature between Earth and Neptune.

64. Find the difference in temperature between Venus and Mars.

65. Find the difference in temperature between the two planets with the lowest temperatures.

66. Find the difference in temperature between Jupiter and Saturn.

67. The coldest temperature ever recorded on Earth was −129°F in Antarctica. The warmest temperature ever recorded was 134°F in Death Valley, California. How many degrees warmer is 134°F than −129°F? (*Source: The World Almanac, 2013*)

68. The coldest temperature ever recorded in the United States was −80°F in Alaska. The warmest temperature ever recorded was 134°F in California. How many degrees warmer is 134°F than −80°F? (*Source: The World Almanac, 2013*)

Solve.

69. Aaron Aiken has $125 in his checking account. He writes a check for $117, makes a deposit of $45, and then writes another check for $69. Find the balance in his account. (Write the amount as an integer.)

70. A woman received a statement of her charge account at Old Navy. She spent $93 on purchases last month. She returned an $18 top because she didn't like the color. She also returned a $26 nightshirt because it was damaged. What does she actually owe on her account?

71. The temperature on a February morning is −6° Celsius at 6 a.m. If the temperature drops 3 degrees by 7 a.m., rises 4 degrees between 7 a.m. and 8 a.m., and then drops 7 degrees between 8 a.m. and 9 a.m., find the temperature at 9 a.m.

72. Mauna Kea in Hawaii has an elevation of 13,796 feet above sea level. The Mid-America Trench in the Pacific Ocean has an elevation of 21,857 feet below sea level. Find the difference in elevation between those two points. (*Source:* National Geographic Society and Defense Mapping Agency)

Some places on Earth lie below sea level, which is the average level of the surface of the oceans. Use this diagram to answer Exercises 73 through 76. (Source: Fantastic Book of Comparisons, Russell Ash)

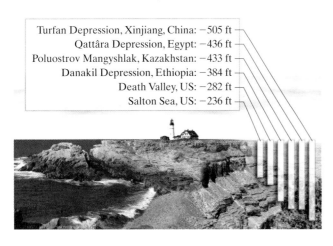

Turfan Depression, Xinjiang, China: −505 ft
Qattâra Depression, Egypt: −436 ft
Poluostrov Mangyshlak, Kazakhstan: −433 ft
Danakil Depression, Ethiopia: −384 ft
Death Valley, US: −282 ft
Salton Sea, US: −236 ft

73. Find the difference in elevation between Death Valley and Quattâra Depression.

74. Find the difference in elevation between Danakil and Turfan Depressions.

75. Find the difference in elevation between the two lowest elevations shown.

76. Find the difference in elevation between the highest elevation shown and the lowest elevation shown.

The bar graph from Section 2.2 shows heights of selected lakes. For Exercises 77 through 80, find the difference in elevation for the lakes listed. (Source: U.S. Geological Survey)

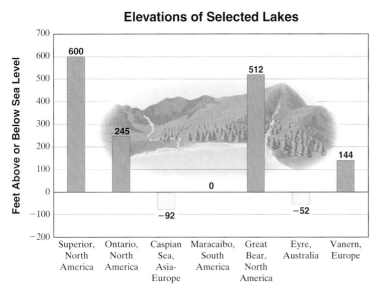

Elevations of Selected Lakes

Feet Above or Below Sea Level

- Superior, North America: 600
- Ontario, North America: 245
- Caspian Sea, Asia-Europe: −92
- Maracaibo, South America: 0
- Great Bear, North America: 512
- Eyre, Australia: −52
- Vanern, Europe: 144

77. Lake Superior and Lake Eyre

78. Great Bear Lake and Caspian Sea

79. Lake Maracaibo and Lake Vanern

80. Lake Eyre and Caspian Sea

Solve.

81. The difference between a country's exports and imports is called the country's *trade balance*. In June 2013, the United States had $191 billion in exports and $225 billion in imports. What was the U.S. trade balance in June 2013? (*Source:* U.S. Department of Commerce)

82. In 2012, the United States exported 1165 million barrels of petroleum products and imported 3878 million barrels of petroleum products. What was the U.S. trade balance for petroleum products in 2012? (*Source:* U.S. Energy Information Administration)

Mixed Practice—Translating (*Sections 2.3, 2.4*) *Translate each phrase to an algebraic expression. Use "x" to represent "a number."*

83. The sum of -5 and a number.

84. The difference of -3 and a number.

85. Subtract a number from -20.

86. Add a number and -36.

Review

Multiply or divide as indicated. See Sections 1.6 and 1.7.

87. $\dfrac{100}{20}$

88. $\dfrac{96}{3}$

89. $\begin{array}{r} 23 \\ \times\ 46 \\ \hline \end{array}$

90. $\begin{array}{r} 51 \\ \times\ 89 \\ \hline \end{array}$

Concept Extensions

91. Name two numbers whose difference is -3.

92. Name two numbers whose difference is -10.

*Each calculation below is **incorrect.** Find the error and correctly calculate. See the Concept Check in this section.*

93. $9 - (-7) \stackrel{?}{=} 2$

94. $-4 - 8 \stackrel{?}{=} 4$

95. $10 - 30 \stackrel{?}{=} 20$

96. $-3 - (-10) \stackrel{?}{=} -13$

Simplify. (Hint: Find the absolute values first.)

97. $|-3| - |-7|$

98. $|-12| - |-5|$

99. $|-6| - |6|$

100. $|-9| - |9|$

101. $|-17| - |-29|$

102. $|-23| - |-42|$

For Exercises 103 and 104, determine whether each statement is true or false.

103. $|-8 - 3| \stackrel{?}{=} 8 - 3$

104. $|-2 - (-6)| \stackrel{?}{=} |-2| - |-6|$

105. In your own words, explain how to subtract one signed number from another.

106. A student explains to you that the first step to simplify $8 + 12 \cdot 5 - 100$ is to add 8 and 12. Is the student correct? Explain why or why not.

Answers

Integers

Represent each quantity by an integer.

1. _____

2. _____

3. _____

4. _____

5. _____

6. _____

7. _____

8. _____

9. _____

10. _____

11. _____

12. _____

13. _____

14. _____

15. _____

16. _____

17. _____

18. _____

19. _____

1. The peak of Mount Everest in Asia is 29,028 feet above sea level. (*Source: U.S. Geological Survey*)

2. The Mariana Trench in the Pacific Ocean is 35,840 feet below sea level. (*Source: The World Almanac*)

3. The deepest hole ever drilled in the Earth's crust is in Russia and its depth is over 7 miles below sea level. (*Source: Fantastic Book of Comparisons*)

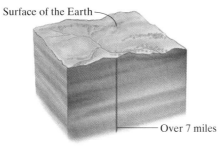

Surface of the Earth

Over 7 miles

4. Graph the signed numbers on the given number line. $-4, 0, -1, 3$

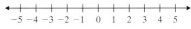

$-5 \quad -4 \quad -3 \quad -2 \quad -1 \quad 0 \quad 1 \quad 2 \quad 3 \quad 4 \quad 5$

Insert < or > between each pair of numbers to make a true statement.

5. $0 \quad -3$ **6.** $-15 \quad -5$ **7.** $-1 \quad 1$ **8.** $-2 \quad -7$

Simplify.

9. $|-1|$ **10.** $-|-4|$ **11.** $|-8|$ **12.** $-(-5)$

Find the opposite of each number.

13. 6 **14.** -3 **15.** 89 **16.** 0

Add or subtract as indicated.

17. $-7 + 12$ **18.** $-9 + (-11)$ **19.** $25 + (-35)$

20. $1 - 3$

21. $26 - (-26)$

22. $-2 - 1$

23. $-18 - (-102)$

24. $-8 + (-6) + 20$

25. $-11 - 7 - (-19)$

26. $-4 + (-8) - 16 - (-9)$

27. Subtract 14 from 26.

28. Subtract -8 from -12.

29. Find the sum of -17 and -27.

Choose all numbers for x from each given list that make each statement true.

30. $|x| > 0$
   a. 0   **b.** 18   **c.** -3 **d.** -21

31. $|x| > -5$
   a. 0   **b.** 3   **c.** -1 **d.** -1000

Evaluate the expressions below for x = −1 and y = 11.

32. $x + y$

33. $x - y$

34. $y - x$

35. $y + x$

36. $5y - x$

37. $x - 3y$

20. _____

21. _____

22. _____

23. _____

24. _____

25. _____

26. _____

27. _____

28. _____

29. _____

30. _____

31. _____

32. _____

33. _____

34. _____

35. _____

36. _____

37. _____

Objectives

A Multiply Integers.

B Divide Integers.

C Evaluate an Algebraic Expression by Multiplying or Dividing.

D Solve Problems by Multiplying and Dividing Integers.

Multiplying and dividing integers is similar to multiplying and dividing whole numbers. One difference is that we need to determine whether the result is a positive number or a negative number.

Objective A Multiplying Integers

Consider the following pattern of products.

First factor decreases by 1 each time.

$$3 \cdot 2 = 6$$
$$2 \cdot 2 = 4$$
$$1 \cdot 2 = 2$$
$$0 \cdot 2 = 0$$

Product decreases by 2 each time.

This pattern can be continued, as follows.

$$-1 \cdot 2 = -2$$
$$-2 \cdot 2 = -4$$
$$-3 \cdot 2 = -6$$

This suggests that the product of a negative number and a positive number is a negative number.

What is the sign of the product of two negative numbers? To find out, we form another pattern of products. Again, we decrease the first factor by 1 each time, but this time the second factor is negative.

$$2 \cdot (-3) = -6$$
$$1 \cdot (-3) = -3$$
$$0 \cdot (-3) = 0$$

Product increases by 3 each time.

This pattern continues as:

$$-1 \cdot (-3) = 3$$
$$-2 \cdot (-3) = 6$$
$$-3 \cdot (-3) = 9$$

This suggests that the product of two negative numbers is a positive number. Thus we can determine the sign of a product when we know the signs of the factors.

Multiplying Numbers

The product of two numbers having the same sign is a positive number.

The product of two numbers having different signs is a negative number.

Product of Like Signs

$$(+)(+) = +$$
$$(-)(-) = +$$

Product of Different Signs

$$(-)(+) = -$$
$$(+)(-) = -$$

Practice 1–4

Multiply.

1. $-3 \cdot 8$ **2.** $-5(-2)$
3. $0 \cdot (-20)$ **4.** $10(-5)$

Answers

1. -24 **2.** 10 **3.** 0 **4.** -50

Examples Multiply.

1. $-7 \cdot 3 = -21$ **2.** $-3(-5) = 15$
3. $0 \cdot (-4) = 0$ **4.** $10(-8) = -80$

Work Practice 1–4

Recall that by the associative and commutative properties for multiplication, we may multiply numbers in any order that we wish. In Example 5, we multiply from left to right.

Examples Multiply.

5. $\overbrace{7(-6)}(-2) = -42(-2)$
$= 84$

6. $\overbrace{(-2)(-3)}(-4) = 6(-4)$
$= -24$

7. $(-1)(-2)(-3)(-4) = -1(-24)$ We have -24 from Example 6.
$= 24$

■ Work Practice 5–7

Practice 5–7

Multiply.
5. $8(-6)(-2)$
6. $(-9)(-2)(-1)$
7. $(-3)(-4)(-5)(-1)$

✓ **Concept Check** What is the sign of the product of five negative numbers? Explain.

Recall from our study of exponents that $2^3 = 2 \cdot 2 \cdot 2 = 8$. We can now work with bases that are negative numbers. For example,

$$(-2)^3 = (-2)(-2)(-2) = -8$$

Example 8 Evaluate: $(-5)^2$

Solution: Remember that $(-5)^2$ means 2 factors of -5.

$$(-5)^2 = (-5)(-5) = 25$$

■ Work Practice 8

Practice 8

Evaluate $(-2)^4$.

Helpful Hint

Have you noticed a pattern when multiplying signed numbers?
If we let $(-)$ represent a negative number and $(+)$ represent a positive number, then

$(-)(-) = (+)$
$(-)(-)(-) = (-)$ ← The product of an **odd** number of negative numbers is a **negative** result.
$(-)(-)(-)(-) = (+)$
$(-)(-)(-)(-)(-) = (-)$

The product of an **even** number of negative numbers is a **positive** result.

Notice in Example 8 the parentheses around -5 in $(-5)^2$. With these parentheses, -5 is the base that is squared. Without parentheses, such as -5^2, only the 5 is squared. In other words, $-5^2 = -(5 \cdot 5) = -25$.

Example 9 Evaluate: -7^2

Solution: Remember that without parentheses, only the 7 is squared.

$$-7^2 = -(7 \cdot 7) = -49$$

■ Work Practice 9

Practice 9

Evaluate: -8^2

Answers
5. 96   **6.** -18 **7.** 60   **8.** 16   **9.** -64

✓**Concept Check Answer**
negative; answers may vary

Helpful Hint

Make sure you understand the difference between Examples 8 and 9.

parentheses, so -5 is squared

$$(-5)^2 = (-5)(-5) = 25$$

no parentheses, so only the 7 is squared

$$-7^2 = -(7 \cdot 7) = -49$$

Objective B Dividing Integers

Division of integers is related to multiplication of integers. The sign rules for division can be discovered by writing a related multiplication problem. For example,

$$\frac{6}{2} = 3 \qquad \text{because } 3 \cdot 2 = 6$$

$$\frac{-6}{2} = -3 \qquad \text{because } -3 \cdot 2 = -6$$

$$\frac{6}{-2} = -3 \qquad \text{because } -3 \cdot (-2) = 6$$

$$\frac{-6}{-2} = 3 \qquad \text{because } 3 \cdot (-2) = -6$$

Helpful Hint Just as for whole numbers, division can be checked by multiplication.

Dividing Numbers

The quotient of two numbers having the same sign is a positive number.

The quotient of two numbers having different signs is a negative number.

Quotient of Like Signs	**Quotient of Different Signs**
$\dfrac{(+)}{(+)} = + \qquad \dfrac{(-)}{(-)} = +$	$\dfrac{(+)}{(-)} = - \qquad \dfrac{(-)}{(+)} = -$

Practice 10–12

Divide.

10. $\dfrac{42}{-7}$

11. $-16 \div (-2)$

12. $\dfrac{-80}{10}$

Examples Divide.

10. $\dfrac{-12}{6} = -2$

11. $-20 \div (-4) = 5$

12. $\dfrac{48}{-3} = -16$

■ Work Practice 10–12

Answers

10. -6 **11.** 8 **12.** -8

✓**Concept Check Answer**

$\dfrac{-36}{-9} = 4$

✓**Concept Check** What is wrong with the following calculation?

$$\frac{-36}{-9} = -4$$

Examples Divide, if possible.

13. $\dfrac{0}{-5} = 0$ because $0 \cdot -5 = 0$

14. $\dfrac{-7}{0}$ is undefined because there is no number that gives a product of -7 when multiplied by 0.

■ Work Practice 13–14

Practice 13–14

Divide, if possible.

13. $\dfrac{-6}{0}$ **14.** $\dfrac{0}{-7}$

Objective C Evaluating Expressions

Next, we practice evaluating expressions given integer replacement values.

Example 15 Evaluate xy for $x = -2$ and $y = 7$.

Solution: Recall that xy means $x \cdot y$.

Replace x with -2 and y with 7.

$$xy = -2 \cdot 7$$
$$= -14$$

■ Work Practice 15

Practice 15

Evaluate xy for $x = 5$ and $y = -8$.

Example 16 Evaluate $\dfrac{x}{y}$ for $x = -24$ and $y = 6$.

Solution: $\dfrac{x}{y} = \dfrac{-24}{6}$ Replace x with -24 and y with 6.

$$= -4$$

■ Work Practice 16

Practice 16

Evaluate $\dfrac{x}{y}$ for $x = -12$ and $y = -3$.

Objective D Solving Problems by Multiplying and Dividing Integers

Many real-life problems involve multiplication and division of signed numbers.

Example 17 Calculating a Total Golf Score

A professional golfer finished seven strokes under par (-7) for each of three days of a tournament. What was his total score for the tournament?

Solution:

1. UNDERSTAND. Read and reread the problem. Although the key word is "total," since this is repeated addition of the same number, we multiply.

2. TRANSLATE.

In words:	golfer's total score	=	number of days	·	score each day
	↓	↓	↓	↓	↓
Translate:	golfer's total	=	3	·	(-7)

3. SOLVE: $3 \cdot (-7) = -21$

4. INTERPRET. Check and state your conclusion: The golfer's total score was -21, or 21 strokes under par.

■ Work Practice 17

Practice 17

A card player had a score of -13 for each of four games. Find her total score.

Answers

13. undefined **14.** 0 **15.** -40
16. 4 **17.** -52

Vocabulary, Readiness & Video Check

Use the choices below to fill in each blank. Each choice may be used more than once.

negative   0
positive   undefined

1. The product of a negative number and a positive number is a(n) _____ number.
2. The product of two negative numbers is a(n) _____ number.
3. The quotient of two negative numbers is a(n) _____ number.
4. The quotient of a negative number and a positive number is a(n) _____ number.
5. The product of a negative number and zero is _____.
6. The quotient of 0 and a negative number is _____.
7. The quotient of a negative number and 0 is _____.

Martin-Gay Interactive Videos   *Watch the section lecture video and answer the following questions.*

See Video 2.5

Objective A   8. Explain the role of parentheses when comparing Examples 3 and 4.

Objective B   9. Complete this statement based on the lecture before Example 6: We can find out about sign rules for division because we know sign rules for _____.

Objective C   10. In Example 10, what are you asked to remember about the algebraic expression *ab*?

Objective D   11. In Example 12, how do we know the example will involve a negative number?

2.5   Exercise Set   MyMathLab®

Objective A   *Multiply. See Examples 1 through 4.*

1. $-6(-2)$ 2. $5(-3)$ 3. $-4(9)$ 4. $-7(-2)$

5. $9(-9)$ 6. $-9(7)$ 7. $0(-11)$ 8. $-6(0)$

Multiply. See Examples 5 through 7.

9. $6(-2)(-4)$ 10. $-2(3)(-7)$ 11. $-1(-3)(-4)$ 12. $-8(-3)(-3)$

13. $-4(4)(-5)$ 14. $2(-5)(-4)$ 15. $10(-5)(0)(-7)$ 16. $3(0)(-4)(-8)$

17. $-5(3)(-1)(-1)$ 18. $-2(-1)(3)(-2)$

Evaluate. See Examples 8 and 9.

19. -3^2

20. -2^4

21. $(-3)^2$

22. $(-1)^4$

23. -6^2

24. -4^3

25. $(-4)^3$

26. $(-3)^3$

Objective B *Find each quotient. See Examples 10 through 14.*

27. $-24 \div 3$

28. $90 \div (-9)$

29. $\dfrac{-30}{6}$

30. $\dfrac{56}{-8}$

31. $\dfrac{-77}{-11}$

32. $\dfrac{-32}{4}$

33. $\dfrac{0}{-21}$

34. $\dfrac{-13}{0}$

35. $\dfrac{-10}{0}$

36. $\dfrac{0}{-15}$

37. $\dfrac{56}{-4}$

38. $\dfrac{-24}{-12}$

Objectives A B **Mixed Practice** *Multiply or divide as indicated. See Examples 1 through 14.*

39. $-14(0)$

40. $0(-100)$

41. $-5(3)$

42. $-6 \cdot 2$

43. $-9 \cdot 7$

44. $-12(13)$

45. $-7(-6)$

46. $-9(-5)$

47. $-3(-4)(-2)$

48. $-7(-5)(-3)$

49. $(-7)^2$

50. $(-6)^2$

51. $-\dfrac{25}{5}$

52. $-\dfrac{30}{5}$

53. $-\dfrac{72}{8}$

54. $-\dfrac{49}{7}$

55. $-18 \div 3$

56. $-15 \div 3$

57. $4(-10)(-3)$

58. $6(-5)(-2)$

59. $-30(6)(-2)(-3)$

60. $-20 \cdot 5 \cdot (-5) \cdot (-3)$

61. $\dfrac{-25}{0}$

62. $\dfrac{0}{-14}$

63. $\dfrac{120}{-20}$

64. $\dfrac{63}{-9}$

65. $280 \div (-40)$

66. $480 \div (-8)$

67. $\dfrac{-12}{-4}$

68. $\dfrac{-36}{-3}$

69. -1^4

70. -2^3

71. $(-2)^5$

72. $(-11)^2$

73. $-2(3)(5)(-6)$

74. $-1(2)(7)(-3)$

75. $(-1)^{32}$

76. $(-1)^{33}$

77. $-2(-3)(-5)$

78. $-2(-2)(-3)(-2)$

79. $-48 \cdot 23$

80. $-56 \cdot 43$

81. $35 \cdot (-82)$

82. $70 \cdot (-23)$

Objective C *Evaluate ab for the given replacement values. See Example 15.*

83. $a = -8$ and $b = 7$

84. $a = 5$ and $b = -1$

85. $a = 9$ and $b = -2$

86. $a = -8$ and $b = 8$

87. $a = -7$ and $b = -5$

88. $a = -9$ and $b = -6$

Evaluate $\dfrac{x}{y}$ for the given replacement values. See Example 16.

89. $x = 5$ and $y = -5$

90. $x = 9$ and $y = -3$

91. $x = -15$ and $y = 0$

92. $x = 0$ and $y = -5$

93. $x = -36$ and $y = -6$

94. $x = -10$ and $y = -10$

Evaluate xy and also $\dfrac{x}{y}$ for the given replacement values. See Examples 15 and 16.

95. $x = -8$ and $y = -2$ **96.** $x = 20$ and $y = -5$ **97.** $x = 0$ and $y = -8$ **98.** $x = -3$ and $y = 0$

Objective D Translating *Translate each phrase; then simplify. See Example 17.*

99. Find the quotient of -54 and 9.

100. Find the quotient of -63 and -3.

101. Find the product of -42 and -6.

102. Find the product of -49 and 5.

Translating *Translate each phrase to an expression. Use x to represent "a number." See Example 17.*

103. The product of -71 and a number

104. The quotient of -8 and a number

105. Subtract a number from -16.

106. The sum of a number and -12

107. -29 increased by a number

108. The difference of a number and -10

109. Divide a number by -33.

110. Multiply a number by -17.

Solve. See Example 17.

111. A football team lost four yards on each of three consecutive plays. Represent the total loss as a product of signed numbers and find the total loss.

112. An investor lost $400 on each of seven consecutive days in the stock market. Represent his total loss as a product of signed numbers and find his total loss.

113. A deep-sea diver must move up or down in the water in short steps in order to keep from getting a physical condition called the "bends." Suppose a diver moves down from the surface in five steps of 20 feet each. Represent his total movement as a product of signed numbers and find the product.

114. A weather forecaster predicts that the temperature will drop five degrees each hour for the next six hours. Represent this drop as a product of signed numbers and find the total drop in temperature.

The graph shows melting points in degrees Celsius of selected elements. Use this graph to answer Exercises 115 through 118.

115. The melting point of nitrogen is 3 times the melting point of radon. Find the melting point of nitrogen.

116. The melting point of rubidium is −1 times the melting point of mercury. Find the melting point of rubidium.

117. The melting point of argon is −3 times the melting point of potassium. Find the melting point of argon.

118. The melting point of strontium is −11 times the melting point of radon. Find the melting point of strontium.

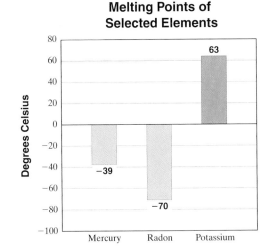

Melting Points of Selected Elements

Solve. See Example 17.

119. For the first quarter of 2013, Wal-Mart, Inc. posted a loss of $33 million in membership and other income. If this trend was consistent for each month of the quarter, how much would you expect this loss to have been for each month? (*Source:* Wal-Mart Stores, Inc.)

120. For the first quarter of 2013, Chrysler Group LLC, maker of Jeep vehicles, posted a loss of about 30,000 Jeep Liberty shipments because they had stopped producing the vehicle in 2012. If this trend was consistent for each month of the quarter, how much would you expect this loss to have been for each month? (*Source:* Chrysler Group, LLC)

121. In 2008, there were 33,319 analog (nondigital) U.S. movie screens. In 2012, this number of screens dropped to 6387. (*Source:* Motion Picture Association: Worldwide Market Research)

   a. Find the change in the number of U.S. analog movie screens from 2008 to 2012.

   b. Find the average change per year in the number of analog movie screens over this period.

122. In 1987, the California Condor was all but extinct in the wild, with about 30 condors in the world. The condors in the wild were captured by the U.S. Fish and Wildlife Service in an aggressive move to rebuild the population by breeding them in captivity and releasing the chicks into the wild. The condor population increased to approximately 405 birds in 2012. (*Source:* Arizona Game and Fish Department)

   a. Find the change in the number of California Condors from 1987 to 2012.

   b. Find the average change per year in the California Condor population over the period in part **a.**

Review

Perform each indicated operation. See Section 1.9.

123. $90 + 12^2 - 5^3$ **124.** $3 \cdot (7 - 4) + 2 \cdot 5^2$ **125.** $12 \div 4 - 2 + 7$ **126.** $12 \div (4 - 2) + 7$

Concept Extensions

Mixed Practice (*Sections 2.3, 2.4, 2.5*) *Perform the indicated operations.*

127. $-57 \div 3$

128. $-9(-11)$

129. $-8 - 20$

130. $-4 + (-3) + 21$

131. $-4 - 15 - (-11)$

132. $-16 - (-2)$

Solve. For Exercises 133 and 134, see the first Concept Check in this section.

133. What is the sign of the product of seven negative numbers?

134. What is the sign of the product of ten negative numbers?

Without actually finding the product, write the list of numbers in Exercises 135 and 136 in order from least to greatest. For help, see a Helpful Hint box in this section.

135. $(-2)^{12}, (-2)^{17}, (-5)^{12}, (-5)^{17}$

136. $(-1)^{50}, (-1)^{55}, 0^{15}, (-7)^{20}, (-7)^{23}$

137. In your own words, explain how to divide two integers.

138. In your own words, explain how to multiply two integers.

2.6 Order of Operations

Objectives

A Simplify Expressions by Using the Order of Operations.

B Evaluate an Algebraic Expression.

C Find the Average of a List of Numbers.

Objective A Simplifying Expressions

We first discussed the order of operations in Chapter 1. In this section, you are given an opportunity to practice using the order of operations when expressions contain signed numbers. The rules for the order of operations from Section 1.9 are repeated here.

Order of Operations

1. Perform all operations within parentheses ( ), brackets [ ], or other grouping symbols such as fraction bars, starting with the innermost set.
2. Evaluate any expressions with exponents.
3. Multiply or divide in order from left to right.
4. Add or subtract in order from left to right.

Before simplifying other expressions, make sure you are confident simplifying Examples 1 through 3.

Examples Find the value of each expression.

1. $(-3)^2 = (-3)(-3) = 9$ The base of the exponent is -3.

2. $-3^2 = -(3)(3) = -9$ The base of the exponent is 3.

3. $2 \cdot 5^2 = 2 \cdot (5 \cdot 5) = 2 \cdot 25 = 50$ The base of the exponent is 5.

■ Work Practice 1–3

Practice 1–3

Find the value of each expression.

1. $(-2)^4$

2. -2^4

3. $3 \cdot 6^2$

Helpful Hint

When simplifying expressions with exponents, remember that parentheses make an important difference.

$(-3)^2$ and -3^2 **do not** mean the same thing.

$(-3)^2$ means $(-3)(-3) = 9$.

-3^2 means the opposite of $3 \cdot 3$, or -9.

Only with parentheses around it is the -3 squared.

Example 4 Simplify: $\dfrac{-6(2)}{-3}$

Solution: First we multiply -6 and 2. Then we divide.

$$\frac{-6(2)}{-3} = \frac{-12}{-3}$$
$$= 4$$

■ Work Practice 4

Practice 4

Simplify: $\dfrac{-25}{5(-1)}$

Example 5 Simplify: $\dfrac{12 - 16}{-1 + 3}$

Solution: We simplify above and below the fraction bar separately. Then we divide.

$$\frac{12 - 16}{-1 + 3} = \frac{-4}{2}$$
$$= -2$$

■ Work Practice 5

Practice 5

Simplify: $\dfrac{-18 + 6}{-3 - 1}$

Example 6 Simplify: $60 + 30 + (-2)^3$

Solution: $60 + 30 + (-2)^3 = 60 + 30 + (-8)$ Write $(-2)^3$ as -8.
$$= 90 + (-8)$$ Add from left to right.
$$= 82$$

■ Work Practice 6

Practice 6

Simplify: $30 + 50 + (-4)^3$

Answers

1. 16 **2.** -16 **3.** 108 **4.** 5

5. 3 **6.** 16

Practice 7

Simplify: $-2^3 + (-4)^2 + 1^5$

Example 7   Simplify: $-4^2 + (-3)^2 - 1^3$

Solution:

$$-4^2 + (-3)^2 - 1^3 = -16 + 9 - 1 \quad \text{Simplify expressions with exponents.}$$
$$= -7 - 1 \quad \text{Add or subtract from left to right.}$$
$$= -8$$

■ Work Practice 7

Practice 8

Simplify:

$2(2 - 9) + (-12) - \sqrt{9}$

Example 8   Simplify: $3(4 - 7) + (-2) - \sqrt{25}$

Solution:

$$3(4 - 7) + (-2) - \sqrt{25} = 3(-3) + (-2) - 5 \quad \begin{array}{l}\text{Simplify inside parentheses}\\ \text{and replace } \sqrt{25} \text{ with 5.}\end{array}$$
$$= -9 + (-2) - 5 \quad \text{Multiply.}$$
$$= -11 - 5 \quad \begin{array}{l}\text{Add or subtract from}\\ \text{left to right.}\end{array}$$
$$= -16$$

■ Work Practice 8

Practice 9

Simplify:

$(-5) \cdot |-8| + (-3) + 2^3$

Example 9   Simplify: $(-3) \cdot |-5| - (-2) + 4^2$

Solution:

$$(-3) \cdot |-5| - (-2) + 4^2 = (-3) \cdot 5 - (-2) + 4^2 \quad \text{Write } |-5| \text{ as 5.}$$
$$= (-3) \cdot 5 - (-2) + 16 \quad \text{Write } 4^2 \text{ as 16.}$$
$$= -15 - (-2) + 16 \quad \text{Multiply.}$$
$$= -13 + 16 \quad \begin{array}{l}\text{Add or subtract from}\\ \text{left to right.}\end{array}$$
$$= 3$$

■ Work Practice 9

Practice 10

Simplify:

$-4[-6 + 5(-3 + 5)] - 7$

Example 10   Simplify: $-2[-3 + 2(-1 + 6)] - 5$

Solution: Here we begin with the innermost set of parentheses.

$$-2[-3 + 2(-1 + 6)] - 5 = -2[-3 + 2(5)] - 5 \quad \text{Write } -1 + 6 \text{ as 5.}$$
$$= -2[-3 + 10] - 5 \quad \text{Multiply.}$$
$$= -2(7) - 5 \quad \text{Add.}$$
$$= -14 - 5 \quad \text{Multiply.}$$
$$= -19 \quad \text{Subtract.}$$

■ Work Practice 10

Answers

7. 9   **8.** -29 **9.** -35 **10.** -23

✓**Concept Check Answer**

false; $-4(3 - 7) - 8(9 - 6) = -8$

✓**Concept Check** True or false? Explain your answer. The result of

$$-4(3 - 7) - 8(9 - 6)$$

is positive because there are four negative signs.

Objective B Evaluating Expressions

Now we practice evaluating expressions.

Example 11 Evaluate x^2 and $-x^2$ for $x = -11$.

Solution: $x^2 = (-11)^2 = (-11)(-11) = 121$

$-x^2 = -(-11)^2 = -(-11)(-11) = -121$

◼ Work Practice 11

Practice 11
Evaluate x^2 and $-x^2$ for $x = -15$.

Example 12 Evaluate $6z^2$ for $z = 2$ and $z = -2$.

Solution: $6z^2 = 6(2)^2 = 6(4) = 24$

$6z^2 = 6(-2)^2 = 6(4) = 24$

◼ Work Practice 12

Practice 12
Evaluate $5y^2$ for $y = 4$ and $y = -4$.

Example 13 Evaluate $x + 2y - z$ for $x = 3, y = -5,$ and $z = -4$.

Solution: Replace x with 3, y with -5, and z with -4, and simplify.

Helpful Hint   Remember to rewrite the subtraction sign.

$x + 2y - z = 3 + 2(-5) - (-4)$ Let $x = 3, y = -5,$ and $z = -4$.

$= 3 + (-10) + 4$ Replace $2(-5)$ with its product, -10.

$= -3$ Add.

◼ Work Practice 13

Practice 13
Evaluate $x - y + 3z$ for $x = -6, y = -3,$ and $z = 12$.

Example 14 Evaluate $7 - x^2$ for $x = -4$.

Solution: Replace x with -4 and simplify carefully!

$7 - x^2 = 7 - (-4)^2$

$= 7 - 16$ $(-4)^2 = (-4)(-4) = 16$

$= -9$ Subtract.

◼ Work Practice 14

Practice 14
Evaluate $4 - x^2$ for $x = -8$.

Objective C Finding Averages

Recall from Chapter 1 that the average of a list of numbers is

$$\text{average} = \frac{\text{sum of numbers}}{\textit{number} \text{ of numbers}}$$

Answers
11. $225; -225$ **12.** $80; 80$ **13.** 33
14. -60

Practice 15

Find the average of the temperatures for the months October through April.

Example 15 The graph shows some monthly normal temperatures for Barrow, Alaska. Use this graph to find the average of the temperatures for the months January through April.

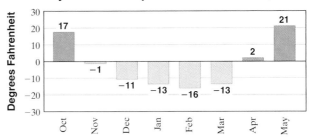

Monthly Normal Temperatures for Barrow, Alaska

Solution: By reading the graph, we have

$$\text{average} = \frac{-13 + (-16) + (-13) + 2}{4} \quad \text{There are 4 months from January through April.}$$

$$= \frac{-40}{4}$$

$$= -10$$

The average of the temperatures is $-10°\text{F}$.

Answer

15. $-5°\text{F}$

■ Work Practice 15

 Calculator Explorations Simplifying an Expression Containing a Fraction Bar

Recall that even though most calculators follow the order of operations, parentheses must sometimes be inserted. For example, to simplify $\dfrac{-8 + 6}{-2}$ on a calculator, enter parentheses around the expression above the fraction bar so that it is simplified separately.

To simplify $\dfrac{-8 + 6}{-2}$, press the keys

(8 +/- + 6) ÷ 2 +/- = or

((−) 8 + 6) ÷ (−) 2 ENTER

The display will read ☐ 1 .

Thus, $\dfrac{-8 + 6}{-2} = 1$.

Use a calculator to simplify.

1. $\dfrac{-120 - 360}{-10}$

2. $\dfrac{4750}{-2 + (-17)}$

3. $\dfrac{-316 + (-458)}{28 + (-25)}$

4. $\dfrac{-234 + 86}{-18 + 16}$

Vocabulary, Readiness & Video Check

Use the choices below to fill in each blank. Not all choices will be used.

average	subtraction	division	$-7 - 3(1)$
addition	multiplication	$-7 - 3(-1)$	

1. To simplify $-2 \div 2 \cdot (3)$, which operation should be performed first? _____

2. To simplify $-9 - 3 \cdot 4$, which operation should be performed first? _____

3. The _____ of a list of numbers is $\dfrac{\text{sum of numbers}}{\text{number of numbers}}$.

4. To simplify $5[-9 + (-3)] \div 4$, which operation should be performed first? _____

5. To simplify $-2 + 3(10 - 12) \cdot (-8)$, which operation should be performed first? _____

6. To evaluate $x - 3y$ for $x = -7$ and $y = -1$, replace x with -7 and y with -1 and evaluate _____.

Identify the bases and exponents of each expression. Do not simplify.

7. -3^2

8. $(-3)^2$

9. $4 \cdot 2^3$

10. $9 \cdot 5^6$

11. $(-7)^5$

12. -9^4

13. $5^7 \cdot 10$

14. $2^8 \cdot 11$

Martin-Gay Interactive Videos   *Watch the section lecture video and answer the following questions.*

Objective A **15.** In Example 1, what two things about the fraction bar are we reminded of?

Objective B **16.** In Example 5, why is it important to place the replacement value for x within parentheses?

Objective C **17.** From the lecture before Example 6, explain why finding the average is a good example of an application for this section.

See Video 2.6

2.6 Exercise Set MyMathLab®

Objective A *Simplify. See Examples 1 through 10.*

1. $(-4)^3$

2. -2^4

3. -4^3

4. $(-2)^4$

5. $6 \cdot 2^2$

6. $5 \cdot 2^3$

7. $3 + (-8) \div 2$

8. $-1(-2) + 1$

9. $9 - 12 - 4$

10. $10 - 23 - 12$

11. $4 + 3(-6)$

12. $-8 + 4(3)$

13. $5(-9) + 2$

14. $7(-6) + 3$

15. $(-10) + 4 \div 2$

16. $(-12) + 6 \div 3$

17. $6 + 7 \cdot 3 - 40$

18. $5 + 9 \cdot 4 - 52$

19. $\dfrac{16 - 13}{-3}$

20. $\dfrac{20 - 15}{-1}$

21. $\dfrac{24}{10 + (-4)}$

22. $\dfrac{88}{-8 - 3}$

23. $5(-3) - (-12)$

24. $7(-4) - (-6)$

25. $(-19) - 12(3)$

26. $(-24) - 14(2)$

27. $-8 + 4^2$

28. $-12 + 3^3$

29. $[8 + (-4)]^2$

30. $[9 + (-2)]^3$

31. $8 \cdot 6 - 3 \cdot 5 + (-20)$

32. $7 \cdot 6 - 6 \cdot 5 + (-10)$

33. $4 - (-3)^4$

34. $20 - (-5)^2$

35. $|5 + 3| \cdot 2^3$

36. $|-3 + 7| \cdot 7^2$

37. $7 \cdot 8^2 + 4$

38. $10 \cdot 5^3 + 7$

39. $5^3 - (4 - 2^3)$

40. $8^2 - (5 - 2)^4$

41. $|3 - 12| \div 3$

42. $|12 - 19| \div 7$

43. $-(-2)^2$

44. $-(-2)^3$

45. $(5 - 9)^2 \div (4 - 2)^2$

46. $(2 - 7)^2 \div (4 - 3)^4$

47. $|8 - 24| \cdot (-2) \div (-2)$

48. $|3 - 15| \cdot (-4) \div (-16)$

49. $(-12 - 20) \div 16 - 25$

50. $(-20 - 5) \div 5 - 15$

51. $5(5 - 2) + (-5)^2 - 6$

52. $3 \cdot (8 - 3) + (-4) - 10$

53. $(2 - 7) \cdot (6 - 19)$

54. $(4 - 12) \cdot (8 - 17)$

55. $2 - 7 \cdot 6 - 19$

56. $4 - 12 \cdot 8 - 17$

57. $(-36 \div 6) - (4 \div 4)$

58. $(-4 \div 4) - (8 \div 8)$

59. $-5^2 - 6^2$

60. $-4^4 - 5^4$

61. $(-5)^2 - 6^2$

62. $(-4)^4 - 5^4$

63. $(10 - 4^2)^2$

64. $(11 - 3^2)^3$

65. $2(8 - 10)^2 - 5(1 - 6)^2$

66. $-3(4 - 8)^2 + 5(14 - 16)^3$

67. $3(-10) \div [5(-3) - 7(-2)]$

68. $12 - [7 - (3 - 6)] + (2 - 3)^3$

69. $\dfrac{(-7)(-3) - (4)(3)}{3[7 \div (3 - 10)]}$

70. $\dfrac{10(-1) - (-2)(-3)}{2[-8 \div (-2 - 2)]}$

71. $-5[4 + 5(-3 + 5)] + 11$

72. $-2[1 + 3(7 - 12)] - 35$

73. $-3[5 + 2(-4 + 9)] + 15$

74. $-2[6 + 4(2 - 8)] - 25$

Objective B *Evaluate each expression for* $x = -2$, $y = 4$, *and* $z = -1$. *See Examples 11 through 14.*

75. $x + y + z$

76. $x - y - z$

77. $2x - 3y - 4z$

78. $5x - y + 4z$

79. $x^2 - y$

80. $x^2 + z$

81. $\dfrac{5y}{z}$

82. $\dfrac{4x}{y}$

Evaluate each expression for $x = -3$ *and* $z = -4$. *See Examples 11 through 14.*

83. x^2

84. z^2

85. $-z^2$

86. $-x^2$

87. $10 - x^2$

88. $3 - z^2$

89. $2x^3 - z$

90. $3z^2 - x$

Objective C *Find the average of each list of numbers. See Example 15.*

91. $-10, 8, -4, 2, 7, -5, -12$

92. $-18, -8, -1, -1, 0, 4$

93. $-17, -26, -20, -13$

94. $-40, -20, -10, -15, -5$

Scores in golf can be 0 (also called par), a positive integer (also called above par), or a negative integer (also called below par). The bar graph shows final scores of selected golfers from a 2013 tournament. Use this graph for Exercises 95 through 100. (Source: LPGA)

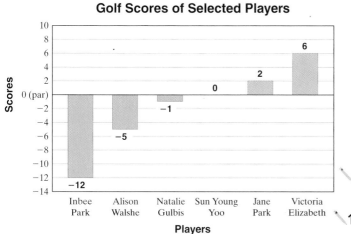

Golf Scores of Selected Players

95. Find the difference between the lowest score shown and the highest score shown.

96. Find the difference between the two lowest scores.

97. Find the average of the scores for Walshe, Gulbis, Yoo, and Jane Park. (*Hint:* Here, the average is the sum of the scores divided by the number of players.)

98. Find the average of the scores for Inbee Park, Walshe, Gulbis, and Elizabeth.

99. Can the average for the scores in Exercise **98** be greater than the highest score, 6? Explain why or why not.

100. Can the average of the scores in Exercise **98** be less than the lowest score, −12? Explain why or why not.

Review

Perform each indicated operation. See Sections 1.3, 1.4, 1.6, and 1.7.

101. $45 \cdot 90$ **102.** $90 \div 45$ **103.** $90 - 45$ **104.** $45 + 90$

Find the perimeter of each figure. See Section 1.3.

105. Square   8 in.

106. Parallelogram   5 cm   3 cm

107. Rectangle   6 ft   9 ft

108. Triangle   17 m   23 m   32 m

Concept Extensions

Insert parentheses where needed so that each expression evaluates to the given number.

109. $2 \cdot 7 - 5 \cdot 3$; evaluates to 12

110. $7 \cdot 3 - 4 \cdot 2$; evaluates to 34

111. $-6 \cdot 10 - 4$; evaluates to −36

112. $2 \cdot 8 \div 4 - 20$; evaluates to −36

113. Are parentheses necessary in the expression $3 + (4 \cdot 5)$? Explain your answer.

114. Are parentheses necessary in the expression $(3 + 4) \cdot 5$? Explain your answer.

115. Discuss the effect parentheses have in an exponential expression. For example, what is the difference between $(-6)^2$ and -6^2?

116. Discuss the effect parentheses have in an exponential expression. For example, what is the difference between $(2 \cdot 4)^2$ and $2 \cdot 4^2$?

Evaluate.

117. $(-12)^4$

118. $(-17)^6$

119. $x^3 - y^2$ for $x = 21$ and $y = -19$

120. $3x^2 + 2x - y$ for $x = -18$ and $y = 2868$

121. $(xy + z)^x$ for $x = 2$, $y = -5$, and $z = 7$

122. $5(ab + 3)^b$ for $a = -2$, $b = 3$

Chapter 2 Group Activity

Magic Squares

Sections 2.2–2.4

A magic square is a set of numbers arranged in a square table so that the sum of the numbers in each column, row, and diagonal is the same. For instance, in the magic square below, the sum of each column, row, and diagonal is 15. Notice that no number is used more than once in the magic square.

2	9	4
7	5	3
6	1	8

The properties of magic squares have been known for a very long time and once were thought to be good luck charms. The ancient Egyptians and Greeks understood their patterns, and a magic square even made it into a famous work of art. The engraving titled *Melencolia I,* created by German artist Albrecht Dürer in 1514, features the following four-by-four magic square on the building behind the central figure.

16	3	2	13
5	10	11	8
9	6	7	12
4	15	14	1

Exercises

1. Verify that what is shown in the Dürer engraving is, in fact, a magic square. What is the common sum of the columns, rows, and diagonals?

2. Negative numbers can also be used in magic squares. Complete the following magic square:

	−1	
0		−4

3. Use the numbers $-16, -12, -8, -4, 0, 4, 8, 12,$ and 16 to form a magic square:

Chapter 2 Vocabulary Check

Fill in each blank with one of the words or phrases listed below.

signed   positive   opposites   negative   absolute value   variable   integers

1. Two numbers that are the same distance from 0 on a number line but are on opposite sides of 0 are called
_____ .

2. Together, positive numbers, negative numbers, and 0 are called _____ numbers.

3. The _____ of a number is that number's distance from 0 on a number line.

4. The _____ are . . . , $-3, -2, -1, 0, 1, 2, 3, \ldots$.

5. A letter used to represent a number is called a(n) _____ .

6. The _____ numbers are numbers less than zero.

7. The _____ numbers are numbers greater than zero.

> **Helpful Hint**
>
> ▶ Are you preparing for your test? Don't forget to take the Chapter 2 Test on page 170. Then check your answers at the back of the text and use the Chapter Test Prep Videos to see the fully worked-out solutions to any of the exercises you want to review.

Chapter Highlights

Definitions and Concepts	Examples
Section 2.1 Introduction to Variables and Algebraic Expressions	
A letter used to represent a number is called a **variable.**	Variables: $x, \quad y, \quad z, \quad a, \quad b$
A combination of operations on variables and numbers is called an **algebraic expression.**	Algebraic expressions: $3 + x, \quad 7y, \quad x^3 + y - 10$
Replacing a variable in an expression by a number, and then finding the value of the expression, is called **evaluating the expression** for the variable.	Evaluate $2x + y$ for $x = 22$ and $y = 4$. $2x + y = 2 \cdot 22 + 4$ Replace x with 22 and y with 4. $\qquad\quad = 44 + 4$ Multiply. $\qquad\quad = 48$ Add.
Section 2.2 Introduction to Integers	
Together, positive numbers, negative numbers, and 0 are called **signed numbers.**	$-432, -10, 0, 15$
The **integers** are . . . , $-3, -2, -1, 0, 1, 2, 3, \ldots$.	
The **absolute value** of a number is that number's distance from 0 on a number line. The symbol for absolute value is $\lvert \ \ \rvert$.	$\lvert -2 \rvert = 2$ $\lvert 2 \rvert = 2$
Two numbers that are the same distance from 0 on a number line but are on opposite sides of 0 are called **opposites.**	5 and -5 are opposites.
If a is a number, then $-(-a) = a$.	$-(-11) = 11$. Do not confuse with $-\lvert -3 \rvert = -3$.

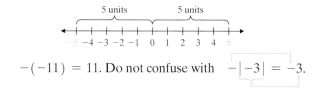

Definitions and Concepts	Examples

Section 2.3   Adding Integers

Adding Two Numbers with the Same Sign

Step 1:   Add their absolute values.

Step 2:   Use their common sign as the sign of the sum.

Adding Two Numbers with Different Signs

Step 1:   Find the larger absolute value minus the smaller absolute value.

Step 2:   Use the sign of the number with the larger absolute value as the sign of the sum.

Add:

$$-3 + (-2) = -5$$
$$-7 + (-15) = -22$$

$$-6 + 4 = -2$$
$$17 + (-12) = 5$$

$$-32 + (-2) + 14 = -34 + 14$$
$$= -20$$

Section 2.4   Subtracting Integers

Subtracting Two Numbers

If a and b are numbers, then $a - b = a + (-b)$.

Subtract:

$$-35 - 4 = -35 + (-4) = -39$$
$$3 - 8 = 3 + (-8) = -5$$
$$-10 - (-12) = -10 + 12 = 2$$
$$7 - 20 - 18 - (-3) = 7 + (-20) + (-18) + (+3)$$
$$= -13 + (-18) + 3$$
$$= -31 + 3$$
$$= -28$$

Section 2.5   Multiplying and Dividing Integers

Multiplying Numbers

The product of two numbers having the same sign is a positive number.
The product of two numbers having different signs is a negative number.

Dividing Numbers

The quotient of two numbers having the same sign is a positive number.
The quotient of two numbers having different signs is a negative number.

Multiply:

$$(-7)(-6) = 42$$
$$9(-4) = -36$$

Evaluate:

$$(-3)^2 = (-3)(-3) = 9$$

Divide:

$$-100 \div (-10) = 10$$

$$\frac{14}{-2} = -7, \quad \frac{0}{-3} = 0, \quad \frac{22}{0} \text{ is undefined.}$$

Section 2.6   Order of Operations

Order of Operations

1. Perform all operations within parentheses ( ), brackets [ ], or other grouping symbols such as fraction bars or square roots, starting with the innermost set.

2. Evaluate any expressions with exponents.

3. Multiply or divide in order from left to right.

4. Add or subtract in order from left to right.

Simplify:

$$3 + 2 \cdot (-5) = 3 + (-10)$$
$$= -7$$

Simplify:

$$\frac{-2(5 - 7)}{-7 + |-3|} = \frac{-2(-2)}{-7 + 3}$$
$$= \frac{4}{-4}$$
$$= -1$$

(2.1) *Evaluate each expression for $x = 5$, $y = 0$, and $z = 2$.*

1. $\dfrac{2x}{z}$

2. $4x - 3$

3. $\dfrac{x + 7}{y}$

4. $\dfrac{y}{5x}$

5. $x^3 - 2z$

6. $\dfrac{7 + x}{3z}$

7. $(y + z)^2$

8. $\dfrac{100}{x} + \dfrac{y}{3}$

Translate each phrase into a variable expression. Use x to represent "a number."

9. Five subtracted from a number

10. Seven more than a number

11. Ten divided by a number

12. The product of 5 and a number

*The map below shows selected cities and their record high and low temperatures in degrees Fahrenheit. Use this map as indicated throughout the rest of this Chapter Review to fill in each missing temperature in the map. The table on the next page may help to insert missing temperatures. Exercise **49** is the first exercise to use this map.*

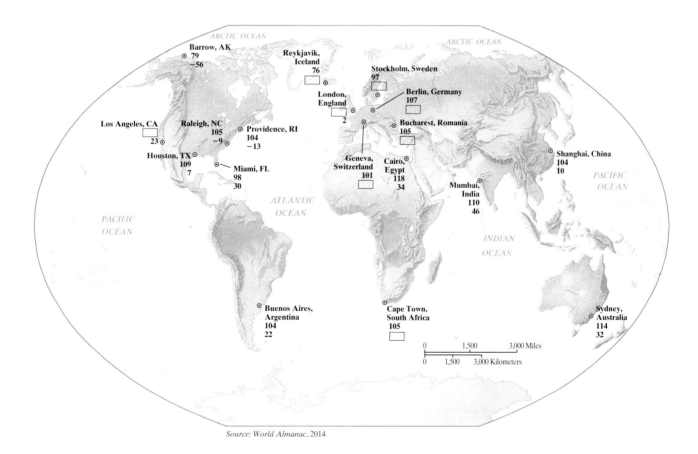

Source: World Almanac, 2014

Record High and Low Temperatures for Selected Locations (in degrees Fahrenheit)					
	Max	**Min**		**Max**	**Min**
Berlin, Germany	107		Barrow, AK	79	−56
Raleigh, NC	105	−9	London, England		2
Houston, TX	109	7	Cairo, Egypt	118	34
Miami, FL	98	30	Sydney, Australia	114	32
Los Angeles, CA		23	Shanghai, China	104	10
Bucharest, Romania	105		Reykjavik, Iceland	76	
Geneva, Switzerland	101		Cape Town, South Africa	105	
Providence, RI	104	−13	Buenos Aires, Argentina	104	22
Stockholm, Sweden	97		Mumbai, India	110	46

(2.2) *Represent each quantity by an integer.*

13. A gold miner is working 1435 feet down in a mine.

14. Mount Hood, in Oregon, has an elevation of 11,239 feet.

Graph each integer in the list on the same number line.

15. $-2, -5, 0, 5$

16. $-7, -1, 0, 7$

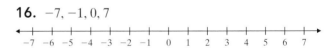

Simplify.

17. $|-12|$

18. $|0|$

19. $-|6|$

20. $-(-9)$

21. $-|-9|$

22. $-(-2)$

Insert $<$ or $>$ between each pair of integers to make a true statement.

23. $-18 \quad -20$

24. $-5 \quad 5$

25. $|-123| \quad -|-198|$

26. $8 - |-12| \quad -|-16|$

Find the opposite of each integer.

27. -12

28. $-(-3)$

Answer true or false for each statement.

29. If $a < b$, then a must be a negative number.

30. The absolute value of an integer is always 0 or a positive number.

31. A negative number is always less than a positive number.

32. If a is a negative number, then $-a$ is a positive number.

(2.3) *Add.*

33. $5 + (-3)$

34. $18 + (-4)$

35. $-12 + 16$

36. $-23 + 40$

37. $-8 + (-15)$

38. $-5 + (-17)$

39. $-24 + 3$

40. $-89 + 19$

41. $15 + (-15)$

42. $-24 + 24$

43. $-43 + (-108)$

44. $-100 + (-506)$

45. The temperature at 5 a.m. on a day in January was $-15°$ Celsius. By 6 a.m. the temperature had fallen 5 degrees. Use a signed number to represent the temperature at 6 a.m.

46. A diver starts out at 127 feet below the surface and then swims downward another 23 feet. Use a signed number to represent the diver's current depth.

47. During the 2014 PGA Masters Tournament, the winner, Bubba Watson, had scores of -3, $+2$, -4, and -3 over four rounds of golf. What was his total score for the tournament? (*Source:* Professional Golfers' Association)

48. During the 2014 Wells Fargo Championship, winning golfer J.B. Holmes had a score of -14. The fourth-place finisher, Jason Bohn, had a score that was 3 points more than the winning score. What was Jason Bohn's score in the Wells Fargo Championship? (*Source:* Professional Golfers' Association)

For Exercises 49 and 50, use the map at the beginning of this Chapter Review.

49. The high temperature for London, England, is 155 degrees greater than the low temperature for Barrow, Alaska. Find the high temperature for London.

50. The high temperature for Los Angeles, California, is 123 degrees greater than the low temperature for Providence, Rhode Island. Find the high temperature for Los Angeles.

(2.4) *Subtract.*

51. $12 - 4$

52. $-12 - 4$

53. $8 - 19$

54. $-8 - 19$

55. $7 - (-13)$

56. $-6 - (-14)$

57. $16 - 16$

58. $-16 - 16$

59. $-12 - (-12)$

60. $|-5| - |-12|$

61. $-(-5) - 12 + (-3)$ **62.** $-8 + |-12| - 10 - |-3|$

Solve.

63. Josh Weidner has $142 in his checking account. He writes a check for $125, makes a deposit of $43, and then writes another check for $85. Represent the final dollar amount in his account by an integer.

64. If the elevation of Lake Superior is 600 feet above sea level and the elevation of the Caspian Sea is 92 feet below sea level, find the difference in the elevations.

For Exercises 65 and 66, use the map at the beginning of this Chapter Review.

65. The low temperature for Reykjavik is 35 degrees less than the low temperature for Sydney, Australia. Find the low temperature for Reykjavik.

66. The low temperature for Berlin, Germany, is 14 degrees less than the low temperature for Shanghai, China. Find the low temperature for Berlin.

Answer true or false for each statement.

67. $|-5| - |-6| = 5 - 6$

68. $|-5 - (-6)| = 5 + 6$

69. If $b > a$, then $b - a$ is a positive number.

70. If $b < a$, then $b - a$ is a negative number.

(2.5) *Multiply.*

71. $-3(-7)$

72. $-6(3)$

73. $-4(16)$

74. $-5(-12)$

75. $(-5)^2$

76. $(-1)^5$

77. $12(-3)(0)$

78. $-1(6)(2)(-2)$

Divide.

79. $-15 \div 3$

80. $\dfrac{-24}{-8}$

81. $\dfrac{0}{-3}$

82. $\dfrac{-46}{0}$

83. $\dfrac{100}{-5}$

84. $\dfrac{-72}{8}$

85. $\dfrac{-38}{-1}$

86. $\dfrac{45}{-9}$

87. A football team lost 5 yards on each of two consecutive plays. Represent the total loss by a product of integers, and find the product.

88. A race horse bettor lost $50 on each of four consecutive races. Represent the total loss by a product of integers, and find the product.

For Exercises 89 through 92, use the map at the beginning of this Chapter Review.

89. The low temperature for Bucharest, Romania, is 2 times the low temperature for Raleigh, North Carolina. Find the low temperature for Bucharest.

90. The low temperature for Geneva, Switzerland, is the same as the low temperature for Miami, Florida, divided by -10. Find the low temperature for Geneva.

91. The low temperature for Cape Town, South Africa, is the same as the low temperature for Barrow, Alaska, divided by -2. Find the low temperature for Cape Town.

92. The low temperature for Stockholm, Sweden, is 2 times the low temperature for Providence, Rhode Island. Find the low temperature for Stockholm.

(2.6) *Simplify.*

93. $(-7)^2$

94. -7^2

95. -2^5

96. $(-2)^5$

97. $5 - 8 + 3$

98. $-3 + 12 + (-7) - 10$

99. $-10 + 3 \cdot (-2)$

100. $5 - 10 \cdot (-3)$

101. $16 \cdot (-2) + 4$

102. $3 \cdot (-12) - 8$

103. $5 + 6 \div (-3)$

104. $-6 + (-10) \div (-2)$

105. $16 + (-3) \cdot 12 \div 4$

106. $-12 + 25 \cdot 1 \div (-5)$

107. $4^3 - (8 - 3)^2$

108. $4^3 - 90$

109. $-(-4) \cdot |-3| - 5$ **110.** $|5 - 1|^2 \cdot (-5)$ **111.** $\dfrac{(-4)(-3) - (-2)(-1)}{-10 + 5}$ **112.** $\dfrac{4(12 - 18)}{-10 \div (-2 - 3)}$

Find the average of each list of numbers.

113. $-18, 25, -30, 7, 0, -2$ **114.** $-45, -40, -30, -25$

Evaluate each expression for $x = -2$ and $y = 1$.

115. $2x - y$ **116.** $y^2 + x^2$ **117.** $\dfrac{3x}{6}$ **118.** $\dfrac{5y - x}{-y}$

119. x^2 **120.** $-x^2$ **121.** $7 - x^2$ **122.** $100 - x^3$

Mixed Review

Perform the indicated operations.

123. $(-4)^2$ **124.** -4^2 **125.** $-6 + (-9)$ **126.** $-16 - 3$

127. $-4(-12)$ **128.** $\dfrac{84}{-4}$ **129.** $-76 - (-97)$ **130.** $-9 + 4$

Elevator shafts in some buildings extend not only above ground, but in many cases below ground to accommodate basements, underground parking, etc. The bar graph shows four such elevators and their shaft distance above and below ground. Use the bar graph to answer Exercises 131 and 132.

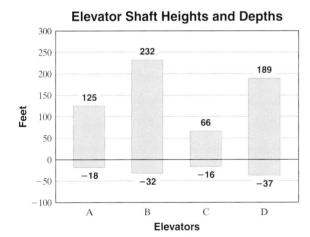

131. Which elevator shaft extends the farthest below ground?

132. Which elevator shaft extends the highest above ground?

133. The top of a mountain has an altitude of 12,923 feet. The bottom of a valley is 195 feet below sea level. Find the difference between these two elevations.

134. Wednesday's lowest temperature was $-18°C$. The cold weather continued and by Friday it had dropped another $9°C$. What was the temperature on Friday?

Simplify.

135. $(3 - 7)^2 \div (6 - 4)^3$ **136.** $(4 + 6)^2 \div (2 - 7)^2$ **137.** $3(4 + 2) + (-6) - 3^2$

138. $4(5 - 3) - (-2) + 3^3$ **139.** $2 - 4 \cdot 3 + \sqrt{25}$ **140.** $4 - 6 \cdot 5 + \sqrt{1}$

141. $\dfrac{-|-14| - 6}{7 + 2(-3)}$ **142.** $5(7 - 6)^3 - 4(2 - 3)^2 + 2^4$

Chapter 2 Test

Step-by-step test solutions are found on the Chapter Test Prep Videos. Where available: MyMathLab® or YouTube

Answers

Simplify each expression.

1. $-5 + 8$

2. $18 - 24$

3. $5 \cdot (-20)$

1. _____

2. _____

3. _____

4. _____

5. _____

6. _____

7. _____

8. _____

9. _____

10. _____

11. _____

12. _____

13. _____

14. _____

15. _____

16. _____

17. _____

18. _____

4. $(-16) \div (-4)$

5. $(-18) + (-12)$

6. $-7 - (-19)$

7. $(-5) \cdot (-13)$

8. $\dfrac{-25}{-5}$

9. $|-25| + (-13)$

10. $14 - |-20|$

11. $|5| \cdot |-10|$

12. $\dfrac{|-10|}{-|-5|}$

13. $(-8) + 9 \div (-3)$

14. $-7 + (-32) - 12 + 5$

15. $(-5)^3 - 24 \div (-3)$

16. $(5 - 9)^2 \cdot (8 - 2)^3$

17. $-(-7)^2 \div 7 \cdot (-4)$

18. $3 - (8 - 2)^3$

170

19. $-6 + (-15) \div (-3)$ **20.** $\dfrac{4}{2} - \dfrac{8^2}{16}$ **21.** $\dfrac{-3(-2) + 12}{-1(-4 - 5)}$

22. $\dfrac{|25 - 30|^2}{2(-6) + 7}$ **23.** $5(-8) - [6 - (2 - 4)] + (12 - 16)^2$

24. $-2^3 - 2^2$

Evaluate each expression for x = 0, y = −3, and z = 2.

25. $3x + y$ **26.** $|y| + |x| + |z|$ **27.** $\dfrac{3z}{2y}$

28. $2y^3$ **29.** $10 - y^2$ **30.** $7x + 3y - 4z$

31. Mary Dunstan, a diver, starts at sea level and then makes 4 successive descents of 22 feet. After the descents, what is her elevation?

32. Aaron Hawn has $129 in his checking account. He writes a check for $79, withdraws $40 from an ATM, and then deposits $35. Represent the new balance in his account by an integer.

33. Mt. Washington in New Hampshire has an elevation of 6288 feet above sea level. The Romanche Gap in the Atlantic Ocean has an elevation of 25,354 feet below sea level. Represent the difference in elevation between these two points by an integer. (*Source:* National Geographic Society and Defense Mapping Agency)

34. Lake Baykal in Siberian Russia is the deepest lake in the world, with a maximum depth of 5315 feet. The elevation of the lake's surface is 1495 feet above sea level. What is the elevation (with respect to sea level) of the deepest point in the lake? (*Source:* U.S. Geological Survey)

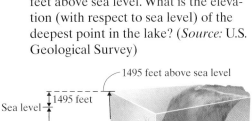
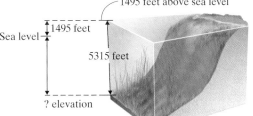

1495 feet above sea level

Sea level — 1495 feet

5315 feet

? elevation

35. Find the average of $-12, -13, 0, 9$.

36. Translate the following phrases into mathematical expressions. Use *x* to represent "a number."

   a. The product of a number and 17

   b. Twice a number subtracted from 20

19. _____

20. _____

21. _____

22. _____

23. _____

24. _____

25. _____

26. _____

27. _____

28. _____

29. _____

30. _____

31. _____

32. _____

33. _____

34. _____

35. _____

36. a. _____

    b. _____

Answers

1. _____

2. _____

3. _____

4. _____

5. _____

6. _____

7. a. _____

   b. _____

   c. _____

8. a. _____

   b. _____

   c. _____

9. _____

10. _____

11. _____

12. _____

13. _____

14. _____

15. _____

16. _____

17. _____

18. _____

19. a. _____

    b. _____

    c. _____

20. a. _____

    b. _____

    c. _____

21. _____

22. _____

23. a. _____

    b. _____

    c. _____

24. a. _____

    b. _____

    c. _____

Find the place value of the digit 3 in each whole number.

1. 396,418

2. 4308

3. 93,192

4. 693,298

5. 534,275,866

6. 267,301,818

7. Insert $<$ or $>$ to make a true statement.
   a. -7 7
   b. 0 -4
   c. -9 -11

8. Insert $<$ or $>$ to make a true statement.
   a. 12 4
   b. 13 31
   c. 82 79

9. Add:
   $13 + 2 + 7 + 8 + 9$

10. Add: $11 + 3 + 9 + 16$

11. Subtract: $7826 - 505$
   Check by adding.

12. Subtract: $3285 - 272$
   Check by adding.

13. The radius of Jupiter is 43,441 miles. The radius of Saturn is 7257 miles less than the radius of Jupiter. Find the radius of Saturn. (*Source:* National Space Science Data Center)

14. C.J. Dufour wants to buy a digital camera. She has $762 in her savings account. If the camera costs $237, how much money will she have in her account after buying the camera?

15. Round 568 to the nearest ten.

16. Round 568 to the nearest hundred.

17. Round each number to the nearest hundred to find an estimated difference.
$$\begin{array}{r} 4725 \\ -2879 \\ \hline \end{array}$$

18. Round each number to the nearest thousand to find an estimated difference.
$$\begin{array}{r} 8394 \\ -2913 \\ \hline \end{array}$$

19. Rewrite each using the distributive property.
   a. $3(4 + 5)$
   b. $10(6 + 8)$
   c. $2(7 + 3)$

20. Rewrite each using the distributive property.
   a. $5(2 + 12)$
   b. $9(3 + 6)$
   c. $4(8 + 1)$

21. Multiply: 631×125

22. Multiply: 299×104

23. Find each quotient. Check by multiplying.
   a. $42 \div 7$
   b. $\dfrac{64}{8}$
   c. $3\overline{)21}$

24. Find each quotient. Check by multiplying.
   a. $\dfrac{35}{5}$
   b. $64 \div 8$
   c. $4\overline{)48}$

25. Divide: $3705 \div 5$
Check by multiplying.

26. Divide: $3648 \div 8$
Check by multiplying.

27. As part of a promotion, an executive receives 238 cards, each good for one free song download. If she wants to share them evenly with 19 friends, how many download cards will each friend receive? How many will be left over?

28. Mrs. Mallory's first-grade class is going to the zoo. She pays a total of $324 for 36 admission tickets. How much does each ticket cost?

Evaluate.

29. 9^2

30. 5^3

31. 6^1

32. 4^1

33. $5 \cdot 6^2$

34. $2^3 \cdot 7$

35. Simplify: $\dfrac{7 - 2 \cdot 3 + 3^2}{5(2 - 1)}$

36. Simplify: $\dfrac{6^2 + 4 \cdot 4 + 2^3}{37 - 5^2}$

37. Evaluate $x + 7$ if x is 8.

38. Evaluate $5 + x$ if x is 9.

39. Simplify:
  a. $|-2|$
  b. $|8|$
  c. $|0|$

40. Simplify:
  a. $|4|$
  b. $|-7|$

41. Add: $-14 + 35$

42. Add: $8 + (-3)$

43. Evaluate $2a - b$ for $a = 8$ and $b = -6$.

44. Evaluate $x - y$ for $x = -2$ and $y = -7$.

45. Multiply: $-7 \cdot 3$

46. Multiply: $5(-2)$

47. Multiply: $0 \cdot (-4)$

48. Multiply: $-6 \cdot 9$

49. Simplify: $3(4 - 7) + (-2) - \sqrt{25}$

50. Simplify: $4 - 8(7 - 3) - (-1)$

25. _____
26. _____
27. _____
28. _____
29. _____
30. _____
31. _____
32. _____
33. _____
34. _____
35. _____
36. _____
37. _____
38. _____
39. a. _____
    b. _____
    c. _____
40. a. _____
    b. _____
41. _____
42. _____
43. _____
44. _____
45. _____
46. _____
47. _____
48. _____
49. _____
50. _____

3

Fractions and Mixed Numbers

Fractions are numbers and, like whole numbers and integers, they can be added, subtracted, multiplied, and divided. Fractions are very useful and appear frequently in everyday language, in common phrases such as "half an hour," "quarter of a pound," and "third of a cup." This chapter reviews the concepts of fractions and mixed numbers and demonstrates how to add, subtract, multiply, and divide these numbers.

What Majors Do College Freshmen Choose?

The following graph is called a circle graph or a pie chart. Each sector (shaped like a piece of pie) shows the fraction of entering college freshmen who choose to major in each discipline shown. Can you find your current choice of major in this graph? In Section 3.2, Exercises 95–98, we show this same circle graph, but in 3-D design. We simplify some of the fractions in it and also study sector size versus fraction value.

College Freshmen Majors

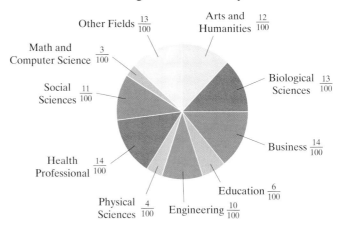

- Other Fields $\frac{13}{100}$
- Arts and Humanities $\frac{12}{100}$
- Math and Computer Science $\frac{3}{100}$
- Biological Sciences $\frac{13}{100}$
- Social Sciences $\frac{11}{100}$
- Business $\frac{14}{100}$
- Health Professional $\frac{14}{100}$
- Education $\frac{6}{100}$
- Physical Sciences $\frac{4}{100}$
- Engineering $\frac{10}{100}$

Source: The Higher Education Research Institute

Objective A Identifying Numerators and Denominators

Whole numbers are used to count whole things or units, such as cars, horses, dollars, and people. To refer to a part of a whole, fractions can be used. Here are some examples of **fractions.** Study these examples for a moment.

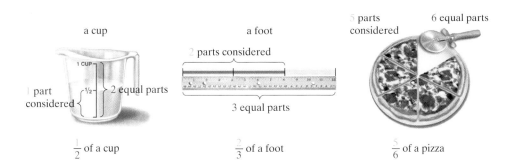

a cup

1 part considered

$\frac{1}{2}$ of a cup

a foot

2 parts considered

3 equal parts

$\frac{2}{3}$ of a foot

5 parts considered

6 equal parts

$\frac{5}{6}$ of a pizza

In a fraction, the top number is called the **numerator** and the bottom number is called the **denominator.** The bar between the numbers is called the **fraction bar.**

Names	Fraction	Meaning
numerator $\longrightarrow$	$\frac{5}{6}$	$\longleftarrow$ number of parts being considered
denominator $\longrightarrow$		$\longleftarrow$ number of equal parts in the whole

Examples Identify the numerator and the denominator of each fraction.

1. $\frac{3}{7}$ $\leftarrow$ numerator
   $\leftarrow$ denominator

2. $\frac{13}{5}$ $\leftarrow$ numerator
   $\leftarrow$ denominator

Work Practice 1–2

Helpful Hint

$\frac{3}{7}$ $\leftarrow$ Remember that the bar in a fraction means division. Since division by 0 is undefined, a fraction with a denominator of 0 is undefined. For example, $\frac{3}{0}$ is undefined.

Objective B Writing Fractions to Represent Parts of Figures or Real-Life Data

One way to become familiar with the concept of fractions is to visualize fractions with shaded figures. We can then write a fraction to represent the shaded area of the figure (or diagram).

Answers
1. numerator = 11, denominator = 2
2. numerator = 10, denominator = 17

Practice 3–4

Write a fraction to represent the shaded part of each figure.

3.

4.

Examples   Write a fraction to represent the shaded part of each figure.

3. In this figure, 2 of the 5 equal parts are shaded. Thus, the fraction is $\frac{2}{5}$.

$$\frac{2}{5} \quad \begin{array}{l} \leftarrow \text{number of parts shaded} \\ \leftarrow \text{number of equal parts} \end{array}$$

4. In this figure, 3 of the 10 rectangles are shaded. Thus, the fraction is $\frac{3}{10}$.

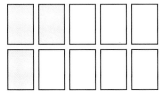

$$\frac{3}{10} \quad \begin{array}{l} \leftarrow \text{number of parts shaded} \\ \leftarrow \text{number of equal parts} \end{array}$$

■ Work Practice 3–4

Practice 5–6

Write a fraction to represent the part of the whole shown.

5. Just consider this part of the syringe

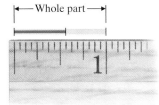

6. |←—Whole part—→|

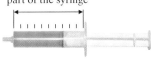

Examples   Write a fraction to represent the shaded part of the diagram.

5.

The fraction is $\frac{3}{10}$.

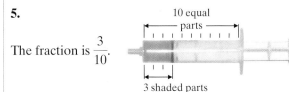

6.

The fraction is $\frac{1}{3}$.

■ Work Practice 5–6

Practice 7

Draw and shade a part of a figure to represent the fraction.

$\frac{2}{3}$ of a figure

Examples   Draw a figure and then shade a part of it to represent each fraction.

7. $\frac{5}{6}$ of a figure

We will use a geometric figure such as a rectangle. Since the denominator is 6, we divide it into 6 equal parts. Then we shade 5 of the equal parts.

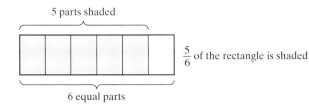

Answers

3. $\frac{3}{8}$ **4.** $\frac{1}{6}$ **5.** $\frac{7}{10}$ **6.** $\frac{9}{16}$

7. answers may vary; for example,

8. $\frac{3}{8}$ of a figure

If you'd like, our figure can consist of 8 triangles of the same size. We will shade 3 of the triangles.

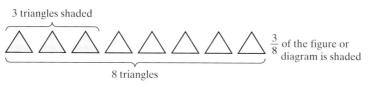

3 triangles shaded

8 triangles

$\frac{3}{8}$ of the figure or diagram is shaded

■ Work Practice 7–8

Practice 8

Draw and shade a part of a figure to represent the fraction.

$\frac{7}{11}$ of a figure

✔**Concept Check** If  represents $\frac{6}{7}$ of a whole diagram, sketch the whole diagram.

Example 9 Writing Fractions from Real-Life Data

Of the eight planets in our solar system (Pluto is now a dwarf planet), three are closer to the sun than Mars is. What fraction of the planets are closer to the sun than Mars is?

Solution: The fraction of planets closer to the sun than Mars is:

$\frac{3}{8}$ ← number of planets closer
   ← number of planets in our solar system

Thus, $\frac{3}{8}$ of the planets in our solar system are closer to the sun than Mars is.

■ Work Practice 9

Practice 9

Of the eight planets in our solar system, five are farther from the sun than Earth is. What fraction of the planets are farther from the sun than Earth is?

Objective C Identifying Proper Fractions, Improper Fractions, and Mixed Numbers ▶

The definitions and statements below apply to positive fractions.

A **proper fraction** is a fraction whose numerator is less than its denominator. Proper fractions are less than 1. For example, the shaded portion of the triangle is represented by $\frac{2}{3}$.

An **improper fraction** is a fraction whose numerator is greater than or equal to its denominator. Improper fractions are greater than or equal to 1.

$\frac{2}{3}$

Answers

8. answers may vary; for example,

9. $\frac{5}{8}$

✔**Concept Check Answer**

The shaded part of the group of circles below is $\frac{9}{4}$. The shaded part of the rectangle is $\frac{6}{6}$. Recall that $\frac{6}{6}$ simplifies to 1 and notice that the entire rectangle (1 whole figure) is shaded below.

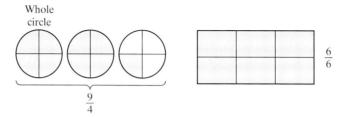

A **mixed number** contains a whole number and a fraction. Mixed numbers are greater than 1. Above, we wrote the shaded part of the group of circles as the improper fraction $\frac{9}{4}$. Now let's write the shaded part of the same group of circles as a mixed number (see below). The shaded part of the group of circles' area is $2\frac{1}{4}$. Read this as "two and one-fourth."

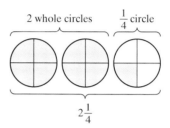

Practice 10

Identify each number as a proper fraction, improper fraction, or mixed number.

a. $\frac{5}{8}$ **b.** $\frac{7}{7}$

c. $\frac{14}{13}$ **d.** $\frac{13}{14}$

e. $5\frac{1}{4}$ **f.** $\frac{100}{49}$

Practice 11–12

Represent the shaded part of each figure group as both an improper fraction and a mixed number.

11.

12.

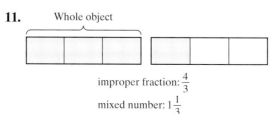

Answers

10. a. proper fraction   **b.** improper fraction   **c.** improper fraction
d. proper fraction   **e.** mixed number
f. improper fraction

11. $\frac{8}{3}$, $2\frac{2}{3}$ **12.** $\frac{5}{4}$, $1\frac{1}{4}$

Example 10

Identify each number as a proper fraction, improper fraction, or mixed number.

a. $\frac{6}{7}$ is a proper fraction.   **b.** $\frac{13}{12}$ is an improper fraction.

c. $\frac{2}{2}$ is an improper fraction.   **d.** $\frac{99}{101}$ is a proper fraction.

e. $1\frac{7}{8}$ is a mixed number.   **f.** $\frac{93}{74}$ is an improper fraction.

Work Practice 10

Helpful Hint

The mixed number $2\frac{1}{4}$ represents $2 + \frac{1}{4}$.

The mixed number $-3\frac{1}{5}$ represents $-\left(3 + \frac{1}{5}\right)$ or $-3 - \frac{1}{5}$. We review this later in this chapter.

Examples

Represent the shaded part of each figure group as both an improper fraction and a mixed number.

11.   Whole object

improper fraction: $\frac{4}{3}$

mixed number: $1\frac{1}{3}$

12.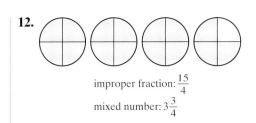

improper fraction: $\frac{15}{4}$

mixed number: $3\frac{3}{4}$

■ Work Practice 11–12

✓**Concept Check** If you were to round $3\frac{3}{4}$, shown in Example 12 above, to the nearest whole number, would you choose 3 or 4? Why?

Objective D Graphing Fractions on a Number Line

Another way to visualize fractions is to graph them on a number line. To do this, think of 1 unit on the number line as a whole. To graph $\frac{2}{5}$, for example, divide the distance from 0 to 1 into 5 equal parts. Then start at 0 and count 2 parts to the right.

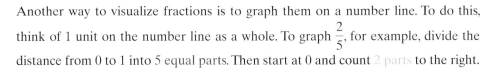

Notice that the graph of $\frac{2}{5}$ lies between 0 and 1. This means

$$0 < \frac{2}{5} \left(\text{or } \frac{2}{5} > 0\right) \text{ and also } \frac{2}{5} < 1$$

Example 13 Graph each fraction on a number line.

a. $\frac{3}{4}$ **b.** $\frac{1}{2}$ **c.** $\frac{3}{6}$ **d.** $\frac{9}{5}$ **e.** $\frac{6}{6}$

Solution:

a. To graph $\frac{3}{4}$, divide the distance from 0 to 1 into 4 parts. Then start at 0 and count over 3 parts.

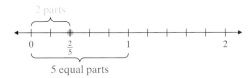

b.

c.

d.

e.

■ Work Practice 13

Practice 13

Graph each fraction on a number line.

a. $\frac{5}{7}$ **b.** $\frac{2}{3}$ **c.** $\frac{4}{6}$

d. $\frac{5}{4}$ **e.** $\frac{7}{7}$

Answers

13. a.

b.

c.

d.

e.

✓**Concept Check Answer**

4; answers may vary

The statements below apply to positive fractions.

The fractions in Example 13, parts a, b, and c, are proper fractions. Notice that the value of each is less than 1. This is always true for proper fractions since the numerator of a proper fraction is less than the denominator.

The fractions in Example 13, parts d and e, are improper fractions. Notice that improper fractions are greater than or equal to 1. This is always true since the numerator of an improper fraction is greater than or equal to the denominator.

Note: We will graph mixed numbers at the end of this chapter.

Objective E Reviewing Division Properties of 0 and 1

Before we continue further, don't forget from Section 1.7 that the fraction bar indicates division. Let's review some division properties of 1 and 0.

$$\frac{9}{9} = 1 \text{ because } 1 \cdot 9 = 9 \qquad \frac{-11}{1} = -11 \text{ because } -11 \cdot 1 = -11$$

$$\frac{0}{6} = 0 \text{ because } 0 \cdot 6 = 0 \qquad \frac{6}{0} \text{ is undefined because there is no number that}$$
$$\text{when multiplied by 0 gives 6.}$$

In general, we can say the following.

Let n be any integer except 0.

$$\frac{n}{n} = 1 \qquad \frac{0}{n} = 0$$

$$\frac{n}{1} = n \qquad \frac{n}{0} \text{ is undefined.}$$

Practice 14–19

Simplify.

14. $\dfrac{9}{9}$ **15.** $\dfrac{-6}{-6}$

16. $\dfrac{0}{-1}$ **17.** $\dfrac{4}{1}$

18. $\dfrac{-13}{0}$ **19.** $\dfrac{-13}{1}$

Examples Simplify.

14. $\dfrac{15}{15} = 1$ **15.** $\dfrac{-2}{-2} = 1$ **16.** $\dfrac{0}{-5} = 0$

17. $\dfrac{-9}{1} = -9$ **18.** $\dfrac{41}{1} = 41$ **19.** $\dfrac{19}{0}$ is undefined

■ Work Practice 14–19

Notice from Example 17 that we can have negative fractions. In fact,

$$\frac{-5}{1} = -5, \qquad \frac{5}{-1} = -5, \qquad \text{and} \quad -\frac{5}{1} = -5$$

Because all of the fractions equal -5, we have

$$\frac{-5}{1} = \frac{5}{-1} = -\frac{5}{1}$$

This means that the negative sign in a fraction can be written in the numerator, in the denominator, or in front of the fraction. Remember this as we work with negative fractions.

Helpful Hint Remember, for example, that

$$-\frac{2}{3} = \frac{-2}{3} = \frac{2}{-3}$$

Answers

14. 1 **15.** 1 **16.** 0 **17.** 4
18. undefined **19.** -13

Vocabulary, Readiness & Video Check

Use the choices below to fill in each blank.

improper	fraction	proper	is undefined	mixed number	$= 0$
≥ 1	denominator	$= 1$	< 1	numerator	

1. The number $\frac{17}{31}$ is called a(n) _____. The number 31 is called its _____ and 17 is called its _____.

2. If we simplify each fraction, $\frac{-9}{-9}$ _____, $\frac{0}{-4}$ _____, and we say $\frac{-4}{0}$ _____.

3. The fraction $\frac{8}{3}$ is called a(n) _____ fraction, the fraction $\frac{3}{8}$ is called a(n) _____ fraction, and $10\frac{3}{8}$ is called a(n) _____.

4. The value of an improper fraction is always _____, and the value of a proper fraction is always _____.

Martin-Gay Interactive Videos

See Video 3.1

Watch the section lecture video and answer the following questions.

Objective A 5. Complete this statement based on ⊞ Example 1: When the numerator is greater than or _____ to the denominator, you have a(n) _____ fraction. ⦿

Objective B 6. In ⊞ Example 4, there are two shapes in the diagram, so why do the representative fractions have a denominator of 3? ⦿

Objective C 7. Based on the lecture during ⊞ Example 4, fill in the blanks: A mixed number has a(n) _____ part and a(n) _____ part. ⦿

Objective D 8. From ⊞ Examples 6 and 7, when graphing a positive fraction on a number line, how does the denominator help? What does the denominator tell you? ⦿

Objective E 9. From ⊞ Example 10, what can you conclude about any fraction where the numerator and denominator are the same nonzero number? ⦿

3.1 **Exercise Set** MyMathLab®

Objectives A C *Identify the numerator and the denominator of each fraction and identify each fraction as proper or improper. See Examples 1, 2, and 10.*

⦿ 1. $\frac{1}{2}$

2. $\frac{1}{4}$

⦿ 3. $\frac{10}{3}$

4. $\frac{53}{21}$

5. $\frac{15}{15}$

6. $\frac{26}{26}$

Objectives B C *Write a proper or improper fraction to represent the shaded part of each diagram. If an improper fraction is appropriate, write the shaded part of the diagram as (a) an improper fraction and (b) a mixed number. (Note to students: If you are familiar with simplifying fractions, note that none of the fractions in this section are simplified.) See Examples 3 through 6 and 11 and 12.*

7.

8.

9.

10.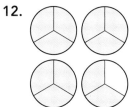

11.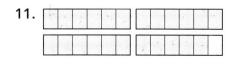

12.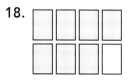

13.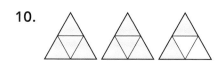

14.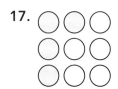

15.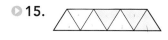

16.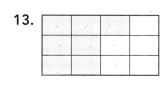

17.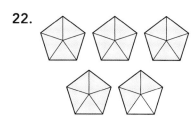

18.

19.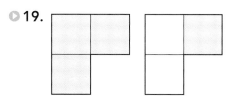

20.

21.

22.

23.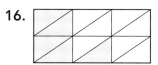

24.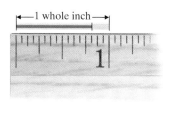

25.

26.

Objective B *Draw and shade a part of a diagram to represent each fraction. See Examples 7 and 8.*

27. $\dfrac{1}{5}$ of a diagram

28. $\dfrac{1}{16}$ of a diagram

29. $\dfrac{6}{7}$ of a diagram

30. $\dfrac{7}{9}$ of a diagram

31. $\dfrac{4}{4}$ of a diagram

32. $\dfrac{6}{6}$ of a diagram

Write each fraction. (Note to students: In case you are familiar with simplifying fractions, note that none of the fractions in this section are simplified.) See Example 9.

33. Of the 131 students at a small private school, 42 are freshmen. What fraction of the students are freshmen?

34. Of the 63 employees at a new biomedical engineering firm, 22 are men. What fraction of the employees are men?

35. Use Exercise 33 to answer **a** and **b.**
   a. How many students are *not* freshmen?
   b. What fraction of the students are *not* freshmen?

36. Use Exercise 34 to answer **a** and **b.**
   a. How many of the employees are women?
   b. What fraction of the employees are women?

37. As of 2014, the United States has had 44 different presidents. A total of seven U.S. presidents were born in the state of Ohio, second only to the state of Virginia in producing U.S. presidents. What fraction of U.S. presidents were born in Ohio? (*Source: World Almanac,* 2014)

38. Of the eight planets in our solar system, four have days that are longer than the 24-hour Earth day. What fraction of the planets have longer days than Earth has? (*Source:* National Space Science Data Center)

39. Hurricane Sandy, which struck the East Coast in October 2012, was the largest Atlantic hurricane ever documented. Sandy was just one of 19 named tropical storms that formed during the 2012 Atlantic hurricane season. A total of 10 of these tropical storms turned into hurricanes. What fraction of the 2012 Atlantic tropical storms escalated into hurricanes? (*Source:* National Oceanic and Atmospheric Administration)

40. There are 12 inches in a foot. What fractional part of a foot does 5 inches represent?

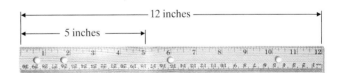

41. There are 31 days in the month of March. What fraction of the month does 11 days represent?

Mon.	Tue.	Wed.	Thu.	Fri.	Sat.	Sun.
					1	2
3	4	5	6	7	8	9
10	11	12	13	14	15	16
17	18	19	20	21	22	23
24	25	26	27	28	29	30
31						

42. There are 60 minutes in an hour. What fraction of an hour does 37 minutes represent?

43. In a prealgebra class containing 31 students, there are 18 freshmen, 10 sophomores, and 3 juniors. What fraction of the class is sophomores?

44. In a sports team with 20 children, there are 9 boys and 11 girls. What fraction of the team is boys?

45. Thirty-three out of the fifty states in the United States contain federal Indian reservations.
   a. What fraction of the states contain federal Indian reservations?
   b. How many states do not contain federal Indian reservations?
   c. What fraction of the states do not contain federal Indian reservations? (*Source:* Tiller Research, Inc., Albuquerque, NM)

46. Consumer fireworks are legal in 46 out of the 50 states in the United States.
   a. In what fraction of the states are consumer fireworks legal?
   b. In how many states are consumer fireworks illegal?
   c. In what fraction of the states are consumer fireworks illegal? (*Source:* United States Fireworks Safety Council)

47. A bag contains 50 red or blue marbles. If 21 marbles are blue,
   a. What fraction of the marbles are blue?
   b. How many marbles are red?
   c. What fraction of the marbles are red?

48. An art dealer is taking inventory. His shop contains a total of 37 pieces, which are all sculptures, watercolor paintings, or oil paintings. If there are 15 watercolor paintings and 17 oil paintings, answer each question.
   a. What fraction of the inventory is watercolor paintings?
   b. What fraction of the inventory is oil paintings?
   c. How many sculptures are there?
   d. What fraction of the inventory is sculptures?

Objective D *Graph each fraction on a number line. See Example 13.*

49. $\frac{1}{4}$

50. $\frac{1}{3}$

51. $\frac{4}{7}$

52. $\frac{5}{6}$

53. $\dfrac{8}{5}$

0

54. $\dfrac{9}{8}$

0

55. $\dfrac{7}{3}$

0

56. $\dfrac{13}{7}$

0

Objective E *Simplify by dividing. See Examples 14 through 19.*

57. $\dfrac{12}{12}$

58. $\dfrac{-3}{-3}$

59. $\dfrac{-5}{1}$

60. $\dfrac{-20}{1}$

61. $\dfrac{0}{-2}$

62. $\dfrac{0}{-8}$

63. $\dfrac{-8}{-8}$

64. $\dfrac{-14}{-14}$

65. $\dfrac{-9}{0}$

66. $\dfrac{-7}{0}$

67. $\dfrac{3}{1}$

68. $\dfrac{5}{1}$

Review

Simplify. See Section 1.9.

69. 3^2

70. 4^3

71. 5^3

72. 3^4

Write each using exponents.

73. $7 \cdot 7 \cdot 7 \cdot 7 \cdot 7$

74. $5 \cdot 5 \cdot 5 \cdot 5$

75. $2 \cdot 2 \cdot 2 \cdot 3$

76. $4 \cdot 4 \cdot 10 \cdot 10 \cdot 10$

Concept Extensions

For Exercises 77–80, write each fraction in two other equivalent ways by inserting the negative sign in different places.

77. $-\dfrac{11}{2} = \quad\quad =$

78. $-\dfrac{21}{4} = \quad\quad =$

79. $\dfrac{-13}{15} = \quad\quad =$

80. $\dfrac{45}{-57} = \quad\quad =$

81. In your own words, explain why $\dfrac{0}{10} = 0$ and $\dfrac{10}{0}$ is undefined.

82. In your own words, explain why $\dfrac{0}{-3} = 0$ and $\dfrac{-3}{0}$ is undefined.

Solve. See the Concept Checks in this section.

83. If ◯◯◯◯ represents $\frac{4}{9}$ of a whole diagram, sketch the whole diagram.

84. If △△ represents $\frac{1}{3}$ of a whole diagram, sketch the whole diagram.

85. Round the mixed number $7\frac{1}{8}$ to the nearest whole number.

86. Round the mixed number $5\frac{11}{12}$ to the nearest whole number.

87. IKEA Group employs workers in four different regions worldwide, as shown on the bar graph. What fraction of IKEA employees work in the North American region?

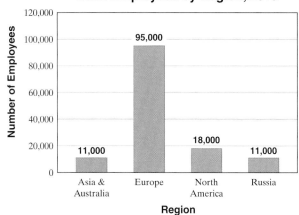

IKEA Employees by Region, 2013

Source: IKEA Group

88. The Public Broadcasting Service (PBS) provides programming to the noncommercial, public TV stations of the United States. The bar graph shows a breakdown of the public television licensees by type. Each licensee operates one or more PBS member TV stations. What fraction of the public television licensees are universities or colleges? (*Source:* The Public Broadcasting Service)

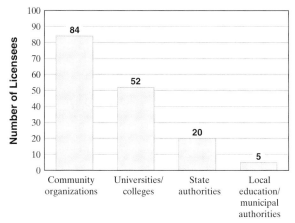

Public Television Licensees

89. Heifer International is a nonprofit world hunger relief organization that focuses on sustainable agriculture programs. Currently, Heifer International is working in 6 North American countries, 5 South American countries, 4 Central/Eastern European countries, 14 African countries, and 7 Asian/South Pacific countries. What fraction of the total countries in which Heifer International is working are located in North America? (*Hint:* First find the total number of countries.) (*Source:* Heifer International)

90. The United States Mint operates six facilities. One facility is the headquarters, one facility is a depository, and four facilities mint coins. What fraction of the United States Mint facilities produce coins? (*Hint:* First find the total number of facilities.) (*Source:* United States Mint)

3.2 Factors and Simplest Form

Objective A Writing a Number as a Product of Prime Numbers

To perform operations on fractions, it is necessary to be able to factor a number. Remember that factoring a number means writing a number as a product. We first practice writing a number as a product of prime numbers.

Recall from Section 1.6 that since $12 = 2 \cdot 6$, the numbers 2 and 6 are called *factors* of 12. A **factor** is any number that divides a number evenly (with a remainder of 0).

Of all the ways to factor a number, one special way is called the **prime factorization.** To help us write prime factorizations, we first review prime and composite numbers.

Prime Numbers

A **prime number** is a natural number that has exactly two different factors, 1 and itself.

The first several prime numbers are

2, 3, 5, 7, 11, 13, 17

It would be helpful to memorize these.

If a natural number other than 1 is not a prime number, it is called a **composite number.**

Composite Numbers

A **composite number** is any natural number, other than 1, that is not prime.

Helpful Hint

The natural number 1 is neither prime nor composite.

Now we are ready to define, and then find, **prime factorizations** of numbers.

Prime Factorization

The **prime factorization** of a number is the factorization in which all the factors are prime numbers.

Earlier, we wrote $12 = 2 \cdot 6$. Although 2 and 6 are factors of 12, the product $2 \cdot 6$ is *not* the prime factorization of 12 because 6 is *not* a prime number.

The prime factorization of 12 is $2 \cdot 2 \cdot 3$ because

$12 = 2 \cdot 2 \cdot 3$

This product is 12 and each
number is a prime number.

Helpful Hint

Don't forget that multiplication is commutative, so the order of the factors is not important. We can write the factorization $2 \cdot 2 \cdot 3$ as $2 \cdot 3 \cdot 2$ or $3 \cdot 2 \cdot 2$. Any of these is called the *prime factorization* of 12.

Every whole number greater than 1 has exactly one prime factorization.

One method for finding the prime factorization of a number is by using a factor tree, as shown in the next example.

Practice 1

Use a factor tree to find the prime factorization of each number.

a. 30   **b.** 56   **c.** 72

Example 1   Write the prime factorization of 45.

Solution: We can begin by writing 45 as the product of two numbers, say, 5 and 9.

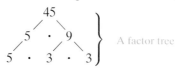

The number 5 is prime but 9 is not, so we write 9 as $3 \cdot 3$.

A factor tree

Each factor is now a prime number, so the prime factorization of 45 is $3 \cdot 3 \cdot 5$ or $3^2 \cdot 5$.

■ Work Practice 1

✓**Concept Check** True or false? Two different numbers can have exactly the same prime factorization. Explain your answer.

Practice 2

Write the prime factorization of 60.

Example 2   Write the prime factorization of 80.

Solution: Write 80 as a product of two numbers. Continue this process until all factors are prime.

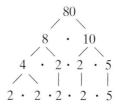

All factors are now prime, so the prime factorization of 80 is

$$2 \cdot 2 \cdot 2 \cdot 2 \cdot 5 \quad \text{or} \quad 2^4 \cdot 5.$$

■ Work Practice 2

Helpful Hint

It makes no difference which factors you start with. The prime factorization of a number will be the same.

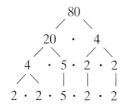

Same factors as in Example 2

Answers

1. a. $2 \cdot 3 \cdot 5$ **b.** $2^3 \cdot 7$ **c.** $2^3 \cdot 3^2$
2. $2^2 \cdot 3 \cdot 5$

✓**Concept Check Answer**

false; answers may vary

There are a few quick **divisibility tests** to determine whether a number is divisible by the primes 2, 3, or 5. (A number is divisible by 2, for example, if 2 divides it evenly so that the remainder is 0.)

Divisibility Tests

A whole number is divisible by:

- 2 if the last digit is 0, 2, 4, 6, or 8.

    132 is divisible by 2 since the last digit is a 2.

- 3 if the sum of the digits is divisible by 3.

    144 is divisible by 3 since $1 + 4 + 4 = 9$, which is divisible by 3.

- 5 if the last digit is 0 or 5.

    1115 is divisible by 5 since the last digit is a 5.

Helpful Hint

Here are a few other divisibility tests you may want to use. A whole number is divisible by:

- 4 if its last two digits are divisible by 4.

    1712 is divisible by 4.

- 6 if it's divisible by 2 and 3.

    9858 is divisible by 6.

- 9 if the sum of its digits is divisible by 9.

    5238 is divisible by 9 since $5 + 2 + 3 + 8 = 18$, which is divisible by 9.

When finding the prime factorization of larger numbers, you may want to use the procedure shown in Example 3.

Example 3   Write the prime factorization of 252.

Solution: For this method, we divide prime numbers into the given number. Since the ones digit of 252 is 2, we know that 252 is divisible by 2.

$$\frac{126}{2\overline{)252}}$$

126 is divisible by 2 also.

$$\frac{63}{2\overline{)126}}$$
$$2\overline{)252}$$

63 is not divisible by 2 but is divisible by 3. Divide 63 by 3 and continue in this same manner until the quotient is a prime number.

$$\frac{7}{3\overline{)\ 21}}$$
$$\frac{}{3\overline{)\ 63}}$$
$$2\overline{)126}$$
$$2\overline{)252}$$

Helpful Hint

The order of choosing prime numbers does not matter. For consistency, we use the order 2, 3, 5, 7, ...

The prime factorization of 252 is $2 \cdot 2 \cdot 3 \cdot 3 \cdot 7$ or $2^2 \cdot 3^2 \cdot 7$.

Practice 3

Write the prime factorization of 297.

Work Practice 3

Answer

3. $3^3 \cdot 11$

In this text, we will write the factorization of a number from the smallest factor to the largest factor.

✓**Concept Check** True or false? The prime factorization of 117 is $9 \cdot 13$. Explain your reasoning.

Objective B Writing Fractions in Simplest Form

Fractions that represent the same portion of a whole or the same point on a number line are called **equivalent fractions.** Study the table below to see two ways to visualize equivalent fractions.

Equivalent Fractions	
Figures	**Number Line**
When we shade $\frac{1}{3}$ and $\frac{2}{6}$ on the same-sized figures,	When we graph $\frac{1}{3}$ and $\frac{2}{6}$ on a number line,
$\frac{1}{3}$ $\frac{2}{6}$	(number line graphic)
notice that both $\frac{1}{3}$ and $\frac{2}{6}$ represent the same portion of a whole. These fractions are called **equivalent fractions,** and we write $\frac{1}{3} = \frac{2}{6}$.	notice that both $\frac{1}{3}$ and $\frac{2}{6}$ correspond to the same point. These fractions are called **equivalent fractions,** and we write $\frac{1}{3} = \frac{2}{6}$.
Thus, $\frac{1}{3} = \frac{2}{6}$ and $\frac{1}{3}$ and $\frac{2}{6}$ are equivalent.	

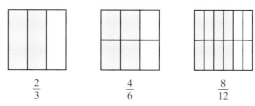

$$\frac{2}{3} \qquad \frac{4}{6} \qquad \frac{8}{12}$$

For example, $\frac{2}{3}, \frac{4}{6},$ and $\frac{8}{12}$ all represent the same shaded portion of the rectangle's area, so they are equivalent fractions. To show that these fractions are equivalent, we place an equal sign between them. In other words,

$$\frac{2}{3} = \frac{4}{6} = \frac{8}{12}$$

There are many equivalent forms of a fraction. A special equivalent form of a fraction is called **simplest form.**

Simplest Form of a Fraction

A fraction is written in **simplest form** or **lowest terms** when the numerator and the denominator have no common factors other than 1.

For example, the fraction $\frac{2}{3}$ *is* in simplest form because 2 and 3 have no common factor other than 1. The fraction $\frac{4}{6}$ *is not* in simplest form because 4 and 6 both

✓**Concept Check Answer**
false; 9 is not prime

have a factor of 2. That is, 2 is a common factor of 4 and 6. The process of writing a fraction in simplest form is called **simplifying** the fraction.

To simplify $\frac{4}{6}$ and write it as $\frac{2}{3}$, let's first study a few properties. Recall from Section 3.1 that any nonzero whole number n divided by itself is 1.

Any nonzero number n divided by itself is 1.

$$\frac{5}{5} = 1, \quad \frac{17}{17} = 1, \quad \frac{24}{24} = 1, \text{ or, in general, } \frac{n}{n} = 1$$

Also, in general, if $\frac{a}{b}$ and $\frac{c}{d}$ are fractions (with b and d not 0), the following is true.

$$\frac{a \cdot c}{b \cdot d} = \frac{a}{b} \cdot \frac{c}{d}*$$

These properties allow us to do the following:

$$\frac{4}{6} = \frac{2 \cdot 2}{2 \cdot 3} = \frac{2}{2} \cdot \frac{2}{3} = 1 \cdot \frac{2}{3} = \frac{2}{3} \qquad \text{When 1 is multiplied by a number, the result is the same number.}$$

This is 1

Example 4 Write in simplest form: $\frac{12}{20}$

Solution: Notice that 12 and 20 have a common factor of 4.

$$\frac{12}{20} = \frac{4 \cdot 3}{4 \cdot 5} = \frac{4}{4} \cdot \frac{3}{5} = 1 \cdot \frac{3}{5} = \frac{3}{5}$$

Since 3 and 5 have no common factors (other than 1), $\frac{3}{5}$ is in simplest form.

■ Work Practice 4

Practice 4

Write in simplest form: $\frac{30}{45}$

If you have trouble finding common factors, write the prime factorization of the numerator and the denominator.

Example 5 Write in simplest form: $\frac{42}{66}$

Solution: Let's write the prime factorizations of 42 and 66.

$$\frac{42}{66} = \frac{2 \cdot 3 \cdot 7}{2 \cdot 3 \cdot 11} = \frac{2}{2} \cdot \frac{3}{3} \cdot \frac{7}{11} = 1 \cdot 1 \cdot \frac{7}{11} = \frac{7}{11}$$

■ Work Practice 5

Practice 5

Write in simplest form: $\frac{39}{51}$

In the example above, you may have saved time by noticing that 42 and 66 have a common factor of 6.

$$\frac{42}{66} = \frac{6 \cdot 7}{6 \cdot 11} = \frac{6}{6} \cdot \frac{7}{11} = 1 \cdot \frac{7}{11} = \frac{7}{11}$$

Helpful Hint

Writing the prime factorizations of the numerator and the denominator is helpful in finding any common factors.

Answers

4. $\frac{2}{3}$ 5. $\frac{13}{17}$

*Note: We will study this concept further in the next section.

The method for simplifying negative fractions is the same as for positive fractions.

Practice 6

Write in simplest form: $-\dfrac{9}{50}$

Example 6 Write in simplest form: $-\dfrac{10}{27}$

Solution:

$$-\frac{10}{27} = -\frac{2 \cdot 5}{3 \cdot 3 \cdot 3} \quad \text{Prime factorizations of 10 and 27}$$

Since 10 and 27 have no common factors, $-\dfrac{10}{27}$ is already in simplest form.

■ Work Practice 6

Practice 7

Write in simplest form: $\dfrac{49}{112}$

Example 7 Write in simplest form: $\dfrac{30}{108}$

Solution:

$$\frac{30}{108} = \frac{2 \cdot 3 \cdot 5}{2 \cdot 2 \cdot 3 \cdot 3 \cdot 3} = \frac{2}{2} \cdot \frac{3}{3} \cdot \frac{5}{2 \cdot 3 \cdot 3} = 1 \cdot 1 \cdot \frac{5}{18} = \frac{5}{18}$$

■ Work Practice 7

We can use a shortcut procedure with common factors when simplifying.

$$\frac{4}{6} = \frac{\overset{1}{\cancel{2}} \cdot 2}{\underset{1}{\cancel{2}} \cdot 3} = \frac{1 \cdot 2}{1 \cdot 3} = \frac{2}{3} \quad \text{Divide out the common factor of 2 in the numerator and denominator.}$$

This procedure is possible because dividing out a common factor in the numerator and denominator is the same as removing a factor of 1 in the product.

Writing a Fraction in Simplest Form

To write a fraction in simplest form, write the prime factorization of the numerator and the denominator and then divide both by all common factors.

Practice 8

Write in simplest form: $-\dfrac{64}{20}$

Example 8 Write in simplest form: $-\dfrac{72}{26}$

Solution:

$$-\frac{72}{26} = -\frac{\overset{1}{\cancel{2}} \cdot 2 \cdot 2 \cdot 3 \cdot 3}{\underset{1}{\cancel{2}} \cdot 13} = -\frac{1 \cdot 2 \cdot 2 \cdot 3 \cdot 3}{1 \cdot 13} = -\frac{36}{13}$$

■ Work Practice 8

✓**Concept Check** Which is the correct way to simplify the fraction $\dfrac{15}{25}$? Or are both correct? Explain.

a. $\dfrac{15}{25} = \dfrac{3 \cdot \overset{1}{\cancel{5}}}{5 \cdot \underset{1}{\cancel{5}}} = \dfrac{3}{5}$

b. $\dfrac{1\overset{1}{\cancel{5}}}{2\underset{1}{\cancel{5}}} = \dfrac{11}{21}$

Answers

6. $-\dfrac{9}{50}$ 7. $\dfrac{7}{16}$ 8. $-\dfrac{16}{5}$

✓**Concept Check Answers**

a. correct   **b.** incorrect

Example 9 Write in simplest form: $\dfrac{6}{60}$

Solution:

$$\frac{6}{60} = \frac{\cancel{2} \cdot \cancel{3}}{\cancel{2} \cdot 2 \cdot \cancel{3} \cdot 5} = \frac{1 \cdot 1}{1 \cdot 2 \cdot 1 \cdot 5} = \frac{1}{10}$$

■ Work Practice 9

Practice 9

Write in simplest form: $\dfrac{8}{56}$

Helpful Hint

Be careful when all factors of the numerator or denominator are divided out. In Example 9, the numerator was $1 \cdot 1 = 1$, so the final result was $\dfrac{1}{10}$.

In the fraction of Example 9, $\dfrac{6}{60}$, you may have immediately noticed that the largest common factor of 6 and 60 is 6. If so, you may simply divide out that common factor.

$$\frac{6}{60} = \frac{\cancel{6}}{\cancel{6} \cdot 10} = \frac{1}{1 \cdot 10} = \frac{1}{10} \qquad \text{Divide out the common factor of 6.}$$

Notice that the result, $\dfrac{1}{10}$, is in simplest form. If it were not, we would repeat the same procedure until the result is in simplest form.

Objective C Determining Whether Two Fractions Are Equivalent

Recall from Objective **B** that two fractions are equivalent if they represent the same part of a whole. One way to determine whether two fractions are equivalent is to see whether they simplify to the same fraction.

Example 10 Determine whether $\dfrac{16}{40}$ and $\dfrac{10}{25}$ are equivalent.

Solution: Simplify each fraction.

$$\frac{16}{40} = \frac{\cancel{8} \cdot 2}{\cancel{8} \cdot 5} = \frac{1 \cdot 2}{1 \cdot 5} = \frac{2}{5}$$

$$\frac{10}{25} = \frac{2 \cdot \cancel{5}}{5 \cdot \cancel{5}} = \frac{2 \cdot 1}{5 \cdot 1} = \frac{2}{5}$$

Since these fractions are the same, $\dfrac{16}{40} = \dfrac{10}{25}$.

■ Work Practice 10

Practice 10

Determine whether $\dfrac{7}{9}$ and $\dfrac{21}{27}$ are equivalent.

There is a shortcut method you may use to check or test whether two fractions are equivalent. In the example above, we learned that the fractions are equivalent, or

$$\frac{16}{40} = \frac{10}{25}$$

In this example above, we call $25 \cdot 16$ and $40 \cdot 10$ **cross products** because they are the products one obtains by multiplying diagonally across the equal sign, as shown below.

Cross Products

$$25 \cdot 16 \qquad\qquad 40 \cdot 10$$

$$\frac{16}{40} = \frac{10}{25}$$

Answers

9. $\dfrac{1}{7}$ **10.** equivalent

Notice that these cross products are equal:

$$25 \cdot 16 = 400, \quad 40 \cdot 10 = 400$$

In general, this is true for equivalent fractions.

Equality of Fractions

$$8 \cdot 6 \qquad\qquad\qquad 24 \cdot 2$$

$$\frac{6}{24} \overset{?}{=} \frac{2}{8}$$

Since the cross products ($8 \cdot 6 = 48$ and $24 \cdot 2 = 48$) are equal, the fractions are equal.

Note: If the cross products are not equal, the fractions are not equal.

Practice 11

Determine whether $\frac{4}{13}$ and $\frac{5}{18}$ are equivalent.

Example 11 Determine whether $\frac{8}{11}$ and $\frac{19}{26}$ are equivalent.

Solution: Let's check cross products.

$$26 \cdot 8 \qquad\qquad\qquad 11 \cdot 19$$
$$= 208 \qquad \frac{8}{11} \overset{?}{=} \frac{19}{26} \qquad = 209$$

Since $208 \neq 209$, then $\frac{8}{11} \neq \frac{19}{26}$.

Helpful Hint
"Not equal to" symbol

■ Work Practice 11

Objective D Solving Problems by Writing Fractions in Simplest Form

Many real-life problems can be solved by writing fractions. To make the answers clearer, these fractions should be written in simplest form.

Practice 12

There are four national historical parks in the state of Virginia. See Example 12 and determine what fraction of the United States' national historical parks can be found in Virginia. Write the fraction in simplest form.

Example 12   Calculating Fraction of Parks in Pennsylvania

There are currently 46 national historical parks in the United States. Two of these historical parks are located in the state of Pennsylvania. What fraction of the United States' national historical parks can be found in Pennsylvania? Write the fraction in simplest form. (*Source:* National Park Service)

Answers

11. not equivalent   **12.** $\frac{2}{23}$

Solution: First we determine the fraction of parks found in Pennsylvania.

$\dfrac{2}{46}$ ← national historical parks in Pennsylvania

   ← total national historical parks

Next we simplify the fraction.

$$\frac{2}{46} = \frac{\cancel{2}}{\cancel{2} \cdot 23} = \frac{1}{1 \cdot 23} = \frac{1}{23}$$

Thus, $\dfrac{1}{23}$ of the United States' national historical parks are in Pennsylvania.

■ Work Practice 12

 Calculator Explorations **Simplifying Fractions**

Scientific Calculator

Many calculators have a fraction key, such as $\boxed{a^b/c}$, that allows you to simplify a fraction on the calculator. For example, to simplify $\dfrac{324}{612}$, enter

$\boxed{324}$ $\boxed{a^b/c}$ $\boxed{612}$ $\boxed{=}$

The display will read

$\boxed{\quad 9 \mid 17 \quad}$

which represents $\dfrac{9}{17}$, the original fraction simplified.

Graphing Calculator

Graphing calculators also allow you to simplify fractions. The fraction option on a graphing calculator may be found under the $\boxed{\text{MATH}}$ menu.

To simplify $\dfrac{324}{612}$, enter

$\boxed{324}$ $\boxed{\div}$ $\boxed{612}$ $\boxed{\text{MATH}}$ $\boxed{\text{ENTER}}$ $\boxed{\text{ENTER}}$

The display will read

$\boxed{324/612 \blacktriangleright \text{Frac } 9/17}$

Helpful Hint The Calculator Explorations boxes in this chapter provide only an introduction to fraction keys on calculators. Any time you use a calculator, there are both advantages and limitations to its use. Never rely solely on your calculator. It is very important that you understand how to perform all operations on fractions by hand in order to progress through later topics. For further information, talk to your instructor.

Use your calculator to simplify each fraction.

1. $\dfrac{128}{224}$ 2. $\dfrac{231}{396}$ 3. $\dfrac{340}{459}$

4. $\dfrac{999}{1350}$ 5. $\dfrac{432}{810}$ 6. $\dfrac{225}{315}$

7. $\dfrac{54}{243}$ 8. $\dfrac{455}{689}$

Vocabulary, Readiness & Video Check

Use the choices below to fill in each blank.

cross products   equivalent   composite

simplest form   prime factorization   prime

1. The number 40 equals $2 \cdot 2 \cdot 2 \cdot 5$. Since each factor is prime, we call $2 \cdot 2 \cdot 2 \cdot 5$ the _____ of 40.

2. A natural number, other than 1, that is not prime is called a(n) _____ number.

3. A natural number that has exactly two different factors, 1 and itself, is called a(n) _____ number.

4. In $\frac{11}{48}$, since 11 and 48 have no common factors other than 1, $\frac{11}{48}$ is in _____.

5. Fractions that represent the same portion of a whole are called _____ fractions.

6. In the statement $\frac{5}{12} = \frac{15}{36}$, $5 \cdot 36$ and $12 \cdot 15$ are called _____.

Without dividing, answer the questions in Exercises 7 and 8.

7. Is 2430 divisible by 2? By 3? By 5?   **8.** Is 1155 divisible by 2? By 3? By 5?

Write the prime factorization of each number.

9. 15   **10.** 10   **11.** 6   **12.** 21

13. 4   **14.** 9

Martin-Gay Interactive Videos

See Video 3.2

Watch the section lecture video and answer the following questions.

Objective A **15.** From ⊞ Example 1, what two things should you check to make sure your prime factorization of a number is correct? ▶

Objective B **16.** From the lecture before ⊞ Example 3, when you have a common factor in the numerator and denominator of a fraction, essentially you have what? ▶

Objective C **17.** Describe another way to solve ⊞ Example 8 besides using cross products. ▶

Objective D **18.** Why isn't $\frac{10}{24}$ the final answer to ⊞ Example 9? What is the final answer? ▶

3.2 Exercise Set MyMathLab®

Objective A *Write the prime factorization of each number. See Examples 1 through 3.*

1. 20

2. 12

▷ 3. 48

4. 75

5. 45

6. 64

7. 162

8. 128

9. 110

10. 130

11. 85

12. 93

▷ 13. 240

14. 836

15. 828

16. 504

Objective B *Write each fraction in simplest form. See Examples 4 through 9.*

17. $\dfrac{3}{12}$

18. $\dfrac{5}{30}$

19. $\dfrac{4}{42}$

20. $\dfrac{9}{48}$

▷ 21. $\dfrac{14}{16}$

22. $\dfrac{22}{34}$

23. $\dfrac{20}{30}$

24. $\dfrac{70}{80}$

25. $\dfrac{35}{50}$

26. $\dfrac{25}{55}$

27. $-\dfrac{63}{81}$

28. $-\dfrac{21}{49}$

▷ 29. $\dfrac{24}{40}$

30. $\dfrac{36}{54}$

31. $\dfrac{27}{64}$

32. $\dfrac{32}{63}$

33. $\dfrac{25}{40}$

34. $\dfrac{36}{42}$

35. $-\dfrac{40}{64}$

36. $-\dfrac{28}{60}$

37. $\dfrac{36}{24}$

38. $\dfrac{60}{36}$

39. $\dfrac{90}{120}$

40. $\dfrac{60}{150}$

▷ 41. $\dfrac{70}{196}$

42. $\dfrac{98}{126}$

43. $\dfrac{66}{308}$

44. $\dfrac{65}{234}$

▷ 45. $-\dfrac{55}{85}$

46. $-\dfrac{78}{90}$

47. $\dfrac{189}{216}$

48. $\dfrac{144}{162}$

49. $\dfrac{224}{16}$

50. $\dfrac{270}{15}$

Objective C *Determine whether each pair of fractions is equivalent. See Examples 10 and 11.*

51. $\dfrac{3}{6}$ and $\dfrac{4}{8}$

52. $\dfrac{3}{9}$ and $\dfrac{2}{6}$

▷ 53. $\dfrac{7}{11}$ and $\dfrac{5}{8}$

54. $\dfrac{2}{5}$ and $\dfrac{4}{11}$

55. $\dfrac{10}{15}$ and $\dfrac{6}{9}$

56. $\dfrac{4}{10}$ and $\dfrac{6}{15}$

▷ 57. $\dfrac{3}{9}$ and $\dfrac{6}{18}$

58. $\dfrac{2}{8}$ and $\dfrac{7}{28}$

59. $\dfrac{10}{13}$ and $\dfrac{12}{15}$

60. $\dfrac{16}{20}$ and $\dfrac{9}{12}$

61. $\dfrac{8}{18}$ and $\dfrac{12}{24}$

62. $\dfrac{6}{21}$ and $\dfrac{14}{35}$

Objective D *Solve. Write each fraction in simplest form. See Example 12.*

63. A work shift for an employee at Starbucks consists of 8 hours. What fraction of the employee's work shift is represented by 2 hours?

64. Two thousand baseball caps were sold one year at the U.S. Open Golf Tournament. What fractional part of this total does 200 caps represent?

65. There are 5280 feet in a mile. What fraction of a mile is represented by 2640 feet?

66. There are 100 centimeters in 1 meter. What fraction of a meter is 20 centimeters?

67. Sixteen out of the total fifty states in the United States have Ritz-Carlton hotels. (*Source:* Ritz-Carlton Hotel Company, LLC)

    a. What fraction of states can claim at least one Ritz-Carlton hotel?

    b. How many states do not have a Ritz-Carlton hotel?

    c. Write the fraction of states without a Ritz-Carlton hotel.

68. There are 75 national monuments in the United States. Ten of these monuments are located in New Mexico. (*Source:* National Park Service)

    a. What fraction of the national monuments in the United States can be found in New Mexico?

    b. How many of the national monuments in the United States are found outside New Mexico?

    c. Write the fraction of national monuments found in states other than New Mexico.

69. The outer wall of the Pentagon is 24 inches thick. Ten inches is concrete, 8 inches is brick, and 6 inches is limestone. What fraction of the wall is concrete?

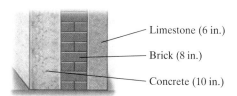

Limestone (6 in.)

Brick (8 in.)

Concrete (10 in.)

70. There are 35 students in a biology class. If 10 students made an A on the first test, what fraction of the students made an A?

71. As Internet usage grows in the United States, more and more state governments are placing services online. Forty-two out of the total fifty states have Web sites that allow residents to file their federal and state income taxes electronically at the same time.

    a. How many states do not have this type of Web site?

    b. What fraction of states do not have this type of Web site? (*Source:* MeF Federal/State Program)

72. Katy Biagini just bought a brand-new 2014 Toyota Camry hybrid for $26,400. Her old car was traded in for $12,000.

    a. How much of her purchase price was not covered by her trade-in?

    b. What fraction of the purchase price was not covered by the trade-in?

73. As of this writing, a total of 464 individuals from the United States are or have been astronauts. Of these, 26 were born in Texas. What fraction of U.S. astronauts were born in Texas? (*Source:* Spacefacts Web site)

74. Worldwide, Hallmark employs nearly 12,000 full-time employees. About 3200 employees work at the Hallmark headquarters in Kansas City, Missouri. What fraction of Hallmark employees work in Kansas City? (*Source:* Hallmark Cards, Inc.)

Review

Multiply. See Section 1.6.

75.
$$\begin{array}{r} 91 \\ \times\ 4 \\ \hline \end{array}$$

76.
$$\begin{array}{r} 73 \\ \times\ 8 \\ \hline \end{array}$$

77.
$$\begin{array}{r} 387 \\ \times\ 6 \\ \hline \end{array}$$

78.
$$\begin{array}{r} 562 \\ \times\ 9 \\ \hline \end{array}$$

79.
$$\begin{array}{r} 72 \\ \times\ 35 \\ \hline \end{array}$$

80.
$$\begin{array}{r} 238 \\ \times\ 26 \\ \hline \end{array}$$

Concept Extensions

81. In your own words, define equivalent fractions.

82. Given a fraction, say, $\frac{3}{8}$, how many fractions are there that are equivalent to it? Explain your answer.

Write each fraction in simplest form.

83. $\dfrac{3975}{6625}$

84. $\dfrac{9506}{12{,}222}$

There are generally considered to be eight basic blood types. The table shows the number of people with the various blood types in a typical group of 100 blood donors. Use the table to answer Exercises 85 through 88. Write each answer in simplest form.

Distribution of Blood Types in Blood Donors	
Blood Type	**Number of People**
O Rh-positive	37
O Rh-negative	7
A Rh-positive	36
A Rh-negative	6
B Rh-positive	9
B Rh-negative	1
AB Rh-positive	3
AB Rh-negative	1
(*Source:* American Red Cross Biomedical Services)	

85. What fraction of blood donors have blood type A Rh-positive?

86. What fraction of blood donors have an O blood type?

87. What fraction of blood donors have an AB blood type?

88. What fraction of blood donors have a B blood type?

Find the prime factorization of each number.

89. 34,020

90. 131,625

Solve.

91. In your own words, define a prime number.

92. The number 2 is a prime number. All other even natural numbers are composite numbers. Explain why.

93. Two students have different prime factorizations for the same number. Is this possible? Explain.

94. Two students work to prime-factor 120. One student starts by writing 120 as 12×10. The other student writes 120 as 24×5. Finish each prime factorization. Are they the same? Why or why not?

*The following graph is called a **circle graph** or **pie chart**. Each sector (shaped like a piece of pie) shows the fraction of entering college freshmen who choose to major in each discipline shown. The whole circle represents the entire class of college freshmen. Use this graph to answer Exercises 95 through 98. Write each fraction answer in simplest form.*

College Freshmen Majors

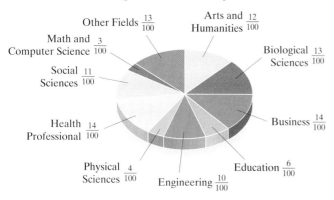

Other Fields $\frac{13}{100}$

Arts and Humanities $\frac{12}{100}$

Math and Computer Science $\frac{3}{100}$

Biological Sciences $\frac{13}{100}$

Social Sciences $\frac{11}{100}$

Health Professional $\frac{14}{100}$

Business $\frac{14}{100}$

Physical Sciences $\frac{4}{100}$

Engineering $\frac{10}{100}$

Education $\frac{6}{100}$

Source: The Higher Education Research Institute

95. What fraction of entering college freshmen plan to major in education?

96. What fraction of entering college freshmen plan to major in engineering?

97. Why is the Business sector the same size as the Health Professional sector?

98. Why is the Physical Sciences sector smaller than the Business sector?

Use this circle graph to answer Exercises 99 through 102.

Areas Maintained by the National Park Service

Preserves/ Reserves $\frac{1}{20}$

Lakeshores/ Seashores $\frac{2}{50}$

Other $\frac{3}{100}$

Battlefields $\frac{3}{50}$

Rivers $\frac{2}{50}$

Memorials $\frac{7}{100}$

Parkways/ Scenic Trails $\frac{1}{50}$

Monuments $\frac{19}{100}$

Parks $\frac{3}{20}$

Recreation Areas $\frac{1}{25}$

Historical Parks and Sites $\frac{31}{100}$

Source: National Park Service

99. What fraction of National Park Service areas are National Memorials?

100. What fraction of National Park Service areas are National Parks?

101. Why is the National Battlefields sector smaller than the National Monuments sector?

102. Why is the National Lakeshores/National Seashores sector the same size as the National Rivers sector?

Use the following numbers for Exercises 103 through 106.

8691   786   1235   2235   85   105   22   222   900   1470

103. List the numbers divisible by both 2 and 3.

104. List the numbers that are divisible by both 3 and 5.

105. The answers to Exercise 103 are also divisible by what number? Tell why.

106. The answers to Exercise 104 are also divisible by what number? Tell why.

Multiplying and Dividing Fractions

Objective A Multiplying Fractions

Let's use a diagram to discover how fractions are multiplied. For example, to multiply $\frac{1}{2}$ and $\frac{3}{4}$, we find $\frac{1}{2}$ of $\frac{3}{4}$. To do this, we begin with a diagram showing $\frac{3}{4}$ of a rectangle's area shaded.

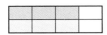
$\frac{3}{4}$ of the rectangle's area is shaded.

To find $\frac{1}{2}$ of $\frac{3}{4}$, we heavily shade $\frac{1}{2}$ of the part that is already shaded.

By counting smaller rectangles, we see that $\frac{3}{8}$ of the larger rectangle is now heavily shaded, so that

$$\frac{1}{2} \text{ of } \frac{3}{4} \text{ is } \frac{3}{8}, \text{ or } \frac{1}{2} \cdot \frac{3}{4} = \frac{3}{8}$$

Notice that $\frac{1}{2} \cdot \frac{3}{4} = \frac{1 \cdot 3}{2 \cdot 4} = \frac{3}{8}$.

Multiplying Fractions

To multiply two fractions, multiply the numerators and multiply the denominators. If $a, b, c,$ and d represent numbers, and b and d are not 0, we have

$$\frac{a}{b} \cdot \frac{c}{d} = \frac{a \cdot c}{b \cdot d}$$

Examples Multiply.

1. $\frac{2}{3} \cdot \frac{5}{11} = \frac{2 \cdot 5}{3 \cdot 11} = \frac{10}{33}$ Multiply numerators.
Multiply denominators.

This fraction is in simplest form since 10 and 33 have no common factors other than 1.

2. $\frac{1}{4} \cdot \frac{1}{2} = \frac{1 \cdot 1}{4 \cdot 2} = \frac{1}{8}$ This fraction is in simplest form.

Work Practice 1–2

Example 3 Multiply and simplify: $\frac{6}{7} \cdot \frac{14}{27}$

Solution:

$$\frac{6}{7} \cdot \frac{14}{27} = \frac{6 \cdot 14}{7 \cdot 27}$$

We can simplify by finding the prime factorizations and using our shortcut procedure of dividing out common factors in the numerator and denominator.

$$\frac{6 \cdot 14}{7 \cdot 27} = \frac{2 \cdot \overset{1}{\cancel{3}} \cdot 2 \cdot \overset{1}{\cancel{7}}}{\underset{1}{\cancel{7}} \cdot \underset{1}{\cancel{3}} \cdot 3 \cdot 3} = \frac{2 \cdot 2}{3 \cdot 3} = \frac{4}{9}$$

Work Practice 3

Objectives

A Multiply Fractions.

B Evaluate Exponential Expressions with Fractional Bases.

C Divide Fractions.

D Multiply or Divide Given Fractional Replacement Values.

E Solve Applications That Require Multiplication of Fractions.

Practice 1–2

Multiply.

1. $\frac{3}{7} \cdot \frac{5}{11}$ **2.** $\frac{1}{3} \cdot \frac{1}{9}$

Practice 3

Multiply and simplify: $\frac{6}{77} \cdot \frac{7}{8}$

Answers

1. $\frac{15}{77}$ **2.** $\frac{1}{27}$ **3.** $\frac{3}{44}$

Helpful Hint

Remember that the shortcut procedure in Example 3 is the same as removing factors of 1 in the product.

$$\frac{6 \cdot 14}{7 \cdot 27} = \frac{2 \cdot 3 \cdot 2 \cdot 7}{7 \cdot 3 \cdot 3 \cdot 3} = \frac{7}{7} \cdot \frac{3}{3} \cdot \frac{2 \cdot 2}{3 \cdot 3} = 1 \cdot 1 \cdot \frac{4}{9} = \frac{4}{9}$$

Practice 4

Multiply and simplify: $\dfrac{4}{27} \cdot \dfrac{3}{8}$

Helpful Hint   Don't forget that we may identify common factors that are not prime numbers.

Example 4   Multiply and simplify: $\dfrac{23}{32} \cdot \dfrac{4}{7}$

Solution: Notice that 4 and 32 have a common factor of 4.

$$\frac{23}{32} \cdot \frac{4}{7} = \frac{23 \cdot 4}{32 \cdot 7} = \frac{23 \cdot 4}{4 \cdot 8 \cdot 7} = \frac{23}{8 \cdot 7} = \frac{23}{56}$$

Work Practice 4

After multiplying two fractions, always check to see whether the product can be simplified.

Practice 5

Multiply.

$$\frac{1}{2} \cdot \left(-\frac{11}{28}\right)$$

Example 5   Multiply: $-\dfrac{1}{4} \cdot \dfrac{1}{2}$

Solution: Recall that the product of a negative number and a positive number is a negative number.

$$-\frac{1}{4} \cdot \frac{1}{2} = -\frac{1 \cdot 1}{4 \cdot 2} = -\frac{1}{8}$$

Work Practice 5

Practice 6–7

Multiply.

6. $\left(-\dfrac{4}{11}\right)\left(-\dfrac{33}{16}\right)$

7. $\dfrac{1}{6} \cdot \dfrac{3}{10} \cdot \dfrac{25}{16}$

Examples   Multiply.

6. $\left(-\dfrac{6}{13}\right)\left(-\dfrac{26}{30}\right) = \dfrac{6 \cdot 26}{13 \cdot 30} = \dfrac{6 \cdot 13 \cdot 2}{13 \cdot 6 \cdot 5} = \dfrac{2}{5}$ The product of two negative numbers is a positive number.

7. $\dfrac{1}{3} \cdot \dfrac{2}{5} \cdot \dfrac{9}{16} = \dfrac{1 \cdot 2 \cdot 9}{3 \cdot 5 \cdot 16} = \dfrac{1 \cdot 2 \cdot 3 \cdot 3}{3 \cdot 5 \cdot 2 \cdot 8} = \dfrac{3}{40}$

Work Practice 6–7

Objective B Evaluating Exponential Expressions with Fractional Bases

The base of an exponential expression can also be a fraction.

$$\left(\frac{1}{3}\right)^4 = \frac{1}{3} \cdot \frac{1}{3} \cdot \frac{1}{3} \cdot \frac{1}{3} = \frac{1 \cdot 1 \cdot 1 \cdot 1}{3 \cdot 3 \cdot 3 \cdot 3} = \frac{1}{81}$$

$\dfrac{1}{3}$ is a factor 4 times.

Answers

4. $\dfrac{1}{18}$ **5.** $-\dfrac{11}{56}$ **6.** $\dfrac{3}{4}$ **7.** $\dfrac{5}{64}$

Example 8 Evaluate.

a. $\left(\dfrac{2}{5}\right)^4 = \dfrac{2}{5} \cdot \dfrac{2}{5} \cdot \dfrac{2}{5} \cdot \dfrac{2}{5} = \dfrac{2 \cdot 2 \cdot 2 \cdot 2}{5 \cdot 5 \cdot 5 \cdot 5} = \dfrac{16}{625}$

b. $\left(-\dfrac{1}{4}\right)^2 = \left(-\dfrac{1}{4}\right) \cdot \left(-\dfrac{1}{4}\right) = \dfrac{1 \cdot 1}{4 \cdot 4} = \dfrac{1}{16}$ The product of two negative numbers is a positive number.

■ Work Practice 8

Practice 8

Evaluate.

a. $\left(\dfrac{3}{4}\right)^3$ **b.** $\left(-\dfrac{4}{5}\right)^2$

Objective C Daniel Dividing Fractions

Before we can divide fractions, we need to know how to find the **reciprocal** of a fraction.

Reciprocal of a Fraction

Two numbers are **reciprocals** of each other if their product is 1. The reciprocal of the fraction $\dfrac{a}{b}$ is $\dfrac{b}{a}$ because $\dfrac{a}{b} \cdot \dfrac{b}{a} = \dfrac{a \cdot b}{b \cdot a} = 1$.

For example,

The reciprocal of $\dfrac{2}{5}$ is $\dfrac{5}{2}$ because $\dfrac{2}{5} \cdot \dfrac{5}{2} = \dfrac{10}{10} = 1$.

The reciprocal of 5 is $\dfrac{1}{5}$ because $5 \cdot \dfrac{1}{5} = \dfrac{5}{1} \cdot \dfrac{1}{5} = \dfrac{5}{5} = 1$.

The reciprocal of $-\dfrac{7}{11}$ is $-\dfrac{11}{7}$ because $-\dfrac{7}{11} \cdot -\dfrac{11}{7} = \dfrac{77}{77} = 1$.

Helpful Hint

Every number has a reciprocal except 0. The number 0 has no reciprocal because there is no number such that $0 \cdot a = 1$.

Division of fractions has the same meaning as division of whole numbers. For example,

$10 \div 5$ means: How many 5s are there in 10?

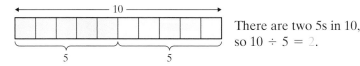

There are two 5s in 10, so $10 \div 5 = 2$.

$\dfrac{3}{4} \div \dfrac{1}{8}$ means: How many $\dfrac{1}{8}$s are there in $\dfrac{3}{4}$?

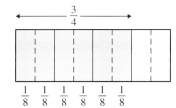

There are six $\dfrac{1}{8}$s in $\dfrac{3}{4}$, so $\dfrac{3}{4} \div \dfrac{1}{8} = 6$.

We use reciprocals to divide fractions.

Answers

8. **a.** $\dfrac{27}{64}$ **b.** $\dfrac{16}{25}$

Dividing Fractions

To divide two fractions, multiply the first fraction by the reciprocal of the second fraction.

If $a, b, c,$ and d represent numbers, and $b, c,$ and d are not 0, then

$$\frac{a}{b} \div \frac{c}{d} = \frac{a}{b} \cdot \frac{d}{c} = \frac{a \cdot d}{b \cdot c}$$

reciprocal

For example,

multiply by reciprocal

$$\frac{3}{4} \div \frac{1}{8} = \frac{3}{4} \cdot \frac{8}{1} = \frac{3 \cdot 8}{4 \cdot 1} = \frac{3 \cdot 2 \cdot \cancel{4}}{\cancel{4} \cdot 1} = \frac{6}{1} \quad \text{or} \quad 6$$

Just as when you are multiplying fractions, *always* check to see whether the result can be simplified when you divide fractions.

Practice 9–11

Divide and simplify.

9. $\dfrac{3}{2} \div \dfrac{14}{5}$ **10.** $\dfrac{8}{7} \div \dfrac{2}{9}$

11. $\dfrac{4}{9} \div \dfrac{1}{2}$

Examples Divide and simplify.

9. $\dfrac{7}{8} \div \dfrac{2}{9} = \dfrac{7}{8} \cdot \dfrac{9}{2} = \dfrac{7 \cdot 9}{8 \cdot 2} = \dfrac{63}{16}$

10. $\dfrac{5}{16} \div \dfrac{3}{4} = \dfrac{5}{16} \cdot \dfrac{4}{3} = \dfrac{5 \cdot 4}{16 \cdot 3} = \dfrac{5 \cdot \cancel{4}}{\cancel{4} \cdot 4 \cdot 3} = \dfrac{5}{12}$

11. $\dfrac{2}{5} \div \dfrac{1}{2} = \dfrac{2}{5} \cdot \dfrac{2}{1} = \dfrac{2 \cdot 2}{5 \cdot 1} = \dfrac{4}{5}$

■ Work Practice 9–11

Helpful Hint

When dividing by a fraction, do not look for common factors to divide out until you rewrite the division as multiplication.

Do not try to divide out these two 2s.

$$\frac{1}{\mathbf{2}} \div \frac{\mathbf{2}}{3} = \frac{1}{2} \cdot \frac{3}{2} = \frac{3}{4}$$

Practice 12

Divide: $-\dfrac{10}{4} \div \dfrac{2}{9}$

Answers

9. $\dfrac{15}{28}$ **10.** $\dfrac{36}{7}$ **11.** $\dfrac{8}{9}$ **12.** $-\dfrac{45}{4}$

Example 12 Divide: $-\dfrac{7}{12} \div -\dfrac{5}{6}$

Solution: Recall that the quotient (or product) of two negative numbers is a positive number.

$$-\frac{7}{12} \div -\frac{5}{6} = -\frac{7}{12} \cdot -\frac{6}{5} = \frac{7 \cdot \cancel{6}}{2 \cdot \cancel{6} \cdot 5} = \frac{7}{10}$$

■ Work Practice 12

✓**Concept Check** Which is the correct way to divide $\frac{3}{5}$ by $\frac{5}{12}$? Or are both correct? Explain.

a. $\frac{3}{5} \div \frac{5}{12} = \frac{3}{5} \cdot \frac{12}{5}$

b. $\frac{3}{5} \div \frac{5}{12} = \frac{5}{3} \cdot \frac{5}{12}$

✓**Concept Check** Which of the following is the correct way to divide $\frac{2}{5}$ by $\frac{3}{4}$? Or are both correct? Explain.

a. $\frac{5}{2} \cdot \frac{3}{4}$

b. $\frac{2}{5} \cdot \frac{4}{3}$

Objective D Multiplying or Dividing with Fractional Replacement Values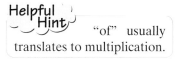

Example 13 If $x = \frac{7}{8}$ and $y = -\frac{1}{3}$, evaluate **(a)** xy and **(b)** $x \div y$.

Solution: Replace x with $\frac{7}{8}$ and y with $-\frac{1}{3}$.

a. $xy = \frac{7}{8} \cdot -\frac{1}{3}$

$= -\frac{7 \cdot 1}{8 \cdot 3}$

$= -\frac{7}{24}$

b. $x \div y = \frac{7}{8} \div -\frac{1}{3}$

$= \frac{7}{8} \cdot -\frac{3}{1}$

$= -\frac{7 \cdot 3}{8 \cdot 1}$

$= -\frac{21}{8}$

⬛ Work Practice 13

Practice 13

If $x = -\frac{3}{4}$ and $y = \frac{9}{2}$, evaluate (a) xy and (b) $x \div y$.

Objective E Solving Problems by Multiplying Fractions ▶

To solve real-life problems that involve multiplying fractions, we use our four problem-solving steps from Chapter 1. In Example 14, a new key word that implies multiplication is used. That key word is "**of.**"

Example 14 Finding the Number of Roller Coasters in an Amusement Park

Cedar Point is an amusement park located in Sandusky, Ohio. Its collection of 72 rides is the largest in the world. Of the rides, $\frac{2}{9}$ are roller coasters. How many roller coasters are in Cedar Point's collection of rides? (*Source:* Wikipedia)

(*Continued on next page*)

Helpful Hint "of" usually translates to multiplication.

Practice 14

Kings Dominion is an amusement park in Doswell, Virginia. Of its 48 rides, $\frac{5}{16}$ of them are roller coasters. How many roller coasters are in Kings Dominion? (*Source:* Cedar Fair Parks)

Answers

13. a. $-\frac{27}{8}$ **b.** $-\frac{1}{6}$

14. 15 roller coasters

✓**Concept Check Answers**

a. correct **b.** incorrect

a. incorrect **b.** correct

Solution:

1. **UNDERSTAND** the problem. To do so, read and reread the problem. We are told that $\frac{2}{9}$ of Cedar Point's rides are roller coasters. The word "of" here means multiplication.

2. **TRANSLATE.**

In words:

number of roller coasters	is	$\frac{2}{9}$	of	total rides at Cedar Point
↓	↓	↓	↓	↓

Translate:

$$\text{number of roller coasters} = \frac{2}{9} \cdot 72$$

3. **SOLVE:** Before we solve, let's estimate a reasonable answer. The fraction $\frac{2}{9}$ is less than $\frac{1}{4}$ (draw a diagram, if needed), and $\frac{1}{4}$ of 72 rides is 18 rides, so the number of roller coasters should be less than 18.

$$\frac{2}{9} \cdot 72 = \frac{2}{9} \cdot \frac{72}{1} = \frac{2 \cdot 72}{9 \cdot 1} = \frac{2 \cdot \cancel{9} \cdot 8}{\cancel{9} \cdot 1} = \frac{16}{1} \quad \text{or} \quad 16$$

4. **INTERPRET.** *Check* your work. From our estimate, our answer is reasonable. *State* your conclusion: The number of roller coasters at Cedar Point is 16.

■ Work Practice 14

Helpful Hint

To help visualize a fractional part of a whole number, look at the diagram below.

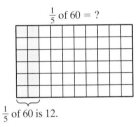

$\frac{1}{5}$ of $60 = ?$

$\frac{1}{5}$ of 60 is 12.

Vocabulary, Readiness & Video Check

Use the choices below to fill in each blank. Not all choices will be used.

multiplication	$\frac{a \cdot d}{b \cdot c}$	$\frac{a \cdot c}{b \cdot d}$	$\frac{2 \cdot 2 \cdot 2}{7}$	$\frac{2}{7} \cdot \frac{2}{7} \cdot \frac{2}{7}$
division	0	reciprocals		

1. To multiply two fractions, we write $\frac{a}{b} \cdot \frac{c}{d} = $ _____ .

2. Two numbers are _____ of each other if their product is 1.

3. The expression $\frac{2^3}{7} = $ _____ whereas $\left(\frac{2}{7}\right)^3 = $ _____ .

4. Every number has a reciprocal except _____.

5. To divide two fractions, we write $\dfrac{a}{b} \div \dfrac{c}{d} =$ _____.

6. The word "of" indicates _____.

Find each product.

7. $\dfrac{1}{3} \cdot \dfrac{2}{5}$

8. $\dfrac{2}{3} \cdot \dfrac{4}{7}$

9. $\dfrac{6}{5} \cdot \dfrac{1}{7}$

10. $\dfrac{7}{3} \cdot \dfrac{2}{3}$

11. $\dfrac{3}{1} \cdot \dfrac{3}{8}$

12. $\dfrac{2}{1} \cdot \dfrac{7}{11}$

Martin-Gay Interactive Videos

See Video 3.3

Watch the section lecture video and answer the following questions.

Objective A **13.** In ▦ Example 1, how do we know that the answer is positive?

Objective B **14.** In ▦ Example 4, does the exponent apply to the negative sign? Why or why not?

Objective C **15.** Complete this statement based on ▦ Example 5: When writing the reciprocal of a fraction, the denominator becomes the _____, and the numerator becomes the _____.

Objective D **16.** In ▦ Example 9a, why don't we write out the prime factorizations of 4 and 9 in the numerator?

Objective E **17.** What formula is used to solve ▦ Example 10?

3.3 **Exercise Set** MyMathLab®

Objective A *Multiply. Write the product in simplest form. See Examples 1 through 7.*

1. $\dfrac{2}{7} \cdot \dfrac{6}{11}$

2. $\dfrac{5}{9} \cdot \dfrac{7}{4}$

3. $-\dfrac{1}{5} \cdot \dfrac{9}{10}$

4. $\dfrac{5}{8} \cdot -\dfrac{1}{3}$

5. $\left(-\dfrac{1}{2}\right)\left(-\dfrac{2}{15}\right)$

6. $\left(-\dfrac{3}{8}\right)\left(-\dfrac{5}{12}\right)$

7. $\dfrac{6}{5} \cdot \dfrac{1}{7}$

8. $\dfrac{7}{3} \cdot \dfrac{1}{4}$

9. $\dfrac{2}{7} \cdot \dfrac{5}{8}$

10. $\dfrac{7}{8} \cdot \dfrac{2}{3}$

11. $\dfrac{5}{28} \cdot \dfrac{2}{25}$

12. $\dfrac{4}{35} \cdot \dfrac{5}{24}$

13. $0 \cdot \dfrac{8}{9}$

14. $\dfrac{11}{12} \cdot 0$

15. $\dfrac{18}{20} \cdot \dfrac{36}{99}$

16. $\dfrac{5}{32} \cdot \dfrac{64}{100}$

17. $\dfrac{11}{20} \cdot \dfrac{1}{7} \cdot \dfrac{5}{22}$

18. $\dfrac{27}{32} \cdot \dfrac{10}{13} \cdot \dfrac{16}{30}$

Objective B *Evaluate. See Example 8.*

19. $\left(\dfrac{1}{5}\right)^3$

20. $\left(-\dfrac{1}{2}\right)^4$

▸ **21.** $\left(-\dfrac{2}{3}\right)^2$

22. $\left(\dfrac{8}{9}\right)^2$

23. $\left(-\dfrac{2}{3}\right)^3 \cdot \dfrac{1}{2}$

24. $\left(-\dfrac{3}{4}\right)^3 \cdot \dfrac{1}{3}$

Objective C *Divide. Write all quotients in simplest form. See Examples 9 through 12.*

▸ **25.** $\dfrac{2}{3} \div \dfrac{5}{6}$

26. $\dfrac{5}{8} \div \dfrac{2}{3}$

27. $-\dfrac{6}{15} \div \dfrac{12}{5}$

28. $-\dfrac{4}{15} \div -\dfrac{8}{3}$

29. $\dfrac{8}{9} \div -\dfrac{1}{2}$

30. $\dfrac{10}{11} \div -\dfrac{4}{5}$

31. $-\dfrac{2}{3} \div 4$

▸ **32.** $-\dfrac{5}{6} \div 10$

33. $\dfrac{1}{10} \div \dfrac{10}{1}$

34. $\dfrac{3}{13} \div \dfrac{13}{3}$

35. $\dfrac{7}{45} \div \dfrac{4}{25}$

36. $\dfrac{14}{52} \div \dfrac{1}{13}$

37. $\dfrac{3}{25} \div \dfrac{27}{40}$

38. $\dfrac{6}{15} \div \dfrac{7}{10}$

▸ **39.** $\dfrac{8}{13} \div 0$

40. $0 \div \dfrac{4}{11}$

▸ **41.** $0 \div \dfrac{7}{8}$

42. $\dfrac{2}{3} \div 0$

Objectives A B C **Mixed Practice** *Perform each indicated operation. See Examples 1 through 12.*

43. $\dfrac{2}{3} \cdot \dfrac{5}{9}$

44. $\dfrac{8}{15} \cdot \dfrac{5}{32}$

45. $-\dfrac{5}{28} \cdot \dfrac{35}{25}$

46. $\dfrac{24}{45} \cdot -\dfrac{5}{8}$

47. $-\dfrac{3}{5} \div -\dfrac{4}{5}$

48. $-\dfrac{11}{16} \div -\dfrac{13}{16}$

49. $\left(-\dfrac{3}{4}\right)^2$

50. $\left(-\dfrac{1}{2}\right)^5$

51. $7 \div \dfrac{2}{11}$

52. $-100 \div \dfrac{1}{2}$

53. $\dfrac{4}{8} \div \dfrac{3}{16}$

54. $\dfrac{9}{2} \div \dfrac{16}{15}$

55. $-\dfrac{1}{8} \cdot \dfrac{3}{7} \cdot \dfrac{14}{27}$

56. $-\dfrac{1}{10} \cdot \dfrac{7}{11} \cdot \dfrac{22}{35}$

Objective D *Given the following replacement values, evaluate (a) xy and (b) x ÷ y. See Example 13.*

57. $x = \dfrac{2}{5}$ and $y = \dfrac{5}{6}$

58. $x = \dfrac{8}{9}$ and $y = \dfrac{1}{4}$

▸ **59.** $x = -\dfrac{4}{5}$ and $y = \dfrac{9}{11}$

60. $x = \dfrac{7}{6}$ and $y = -\dfrac{1}{2}$

Objective E Translating *Solve. Write each answer in simplest form. For Exercises 61 through 64, recall that "of" translates to multiplication. See Example 14.*

61. Find $\frac{1}{4}$ of 200.   **62.** Find $\frac{1}{5}$ of 200.   **63.** Find $\frac{5}{6}$ of 24.   **64.** Find $\frac{5}{8}$ of 24.

Solve. For Exercises 65 and 66, the solutions have been started for you. See Example 14.

65. In the United States, $\frac{7}{50}$ of college freshmen major in business. A community college in Pennsylvania has a freshmen enrollment of approximately 800 students. How many of these freshmen might we project are majoring in business?

Start the solution:

1. UNDERSTAND the problem. Reread it as many times as needed.
2. TRANSLATE into an equation. (Fill in the blank below.)

freshmen majoring in business	is	$\frac{7}{50}$	of	community college freshmen enrollment
↓	↓	↓	↓	↓

$$\text{freshmen majoring in business} = \frac{7}{50} \cdot \underline{\qquad}$$

Finish with:

3. SOLVE
4. INTERPRET

66. A patient was told that, at most, $\frac{1}{5}$ of his calories should come from fat. If his diet consists of 3000 calories a day, find the maximum number of calories that can come from fat.

Start the solution:

1. UNDERSTAND the problem. Reread it as many times as needed.
2. TRANSLATE into an equation. (Fill in the blank below.)

patient's fat calories	is	$\frac{1}{5}$	of	his daily calories
↓	↓	↓	↓	↓

$$\text{patient's fat calories} = \frac{1}{5} \cdot \underline{\qquad}$$

Finish with:

3. SOLVE
4. INTERPRET

67. In 2013, there were approximately 230 million moviegoers in the United States and Canada. Of these, about $\frac{3}{10}$ viewed at least one 3-D movie. Find the approximate number of people who viewed at least one 3-D movie. (*Source:* Motion Picture Association of America)

68. In 2013, cinemas in the United States and Canada sold about 1300 million movie tickets. About $\frac{11}{100}$ of these tickets were purchased by frequent moviegoers, who go to the cinema once or more per month. Find the number of tickets purchased by frequent moviegoers in 2013. (*Source:* Motion Picture Association of America)

69. The Oregon National Historic Trail is 2170 miles long. It begins in Independence, Missouri, and ends in Oregon City, Oregon. Manfred Coulon has hiked $\frac{2}{5}$ of the trail before. How many miles has he hiked? (*Source:* National Park Service)

70. Each turn of a screw sinks it $\frac{3}{16}$ of an inch deeper into a piece of wood. Find how deep the screw is after 8 turns.

71. The radius of a circle is one-half of its diameter, as shown. If the diameter of a circle is $\frac{3}{8}$ of an inch, what is its radius?

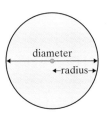

diameter
←radius→

72. The diameter of a circle is twice its radius, as shown in the Exercise 71 illustration. If the radius of a circle is $\frac{7}{20}$ of a foot, what is its diameter?

73. A special on a cruise to the Bahamas is advertised to be $\frac{2}{3}$ of the regular price. If the regular price is $2757, what is the sale price?

74. A family recently sold their house for $102,000, but $\frac{3}{50}$ of this amount goes to the real estate companies that helped them sell their house. How much money does the family pay to the real estate companies?

75. The state of Mississippi houses $\frac{1}{184}$ of the total U.S. libraries. If there are about 9200 libraries in the United States, how many libraries are in Mississippi?

76. There have been about 410 different contestants on the reality television show *Survivor* over 27 seasons. Some of these contestants appeared in two different series and/or seasons. If the number of repeat contestants was $\frac{6}{41}$ of the total number of participants in the first 27 seasons, how many contestants participated more than once? (*Source:* Survivor.com)

Find the area of each rectangle. Recall that area = length · width.

77.

$\frac{1}{5}$ foot

$\frac{5}{14}$ foot

78.

$\frac{1}{2}$ mile

$\frac{3}{8}$ mile

*Recall from Section 3.2 that the following graph is called a **circle graph** or **pie chart.** Each sector (shaped like a piece of pie) shows the fractional part of a car's total mileage that falls into a particular category. The whole circle represents a car's total mileage.*

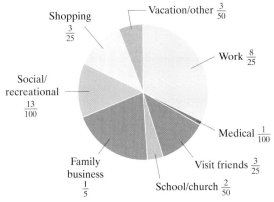

Shopping $\frac{3}{25}$

Vacation/other $\frac{3}{50}$

Social/recreational $\frac{13}{100}$

Work $\frac{8}{25}$

Family business $\frac{1}{5}$

School/church $\frac{2}{50}$

Visit friends $\frac{3}{25}$

Medical $\frac{1}{100}$

Source: The American Automobile Manufacturers Association and The National Automobile Dealers Association

In one year, a family drove 12,000 miles in the family car. Use the circle graph to determine how many of these miles might be expected to have fallen in the categories shown in Exercises 79 through 82.

79. Work

80. Shopping

81. Family business

82. Medical

Review

Perform each indicated operation. See Sections 1.3 and 1.4.

83.
$$\begin{array}{r} 27 \\ 76 \\ + \ 98 \\ \hline \end{array}$$

84.
$$\begin{array}{r} 811 \\ 42 \\ + \ 69 \\ \hline \end{array}$$

85.
$$\begin{array}{r} 968 \\ - \ 772 \\ \hline \end{array}$$

86.
$$\begin{array}{r} 882 \\ - \ 773 \\ \hline \end{array}$$

Concept Extensions

87. In your own words, describe how to divide fractions.

88. In your own words, explain how to multiply fractions.

Simplify.

89. $\left(\dfrac{1}{2} \cdot \dfrac{2}{3}\right) \div \dfrac{5}{6}$

90. $\left(\dfrac{3}{4} \cdot \dfrac{8}{9}\right) \div \dfrac{2}{5}$

91. $\dfrac{42}{25} \cdot \dfrac{125}{36} \div \dfrac{7}{6}$

92. $\left(\dfrac{8}{13} \cdot \dfrac{39}{16} \cdot \dfrac{8}{9}\right)^2 \div \dfrac{1}{2}$

Solve.

93. Approximately $\dfrac{1}{8}$ of the U.S. population lives in the state of California. If the U.S. population is approximately 313,914,000, find the approximate population of California. (*Source:* U.S. Census Bureau)

94. The estimated population of New Zealand was 4,433,000 in 2012. About $\dfrac{3}{20}$ of New Zealand's population is of Māori descent. How many Māori lived in New Zealand in 2012? (*Source:* Statistics New Zealand)

95. In 2013, a survey found that about $\dfrac{11}{20}$ of all adults in the United States owned a smartphone. There were roughly 240,000,000 U.S. adults at that time. How many U.S. adults owned a smartphone in 2013? (*Sources:* Pew Research Center, U.S. Census Bureau)

96. If $\dfrac{3}{4}$ of 36 students on a first bus are girls and $\dfrac{2}{3}$ of the 30 students on a second bus are *boys*, how many students on the two buses are girls?

Objectives

A Add or Subtract Like Fractions.

B Add or Subtract Given Fractional Replacement Values.

C Solve Problems by Adding or Subtracting Like Fractions.

D Find the Least Common Denominator of a List of Fractions.

E Write Equivalent Fractions.

Fractions with the same denominator are called **like fractions.** Fractions that have different denominators are called **unlike fractions.**

Like Fractions	Unlike Fractions
$\frac{2}{5}$ and $\frac{3}{5}$ same denominator	$\frac{2}{5}$ and $\frac{3}{4}$ different denominators
$\frac{5}{21}, \frac{16}{21}$, and $\frac{7}{21}$ same denominator	$\frac{5}{7}$ and $\frac{5}{9}$ different denominators

Objective A Adding or Subtracting Like Fractions

To see how we add like fractions (fractions with the same denominator), study one or both illustrations below.

Add: $\dfrac{1}{5} + \dfrac{3}{5}$	
Figures	**Number Line**
$\frac{1}{5}$ + $\frac{3}{5}$ $\dfrac{1}{5} + \dfrac{3}{5} = \dfrac{4}{5}$	To add $\dfrac{1}{5} + \dfrac{3}{5}$, start at 0 and draw an arrow $\dfrac{1}{5}$ of a unit long pointing to the right. From the tip of this arrow, draw an arrow $\dfrac{3}{5}$ of a unit long, also pointing to the right. The tip of the second arrow ends at their sum, $\dfrac{4}{5}$. Start End 0 $\frac{4}{5}$ 1 $\dfrac{1}{5} + \dfrac{3}{5} = \dfrac{4}{5}$
Thus, $\dfrac{1}{5} + \dfrac{3}{5} = \dfrac{4}{5}$.	

Notice that the numerator of the sum is the sum of the numerators. Also, the denominator of the sum is the **common denominator.** This is how we add fractions. Similar illustrations can be shown for subtracting fractions.

Adding or Subtracting Like Fractions (Fractions with the Same Denominator)

If a, b, and c are numbers and b is not 0, then

$$\frac{a}{b} + \frac{c}{b} = \frac{a + c}{b} \qquad \text{and also} \qquad \frac{a}{b} - \frac{c}{b} = \frac{a - c}{b}$$

In other words, to add or subtract fractions with the same denominator, add or subtract their numerators and write the sum or difference over the **common** denominator.

For example,

$$\frac{1}{4} + \frac{2}{4} = \frac{1+2}{4} = \frac{3}{4} \qquad \text{Add the numerators.}$$
Keep the common denominator.

$$\frac{4}{5} - \frac{2}{5} = \frac{4-2}{5} = \frac{2}{5} \qquad \text{Subtract the numerators.}$$
Keep the common denominator.

Helpful Hint

As usual, don't forget to write all answers in simplest form.

Examples · Add and simplify.

1. $\dfrac{2}{7} + \dfrac{3}{7} = \dfrac{2+3}{7} = \dfrac{5}{7}$ ← Add the numerators.
                                ← Keep the common denominator.

2. $\dfrac{3}{16} + \dfrac{7}{16} = \dfrac{3+7}{16} = \dfrac{10}{16} = \dfrac{\cancel{2} \cdot 5}{\cancel{2} \cdot 8} = \dfrac{5}{8}$

3. $\dfrac{7}{8} + \dfrac{6}{8} + \dfrac{3}{8} = \dfrac{7+6+3}{8} = \dfrac{16}{8}$ or 2

■ Work Practice 1–3

✔ Concept Check · Find and correct the error in the following:

$$\frac{1}{5} + \frac{1}{5} = \frac{2}{10}$$

Examples · Subtract and simplify.

4. $\dfrac{8}{9} - \dfrac{1}{9} = \dfrac{8-1}{9} = \dfrac{7}{9}$ ← Subtract the numerators.
                                ← Keep the common denominator.

5. $\dfrac{7}{8} - \dfrac{5}{8} = \dfrac{7-5}{8} = \dfrac{2}{8} = \dfrac{\cancel{2}}{\cancel{2} \cdot 4} = \dfrac{1}{4}$

■ Work Practice 4–5

From our earlier work, we know that

$$\frac{-12}{6} = \frac{12}{-6} = -\frac{12}{6} \quad \text{since these all simplify to } -2.$$

In general, the following is true:

$$\frac{-a}{b} = \frac{a}{-b} = -\frac{a}{b} \quad \text{as long as } b \text{ is not 0.}$$

Example 6 · Add: $-\dfrac{11}{8} + \dfrac{6}{8}$

Solution: $-\dfrac{11}{8} + \dfrac{6}{8} = \dfrac{-11+6}{8}$

$$= \frac{-5}{8} \quad \text{or} \quad -\frac{5}{8}$$

■ Work Practice 6

Practice 1–3

Add and simplify.

1. $\dfrac{6}{13} + \dfrac{2}{13}$

2. $\dfrac{5}{8} + \dfrac{1}{8}$

3. $\dfrac{20}{11} + \dfrac{6}{11} + \dfrac{7}{11}$

Practice 4–5

Subtract and simplify.

4. $\dfrac{11}{12} - \dfrac{6}{12}$

5. $\dfrac{7}{15} - \dfrac{2}{15}$

Practice 6

Add: $-\dfrac{8}{17} + \dfrac{4}{17}$

Answers

1. $\dfrac{8}{13}$ **2.** $\dfrac{3}{4}$ **3.** 3   **4.** $\dfrac{5}{12}$

5. $\dfrac{1}{3}$ **6.** $-\dfrac{4}{17}$

✔ Concept Check Answer

We don't add denominators together; correction: $\dfrac{1}{5} + \dfrac{1}{5} = \dfrac{2}{5}$.

Practice 7

Subtract: $\dfrac{2}{5} - \dfrac{7}{5}$

Example 7 Subtract: $\dfrac{3}{4} - \dfrac{7}{4}$

Solution: $\dfrac{3}{4} - \dfrac{7}{4} = \dfrac{3 - 7}{4} = \dfrac{3 + (-7)}{4} = \dfrac{-4}{4} = -1$

■ Work Practice 7

Practice 8

Subtract: $\dfrac{4}{11} - \dfrac{6}{11} - \dfrac{3}{11}$

Example 8 Subtract: $\dfrac{3}{7} - \dfrac{6}{7} - \dfrac{3}{7}$

Solution: $\dfrac{3}{7} - \dfrac{6}{7} - \dfrac{3}{7} = \dfrac{3 - 6 - 3}{7} = \dfrac{-6}{7}$ or $-\dfrac{6}{7}$

■ Work Practice 8

Helpful Hint

Recall that $\dfrac{-6}{7} = -\dfrac{6}{7}$ $\left(\text{Also, } \dfrac{6}{-7} = -\dfrac{6}{7}, \text{if needed.}\right)$

Objective B Adding or Subtracting Given Fractional Replacement Values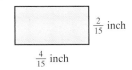

Practice 9

Evaluate $x + y$ if $x = -\dfrac{10}{12}$ and $y = \dfrac{5}{12}$.

Example 9 Evaluate $y - x$ if $x = -\dfrac{3}{10}$ and $y = -\dfrac{8}{10}$.

Solution: Be very careful when replacing x and y with replacement values.

$$y - x = -\dfrac{8}{10} - \left(-\dfrac{3}{10}\right) \quad \text{Replace } x \text{ with } -\dfrac{3}{10} \text{ and } y \text{ with } -\dfrac{8}{10}.$$
$$= \dfrac{-8 - (-3)}{10}$$
$$= \dfrac{-5}{10} = \dfrac{-1 \cdot 5}{2 \cdot 5} = \dfrac{-1}{2} \text{ or } -\dfrac{1}{2}$$

■ Work Practice 9

Objective C Solving Problems by Adding or Subtracting Like Fractions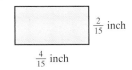

Practice 10

Find the perimeter of the square.

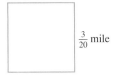

$\frac{3}{20}$ mile

Many real-life problems involve finding the perimeters of square- or rectangular-shaped figures such as pastures, swimming pools, and so on. We can use our knowledge of adding fractions to find perimeters.

△ **Example 10** Find the perimeter of the rectangle.

$\frac{2}{15}$ inch

$\frac{4}{15}$ inch

Answers

7. -1　8. $-\dfrac{5}{11}$　9. $-\dfrac{5}{12}$　10. $\dfrac{3}{5}$ mi

Solution: Recall that perimeter means distance around and that opposite sides of a rectangle are the same length.

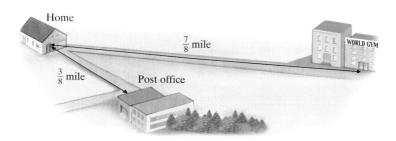

$$\text{Perimeter} = \frac{2}{15} + \frac{4}{15} + \frac{2}{15} + \frac{4}{15} = \frac{2 + 4 + 2 + 4}{15}$$

$$= \frac{12}{15} = \frac{3 \cdot 4}{3 \cdot 5} = \frac{4}{5}$$

The perimeter of the rectangle is $\frac{4}{5}$ inch.

■ Work Practice 10

We can combine our skills in adding and subtracting fractions with our four problem-solving steps from Chapter 1 to solve many kinds of real-life problems.

Example 11 Calculating Distance

The distance from home to the World Gym is $\frac{7}{8}$ of a mile and from home to the post office is $\frac{3}{8}$ of a mile. How much farther is it from home to the World Gym than from home to the post office?

Home

$\frac{7}{8}$ mile

WORLD GYM

$\frac{3}{8}$ mile   Post office

Practice 11

A jogger ran $\frac{13}{4}$ miles on Monday and $\frac{11}{4}$ miles on Wednesday. How much farther did he run on Monday than on Wednesday?

Solution:

1. UNDERSTAND. Read and reread the problem. The phrase "How much farther" tells us to subtract distances.

2. TRANSLATE.

In words:	distance farther	is	home to World Gym distance	minus	home to post office distance
	↓	↓	↓	↓	↓
Translate:	distance farther	=	$\frac{7}{8}$	−	$\frac{3}{8}$

3. SOLVE: $\dfrac{7}{8} - \dfrac{3}{8} = \dfrac{7 - 3}{8} = \dfrac{4}{8} = \dfrac{4}{2 \cdot 4} = \dfrac{1}{2}$

4. INTERPRET. *Check* your work. *State* your conclusion: The distance from home to the World Gym is $\frac{1}{2}$ mile farther than from home to the post office.

■ Work Practice 11

Answer

11. $\frac{1}{2}$ mi

Objective D Finding the Least Common Denominator ▶

In the next section, we will add and subtract fractions that have different, or unlike, denominators. To do so, we first write them as equivalent fractions with a common denominator.

Although any common denominator can be used to add or subtract unlike fractions, we will use the **least common denominator (LCD).** The LCD of a list of fractions is the same as the **least common multiple (LCM)** of the denominators. Why do we use this number as the common denominator? Since the LCD is the *smallest* of all common denominators, operations are usually less tedious with this number.

> The **least common denominator (LCD)** of a list of fractions is the smallest positive number divisible by all the denominators in the list. (The least common denominator is also the **least common multiple (LCM) of the denominators**).

For example, the LCD of $\frac{1}{4}$ and $\frac{3}{10}$ is 20 because 20 is the smallest positive number divisible by both 4 and 10.

Finding the LCD: Method 1

One way to find the LCD is to see whether the larger denominator is divisible by the smaller denominator. If so, the larger number is the LCD. If not, then check consecutive multiples of the larger denominator until the LCD is found.

> **Method 1: Finding the LCD of a List of Fractions Using Multiples of the Largest Number**
>
> **Step 1:** Write the multiples of the largest denominator (starting with the number itself) until a multiple common to all denominators in the list is found.
>
> **Step 2:** The multiple found in Step 1 is the LCD.

Practice 12

Find the LCD of $\frac{7}{8}$ and $\frac{11}{16}$.

Example 12 Find the LCD of $\frac{3}{7}$ and $\frac{5}{14}$.

Solution: The denominators are 7 and 14. We write the multiples of 14 until we find one that is also a multiple of 7.

$14 \cdot 1 = 14$ A multiple of 7

The LCD is 14.

■ Work Practice 12

Practice 13

Find the LCD of $\frac{23}{25}$ and $\frac{1}{30}$.

Example 13 Find the LCD of $\frac{11}{12}$ and $\frac{7}{20}$.

Solution: We write the multiples of the larger denominator, 20, until we find one that is also a multiple of 12.

$20 \cdot 1 = 20$ Not a multiple of 12
$20 \cdot 2 = 40$ Not a multiple of 12
$20 \cdot 3 = 60$ A multiple of 12

The LCD is 60.

■ Work Practice 13

Answers

12. 16 **13.** 150

Method 1 for finding multiples works fine for smaller numbers, but may get tedious for larger numbers. For this reason, let's study a second method, which uses prime factorization.

Finding the LCD: Method 2

For example, to find the LCD of $\frac{11}{12}$ and $\frac{7}{20}$, such as in Example 13, let's look at the prime factorization of each denominator.

$$12 = 2 \cdot 2 \cdot 3$$
$$20 = 2 \cdot 2 \cdot 5$$

Recall that the LCD must be a multiple of both 12 and 20. Thus, to build the LCD, we will circle the greatest number of factors for each different prime number. The LCD is the product of the circled factors.

Prime Number Factors

$12 = \boxed{2 \cdot 2} \cdot \boxed{3}$
$20 = 2 \cdot 2 \cdot \boxed{5}$ *Circle either pair of 2s, but not both.*
$\text{LCD} = 2 \cdot 2 \cdot 3 \cdot 5 = 60$

The number 60 is the smallest number that both 12 and 20 divide into evenly. This method is summarized below:

Method 2: Finding the LCD of a List of Denominators Using Prime Factorization

Step 1: Write the prime factorization of each denominator.

Step 2: For each different prime factor in Step 1, circle the *greatest* number of times that factor occurs in any one factorization.

Step 3: The LCD is the product of the circled factors.

Example 14   Find the LCD of $-\frac{23}{72}$ and $\frac{17}{60}$.

Solution: First we write the prime factorization of each denominator.

$$72 = 2 \cdot 2 \cdot 2 \cdot 3 \cdot 3$$
$$60 = 2 \cdot 2 \cdot 3 \cdot 5$$

For the prime factors shown, we circle the greatest number of factors found in either factorization.

$72 = \boxed{2 \cdot 2 \cdot 2} \cdot \boxed{3 \cdot 3}$
$60 = 2 \cdot 2 \cdot 3 \cdot \boxed{5}$

The LCD is the product of the circled factors.

$$\text{LCD} = 2 \cdot 2 \cdot 2 \cdot 3 \cdot 3 \cdot 5 = 360$$

The LCD is 360.

Work Practice 14

Practice 14

Find the LCD of $-\frac{3}{40}$ and $\frac{11}{108}$.

Helpful Hint   If you prefer working with exponents, circle the factor with the greatest exponent.

Example 14:

$72 = \boxed{2^3} \cdot \boxed{3^2}$
$60 = 2^2 \cdot 3 \cdot \boxed{5}$
$\text{LCD} = 2^3 \cdot 3^2 \cdot 5 = 360$

Answer

14. 1080

Copyright 2016 Pearson Education, Inc.

Helpful Hint

If the number of factors of a prime number is equal, circle either one, but not both. For example,

$$12 = \boxed{2 \cdot 2} \cdot \boxed{3}$$
$$15 = 3 \cdot \boxed{5} \qquad \text{— Circle either 3 but not both.}$$

The LCD is $2 \cdot 2 \cdot 3 \cdot 5 = 60$.

Practice 15

Find the LCD of $\dfrac{7}{20}, \dfrac{1}{24},$ and $\dfrac{13}{45}$.

Example 15   Find the LCD of $\dfrac{1}{15}, \dfrac{5}{18},$ and $\dfrac{53}{54}$.

Solution: $15 = 3 \cdot \boxed{5}$
$18 = \boxed{2} \cdot 3 \cdot 3$
$54 = 2 \cdot \boxed{3 \cdot 3 \cdot 3}$

The LCD is $2 \cdot 3 \cdot 3 \cdot 3 \cdot 5$ or 270.

■ Work Practice 15

✓**Concept Check**   True or false? The LCD of the fractions $\dfrac{1}{6}$ and $\dfrac{1}{8}$ is 48.

Objective E   Writing Equivalent Fractions

To add or subtract unlike fractions in the next section, we first write equivalent fractions with the LCD as the denominator.

To write $\dfrac{1}{3}$ as an equivalent fraction with a denominator of 6, we multiply by 1 in the form of $\dfrac{2}{2}$. Why? Because $3 \cdot 2 = 6$, so the new denominator will become 6, as shown below.

$$\frac{1}{3} = \frac{1}{3} \cdot 1 = \frac{1}{3} \cdot \frac{2}{2} = \frac{1 \cdot 2}{3 \cdot 2} = \frac{2}{6}$$
$$\frac{2}{2} = 1$$

So $\dfrac{1}{3} = \dfrac{2}{6}$.

To write an equivalent fraction,

$$\frac{a}{b} = \frac{a}{b} \cdot \frac{c}{c} = \frac{a \cdot c}{b \cdot c}$$

where $a, b,$ and c are nonzero numbers.

✓**Concept Check**   Which of the following is *not* equivalent to $\dfrac{3}{4}$?

a. $\dfrac{6}{8}$ **b.** $\dfrac{18}{24}$ **c.** $\dfrac{9}{14}$ **d.** $\dfrac{30}{40}$

Answer
15. 360

✓Concept Check Answers
false; it is 24
c

Example 16 Write $\frac{3}{4}$ as an equivalent fraction with a denominator of 20.

$$\frac{3}{4} = \frac{}{20}$$

Solution: In the denominators, since $4 \cdot 5 = 20$, we will multiply by 1 in the form of $\frac{5}{5}$.

$$\frac{3}{4} = \frac{3}{4} \cdot \frac{5}{5} = \frac{3 \cdot 5}{4 \cdot 5} = \frac{15}{20}$$

Thus, $\frac{3}{4} = \frac{15}{20}$.

■ Work Practice 16

Practice 16

Write $\frac{7}{8}$ as an equivalent fraction with a denominator of 56.

$$\frac{7}{8} = \frac{}{56}$$

Helpful Hint

To check Example 16, write $\frac{15}{20}$ in simplest form.

$$\frac{15}{20} = \frac{3 \cdot \cancel{5}}{4 \cdot \cancel{5}} = \frac{3}{4}, \text{ the original fraction.}$$

If the original fraction is in lowest terms, we can check our work by writing the new, equivalent fraction in simplest form. This form should be the original fraction.

✔**Concept Check** True or false? When the fraction $\frac{2}{9}$ is rewritten as an equivalent fraction with 27 as the denominator, the result is $\frac{2}{27}$.

Example 17 Write an equivalent fraction with the given denominator.

$$\frac{2}{5} = \frac{}{15}$$

Solution: Since $5 \cdot 3 = 15$, we multiply by 1 in the form of $\frac{3}{3}$.

$$\frac{2}{5} = \frac{2}{5} \cdot \frac{3}{3} = \frac{2 \cdot 3}{5 \cdot 3} = \frac{6}{15}$$

Then $\frac{2}{5}$ is equivalent to $\frac{6}{15}$. They both represent the same part of a whole.

■ Work Practice 17

Practice 17

Write an equivalent fraction with the given denominator.

$$\frac{1}{4} = \frac{}{44}$$

Example 18 Write an equivalent fraction with the given denominator.

$$3 = \frac{}{7}$$

Solution: Recall that $3 = \frac{3}{1}$. Since $1 \cdot 7 = 7$, multiply by 1 in the form $\frac{7}{7}$.

$$\frac{3}{1} = \frac{3}{1} \cdot \frac{7}{7} = \frac{3 \cdot 7}{1 \cdot 7} = \frac{21}{7}$$

■ Work Practice 18

Practice 18

Write an equivalent fraction with the given denominator.

$$4 = \frac{}{6}$$

Answers

16. $\frac{49}{56}$ **17.** $\frac{11}{44}$ **18.** $\frac{24}{6}$

✔**Concept Check Answers**

false; the correct result is $\frac{6}{27}$

answers may vary

✔**Concept Check** What is the first step in writing $\frac{3}{10}$ as an equivalent fraction whose denominator is 100?

Vocabulary, Readiness & Video Check

Use the choices below to fill in each blank. Not all choices will be used.

least common denominator (LCD)   like $-\dfrac{a}{b}$ $\dfrac{a - c}{b}$ $\dfrac{a + c}{b}$ $-\dfrac{a}{-b}$

perimeter   unlike   equivalent

1. The fractions $\dfrac{9}{11}$ and $\dfrac{13}{11}$ are called _____ fractions while $\dfrac{3}{4}$ and $\dfrac{1}{3}$ are called _____ fractions.

2. $\dfrac{a}{b} + \dfrac{c}{b} =$ _____ and $\dfrac{a}{b} - \dfrac{c}{b} =$ _____.

3. As long as b is not 0, $\dfrac{-a}{b} = \dfrac{a}{-b} =$ _____.

4. The distance around a figure is called its _____.

5. The smallest positive number divisible by all the denominators of a list of fractions is called the _____.

6. Fractions that represent the same portion of a whole are called _____ fractions.

State whether the fractions in each list are like or unlike fractions.

7. $\dfrac{7}{8}, \dfrac{7}{10}$

8. $\dfrac{2}{3}, \dfrac{4}{9}$

9. $\dfrac{9}{10}, \dfrac{1}{10}$

10. $\dfrac{8}{11}, \dfrac{2}{11}$

11. $\dfrac{2}{31}, \dfrac{30}{31}, \dfrac{19}{31}$

12. $\dfrac{3}{10}, \dfrac{3}{11}, \dfrac{3}{13}$

13. $\dfrac{5}{12}, \dfrac{7}{12}, \dfrac{12}{11}$

14. $\dfrac{1}{5}, \dfrac{2}{5}, \dfrac{4}{5}$

Martin-Gay Interactive Videos

See Video 3.4

Watch the section lecture video and answer the following questions.

Objective A 15. Complete this statement based on the lecture before ▥ Example 1: To add like fractions, we add the _____ and keep the same _____ ▶.

Objective B 16. In ▥ Example 8, why are we told to be careful when substituting the replacement value for y? ▶

Objective C 17. What is the perimeter equation used to solve ▥ Example 9? What is the final answer? ▶

Objective D 18. In ▥ Example 10, the LCD is found to be 45. What does this mean in terms of the specific fractions in the problem? ▶

Objective E 19. From ▥ Example 12, why can we multiply a fraction by a form of 1 to get an equivalent fraction? ▶

3.4 Exercise Set MyMathLab®

Objective A *Add and simplify. See Examples 1 through 3, and 6.*

1. $\dfrac{1}{7} + \dfrac{2}{7}$

2. $\dfrac{9}{17} + \dfrac{2}{17}$

3. $\dfrac{1}{10} + \dfrac{1}{10}$

4. $\dfrac{1}{4} + \dfrac{1}{4}$

5. $\dfrac{2}{9} + \dfrac{4}{9}$

6. $\dfrac{3}{10} + \dfrac{2}{10}$

7. $-\dfrac{6}{20} + \dfrac{1}{20}$

8. $-\dfrac{1}{8} + \dfrac{3}{8}$

9. $-\dfrac{3}{14} + \left(-\dfrac{4}{14}\right)$

10. $-\dfrac{5}{24} + \left(-\dfrac{7}{24}\right)$

11. $\dfrac{10}{11} + \dfrac{3}{11}$

12. $\dfrac{13}{17} + \dfrac{9}{17}$

13. $\dfrac{4}{13} + \dfrac{2}{13} + \dfrac{1}{13}$

14. $\dfrac{5}{11} + \dfrac{1}{11} + \dfrac{2}{11}$

15. $-\dfrac{7}{18} + \dfrac{3}{18} + \dfrac{2}{18}$

16. $-\dfrac{7}{15} + \dfrac{3}{15} + \dfrac{1}{15}$

Subtract and simplify. See Examples 4, 5, and 7.

17. $\dfrac{10}{11} - \dfrac{4}{11}$

18. $\dfrac{9}{13} - \dfrac{5}{13}$

19. $\dfrac{4}{5} - \dfrac{1}{5}$

20. $\dfrac{7}{8} - \dfrac{4}{8}$

21. $\dfrac{7}{4} - \dfrac{3}{4}$

22. $\dfrac{18}{5} - \dfrac{3}{5}$

23. $\dfrac{7}{8} - \dfrac{1}{8}$

24. $\dfrac{5}{6} - \dfrac{1}{6}$

25. $-\dfrac{25}{12} - \dfrac{15}{12}$

26. $-\dfrac{30}{20} - \dfrac{15}{20}$

27. $\dfrac{11}{10} - \dfrac{3}{10}$

28. $\dfrac{14}{15} - \dfrac{4}{15}$

29. $-\dfrac{27}{33} - \left(-\dfrac{8}{33}\right)$

30. $-\dfrac{37}{45} - \left(-\dfrac{18}{45}\right)$

Mixed Practice *Perform the indicated operation. See Examples 1 through 8.*

31. $\dfrac{8}{21} + \dfrac{5}{21}$

32. $\dfrac{7}{37} + \dfrac{9}{37}$

33. $\dfrac{99}{100} - \dfrac{9}{100}$

34. $\dfrac{85}{200} - \dfrac{15}{200}$

35. $-\dfrac{13}{28} - \dfrac{13}{28}$

36. $-\dfrac{15}{26} - \dfrac{15}{26}$

37. $-\dfrac{3}{16} + \left(-\dfrac{7}{16}\right) + \left(-\dfrac{2}{16}\right)$

38. $-\dfrac{5}{18} + \left(-\dfrac{1}{18}\right) + \left(-\dfrac{6}{18}\right)$

Objective B *Evaluate each expression for the given replacement values. See Example 9.*

39. $x + y; x = \dfrac{3}{4}, y = \dfrac{2}{4}$

40. $x - y; x = \dfrac{7}{8}, y = \dfrac{9}{8}$

41. $x - y; x = -\dfrac{1}{5}, y = -\dfrac{3}{5}$

42. $x + y; x = -\dfrac{1}{6}, y = \dfrac{5}{6}$

Objective C *Find the perimeter of each figure. (Hint: Recall that perimeter means distance around.) See Example 10.*

△ **43.**

$\frac{4}{20}$ inch $\frac{7}{20}$ inch

Triangle

$\frac{9}{20}$ inch

△ **44.**

$\frac{3}{13}$ foot

$\frac{2}{13}$ foot   Pentagon $\frac{6}{13}$ foot

$\frac{3}{13}$ foot

$\frac{4}{13}$ foot

45.

$\frac{5}{12}$ meter | Rectangle

$\frac{7}{12}$ meter

△ **46.**

Square | $\frac{1}{6}$ centimeter

Solve. For Exercises 47 and 48, the solutions have been started for you. Write each answer in simplest form. See Example 11.

47. A railroad inspector must inspect $\frac{19}{20}$ of a mile of railroad track. If she has already inspected $\frac{5}{20}$ of a mile, how much more does she need to inspect?

Start the solution:

1. UNDERSTAND the problem. Reread it as many times as needed.

2. TRANSLATE into an equation. (Fill in the blanks.)

distance left to inspect	is	distance needed to inspect	minus	distance already inspected
↓	↓	↓	↓	↓
distance left to inspect	=	_____	−	_____

Finish with:

3. SOLVE.
4. INTERPRET.

48. Scott Davis has run $\frac{11}{8}$ miles already and plans to complete $\frac{16}{8}$ miles. To do this, how much farther must he run?

Start the solution:

1. UNDERSTAND the problem. Reread it as many times as needed.

2. TRANSLATE into an equation. (Fill in the blanks.)

distance left to run	is	distance planned to run	minus	distance already run
↓	↓	↓	↓	↓
distance left to run	=	_____	−	_____

Finish with:

3. SOLVE.
4. INTERPRET.

49. As of 2013, the fraction of states in the United States with maximum interstate highway speed limits up to and including 70 mph was $\frac{33}{50}$. The fraction of states with 70 mph speed limits was $\frac{20}{50}$. What fraction of states had speed limits that were less than 70 mph? (*Source:* Insurance Institute for Highway Safety)

50. When people take aspirin, $\frac{31}{50}$ of the time it is used to treat some type of pain. Approximately $\frac{7}{50}$ of all aspirin use is for treating headaches. What fraction of aspirin use is for treating pain other than headaches? (*Source:* Bayer Market Research)

The map of the world below shows the fraction of the world's surface land area taken up by each continent. In other words, the continent of Africa, for example, makes up $\frac{20}{100}$ of the land in the world. Use this map to solve Exercises 51 through 54. Write answers in simplest form.

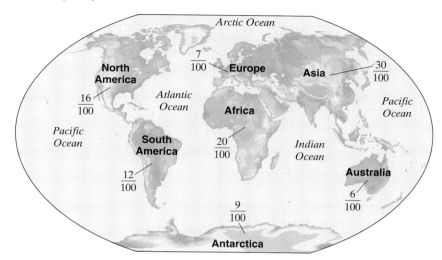

51. Find the fractional part of the world's land area within the continents of North America and South America.

52. Find the fractional part of the world's land area within the continents of Asia and Africa.

53. How much greater is the fractional part of the continent of Antarctica than the fractional part of the continent of Europe?

54. How much greater is the fractional part of the continent of Asia than the fractional part of the continent of Australia?

Objective D   *Find the LCD of each list of fractions. See Examples 12 through 15.*

55. $\frac{2}{9}, \frac{6}{15}$

56. $\frac{7}{12}, \frac{3}{20}$

57. $-\frac{1}{36}, \frac{1}{24}$

58. $-\frac{1}{15}, \frac{1}{90}$

59. $\frac{2}{25}, \frac{3}{15}, \frac{5}{6}$

60. $\frac{3}{4}, \frac{1}{14}, \frac{13}{20}$

61. $-\frac{7}{24}, -\frac{5}{7}$

62. $-\frac{11}{3}, -\frac{13}{64}$

63. $\frac{23}{18}, \frac{1}{21}$

64. $\frac{45}{24}, \frac{2}{45}$

65. $\frac{4}{3}, \frac{8}{21}, \frac{3}{56}$

66. $\frac{12}{11}, \frac{20}{33}, \frac{12}{121}$

Objective E   *Write each fraction as an equivalent fraction with the given denominator. See Examples 16 through 18.*

67. $\frac{2}{3} = \frac{}{21}$

68. $\frac{5}{6} = \frac{}{24}$

69. $\frac{4}{7} = \frac{}{35}$

70. $\frac{3}{5} = \frac{}{100}$

71. $\frac{1}{2} = \frac{}{50}$

72. $\frac{1}{5} = \frac{}{50}$

73. $\frac{14}{17} = \frac{}{68}$

74. $\frac{19}{21} = \frac{}{126}$

75. $\frac{2}{3} = \frac{}{12}$

76. $\frac{3}{2} = \frac{}{12}$

77. $\frac{5}{9} = \frac{}{36}$

78. $\frac{7}{6} = \frac{}{36}$

A nonstore retailer is a mail-order business that sells goods via catalogs, toll-free telephone numbers, or online media. The table shows the fraction of nonstore retailers' goods that were sold online in a recent year by types of goods. Use this table to answer Exercises 79 through 82. (Hint for Exercises 80, 81, and 82: To compare fractions with the same denominator, simply compare their numerators.)

Types of Goods Sold by Nonstore Retailers	Fraction of Goods That Were Sold Online	Equivalent Fraction with a Denominator of 100
Books and magazines	$\frac{43}{50}$	
Clothing and accessories	$\frac{4}{5}$	
Computer hardware	$\frac{29}{50}$	
Computer software	$\frac{17}{25}$	
Drugs, health and beauty aids	$\frac{3}{25}$	
Electronics and appliances	$\frac{21}{25}$	
Food, beer, and wine	$\frac{17}{25}$	
Home furnishings	$\frac{81}{100}$	
Music and videos	$\frac{9}{10}$	
Office equipment and supplies	$\frac{79}{100}$	
Sporting goods	$\frac{37}{50}$	
Toys, hobbies, and games	$\frac{39}{50}$	

(*Source:* U.S. Census Bureau)

79. Complete the table by writing each fraction as an equivalent fraction with a denominator of 100.

80. Which of these types of goods had the largest fraction sold online?

81. Which of these types of goods had the smallest fraction sold online?

82. Which of the types of goods had **more than** $\frac{4}{5}$ of the goods sold online? (*Hint:* Write $\frac{4}{5}$ as an equivalent fraction with a denominator of 100.)

Review

Simplify. See Section 1.9.

83. 3^2

84. 4^3

85. 3^4

86. 7^2

87. $2^3 \cdot 3$

88. $4^2 \cdot 5$

Concept Extensions

Find and correct the error. See the first Concept Check in this section.

89. $\dfrac{2}{7} + \dfrac{9}{7} = \dfrac{11}{14}$

90. $\dfrac{3}{4} - \dfrac{1}{4} = \dfrac{2}{8} = \dfrac{1}{4}$

Solve.

91. In your own words, explain how to add like fractions.

92. In your own words, explain how to subtract like fractions.

93. Use the map of the world for Exercises 51 through 54 and find the sum of all the continents' fractions. Explain your answer.

94. Mike Cannon jogged $\dfrac{3}{8}$ of a mile from home and then rested. Then he continued jogging farther from home for another $\dfrac{3}{8}$ of a mile until he discovered his watch had fallen off. He walked back along the same path for $\dfrac{4}{8}$ of a mile until he found his watch. Find how far he was from his home.

Write each fraction as an equivalent fraction with the indicated denominator.

95. $\dfrac{37}{165} = \dfrac{}{3630}$

96. $\dfrac{108}{215} = \dfrac{}{4085}$

97. In your own words, explain how to find the LCD of two fractions.

98. In your own words, explain how to write a fraction as an equivalent fraction with a given denominator.

Solve. See the second through fourth Concept Checks in this section.

99. Which of the following are equivalent to $\dfrac{2}{3}$?

    a. $\dfrac{10}{15}$ **b.** $\dfrac{40}{60}$

    c. $\dfrac{16}{20}$ **d.** $\dfrac{200}{300}$

100. True or false? When the fraction $\dfrac{7}{12}$ is rewritten with a denominator of 48, the result is $\dfrac{11}{48}$. If false, give the correct fraction.

Answers

1. _____

2. _____

3. _____

4. _____

5. _____

6. _____

7. _____

8. _____

9. _____

10. _____

11. _____

12. _____

13. _____

14. _____

15. _____

16. _____

17. _____

18. _____

19. _____

20. _____

21. _____

22. _____

23. a.   b.   c. _____

24. a.   b.   c. _____

226

Summary on Fractions and Operations on Fractions

Use a fraction to represent the shaded area of each figure. If the fraction is improper, also write the fraction as a mixed number.

1.

2.

Solve.

3. In a survey, 73 people out of 85 get less than 8 hours of sleep each night. What fraction of people in the survey get less than 8 hours of sleep?

4. Sketch a diagram to represent $\dfrac{9}{13}$.

Simplify.

5. $\dfrac{11}{-11}$

6. $\dfrac{17}{1}$

7. $\dfrac{0}{-3}$

8. $\dfrac{7}{0}$

Write the prime factorization of each composite number. Write any repeated factors using exponents.

9. 65

10. 70

11. 315

12. 441

Write each fraction in simplest form.

13. $\dfrac{2}{14}$

14. $\dfrac{24}{20}$

15. $-\dfrac{56}{60}$

16. $-\dfrac{72}{80}$

17. $\dfrac{54}{135}$

18. $\dfrac{90}{240}$

19. $\dfrac{165}{210}$

20. $\dfrac{245}{385}$

Determine whether each pair of fractions is equivalent.

21. $\dfrac{7}{8}$ and $\dfrac{9}{10}$

22. $\dfrac{10}{12}$ and $\dfrac{15}{18}$

23. Of the 50 states, 2 states are not adjacent to any other states.

    a. What fraction of the states are not adjacent to other states?

    b. How many states are adjacent to other states?

    c. What fraction of the states are adjacent to other states?

24. In a recent year, 540 new films were released and rated. Of these, 145 were rated PG-13. (*Source:* Nash Information, LLC)

    a. What fraction were rated PG-13?

    b. How many films were rated other than PG-13?

    c. What fraction of films were rated other than PG-13?

Find the LCD of each list of fractions.

25. $\dfrac{3}{5}, \dfrac{1}{6}$

26. $\dfrac{1}{2}, \dfrac{11}{14}$

27. $\dfrac{5}{6}, \dfrac{13}{18}, \dfrac{17}{30}$

Write each fraction as an equivalent fraction with the indicated denominator.

28. $\dfrac{7}{9} = \dfrac{}{36}$

29. $\dfrac{11}{15} = \dfrac{}{75}$

30. $\dfrac{5}{6} = \dfrac{}{48}$

The following summary will help you with the following review of operations on fractions.

Operations on Fractions

Let a, b, c, and d be integers.

Addition: $\dfrac{a}{b} + \dfrac{c}{b} = \dfrac{a+c}{b}$

$(b \neq 0)$ ↑   ↑
common denominator

Subtraction: $\dfrac{a}{b} - \dfrac{c}{b} = \dfrac{a-c}{b}$

$(b \neq 0)$ ↑   ↑
common denominator

Multiplication: $\dfrac{a}{b} \cdot \dfrac{c}{d} = \dfrac{a \cdot c}{b \cdot d}$

$(b \neq 0, d \neq 0)$

Division: $\dfrac{a}{b} \div \dfrac{c}{d} = \dfrac{a}{b} \cdot \dfrac{d}{c} = \dfrac{a \cdot d}{b \cdot c}$

$(b \neq 0, d \neq 0, c \neq 0)$

Perform each indicated operation.

31. $\dfrac{9}{10} + \dfrac{3}{10}$

32. $\dfrac{9}{10} - \dfrac{3}{10}$

33. $\dfrac{9}{10} \cdot \dfrac{2}{3}$

34. $\dfrac{9}{10} \div \dfrac{2}{3}$

35. $\dfrac{21}{70} - \dfrac{3}{70}$

36. $\dfrac{21}{70} + \dfrac{3}{70}$

37. $\dfrac{21}{25} \div \dfrac{3}{70}$

38. $\dfrac{21}{25} \cdot \dfrac{3}{70}$

39. $-\dfrac{7}{9} \cdot \dfrac{4}{5}$

40. $-\dfrac{3}{11} \div \left(-\dfrac{3}{10}\right)$

41. $-\dfrac{14}{27} - \dfrac{4}{27}$

42. $-\dfrac{8}{45} + \dfrac{6}{45}$

Solve.

43. A contractor is using 18 acres of his land to sell $\dfrac{3}{4}$-acre lots. How many lots can he sell?

44. Suppose that the cross-section of a piece of pipe looks like the diagram shown. What is the inner diameter?

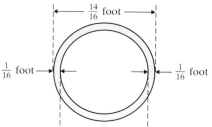

$\frac{14}{16}$ foot

$\frac{1}{16}$ foot ← → $\frac{1}{16}$ foot

inner diameter

25. _____

26. _____

27. _____

28. _____

29. _____

30. _____

31. _____

32. _____

33. _____

34. _____

35. _____

36. _____

37. _____

38. _____

39. _____

40. _____

41. _____

42. _____

43. _____

44. _____

Objectives

A Add or Subtract Unlike Fractions.

B Write Fractions in Order.

C Evaluate Expressions Given Fractional Replacement Values.

D Solve Problems by Adding or Subtracting Unlike Fractions.

Objective **A** Adding or Subtracting Unlike Fractions

In this section we add and subtract fractions with unlike denominators. To add or subtract these unlike fractions, we first write the fractions as equivalent fractions with a common denominator and then add or subtract the like fractions. Recall from the previous section that the common denominator we use is called the **least common denominator (LCD)**.

To begin, let's add the unlike fractions $\frac{3}{4} + \frac{1}{6}$.

The LCD of these fractions is 12. So we write each fraction as an equivalent fraction with a denominator of 12, and then add as usual. This addition process is shown next and is also illustrated by figures.

Add: $\frac{3}{4} + \frac{1}{6}$	The LCD is 12.
Figures	**Algebra**
$\frac{3}{4}$ + $\frac{1}{6}$	$\frac{3}{4} = \frac{3}{4} \cdot \frac{3}{3} = \frac{9}{12}$ and $\frac{1}{6} = \frac{1}{6} \cdot \frac{2}{2} = \frac{2}{12}$
	Remember, $\frac{3}{3} = 1$ and $\frac{2}{2} = 1$.
$\frac{9}{12}$ + $\frac{2}{12}$	Now we can add just as we did in Section 3.4.
$\frac{9}{12} + \frac{2}{12} = \frac{11}{12}$	$\frac{3}{4} + \frac{1}{6} = \frac{9}{12} + \frac{2}{12} = \frac{11}{12}$
Thus, the sum is $\frac{11}{12}$.	

Adding or Subtracting Unlike Fractions

Step 1: Find the least common denominator (LCD) of the fractions.

Step 2: Write each fraction as an equivalent fraction whose denominator is the LCD.

Step 3: Add or subtract the like fractions.

Step 4: Write the sum or difference in simplest form.

Example 1 Add: $\dfrac{2}{5} + \dfrac{4}{15}$

Practice 1
Add: $\dfrac{2}{7} + \dfrac{8}{21}$

Solution:

Step 1: The LCD of the fractions is 15. In later examples, we will simply say, for example, that the LCD of the denominators 5 and 15 is 15.

Step 2: $\dfrac{2}{5} = \dfrac{2}{5} \cdot \dfrac{3}{3} = \dfrac{6}{15}$ $\dfrac{4}{15} = \dfrac{4}{15}$ ← This fraction already has a denominator of 15.

⌞— Multiply by 1 in the form $\dfrac{3}{3}$.

Step 3: $\dfrac{2}{5} + \dfrac{4}{15} = \dfrac{6}{15} + \dfrac{4}{15} = \dfrac{10}{15}$

Step 4: Write in simplest form.

$$\dfrac{10}{15} = \dfrac{2 \cdot 5}{3 \cdot 5} = \dfrac{2}{3}$$

■ Work Practice 1

Example 2 Add: $\dfrac{2}{15} + \dfrac{3}{10}$

Practice 2
Add: $\dfrac{5}{6} + \dfrac{2}{9}$

Solution:

Step 1: The LCD of the denominators 15 and 10 is 30.

Step 2: $\dfrac{2}{15} = \dfrac{2}{15} \cdot \dfrac{2}{2} = \dfrac{4}{30}$ $\dfrac{3}{10} = \dfrac{3}{10} \cdot \dfrac{3}{3} = \dfrac{9}{30}$

Step 3: $\dfrac{2}{15} + \dfrac{3}{10} = \dfrac{4}{30} + \dfrac{9}{30} = \dfrac{13}{30}$

Step 4: $\dfrac{13}{30}$ is in simplest form.

■ Work Practice 2

Example 3 Add: $-\dfrac{1}{6} + \dfrac{1}{2}$

Practice 3
Add: $-\dfrac{1}{5} + \dfrac{9}{20}$

Solution: The LCD of the denominators 6 and 2 is 6.

$$-\dfrac{1}{6} + \dfrac{1}{2} = \dfrac{-1}{6} + \dfrac{1 \cdot 3}{2 \cdot 3}$$
$$= \dfrac{-1}{6} + \dfrac{3}{6}$$
$$= \dfrac{2}{6}$$

Helpful Hint Recall that
$$-\dfrac{1}{6} = \dfrac{-1}{6} = \dfrac{1}{-6}$$

Next, simplify $\dfrac{2}{6}$.

$$\dfrac{2}{6} = \dfrac{2}{2 \cdot 3} = \dfrac{1}{3}$$

■ Work Practice 3

Answers
1. $\dfrac{2}{3}$ **2.** $\dfrac{19}{18}$ **3.** $\dfrac{1}{4}$

✓**Concept Check** Find and correct the error in the following:

$$\frac{2}{9} + \frac{4}{11} \cancel{=} \frac{6}{20} = \frac{3}{10}$$

Practice 4

Subtract: $\dfrac{7}{12} - \dfrac{5}{24}$

Example 4 Subtract: $\dfrac{2}{5} - \dfrac{3}{20}$

Solution:

Step 1: The LCD of the denominators 5 and 20 is 20.

Step 2: $\dfrac{2}{5} = \dfrac{2}{5} \cdot \dfrac{4}{4} = \dfrac{8}{20}$ $\dfrac{3}{20} = \dfrac{3}{20}$ ← This fraction already has a denominator of 20.

Step 3: $\dfrac{2}{5} - \dfrac{3}{20} = \dfrac{8}{20} - \dfrac{3}{20} = \dfrac{5}{20}$

Step 4: Write in simplest form.

$$\frac{5}{20} = \frac{\cancel{5}}{\cancel{5} \cdot 4} = \frac{1}{4}$$

■ Work Practice 4

Practice 5

Subtract: $\dfrac{5}{7} - \dfrac{9}{10}$

Example 5 Subtract: $\dfrac{2}{3} - \dfrac{10}{11}$

Solution:

Step 1: The LCD of the denominators 3 and 11 is 33.

Step 2: $\dfrac{2}{3} = \dfrac{2}{3} \cdot \dfrac{11}{11} = \dfrac{22}{33}$ $\dfrac{10}{11} = \dfrac{10}{11} \cdot \dfrac{3}{3} = \dfrac{30}{33}$

Step 3: $\dfrac{2}{3} - \dfrac{10}{11} = \dfrac{22}{33} - \dfrac{30}{33} = \dfrac{-8}{33}$ or $-\dfrac{8}{33}$

Step 4: $-\dfrac{8}{33}$ is in simplest form.

■ Work Practice 5

Practice 6

Find: $\dfrac{5}{8} - \dfrac{1}{3} - \dfrac{1}{12}$

Example 6 Find: $-\dfrac{3}{4} - \dfrac{1}{14} + \dfrac{6}{7}$

Solution: The LCD of the denominators 4, 14, and 7 is 28.

$$-\frac{3}{4} - \frac{1}{14} + \frac{6}{7} = -\frac{3 \cdot 7}{4 \cdot 7} - \frac{1 \cdot 2}{14 \cdot 2} + \frac{6 \cdot 4}{7 \cdot 4}$$

$$= -\frac{21}{28} - \frac{2}{28} + \frac{24}{28}$$

$$= \frac{1}{28}$$

■ Work Practice 6

Answers

4. $\dfrac{3}{8}$ 5. $-\dfrac{13}{70}$ 6. $\dfrac{5}{24}$

✓**Concept Check Answers**

When adding or subtracting fractions, we don't add or subtract the denominators. Correct solutions:

$\dfrac{2}{9} + \dfrac{4}{11} = \dfrac{22}{99} + \dfrac{36}{99} = \dfrac{58}{99}$

$\dfrac{7}{12} - \dfrac{3}{4} = \dfrac{7}{12} - \dfrac{9}{12} = -\dfrac{2}{12} = -\dfrac{1}{6}$

✓**Concept Check** Find and correct the error in the following:

$$\frac{7}{12} - \frac{3}{4} \cancel{=} \frac{4}{8} = \frac{1}{2}$$

Objective B Writing Fractions in Order

One important application of the least common denominator is using the LCD to help order or compare fractions.

Example 7   Insert $<$ or $>$ to form a true sentence.

$$\frac{3}{4} \qquad \frac{9}{11}$$

Solution: The LCD for these fractions is 44. Let's write each fraction as an equivalent fraction with a denominator of 44.

$$\frac{3}{4} = \frac{3 \cdot 11}{4 \cdot 11} = \frac{33}{44} \qquad\qquad \frac{9}{11} = \frac{9 \cdot 4}{11 \cdot 4} = \frac{36}{44}$$

Since $33 < 36$, then

$$\frac{33}{44} < \frac{36}{44} \text{ or}$$

$$\frac{3}{4} < \frac{9}{11}$$

■ Work Practice 7

Practice 7

Insert $<$ or $>$ to form a true sentence.

$$\frac{5}{8} \qquad \frac{11}{20}$$

Example 8   Insert $<$ or $>$ to form a true sentence.

$$-\frac{2}{7} \qquad -\frac{1}{3}$$

Solution: The LCD of the denominators 7 and 3 is 21.

$$-\frac{2}{7} = -\frac{2 \cdot 3}{7 \cdot 3} = -\frac{6}{21} \text{ or } \frac{-6}{21} \qquad -\frac{1}{3} = -\frac{1 \cdot 7}{3 \cdot 7} = -\frac{7}{21} \text{ or } \frac{-7}{21}$$

Since $-6 > -7$, then

$$-\frac{6}{21} > -\frac{7}{21} \text{ or}$$

$$-\frac{2}{7} > -\frac{1}{3}$$

■ Work Practice 8

Practice 8

Insert $<$ or $>$ to form a true sentence.

$$-\frac{17}{20} \qquad -\frac{4}{5}$$

Objective C Evaluating Expressions Given Fractional Replacement Values

Example 9   Evaluate $x - y$ if $x = \frac{7}{18}$ and $y = \frac{2}{9}$.

Solution: Replace x with $\frac{7}{18}$ and y with $\frac{2}{9}$ in the expression $x - y$.

$$x - y = \frac{7}{18} - \frac{2}{9}$$

(Continued on next page)

Practice 9

Evaluate $x - y$ if $x = \frac{10}{11}$ and $y = \frac{9}{22}$.

Answers

7. $>$ **8.** $<$ **9.** $\frac{1}{2}$

The LCD of the denominators 18 and 9 is 18. Then

$$\frac{7}{18} - \frac{2}{9} = \frac{7}{18} - \frac{2 \cdot 2}{9 \cdot 2}$$

$$= \frac{7}{18} - \frac{4}{18}$$

$$= \frac{3}{18} = \frac{\cancel{3}}{\cancel{3} \cdot 6} = \frac{1}{6} \quad \text{Simplified}$$

■ Work Practice 9

Objective D Solving Problems by Adding or Subtracting Unlike Fractions ▶

Very often, real-world problems involve adding or subtracting unlike fractions.

Example 10 Finding Total Weight

A freight truck has $\frac{1}{4}$ ton of computers, $\frac{1}{3}$ ton of televisions, and $\frac{3}{8}$ ton of small appliances. Find the total weight of its load.

Practice 10

To repair her sidewalk, a homeowner must pour cement in three different locations. She needs $\frac{3}{5}$ of a cubic yard, $\frac{3}{10}$ of a cubic yard, and $\frac{1}{15}$ of a cubic yard for these locations. Find the total amount of cement the homeowner needs.

Solution:

1. **UNDERSTAND.** Read and reread the problem. The phrase "total weight" tells us to add.

2. **TRANSLATE.**

In words:	total weight	is	weight of computers	plus	weight of televisions	plus	weight of appliances
	↓	↓	↓	↓	↓	↓	↓
Translate:	total weight	=	$\frac{1}{4}$	+	$\frac{1}{3}$	+	$\frac{3}{8}$

3. **SOLVE:** The LCD is 24.

$$\frac{1}{4} + \frac{1}{3} + \frac{3}{8} = \frac{1}{4} \cdot \frac{6}{6} + \frac{1}{3} \cdot \frac{8}{8} + \frac{3}{8} \cdot \frac{3}{3}$$

$$= \frac{6}{24} + \frac{8}{24} + \frac{9}{24}$$

$$= \frac{23}{24}$$

4. **INTERPRET.** *Check* the solution. *State* your conclusion: The total weight of the truck's load is $\frac{23}{24}$ ton.

■ Work Practice 10

Answer

10. $\frac{29}{30}$ cu yd

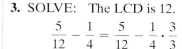 **Example 11** Calculating Flight Time

A flight from Tucson to Phoenix, Arizona, requires $\frac{5}{12}$ of an hour. If the plane has been flying $\frac{1}{4}$ of an hour, find how much time remains before landing.

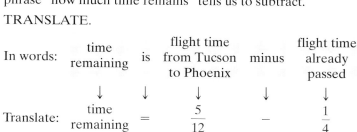

Solution:

1. **UNDERSTAND.** Read and reread the problem. The phrase "how much time remains" tells us to subtract.

2. **TRANSLATE.**

In words:	time remaining	is	flight time from Tucson to Phoenix	minus	flight time already passed
	↓	↓	↓	↓	↓
Translate:	time remaining	=	$\frac{5}{12}$	−	$\frac{1}{4}$

3. **SOLVE:** The LCD is 12.

$$\frac{5}{12} - \frac{1}{4} = \frac{5}{12} - \frac{1}{4} \cdot \frac{3}{3}$$

$$= \frac{5}{12} - \frac{3}{12}$$

$$= \frac{2}{12} = \frac{\cancel{2}}{\cancel{2} \cdot 6} = \frac{1}{6}$$

4. **INTERPRET.** *Check* the solution. *State* your conclusion: The remaining flight time is $\frac{1}{6}$ of an hour.

■ Work Practice 11

Practice 11

Find the difference in length of two boards if one board is $\frac{3}{4}$ of a foot long and the other is $\frac{2}{3}$ of a foot long.

Answer

11. $\frac{1}{12}$ ft

 Calculator Explorations **Performing Operations on Fractions**

Scientific Calculator

Many calculators have a fraction key, such as $\boxed{a^b/c}$, that allows you to enter fractions and perform operations on fractions and will give the result as a fraction. If your calculator has a fraction key, use it to calculate

$$\frac{3}{5} + \frac{4}{7}$$

Enter the keystrokes

$\boxed{3}\ \boxed{a^b/c}\ \boxed{5}\ \boxed{+}\ \boxed{4}\ \boxed{a^b/c}\ \boxed{7}\ \boxed{=}$

The display should read $\boxed{1\_6\ |\ 35}$

which represents the mixed number $1\frac{6}{35}$. Let's write the result as a fraction. To convert from mixed number notation to fractional notation, press

$\boxed{2^{nd}}\ \boxed{d/c}$

The display now reads $\boxed{41\ |\ 35}$

which represents $\frac{41}{35}$, the sum in fractional notation.

Graphing Calculator

Graphing calculators also allow you to perform operations on fractions and will give exact fractional results. The fraction option on a graphing calculator may be found under the $\boxed{\text{MATH}}$ menu. To perform the addition in the left column, try the keystrokes

$\boxed{3}\ \boxed{÷}\ \boxed{5}\ \boxed{+}\ \boxed{4}\ \boxed{÷}\ \boxed{7}\ \boxed{\text{MATH}}\ \boxed{\text{ENTER}}$
$\boxed{\text{ENTER}}$

The display should read

$\boxed{3/5 + 4/7 \blacktriangleright \text{Frac } 41/35}$

Use a calculator to add the following fractions. Give each sum as a fraction.

1. $\frac{1}{16} + \frac{2}{5}$ **2.** $\frac{3}{20} + \frac{2}{25}$ **3.** $\frac{4}{9} + \frac{7}{8}$

4. $\frac{9}{11} + \frac{5}{12}$ **5.** $\frac{10}{17} + \frac{12}{19}$ **6.** $\frac{14}{31} + \frac{15}{21}$

Vocabulary, Readiness & Video Check

Use the choices below to fill in each blank. Any numerical answers are not listed.

expression   least common denominator   >

equivalent   <

1. To add or subtract unlike fractions, we first write the fractions as _____ fractions with a common denominator. The common denominator we use is called the _____.

2. The LCD for $\dfrac{1}{6}$ and $\dfrac{5}{8}$ is _____.

3. $\dfrac{1}{6} + \dfrac{5}{8} = \dfrac{1}{6} \cdot \dfrac{4}{4} + \dfrac{5}{8} \cdot \dfrac{3}{3} = \dfrac{\quad}{\quad} + \dfrac{\quad}{\quad} = \dfrac{\quad}{\quad}$.

4. $\dfrac{1}{6} - \dfrac{5}{8} = \dfrac{1}{6} \cdot \dfrac{4}{4} - \dfrac{5}{8} \cdot \dfrac{3}{3} = \dfrac{\quad}{\quad} - \dfrac{\quad}{\quad} = \dfrac{\quad}{\quad}$.

5. $x - y$ is a(n) _____.

6. Since $-10 < -1$, we know that $-\dfrac{10}{13}$ _____ $-\dfrac{1}{13}$.

Write the LCD of each pair of fractions.

7. $\dfrac{1}{2}, \dfrac{2}{3}$

8. $\dfrac{1}{2}, \dfrac{3}{4}$

9. $\dfrac{1}{6}, \dfrac{5}{12}$

10. $\dfrac{2}{5}, \dfrac{7}{10}$

11. $\dfrac{4}{7}, \dfrac{1}{8}$

12. $\dfrac{23}{24}, \dfrac{1}{3}$

13. $\dfrac{11}{12}, \dfrac{3}{4}$

14. $\dfrac{2}{3}, \dfrac{3}{11}$

Martin-Gay Interactive Videos

See Video 3.5

Watch the section lecture video and answer the following questions.

Objective A **15.** In ▭ Example 4, why can't we add the two terms in the numerators? ▶

Objective B **16.** In ▭ Example 6, when comparing two fractions, how does writing each fraction with the same denominator help? ▶

Objective C **17.** In ▭ Example 7, if we had chosen to simplify the first fraction before adding, what would our addition problem have become and what would our LCD have been? ▶

Objective D **18.** Is the answer to ▭ Example 8 a proper or an improper fraction? ▶

3.5 **Exercise Set** MyMathLab®

Objective A *Add or subtract as indicated. See Examples 1 through 6.*

1. $\dfrac{2}{3} + \dfrac{1}{6}$

2. $\dfrac{5}{6} + \dfrac{1}{12}$

3. $\dfrac{1}{2} - \dfrac{1}{3}$

4. $\dfrac{2}{3} - \dfrac{1}{4}$

▶5. $-\dfrac{2}{11} + \dfrac{2}{33}$

6. $-\dfrac{5}{9} + \dfrac{1}{3}$

7. $\dfrac{3}{14} - \dfrac{3}{7}$

8. $\dfrac{2}{15} - \dfrac{2}{5}$

9. $\frac{11}{35} + \frac{2}{7}$

10. $\frac{2}{5} + \frac{3}{25}$

11. $2 - \frac{5}{12}$

12. $5 - \frac{3}{20}$

13. $\frac{5}{12} - \frac{1}{9}$

14. $\frac{7}{12} - \frac{5}{18}$

15. $\frac{5}{7} + 1$

16. $-10 + \frac{7}{10}$

17. $\frac{5}{11} + \frac{4}{9}$

18. $\frac{7}{18} + \frac{2}{9}$

19. $\frac{2}{3} - \frac{1}{6}$

20. $\frac{5}{6} - \frac{1}{12}$

21. $\frac{1}{3} + \frac{1}{9} + \frac{1}{27}$

22. $\frac{1}{4} + \frac{1}{16} + \frac{1}{64}$

23. $-\frac{2}{11} - \frac{2}{33}$

24. $-\frac{5}{9} - \frac{1}{3}$

25. $\frac{9}{14} - \frac{3}{7}$

26. $\frac{4}{5} - \frac{2}{15}$

27. $\frac{11}{35} - \frac{2}{7}$

28. $\frac{2}{5} - \frac{3}{25}$

29. $\frac{1}{9} - \frac{5}{12}$

30. $\frac{5}{18} - \frac{7}{12}$

31. $\frac{7}{15} - \frac{5}{12}$

32. $\frac{5}{8} - \frac{3}{20}$

33. $\frac{5}{7} - \frac{1}{8}$

34. $\frac{10}{13} - \frac{7}{10}$

35. $\frac{7}{8} + \frac{3}{16}$

36. $-\frac{7}{18} - \frac{2}{9}$

37. $\frac{5}{9} + \frac{3}{9}$

38. $\frac{4}{13} - \frac{1}{13}$

39. $-\frac{2}{5} + \frac{1}{3} - \frac{3}{10}$

40. $-\frac{1}{3} - \frac{1}{4} + \frac{2}{5}$

41. $-\frac{5}{6} - \frac{3}{7}$

42. $\frac{1}{2} - \frac{3}{29}$

43. $\frac{7}{9} - \frac{1}{6}$

44. $\frac{9}{16} - \frac{3}{8}$

45. $\frac{5}{11} + \frac{3}{13}$

46. $\frac{3}{7} + \frac{9}{17}$

47. $\frac{7}{30} - \frac{5}{12}$

48. $\frac{7}{30} - \frac{3}{20}$

49. $\frac{6}{5} - \frac{3}{4} + \frac{1}{2}$

50. $\frac{6}{5} + \frac{3}{4} - \frac{1}{2}$

51. $\frac{4}{5} + \frac{4}{9}$

52. $\frac{11}{12} - \frac{7}{24}$

53. $-\frac{9}{12} + \frac{17}{24} - \frac{1}{6}$

54. $-\frac{5}{14} + \frac{3}{7} - \frac{1}{2}$

55. $\frac{1}{1000} - \frac{1}{100}$

56. $\frac{1}{500} - \frac{1}{50}$

Objective B Insert < or > to form a true sentence. See Examples 7 and 8.

57. $\frac{2}{7} \quad \frac{3}{10}$

58. $\frac{5}{9} \quad \frac{6}{11}$

59. $\frac{5}{6} \quad -\frac{13}{15}$

60. $-\frac{7}{8} \quad \frac{5}{6}$

61. $-\frac{3}{4} \quad -\frac{11}{14}$

62. $-\frac{2}{9} \quad -\frac{3}{13}$

Objective C and Section 3.3 Mixed Practice Evaluate each expression if $x = \frac{1}{3}$ and $y = \frac{3}{4}$. See Example 9.

63. $x + y$

64. $x - y$

65. xy

66. $x \div y$

67. $2y + x$

68. $2x + y$

Objective D *Find the perimeter of each geometric figure. (Hint: Recall that perimeter means distance around.)*

69.
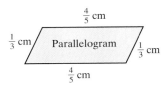

70.
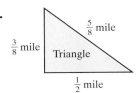

71.
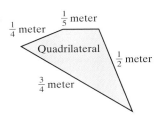

72.
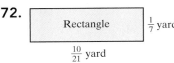

Translating *Translate each phrase to an algebraic expression. Use "x" to represent "a number." See Examples 10 and 11.*

73. The sum of a number and $\dfrac{1}{2}$

74. A number increased by $-\dfrac{2}{5}$

75. A number subtracted from $-\dfrac{3}{8}$

76. The difference of a number and $\dfrac{7}{20}$

Solve. For Exercises 77 and 78, the solutions have been started for you. See Examples 10 and 11.

77. The slowest mammal is the three-toed sloth from South America. The sloth has an average ground speed of $\dfrac{1}{10}$ mph. In the trees, it can accelerate to $\dfrac{17}{100}$ mph. How much faster can a sloth travel in the trees? (*Source: The Guinness Book of World Records*)

Start the solution:

1. UNDERSTAND the problem. Reread it as many times as needed.
2. TRANSLATE into an equation. (Fill in the blanks.)

how much faster sloth travels in trees	is	sloth speed in trees	minus	sloth speed on ground
↓	↓	↓	↓	↓
how much faster sloth travels in trees	=	_____	−	_____

Finish with:
3. SOLVE.
4. INTERPRET.

78. Killer bees have been known to chase people for up to $\dfrac{1}{4}$ of a mile, while domestic European honeybees will normally chase a person for no more than 100 feet, or $\dfrac{5}{264}$ of a mile. How much farther will a killer bee chase a person than a domestic honeybee? (*Source: Coachella Valley Mosquito & Vector Control District*)

Start the solution:

1. UNDERSTAND the problem. Reread it as many times as needed.
2. TRANSLATE into an equation. (Fill in the blanks.)

how much farther killer bee will chase than honeybee	is	distance killer bee chases	minus	distance honeybee chases
↓	↓	↓	↓	↓
how much farther killer bee will chase than honeybee	=	_____	−	_____

Finish with:
3. SOLVE.
4. INTERPRET.

79. Find the inner diameter of the washer.

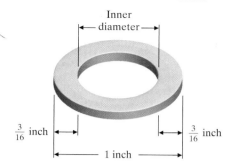

Inner
← diameter →

$\frac{3}{16}$ inch ←→   ←→ $\frac{3}{16}$ inch

←——— 1 inch ———→

80. Find the inner diameter of the tubing.

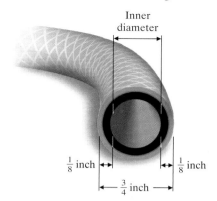

Inner
diameter

$\frac{1}{8}$ inch ←→   ←→ $\frac{1}{8}$ inch

←— $\frac{3}{4}$ inch —→

81. About $\frac{13}{20}$ of American students ages 10 to 17 name math, science, or art as their favorite subject in school. Art is the favorite subject for about $\frac{4}{25}$ of the American students ages 10 to 17. For what fraction of students this age is math or science their favorite subject? (*Source:* Peter D. Hart Research Associates for the National Science Foundation)

82. After 14 races in the 2014 FIA Formula One World Championship, the Mercedes AMG Petronas F1 team won $\frac{11}{14}$ of the races. If Mercedes driver Lewis Hamilton won $\frac{1}{2}$ of the races, what fraction did the other Mercedes driver, Nico Rosberg, win? (*Source:* Formula1.com)

83. Given the following diagram, find its total length.

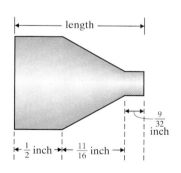

←——— length ———→

$\frac{9}{32}$ inch

←| $\frac{1}{2}$ inch |→←— $\frac{11}{16}$ inch —→

84. Given the following diagram, find its total width.

width

$\frac{11}{16}$ inch

$\frac{5}{8}$ inch

$\frac{11}{16}$ inch

The table below shows the fraction of the Earth's water area taken up by each ocean. Use this table for Exercises 85 and 86.

Fraction of Earth's Water Area per Ocean	
Ocean	**Fraction**
Arctic	$\frac{1}{25}$
Atlantic	$\frac{13}{50}$
Pacific	$\frac{1}{2}$
Indian	$\frac{1}{5}$

85. What fraction of the world's water surface area is accounted for by the Pacific and Atlantic Oceans?

86. What fraction of the world's water surface area is accounted for by the Arctic and Indian Oceans?

Use this circle graph to answer Exercises 87 through 90.

Areas Maintained by the National Park Service

Preserves/ $\frac{1}{20}$
Reserves

Other $\frac{3}{100}$

Lakeshores/ $\frac{2}{50}$
Seashores

Battlefields $\frac{3}{50}$

Rivers $\frac{2}{50}$

Memorials $\frac{7}{100}$

Parkways/ $\frac{1}{50}$
Scenic Trails

Monuments $\frac{19}{100}$

Parks $\frac{3}{20}$

Recreation Areas $\frac{1}{25}$

Historical Parks and Sites $\frac{31}{100}$

Source: National Park Service

87. What fraction of areas maintained by the National Park Service in 2013 were designated as either National Parks or National Monuments?

88. What fraction of areas maintained by the National Park Service in 2013 were designated as either National Memorials or National Battlefields?

89. What fraction of areas maintained by the National Park Service in 2013 were NOT National Monuments?

90. What fraction of areas maintained by the National Park Service in 2013 were NOT National Parkways or Scenic Trails?

Review

Use order of operations to simplify. See Section 1.9.

91. $50 \div 5 \cdot 2$

92. $8 - 6 \cdot 4 - 7$

93. $(8 - 6) \cdot (4 - 7)$

94. $50 \div (5 \cdot 2)$

Concept Extensions

For Exercises 95 and 96 below, do the following:

a. *Draw three rectangles of the same size and represent each fraction in the sum or difference, one fraction per rectangle, by shading.*

b. *Using these rectangles as estimates, determine whether there is an error in the sum or difference.*

c. *If there is an error, correctly calculate the sum or difference.*

See the Concept Checks in this section.

95. $\frac{3}{5} + \frac{4}{5} \stackrel{?}{=} \frac{7}{10}$

96. $\frac{5}{8} - \frac{3}{4} \stackrel{?}{=} \frac{2}{4}$

Subtract from left to right.

97. $\frac{2}{3} - \frac{1}{4} - \frac{2}{540}$

98. $\frac{9}{10} - \frac{7}{200} - \frac{1}{3}$

Perform each indicated operation.

99. $\frac{30}{55} + \frac{1000}{1760}$

100. $\frac{19}{26} - \frac{968}{1352}$

101. In your own words, describe how to add or subtract two fractions with different denominators.

102. Find the sum of the fractions in the circle graph above. Did the sum surprise you? Why or why not?

103. In 2012, about $\frac{69}{160}$ of the total number of pieces of mail delivered by the United States Postal Service was first-class mail. That same year, about $\frac{1}{2}$ of the total number of pieces of mail delivered by the United States Postal Service was standard mail. Which of these two categories accounts for a greater portion of the mail handled by volume? (*Source:* United States Postal Service)

Complex Fractions, Order of Operations, and Mixed Numbers

Objective A Simplifying Complex Fractions

Thus far, we have studied operations on fractions. We now practice simplifying fractions whose numerators or denominators themselves contain fractions. These fractions are called **complex fractions.**

Complex Fraction

A fraction whose numerator or denominator or both numerator and denominator contain fractions is called a **complex fraction.**

Examples of complex fractions are

$$\dfrac{\dfrac{1}{4}}{\dfrac{3}{2}} \qquad \dfrac{\dfrac{1}{2}+\dfrac{3}{8}}{\dfrac{3}{4}-\dfrac{1}{6}} \qquad \dfrac{\dfrac{4}{5}-2}{\dfrac{3}{10}}$$

Objectives

A Simplify Complex Fractions.

B Review the Order of Operations.

C Evaluate Expressions Given Replacement Values.

D Write Mixed Numbers as Improper Fractions.

E Write Improper Fractions as Mixed Numbers or Whole Numbers.

Method 1 for Simplifying Complex Fractions

Two methods are presented to simplify complex fractions. The first method makes use of the fact that a fraction bar means division.

Example 1 Simplify: $\dfrac{\dfrac{1}{4}}{\dfrac{3}{2}}$

Solution: Since a fraction bar means division, the complex fraction $\dfrac{\dfrac{1}{4}}{\dfrac{3}{2}}$ can be written as $\dfrac{1}{4} \div \dfrac{3}{2}$. Then divide as usual to simplify.

$$\dfrac{1}{4} \div \dfrac{3}{2} = \dfrac{1}{4} \cdot \dfrac{2}{3} \qquad \text{Multiply by the reciprocal.}$$

$$= \dfrac{1 \cdot \cancel{2}}{\cancel{2} \cdot 2 \cdot 3}$$

$$= \dfrac{1}{6}$$

Practice 1

Simplify: $\dfrac{\dfrac{7}{10}}{\dfrac{1}{5}}$

■ Work Practice 1

Example 2 Simplify: $\dfrac{\dfrac{1}{2}+\dfrac{3}{8}}{\dfrac{3}{4}-\dfrac{1}{6}}$

Solution: Recall the order of operations. Since the fraction bar is considered a grouping symbol, we simplify the numerator and the denominator of the complex fraction separately. Then we divide.

$$\dfrac{\dfrac{1}{2}+\dfrac{3}{8}}{\dfrac{3}{4}-\dfrac{1}{6}} = \dfrac{\dfrac{1\cdot 4}{2\cdot 4}+\dfrac{3}{8}}{\dfrac{3\cdot 3}{4\cdot 3}-\dfrac{1\cdot 2}{6\cdot 2}} = \dfrac{\dfrac{4}{8}+\dfrac{3}{8}}{\dfrac{9}{12}-\dfrac{2}{12}} = \dfrac{\dfrac{7}{8}}{\dfrac{7}{12}}$$

Practice 2

Simplify: $\dfrac{\dfrac{1}{2}+\dfrac{1}{6}}{\dfrac{3}{4}-\dfrac{2}{3}}$

Answers

1. $\dfrac{7}{2}$ **2.** $\dfrac{8}{1}$ or 8

(Continued on next page)

239

Thus,

$$\frac{\dfrac{1}{2} + \dfrac{3}{8}}{\dfrac{3}{4} - \dfrac{1}{6}} = \frac{\dfrac{7}{8}}{\dfrac{7}{12}}$$

$$= \frac{7}{8} \div \frac{7}{12} \qquad \text{Rewrite the quotient using the } \div \text{ sign.}$$

$$= \frac{7}{8} \cdot \frac{12}{7} \qquad \text{Multiply by the reciprocal.}$$

$$= \frac{\cancel{7} \cdot 3 \cdot \cancel{4}}{2 \cdot \cancel{4} \cdot \cancel{7}} \qquad \text{Multiply.}$$

$$= \frac{3}{2} \qquad \text{Simplify.}$$

■ Work Practice 2

Method 2 for Simplifying Complex Fractions

The second method for simplifying complex fractions is to multiply the numerator and the denominator of the complex fraction by the LCD of all the fractions in its numerator and its denominator. This has the effect of leaving sums and differences of integers in the numerator and the denominator, as we shall see in the example below.

Let's use this second method to simplify the complex fraction in Example 2 again.

Practice 3

Use Method 2 to simplify:

$$\frac{\dfrac{1}{2} + \dfrac{1}{6}}{\dfrac{3}{4} - \dfrac{2}{3}}$$

Example 3　Simplify: $\dfrac{\dfrac{1}{2} + \dfrac{3}{8}}{\dfrac{3}{4} - \dfrac{1}{6}}$

Solution: The complex fraction contains fractions with denominators 2, 8, 4, and 6. The LCD is 24. By what we can call the fundamental property of fractions, recall that we can multiply the numerator and the denominator of the complex fraction by 24. Notice below that by the distributive property, this means that we multiply each term in the numerator and denominator by 24.

$$\frac{\dfrac{1}{2} + \dfrac{3}{8}}{\dfrac{3}{4} - \dfrac{1}{6}} = \frac{24\left(\dfrac{1}{2} + \dfrac{3}{8}\right)}{24\left(\dfrac{3}{4} - \dfrac{1}{6}\right)}$$

$$= \frac{\left(\overset{12}{\cancel{24}} \cdot \dfrac{1}{2}\right) + \left(\overset{3}{\cancel{24}} \cdot \dfrac{3}{8}\right)}{\left(\overset{6}{\cancel{24}} \cdot \dfrac{3}{4}\right) - \left(\overset{4}{\cancel{24}} \cdot \dfrac{1}{6}\right)} \qquad \begin{array}{l}\text{Apply the distributive property. Then divide out} \\ \text{common factors to aid in multiplying.}\end{array}$$

$$= \frac{12 + 9}{18 - 4} \qquad \text{Multiply.}$$

$$= \frac{21}{14}$$

$$= \frac{\cancel{7} \cdot 3}{\cancel{7} \cdot 2} = \frac{3}{2} \qquad \text{Simplify.}$$

■ Work Practice 3

Answer

3. $\dfrac{8}{1}$ or 8

The simplified result is the same, of course, no matter which method is used.

Example 4 Simplify: $\dfrac{\dfrac{4}{5} - 2}{\dfrac{3}{10}}$

Solution: Use the second method and multiply the numerator and the denominator of the complex fraction by the LCD of all fractions. Recall that $2 = \dfrac{2}{1}$. The LCD of the denominators 5, 1, and 10 is 10.

$$\frac{\dfrac{4}{5} - \dfrac{2}{1}}{\dfrac{3}{10}} = \frac{10\left(\dfrac{4}{5} - \dfrac{2}{1}\right)}{10\left(\dfrac{3}{10}\right)}$$ Multiply the numerator and denominator by 10.

$$= \frac{\left(\overset{2}{\cancel{10}} \cdot \dfrac{4}{\cancel{5}}\right) - \left(10 \cdot \dfrac{2}{1}\right)}{\overset{1}{\cancel{10}} \cdot \dfrac{3}{\cancel{10}}}$$ Apply the distributive property. Then divide out common factors to aid in multiplying.

$$= \frac{8 - 20}{3}$$ Multiply.

$$= \frac{-12}{3} = -4$$ Simplify.

■ Work Practice 4

Practice 4

Simplify: $\dfrac{\dfrac{3}{4}}{\dfrac{3}{5} - 1}$

Helpful Hint   Don't forget to multiply the numerator and the denominator of the complex fraction by the same number—the LCD.

Objective B Reviewing the Order of Operations ▶

At this time, it is probably a good idea to review the order of operations on expressions containing fractions. Before we do so, let's review how we perform operations on fractions.

Review of Operations on Fractions		
Operation	**Procedure**	**Example**
Multiply	Multiply the numerators and multiply the denominators.	$\dfrac{5}{9} \cdot \dfrac{1}{2} = \dfrac{5 \cdot 1}{9 \cdot 2} = \dfrac{5}{18}$
Divide	Multiply the first fraction by the reciprocal of the second fraction.	$\dfrac{2}{3} \div \dfrac{11}{13} = \dfrac{2}{3} \cdot \dfrac{13}{11} = \dfrac{2 \cdot 13}{3 \cdot 11} = \dfrac{26}{33}$
Add or Subtract	**1.** Write each fraction as an equivalent fraction whose denominator is the LCD. **2.** Add or subtract numerators and write the result over the common denominator.	$\dfrac{3}{4} + \dfrac{1}{8} = \dfrac{3}{4} \cdot \dfrac{2}{2} + \dfrac{1}{8} = \dfrac{6}{8} + \dfrac{1}{8} = \dfrac{7}{8}$

Now let's review the order of operations.

Order of Operations

1. Perform all operations within parentheses ( ), brackets [ ], or other grouping symbols such as fraction bars or square roots, starting with the innermost set.
2. Evaluate any expressions with exponents.
3. Multiply or divide in order from left to right.
4. Add or subtract in order from left to right.

Example 5 Simplify: $\left(\dfrac{4}{5}\right)^2 - 1$

Solution: According to the order of operations, first evaluate $\left(\dfrac{4}{5}\right)^2$.

$$\left(\dfrac{4}{5}\right)^2 - 1 = \dfrac{16}{25} - 1$$ Write $\left(\dfrac{4}{5}\right)^2$ as $\dfrac{16}{25}$.

(Continued on next page)

Practice 5

Simplify: $\left(\dfrac{2}{3}\right)^3 - 2$

Answers

4. $-\dfrac{15}{8}$ **5.** $-\dfrac{46}{27}$

Next, combine the fractions. The LCD of 25 and 1 is 25.

$$\frac{16}{25} - 1 = \frac{16}{25} - \frac{25}{25} \qquad \text{Write 1 as } \frac{25}{25}.$$

$$= \frac{-9}{25} \text{ or } -\frac{9}{25} \qquad \text{Subtract.}$$

■ Work Practice 5

Practice 6

Simplify: $\left(-\frac{1}{2} + \frac{1}{5}\right)\left(\frac{7}{8} + \frac{1}{8}\right)$

Example 6 Simplify: $\left(\frac{1}{4} + \frac{2}{3}\right)\left(\frac{11}{12} + \frac{1}{4}\right)$

Solution: First perform operations inside parentheses. Then multiply.

$$\left(\frac{1}{4} + \frac{2}{3}\right)\left(\frac{11}{12} + \frac{1}{4}\right) = \left(\frac{1 \cdot 3}{4 \cdot 3} + \frac{2 \cdot 4}{3 \cdot 4}\right)\left(\frac{11}{12} + \frac{1 \cdot 3}{4 \cdot 3}\right) \quad \text{Each LCD is 12.}$$

$$= \left(\frac{3}{12} + \frac{8}{12}\right)\left(\frac{11}{12} + \frac{3}{12}\right)$$

$$= \left(\frac{11}{12}\right)\left(\frac{14}{12}\right) \qquad \text{Add.}$$

$$= \frac{11 \cdot \overset{1}{\cancel{2}} \cdot 7}{\cancel{2} \cdot 6 \cdot 12} \qquad \text{Multiply.}$$

$$= \frac{77}{72} \qquad \text{Simplify.}$$

■ Work Practice 6

Helpful Hint If you find it difficult replacing a variable with a number, try the following. First, replace the variable with a set of parentheses, and then place the replacement number between the parentheses.

If $x = \frac{4}{5}$, find $2x + x^2$.

$2x + x^2 = 2(\ ) + (\ )^2$

$= 2\left(\frac{4}{5}\right) + \left(\frac{4}{5}\right)^2 \dots$

and then continue simplifying.

✓**Concept Check** What should be done first to simplify the expression

$$\frac{1}{5} \cdot \frac{5}{2} - \left(\frac{2}{3} + \frac{4}{5}\right)^2 ?$$

Objective C Evaluating Algebraic Expressions

Practice 7

Evaluate $-\frac{3}{5} - xy$ if $x = \frac{3}{10}$ and $y = \frac{2}{3}$.

Example 7 Evaluate $2x + y^2$ if $x = -\frac{1}{2}$ and $y = \frac{1}{3}$.

Solution: Replace x and y with the given values and simplify.

$$2x + y^2 = 2\left(-\frac{1}{2}\right) + \left(\frac{1}{3}\right)^2 \quad \text{Replace } x \text{ with } -\frac{1}{2} \text{ and } y \text{ with } \frac{1}{3}.$$

$$= 2\left(-\frac{1}{2}\right) + \frac{1}{9} \qquad \text{Write } \left(\frac{1}{3}\right)^2 \text{ as } \frac{1}{9}.$$

$$= -1 + \frac{1}{9} \qquad \text{Multiply.}$$

$$= -\frac{9}{9} + \frac{1}{9} \qquad \text{The LCD is 9.}$$

$$= -\frac{8}{9} \qquad \text{Add.}$$

■ Work Practice 7

Answers

6. $-\frac{3}{10}$ 7. $-\frac{4}{5}$

✓**Concept Check Answer**

Add inside parentheses.

Objective D Writing Mixed Numbers as Improper Fractions

In Section 3.1, mixed numbers and improper fractions were both used to represent the shaded part of figure groups. For example,

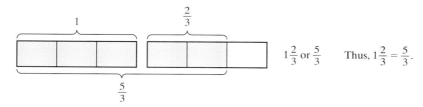

$1\frac{2}{3}$ or $\frac{5}{3}$ Thus, $1\frac{2}{3} = \frac{5}{3}$.

The following steps may be used to write a mixed number as an improper fraction:

Writing a Mixed Number as an Improper Fraction

To write a mixed number as an improper fraction:

Step 1: Multiply the denominator of the fraction by the whole number.

Step 2: Add the numerator of the fraction to the product from Step 1.

Step 3: Write the sum from Step 2 as the numerator of the improper fraction over the original denominator.

For example,

Step 1 Step 2

$$1\frac{2}{3} = \frac{3 \cdot 1 + 2}{3} = \frac{3 + 2}{3} = \frac{5}{3} \quad \text{or} \quad 1\frac{2}{3} = \frac{5}{3}, \text{ as stated above.}$$

Step 3

Example 8 Write each as an improper fraction.

a. $4\frac{2}{9} = \frac{9 \cdot 4 + 2}{9} = \frac{36 + 2}{9} = \frac{38}{9}$

b. $1\frac{8}{11} = \frac{11 \cdot 1 + 8}{11} = \frac{11 + 8}{11} = \frac{19}{11}$

■ Work Practice 8

Practice 8

Write each as an improper fraction.

a. $5\frac{2}{7}$ **b.** $6\frac{2}{3}$ **c.** $10\frac{9}{10}$ **d.** $4\frac{1}{5}$

Objective E Writing Improper Fractions as Mixed Numbers or Whole Numbers

Just as there are times when an improper fraction is preferred, sometimes a mixed or a whole number better suits a situation. To write improper fractions as mixed or whole numbers, we use division. Recall once again from Section 1.7 that the fraction bar means division. This means that the fraction

$\frac{5}{3}$ numerator
denominator

means $3\overline{)5}$
numerator
denominator

Answers

8. a. $\frac{37}{7}$ **b.** $\frac{20}{3}$ **c.** $\frac{109}{10}$ **d.** $\frac{21}{5}$

Writing an Improper Fraction as a Mixed Number or a Whole Number

To write an improper fraction as a mixed number or a whole number:

Step 1: Divide the denominator into the numerator.

Step 2: The whole number part of the mixed number is the quotient. The fraction part of the mixed number is the remainder over the original denominator.

$$\text{quotient}\,\frac{\text{remainder}}{\text{original denominator}}$$

For example,

$$\frac{5}{3}: \quad 3\overline{)5} \quad \frac{-3}{\;\;2} \qquad \frac{5}{3} = 1\frac{2}{3} \;\leftarrow \text{remainder} \\ \qquad\qquad\qquad\qquad\quad \leftarrow \text{original denominator}$$

Step 1 · quotient

Practice 9

Write each as a mixed number or a whole number.

a. $\dfrac{9}{5}$ **b.** $\dfrac{23}{9}$ **c.** $\dfrac{48}{4}$

d. $\dfrac{62}{13}$ **e.** $\dfrac{51}{7}$ **f.** $\dfrac{21}{20}$

Answers

9. a. $1\dfrac{4}{5}$ **b.** $2\dfrac{5}{9}$ **c.** 12 **d.** $4\dfrac{10}{13}$

e. $7\dfrac{2}{7}$ **f.** $1\dfrac{1}{20}$

Example 9 Write each as a mixed number or a whole number.

a. $\dfrac{30}{7}$ **b.** $\dfrac{16}{15}$ **c.** $\dfrac{84}{6}$

Solution:

a. $\dfrac{30}{7}$: $\quad 7\overline{)30} \quad \dfrac{-28}{\;\;\;2}$ $\qquad \dfrac{30}{7} = 4\dfrac{2}{7}$

b. $\dfrac{16}{15}$: $\quad 15\overline{)16} \quad \dfrac{-15}{\;\;\;1}$ $\qquad \dfrac{16}{15} = 1\dfrac{1}{15}$

c. $\dfrac{84}{6}$: $\quad 6\overline{)84} \quad \dfrac{-6}{\;24} \quad \dfrac{-24}{\;\;\;0}$ $\qquad \dfrac{84}{6} = 14$ Since the remainder is 0, the result is the whole number 14.

Helpful Hint When the remainder is 0, the improper fraction is a whole number. For example, $\dfrac{92}{4} = 23$.

$$4\overline{)92} \quad \dfrac{-8}{\;12} \quad \dfrac{-12}{\;\;\;0}$$

■ Work Practice 9

Vocabulary, Readiness & Video Check

Use the choices below to fill in each blank.

addition	multiplication	evaluate the exponential expression
subtraction	division	complex

1. A fraction whose numerator or denominator or both numerator and denominator contain fractions is called a(n) _____ fraction.

2. To simplify $-\dfrac{1}{2} + \dfrac{2}{3} \cdot \dfrac{7}{8}$, which operation do we perform first? _____

3. To simplify $-\dfrac{1}{2} \div \dfrac{2}{3} \cdot \dfrac{7}{8}$, which operation do we perform first? _____

4. To simplify $\dfrac{7}{8} \cdot \left(\dfrac{1}{2} - \dfrac{2}{3} \right)$, which operation do we perform first? _____

5. To simplify $\dfrac{1}{3} \div \dfrac{1}{4} \cdot \left(\dfrac{9}{11} + \dfrac{3}{8} \right)^3$, which operation do we perform first? _____

6. To simplify $9 - \left(-\dfrac{3}{4} \right)^2$, which operation do we perform first? _____

Martin-Gay Interactive Videos   Watch the section lecture video and answer the following questions.

See Video 3.6

Objective A   **7.** In ▣ Example 3, what property is used to simplify the denominator of the complex fraction? ◯

Objective B   **8.** In ▣ Example 4, why can we add the fractions in the first set of parentheses right away? ◯

Objective C   **9.** In ▣ Example 5, why do we use parentheses when substituting the replacement value for x? What would happen if we didn't use parentheses? ◯

Objective D **10.** Complete this statement based on the lecture before ▣ Example 6: The operation of _____ is understood in a mixed number notation; for example, $1\dfrac{1}{3}$ means 1 _____ $\dfrac{1}{3}$. ◯

Objective E **11.** From the lecture before ▣ Example 9, what operation is used to write an improper fraction as a mixed number? ◯

3.6 **Exercise Set** MyMathLab®

Objective A *Simplify each complex fraction. See Examples 1 through 4.*

1. $\dfrac{\frac{1}{8}}{\frac{3}{4}}$

2. $\dfrac{\frac{5}{12}}{\frac{15}{12}}$

3. $\dfrac{\frac{2}{3}}{\frac{2}{7}}$

4. $\dfrac{\frac{9}{25}}{\frac{6}{25}}$

5. $\dfrac{\frac{2}{27}}{\frac{4}{9}}$

6. $\dfrac{\frac{3}{11}}{\frac{1}{2}}$

7. $\dfrac{\frac{3}{4} + \frac{2}{5}}{\frac{1}{2} + \frac{3}{5}}$

8. $\dfrac{\frac{7}{6} + \frac{2}{3}}{\frac{3}{2} - \frac{8}{9}}$

9. $\dfrac{\frac{3}{4}}{5 - \frac{1}{8}}$

10. $\dfrac{\frac{3}{10} + 2}{\frac{2}{5}}$

Objective B *Use the order of operations to simplify each expression. See Examples 5 and 6.*

11. $\dfrac{1}{5} + \dfrac{1}{3} \cdot \dfrac{1}{4}$

12. $\dfrac{1}{2} + \dfrac{1}{6} \cdot \dfrac{1}{3}$

13. $\dfrac{5}{6} \div \dfrac{1}{3} \cdot \dfrac{1}{4}$

14. $\dfrac{7}{8} \div \dfrac{1}{4} \cdot \dfrac{1}{7}$

15. $2^2 - \left(\dfrac{1}{3}\right)^2$

16. $3^2 - \left(\dfrac{1}{2}\right)^2$

▶**17.** $\left(\dfrac{2}{9} + \dfrac{4}{9}\right)\left(\dfrac{1}{3} - \dfrac{9}{10}\right)$

18. $\left(\dfrac{1}{5} - \dfrac{1}{10}\right)\left(\dfrac{1}{5} + \dfrac{1}{10}\right)$

19. $\left(\dfrac{7}{8} - \dfrac{1}{2}\right) \div \dfrac{3}{11}$

20. $\left(-\dfrac{2}{3} - \dfrac{7}{3}\right) \div \dfrac{4}{9}$

21. $2 \cdot \left(\dfrac{1}{4} + \dfrac{1}{5}\right) + 2$

22. $\dfrac{2}{5} \cdot \left(5 - \dfrac{1}{2}\right) - 1$

23. $\left(\dfrac{3}{4}\right)^2 \div \left(\dfrac{3}{4} - \dfrac{1}{12}\right)$

24. $\left(\dfrac{8}{9}\right)^2 \div \left(2 - \dfrac{2}{3}\right)$

25. $\left(\dfrac{2}{5} - \dfrac{3}{10}\right)^2$

26. $\left(\dfrac{3}{2} - \dfrac{4}{3}\right)^3$

27. $\left(\dfrac{3}{4} + \dfrac{1}{8}\right)^2 - \left(\dfrac{1}{2} + \dfrac{1}{8}\right)$

28. $\left(\dfrac{1}{6} + \dfrac{1}{3}\right)^3 + \left(\dfrac{2}{5} \cdot \dfrac{3}{4}\right)^2$

Objective C *Evaluate each expression if $x = -\dfrac{1}{3}$, $y = \dfrac{2}{5}$, and $z = \dfrac{5}{6}$. See Example 7.*

29. $5y - z$

30. $2z - x$

31. $\dfrac{x}{z}$

32. $\dfrac{y + x}{z}$

▶**33.** $x^2 - yz$

34. $x^2 - z^2$

35. $(1 + x)(1 + z)$

36. $(1 - x)(1 - z)$

Objectives A B Mixed Practice *Simplify the following. See Examples 1 through 6.*

37. $\dfrac{\dfrac{5}{24}}{\dfrac{1}{12}}$

38. $\dfrac{\dfrac{7}{10}}{\dfrac{14}{25}}$

39. $\left(\dfrac{3}{2}\right)^3 + \left(\dfrac{1}{2}\right)^3$

40. $\left(\dfrac{5}{21} \div \dfrac{1}{2}\right) + \left(\dfrac{1}{7} \cdot \dfrac{1}{3}\right)$

41. $\left(-\dfrac{1}{2}\right)^2 + \dfrac{1}{5}$

42. $\left(-\dfrac{3}{4}\right)^2 + \dfrac{3}{8}$

43. $\dfrac{2 + \dfrac{1}{6}}{1 - \dfrac{4}{3}}$

44. $\dfrac{3 - \dfrac{1}{2}}{4 + \dfrac{1}{5}}$

45. $\left(1 - \dfrac{2}{5}\right)^2$

46. $\left(-\dfrac{1}{2}\right)^2 - \left(\dfrac{3}{4}\right)^2$

47. $\left(\dfrac{3}{4} - 1\right)\left(\dfrac{1}{8} + \dfrac{1}{2}\right)$

48. $\left(\dfrac{1}{10} + \dfrac{3}{20}\right)\left(\dfrac{1}{5} - 1\right)$

49. $\left(-\dfrac{2}{9} - \dfrac{7}{9}\right)^4$

50. $\left(\dfrac{5}{9} - \dfrac{2}{3}\right)^2$

51. $\dfrac{\dfrac{1}{3} - \dfrac{5}{6}}{\dfrac{3}{4} + \dfrac{1}{2}}$

52. $\dfrac{\dfrac{7}{10} + \dfrac{1}{2}}{\dfrac{4}{5} + \dfrac{3}{4}}$

53. $\left(\dfrac{3}{4} \div \dfrac{6}{5}\right) - \left(\dfrac{3}{4} \cdot \dfrac{6}{5}\right)$

54. $\left(\dfrac{1}{2} \cdot \dfrac{2}{7}\right) - \left(\dfrac{1}{2} \div \dfrac{2}{7}\right)$

Objective D *Write each mixed number as an improper fraction. See Example 8.*

○ 55. $2\dfrac{1}{3}$

56. $1\dfrac{13}{17}$

○ 57. $3\dfrac{3}{5}$

58. $2\dfrac{5}{9}$

59. $6\dfrac{5}{8}$

60. $7\dfrac{3}{8}$

61. $11\dfrac{6}{7}$

62. $12\dfrac{2}{5}$

○ 63. $9\dfrac{7}{20}$

64. $10\dfrac{14}{27}$

65. $166\dfrac{2}{3}$

66. $114\dfrac{2}{7}$

Objective E *Write each improper fraction as a mixed number or a whole number. See Example 9.*

○ 67. $\dfrac{17}{5}$

68. $\dfrac{13}{7}$

○ 69. $\dfrac{37}{8}$

70. $\dfrac{64}{9}$

71. $\dfrac{47}{15}$

72. $\dfrac{65}{12}$

73. $\dfrac{225}{15}$

74. $\dfrac{196}{14}$

75. $\dfrac{182}{175}$

76. $\dfrac{149}{143}$

77. $\dfrac{737}{112}$

78. $\dfrac{901}{123}$

Review

Perform each indicated operation. If the result is an improper fraction, also write the improper fraction as a mixed number. See Section 3.5.

79. $3 + \dfrac{1}{2}$

80. $2 + \dfrac{2}{3}$

81. $9 - \dfrac{5}{6}$

82. $4 - \dfrac{1}{5}$

Concept Extensions

83. In your own words, explain how to write an improper fraction as a mixed number.

84. In your own words, explain how to write a mixed number as an improper fraction.

85. Calculate $\dfrac{2^3}{3}$ and $\left(\dfrac{2}{3}\right)^3$. Do both of these expressions simplify to the same number? Explain why or why not.

86. Calculate $\left(\dfrac{1}{2}\right)^2 \cdot \left(\dfrac{3}{4}\right)^2$ and $\left(\dfrac{1}{2} \cdot \dfrac{3}{4}\right)^2$. Do both of these expressions simplify to the same number? Explain why or why not.

Recall that to find the average of two numbers, we find their sum and divide by 2. For example, the average of $\frac{1}{2}$ and $\frac{3}{4}$ is $\dfrac{\frac{1}{2} + \frac{3}{4}}{2}$. Find the average of each pair of numbers.

87. $\dfrac{1}{2}, \dfrac{3}{4}$

88. $\dfrac{3}{5}, \dfrac{9}{10}$

89. $\dfrac{1}{4}, \dfrac{2}{14}$

90. $\dfrac{5}{6}, \dfrac{7}{9}$

Solve.

91. Two positive numbers, a and b, are graphed below. Where should the graph of their average lie?

92. Study Exercise 91. Without calculating, can $\dfrac{1}{3}$ be the average of $\dfrac{1}{2}$ and $\dfrac{8}{9}$? Explain why or why not.

For Exercises 93–98, answer true or false for each statement.

93. It is possible for the average of two numbers to be greater than both numbers.

94. It is possible for the average of two numbers to be less than both numbers.

95. The sum of two negative fractions is always a negative number.

96. The sum of a negative fraction and a positive fraction is always a positive number.

97. It is possible for the sum of two fractions to be a whole number.

98. It is possible for the difference of two fractions to be a whole number.

99. What operation should be performed first to simplify

$$\frac{1}{5} \cdot \frac{5}{2} - \left(\frac{2}{3} + \frac{4}{5}\right)^2 ?$$

Explain your answer.

100. A student is to evaluate $x - y$ when $x = \dfrac{1}{5}$ and $y = -\dfrac{1}{7}$. This student is asking you if he should evaluate $\dfrac{1}{5} - \dfrac{1}{7}$. What do you tell this student and why?

Each expression contains one addition, one subtraction, one multiplication, and one division. Write the operations in the order that they should be performed. Do not actually simplify. See the Concept Check in this section.

101. $[9 + 3(4 - 2)] \div \dfrac{10}{21}$

102. $[30 - 4(3 + 2)] \div \dfrac{5}{2}$

103. $\dfrac{1}{3} \div \left(\dfrac{2}{3}\right)\left(\dfrac{4}{5}\right) - \dfrac{1}{4} + \dfrac{1}{2}$

104. $\left(\dfrac{5}{6} - \dfrac{1}{3}\right) \cdot \dfrac{1}{3} + \dfrac{1}{2} \div \dfrac{9}{8}$

Evaluate each expression if $x = \dfrac{3}{4}$ and $y = -\dfrac{4}{7}$.

105. $\dfrac{2 + x}{y}$

106. $4x + y$

107. $x^2 + 7y$

108. $\dfrac{\frac{9}{14}}{x + y}$

Objective A Graphing Fractions and Mixed Numbers

Let's review graphing fractions and practice graphing mixed numbers on a number line. This will help us visualize rounding and estimating operations with mixed numbers.

Recall that $5\frac{2}{3}$ means $5 + \frac{2}{3}$ and that

$$-4\frac{1}{6} \text{ means } -4 - \frac{1}{6} \quad \text{or} \quad -4 + \left(-\frac{1}{6}\right)$$

Example 1 Graph the numbers on a number line:

$$\frac{1}{2}, -\frac{3}{4}, 2\frac{2}{3}, -3, -3\frac{1}{8}$$

Solution: Remember that $2\frac{2}{3}$ means $2 + \frac{2}{3}$.

Also, $-3\frac{1}{8}$ means $-3 - \frac{1}{8}$, so $-3\frac{1}{8}$ lies to the left of -3.

Work Practice 1

✓**Concept Check** Which of the following are equivalent to 9?

a. $7\frac{6}{3}$ **b.** $8\frac{4}{4}$ **c.** $8\frac{9}{9}$ **d.** $\frac{18}{2}$ **e.** all of these

Objective B Multiplying or Dividing with Mixed Numbers or Whole Numbers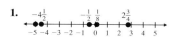

When multiplying or dividing a fraction and a mixed or a whole number, remember that mixed and whole numbers can be written as improper fractions.

Multiplying or Dividing Fractions and Mixed Numbers or Whole Numbers

To multiply or divide with mixed numbers or whole numbers, first write any mixed or whole numbers as improper fractions and then multiply or divide as usual.

(*Note:* If an exercise contains a mixed number, we will write the answer as a mixed number, if possible.)

Objectives

A Graph Positive and Negative Fractions and Mixed Numbers.

B Multiply or Divide Mixed or Whole Numbers.

C Add or Subtract Mixed Numbers.

D Solve Problems Containing Mixed Numbers.

E Perform Operations on Negative Mixed Numbers.

Practice 1

Graph the numbers on a number line.

$$-5, -4\frac{1}{2}, 2\frac{3}{4}, \frac{1}{8}, -\frac{1}{2}$$

Answer

1.

✓**Concept Check Answer**

e

Practice 2

Multiply and simplify: $1\dfrac{2}{3}\cdot\dfrac{11}{15}$

Example 2 Multiply: $3\dfrac{1}{3}\cdot\dfrac{7}{8}$

Solution: Recall from Section 3.6 that the mixed number $3\dfrac{1}{3}$ can be written as the fraction $\dfrac{10}{3}$. Then

$$3\dfrac{1}{3}\cdot\dfrac{7}{8}=\dfrac{10}{3}\cdot\dfrac{7}{8}=\dfrac{\overset{1}{\cancel{2}}\cdot 5\cdot 7}{3\cdot\underset{1}{\cancel{2}}\cdot 4}=\dfrac{35}{12}\quad\text{or}\quad 2\dfrac{11}{12}$$

■ Work Practice 2

Don't forget that a whole number can be written as a fraction by writing the whole number over 1. For example,

$$20=\dfrac{20}{1}\qquad\text{and}\qquad 7=\dfrac{7}{1}$$

Practice 3

Multiply: $\dfrac{5}{6}\cdot 18$

Example 3 Multiply: $\dfrac{3}{4}\cdot 20$

Solution: $\dfrac{3}{4}\cdot 20=\dfrac{3}{4}\cdot\dfrac{20}{1}=\dfrac{3\cdot 20}{4\cdot 1}=\dfrac{3\cdot\overset{5}{\cancel{4}}\cdot 5}{\underset{1}{\cancel{4}}\cdot 1}=\dfrac{15}{1}\quad\text{or}\quad 15$

■ Work Practice 3

When both numbers to be multiplied are mixed or whole numbers, it is a good idea to estimate the product to see if your answer is reasonable. To do this, we first practice rounding mixed numbers to the nearest whole. If the fraction part of the mixed number is $\dfrac{1}{2}$ or greater, we round the whole number part up. If the fraction part of the mixed number is less than $\dfrac{1}{2}$, then we do not round the whole number part up. Study the table below for examples.

Mixed Number	Rounding
$5\dfrac{1}{4}$ $\dfrac{1}{4}$ is less than $\dfrac{1}{2}$. $\dfrac{1}{4}$ $\dfrac{1}{2}$	Thus, $5\dfrac{1}{4}$ rounds to 5.
$3\dfrac{9}{16}$ ← 9 is greater than 8 → Half of 16 is 8.	Thus, $3\dfrac{9}{16}$ rounds to 4.
$1\dfrac{3}{7}$ ← 3 is less than $3\dfrac{1}{2}$. → Half of 7 is $3\dfrac{1}{2}$.	Thus, $1\dfrac{3}{7}$ rounds to 1.

Practice 4

Multiply. Check by estimating.
$3\dfrac{1}{5}\cdot 2\dfrac{3}{4}$

Example 4 Multiply $1\dfrac{2}{3}\cdot 2\dfrac{1}{4}$. Check by estimating.

Solution: $1\dfrac{2}{3}\cdot 2\dfrac{1}{4}=\dfrac{5}{3}\cdot\dfrac{9}{4}=\dfrac{5\cdot 9}{3\cdot 4}=\dfrac{5\cdot\overset{3}{\cancel{9}}\cdot 3}{\underset{1}{\cancel{3}}\cdot 4}=\dfrac{15}{4}\text{ or }3\dfrac{3}{4}$ Exact

Let's check by estimating.

$$1\dfrac{2}{3}\text{ rounds to 2, }2\dfrac{1}{4}\text{ rounds to 2, and }2\cdot 2=4.\quad\text{Estimate}$$

The estimate is close to the exact value, so our answer is reasonable.

Answers

2. $1\dfrac{2}{9}$ 3. 15   4. $8\dfrac{4}{5}$

■ Work Practice 4

Example 5 Multiply: $7 \cdot 2\frac{11}{14}$. Check by estimating.

Solution: $7 \cdot 2\frac{11}{14} = \frac{7}{1} \cdot \frac{39}{14} = \frac{7 \cdot 39}{1 \cdot 14} = \frac{\overset{1}{\cancel{7}} \cdot 39}{1 \cdot 2 \cdot \cancel{7}} = \frac{39}{2}$ or $19\frac{1}{2}$ Exact

To estimate,

$2\frac{11}{14}$ rounds to 3 and $7 \cdot 3 = 21$. Estimate

The estimate is close to the exact value, so our answer is reasonable.

■ Work Practice 5

✔**Concept Check** Find the error.

$2\frac{1}{4} \cdot \frac{1}{2} = 2\frac{1 \cdot 1}{4 \cdot 2} = 2\frac{1}{8}$

Examples Divide.

6. $\frac{3}{4} \div 5 = \frac{3}{4} \div \frac{5}{1} = \frac{3}{4} \cdot \frac{1}{5} = \frac{3 \cdot 1}{4 \cdot 5} = \frac{3}{20}$

7. $\frac{11}{18} \div 2\frac{5}{6} = \frac{11}{18} \div \frac{17}{6} = \frac{11}{18} \cdot \frac{6}{17} = \frac{11 \cdot 6}{18 \cdot 17} = \frac{11 \cdot \overset{1}{\cancel{6}}}{\underset{1}{\cancel{6}} \cdot 3 \cdot 17} = \frac{11}{51}$

8. $5\frac{2}{3} \div 2\frac{5}{9} = \frac{17}{3} \div \frac{23}{9} = \frac{17}{3} \cdot \frac{9}{23} = \frac{17 \cdot 9}{3 \cdot 23} = \frac{17 \cdot \overset{3}{\cancel{3}} \cdot 3}{\underset{1}{\cancel{3}} \cdot 23} = \frac{51}{23}$ or $2\frac{5}{23}$

■ Work Practice 6–8

Objective C Adding or Subtracting Mixed Numbers ▶

We can add or subtract mixed numbers, too, by first writing each mixed number as an improper fraction. But it is often easier to add or subtract the whole-number parts and add or subtract the proper-fraction parts vertically.

Adding or Subtracting Mixed Numbers

To add or subtract mixed numbers, add or subtract the fraction parts and then add or subtract the whole number parts.

Example 9 Add: $2\frac{1}{3} + 5\frac{3}{8}$. Check by estimating.

Solution: The LCD of the denominators 3 and 8 is 24.

$$2\frac{1 \cdot 8}{3 \cdot 8} = 2\frac{8}{24}$$
$$+5\frac{3 \cdot 3}{8 \cdot 3} = 5\frac{9}{24}$$
$$\overline{7\frac{17}{24}} \quad \leftarrow \text{Add the fractions.}$$
$$\text{Add the whole numbers.}$$

To check by estimating, we round as usual. The fraction $2\frac{1}{3}$ rounds to 2, $5\frac{3}{8}$ rounds to 5, and $2 + 5 = 7$, our estimate.

Our exact answer is close to 7, so our answer is reasonable.

■ Work Practice 9

Helpful Hint

When adding or subtracting mixed numbers and whole numbers, it is a good idea to estimate to see if your answer is reasonable.

For the rest of this section, we leave most of the checking by estimating to you.

Practice 10

Add: $3\dfrac{5}{14} + 2\dfrac{6}{7}$

Example 10   Add: $3\dfrac{4}{5} + 1\dfrac{4}{15}$

Solution: The LCD of the denominators 5 and 15 is 15.

$$
\begin{array}{rcl}
3\dfrac{4}{5} &=& 3\dfrac{12}{15}\\[2mm]
+1\dfrac{4}{15} &=& 1\dfrac{4}{15}\\[2mm]
\hline
&& 4\dfrac{16}{15}
\end{array}
$$

Add the fractions; then add the whole numbers.

Notice that the fraction part is improper.

Since $\dfrac{16}{15}$ is $1\dfrac{1}{15}$, we can write the sum as

$$4\dfrac{16}{15} = 4 + 1\dfrac{1}{15} = 5\dfrac{1}{15}$$

■ Work Practice 10

✔**Concept Check**   Explain how you could estimate the following sum:

$$5\dfrac{1}{9} + 14\dfrac{10}{11}.$$

Practice 11

Add: $12 + 3\dfrac{6}{7} + 2\dfrac{1}{5}$

Example 11   Add: $2\dfrac{4}{5} + 5 + 1\dfrac{1}{2}$

Solution: The LCD of the denominators 5 and 2 is 10.

$$
\begin{array}{rcl}
2\dfrac{4}{5} &=& 2\dfrac{8}{10}\\[2mm]
5 &=& 5\\[2mm]
+1\dfrac{1}{2} &=& 1\dfrac{5}{10}\\[2mm]
\hline
&& 8\dfrac{13}{10} = 8 + 1\dfrac{3}{10} = 9\dfrac{3}{10}
\end{array}
$$

■ Work Practice 11

Practice 12

Subtract: $32\dfrac{7}{9} - 16\dfrac{5}{18}$. Check by estimating.

Example 12   Subtract: $8\dfrac{3}{7} - 5\dfrac{2}{21}$. Check by estimating.

Solution: The LCD of the denominators 7 and 21 is 21.

$$
\begin{array}{rcl}
8\dfrac{3}{7} &=& 8\dfrac{9}{21}\\[2mm]
-5\dfrac{2}{21} &=& -5\dfrac{2}{21}\\[2mm]
\hline
&& 3\dfrac{7}{21}
\end{array}
$$

← The LCD of 7 and 21 is 21.

← Subtract the fractions.

Subtract the whole numbers.

Answers

10. $6\dfrac{3}{14}$ **11.** $18\dfrac{2}{35}$ **12.** $16\dfrac{1}{2}$

✔**Concept Check Answer**

Round each mixed number to the nearest whole number and add. $5\dfrac{1}{9}$ rounds to 5 and $14\dfrac{10}{11}$ rounds to 15, and the estimated sum is $5 + 15 = 20$.

Then $3\dfrac{7}{21}$ simplifies to $3\dfrac{1}{3}$. The difference is $3\dfrac{1}{3}$.

To check, $8\dfrac{3}{7}$ rounds to 8, $5\dfrac{2}{21}$ rounds to 5, and $8 - 5 = 3$, our estimate.

Our exact answer is close to 3, so our answer is reasonable.

■ Work Practice 12

When subtracting mixed numbers, borrowing may be needed, as shown in the next example.

Example 13　Subtract:　$7\dfrac{3}{14} - 3\dfrac{6}{7}$

Solution:　The LCD of the denominators 7 and 14 is 14.

$$7\dfrac{3}{14} = 7\dfrac{3}{14} \qquad \text{Notice that we cannot subtract } \dfrac{12}{14} \text{ from } \dfrac{3}{14}, \text{ so we borrow}$$
$$-3\dfrac{6}{7} = -3\dfrac{12}{14} \qquad \text{from the whole number, 7.}$$

borrow 1 from 7

$$7\dfrac{3}{14} = 6 + 1\dfrac{3}{14} = 6 + \dfrac{17}{14} \text{ or } 6\dfrac{17}{14}$$

Now subtract.

$$7\dfrac{3}{14} = 7\dfrac{3}{14} = 6\dfrac{17}{14}$$
$$-3\dfrac{6}{7} = -3\dfrac{12}{14} = -3\dfrac{12}{14}$$
$$\overline{} \qquad \overline{} \qquad 3\dfrac{5}{14} \leftarrow \text{Subtract the fractions.}$$
$$\uparrow$$
$$\text{Subtract the whole numbers.}$$

■ Work Practice 13

Practice 13

Subtract:　$9\dfrac{7}{15} - 4\dfrac{3}{5}$

✓**Concept Check** In the subtraction problem $5\dfrac{1}{4} - 3\dfrac{3}{4}, 5\dfrac{1}{4}$ must be rewritten because $\dfrac{3}{4}$ cannot be subtracted from $\dfrac{1}{4}$. Why is it incorrect to rewrite $5\dfrac{1}{4}$ as $5\dfrac{5}{4}$?

Example 14　Subtract:　$14 - 8\dfrac{3}{7}$

Solution:
$$14 = 13\dfrac{7}{7} \qquad \text{Borrow 1 from 14 and write it as } \dfrac{7}{7}.$$
$$-8\dfrac{3}{7} = -8\dfrac{3}{7}$$
$$\overline{} \qquad 5\dfrac{4}{7} \leftarrow \text{Subtract the fractions.}$$
$$\uparrow$$
$$\text{Subtract the whole numbers.}$$

■ Work Practice 14

Practice 14

Subtract:　$25 - 10\dfrac{2}{9}$

Answers

13. $4\dfrac{13}{15}$　**14.** $14\dfrac{7}{9}$

✓**Concept Check Answer**

Rewrite $5\dfrac{1}{4}$ as $4\dfrac{5}{4}$ by borrowing from the 5.

Objective D Solving Problems Containing Mixed Numbers

Now that we know how to perform operations on mixed numbers, we can solve real-life problems.

Practice 15

The measurement around the trunk of a tree just below shoulder height is called its girth. The largest known American beech tree in the United States has a girth of $24\frac{1}{6}$ feet. The largest known sugar maple tree in the United States has a girth of $19\frac{5}{12}$ feet. How much larger is the girth of the largest known American beech tree than the girth of the largest known sugar maple tree? (*Source: American Forests*)

Girth

| **Example 15** | Finding Legal Lobster Size |

Lobster fishermen must measure the upper body shells of the lobsters they catch. Lobsters that are too small are thrown back into the ocean. Each state has its own size standard for lobsters to help control the breeding stock. Massachusetts divides its waters into four Lobster Conservation Management Areas, with a different minimum lobster size permitted in each area. In area three, the legal lobster size increased from $3\frac{13}{32}$ inches to $3\frac{1}{2}$ inches. How much of an increase was this?

(*Source:* Massachusetts Division of Marine Fisheries)

Solution:

1. UNDERSTAND. Read and reread the problem carefully. The word "increase" found in the problem might make you think that we add to solve the problem. But the phrase "how much of an increase" tells us to subtract to find the increase.

2. TRANSLATE.

In words:	increase	is	new lobster size	minus	old lobster size
	↓	↓	↓	↓	↓
Translate:	increase	=	$3\frac{1}{2}$	−	$3\frac{13}{32}$

3. SOLVE: Before we solve, let's estimate by rounding to the nearest wholes. The fraction $3\frac{1}{2}$ can be rounded up to 4, $3\frac{13}{32}$ rounds to 3, and $4 - 3 = 1$. The increase is not 1 but will be smaller since we rounded $3\frac{1}{2}$ up and rounded $3\frac{13}{32}$ down.

$$
\begin{array}{r}
3\frac{1}{2} = 3\frac{16}{32} \\
-3\frac{13}{32} = -3\frac{13}{32} \\
\hline
\frac{3}{32}
\end{array}
$$

4. INTERPRET. *Check* your work. Our estimate tells us that the exact increase of $\frac{3}{32}$ is reasonable. *State* your conclusion: The increase in lobster size was $\frac{3}{32}$ of an inch.

Work Practice 15

Answer

15. $4\frac{3}{4}$ ft

Practice 16

A designer of women's clothing designs a woman's dress that requires $3\frac{1}{7}$ yards of material. How many dresses can be made from a 44-yard bolt of material?

Example 16 Calculating Manufacturing Materials Needed

In a manufacturing process, a metal-cutting machine cuts strips $1\frac{3}{5}$ inches long from a piece of metal stock. How many such strips can be cut from a 48-inch piece of stock?

Solution:

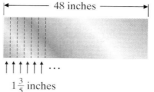

48 inches

$1\frac{3}{5}$ inches

1. UNDERSTAND the problem. To do so, read and reread the problem. Then draw a diagram:

   We want to know how many $1\frac{3}{5}$s there are in 48.

2. TRANSLATE.

   In words:
   | number of strips | is | 48 | divided by | $1\frac{3}{5}$ |
   | ↓ | ↓ | ↓ | ↓ | ↓ |

   Translate:
   | number of strips | = | 48 | ÷ | $1\frac{3}{5}$ |

3. SOLVE: Let's estimate a reasonable answer. The mixed number $1\frac{3}{5}$ rounds to 2 and 48 ÷ 2 = 24.

$$48 \div 1\frac{3}{5} = 48 \div \frac{8}{5} = \frac{48}{1} \cdot \frac{5}{8} = \frac{48 \cdot 5}{1 \cdot 8} = \frac{\cancel{8} \cdot 6 \cdot 5}{1 \cdot \cancel{8}} = \frac{30}{1} \quad \text{or} \quad 30$$

4. INTERPRET. *Check* your work. Since the exact answer of 30 is close to our estimate of 24, our answer is reasonable. *State* your conclusion: Thirty strips can be cut from the 48-inch piece of stock.

Work Practice 16

Objective E Operating on Negative Mixed Numbers

To perform operations on negative mixed numbers, let's first practice writing these numbers as negative fractions and negative fractions as negative mixed numbers.

To understand negative mixed numbers, we simply need to know that, for example,

$$-3\frac{2}{5} \text{ means } -\left(3\frac{2}{5}\right)$$

Thus, to write a negative mixed number as a fraction, we do the following.

$$-3\frac{2}{5} = -\left(3\frac{2}{5}\right) = -\left(\frac{5 \cdot 3 + 2}{5}\right) = -\left(\frac{17}{5}\right) \quad \text{or} \quad -\frac{17}{5}$$

Examples Write each as a fraction.

17. $-1\frac{7}{8} = -\frac{8 \cdot 1 + 7}{8} = -\frac{15}{8}$ Write $1\frac{7}{8}$ as an improper fraction and keep the negative sign.

18. $-23\frac{1}{2} = -\frac{2 \cdot 23 + 1}{2} = -\frac{47}{2}$ Write $23\frac{1}{2}$ as an improper fraction and keep the negative sign.

Work Practice 17–18

Practice 17–18

Write each as a fraction.

17. $-9\frac{3}{7}$ 18. $-5\frac{10}{11}$

Answers

16. 14 dresses   17. $-\frac{66}{7}$ 18. $-\frac{65}{11}$

To write a negative fraction as a negative mixed number, we use a similar procedure. We simply disregard the negative sign, convert the improper fraction to a mixed number, and then reinsert the negative sign.

Practice 19–20

Write each as a mixed number.

19. $-\dfrac{37}{8}$ **20.** $-\dfrac{46}{5}$

Examples Write each as a mixed number.

19. $-\dfrac{22}{5} = -4\dfrac{2}{5}$

$5\overline{)22} \quad \dfrac{22}{5} = 4\dfrac{2}{5}$
$\underline{-20}$
$\quad 2$

20. $-\dfrac{9}{4} = -2\dfrac{1}{4}$

$4\overline{)9} \quad \dfrac{9}{4} = 2\dfrac{1}{4}$
$\underline{-8}$
$\quad 1$

■ Work Practice 19–20

We multiply or divide with negative mixed numbers the same way that we multiply or divide with positive mixed numbers. We first write each mixed number as a fraction.

Practice 21–22

21. $2\dfrac{3}{4} \cdot \left(-3\dfrac{3}{5}\right)$

22. $-4\dfrac{2}{7} \div 1\dfrac{1}{4}$

Examples Perform the indicated operations.

21. $-4\dfrac{2}{5} \cdot 1\dfrac{3}{11} = -\dfrac{22}{5} \cdot \dfrac{14}{11} = -\dfrac{22 \cdot 14}{5 \cdot 11} = -\dfrac{2 \cdot 11 \cdot 14}{5 \cdot 11} = -\dfrac{28}{5}$ or $-5\dfrac{3}{5}$

22. $-2\dfrac{1}{3} \div \left(-2\dfrac{1}{2}\right) = -\dfrac{7}{3} \div \left(-\dfrac{5}{2}\right) = -\dfrac{7}{3} \cdot \left(-\dfrac{2}{5}\right) = \dfrac{7 \cdot 2}{3 \cdot 5} = \dfrac{14}{15}$

■ Work Practice 21–22

Helpful Hint Recall that $(-) \cdot (-) = +$

To add or subtract with negative mixed numbers, we must be very careful! Recall that

$$-3\dfrac{2}{5} \text{ means } -\left(3\dfrac{2}{5}\right)$$

This means that

$$-3\dfrac{2}{5} = -\left(3\dfrac{2}{5}\right) = -\left(3 + \dfrac{2}{5}\right) = -3 - \dfrac{2}{5} \quad \text{This can sometimes be easily overlooked.}$$

To avoid problems, we will add or subtract negative mixed numbers by rewriting them as addition and recalling how to add signed numbers.

Answers

19. $-4\dfrac{5}{8}$ **20.** $-9\dfrac{1}{5}$ **21.** $-9\dfrac{9}{10}$

22. $-3\dfrac{3}{7}$

Example 23 Add: $6\dfrac{3}{5} + \left(-9\dfrac{7}{10}\right)$

Solution: Here we are adding two numbers with different signs. Recall that we then subtract the absolute values and keep the sign of the larger absolute value.

Since $-9\dfrac{7}{10}$ has the larger absolute value, the answer is negative.

First, subtract absolute values:

$$
\begin{array}{r}
9\dfrac{7}{10} = 9\dfrac{7}{10} \\[2mm]
-6\dfrac{3 \cdot 2}{5 \cdot 2} = -6\dfrac{6}{10} \\[2mm]
\hline
3\dfrac{1}{10}
\end{array}
$$

Thus,

$$6\dfrac{3}{5} + \left(-9\dfrac{7}{10}\right) = -3\dfrac{1}{10} \quad \text{The result is negative since } -9\dfrac{7}{10} \text{ has the larger absolute value.}$$

■ Work Practice 23

Practice 23

Add: $6\dfrac{2}{3} + \left(-12\dfrac{3}{4}\right)$

Example 24 Subtract: $-11\dfrac{5}{6} - 20\dfrac{4}{9}$

Solution: Let's write as an equivalent addition: $-11\dfrac{5}{6} + \left(-20\dfrac{4}{9}\right)$. Here, we are adding two numbers with like signs. Recall that we add their absolute values and keep the common negative sign.

First, add absolute values:

$$
\begin{array}{r}
11\dfrac{5 \cdot 3}{6 \cdot 3} = 11\dfrac{15}{18} \\[2mm]
+20\dfrac{4 \cdot 2}{9 \cdot 2} = +20\dfrac{8}{18} \\[2mm]
\hline
31\dfrac{23}{18} \text{ or } 32\dfrac{5}{18}
\end{array}
$$

Since $\dfrac{23}{18} = 1\dfrac{5}{18}$

Thus,

$$-11\dfrac{5}{6} - 20\dfrac{4}{9} = -32\dfrac{5}{18}$$

Keep the common sign.

■ Work Practice 24

Practice 24

Subtract: $-9\dfrac{2}{7} - 30\dfrac{11}{14}$

Answers

23. $-6\dfrac{1}{12}$ **24.** $-40\dfrac{1}{14}$

Calculator Explorations **Converting Between Mixed Number and Fraction Notation**

If your calculator has a fraction key, such as $\boxed{a^b/_c}$, you can use it to convert between mixed number notation and fraction notation.

To write $13\dfrac{7}{16}$ as an improper fraction, press

$\boxed{13}$ $\boxed{a^b/_c}$ $\boxed{7}$ $\boxed{a^b/_c}$ $\boxed{16}$ $\boxed{2nd}$ $\boxed{d/c}$

The display will read

$\boxed{215\,|\,16}$

which represents $\dfrac{215}{16}$. Thus $13\dfrac{7}{16} = \dfrac{215}{16}$.

To convert $\dfrac{190}{13}$ to a mixed number, press

$\boxed{190}$ $\boxed{a^b/_c}$ $\boxed{13}$ $\boxed{=}$

The display will read

$\boxed{14\_8/13}$

which represents $14\dfrac{8}{13}$. Thus $\dfrac{190}{13} = 14\dfrac{8}{13}$.

Write each mixed number as a fraction and each fraction as a mixed number.

1. $25\dfrac{5}{11}$ **2.** $67\dfrac{14}{15}$ **3.** $107\dfrac{31}{35}$

4. $186\dfrac{17}{21}$ **5.** $\dfrac{365}{14}$ **6.** $\dfrac{290}{13}$

7. $\dfrac{2769}{30}$ **8.** $\dfrac{3941}{17}$

Vocabulary, Readiness & Video Check

Use the choices below to fill in each blank.

round   fraction   whole number
improper   mixed number

1. The number $5\dfrac{3}{4}$ is called a(n) _____.

2. For $5\dfrac{3}{4}$, the 5 is called the _____ part and $\dfrac{3}{4}$ is called the _____ part.

3. To estimate operations on mixed numbers, we _____ mixed numbers to the nearest whole number.

4. The mixed number $2\dfrac{5}{8}$ written as a(n) _____ fraction is $\dfrac{21}{8}$.

Martin-Gay Interactive Videos

See Video 3.7 🔴

Watch the section lecture video and answer the following questions.

Objective A **5.** In ⊞ Example 1, why is the unit distance between −4 and −3 on the number line split into 5 equal parts? 🔘

Objective B **6.** Why do we need to know how to multiply fractions to solve ⊞ Example 2? 🔘

Objective C **7.** In ⊞ Example 4, why is the first form of the answer not an appropriate form? 🔘

Objective D **8.** Why do we need to know how to subtract fractions to solve ⊞ Example 5? 🔘

Objective E **9.** In ⊞ Example 6, how is it determined whether the answer is positive or negative? 🔘

3.7 **Exercise Set** MyMathLab®

Objective A *Graph each list of numbers on the given number line. See Example 1.*

1. $-2, -2\frac{2}{3}, 0, \frac{7}{8}, -\frac{1}{3}$

2. $-1, -1\frac{1}{4}, -\frac{1}{4}, 3\frac{1}{4}, 3$

3. $4, \frac{1}{3}, -3, -3\frac{4}{5}, 1\frac{1}{3}$

4. $3, \frac{3}{8}, -4, -4\frac{1}{3}, -\frac{9}{10}$

Objective B *Choose the best estimate for each product or quotient. See Examples 4 and 5.*

5. $2\frac{11}{12} \cdot 1\frac{1}{4}$

   a. 2   **b.** 3   **c.** 1   **d.** 12

6. $5\frac{1}{6} \cdot 3\frac{5}{7}$

   a. 9   **b.** 15   **c.** 8   **d.** 20

7. $12\frac{2}{11} \div 3\frac{9}{10}$

   a. 3   **b.** 4   **c.** 36   **d.** 9

8. $20\frac{3}{14} \div 4\frac{8}{11}$

   a. 5   **b.** 80   **c.** 4   **d.** 16

Multiply or divide. For Exercises 13 through 16, find an exact answer and an estimated answer. See Examples 2 through 8.

9. $2\frac{2}{3} \cdot \frac{1}{7}$

10. $\frac{5}{9} \cdot 4\frac{1}{5}$

11. $7 \div 1\frac{3}{5}$

12. $9 \div 1\frac{2}{3}$

13. $2\frac{1}{5} \cdot 3\frac{1}{2}$

Exact:

Estimate:

14. $2\frac{1}{4} \cdot 7\frac{1}{8}$

Exact:

Estimate:

15. $3\frac{4}{5} \cdot 6\frac{2}{7}$

Exact:

Estimate:

16. $5\frac{5}{6} \cdot 7\frac{3}{5}$

Exact:

Estimate:

17. $5 \cdot 2\frac{1}{2}$

18. $6 \cdot 3\frac{1}{3}$

19. $3\frac{2}{3} \cdot 1\frac{1}{2}$

20. $2\frac{4}{5} \cdot 2\frac{5}{8}$

21. $2\frac{2}{3} \div \frac{1}{7}$

22. $\frac{5}{9} \div 4\frac{1}{5}$

Objective C *Choose the best estimate for each sum or difference. See Examples 9 and 12.*

23. $3\frac{7}{8} + 2\frac{1}{5}$

    a. 6   **b.** 5   **c.** 1   **d.** 2

24. $3\frac{7}{8} - 2\frac{1}{5}$

    a. 6   **b.** 5   **c.** 1   **d.** 2

25. $8\frac{1}{3} + 1\frac{1}{2}$

    a. 4   **b.** 10   **c.** 6   **d.** 16

26. $8\frac{1}{3} - 1\frac{1}{2}$

    a. 4   **b.** 10   **c.** 6   **d.** 16

Add. For Exercises 27 through 30, find an exact sum and an estimated sum. See Examples 9 through 11.

27. $4\frac{7}{12}$
$+2\frac{1}{12}$
Exact:
Estimate:

28. $7\frac{4}{11}$
$+3\frac{2}{11}$
Exact:
Estimate:

29. $10\frac{3}{14}$
$+\ 3\frac{4}{7}$
Exact:
Estimate:

30. $12\frac{5}{12}$
$+\ 4\frac{1}{6}$
Exact:
Estimate:

31. $9\frac{1}{5}$
$+8\frac{2}{25}$

32. $6\frac{2}{13}$
$+8\frac{7}{26}$

33. $12\frac{3}{14}$
10
$+25\frac{5}{12}$

34. $8\frac{2}{9}$
32
$+\ 9\frac{10}{21}$

35. $15\frac{4}{7}$
$+9\frac{11}{14}$

36. $23\frac{3}{5}$
$+8\frac{8}{15}$

37. $3\frac{5}{8}$
$2\frac{1}{6}$
$+7\frac{3}{4}$

38. $4\frac{2}{3}$
$9\frac{2}{5}$
$+3\frac{1}{6}$

Subtract. For Exercises 39 through 42, find an exact difference and an estimated difference. See Examples 12 through 14.

39. $4\frac{7}{10}$
$-2\frac{1}{10}$
Exact:
Estimate:

40. $7\frac{4}{9}$
$-3\frac{2}{9}$
Exact:
Estimate:

41. $10\frac{13}{14}$
$-\ 3\frac{4}{7}$
Exact:
Estimate:

42. $12\frac{5}{12}$
$-\ 4\frac{1}{6}$
Exact:
Estimate:

43. $9\dfrac{1}{5}$

$\quad -8\dfrac{6}{25}$

44. $5\dfrac{2}{13}$

$\quad -4\dfrac{7}{26}$

45. 6

$\quad -2\dfrac{4}{9}$

46. 8

$\quad -1\dfrac{7}{10}$

47. $63\dfrac{1}{6}$

$\quad -47\dfrac{5}{12}$

48. $86\dfrac{2}{15}$

$\quad -27\dfrac{3}{10}$

Objectives B C Mixed Practice *Perform each indicated operation. See Examples 2 through 14.*

49. $2\dfrac{3}{4}$

$\quad +1\dfrac{1}{4}$

50. $5\dfrac{5}{8}$

$\quad +2\dfrac{3}{8}$

51. $15\dfrac{4}{7}$

$\quad -9\dfrac{11}{14}$

52. $23\dfrac{8}{15}$

$\quad -8\dfrac{3}{5}$

53. $3\dfrac{1}{9} \cdot 2$

54. $4\dfrac{1}{2} \cdot 3$

55. $1\dfrac{2}{3} \div 2\dfrac{1}{5}$

56. $5\dfrac{1}{5} \div 3\dfrac{1}{4}$

57. $22\dfrac{4}{9} + 13\dfrac{5}{18}$

58. $15\dfrac{3}{25} + 5\dfrac{2}{5}$

59. $5\dfrac{2}{3} - 3\dfrac{1}{6}$

60. $5\dfrac{3}{8} - 2\dfrac{3}{16}$

61. $15\dfrac{4}{5}$

$\quad 20\dfrac{3}{10}$

$\quad +37\dfrac{2}{15}$

62. $3\dfrac{7}{16}$

$\quad 6\dfrac{1}{2}$

$\quad +9\dfrac{3}{8}$

63. $6\dfrac{4}{7} - 5\dfrac{11}{14}$

64. $47\dfrac{5}{12} - 23\dfrac{19}{24}$

65. $4\dfrac{2}{7} \cdot 1\dfrac{3}{10}$

66. $6\dfrac{2}{3} \cdot 2\dfrac{3}{4}$

67. $6\dfrac{2}{11}$

$\quad 3$

$\quad +4\dfrac{10}{33}$

68. $7\dfrac{3}{7}$

$\quad 15$

$\quad +20\dfrac{1}{2}$

Objective D **Translating** *Translate each phrase to an algebraic expression. Use x to represent "a number." See Examples 15 and 16.*

69. $-5\frac{2}{7}$ decreased by a number

70. The sum of $8\frac{3}{4}$ and a number

71. Multiply $1\frac{9}{10}$ by a number.

72. Divide a number by $-6\frac{1}{11}$.

Solve. For Exercises 73 and 74, the solutions have been started for you. Write each answer in simplest form. See Examples 15 and 16.

73. A heart attack patient in rehabilitation walked on a treadmill $12\frac{3}{4}$ miles over 4 days. How many miles is this per day?

Start the solution:

1. UNDERSTAND the problem. Reread it as many times as needed.
2. TRANSLATE into an equation. (Fill in the blanks.)

miles per day	is	total miles	divided by	number of days
↓	↓	↓	↓	↓

$$\text{miles per day} = \underline{\quad} \div \underline{\quad}$$

Finish with:

3. SOLVE. and 4. INTERPRET.

74. A local restaurant is selling hamburgers from a booth on Memorial Day. A total of $27\frac{3}{4}$ pounds of hamburger have been ordered. How many quarter-pound hamburgers can this make?

Start the solution:

1. UNDERSTAND the problem. Reread it as many times as needed.
2. TRANSLATE into an equation. (Fill in the blanks.)

how many quarter-pound hamburgers	is	total pounds of hamburger	divided by	a quarter-pound
↓	↓	↓	↓	↓

$$\text{how many quarter-pound hamburgers} = \underline{\quad} \div \underline{\quad}$$

Finish with:

3. SOLVE. and 4. INTERPRET.

75. The Gauge Act of 1846 set the standard gauge for U.S. railroads at $56\frac{1}{2}$ inches. (See figure.) If the standard gauge in Spain is $65\frac{9}{10}$ inches, how much wider is Spain's standard gauge than the U.S. standard gauge? (*Source:* San Diego Railroad Museum)

76. The standard railroad track gauge (see figure) in Spain is $65\frac{9}{10}$ inches, while in neighboring Portugal it is $65\frac{11}{20}$ inches. Which gauge is wider and by how much? (*Source:* San Diego Railroad Museum)

Track gauge (U.S. $56\frac{1}{2}$ inches)

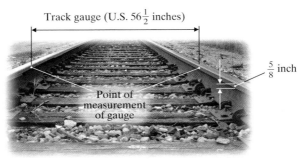

$\frac{5}{8}$ inch

Point of measurement of gauge

Section 3.7 | Operations on Mixed Numbers

77. If Tucson's average rainfall is $11\frac{1}{4}$ inches and Yuma's is $3\frac{3}{5}$ inches, how much more rain, on the average, does Tucson get than Yuma?

78. A pair of crutches needs adjustment. One crutch is 43 inches and the other is $41\frac{5}{8}$ inches. Find how much the shorter crutch should be lengthened to make both crutches the same length.

For Exercises 79 and 80, find the area of each figure.

79.

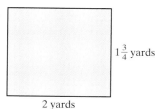

$1\frac{3}{4}$ yards

2 yards

80.

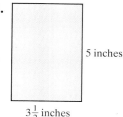

5 inches

$3\frac{1}{2}$ inches

81. A model for a proposed computer chip measures $\frac{3}{4}$ inch by $1\frac{1}{4}$ inches. Find its area.

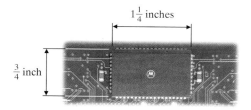

$1\frac{1}{4}$ inches

$\frac{3}{4}$ inch

82. The Saltalamachios are planning to build a deck that measures $4\frac{1}{2}$ yards by $6\frac{1}{3}$ yards. Find the area of their proposed deck.

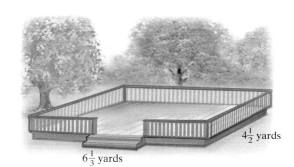

$4\frac{1}{2}$ yards

$6\frac{1}{3}$ yards

For Exercises 83 and 84, find the perimeter of each figure.

83.

$5\frac{1}{3}$ meters

3 meters

5 meters

$7\frac{7}{8}$ meters

84.

$3\frac{1}{4}$ yards $3\frac{1}{4}$ yards

$3\frac{1}{4}$ yards $3\frac{1}{4}$ yards

$3\frac{1}{4}$ yards

85. A homeowner has $15\frac{2}{3}$ feet of plastic pipe. She cuts off a $2\frac{1}{2}$-foot length and then a $3\frac{1}{4}$-foot length. If she now needs a 10-foot piece of pipe, will the remaining piece do? If not, by how much will the piece be short?

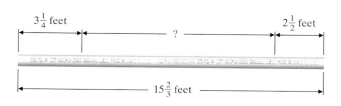

$3\frac{1}{4}$ feet

?

$2\frac{1}{2}$ feet

$15\frac{2}{3}$ feet

86. A trim carpenter cuts a board $3\frac{3}{8}$ feet long from one 6 feet long. How long is the remaining piece?

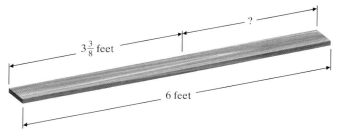

?

$3\frac{3}{8}$ feet

6 feet

△ **87.** The area of the rectangle below is 12 square meters. If its width is $2\frac{4}{7}$ meters, find its length.

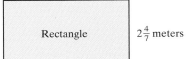

Rectangle $2\frac{4}{7}$ meters

△ **88.** The perimeter of the square below is $23\frac{1}{2}$ feet. Find the length of each side.

Square

The following table lists three upcoming total eclipses of the sun that will be visible from North America. The duration of each eclipse is listed in the table. Use the table to answer Exercises 89 through 92.

Total Solar Eclipses Visible from North America	
Date of Eclipse	**Duration (in Minutes)**
August 21, 2017	$2\frac{2}{3}$
April 8, 2024	$4\frac{7}{15}$
March 30, 2033	$2\frac{37}{60}$
(*Source:* NASA/Goddard Space Flight Center)	

89. What is the total duration for the three eclipses?

90. What is the total duration for the two eclipses occurring in odd-numbered years?

91. How much longer will the April 8, 2024, eclipse be than the August 21, 2017, eclipse?

92. How much longer will the April 8, 2024, eclipse be than the March 30, 2033, eclipse?

Objective **E** *Perform the indicated operations. See Examples 17 through 24.*

93. $-4\frac{2}{5} \cdot 2\frac{3}{10}$

94. $-3\frac{5}{6} \div \left(-3\frac{2}{3}\right)$

95. $-5\frac{1}{8} - 19\frac{3}{4}$

96. $17\frac{5}{9} + \left(-14\frac{2}{3}\right)$

▶ **97.** $-31\frac{2}{15} + 17\frac{3}{20}$

98. $-31\frac{7}{8} - \left(-26\frac{5}{12}\right)$

99. $-1\frac{5}{7} \cdot \left(-2\frac{1}{2}\right)$

100. $1\frac{3}{4} \div \left(-3\frac{1}{2}\right)$

101. $11\frac{7}{8} - 13\frac{5}{6}$

102. $-20\frac{2}{5} + \left(-30\frac{3}{10}\right)$

103. $-7\frac{3}{10} \div (-100)$

104. $-4\frac{1}{4} \div 2\frac{3}{8}$

Review

Evaluate each expression. See Section 1.9.

105. $20 \div 10 \cdot 2$ **106.** $36 - 5 \cdot 6 + 10$ **107.** $2 + 3(8 \cdot 7 - 1)$ **108.** $2(10 - 2 \cdot 5) + 13$

Concept Extensions

Solve. See the first Concept Check in this section.

109. Which of the following are equivalent to 10?

a. $9\frac{5}{5}$ **b.** $9\frac{100}{100}$ **c.** $6\frac{44}{11}$ **d.** $8\frac{13}{13}$

110. Which of the following are equivalent to $7\frac{3}{4}$?

a. $6\frac{7}{4}$ **b.** $5\frac{11}{4}$ **c.** $7\frac{12}{16}$ **d.** all of them

Solve. See the second Concept Check in this section.

111. A student asked you to check her work below. Is it correct? If not, where is the error?

$$20\frac{2}{3} \div 10\frac{1}{2} \stackrel{?}{=} 2\frac{1}{3}$$

112. A student asked you to check his work below. Is it correct? If not, where is the error?

$$3\frac{2}{3} \cdot 1\frac{1}{7} \stackrel{?}{=} 3\frac{2}{21}$$

113. In your own words, describe how to divide mixed numbers.

114. In your own words, explain how to multiply
a. fractions
b. mixed numbers

Solve. See the third Concept Check in this section.

115. In your own words, explain how to round a mixed number to the nearest whole number.

116. Use rounding to estimate the best sum for $11\frac{19}{20} + 9\frac{1}{10}$.

a. 2 **b.** 3 **c.** 20 **d.** 21

Solve.

117. Explain in your own words why $9\frac{13}{9}$ is equal to $10\frac{4}{9}$.

118. In your own words, explain
a. when to borrow when subtracting mixed numbers, and
b. how to borrow when subtracting mixed numbers.

Chapter 3 Group Activity

Lobster Classification Sections 3.1, 3.6, 3.7

This activity may be completed by working in groups or individually.

Lobsters are normally classified by weight. Use the weight classification table to answer the questions in this activity.

Classification of Lobsters

Class	Weight (in Pounds)
Chicken	1 to $1\frac{1}{8}$
Quarter	$1\frac{1}{4}$
Half	$1\frac{1}{2}$ to $1\frac{3}{4}$
Select	$1\frac{3}{4}$ to $2\frac{1}{2}$
Large select	$2\frac{1}{2}$ to $3\frac{1}{2}$
Jumbo	Over $3\frac{1}{2}$

(*Source:* The Maine Lobster Promotion Council)

1. A lobster fisher has kept four lobsters from a lobster trap. Classify each lobster if they have the following weights:

   a. $1\frac{7}{8}$ pounds

   b. $1\frac{9}{16}$ pounds

   c. $2\frac{3}{4}$ pounds

   d. $2\frac{3}{8}$ pounds

2. A recipe requires 5 pounds of lobster. Using the minimum weight for each class, decide whether a chicken, half, and select lobster will be enough for the recipe, and explain your reasoning. If not, suggest a better choice of lobsters to meet the recipe requirements.

3. A lobster market customer has selected two chickens, a select, and a large select. What is the most that these four lobsters could weigh? What is the least that these four lobsters could weigh?

4. A lobster market customer wishes to buy three quarters. If lobsters sell for $7 per pound, how much will the customer owe for her purchase?

5. Why do you think there is no classification for lobsters weighing under 1 pound?

Chapter 3 Vocabulary Check

Fill in each blank with one of the words or phrases listed below.

mixed number	complex fraction	like	numerator	prime factorization
composite number	equivalent	cross products	least common denominator	denominator
prime number	improper fraction	simplest form	undefined	0
reciprocals	proper fraction			

1. Two numbers are _____ of each other if their product is 1.

2. A(n) _____ is a natural number greater than 1 that is not prime.

3. Fractions that represent the same portion of a whole are called _____ fractions.

4. A(n) _____ is a fraction whose numerator is greater than or equal to its denominator.

5. A(n) _____ is a natural number greater than 1 whose only factors are 1 and itself.

6. A fraction is in _____ when the numerator and the denominator have no factors in common other than 1.

7. A(n) _____ is one whose numerator is less than its denominator.

8. A(n) _____ contains a whole number part and a fraction part.

9. In the fraction $\frac{7}{9}$, the 7 is called the _____ and the 9 is called the _____.

10. The _____ of a number is the factorization in which all the factors are prime numbers.

11. The fraction $\frac{3}{0}$ is _____.

12. The fraction $\frac{0}{5} =$ _____.

13. Fractions that have the same denominator are called _____ fractions.

14. The LCM of the denominators in a list of fractions is called the _____.

15. A fraction whose numerator or denominator or both numerator and denominator contain fractions is called a(n) _____.

16. In $\frac{a}{b} = \frac{c}{d}$, $a \cdot d$ and $b \cdot c$ are called _____.

Helpful Hint ▸ Are you preparing for your test? Don't forget to take the Chapter 3 Test on page 277. Then check your answers at the back of the text and use the Chapter Test Prep Videos to see the fully worked-out solutions to any of the exercises you want to review.

3 Chapter Highlights

Definitions and Concepts	Examples
Section 3.1 Introduction to Fractions and Mixed Numbers	

Definitions and Concepts	Examples
A **fraction** is of the form $\dfrac{\text{numerator}}{\text{denominator}}$ ← number of parts being considered ← number of equal parts in the whole	Write a fraction to represent the shaded part of the figure. $\dfrac{3}{8}$ ← number of parts shaded ← number of equal parts
A fraction is called a **proper fraction** if its numerator is less than its denominator. A fraction is called an **improper fraction** if its numerator is greater than or equal to its denominator. A **mixed number** contains a whole number and a fraction.	Proper Fractions: $\dfrac{1}{3}, \dfrac{2}{5}, \dfrac{7}{8}, \dfrac{100}{101}$ Improper Fractions: $\dfrac{5}{4}, \dfrac{2}{2}, \dfrac{9}{7}, \dfrac{101}{100}$ Mixed Numbers: $1\dfrac{1}{2}, 5\dfrac{7}{8}, 25\dfrac{9}{10}$

Definitions and Concepts	Examples

Section 3.2   Factors and Simplest Form

A **prime number** is a natural number that has exactly two different factors, 1 and itself.	$2, 3, 5, 7, 11, 13, 17, \ldots$
A **composite number** is any natural number other than 1 that is not prime.	$4, 6, 8, 9, 10, 12, 14, 15, 16, \ldots$
The **prime factorization** of a number is the factorization in which all the factors are prime numbers.	Write the prime factorization of 60. $$60 = 6 \cdot 10$$ $$= 2 \cdot 3 \cdot 2 \cdot 5 \quad \text{or} \quad 2^2 \cdot 3 \cdot 5$$

Fractions that represent the same portion of a whole are called **equivalent fractions.**

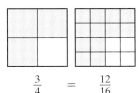

$$\frac{3}{4} \quad = \quad \frac{12}{16}$$

A fraction is in **simplest form** or **lowest terms** when the numerator and the denominator have no common factors other than 1.	The fraction $\frac{2}{3}$ is in simplest form.
To write a fraction in simplest form, write the prime factorizations of the numerator and the denominator and then divide both by all common factors.	Write in simplest form: $\dfrac{30}{36}$

$$\frac{30}{36} = \frac{2 \cdot 3 \cdot 5}{2 \cdot 2 \cdot 3 \cdot 3} = \frac{2}{2} \cdot \frac{3}{3} \cdot \frac{5}{2 \cdot 3} = 1 \cdot 1 \cdot \frac{5}{6} = \frac{5}{6}$$

$$\text{or} \quad \frac{30}{36} = \frac{\cancel{2} \cdot \cancel{3} \cdot 5}{\cancel{2} \cdot 2 \cdot \cancel{3} \cdot 3} = \frac{5}{6}$$

Two fractions are equivalent if

Method 1: They simplify to the same fraction.

Method 2: Their cross products are equal.

$$\begin{array}{ccc} 24 \cdot 7 & & 8 \cdot 21 \\ = 168 & \dfrac{7}{8} = \dfrac{21}{24} & = 168 \end{array}$$

Since $168 = 168, \dfrac{7}{8} = \dfrac{21}{24}.$

Determine whether $\dfrac{7}{8}$ and $\dfrac{21}{24}$ are equivalent.

$\dfrac{7}{8}$ is in simplest form.

$$\frac{21}{24} = \frac{\cancel{3} \cdot 7}{\cancel{3} \cdot 8} = \frac{1 \cdot 7}{1 \cdot 8} = \frac{7}{8}$$

Since both simplify to $\dfrac{7}{8}$, then $\dfrac{7}{8} = \dfrac{21}{24}.$

Section 3.3   Multiplying and Dividing Fractions

To multiply two fractions, multiply the numerators and multiply the denominators.	Multiply.

$$\frac{2}{3} \cdot \frac{5}{7} = \frac{2 \cdot 5}{3 \cdot 7} = \frac{10}{21}$$

$$\frac{3}{4} \cdot \frac{1}{6} = \frac{3 \cdot 1}{4 \cdot 6} = \frac{\cancel{3} \cdot 1}{4 \cdot \cancel{3} \cdot 2} = \frac{1}{8}$$

To find the **reciprocal** of a fraction, interchange its numerator and denominator.	The reciprocal of $\dfrac{3}{5}$ is $\dfrac{5}{3}.$
To divide two fractions, multiply the first fraction by the reciprocal of the second fraction.	Divide.

$$-\frac{3}{10} \div \frac{7}{9} = -\frac{3}{10} \cdot \frac{9}{7} = -\frac{3 \cdot 9}{10 \cdot 7} = -\frac{27}{70}$$

Definitions and Concepts	Examples

Section 3.4 Adding and Subtracting Like Fractions, Least Common Denominator, and Equivalent Fractions

Fractions that have the same denominator are called **like fractions**.

To add or subtract like fractions, combine the numerators and place the sum or difference over the common denominator.

$-\dfrac{1}{3}$ and $\dfrac{2}{3}$; $\dfrac{5}{7}$ and $\dfrac{6}{7}$

$\dfrac{2}{7} + \dfrac{3}{7} = \dfrac{5}{7}$ ← Add the numerators.
  ← Keep the common denominator.

$\dfrac{7}{8} - \dfrac{4}{8} = \dfrac{3}{8}$ ← Subtract the numerators.
  ← Keep the common denominator.

The **least common denominator (LCD)** of a list of fractions is the smallest positive number divisible by all the denominators in the list.

The LCD of $\dfrac{1}{2}$ and $\dfrac{5}{6}$ is 6 because 6 is the smallest positive number that is divisible by both 2 and 6.

Method 1 for Finding the LCD of a List of Fractions Using Multiples

Step 1: Write the multiples of the largest denominator (starting with the number itself) until a multiple common to all denominators in the list is found.

Step 2: The multiple found in Step 1 is the LCD.

Find the LCD of $\dfrac{1}{4}$ and $\dfrac{5}{6}$ using Method 1.

$6 \cdot 1 = 6$ Not a multiple of 4
$6 \cdot 2 = 12$ A multiple of 4

The LCD is 12.

Method 2 for Finding the LCD of a List of a Fractions Using Prime Factorization

Step 1: Write the prime factorization of each denominator.

Step 2: For each different prime factor in Step 1, circle the greatest number of times that factor occurs in any one factorization.

Step 3: The LCD is the product of the circled factors.

Find the LCD of $\dfrac{5}{6}$ and $\dfrac{11}{20}$ using Method 2.

$6 = 2 \cdot ③$
$20 = ②\cdot②\cdot⑤$

The LCD is

$2 \cdot 2 \cdot 3 \cdot 5 = 60$

Equivalent fractions represent the same portion of a whole.

Write an equivalent fraction with the indicated denominator.

$\dfrac{2}{8} = \dfrac{}{16}$

$\dfrac{2 \cdot 2}{8 \cdot 2} = \dfrac{4}{16}$

Section 3.5 Adding and Subtracting Unlike Fractions

To Add or Subtract Fractions with Unlike Denominators

Step 1: Find the LCD.

Step 2: Write each fraction as an equivalent fraction whose denominator is the LCD.

Step 3: Add or subtract the like fractions.

Step 4: Write the sum or difference in simplest form.

Add: $\dfrac{3}{20} + \dfrac{2}{5}$

Step 1: The LCD of the denominators 20 and 5 is 20.

Step 2: $\dfrac{3}{20} = \dfrac{3}{20}; \dfrac{2}{5} = \dfrac{2}{5} \cdot \dfrac{4}{4} = \dfrac{8}{20}$

Step 3: $\dfrac{3}{20} + \dfrac{2}{5} = \dfrac{3}{20} + \dfrac{8}{20} = \dfrac{11}{20}$

Step 4: $\dfrac{11}{20}$ is in simplest form.

Definitions and Concepts	Examples

Section 3.6   Complex Fractions, Order of Operations, and Mixed Numbers

A fraction whose numerator or denominator or both contain fractions is called a **complex fraction.**

Complex Fractions:

$$\frac{\dfrac{11}{4}}{\dfrac{7}{10}}, \quad \frac{\dfrac{1}{6} - 11}{\dfrac{4}{3}}$$

One method for simplifying complex fractions is to multiply the numerator and the denominator of the complex fraction by the LCD of all fractions in its numerator and its denominator.

$$\frac{\dfrac{1}{6} - 11}{\dfrac{4}{3}} = \frac{6\left(\dfrac{1}{6} - 11\right)}{6\left(\dfrac{4}{3}\right)} = \frac{6\left(\dfrac{1}{6}\right) - 6(11)}{6\left(\dfrac{4}{3}\right)}$$

$$= \frac{1 - 66}{8} = \frac{-65}{8} = -\frac{65}{8}$$

To Write a Mixed Number as an Improper Fraction

1. Multiply the denominator of the fraction by the whole number.

2. Add the numerator of the fraction to the product from Step 1.

3. Write the sum from Step 2 as the numerator of the improper fraction over the original denominator.

$$5\frac{2}{7} = \frac{7 \cdot 5 + 2}{7} = \frac{35 + 2}{7} = \frac{37}{7}$$

To Write an Improper Fraction as a Mixed Number or a Whole Number

1. Divide the denominator into the numerator.

2. The whole number part of the mixed number is the quotient. The fraction is the remainder over the original denominator.

$$\text{quotient}\,\frac{\text{remainder}}{\text{original denominator}}$$

$$\frac{17}{3} = 5\frac{2}{3}$$

$$\begin{array}{r} 5 \\ 3\overline{)17} \\ -15 \\ \hline 2 \end{array}$$

Section 3.7   Operations on Mixed Numbers

To multiply with mixed numbers or whole numbers, first write any mixed or whole numbers as improper fractions and then multiply as usual.

To divide with mixed numbers or whole numbers, first write any mixed or whole numbers as fractions and then divide as usual.

$$2\frac{1}{3} \cdot \frac{1}{9} = \frac{7}{3} \cdot \frac{1}{9} = \frac{7 \cdot 1}{3 \cdot 9} = \frac{7}{27}$$

$$2\frac{5}{8} \div 3\frac{7}{16} = \frac{21}{8} \div \frac{55}{16} = \frac{21}{8} \cdot \frac{16}{55} = \frac{21 \cdot 16}{8 \cdot 55}$$

$$= \frac{21 \cdot 2 \cdot \overset{1}{\cancel{8}}}{\cancel{8} \cdot 55} = \frac{42}{55}$$

To add or subtract with mixed numbers, add or subtract the fractions and then add or subtract the whole numbers.

Add: $2\dfrac{1}{2} + 5\dfrac{7}{8}$

$$2\frac{1}{2} = 2\frac{4}{8}$$

$$\begin{array}{r} + 5\frac{7}{8} = 5\frac{7}{8} \\ \hline 7\frac{11}{8} = 7 + 1\frac{3}{8} = 8\frac{3}{8} \end{array}$$

(3.1) *Determine whether each number is an improper fraction, a proper fraction, or a mixed number.*

1. $\dfrac{11}{23}$

2. $\dfrac{9}{8}$

3. $\dfrac{1}{2}$

4. $2\dfrac{1}{4}$

Write a fraction to represent the shaded area. If the fraction is improper, write the shaded area as a mixed number also. (For Exercises 5–8, do not simplify answers.)

5.

6.

7.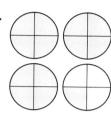

8.

9. A basketball player made 11 free throws out of 12 attempts during a game. What fraction of free throws did the player make?

10. A new car lot contains 23 blue cars out of a total of 131 cars.

    a. How many cars on the lot are not blue?

    b. What fraction of cars on the lot are not blue?

Simplify by dividing.

11. $\dfrac{3}{-3}$

12. $\dfrac{-20}{-20}$

13. $\dfrac{0}{-1}$

14. $\dfrac{4}{0}$

Graph each fraction on a number line.

15. $\dfrac{7}{9}$

16. $\dfrac{4}{7}$

17. $\dfrac{5}{4}$

18. $\dfrac{7}{5}$

(3.2) *Find the prime factorization of each number.*

19. 68

20. 90

21. 785

22. 255

Write each fraction in simplest form.

23. $\dfrac{12}{28}$

24. $\dfrac{15}{27}$

25. $-\dfrac{25}{75}$

26. $-\dfrac{36}{72}$

27. $\dfrac{29}{32}$ **28.** $\dfrac{18}{23}$ **29.** $\dfrac{48}{6}$ **30.** $\dfrac{54}{9}$

31. There are 12 inches in a foot. What fractional part of a foot does 8 inches represent?

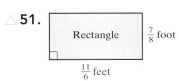

32. Six out of 15 cars are white. What fraction of the cars are *not* white?

Determine whether each two fractions are equivalent.

33. $\dfrac{10}{34}$ and $\dfrac{4}{14}$

34. $\dfrac{30}{50}$ and $\dfrac{9}{15}$

(3.3) *Multiply. Write each answer in simplest form.*

35. $-\dfrac{3}{5} \cdot \dfrac{1}{2}$ **36.** $\dfrac{6}{7} \cdot \dfrac{5}{12}$ **37.** $-\dfrac{24}{5} \cdot \left(-\dfrac{15}{8}\right)$ **38.** $\dfrac{39}{3} \cdot \dfrac{7}{13} \cdot \dfrac{5}{21}$

39. $\left(-\dfrac{1}{3}\right)^3$ **40.** $\left(-\dfrac{5}{12}\right)^2$ **41.** Evaluate xy if $x = \dfrac{2}{3}$ and $y = \dfrac{1}{5}$. **42.** Evaluate ab if $a = -7$ and $b = \dfrac{9}{10}$.

Find the reciprocal of each number.

43. 7

44. $\dfrac{14}{23}$

Divide. Write each answer in simplest form.

45. $-\dfrac{3}{4} \div \dfrac{3}{8}$ **46.** $\dfrac{21}{4} \div \dfrac{7}{5}$ **47.** $\dfrac{5}{3} \div (-2)$

48. $-\dfrac{9}{2} \div -\dfrac{1}{3}$ **49.** Evaluate $x \div y$ if $x = \dfrac{9}{7}$ and $y = \dfrac{3}{4}$. **50.** Evaluate $a \div b$ if $a = -5$ and $b = \dfrac{2}{3}$.

Find the area of each figure.

△ **51.**
Rectangle $\dfrac{7}{8}$ foot
$\dfrac{11}{6}$ feet

△ **52.**
Square $\dfrac{2}{3}$ meter

(3.4) *Add or subtract as indicated.*

53. $\dfrac{7}{11} + \dfrac{3}{11}$

54. $\dfrac{4}{9} + \dfrac{2}{9}$

55. $\dfrac{1}{12} - \dfrac{5}{12}$

56. $\dfrac{11}{15} + \dfrac{1}{15}$

57. $\dfrac{4}{21} - \dfrac{3}{21}$

58. $\dfrac{4}{15} - \dfrac{3}{15} - \dfrac{2}{15}$

Find the LCD of each list of fractions.

59. $\dfrac{2}{3}, \dfrac{5}{7}$

60. $\dfrac{3}{4}, \dfrac{3}{8}, \dfrac{7}{12}$

Write each fraction as an equivalent fraction with the given denominator.

61. $\dfrac{2}{3} = \dfrac{?}{30}$

62. $\dfrac{5}{8} = \dfrac{?}{56}$

63. $\dfrac{7}{6} = \dfrac{?}{42}$

64. $\dfrac{9}{4} = \dfrac{?}{20}$

65. $\dfrac{4}{5} = \dfrac{?}{50}$

66. $\dfrac{5}{9} = \dfrac{?}{18}$

Solve.

67. One evening Mark Alorenzo did $\dfrac{3}{8}$ of his homework before supper, another $\dfrac{2}{8}$ of it while his children did their homework, and $\dfrac{1}{8}$ after his children went to bed. What part of his homework did he do that evening?

△ **68.** The Simpsons will be fencing in their land, which is in the shape of a rectangle. In order to do this, they need to find its perimeter. Find the perimeter of their land.

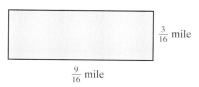

$\frac{3}{16}$ mile

$\frac{9}{16}$ mile

(3.5) *Add or subtract as indicated.*

69. $\dfrac{7}{18} + \dfrac{2}{9}$

70. $\dfrac{4}{13} - \dfrac{1}{26}$

71. $-\dfrac{1}{3} + \dfrac{1}{4}$

72. $-\dfrac{2}{3} + \dfrac{1}{4}$

73. $-\dfrac{9}{14} - \dfrac{3}{7}$

74. $\dfrac{5}{12} - \dfrac{2}{9}$

75. $\dfrac{4}{25} + \dfrac{23}{75} + \dfrac{7}{50}$

76. $\dfrac{2}{3} - \dfrac{2}{9} - \dfrac{1}{6}$

Find the perimeter of each figure.

△ **77.**

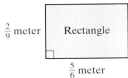

$\frac{2}{9}$ meter ☐ Rectangle

$\frac{5}{6}$ meter

△ **78.**

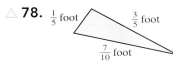

$\frac{1}{5}$ foot $\frac{3}{5}$ foot

$\frac{7}{10}$ foot

79. In a group of 100 blood donors, typically $\frac{9}{25}$ have type A Rh-positive blood and $\frac{3}{50}$ have type A Rh-negative blood. What fraction have type A blood?

80. Find the difference in length of two scarves if one scarf is $\frac{5}{12}$ of a yard long and the other is $\frac{2}{3}$ of a yard long.

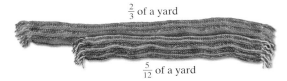

$\frac{2}{3}$ of a yard

$\frac{5}{12}$ of a yard

Insert < or > to form a true statement.

81. $\frac{5}{11}$ $\frac{6}{11}$

82. $\frac{4}{35}$ $\frac{3}{35}$

83. $-\frac{5}{14}$ $-\frac{16}{42}$

84. $-\frac{6}{35}$ $-\frac{17}{105}$

(3.6) *Simplify each complex fraction.*

85. $\dfrac{\frac{2}{5}}{\frac{7}{10}}$

86. $\dfrac{\frac{3}{7}}{\frac{11}{7}}$

87. $\dfrac{\frac{2}{5} - \frac{1}{2}}{\frac{3}{4} - \frac{7}{10}}$

88. $\dfrac{\frac{5}{6} - \frac{1}{4}}{-\frac{1}{12}}$

Evaluate each expression if $x = \frac{1}{2}$, $y = -\frac{2}{3}$, and $z = \frac{4}{5}$.

89. y^2

90. $x - z$

Write each improper fraction as a mixed number or a whole number.

91. $\frac{15}{4}$

92. $\frac{39}{13}$

93. $\frac{7}{7}$

94. $\frac{125}{4}$

Write each mixed number as an improper fraction.

95. $2\frac{1}{5}$

96. $3\frac{8}{9}$

Evaluate each expression. Use the order of operations to simplify.

97. $\dfrac{5}{13} \div \dfrac{1}{2} \cdot \dfrac{4}{5}$

98. $\dfrac{2}{27} - \left(\dfrac{1}{3}\right)^2$

99. $\dfrac{9}{10} \cdot \dfrac{1}{3} - \dfrac{2}{5} \cdot \dfrac{1}{11}$

100. $-\dfrac{2}{7} \cdot \left(\dfrac{1}{5} + \dfrac{3}{10}\right)$

(3.7) *Perform operations as indicated. Simplify your answers. Estimate where indicated.*

101. $31\dfrac{2}{7} + 14\dfrac{10}{21}$

102. $\begin{array}{r} 7\dfrac{3}{8} \\ 9\dfrac{5}{6} \\ + 3\dfrac{1}{12} \\ \hline \end{array}$

103. $\begin{array}{r} 9\dfrac{1}{7} \\ - 4\dfrac{3}{5} \\ \hline \end{array}$

104. $\begin{array}{r} 8\dfrac{1}{5} \\ - 5\dfrac{3}{11} \\ \hline \end{array}$

105. $1\dfrac{5}{8} \cdot 3\dfrac{1}{5}$
Exact:
Estimate:

106. $3\dfrac{6}{11} \cdot 1\dfrac{7}{13}$
Exact:
Estimate:

107. $6\dfrac{3}{4} \div 1\dfrac{2}{7}$

108. $5\dfrac{1}{2} \div 2\dfrac{1}{11}$

109. A truck traveled 341 miles on $15\dfrac{1}{2}$ gallons of gas. How many miles might we expect the truck to travel on 1 gallon of gas?

110. There are $7\dfrac{1}{3}$ grams of fat in each ounce of hamburger. How many grams of fat are in a 5-ounce hamburger patty?

Find the unknown measurements.

111.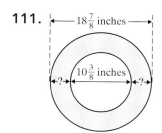

112.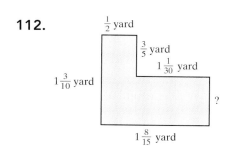

Perform the indicated operations.

113. $-12\dfrac{1}{7} + \left(-15\dfrac{3}{14}\right)$

114. $23\dfrac{7}{8} - 24\dfrac{7}{10}$

115. $-3\dfrac{1}{5} \div \left(-2\dfrac{7}{10}\right)$

116. $-2\dfrac{1}{4} \cdot 1\dfrac{3}{4}$

Mixed Review

Perform indicated operations. Write each answer in simplest form. Estimate where indicated.

117. $\dfrac{7}{8} \cdot \dfrac{2}{3}$

118. $\dfrac{6}{15} \cdot \dfrac{5}{8}$

119. $\dfrac{18}{5} \div \dfrac{2}{5}$

120. $\dfrac{9}{2} \div \dfrac{1}{3}$

121. $\dfrac{5}{12} - \dfrac{3}{12}$

122. $\dfrac{3}{10} - \dfrac{1}{10}$

123. $\dfrac{2}{3} + \dfrac{1}{4}$

124. $\dfrac{5}{11} + \dfrac{2}{55}$

125. $4\dfrac{1}{6} \cdot 2\dfrac{2}{5}$

Exact:

Estimate:

126. $5\dfrac{2}{3} \cdot 2\dfrac{1}{4}$

Exact:

Estimate:

127. $\dfrac{7}{2} \div 1\dfrac{1}{2}$

128. $1\dfrac{3}{5} \div \dfrac{1}{4}$

129. $\begin{array}{r} 7\frac{3}{4} \\ +5\frac{2}{3} \\ \hline \end{array}$

130. $\begin{array}{r} 2\frac{7}{8} \\ +9\frac{1}{2} \\ \hline \end{array}$

131. $\begin{array}{r} 12\frac{3}{5} \\ -9\frac{1}{7} \\ \hline \end{array}$

132. $\begin{array}{r} 32\frac{10}{21} \\ -24\frac{3}{7} \\ \hline \end{array}$

Evaluate each expression. Use the order of operations to simplify.

133. $\dfrac{2}{5} + \left(\dfrac{2}{5}\right)^2 - \dfrac{3}{25}$

134. $\left(\dfrac{5}{6} - \dfrac{3}{4}\right)^2$

135. $-\dfrac{3}{8} \cdot \left(\dfrac{2}{3} - \dfrac{4}{9}\right)$

Solve.

136. Two packages to be mailed weigh $3\dfrac{3}{4}$ pounds and $2\dfrac{3}{5}$ pounds. Find their combined weight.

137. An area of Mississippi received $23\dfrac{1}{2}$ inches of rain in $30\dfrac{1}{2}$ hours. How many inches per 1 hour is this?

138. A ribbon $5\dfrac{1}{2}$ yards long is cut from a reel of ribbon with 50 yards on it. Find the length of the piece remaining on the reel.

△ **139.** A slab of natural granite is purchased and a rectangle with length $7\dfrac{4}{11}$ feet and width $5\dfrac{1}{2}$ feet is cut from it. Find the area of the rectangle.

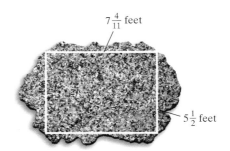

$7\frac{4}{11}$ feet

$5\frac{1}{2}$ feet

Write a fraction to represent the shaded area.

1.

Answers

Write the mixed number as an improper fraction.

2. $7\dfrac{2}{3}$

1. _____

2. _____

Write the improper fraction as a mixed number.

3. _____

3. $\dfrac{75}{4}$

4. _____

5. _____

Write each fraction in simplest form.

6. _____

4. $\dfrac{24}{210}$ **5.** $-\dfrac{42}{70}$

7. _____

8. _____

Determine whether these fractions are equivalent.

9. _____

6. $\dfrac{5}{7}$ and $\dfrac{8}{11}$ **7.** $\dfrac{6}{27}$ and $\dfrac{14}{63}$

10. _____

11. _____

Find the prime factorization of each number.

12. _____

8. 84 **9.** 495 **10.** Find the LCM of 8, 9, and 12.

13. _____

14. _____

Perform each indicated operation. Simplify your answers.

15. _____

16. _____

11. $\dfrac{7}{9} + \dfrac{1}{9}$ **12.** $-\dfrac{8}{15} - \dfrac{2}{15}$ **13.** $\dfrac{4}{4} \div \dfrac{3}{4}$ **14.** $-\dfrac{4}{3} \cdot \dfrac{4}{4}$

17. _____

18. _____

15. $\dfrac{1}{6} + \dfrac{3}{14}$ **16.** $\dfrac{7}{8} - \dfrac{1}{3}$ **17.** $-\dfrac{2}{3} \cdot -\dfrac{8}{15}$ **18.** $8 \div \dfrac{1}{2}$

19. _____

20. _____

19. $\dfrac{6}{21} - \dfrac{1}{7}$ **20.** $\dfrac{16}{25} - \dfrac{1}{2}$ **21.** $\dfrac{3}{8} \cdot \dfrac{16}{6} \cdot \dfrac{4}{11}$ **22.** $5\dfrac{1}{4} \div \dfrac{7}{12}$

21. _____

22. _____

23. _____

24. _____

25. _____

26. _____

27. _____

28. _____

29. _____

30. _____

31. _____

32. _____

33. _____

34. _____

35. _____

36. _____

37. _____

38. _____

39. _____

40. _____

23. $\dfrac{11}{12} - \dfrac{3}{8} + \dfrac{5}{24}$

24.
$$3\dfrac{7}{8}$$
$$7\dfrac{2}{5}$$
$$+2\dfrac{3}{4}$$

25.
$$19$$
$$-2\dfrac{3}{11}$$

26. $-\dfrac{16}{3} \div -\dfrac{3}{12}$

27. $3\dfrac{1}{3} \cdot 6\dfrac{3}{4}$

28. $-\dfrac{2}{7} \cdot \left(6 - \dfrac{1}{6}\right)$

29. $\dfrac{1}{2} \div \dfrac{2}{3} \cdot \dfrac{3}{4}$

30. $\left(-\dfrac{3}{4}\right)^2 \div \left(\dfrac{2}{3} + \dfrac{5}{6}\right)$

31. Find the average of $\dfrac{5}{6}, \dfrac{4}{3},$ and $\dfrac{7}{12}.$

Simplify complex fraction.

32. $\dfrac{5 + \dfrac{3}{7}}{2 - \dfrac{1}{2}}$

Evaluate the expression for the given replacement value.

33. $-5x; x = -\dfrac{1}{2}$

34. $x \div y; x = \dfrac{1}{2}, y = 3\dfrac{7}{8}$

Solve.

35. A carpenter cuts a piece $2\dfrac{3}{4}$ feet long from a cedar plank that is $6\dfrac{1}{2}$ feet long. How long is the remaining piece?

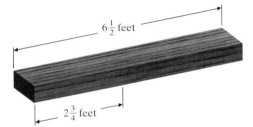
$6\frac{1}{2}$ feet

$2\frac{3}{4}$ feet

The circle graph below shows us how the average consumer spends money. For example, $\dfrac{7}{50}$ of your spending goes for food. Use this information for Exercises 36 through 38.

Consumer Spending

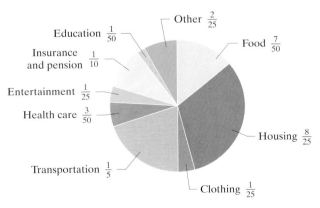

Other $\frac{2}{25}$

Education $\frac{1}{50}$

Insurance and pension $\frac{1}{10}$

Entertainment $\frac{1}{25}$

Health care $\frac{3}{50}$

Transportation $\frac{1}{5}$

Clothing $\frac{1}{25}$

Food $\frac{7}{50}$

Housing $\frac{8}{25}$

Source: U.S. Bureau of Labor Statistics; based on survey

36. What fraction of spending goes for housing and food combined?

37. What fraction of spending goes for education, transportation, and clothing?

38. Suppose your family spent $47,000 on the items in the graph. How much might we expect was spent on health care?

Find the perimeter and area of the figure.

△ **39.**

| Rectangle | $\frac{2}{3}$ foot |

1 foot

40. During a 258-mile trip, a car used $10\dfrac{3}{4}$ gallons of gas. How many miles would we expect the car to travel on 1 gallon of gas?

Write each number in words.

1. 126

2. 115

3. 27,034

4. 6573

5. Add: 23 + 136

6. Add: 587 + 44

7. Subtract: 543 − 29. Check by adding.

8. Subtract: 995 − 62. Check by adding.

9. Round 278,362 to the nearest thousand.

10. Round 1436 to the nearest ten.

11. A digital video disc (DVD) can hold about 4800 megabytes (MB) of information. How many megabytes can 12 DVDs hold?

12. On a trip across the country, Daniel Daunis travels 435 miles per day. How many total miles does he travel in 3 days?

13. Divide and check: 56,717 ÷ 8

14. Divide and check: 4558 ÷ 12

Write using exponential notation.

15. 7 · 7 · 7

16. 7 · 7

17. 3 · 3 · 3 · 3 · 17 · 17 · 17

18. 9 · 9 · 9 · 9 · 5 · 5

19. Evaluate $2(x - y)$ for $x = 8$ and $y = 4$.

20. Evaluate $8a + 3(b - 5)$ for $a = 5$ and $b = 9$.

21. The world's deepest cave is Krubera (or Voronja), in the country of Georgia, located by the Black Sea in Asia. It has been explored to a depth of 7188 feet below the surface of the Earth. Represent this position using an integer.

22. The temperature on a cold day in Minneapolis, MN, is 21°F below zero. Represent this temperature using an integer.

Answers

1. _____

2. _____

3. _____

4. _____

5. _____

6. _____

7. _____

8. _____

9. _____

10. _____

11. _____

12. _____

13. _____

14. _____

15. _____

16. _____

17. _____

18. _____

19. _____

20. _____

21. _____

22. _____

23. Add using a number line: $-7 + 3$

24. Add using a number line: $-3 + 8$

25. Simplify: $7 - 8 - (-5) - 1$

26. Simplify: $6 + (-8) - (-9) + 3$

27. Evaluate: $(-5)^2$

28. Evaluate: -2^4

29. Simplify: $\dfrac{12 - 16}{-1 + 3}$

30. Simplify: $(20 - 5^2)^2$

31. Write the prime factorization of 45.

32. Write the prime factorization of 92.

Multiply.

33. $\dfrac{2}{3} \cdot \dfrac{5}{11}$

34. $\dfrac{1}{7} \cdot \dfrac{2}{5}$

35. $\dfrac{1}{4} \cdot \dfrac{1}{2}$

36. $\dfrac{3}{5} \cdot \dfrac{1}{5}$

37. Write a fraction to represent the shaded part of the figure.

38. Write the prime factorization of 156.

39. Write each as an improper fraction.

    a. $4\dfrac{2}{9}$ **b.** $1\dfrac{8}{11}$

40. Write $7\dfrac{4}{5}$ as an improper fraction.

41. Write in simplest form: $\dfrac{42}{66}$

42. Write in simplest form: $\dfrac{70}{105}$

43. Multiply: $3\dfrac{1}{3} \cdot \dfrac{7}{8}$

44. Multiply: $\dfrac{2}{3} \cdot 4$

45. Divide and simplify: $\dfrac{5}{16} \div \dfrac{3}{4}$

46. Divide: $1\dfrac{1}{10} \div 5\dfrac{3}{5}$

23. _____

24. _____

25. _____

26. _____

27. _____

28. _____

29. _____

30. _____

31. _____

32. _____

33. _____

34. _____

35. _____

36. _____

37. _____

38. _____

39. a. _____

   b. _____

40. _____

41. _____

42. _____

43. _____

44. _____

45. _____

46. _____

Decimals

On Average, What Age Group Texts the Most?

The graph below shows the age group distribution for average daily texting. (Check your age group and see if the data are accurate based on your own experiences.) While we have practiced calculating averages before in this text, an average certainly does not usually simplify to a whole number. While fractions are useful, decimals are also an important system of numbers that can be used to show values between whole numbers. Data are easy to round when they are in the form of a decimal. In Section 4.2, Exercises 77 and 78, we study email traffic and how it is predicted to increase each year.

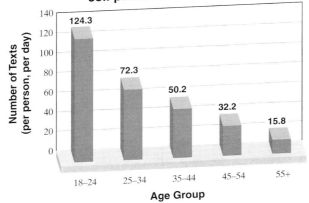

Average Number of Texts Sent/Received per Day
(based on adults who use
cell phone text messaging)

Source: Experian, March 2013

Decimal numbers represent parts of a whole, just like fractions. For example, one penny is 0.01 or $\frac{1}{100}$ of a dollar. In this chapter, we learn to perform arithmetic operations on decimals and to analyze the relationship between fractions and decimals. We also learn how decimals are used in the real world.

Objectives

A Know the Meaning of Place Value for a Decimal Number and Write Decimals in Words.

B Write Decimals in Standard Form.

C Write Decimals as Fractions.

D Compare Decimals.

E Round Decimals to Given Place Values.

Objective A Decimal Notation and Writing Decimals in Words

Like fractional notation, decimal notation is used to denote a part of a whole. Numbers written in decimal notation are called **decimal numbers,** or simply **decimals.** The decimal 17.758 has three parts.

In Section 1.2, we introduced place value for whole numbers. Place names and place values for the whole number part of a decimal number are exactly the same. Place names and place values for the decimal part are shown below.

Helpful Hint Notice that place values to the left of the decimal point end in "s." Place values to the right of the decimal point end in "ths."

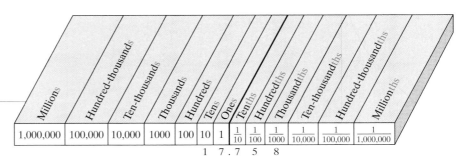

Notice that the value of each place is $\frac{1}{10}$ of the value of the place to its left. For example,

$$1 \cdot \frac{1}{10} = \frac{1}{10} \quad \text{and} \quad \frac{1}{10} \cdot \frac{1}{10} = \frac{1}{100}$$

ones   tenths   tenths   hundredths

The decimal number 17.758 means

1 ten + 7 ones + 7 tenths + 5 hundredths + 8 thousandths

or $1 \cdot 10 + 7 \cdot 1 + 7 \cdot \frac{1}{10} + 5 \cdot \frac{1}{100} + 8 \cdot \frac{1}{1000}$

or $10 + 7 + \frac{7}{10} + \frac{5}{100} + \frac{8}{1000}$

Writing (or Reading) a Decimal in Words

Step 1: Write the whole number part in words.

Step 2: Write "and" for the decimal point.

Step 3: Write the decimal part in words as though it were a whole number, followed by the place value of the last digit.

282

Example 1 Write each decimal in words.

a. 0.3   **b.** −50.82   **c.** 21.093

Solution:

a. Three tenths

b. Negative fifty and eighty-two hundredths

c. Twenty-one and ninety-three thousandths

▪ Work Practice 1

Practice 1

Write each decimal in words.
a. 0.08   **b.** −500.025
c. 0.0329

Example 2 Write the decimal in the following sentence in words: The Golden Jubilee Diamond is a 545.67-carat cut diamond. (*Source: The Guinness Book of Records*)

Solution: five hundred forty-five and sixty-seven hundredths

▪ Work Practice 2

Practice 2

Write the decimal 97.28 in words.

Example 3 Write the decimal in the following sentence in words: The oldest known fragments of the Earth's crust are Zircon crystals; they were discovered in Australia and are thought to be 4.276 billion years old. (*Source: The Guinness Book of Records*)

Solution: four and two hundred seventy-six thousandths

▪ Work Practice 3

Practice 3

Write the decimal 72.1085 in words.

Suppose that you are paying for a purchase of $368.42 at Circuit City by writing a check. Checks are usually written using the following format.

Elayn Martin-Gay

60–8124/7233
1000613331

1403

DATE *(Current date)*

PAY TO THE ORDER OF *Circuit City* | $ *368.42*

Three hundred sixty-eight and $\frac{42}{100}$ ——— DOLLARS

FIRST STATE BANK
OF FARTHINGTON
FARTHINGTON, IL 64422

MEMO

Elayn Martin-Gay

⑆621497260⑆ 1000613331⑈ 1403

Answers

1. a. eight hundredths   **b.** negative five hundred and twenty-five thousandths **c.** three hundred twenty-nine ten-thousandths   **2.** ninety-seven and twenty-eight hundredths   **3.** seventy-two and one thousand eighty-five ten-thousandths

Practice 4

Fill in the check to CLECO (Central Louisiana Electric Company) to pay for your monthly electric bill of $207.40.

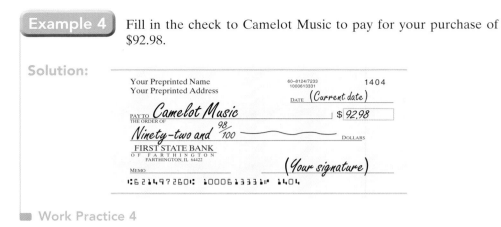

Example 4 Fill in the check to Camelot Music to pay for your purchase of $92.98.

Solution:

Work Practice 4

Objective B Writing Decimals in Standard Form

A decimal written in words can be written in standard form by reversing the procedure in Objective A.

Examples Write each decimal in standard form.

5. Forty-eight and twenty-six hundredths is

48.26
hundredths place

6. Six and ninety-five thousandths is

6.095
thousandths place

Work Practice 5–6

Practice 5–6

Write each decimal in standard form.

5. Three hundred and ninety-six hundredths

6. Thirty-nine and forty-two thousandths

Helpful Hint

When converting a decimal from words to decimal notation, make sure the last digit is in the correct place by inserting 0s if necessary. For example,

Two and thirty-eight thousandths is 2.038
thousandths place

Objective C Writing Decimals as Fractions

Once you master reading and writing decimals, writing a decimal as a fraction follows naturally.

Decimal	In Words	Fraction
0.7	seven tenths	$\dfrac{7}{10}$
0.51	fifty-one hundredths	$\dfrac{51}{100}$
0.009	nine thousandths	$\dfrac{9}{1000}$
0.05	five hundredths	$\dfrac{5}{100} = \dfrac{1}{20}$

Answers

4.

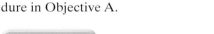

5. 300.96   **6.** 39.042

Notice that the number of decimal places in a decimal number is the same as the number of zeros in the denominator of the equivalent fraction. We can use this fact to write decimals as fractions.

$$0.31 = \frac{31}{100} \qquad 0.007 = \frac{7}{1000}$$

2 decimal places   2 zeros   3 decimal places   3 zeros

 **Example 7**   Write 0.43 as a fraction.

Solution: $0.43 = \frac{43}{100}$

2 decimal places   2 zeros

■ Work Practice 7

Practice 7

Write 0.037 as a fraction.

Example 8   Write 5.7 as a mixed number.

Solution: $5.7 = 5\frac{7}{10}$

1 decimal place   1 zero

■ Work Practice 8

Practice 8

Write 14.97 as a mixed number.

Examples   Write each decimal as a fraction or a mixed number. Write your answer in simplest form.

9. $0.125 = \frac{125}{1000} = \frac{\cancel{125}}{8 \cdot \cancel{125}} = \frac{1}{8}$

10. $23.5 = 23\frac{5}{10} = 23\frac{\cancel{5}}{2 \cdot \cancel{5}} = 23\frac{1}{2 \cdot 1} = 23\frac{1}{2}$

11. $-105.083 = -105\frac{83}{1000}$

■ Work Practice 9–11

Practice 9–11

Write each decimal as a fraction or mixed number. Write your answer in simplest form.
 9. 0.12
10. 57.8
11. −209.086

Later in the chapter, we write fractions as decimals. If you study Examples 7–11, you already know how to write fractions with denominators of 10, 100, 1000, and so on, as decimals.

Objective D Comparing Decimals

One way to compare positive decimals is by comparing digits in corresponding places. To see why this works, let's compare 0.5 or $\frac{5}{10}$ and 0.8 or $\frac{8}{10}$. We know

$$\frac{5}{10} < \frac{8}{10} \quad \text{since } 5 < 8, \text{ so}$$

$$0.5 < 0.8 \quad \text{since } 5 < 8$$

This leads to the following.

Answers

7. $\frac{37}{1000}$ **8.** $14\frac{97}{100}$ **9.** $\frac{3}{25}$

10. $57\frac{4}{5}$ **11.** $-209\frac{43}{500}$

Comparing Two Positive Decimals

Compare digits in the same places from left to right. When two digits are not equal, the number with the larger digit is the larger decimal. If necessary, insert 0s after the last digit to the right of the decimal point to continue comparing.

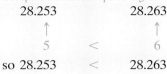

Compare hundredths place digits

28.253   28.263

↑   ↑

5   <   6

so 28.253   <   28.263

Helpful Hint

For any decimal, writing 0s after the last digit to the right of the decimal point does not change the value of the number.

7.6 = 7.60 = 7.600, and so on

When a whole number is written as a decimal, the decimal point is placed to the right of the ones digit.

25 = 25.0 = 25.00, and so on

Practice 12

Insert $<$, $>$, or $=$ to form a true statement.

13.208   13.28

Example 12 Insert $<$, $>$, or $=$ to form a true statement.

0.378   0.368

Solution:

0.3 78   0.3 68   The tenths places are the same.

0.3 7 8   0.3 6 8   The hundredths places are different.

Since 7 > 6, then 0.378 > 0.368.

Work Practice 12

Practice 13

Insert $<$, $>$, or $=$ to form a true statement.

0.12   0.086

Example 13 Insert $<$, $>$, or $=$ to form a true statement.

0.052   0.236

Solution: 0. 0 52 < 0. 2 36   0 is smaller than 2 in the tenths place.

↑   ↑

Work Practice 13

We can also use a number line to compare decimals. This is especially helpful when comparing negative decimals. Remember, the number whose graph is to the left is smaller, and the number whose graph is to the right is larger.

-1.7 -1.2   0.5 0.8

-2   -1   0   1   2

-1.7 < -1.2   0.5 < 0.8

12. $<$ **13.** $>$

Helpful Hint

If you have trouble comparing two negative decimals, try the following: Compare their absolute values. Then to correctly compare the negative decimals, reverse the direction of the inequality symbol.

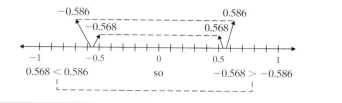

$$0.568 < 0.586 \qquad \text{so} \qquad -0.568 > -0.586$$

Example 14   Insert $<$, $>$, or $=$ to form a true statement.

    -0.0101 -0.00109

Solution: Since $0.0101 > 0.00109$, then $-0.0101 \ < \ -0.00109$.

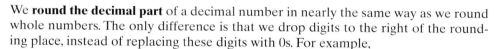

■ Work Practice 14

Practice 14

Insert $<$, $>$, or $=$ to form a true statement.

    -0.029 -0.0209

Objective E   Rounding Decimals ▶

We **round the decimal part** of a decimal number in nearly the same way as we round whole numbers. The only difference is that we drop digits to the right of the rounding place, instead of replacing these digits with 0s. For example,

    36.954   rounded to the nearest hundredth is   36.95

Rounding Decimals to a Place Value to the Right of the Decimal Point

Step 1: Locate the digit to the right of the given place value.

Step 2: If this digit is 5 or greater, add 1 to the digit in the given place value and drop all digits to its right. If this digit is less than 5, drop all digits to the right of the given place.

Example 15   Round 736.2359 to the nearest tenth.

Solution:

Step 1: We locate the digit to the right of the tenths place.

        tenths place

    736.2 3 59

        digit to the right

Step 2: Since the digit to the right is less than 5, we drop it and all digits to its right.

Thus, 736.2359 rounded to the nearest tenth is 736.2.

■ Work Practice 15

The same steps for rounding can be used when the decimal is negative.

Practice 15

Round 123.7815 to the nearest thousandth.

Answers

14. $<$ **15.** 123.782

Practice 16

Round −0.072 to the nearest hundredth.

Example 16   Round −0.027 to the nearest hundredth.

Solution:

Step 1: Locate the digit to the right of the hundredths place.

$$-0.02\ 7$$

Step 2: Since the digit to the right is 5 or greater, we add 1 to the hundredths digit and drop all digits to its right.

Thus, −0.027 is −0.03 rounded to the nearest hundredth.

Work Practice 16

The following number line illustrates the rounding of negative decimals.

$$-0.03 \quad -0.027 \quad -0.025 \qquad -0.02$$

In Section 4.3, we will introduce a formula for the distance around a circle. The distance around a circle is given the special name **circumference.**

The symbol π is the Greek letter pi, pronounced "pie." We use π to denote the following constant:

$$\pi = \frac{\text{circumference of a circle}}{\text{diameter of a circle}}$$

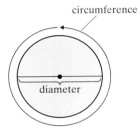

The value π is an **irrational number.** This means if we try to write it as a decimal, it neither ends nor repeats in a pattern.

Practice 17

$\pi \approx 3.14159265$. Round π to the nearest ten-thousandth.

Example 17 $\pi \approx 3.14159265$. Round π to the nearest hundredth.

Solution:

hundredths place ⎯⎯⎯⎯⎯⎯⎯ 1 is less than 5.

$$3.14159265$$

Delete these digits.

Thus, 3.14159265 rounded to the nearest hundredth is 3.14. In other words, $\pi \approx 3.14$.

Work Practice 17

Rounding often occurs with money amounts. Since there are 100 cents in a dollar, each cent is $\frac{1}{100}$ of a dollar. This means that if we want to round to the nearest cent, we round to the nearest hundredth of a dollar.

Answers
16. −0.07   **17.** $\pi \approx 3.1416$

✓ Concept Check Answer
c

✓ Concept Check   1756.0894 rounded to the nearest *ten* is

a. 1756.1   **b.** 1760.0894   **c.** 1760   **d.** 1750

Example 18 Determining State Taxable Income

A high school teacher's taxable income is $41,567.72. The tax tables in the teacher's state use amounts rounded to the nearest dollar. Round the teacher's income to the nearest whole dollar.

Solution: Rounding to the nearest whole dollar means rounding to the ones place.

ones place ———┐   ┌——— 7 is greater than 5.

$41,567.72

Add 1. ———┘   └——— Delete these digits.

Thus, the teacher's income rounded to the nearest dollar is $41,568.

■ Work Practice 18

Practice 18

Water bills in Mexia are always rounded to the nearest dollar. Round a water bill of $24.62 to the nearest dollar.

Answer

18. $25

Vocabulary, Readiness & Video Check

Use the choices below to fill in each blank.

words	decimals	tenths	after
tens	circumference	and	standard form

1. The number "twenty and eight hundredths" is written in _____ and "20.08" is written in _____.

2. Another name for the distance around a circle is its _____.

3. Like fractions, _____ are used to denote part of a whole.

4. When writing a decimal number in words, the decimal point is written as "_____".

5. The place value _____ is to the right of the decimal point while _____ is to the left of the decimal point.

6. The decimal point in a whole number is _____ the last digit.

Determine the place value for the digit 7 in each number.

7. 70   **8.** 700   **9.** 0.7   **10.** 0.07

Martin-Gay Interactive Videos   *Watch the section lecture video and answer the following questions.*

See Video 4.1

Objective A **11.** In ▣ Example 1, how is the decimal point written? ⊙

Objective B **12.** Why is 9.8 not the correct answer to ▣ Example 3? What is the correct answer? ⊙

Objective C **13.** From ▣ Example 5, why does reading a decimal number correctly help you write it as an equivalent fraction? ⊙

Objective D **14.** In ▣ Example 7, we compare place value by place value in which direction? ⊙

Objective E **15.** ▣ Example 8 is being rounded to the nearest tenth, so why is the digit 7, which is not in the tenths place, looked at? ⊙

4.1 Exercise Set MyMathLab®

Objective A *Write each decimal number in words. See Examples 1 through 3.*

1. 6.52

2. 7.59

3. 16.23

4. 47.65

5. −0.205

6. −0.495

7. 167.009

8. 233.056

9. 3000.04

10. 5000.02

11. 105.6

12. 410.3

13. The Akashi Kaikyo Bridge, between Kobe and Awaji-Shima, Japan, is approximately 2.43 miles long.

14. The English Channel Tunnel is a 31.04-mile-long undersea rail tunnel connecting England and France. (*Source: Railway Directory & Year Book*)

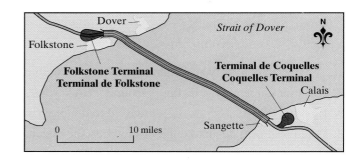

Fill in each check for the described purchase. See Example 4.

15. Your monthly car loan of $321.42 to R.W. Financial.

16. Your part of the monthly apartment rent, which is $213.70. You pay this to Amanda Dupre.

17. Your bill of $91.68 to Verizon wireless.

18. Your grocery bill of $387.49 at Kroger.

Objective B *Write each decimal number in standard form. See Examples 5 and 6.*

19. Six and five tenths

20. Three and nine tenths

21. Nine and eight hundredths

22. Twelve and six hundredths

23. Negative seven hundred five and six hundred twenty-five thousandths

24. Negative eight hundred four and three hundred ninety-nine thousandths

25. Forty-six ten-thousandths

26. Eighty-three ten-thousandths

Objective C *Write each decimal as a fraction or a mixed number. Write your answer in simplest form. See Examples 7 through 11.*

27. 0.3   **28.** 0.9   **29.** 0.27   **30.** 0.39

31. 0.4   **32.** 0.8   **33.** 5.4   **34.** 6.8

35. −0.058   **36.** −0.024   **37.** 7.008   **38.** 9.005

39. 15.802   **40.** 11.406   **41.** 0.3005   **42.** 0.2006

Objectives A B C **Mixed Practice** *Fill in the chart. The first row is completed for you. See Examples 1 through 11.*

	Decimal Number in Standard Form	In Words	Fraction
	0.37	thirty-seven hundredths	$\dfrac{37}{100}$
43.		eight tenths	
44.		five tenths	
45.	0.077		
46.	0.019		

Objective D *Insert* $<, >,$ *or* $=$ *between each pair of numbers to form a true statement. See Examples 12 through 14.*

47. 0.15   0.16   **48.** 0.12   0.15   **49.** −0.57   −0.54   **50.** −0.59   −0.52

51. 0.098   0.1   **52.** 0.0756   0.2   **53.** 0.54900   0.549   **54.** 0.98400   0.984

55. 167.908   167.980   **56.** 519.3405   519.3054   **57.** −1.062   −1.07   **58.** −18.1   −18.01

59. −7.052   7.0052   **60.** 0.01   −0.1   **61.** −0.023   −0.024   **62.** −0.562   −0.652

Objective E *Round each decimal to the given place value. See Examples 15 through 18.*

63. 0.57, nearest tenth   **64.** 0.68, nearest tenth   **65.** 98,207.23, nearest ten

66. 68,935.543, nearest ten   **67.** −0.234, nearest hundredth   **68.** −0.892, nearest hundredth

69. 0.5942, nearest thousandth   **70.** 63.4523, nearest thousandth

Recall that the number π, written as a decimal, neither ends nor repeats in a pattern. Given that π ≈ 3.14159265, round π to the given place values below. (We study π further in Section 4.3.) See Example 17.

71. tenth   **72.** one

73. thousandth   **74.** hundred-thousandth

Round each monetary amount to the nearest cent or dollar as indicated. See Example 18.

75. $26.95, to the nearest dollar   **76.** $14,769.52, to the nearest dollar

77. $0.1992, to the nearest cent   **78.** $0.7633, to the nearest cent

Round each number to the given place value. See Example 18.

79. At the time of this writing, the Apple MacBook Air is the thinnest Mac in production. At its thickest point, it measures 0.68 in. Round this number to the nearest tenth. (*Source:* Apple, Inc.)

80. A large tropical cockroach of the family Dictyoptera is the fastest-moving insect. This insect was clocked at a speed of 3.36 miles per hour. Round this number to the nearest tenth. (*Source:* University of California, Berkeley)

13-inch MacBook Air

0.68 in.

81. Missy Franklin of the United States won the gold medal for the 200 m backstroke in the 2012 London Summer Olympics with a record time of 2.0677 minutes. Round this time to the nearest hundredth of a minute.

82. The population density of the state of Utah is 34.745 people per square mile. Round this population density to the nearest tenth. (*Source:* U.S. Census Bureau)

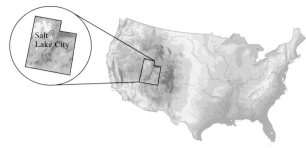

83. A used biology textbook is priced at $67.89. Round this price to the nearest dollar.

84. A used office desk is advertised at $19.95 by Drawley's Office Furniture. Round this price to the nearest dollar.

85. Venus makes a complete orbit around the Sun every 224.695 days. Round this figure to the nearest whole day. (*Source:* National Space Science Data Center)

86. The length of a day on Mars, a full rotation about its axis, is 24.6229 hours. Round this figure to the nearest thousandth. (*Source:* National Space Science Data Center)

Review

Perform each indicated operation. See Sections 1.3 and 1.4.

87. $3452 + 2314$

88. $8945 + 4536$

89. $82 - 47$

90. $4002 - 3897$

Concept Extensions

Solve. See the Concept Check in this section.

91. 2849.1738 rounded to the nearest hundred is
 a. 2849.17 **b.** 2800 **c.** 2850 **d.** 2849.174

92. 146.059 rounded to the nearest ten is
 a. 146.0 **b.** 146.1 **c.** 140 **d.** 150

93. 2849.1738 rounded to the nearest hundredth is
 a. 2849.17 **b.** 2800 **c.** 2850 **d.** 2849.18

94. 146.059 rounded to the nearest tenth is
 a. 146.0 **b.** 146.1 **c.** 140 **d.** 150

Solve.

95. In your own words, describe how to write a decimal as a fraction or a mixed number.

96. Explain how to identify the value of the 9 in the decimal 486.3297.

97. Write $7\dfrac{12}{100}$ as a decimal.

98. Write $17\dfrac{268}{1000}$ as a decimal.

99. Write 0.00026849577 as a fraction.

100. Write 0.00026849577 in words.

101. Write a 5-digit number that rounds to 1.7.

102. Write a 4-digit number that rounds to 26.3.

103. Write a decimal number that is greater than 8 but less than 9.

104. Write a decimal number that is greater than 48.1 but less than 48.2.

105. Which number(s) rounds to 0.26?
0.26559   0.26499   0.25786   0.25186

106. Which number(s) rounds to 0.06?
0.0612   0.066   0.0586   0.0506

For Exercises 107 and 108, write the numbers from smallest to largest.

107. 0.9
0.1038
0.10299
0.1037

108. 0.01
0.0839
0.09
0.1

109. The all-time top six movies (those that have earned the most money in the United States) along with the approximate amount of money they have earned are listed in the table. Estimate the total amount of money that these movies have earned by first rounding each earning to the nearest hundred-million. (*Source:* The Internet Movie Database)

All-Time Top American Movies	
Movie	**Gross Domestic Earnings**
Avatar (2009)	$760.5 million
Titanic (1997)	$658.7 million
The Avengers (2012)	$623.3 million
The Dark Knight (2008)	$533.3 million
Star Wars: The Phantom Menace (1999)	$474.5 million
Star Wars (1977)	$460.9 million

110. In 2013, there were 1328.9 million singles downloaded at an average price of $1.20 each. Find an estimate of the total revenue from downloaded singles by answering parts **a**–**c**. (*Source:* Recording Industry Association of America)

a. Round 1328.9 million to the nearest ten million.

b. Multiply the rounded value in part **a** by 12.

c. Move the decimal point in the product from part **b** one place to the left. This number is the total revenue in million dollars.

Adding and Subtracting Decimals

Objective A Adding Decimals

Adding decimals is similar to adding whole numbers. We add digits in corresponding place values from right to left, carrying if necessary. To make sure that digits in corresponding place values are added, we line up the decimal points vertically.

As we shall see later in this section, subtracting decimals is similar to subtracting whole numbers also.

Objectives

A Add Decimals.

B Subtract Decimals.

C Estimate when Adding or Subtracting Decimals.

D Evaluate Expressions with Decimal Replacement Values.

E Solve Problems That Involve Adding or Subtracting Decimals.

Adding (or Subtracting) Decimals

Step 1: Write the decimals so that the decimal points line up vertically.

Step 2: Add or subtract as with whole numbers.

Step 3: Place the decimal point in the sum or difference so that it lines up vertically with the decimal points in the problem.

In this section, we will insert zeros in decimal numbers so that place value digits line up neatly. This is shown in Example 1.

Example 1   Add: $23.85 + 1.604$

Solution: First we line up the decimal points vertically.

$$\begin{array}{r} 23.850 \\ +\ 1.604 \\ \hline \end{array}$$

Insert one 0 so that digits line up neatly.

Line up decimal points.

Then we add the digits from right to left as for whole numbers.

$$\begin{array}{r} 23.850 \\ +\ 1.604 \\ \hline 25.454 \end{array}$$

Place the decimal point in the sum so that all decimal points line up.

■ Work Practice 1

Practice 1

Add.

a. $15.52 + 2.371$

b. $20.06 + 17.612$

c. $0.125 + 122.8$

Helpful Hint

Recall that 0s may be placed after the last digit to the right of the decimal point without changing the value of the decimal. This may be used to help line up place values when adding decimals.

$$\begin{array}{rcl} 3.2 & \text{becomes} & 3.200 \\ 15.567 & & 15.567 \\ +\ 0.11 & & +\ 0.110 \\ \hline & & 18.877 \end{array}$$

Insert two 0s.

Insert one 0.

Add.

Answers

1. a. 17.891   **b.** 37.672   **c.** 122.925

Practice 2

Add.

a. $34.567 + 129.43 + 2.8903$

b. $11.21 + 46.013 + 362.526$

Example 2 Add: $763.7651 + 22.001 + 43.89$

Solution: First we line up the decimal points.

$$
\begin{array}{r}
763.7651 \\
22.001\textcolor{gray}{0} \quad \text{\small Insert one 0.} \\
+\quad 43.89\textcolor{gray}{00} \quad \text{\small Insert two 0s.} \\
\hline
829.6561 \quad \text{\small Add.}
\end{array}
$$

■ Work Practice 2

> **Helpful Hint**
>
> Don't forget that the decimal point in a whole number is positioned after the last digit.

Practice 3

Add: $119 + 26.072$

Example 3 Add: $45 + 2.06$

Solution:

$$
\begin{array}{r}
45.00 \quad \text{\small Insert a decimal point and two 0s.} \\
+\quad 2.06 \quad \text{\small Line up decimal points.} \\
\hline
47.06 \quad \text{\small Add.}
\end{array}
$$

■ Work Practice 3

✓**Concept Check** What is wrong with the following calculation of the sum of 7.03, 2.008, 19.16, and 3.1415?

$$
\begin{array}{r}
7.03 \\
2.008 \\
19.16 \\
+\,3.1415 \\
\hline
3.6042
\end{array}
$$

Practice 4

Add: $8.1 + (-99.2)$

Example 4 Add: $3.62 + (-4.78)$

Solution: Recall from Chapter 2 that to add two numbers with different signs, we find the difference of the larger absolute value and the smaller absolute value. The sign of the answer is the same as the sign of the number with the larger absolute value.

$$
\begin{array}{r}
4.78 \\
-\,3.62 \\
\hline
1.16 \quad \text{\small Subtract the absolute values.}
\end{array}
$$

Thus, $3.62 + (-4.78) = -1.16$

The sign of the number with the larger absolute value; -4.78 has the larger absolute value.

■ Work Practice 4

Answers

2. a. 166.8873 **b.** 419.749
3. 145.072 **4.** −91.1

✓**Concept Check Answer**

The decimal places are not lined up properly.

Objective B Subtracting Decimals

Subtracting decimals is similar to subtracting whole numbers. We line up digits and subtract from right to left, borrowing when needed.

Example 5 Subtract: $3.5 - 0.068$. Check your answer.

Solution:

$$
\begin{array}{r}
3.5\overset{9}{\cancel{\underset{4}{5}}}\overset{10}{\cancel{0}}0 \quad \text{Insert two 0s.} \\
-0.068 \quad \text{Line up decimal points.} \\
\hline
3.432 \quad \text{Subtract.}
\end{array}
$$

Check: Recall that we can check a subtraction problem by adding.

$$
\begin{array}{r}
3.432 \quad \text{Difference} \\
+0.068 \quad \text{Subtrahend} \\
\hline
3.500 \quad \text{Minuend}
\end{array}
$$

■ Work Practice 5

Example 6 Subtract: $85 - 17.31$. Check your answer.

Solution:

$$
\begin{array}{r}
8\overset{14}{\cancel{5}}.\overset{9}{\cancel{0}}\overset{10}{\cancel{0}} \\
-17.31 \\
\hline
67.69
\end{array}
$$

Check:
$$
\begin{array}{r}
67.69 \quad \text{Difference} \\
+17.31 \quad \text{Subtrahend} \\
\hline
85.00 \quad \text{Minuend}
\end{array}
$$

■ Work Practice 6

Practice 6

Subtract. Check your answers.
a. $53 - 29.31$
b. $120 - 68.22$

Example 7 Subtract 3 from 6.98. Check your answer.

Solution:

$$
\begin{array}{r}
6.98 \\
-3.00 \quad \text{Insert two 0s.} \\
\hline
3.98
\end{array}
$$

Check:
$$
\begin{array}{r}
3.98 \quad \text{Difference} \\
+3.00 \quad \text{Subtrahend} \\
\hline
6.98 \quad \text{Minuend}
\end{array}
$$

■ Work Practice 7

Practice 7

Subtract: 18 from 26.99

Example 8 Subtract: $-5.8 - 1.7$

Solution: Recall from Chapter 2 that to subtract 1.7, we add the opposite of 1.7, or -1.7. Thus

$-5.8 - 1.7 = -5.8 + (-1.7)$ To subtract, add the opposite of 1.7 which is -1.7.

   Add the absolute values.

$= -7.5$

   Use the common negative sign.

■ Work Practice 8

Practice 8

Subtract: $-3.4 - 9.6$

Example 9 Subtract: $-2.56 - (-4.01)$

Solution: $-2.56 - (-4.01) = -2.56 + 4.01$ To subtract, add the opposite of -4.01, which is 4.01.

   Subtract the absolute values.

$= 1.45$

   The answer is positive since 4.01 has the larger absolute value.

■ Work Practice 9

Practice 9

Subtract: $-1.05 - (-7.23)$

Objective C Estimating when Adding or Subtracting Decimals

To help avoid errors, we can also estimate to see if our answer is reasonable when adding or subtracting decimals. Although only one estimate is needed per operation, we show two for variety.

Practice 10

Add or subtract as indicated. Then estimate to see if the answer is reasonable by rounding the given numbers and adding or subtracting the rounded numbers.

a. $48.1 + 326.97$

b. $18.09 - 0.746$

Example 10 Add or subtract as indicated. Then estimate to see if the answer is reasonable by rounding the given numbers and adding or subtracting the rounded numbers.

a. $27.6 + 519.25$

Exact		Estimate 1		Estimate 2
27.60	rounds to	30		30
+ 519.25	rounds to	+500	or	+ 520
546.85		530		550

Since the exact answer is close to either estimate, it is reasonable. (In the first estimate, each number is rounded to the place value of the leftmost digit. In the second estimate, each number is rounded to the nearest ten.)

b. $11.01 - 0.862$

Exact		Estimate 1		Estimate 2
$1\cancel{1}.\cancel{0}\cancel{1}\cancel{0}$	rounds to	10		11
$- 0.862$	rounds to	-1	or	-1
10.148		9		10

In the first estimate, we rounded the first number to the nearest ten and the second number to the nearest one. In the second estimate, we rounded both numbers to the nearest one. Both estimates show us that our answer is reasonable.

■ Work Practice 10

Helpful Hint Remember that estimates are used for our convenience to quickly check the reasonableness of an answer.

✓**Concept Check** Why shouldn't the sum $21.98 + 42.36$ be estimated as $30 + 50 = 80$?

Objective D Using Decimals as Replacement Values

Let's review evaluating expressions with given replacement values. This time the replacement values are decimals.

Practice 11

Evaluate $y - z$ for $y = 11.6$ and $z = 10.8$.

Example 11 Evaluate $x - y$ for $x = 2.8$ and $y = 0.92$.

Solution: Replace x with 2.8 and y with 0.92 and simplify.

$$x - y = 2.8 - 0.92$$
$$= 1.88$$

$$\begin{array}{r} 2.80 \\ -0.92 \\ \hline 1.88 \end{array}$$

■ Work Practice 11

Answers

10. a. Exact: 375.07 **b.** Exact: 17.344
11. 0.8

✓**Concept Check Answer**

Each number is rounded incorrectly. The estimate is too high.

Objective E Solving Problems by Adding or Subtracting Decimals ▶

Decimals are very common in real-life problems.

Example 12 Calculating the Cost of Owning an Automobile

Find the total monthly cost of owning and operating a certain automobile given the expenses shown.

Monthly car payment:	$256.63
Monthly insurance cost:	$47.52
Average gasoline bill per month:	$195.33

Solution:

1. **UNDERSTAND.** Read and reread the problem. The phrase "total monthly cost" tells us to add.

2. **TRANSLATE.**

In words:

total monthly cost	is	car payment	plus	insurance cost	plus	gasoline bill
↓	↓	↓	↓	↓	↓	↓

Translate:

| total monthly cost | = | $256.63 | + | $47.52 | + | $195.33 |

3. **SOLVE:** Let's also estimate by rounding each number to the nearest ten.

$$
\begin{array}{r}
\overset{1\ 1\ 1}{256.63} \quad \text{rounds to} \quad 260 \\
47.52 \quad \text{rounds to} \quad 50 \\
+\ 195.33 \quad \text{rounds to} \quad 200 \\
\hline
499.48 \quad \text{Exact} \qquad 510 \quad \text{Estimate}
\end{array}
$$

4. **INTERPRET.** *Check* your work. Since our estimate is close to our exact answer, our answer is reasonable. *State* your conclusion: The total monthly cost is $499.48.

▨ Work Practice 12

Practice 12

Find the total monthly cost of owning and operating a certain automobile given the expenses shown.

Monthly car payment:	$563.52
Monthly insurance cost:	$52.68
Average gasoline bill per month:	$127.50

The bar graph on the next page has horizontal bars. To visualize the value represented by a bar, see how far it extends to the right. For this horizontal bar graph, the value of each bar is labeled. We will study bar graphs further in a later chapter.

Example 13 Comparing Average Heights

The bar graph shows the current average heights for adults in various countries. How much greater is the average height in Denmark than the average height in the United States?

(Continued on next page)

Practice 13

Use the bar graph in Example 13. How much greater is the average height in the Netherlands than the average height in Czechoslovakia?

Answers
12. $743.70   **13.** 1.8 in.

Average Adult Height

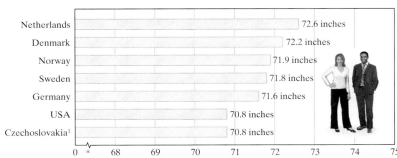

Netherlands	72.6 inches
Denmark	72.2 inches
Norway	71.9 inches
Sweden	71.8 inches
Germany	71.6 inches
USA	70.8 inches
Czechoslovakia[1]	70.8 inches

0   *   68   69   70   71   72   73   74   75

[1]Average for Czech Republic, Slovakia
Source: USA Today

* The ∿ means that some numbers are purposefully missing from the axis.

Solution:

1. UNDERSTAND. Read and reread the problem. Since we want to know "how much greater," we subtract.

2. TRANSLATE.

In words:	How much greater	is	Denmark's average height	minus	U.S. average height
	↓	↓	↓	↓	↓
Translate:	How much greater	=	72.2	−	70.8

3. SOLVE: We estimate by rounding each number to the nearest whole.

$$
\begin{array}{r}
\overset{\overset{1}{7}\overset{12}{\cancel{2}}}{.\cancel{2}} \\
-\ 7\ 0\ .\ 8 \\
\hline
1\ .\ 4
\end{array}
$$

7̶2̶.2̶   rounds to   72
− 7 0 . 8   rounds to   71
      1 . 4   Exact   1   Estimate

4. INTERPRET. *Check* your work. Since our estimate is close to our exact answer, 1.4 inches is reasonable. *State* your conclusion: The average height in Denmark is 1.4 inches greater than the average U.S. height.

■ Work Practice 13

🖩 Calculator Explorations   Decimals

Entering Decimal Numbers

To enter a decimal number, find the key marked ⟦ · ⟧.
To enter the number 2.56, for example, press the keys
⟦2⟧⟦·⟧⟦56⟧.
The display will read ⟦ 2.56 ⟧.

Operations on Decimal Numbers

Operations on decimal numbers are performed in the same way as operations on whole or signed numbers. For example, to find 8.625 − 4.29, press the keys
⟦8.625⟧⟦−⟧⟦4.29⟧ and then ⟦=⟧ or ⟦ENTER⟧.

The display will read ⟦ 4.335 ⟧. (Although entering 8.625, for example, requires pressing more than one key, we group numbers together here for easier reading.)

Use a calculator to perform each indicated operation.

1. 315.782 + 12.96

2. 29.68 + 85.902

3. 6.249 − 1.0076

4. 5.238 − 0.682

5.
```
   12.555
  224.987
    5.2
+ 622.65
```

6.
```
   47.006
    0.17
  313.259
+ 139.088
```

Vocabulary, Readiness & Video Check

Use the choices below to fill in each blank. Not all choices will be used and some may be used more than once.

minuend	vertically	like	true	horizontally
difference	subtrahend	last	false	

1. The decimal point in a whole number is positioned after the _____ digit.

2. In $89.2 - 14.9 = 74.3$, the number 74.3 is called the _____, 89.2 is the _____, and 14.9 is the _____.

3. To add or subtract decimals, we line up the decimal points _____.

4. True or false: If we replace x with 11.2 and y with -8.6 in the expression $x - y$, we have $11.2 - 8.6$. _____

Mentally find the sum or difference.

5. $\begin{array}{r} 0.3 \\ +0.2 \\ \hline \end{array}$

6. $\begin{array}{r} 0.4 \\ +0.5 \\ \hline \end{array}$

7. $\begin{array}{r} 1.00 \\ +0.26 \\ \hline \end{array}$

8. $\begin{array}{r} 3.00 \\ +0.19 \\ \hline \end{array}$

9. $\begin{array}{r} 7.6 \\ +1.3 \\ \hline \end{array}$

10. $\begin{array}{r} 4.5 \\ +3.2 \\ \hline \end{array}$

11. $\begin{array}{r} 0.9 \\ -0.3 \\ \hline \end{array}$

12. $\begin{array}{r} 0.6 \\ -0.2 \\ \hline \end{array}$

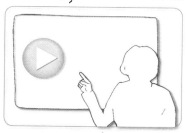

Martin-Gay Interactive Videos   Watch the section lecture video and answer the following questions.

See Video 4.2

Objectives A B **13.** From ▣ Examples 1–3, why do you think we line up decimal points?

Objective C **14.** In ▣ Example 4, estimating is used to check whether the answer to the subtraction problem is reasonable, but what is the best way to fully check?

Objective D **15.** In ▣ Example 5, why is the actual subtraction performed to the side?

Objective E **16.** In ▣ Example 6, to calculate the amount of border material needed, we are actually calculating the _____ of the triangle.

4.2 Exercise Set MyMathLab®

Objectives A C Mixed Practice *Add. See Examples 1 through 4, and 10. For those exercises marked, also estimate to see if the answer is reasonable.*

1. $1.3 + 2.2$

2. $2.5 + 4.1$

3. $5.7 + 1.13$

4. $2.31 + 6.4$

 5. $24.6 + 2.39 + 0.0678$

6. $32.4 + 1.58 + 0.0934$

7. $\begin{array}{r} 6.014 \\ 9.009 \\ +\ 76.181 \\ \hline \end{array}$

8. $\begin{array}{r} 3.003 \\ 8.098 \\ +\ 71.486 \\ \hline \end{array}$

9. $-2.6 + (-5.97)$

10. $-18.2 + (-10.8)$

11. $15.78 + (-4.62)$

12. $6.91 + (-7.03)$

13.
$$
\begin{array}{r}
234.89 \\
+\ 230.67 \\
\end{array}
$$
Exact: _____ Estimate: _____

14.
$$
\begin{array}{r}
734.89 \\
+\ 640.56 \\
\end{array}
$$
Exact: _____ Estimate: _____

15.
$$
\begin{array}{r}
100.009 \\
6.08 \\
+\ \ \ 9.034 \\
\end{array}
$$
Exact: _____ Estimate: _____

16.
$$
\begin{array}{r}
200.89 \\
7.49 \\
+\ \ 62.83 \\
\end{array}
$$
Exact: _____ Estimate: _____

17. Find the sum of 45.023, 3.006, and 8.403.

18. Find the sum of 65.0028, 5.0903, and 6.9003.

Objectives B C Mixed Practice *Subtract and check. See Examples 5 through 10. For those exercises marked, also estimate to see if the answer is reasonable.*

19. $8.8 - 2.3$

20. $7.6 - 2.1$

21. $18 - 2.7$

22. $28 - 3.3$

▶ 23.
$$
\begin{array}{r}
654.9 \\
-\ \ 56.67 \\
\end{array}
$$

24.
$$
\begin{array}{r}
863.23 \\
-\ \ 39.453 \\
\end{array}
$$

25. $5.9 - 4.07$
Exact: _____
Estimate: _____

26. $6.4 - 3.04$
Exact: _____
Estimate: _____

27. $923.5 - 61.9$

28. $845.93 - 45.8$

▶ 29.
$$
\begin{array}{r}
1000 \\
-\ \ 123.4 \\
\end{array}
$$
Exact: _____

Estimate: _____

30.
$$
\begin{array}{r}
2000 \\
-\ \ 327.47 \\
\end{array}
$$
Exact: _____

Estimate: _____

31. $200 - 5.6$

32. $800 - 8.9$

▶ 33. $-1.12 - 5.2$

34. $-8.63 - 5.6$

35. $7.7 - 14.1$

36. $10.25 - 21.76$

37. $-2.6 - (-5.7)$

38. $-9.4 - (-10.4)$

39. $3 - 0.0012$

40. $7 - 0.097$

41. Subtract 6.7 from 23.

42. Subtract 9.2 from 45.

Objectives A B Mixed Practice *Perform the indicated operation. See Examples 1 through 9.*

43. $0.9 + 2.2$

44. $0.7 + 3.4$

45. $-6.06 + 0.44$

46. $-5.05 + 0.88$

47. $900.34 - 123.45$

48. $800.74 - 463.98$

49. $50.2 - 600$

50. $40.3 - 700$

51. Subtract 61.9 from 923.5.

52. Subtract 45.8 from 845.9.

53. Add 500.008 and 4.06 and 8.033.

54. Add 400.009 and 7.05 and 6.088.

55. $-0.003 + 0.091$

56. $-0.004 + 0.085$

57. $-102.4 - 78.04$ **58.** $-36.2 - 10.02$ **59.** $-2.9 - (-1.8)$ **60.** $-6.5 - (-3.3)$

Objective D *Evaluate each expression for* $x = 3.6, y = 5,$ *and* $z = 0.21.$ *See Example 11.*

61. $x + z$

62. $y + x$

63. $x - z$

64. $y - z$

65. $y - x + z$

66. $x + y + z$

Objective E *Solve. For Exercises 67 and 68, the solutions have been started for you. See Examples 12 and 13.*

67. Ann-Margaret Tober bought a book for $32.48. If she paid with two $20 bills, what was her change?

Start the solution:

1. UNDERSTAND the problem. Reread it as many times as needed.
2. TRANSLATE into an equation. (Fill in the blank.)

change	is	two $20 bills	minus	cost of book
↓	↓	↓	↓	↓
change	=	40	−	_____

Finish with
3. SOLVE and 4. INTERPRET.

68. Phillip Guillot bought a car part for $18.26. If he paid with two $10 bills, what was his change?

Start the solution:

1. UNDERSTAND the problem. Reread it as many times as needed.
2. TRANSLATE into an equation. (Fill in the blank.)

change	is	two $10 bills	minus	cost of car part
↓	↓	↓	↓	↓
change	=	20	−	_____

Finish with
3. SOLVE and 4. INTERPRET.

69. Microsoft stock opened the day at $35.17 per share, and the closing price the same day was $34.75. By how much did the price of each share change?

70. A pair of eyeglasses costs a total of $347.89. The frames of the glasses are $97.23. How much do the lenses of the eyeglasses cost?

71. Find the perimeter.

Square | 7.14 meters

72. Find the perimeter.

4.2 in.   5.78 in.   7.8 in.

73. The Apple iPhone 5 was released in 2012. It measures 4.87 inches by 2.31 inches rounded. Find the approximate perimeter of this phone. (*Source:* Apple.com)

74. The Google Nexus 4, released in 2012, is the newest Google phone (at this writing). It measures 5.27 inches by 2.7 inches. Find the perimeter of the phone. (*Source:* Google.com)

75. The average wind speed at the weather station on Mt. Washington in New Hampshire is 35.2 miles per hour. The highest speed ever recorded at the station is 231.0 miles per hour. How much faster is the highest speed than the average wind speed? (*Source:* National Climatic Data Center)

76. The average annual rainfall in Omaha, Nebraska, is 30.08 inches. The average annual rainfall in New Orleans, Louisiana, is 64.16 inches. On average, how much more rain does New Orleans receive annually than Omaha? (*Source:* National Climatic Data Center)

This bar graph shows the predicted increase in the total number of worldwide emails sent or received per day. Use this graph for Exercises 77 and 78. (Source: The Radicati Group, Inc.)

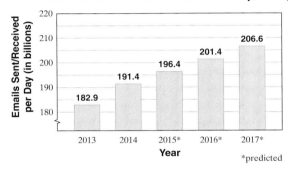

Worldwide Emails Sent/Received per Day

*predicted

77. Find the predicted increase in the number of emails sent or received from 2016 to 2017.

78. Find the predicted increase in the number of emails sent or received per day from 2013 to 2017.

79. As of this writing, the top three U.S. movies that have made the most money through movie ticket sales are *Avatar* (2009), $760.5 million; *Titanic* (1997), $658.7 million; and *The Avengers* (2012), $623.3 million. What is the total amount of ticket sales for these three movies? (*Source:* MovieWeb)

80. In 2011, the average credit card late fee was $23.15. In 2012, the average credit card late fee had increased by about $3.69. Find the average credit card late fee in 2012. (*Source:* Consumer Financial Protection Bureau)

81. The snowiest city in the United States is Valdez, AK, which receives an average of 110.5 more inches of snow than the second snowiest city. The second snowiest city in the United States is Crested Butte, CO. Crested Butte receives an average of 215.8 inches annually. How much snow does Valdez receive on average each year? (*Source:* The Weather Channel)

82. The driest place in the world is the Atacama Desert in Chile, which receives an average of only 0.004 inch of rain per year. Yuma, Arizona, is the driest city in the United States. Yuma receives an average of 3.006 more inches of rain each year than the Atacama Desert. What is the average annual rainfall in Yuma? (*Source:* National Climatic Data Center)

83. A landscape architect is planning a border for a flower garden shaped like a triangle. The sides of the garden measure 12.4 feet, 29.34 feet, and 25.7 feet. Find the amount of border material needed.

84. A contractor purchased enough railing to completely enclose the newly built deck shown below. Find the amount of railing purchased.

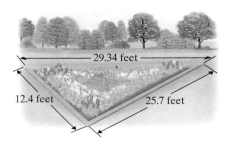

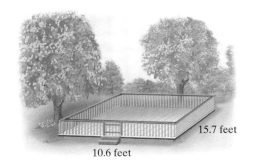

The table shows the average speeds for the Daytona 500 winners for the years shown. Use this table to answer Exercises 85 and 86. (Source: Daytona International Speedway)

Daytona 500 Winners		
Year	**Winner**	**Average Speed**
1978	Bobby Allison	159.73
1988	Bobby Allison	137.531
1998	Dale Earnhardt	172.712
2008	Ryan Newman	152.672
2013	Jimmie Johnson	159.250

85. How much slower was the average Daytona 500 winning speed in 2013 than in 1998?

86. How much faster was Bobby Allison's average Daytona 500 winning speed in 1978 than his average Daytona 500 winning speed in 1988?

The bar graph shows the top five chocolate-consuming nations in the world. Use this graph to answer Exercises 87 through 91.

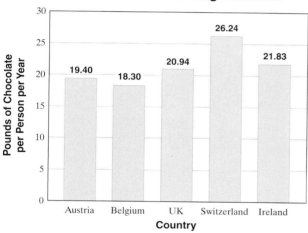

The World's Top Chocolate-Consuming Countries

Source: Confectionary News and Leatherhead Food Research

87. Which country in the graph has the greatest chocolate consumption per person?

88. Which country in the graph has the least chocolate consumption per person?

89. How much more is the greatest chocolate consumption than the least chocolate consumption shown in the graph?

90. How much more chocolate does the average person in Ireland consume per year than the average person in Austria?

91. Make a table listing the countries and their corresponding chocolate consumptions in order from greatest to least.

92. In your opinion, which listing of the data is easier for you to use—the bar graph or your table? Tell why.

Review

Multiply. See Sections 1.6 and 3.3.

93. $23 \cdot 2$

94. $46 \cdot 3$

95. $\left(\dfrac{2}{3}\right)^2$

96. $\left(\dfrac{1}{5}\right)^3$

Concept Extensions

A friend asks you to check his calculations for Exercises 97 and 98. Are they correct? If not, explain your friend's errors and correct the calculations. See the first Concept Check in this section.

97.
$$\begin{array}{r} \overset{1}{9}.2 \\ \overset{1}{8}.63 \\ +\,4.005 \\ \hline 4.960 \end{array}$$

98.
$$\begin{array}{r} \overset{8\,9\,9\,9}{900.0} \\ -\;96.4 \\ \hline 803.5 \end{array}$$

Find the unknown length in each figure.

△**99.**

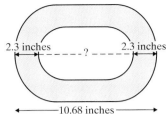

2.3 inches   ?   2.3 inches

10.68 inches

△**100.**

5.26 meters   7.82 meters   ? meters

17.67 meters

Let's review the values of these common U.S. coins in order to answer the following exercises.

Penny   Nickel   Dime   Quarter

$0.01 $0.05 $0.10 $0.25

For Exercises 101 and 102, write the value of each group of coins. To do so, it is usually easiest to start with the coin(s) of greatest value and end with the coin(s) of least value.

101.

102.

103. Name the different ways that coins can have a value of $0.17 given that you may use no more than 10 coins.

104. Name the different ways that coin(s) can have a value of $0.25 given that there are no pennies.

Solve.

105. Why shouldn't the sum

$$82.95 + 51.26$$

be estimated as $90 + 60 = 150$?
See the second Concept Check in this section.

106. Laser beams can be used to measure the distance to the moon. One measurement showed the distance to the moon to be 256,435.235 miles. A later measurement showed that the distance is 256,436.012 miles. Find how much farther away the moon is in the second measurement compared to the first.

107. Explain how adding or subtracting decimals is similar to adding or subtracting whole numbers.

108. Can the sum of two negative decimals ever be a positive decimal? Why or why not?

4.3 # Multiplying Decimals and Circumference of a Circle

Objective A Multiplying Decimals

Multiplying decimals is similar to multiplying whole numbers. The only difference is that we place a decimal point in the product. To discover where a decimal point is placed in the product, let's multiply 0.6×0.03. We first write each decimal as an equivalent fraction and then multiply.

$$\underset{\substack{\uparrow \\ \text{1 decimal} \\ \text{place}}}{0.6} \quad \times \quad \underset{\substack{\uparrow \\ \text{2 decimal} \\ \text{places}}}{0.03} \quad = \frac{6}{10} \times \frac{3}{100} = \frac{18}{1000} = \underset{\substack{\uparrow \\ \text{3 decimal} \\ \text{places}}}{0.018}$$

Notice that $1 + 2 = 3$, the number of decimal places in the product. Now let's multiply 0.03×0.002.

$$\underset{\substack{\uparrow \\ \text{2 decimal} \\ \text{places}}}{0.03} \quad \times \quad \underset{\substack{\uparrow \\ \text{3 decimal} \\ \text{places}}}{0.002} \quad = \frac{3}{100} \times \frac{2}{1000} = \frac{6}{100,000} = \underset{\substack{\uparrow \\ \text{5 decimal} \\ \text{places}}}{0.00006}$$

Again, we see that $2 + 3 = 5$, the number of decimal places in the product.

Instead of writing decimals as fractions each time we want to multiply, we notice a pattern from these examples and state a rule that we can use:

Multiplying Decimals

Step 1: Multiply the decimals as though they are whole numbers.

Step 2: The decimal point in the product is placed so that the number of decimal places in the product is equal to the *sum* of the number of decimal places in the factors.

Example 1 Multiply: 23.6×0.78

Solution:
$$\begin{array}{r} 23.6 \\ \times\ 0.78 \\ \hline 1888 \\ 16520 \\ \hline 18.408 \end{array}$$

23.6 1 decimal place
$\times$ 0.78 2 decimal places

18.408 Since $1 + 2 = 3$, insert the decimal point in the product so that there are 3 decimal places.

■ Work Practice 1

Example 2 Multiply: 0.0531×16

Solution:
$$\begin{array}{r} 0.0531 \\ \times\ \ \ \ \ 16 \\ \hline 3186 \\ 5310 \\ \hline 0.8496 \end{array}$$

0.0531 4 decimal places
$\times$ 16 0 decimal places

0.8496 4 decimal places $(4 + 0 = 4)$

■ Work Practice 2

Objectives

A Multiply Decimals.

B Estimate when Multiplying Decimals.

C Multiply Decimals by Powers of 10.

D Evaluate Expressions with Decimal Replacement Values.

E Find the Circumference of Circles.

F Solve Problems by Multiplying Decimals.

Practice 1
Multiply: 45.9×0.42

Practice 2
Multiply: 0.0721×48

Answers
1. 19.278 **2.** 3.4608

✓**Concept Check** True or false? The number of decimal places in the product of 0.261 and 0.78 is 6. Explain.

Practice 3

Multiply: $(5.4)(-1.3)$

Example 3 Multiply: $(-2.6)(0.8)$

Solution: Recall that the product of a negative number and a positive number is a negative number.

$$(-2.6)(0.8) = -2.08$$

▨ Work Practice 3

Objective B Estimating when Multiplying Decimals

Just as for addition and subtraction, we can estimate when multiplying decimals to check the reasonableness of our answer.

Practice 4

Multiply: 30.26×2.98. Then estimate to see whether the answer is reasonable.

Example 4 Multiply: 28.06×1.95. Then estimate to see whether the answer is reasonable by rounding each factor and then multiplying the rounded numbers.

Solution:

Exact	Estimate 1		Estimate 2
28.06	28 Rounded to ones		30 Rounded to tens
$\times$ 1.95	$\times$ 2	or	$\times$ 2
14030	56		60
252540			
280600			
54.7170			

The answer 54.7170 is reasonable.

▨ Work Practice 4

As shown in Example 4, estimated results will vary depending on what estimates are used. Notice that estimating results is a good way to see whether the decimal point has been correctly placed.

Objective C Multiplying Decimals by Powers of 10

There are some patterns that occur when we multiply a number by a power of 10 such as 10, 100, 1000, 10,000, and so on.

$23.6951 \times 10 = 236.951$ Move the decimal point *1 place* to the *right*.
            1 zero

$23.6951 \times 100 = 2369.51$ Move the decimal point *2 places* to the *right*.
            2 zeros

$23.6951 \times 100,000 = 2,369,510.$ Move the decimal point *5 places* to the *right* (insert a 0).
            5 zeros

Notice that we move the decimal point the same number of places as there are zeros in the power of 10.

Answers

3. -7.02
4. Exact: 90.1748

✓**Concept Check Answer**

false: 3 decimal places and 2 decimal places means 5 decimal places in the product

Multiplying Decimals by Powers of 10 Such as 10, 100, 1000, 10,000

Move the decimal point to the *right* the same number of places as there are *zeros* in the power of 10.

Examples Multiply.

5. $7.68 \times 10 = 76.8$ 7.68

6. $23.702 \times 100 = 2370.2$ 23.702

7. $(-76.3)(1000) = -76,300$ 76.300

■ Work Practice 5–7

Practice 5–7

Multiply.

5. 46.8×10

6. 203.004×100

7. $(-2.33)(1000)$

There are also powers of 10 that are less than 1. The decimals 0.1, 0.01, 0.001, 0.0001, and so on, are examples of powers of 10 less than 1. Notice the pattern when we multiply by these powers of 10:

$569.2 \times 0.1 = 56.92$ Move the decimal point *1 place* to the *left*.

1 decimal place

$569.2 \times 0.01 = 5.692$ Move the decimal point *2 places* to the *left*.

2 decimal places

$569.2 \times 0.0001 = 0.05692$ Move the decimal point *4 places* to the *left* (insert one 0).

4 decimal places

Multiplying Decimals by Powers of 10 Such as 0.1, 0.01, 0.001, 0.0001

Move the decimal point to the *left* the same number of places as there are *decimal places* in the power of 10.

Examples Multiply.

8. $42.1 \times 0.1 = 4.21$ 42.1

9. $76,805 \times 0.01 = 768.05$ 76,805.

10. $(-9.2)(-0.001) = 0.0092$ 0009.2

■ Work Practice 8–10

Practice 8–10

Multiply.

8. 6.94×0.1

9. 3.9×0.01

10. $(-7682)(-0.001)$

Many times we see large numbers written, for example, in the form 279.9 million rather than in the longer standard notation. The next example shows us how to interpret these numbers.

Answers

5. 468   **6.** 20,300.4   **7.** −2330

8. 0.694   **9.** 0.039   **10.** 7.682

Practice 11

There were 59.2 million married couples in the United States in 2013. Write this number in standard notation. (*Source:* U.S. Census Bureau)

Example 11 In 2050, the population of the United States is projected to be 420.3 million. Write this number in standard notation. (*Source:* U.S. Census Bureau)

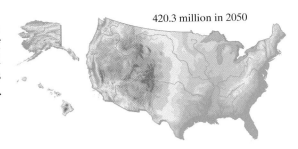

420.3 million in 2050

Solution: 420.3 million = 420.3 × 1 million

= 420.3 × 1,000,000 = 420,300,000

■ Work Practice 11

Objective D Using Decimals as Replacement Values

Now let's practice working with variables.

Practice 12

Evaluate $7y$ for $y = -0.028$.

Example 12 Evaluate xy for $x = 2.3$ and $y = 0.44$.

Solution: Recall that xy means $x \cdot y$.

$$xy = (2.3)(0.44)$$

$$\begin{array}{r} 2.3 \\ \times\ 0.44 \\ \hline 92 \\ 920 \\ \hline 1.012 \end{array}$$

$$= 1.012 \longleftarrow 1.012$$

■ Work Practice 12

Objective E Finding the Circumference of a Circle

Recall from Section 1.3 that the distance around a polygon is called its perimeter. The distance around a circle is given the special name **circumference,** and this distance depends on the radius or the diameter of the circle.

Circumference of a Circle

Circumference = $2 \cdot \pi \cdot$ **r**adius or Circumference = $\pi \cdot$ **d**iameter

$$C = 2\pi r \quad \text{or} \quad C = \pi d$$

In Section 4.1, we learned about the symbol π as the Greek letter pi, pronounced "pie." It is a constant between 3 and 4. A decimal approximation for π is 3.14. Also, a fraction approximation for π is $\dfrac{22}{7}$.

Answers

11. 59,200,000   **12.** −0.196

Example 13 Circumference of a Circle

Find the circumference of a circle whose radius is 5 inches. Then use the approximation 3.14 for π to approximate the circumference.

Solution: Let $r = 5$ in the formula $C = 2\pi r$.

$$C = 2\pi r$$
$$= 2\pi \cdot 5$$
$$= 10\pi$$

5 inches

Next, replace π with the approximation 3.14.

$$C = 10\pi$$
(is approximately) $\longrightarrow$ $\approx 10(3.14)$
$$= 31.4$$

The **exact** circumference or distance around the circle is 10π inches, which is **approximately** 31.4 inches.

■ Work Practice 13

Practice 13

Find the circumference of a circle whose radius is 11 meters. Then use the approximation 3.14 for π to approximate this circumference.

Objective F Solving Problems by Multiplying Decimals

The solutions to many real-life problems are found by multiplying decimals. We continue using our four problem-solving steps to solve such problems.

Example 14 Finding the Total Cost of Materials for a Job

A college student is hired to paint a billboard with paint costing $2.49 per quart. If the job requires 3 quarts of paint, what is the total cost of the paint?

Solution:

1. **UNDERSTAND.** Read and reread the problem. The phrase "total cost" might make us think addition, but since this problem requires repeated addition, let's multiply.

2. **TRANSLATE.**

In words:	total cost	is	cost per quart of paint	times	number of quarts
	↓	↓	↓	↓	↓
Translate:	total cost	=	2.49	×	3

3. **SOLVE.** We can estimate to check our calculations. The number 2.49 rounds to 2 and $2 \times 3 = 6$.

$$\begin{array}{r} 2.49 \\ \times \quad 3 \\ \hline 7.47 \end{array}$$

4. **INTERPRET.** *Check* your work. Since 7.47 is close to our estimate of 6, our answer is reasonable. *State* your conclusion: The total cost of the paint is $7.47.

■ Work Practice 14

Practice 14

A biology major is fertilizing her personal garden. She uses 5.6 ounces of fertilizer per square yard. The garden measures 60.5 square yards. How much fertilizer does she need?

Answers
13. 22π m ≈ 69.08 m
14. 338.8 oz

Vocabulary, Readiness & Video Check

Use the choices below to fill in each blank.

circumference	left	sum	zeros
decimal places	right	product	factor

1. When multiplying decimals, the number of decimal places in the product is equal to the _____ of the number of decimal places in the factors.

2. In $8.6 \times 5 = 43$, the number 43 is called the _____.

3. The distance around a circle is called its _____.

4. In $8.6 \times 5 = 43$, the numbers 8.6 and 5 are each called a _____.

5. When multiplying a decimal number by powers of 10 such as 10, 100, 1000, and so on, we move the decimal point in the number to the _____ the same number of places as there are _____ in the power of 10.

6. When multiplying a decimal number by powers of 10 such as 0.1, 0.01, and so on, we move the decimal point in the number to the _____ the same number of places as there are _____ in the power of 10.

Do not multiply. Just give the number of decimal places in the product. See the Concept Check in this section.

7.
$$\begin{array}{r} 0.46 \\ \times\ 0.81 \\ \hline \end{array}$$

8.
$$\begin{array}{r} 57.9 \\ \times\ 0.36 \\ \hline \end{array}$$

9.
$$\begin{array}{r} 0.0073 \\ \times\ \ \ \ 21 \\ \hline \end{array}$$

10.
$$\begin{array}{r} 0.428 \\ \times\ \ \ 0.2 \\ \hline \end{array}$$

11.
$$\begin{array}{r} 5.1296 \\ \times 7.3987 \\ \hline \end{array}$$

12.
$$\begin{array}{r} 0.028 \\ \times\ 1.36 \\ \hline \end{array}$$

Martin-Gay Interactive Videos

See Video 4.3

Watch the section lecture video and answer the following questions.

Objective A 13. From the lecture before ⊞ Example 1, what's the main difference between multiplying whole numbers and multiplying decimal numbers?

Objective B 14. From ⊞ Example 3, what does estimating especially help us with?

Objective C 15. Why don't we do any actual multiplying in ⊞ Example 5?

Objective D 16. In ⊞ Example 8, once all replacement values are inserted in the variable expression, what is the resulting expression to evaluate?

Objective E 17. Why is 31.4 cm not the exact answer to ⊞ Example 9?

Objective F 18. In ⊞ Example 10, why is 24.8 not the complete answer? What is the complete answer?

4.3 Exercise Set MyMathLab®

Objectives **A** **B** **Mixed Practice** *Multiply. See Examples 1 through 4. For those exercises marked, also estimate to see if the answer is reasonable.*

1. 0.17×8

2. 0.19×6

3.
$$\begin{array}{r} 1.2 \\ \times 0.5 \\ \hline \end{array}$$

4.
$$\begin{array}{r} 6.8 \\ \times 0.3 \\ \hline \end{array}$$

5. $(-2.3)(7.65)$

6. $(4.7)(-9.02)$

7. $(-5.73)(-9.6)$

8. $(-7.84)(-3.5)$

9. 6.8×4.2
Exact:
Estimate:

10. 8.3×2.7
Exact:
Estimate:

11.
$$\begin{array}{r} 0.347 \\ \times \ \ 0.3 \\ \hline \end{array}$$

12.
$$\begin{array}{r} 0.864 \\ \times \ \ 0.4 \\ \hline \end{array}$$

13.
$$\begin{array}{r} 1.0047 \\ \times \ \ \ 8.2 \\ \hline \end{array}$$
Exact: Estimate:

14.
$$\begin{array}{r} 2.0005 \\ \times \ \ \ 5.5 \\ \hline \end{array}$$
Exact: Estimate:

15.
$$\begin{array}{r} 490.2 \\ \times 0.023 \\ \hline \end{array}$$

16.
$$\begin{array}{r} 300.9 \\ \times 0.032 \\ \hline \end{array}$$

Objective C   *Multiply. See Examples 5 through 10.*

17. 6.5×10

18. 7.2×100

19. 8.3×0.1

20. 23.4×0.1

21. $(-7.093)(1000)$

22. $(-1.123)(1000)$

23. 0.7×100

24. 0.5×100

25. $(-9.83)(-0.01)$

26. $(-4.72)(-0.01)$

27. 25.23×0.001

28. 36.41×0.001

Objectives A B C   Mixed Practice *Multiply. See Examples 1 through 10.*

29. 0.123×0.4

30. 0.216×0.3

31. $(147.9)(100)$

32. $(345.2)(100)$

33. 8.6×0.15

34. 0.42×5.7

35. $(937.62)(-0.01)$

36. $(-0.001)(562.01)$

37. 562.3×0.001

38. 993.5×0.001

39.
$$\begin{array}{r} 6.32 \\ \times \ 5.7 \\ \hline \end{array}$$

40.
$$\begin{array}{r} 9.21 \\ \times \ 3.8 \\ \hline \end{array}$$

Write each number in standard notation. See Example 11.

41. The cost of the Hubble Space Telescope at launch was \$1.5 billion. (*Source:* NASA)

42. About 56.7 million American households own at least one dog. (*Source:* American Pet Products Manufacturers Association)

43. The Blue Streak is the oldest roller coaster at Cedar Point, an amusement park in Sandusky, Ohio. Since 1964, it has given more than 49.8 million rides. (*Source:* Cedar Fair, L.P.)

44. In 2013, the restaurant industry had projected sales of \$660.5 billion. (*Source:* National Restaurant Association)

Objective D   *Evaluate each expression for $x = 3$, $y = -0.2$, and $z = 5.7$. See Example 12.*

45. xy

46. yz

47. $xz - y$

48. $-5y + z$

Objective E   *Find the circumference of each circle. Then use the approximation 3.14 for π and approximate each circumference. See Example 13.*

49.
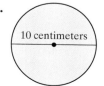
10 centimeters

50.

22 inches

51.

9.1 yards

52.
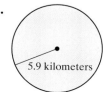
5.9 kilometers

Objectives E F Mixed Practice *Solve. For Exercises 53 and 54, the solutions have been started for you. See Examples 13 and 14. For circumference applications, find the exact circumference and then use 3.14 for π to approximate the circumference.*

53. An electrician for Central Power and Light worked 40 hours last week. Calculate his pay before taxes for last week if his hourly wage is $17.88.

Start the solution:

1. UNDERSTAND the problem. Reread it as many times as needed.
2. TRANSLATE into an equation. (Fill in the blanks.)

pay before taxes	is	hourly wage	times	hours worked
↓	↓	↓	↓	↓

pay before taxes = _____ × _____

Finish with:

3. SOLVE and 4. INTERPRET.

54. An assembly line worker worked 20 hours last week. Her hourly rate is $19.52 per hour. Calculate her pay before taxes.

Start the solution:

1. UNDERSTAND the problem. Reread it as many times as needed.
2. TRANSLATE into an equation. (Fill in the blanks.)

pay before taxes	is	hourly rate	times	hours worked
↓	↓	↓	↓	↓

pay before taxes = _____ × _____

Finish with:

3. SOLVE and 4. INTERPRET.

55. A 1-ounce serving of cream cheese contains 6.2 grams of saturated fat. How much saturated fat is in 4 ounces of cream cheese? (*Source: Home and Garden Bulletin No. 72;* U.S. Department of Agriculture)

56. A 3.5-ounce serving of lobster meat contains 0.1 gram of saturated fat. How much saturated fat do 3 servings of lobster meat contain? (*Source:* The National Institutes of Health)

57. Recall that the face of the Apple iPhone 5 (see Section 4.2) measures 4.87 inches by 2.3 inches rounded. Find the approximate area of the face of the Apple iPhone 5.

58. Recall that the face of the Google Nexus 4 (see Section 4.2) measures 5.27 inches by 2.7 inches. Find the area of the face of the Google Nexus 4.

59. In 1893, the first ride called a Ferris wheel was constructed by Washington Gale Ferris. Its diameter was 250 feet. Find its circumference. Give the exact answer and an approximation using 3.14 for π. (*Source: The Handy Science Answer Book,* Visible Ink Press, 1994)

60. The radius of Earth is approximately 3950 miles. Find the distance around Earth at the equator. Give the exact answer and an approximation using 3.14 for π. (*Hint:* Find the circumference of a circle with radius 3950 miles.)

61. The London Eye, built for the Millennium celebration in London, resembles a gigantic Ferris wheel with a diameter of 135 meters. If Adam Hawn rides the Eye for one revolution, find how far he travels. Give the exact answer and an approximation using 3.14 for π. (*Source:* Londoneye.com)

62. The world's longest suspension bridge is the Akashi Kaikyo Bridge in Japan. This bridge has two circular caissons, which are underwater foundations. If the diameter of a caisson is 80 meters, find its circumference. Give the exact answer and an approximation using 3.14 for π. (*Source: Scientific American;* How Things Work Today)

80 meters
Caisson

63. A meter is a unit of length in the metric system that is approximately equal to 39.37 inches. Sophia Wagner is 1.65 meters tall. Find her approximate height in inches.

64. The doorway to a room is 2.15 meters tall. Approximate this height in inches. (*Hint:* See Exercise 63.)

△**65. a.** Approximate the circumference of each circle.

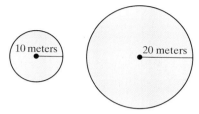

b. If the radius of a circle is doubled, is its corresponding circumference doubled?

△**66. a.** Approximate the circumference of each circle.

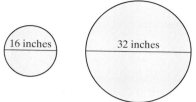

b. If the diameter of a circle is doubled, is its corresponding circumference doubled?

67. On one day in 2014, the price of wheat was $7.08 per bushel. How much would 100 bushels of wheat cost at this price? (*Source:* National Agricultural Statistics Service)

68. On one day in 2014, the price of soybeans was $14.98 per bushel. How much would a company pay for 10,000 bushels of soybeans? (*Source:* National Agricultural Statistics Service)

The table shows currency exchange rates for various countries on a particular day in 2013. To find the amount of foreign currency equivalent to an amount of U.S. dollars, multiply the U.S. dollar amount by the exchange rate listed in the table. Use this table to answer Exercises 69 through 72. Round answers for Exercises 70 through 72 to the nearest hundredth.

Foreign Currency Exchange Rates	
Foreign Currency	**Exchange Rate**
Canadian dollar	1.04920
European Union euro	0.74697
New Zealand dollar	1.28084
Chinese yuan	6.15231
Japanese yen	98.68
Swiss franc	0.92090

69. How many Canadian dollars were equivalent to $750 U.S.?

70. Suppose you exchanged 300 American dollars for Chinese yuan. How much money, in Chinese yuan, did you receive?

71. The Scarpulla family traveled to New Zealand. How many New Zealand dollars did they "buy" with 800 U.S. dollars?

72. A French tourist to the United States spent $130 for souvenirs at the *Head of the Charles Regatta* in Boston. How much money did he spend in euros?

Review

Divide. See Sections 1.7 and 3.3.

73. $2916 \div 6$

74. $2920 \div 365$

75. $-\dfrac{24}{7} \div \dfrac{8}{21}$

76. $\dfrac{162}{25} \div -\dfrac{9}{75}$

Concept Extensions

Mixed Practice (*Sections 4.2, 4.3*) *Perform the indicated operations.*

77. $3.6 + 0.04$

78. $7.2 + 0.14 + 98.6$

79. $3.6 - 0.04$

80. $100 - 48.6$

81. -0.221×0.5

82. -3.6×0.04

Solve.

83. Find how far radio waves travel in 20.6 seconds. (Radio waves travel at a speed of $1.86 \times 100,000$ miles per second.)

84. If it takes radio waves approximately 8.3 minutes to travel from the Sun to the Earth, find approximately how far it is from the Sun to the Earth. (*Hint:* See Exercise 83.)

85. In your own words, explain how to find the number of decimal places in a product of decimal numbers.

86. In your own words, explain how to multiply by a power of 10.

87. Write down two decimal numbers whose product will contain 5 decimal places. Without multiplying, explain how you know your answer is correct.

4.4 Dividing Decimals

Objectives

A Divide Decimals.

B Estimate when Dividing Decimals.

C Divide Decimals by Powers of 10.

D Evaluate Expressions with Decimal Replacement Values.

E Solve Problems by Dividing Decimals.

Practice 1

Divide: $517.2 \div 6$. Check your answer.

Objective A Dividing Decimals

Dividing decimal numbers is similar to dividing whole numbers. The only difference is that we place a decimal point in the quotient. If the divisor is a whole number, we place the decimal point in the quotient directly above the decimal point in the dividend, and then we divide as with whole numbers. Recall that division can be checked by multiplication.

Example 1 Divide: $270.2 \div 7$. Check your answer.

Solution: We divide as usual. The decimal point in the quotient is directly above the decimal point in the dividend.

```
                  ┌─ Write the decimal point.
            38.6  ← quotient
divisor → 7)270.2  ← dividend
          −21↓ │
            60 │
          −56↓
            4 2
          −4 2
              0
```

Check:
```
       6 4
      38.6
    ×    7
    ─────────
     270.2
```

The quotient is 38.6.

Work Practice 1

Answer
1. 86.2

Example 2 Divide: $32\overline{)8.32}$. Check your answer.

Solution: We divide as usual. The decimal point in the quotient is directly above the decimal point in the dividend.

$$
\begin{array}{r}
0.26 \leftarrow \text{quotient} \\
\text{divisor} \rightarrow 32\overline{)8.32} \leftarrow \text{dividend} \\
-6\,4 \\
\hline
1\,92 \\
-1\,92 \\
\hline
0
\end{array}
$$

Check:
$$
\begin{array}{r}
0.26 \quad \text{quotient} \\
\times\ 32 \quad \text{divisor} \\
\hline
52 \\
7\,80 \\
\hline
8.32 \quad \text{dividend}
\end{array}
$$

Work Practice 2

Sometimes to continue dividing we need to insert zeros after the last digit in the dividend.

Example 3 Divide: $-5.98 \div 115$

Solution: Recall that a negative number divided by a positive number gives a negative quotient.

$$
\begin{array}{r}
0.052 \\
115\overline{)5.980} \leftarrow \text{Insert one 0.} \\
-5\,75 \\
\hline
230 \\
-230 \\
\hline
0
\end{array}
$$

Thus $-5.98 \div 115 = -0.052$.

Work Practice 3

If the divisor is not a whole number, before we divide we need to move the decimal point to the right until the divisor is a whole number.

$$1.5\overline{)64.85}$$

divisor ⌐   ⌐ dividend

To understand how this works, let's rewrite

$1.5\overline{)64.85}$ as $\dfrac{64.85}{1.5}$

and then multiply by 1 in the form of $\dfrac{10}{10}$. We use the form $\dfrac{10}{10}$ so that the denominator (divisor) becomes a whole number.

$$\frac{64.85}{1.5} = \frac{64.85}{1.5} \cdot 1 = \frac{64.85}{1.5} \cdot \frac{10}{10} = \frac{64.85 \cdot 10}{1.5 \cdot 10} = \frac{648.5}{15},$$

which can be written as $15.\overline{)648.5}$. Notice that

$1.5\overline{)64.85}$ is equivalent to $15.\overline{)648.5}$.

Practice 2
Divide: $48\overline{)34.08}$. Check your answer.

Practice 3
Divide and check.
a. $-13.62 \div 12$
b. $-2.808 \div (-104)$

Answers
2. 0.71 **3. a.** −1.135 **b.** 0.027

The decimal points in the dividend and the divisor were both moved one place to the right, and the divisor is now a whole number. This procedure is summarized next:

Dividing by a Decimal

Step 1: Move the decimal point in the divisor to the right until the divisor is a whole number.

Step 2: Move the decimal point in the dividend to the right the *same number of places* as the decimal point was moved in Step 1.

Step 3: Divide. Place the decimal point in the quotient directly over the moved decimal point in the dividend.

Practice 4
Divide: 166.88 ÷ 5.6

Example 4 Divide: 10.764 ÷ 2.3

Solution: We move the decimal points in the divisor and the dividend one place to the right so that the divisor is a whole number.

$$
2.3\overline{)10.764} \qquad \text{becomes} \qquad
\begin{array}{r}
4.68 \\
23\overline{)107.64} \\
-92 \\
\hline
15\,6 \\
-13\,8 \\
\hline
1\,84 \\
-1\,84 \\
\hline
0
\end{array}
$$

■ Work Practice 4

Practice 5
Divide: 1.976 ÷ 0.16

Example 5 Divide: 5.264 ÷ 0.32

Solution:

$$
0.32\overline{)5.264} \qquad \text{becomes} \qquad
\begin{array}{r}
16.45 \\
32\overline{)526.40} \quad \text{Insert one 0.} \\
-32 \\
\hline
206 \\
-192 \\
\hline
14\,4 \\
-12\,8 \\
\hline
1\,60 \\
-1\,60 \\
\hline
0
\end{array}
$$

■ Work Practice 5

✔**Concept Check** Is it always true that the number of decimal places in a quotient equals the sum of the decimal places in the dividend and divisor?

Copyright 2016 Pearson Education, Inc.

Answers

4. 29.8 **5.** 12.35

✔**Concept Check Answer**

no

Example 6 Divide: $17.5 \div 0.48$. Round the quotient to the nearest hundredth.

Solution: First we move the decimal points in the divisor and the dividend two places. Then we divide and round the quotient to the nearest hundredth.

```
                      hundredths place
        36.458  ≈  36.46
  48.)1750.000
     −144
       310
      −288
        22 0
       −19 2
         2 80
        −2 40
           400
          −384
            16
```

When rounding to the nearest hundredth, carry the division process out to one more decimal place, the thousandths place.

"is approximately"

Practice 6

Divide: $23.4 \div 0.57$.
Round the quotient to the nearest hundredth.

◼ Work Practice 6

✓**Concept Check** If a quotient is to be rounded to the nearest thousandth, to what place should the division be carried out? (Assume that the division carries out to your answer.)

Objective B Estimating when Dividing Decimals

Just as for addition, subtraction, and multiplication of decimals, we can estimate when dividing decimals to check the reasonableness of our answer.

Example 7 Divide: $272.356 \div 28.4$. Then estimate to see whether the proposed result is reasonable.

Solution:

Exact:	**Estimate 1**		**Estimate 2**

```
        9.59              9                    10
 284.)2723.56       30)270        or     30)300
     −2556
       167 5
      −142 0
        25 56
       −25 56
            0
```

The estimate is 9 or 10, so 9.59 is reasonable.

Practice 7

Divide: $713.7 \div 91.5$. Then estimate to see whether the proposed answer is reasonable.

◼ Work Practice 7

Answers
6. 41.05 **7.** Exact: 7.8

✓**Concept Check Answer**
ten-thousandths place

Objective C Dividing Decimals by Powers of 10

As with multiplication, there are patterns that occur when we divide decimals by powers of 10 such as 10, 100, 1000, and so on.

$$\frac{569.2}{10} = 56.92$$ Move the decimal point *1 place* to the *left.*

         └── 1 zero

$$\frac{569.2}{10,000} = 0.05692$$ Move the decimal point *4 places* to the *left.*

         └── 4 zeros

This pattern suggests the following rule:

Dividing Decimals by Powers of 10 Such as 10, 100, or 1000

Move the decimal point of the dividend to the *left* the same number of places as there are *zeros* in the power of 10.

Practice 8–9

Divide.

8. $\dfrac{128.3}{1000}$ **9.** $\dfrac{-0.56}{10}$

Examples Divide.

8. $\dfrac{786.1}{1000} = 0.7861$ Move the decimal point *3 places* to the *left.*

        └── 3 zeros

9. $\dfrac{-0.12}{10} = -0.012$ Move the decimal point *1 place* to the *left.*

        └── 1 zero

■ Work Practice 8–9

Objective D Using Decimals as Replacement Values

Practice 10

Evaluate $x \div y$ for $x = 0.035$ and $y = 0.02$.

Example 10 Evaluate $x \div y$ for $x = 2.5$ and $y = 0.05$.

Solution: Replace x with 2.5 and y with 0.05.

$$x \div y = 2.5 \div 0.05 \quad 0.05\overline{)2.5} \quad \text{becomes} \quad 5\overline{)250} \;(= 50)$$
$$= 50$$

■ Work Practice 10

Practice 11

A bag of fertilizer covers 1250 square feet of lawn. Tim Parker's lawn measures 14,800 square feet. How many bags of fertilizer does he need? If he can buy only whole bags of fertilizer, how many whole bags does he need?

Objective E Solving Problems by Dividing Decimals

Many real-life problems involve dividing decimals.

△ **Example 11** Calculating Materials Needed for a Job

A gallon of paint covers a 250-square-foot area. If Betty Adkins wishes to paint a wall that measures 1450 square feet, how many gallons of paint does she need? If she can buy only gallon containers of paint, how many gallon containers does she need?

Answers

8. 0.1283 **9.** −0.056 **10.** 1.75
11. 11.84 bags; 12 bags

Solution:

1. **UNDERSTAND.** Read and reread the problem. We need to know how many 250s are in 1450, so we divide.

2. **TRANSLATE.**

In words:

number of gallons	is	square feet	divided by	square feet per gallon
↓	↓	↓	↓	↓

Translate:

number of gallons	=	1450	÷	250

3. **SOLVE.** Let's see if our answer is reasonable by estimating. The dividend 1450 rounds to 1500 and the divisor 250 rounds to 300. Then $1500 \div 300 = 5$.

$$
\begin{array}{r}
5.8 \\
250 \overline{)1450.0} \\
\underline{-1250} \\
200\ 0 \\
\underline{-200\ 0} \\
0
\end{array}
$$

4. **INTERPRET.** *Check* your work. Since our estimate is close to our answer of 5, our answer is reasonable. *State* your conclusion: Betty needs 5.8 gallons of paint. If she can buy only gallon containers of paint, she needs 6 gallon containers of paint to complete the job.

■ Work Practice 11

🖳 Calculator Explorations　**Estimation**

Calculator errors can easily be made by pressing an incorrect key or by not pressing a correct key hard enough. Estimation is a valuable tool that can be used to check calculator results.

Example　Use estimation to determine whether the calculator result is reasonable or not. (For example, a result that is not reasonable can occur if proper keys are not pressed.)

Simplify:　$82.064 \div 23$

Calculator display: 　　| 35.68 |

Solution:　Round each number to the nearest 10. Since $80 \div 20 = 4$, the calculator display 35.68 is not reasonable.

Use estimation to determine whether each result is reasonable or not.

1. 102.62×41.8 Result: 428.9516

2. $174.835 \div 47.9$ Result: 3.65

3. $1025.68 - 125.42$ Result: 900.26

4. $562.781 + 2.96$ Result: 858.781

Vocabulary, Readiness & Video Check

Use the choices below to fill in each blank. Some choices may be used more than once and some not used at all.

dividend	divisor	quotient	true
zeros	left	right	false

1. In $6.5 \div 5 = 1.3$, the number 1.3 is called the _____, 5 is the _____, and 6.5 is the _____.

2. To check a division exercise, we can perform the following multiplication: quotient · _____ = _____.

3. To divide a decimal number by a power of 10 such as 10, 100, 1000, and so on, we move the decimal point in the number to the _____ the same number of places as there are _____ in the power of 10.

4. True or false: If we replace x with -12.6 and y with 0.3 in the expression $y \div x$, we have $0.3 \div (-12.6)$. _____

Recall properties of division and simplify without the use of paper and pencil.

5. $\dfrac{5.9}{1}$

6. $\dfrac{0.7}{0.7}$

7. $\dfrac{0}{9.86}$

8. $\dfrac{2.36}{0}$

9. $\dfrac{7.261}{7.261}$

10. $\dfrac{8.25}{1}$

11. $\dfrac{11.1}{0}$

12. $\dfrac{0}{89.96}$

Martin-Gay Interactive Videos

See Video 4.4

Watch the section lecture video and answer the following questions.

Objective A 13. From the lecture before ▣ Example 1, what must we make sure the divisor is before dividing decimals? ▷

Objective B 14. From ▣ Example 4, what does estimating especially help us with? ▷

Objective C 15. Why don't we do any actual dividing in ▣ Example 6? ▷

Objective D 16. In ▣ Example 7, 8 does not divide into 5. How does this affect the quotient? ▷

Objective E 17. In ▣ Example 8, why is the division carried to the hundredths place? ▷

4.4   **Exercise Set** MyMathLab® ▷

Objectives A B Mixed Practice *Divide. See Examples 1 through 5 and 7. For those exercises marked, also estimate to see if the answer is reasonable.*

1. $3\overline{)13.8}$

2. $2\overline{)11.8}$

3. $5\overline{)0.47}$

4. $6\overline{)0.51}$

5. $0.06\overline{)18}$

6. $0.04\overline{)20}$

7. $0.54\overline{)1.404}$

8. $0.73\overline{)4.526}$

▷ 9. $5.5\overline{)36.3}$
Exact:
Estimate:

10. $2.2\overline{)21.78}$
Exact:
Estimate:

11. $7.434 \div 18$ **12.** $8.304 \div 16$ **13.** $36 \div (-0.06)$ **14.** $36 \div (-0.04)$ **15.** Divide -4.2 by -0.6.

16. Divide -3.6 by -0.9.   **17.** $0.27\overline{)1.296}$ **18.** $0.34\overline{)2.176}$ **19.** $0.02\overline{)42}$ **20.** $0.03\overline{)24}$

21. $4.756 \div 0.82$ **22.** $3.312 \div 0.92$ **23.** $-36.3 \div (-6.6)$ **24.** $-21.78 \div (-9.9)$ **25.** $7.2\overline{)70.56}$
Exact:
Estimate:

26. $6.3\overline{)54.18}$
Exact:
Estimate: **27.** $5.4\overline{)51.84}$ **28.** $7.7\overline{)33.88}$ **29.** $\dfrac{1.215}{0.027}$ **30.** $\dfrac{3.213}{0.051}$

31. $0.25\overline{)13.648}$ **32.** $0.75\overline{)49.866}$ **33.** $3.78\overline{)0.02079}$ **34.** $2.96\overline{)0.01332}$

Divide. Round the quotients as indicated. See Example 6.

35. Divide: 429.34 by 2.4. Round the quotient to the nearest whole number.

36. Divide: 54.8 by 2.6. Round the quotient to the nearest whole number.

37. Divide: $0.549 \div 0.023$. Round the quotient to the nearest hundredth.

38. Divide: $0.0453 \div 0.98$. Round the quotient to the nearest thousandth.

39. Divide: $68.39 \div 0.6$. Round the quotient to the nearest tenth.

40. Divide: $98.83 \div 3.5$. Round the quotient to the nearest tenth.

Objective C   *Divide. See Examples 8 and 9.*

41. $\dfrac{54.982}{100}$ **42.** $\dfrac{342.54}{100}$ **43.** $\dfrac{26.87}{10}$ **44.** $\dfrac{13.49}{10}$ **45.** $-12.9 \div 1000$ **46.** $-13.49 \div 10{,}000$

Objectives A C   Mixed Practice   *Divide. See Examples 1 through 5, 8, and 9.*

47. $7\overline{)88.2}$ **48.** $9\overline{)130.5}$ **49.** $\dfrac{13.1}{10}$ **50.** $\dfrac{17.7}{10}$

51. $\dfrac{456.25}{10{,}000}$ **52.** $\dfrac{986.11}{10{,}000}$ **53.** $1.239 \div 3$ **54.** $0.54 \div 12$

55. Divide 4.8 by -0.6.   **56.** Divide 4.9 by -0.7.   **57.** $-1.224 \div 0.17$ **58.** $-1.344 \div 0.42$

59. Divide 42 by 0.03.   **60.** Divide 27 by 0.03.   **61.** Divide -18 by -0.6.   **62.** Divide 20 by 0.4.

63. Divide 87 by -0.0015.   **64.** Divide 35 by -0.0007.   **65.** $-1.104 \div 1.6$ **66.** $-2.156 \div 0.98$

67. $-2.4 \div (-100)$ **68.** $-86.79 \div (-1000)$ **69.** $\dfrac{4.615}{0.071}$ **70.** $\dfrac{23.8}{0.035}$

Objective D *Evaluate each expression for $x = 5.65$, $y = -0.8$, and $z = 4.52$. See Example 10.*

71. $z \div y$ **72.** $z \div x$ ▶**73.** $x \div y$ **74.** $y \div 2$

Objective E *Solve. For Exercises 75 and 76, the solutions have been started for you. See Example 11.*

△ **75.** A new homeowner is painting the walls of a room. The walls have a total area of 546 square feet. A quart of paint covers 52 square feet. If he must buy paint in whole quarts, how many quarts does he need?

Start the solution:

1. UNDERSTAND the problem. Reread it as many times as needed.
2. TRANSLATE into an equation. (Fill in the blanks.)

number of quarts	is	square feet	divided by	square feet per quart
↓	↓	↓	↓	↓

number of quarts $=$ _____ $\div$ _____

3. SOLVE. Don't forget to round up your quotient.
4. INTERPRET.

76. A shipping box can hold 36 books. If 486 books must be shipped, how many boxes are needed?

Start the solution:

1. UNDERSTAND the problem. Reread it as many times as needed.
2. TRANSLATE into an equation. (Fill in the blanks.)

number of boxes	is	number of books	divided by	books per box
↓	↓	↓	↓	↓

number of boxes $=$ _____ $\div$ _____

3. SOLVE. Don't forget to round up your quotient.
4. INTERPRET.

77. There are approximately 39.37 inches in 1 meter. How many meters, to the nearest tenth of a meter, are there in 200 inches?

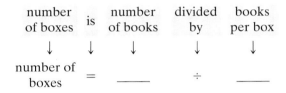

←———1 meter———→

←—≈39.37 inches—→

78. There are 2.54 centimeters in 1 inch. How many inches are there in 50 centimeters? Round to the nearest tenth.

←——— 1 inch ———→

←——— 2.54 cm ———→

▶**79.** In the United States, an average child will wear down 730 crayons by his or her tenth birthday. Find the number of boxes of 64 crayons this is equivalent to. Round to the nearest tenth. (*Source:* Binney & Smith Inc.)

80. In 2013, American farmers received an average of $66.50 per hundred pounds of turkey. What was the average price per pound for turkeys? Round to the nearest cent. (*Source:* National Agricultural Statistics Service)

A child is to receive a dose of 0.5 teaspoon of cough medicine every 4 hours. If the bottle contains 4 fluid ounces, answer Exercises 81 through 84.

81. A fluid ounce equals 6 teaspoons. How many teaspoons are in 4 fluid ounces?

82. The bottle of medicine contains how many doses for the child? (*Hint:* See Exercise 81.)

83. If the child takes a dose every four hours, how many days will the medicine last?

84. If the child takes a dose every six hours, how many days will the medicine last?

Solve.

85. Americans ages 16–19 drive, on average, 7624 miles per year. About how many miles each week is that? Round to the nearest tenth. (*Note:* There are 52 weeks in a year.) (*Source:* Federal Highway Administration)

86. Drake Saucier was interested in the gas mileage on his "new" used car. He filled the tank, drove 423.8 miles, and filled the tank again. When he refilled the tank, it took 19.35 gallons of gas. Calculate the miles per gallon for Drake's car. Round to the nearest tenth.

87. The leading money winner in men's professional golf in 2014 was Rory McIlroy. He earned approximately $8,280,000. Suppose he had earned this working 40 hours each week for a year. Determine his hourly wage to the nearest cent. (*Note:* There are 52 weeks in a year.) (*Source:* Professional Golfers' Association)

88. The book *Harry Potter and the Deathly Hallows* was released to the public on July 21, 2007. Booksellers in the United States sold approximately 8292 thousand copies in the first 24 hours after release. If the same number of books was sold each hour, calculate the number of books sold each hour in the United States for that first day.

Review

Write each decimal as a fraction. See Section 4.1.

89. 0.9

90. 0.7

91. 0.05

92. 0.08

Concept Extensions

Mixed Practice (Sections 4.2, 4.3, 4.4) *Perform the indicated operation.*

93. $1.278 \div 0.3$

94. 1.278×0.3

95. $1.278 + 0.3$

96. $1.278 - 0.3$

97. $(-8.6)(3.1)$

98. $7.2 + 0.05 + 49.1$

99. $\begin{array}{r} 1000 \\ -\ 95.71 \\ \hline \end{array}$

100. $\dfrac{87.2}{-10,000}$

Choose the best estimate.

101. 8.62×41.7
a. 36
b. 32
c. 360
d. 3.6

102. $1.437 + 20.69$
a. 34
b. 22
c. 3.4
d. 2.2

103. $78.6 \div 97$
a. 7.86
b. 0.786
c. 786
d. 7860

104. $302.729 - 28.697$
a. 270
b. 20
c. 27
d. 300

Recall from Section 1.7 that the average of a list of numbers is their total divided by how many numbers there are in the list. Use this procedure to find the average of the test scores listed in Exercises 105 and 106. If necessary, round to the nearest tenth.

105. 86, 78, 91, 87

106. 56, 75, 80

△**107.** The area of a rectangle is 38.7 square feet. If its width is 4.5 feet, find its length.

△**108.** The perimeter of a square is 180.8 centimeters. Find the length of a side.

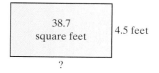

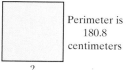

109. When dividing decimals, describe the process you use to place the decimal point in the quotient.

110. In your own words, describe how to quickly divide a number by a power of 10 such as 10, 100, 1000, etc.

To convert wind speeds in miles per hour to knots, divide by 1.15. Use this information and the Saffir-Simpson Hurricane Intensity chart below to answer Exercises 111 and 112. Round to the nearest tenth.

Saffir-Simpson Hurricane Intensity Scale				
Category	**Wind Speed**	**Barometric Pressure [inches of mercury (Hg)]**	**Storm Surge**	**Damage Potential**
1 (Weak)	75–95 mph	≥28.94 in.	4–5 ft	Minimal damage to vegetation
2 (Moderate)	96–110 mph	28.50–28.93 in.	6–8 ft	Moderate damage to houses
3 (Strong)	111–130 mph	27.91–28.49 in.	9–12 ft	Extensive damage to small buildings
4 (Very Strong)	131–155 mph	27.17–27.90 in.	13–18 ft	Extreme structural damage
5 (Devastating)	>155 mph	<27.17 in.	>18 ft	Catastrophic building failures possible

111. The chart gives wind speeds in miles per hour. What is the range of wind speeds for a Category 1 hurricane in knots?

112. What is the range of wind speeds for a Category 4 hurricane in knots?

Solve.

113. A rancher is building a horse corral that's shaped like a rectangle with a width of 24.3 meters. He plans to make a four-wire fence; that is, he will string four wires around the corral. If he plans to use all of his 412.8 meters of wire, find the length of the corral he can construct.

114. A college student signed up for a new credit card that guarantees her no interest charges on transferred balances for a year. She transferred over a $2523.86 balance from her old credit card. Her minimum payment is $185.35 per month. If she pays only the minimum, will she pay off her balance before interest charges start again?

Operations on Decimals

Perform the indicated operations.

1. $1.6 + 0.97$ **2.** $3.2 + 0.85$ **3.** $9.8 - 0.9$ **4.** $10.2 - 6.7$

5. $\begin{array}{r} 0.8 \\ \times\, 0.2 \\ \hline \end{array}$ **6.** $\begin{array}{r} 0.6 \\ \times\, 0.4 \\ \hline \end{array}$ **7.** $8\overline{)2.16}$ **8.** $6\overline{)3.12}$

9. $(9.6)(-0.5)$ **10.** $(-8.7)(-0.7)$ **11.** $\begin{array}{r} 123.6 \\ -\,48.04 \\ \hline \end{array}$ **12.** $\begin{array}{r} 325.2 \\ -\,36.08 \\ \hline \end{array}$

13. $-25 + 0.026$ **14.** $0.125 + (-44)$ **15.** $29.24 \div (-3.4)$ **16.** $-10.26 \div (-1.9)$

17. -2.8×100 **18.** 1.6×1000 **19.** $\begin{array}{r} 96.21 \\ 7.028 \\ +\,121.7 \\ \hline \end{array}$ **20.** $\begin{array}{r} 0.268 \\ 1.93 \\ +\,142.881 \\ \hline \end{array}$

21. $-25.76 \div (-46)$ **22.** $-27.09 \div 43$ **23.** $\begin{array}{r} 12.004 \\ \times\, 2.3 \\ \hline \end{array}$ **24.** $\begin{array}{r} 28.006 \\ \times\, 5.2 \\ \hline \end{array}$

1. _____
2. _____
3. _____
4. _____
5. _____
6. _____
7. _____
8. _____
9. _____
10. _____
11. _____
12. _____
13. _____
14. _____
15. _____
16. _____
17. _____
18. _____
19. _____
20. _____
21. _____
22. _____
23. _____
24. _____

25. _____

26. _____

27. _____

28. _____

29. _____

30. _____

31. _____

32. _____

33. _____

34. _____

35. _____

36. _____

37. _____

38. _____

39. _____

25. Subtract 4.6 from 10.

26. Subtract 18 from 0.26.

27. $-268.19 - 146.25$

28. $-860.18 - 434.85$

29. $\dfrac{2.958}{-0.087}$

30. $\dfrac{-1.708}{0.061}$

31. $160 - 43.19$

32. $120 - 101.21$

33. 15.62×10

34. $15.62 \div 10$

35. $15.62 + 10$

36. $15.62 - 10$

37. Estimate the distance in miles between Garden City, Kansas, and Wichita, Kansas, by rounding each given distance to the nearest ten.

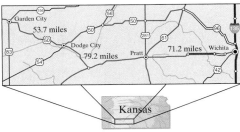

38. It costs $7.29 to send a 5-pound package locally via parcel post at a U.S. Post Office. To send the same package Priority Mail costs $8.10. How much more does it cost to send a package as Priority Mail? (_Source:_ United States Postal Service)

39. In 2013, sales of Blu-ray Discs were $2.2 billion, but DVD sales dropped to $5.2 billion. Find the total spent to buy Blu-ray Discs or DVDs in 2013. Write the total in billions of dollars and also in standard notation. (_Source: USA Today_)

4.5 Fractions, Decimals, and Order of Operations

Objective A Writing Fractions as Decimals

To write a fraction as a decimal, we interpret the fraction bar to mean division and find the quotient.

Objectives

A Write Fractions as Decimals.

B Compare Decimals and Fractions.

C Simplify Expressions Containing Decimals and Fractions Using Order of Operations.

D Solve Area Problems Containing Fractions and Decimals.

E Evaluate Expressions Given Decimal Replacement Values.

Writing Fractions as Decimals

To write a fraction as a decimal, divide the numerator by the denominator.

Example 1 Write $\frac{1}{4}$ as a decimal.

Solution: $\frac{1}{4} = 1 \div 4$

$$
\begin{array}{r}
0.25 \\
4\overline{)1.00} \\
-8 \\
\hline
20 \\
-20 \\
\hline
0
\end{array}
$$

Thus, $\frac{1}{4}$ written as a decimal is 0.25.

■ Work Practice 1

Example 2 Write $-\frac{5}{8}$ as a decimal.

Solution: $-\frac{5}{8} = -(5 \div 8) = -0.625$

$$
\begin{array}{r}
0.625 \\
8\overline{)5.000} \\
-4\,8 \\
\hline
20 \\
-16 \\
\hline
40 \\
-40 \\
\hline
0
\end{array}
$$

■ Work Practice 2

Example 3 Write $\frac{2}{3}$ as a decimal.

Solution:

$$
\begin{array}{r}
0.666\ldots \\
3\overline{)2.000} \\
-1\,8 \\
\hline
20 \\
-18 \\
\hline
20 \\
-18 \\
\hline
2
\end{array}
$$

This pattern will continue because $\frac{2}{3} = 0.6666\ldots$

Remainder is 2, and then 0 is brought down.

Remainder is 2, and then 0 is brought down.

Remainder is 2.

Practice 1

a. Write $\frac{2}{5}$ as a decimal.

b. Write $\frac{9}{40}$ as a decimal.

Practice 2

Write $-\frac{3}{8}$ as a decimal.

Practice 3

a. Write $\frac{5}{6}$ as a decimal.

b. Write $\frac{2}{9}$ as a decimal.

Answers

1. a. 0.4 **b.** 0.225 **2.** −0.375
3. a. 0.8$\overline{3}$ **b.** 0.$\overline{2}$

(Continued on next page)

Notice that the digit 2 keeps occurring as the remainder. This will continue, and the digit 6 will keep repeating in the quotient. We place a bar over the digit 6 to indicate that it repeats.

$$\frac{2}{3} = 0.666\ldots = 0.\overline{6}$$

We can also write a decimal approximation for $\frac{2}{3}$. For example, $\frac{2}{3}$ rounded to the nearest hundredth is 0.67. This can be written as $\frac{2}{3} \approx 0.67$.

■ Work Practice 3

Practice 4

Write $\frac{28}{13}$ as a decimal. Round to the nearest thousandth.

Example 4 Write $\frac{22}{7}$ as a decimal. (Recall that the fraction $\frac{22}{7}$ is an approximation for π.) Round to the nearest hundredth.

Solution:

$$\begin{array}{r} 3.142 \approx 3.14 \\ 7\overline{)22.000} \\ \underline{-21} \\ 1\,0 \\ \underline{-7} \\ 30 \\ \underline{-28} \\ 20 \\ \underline{-14} \\ 6 \end{array}$$ Carry the division out to the thousandths place.

The fraction $\frac{22}{7}$ in decimal form is approximately 3.14.

■ Work Practice 4

Practice 5

Write $3\frac{5}{16}$ as a decimal.

Example 5 Write $2\frac{3}{16}$ as a decimal.

Solution:

Option 1. Write the fractional part only as a decimal.

$$\frac{3}{16} \longrightarrow \begin{array}{r} 0.1875 \\ 16\overline{)3.0000} \\ \underline{-1\,6} \\ 1\,40 \\ \underline{-1\,28} \\ 120 \\ \underline{-112} \\ 80 \\ \underline{-80} \\ 0 \end{array}$$

Thus $2\frac{3}{16} = 2.1875$

Option 2. Write $2\frac{3}{16}$ as an improper fraction, and divide.

$$2\frac{3}{16} = \frac{35}{16} \longrightarrow \begin{array}{r} 2.1875 \\ 16\overline{)35.0000} \\ \underline{-32} \\ 3\,0 \\ \underline{-1\,6} \\ 1\,40 \\ \underline{-1\,28} \\ 120 \\ \underline{-112} \\ 80 \\ \underline{-80} \\ 0 \end{array}$$

■ Work Practice 5

Answers

4. 2.154   **5.** 3.3125

Some fractions may be written as decimals using our knowledge of decimals. From Section 4.1, we know that if the denominator of a fraction is 10, 100, 1000, or so on, we can immediately write the fraction as a decimal. For example,

$$\frac{4}{10} = 0.4, \qquad \frac{12}{100} = 0.12, \text{ and so on}$$

Example 6 Write $\frac{4}{5}$ as a decimal.

Solution: Let's write $\frac{4}{5}$ as an equivalent fraction with a denominator of 10.

$$\frac{4}{5} = \frac{4}{5} \cdot \frac{2}{2} = \frac{8}{10} = 0.8$$

■ Work Practice 6

Practice 6
Write $\frac{3}{5}$ as a decimal.

Example 7 Write $\frac{1}{25}$ as a decimal.

Solution: $\frac{1}{25} = \frac{1}{25} \cdot \frac{4}{4} = \frac{4}{100} = 0.04$

■ Work Practice 7

Practice 7
Write $\frac{3}{50}$ as a decimal.

✓**Concept Check** Suppose you are writing the fraction $\frac{9}{16}$ as a decimal. How do you know you have made a mistake if your answer is 1.735?

Objective B Comparing Decimals and Fractions

Now we can compare decimals and fractions by writing fractions as equivalent decimals.

Example 8 Insert $<$, $>$, or $=$ to form a true statement.

$$\frac{1}{8} \qquad 0.12$$

Solution: First we write $\frac{1}{8}$ as an equivalent decimal. Then we compare decimal places.

$$\begin{array}{r} 0.125 \\ 8\overline{)1.000} \\ \underline{-8} \\ 20 \\ \underline{-16} \\ 40 \\ \underline{-40} \\ 0 \end{array}$$

Original numbers	$\frac{1}{8}$	0.12
Decimals	0.125	0.120
Compare	0.125 $>$ 0.12	

Thus, $\frac{1}{8} > 0.12$

■ Work Practice 8

Practice 8
Insert $<$, $>$, or $=$ to form a true statement.

$$\frac{1}{5} \qquad 0.25$$

Answers
6. 0.6 **7.** 0.06 **8.** $<$

✓**Concept Check Answer**
$\frac{9}{16}$ is less than 1 while 1.735 is greater than 1.

Practice 9

Insert <, >, or = to form a true statement.

a. $\dfrac{1}{2}$ 0.54 **b.** $0.\overline{4}$ $\dfrac{4}{9}$

c. $\dfrac{5}{7}$ 0.72

Example 9 Insert <, >, or = to form a true statement.

$$0.\overline{7} \qquad \frac{7}{9}$$

Solution: We write $\dfrac{7}{9}$ as a decimal and then compare.

$$
\begin{array}{r}
0.77\ldots = 0.\overline{7} \\
9\overline{)7.00} \\
-6\,3 \\
\hline
70 \\
-63 \\
\hline
7
\end{array}
$$

Original numbers	$0.\overline{7}$	$\dfrac{7}{9}$
Decimals	$0.\overline{7}$	$0.\overline{7}$
Compare	$0.\overline{7} = 0.\overline{7}$	

Thus, $0.\overline{7} = \dfrac{7}{9}$

■ Work Practice 9

Practice 10

Write the numbers in order from smallest to largest.

a. $\dfrac{1}{3}, 0.302, \dfrac{3}{8}$

b. $1.26, 1\dfrac{1}{4}, 1\dfrac{2}{5}$

c. $0.4, 0.41, \dfrac{3}{7}$

Example 10 Write the numbers in order from smallest to largest.

$$\frac{9}{20}, \frac{4}{9}, 0.456$$

Solution:

Original numbers	$\dfrac{9}{20}$	$\dfrac{4}{9}$	0.456
Decimals	0.450	0.444 . . .	0.456
Compare in order	2nd	1st	3rd

Written in order, we have

$$
\begin{array}{ccc}
\text{1st} & \text{2nd} & \text{3rd} \\
\downarrow & \downarrow & \downarrow \\
\dfrac{4}{9}, & \dfrac{9}{20}, & 0.456
\end{array}
$$

■ Work Practice 10

Objective C Simplifying Expressions with Decimals and Fractions ▶

In the remaining examples, we will review the order of operations by simplifying expressions that contain decimals and fractions.

Order of Operations

1. Perform all operations within parentheses ( ), brackets [ ], or other grouping symbols such as fraction bars or square roots.
2. Evaluate any expressions with exponents.
3. Multiply or divide in order from left to right.
4. Add or subtract in order from left to right.

Answers

9. a. < **b.** = **c.** <

10. a. $0.302, \dfrac{1}{3}, \dfrac{3}{8}$ **b.** $1\dfrac{1}{4}, 1.26, 1\dfrac{2}{5}$

c. $0.4, 0.41, \dfrac{3}{7}$

Example 11 Simplify: $723.6 \div 1000 \times 10$

Solution: Multiply or divide in order from left to right.

$$723.6 \div 1000 \times 10 = 0.7236 \times 10 \quad \text{Divide.}$$
$$= 7.236 \quad \text{Multiply.}$$

■ Work Practice 11

Practice 11
Simplify: $897.8 \div 100 \times 10$

Example 12 Simplify: $-0.5(8.6 - 1.2)$

Solution: According to the order of operations, we simplify inside the parentheses first.

$$-0.5(8.6 - 1.2) = -0.5(7.4) \quad \text{Subtract.}$$
$$= -3.7 \quad \text{Multiply.}$$

■ Work Practice 12

Practice 12
Simplify: $-8.69(3.2 - 1.8)$

Example 13 Simplify: $\left(-\dfrac{13}{10}\right)^2 + 2.4$

Solution: Recall the meaning of an exponent.

$$\left(-\frac{13}{10}\right)^2 + 2.4 = \left(-\frac{13}{10}\right)\left(-\frac{13}{10}\right) + 2.4 \quad \text{Use the definition of an exponent.}$$
$$= \frac{169}{100} + 2.4 \quad \text{Multiply. The product of two negative numbers is a positive number.}$$
$$= 1.69 + 2.4 \quad \text{Write } \frac{169}{100} \text{ as a decimal.}$$
$$= 4.09 \quad \text{Add.}$$

■ Work Practice 13

Practice 13
Simplify: $\left(-\dfrac{7}{10}\right)^2 + 2.1$

Example 14 Simplify: $\dfrac{5.68 + (0.9)^2 \div 100}{0.2}$

Solution: First we simplify the numerator of the fraction. Then we divide.

$$\frac{5.68 + (0.9)^2 \div 100}{0.2} = \frac{5.68 + 0.81 \div 100}{0.2} \quad \text{Simplify } (0.9)^2.$$
$$= \frac{5.68 + 0.0081}{0.2} \quad \text{Divide.}$$
$$= \frac{5.6881}{0.2} \quad \text{Add.}$$
$$= 28.4405 \quad \text{Divide.}$$

■ Work Practice 14

Practice 14
Simplify: $\dfrac{20.06 - (1.2)^2 \div 10}{0.02}$

Answers
11. 89.78   **12.** -12.166 **13.** 2.59
14. 995.8

Objective D Solving Area Problems Containing Fractions and Decimals

Sometimes real-life problems contain both fractions and decimals. In the next example, we review the area of a triangle, and when values are substituted, the result may be an expression containing both fractions and decimals.

Practice 15

Find the area of the triangle.

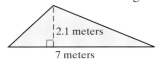

2.1 meters

7 meters

Example 15 The area of a triangle is Area $= \frac{1}{2} \cdot$ base $\cdot$ height. Find the area of the triangle shown.

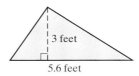

3 feet

5.6 feet

Solution:

$$\text{Area} = \frac{1}{2} \cdot \text{base} \cdot \text{height}$$

$$= \frac{1}{2} \cdot 5.6 \cdot 3$$

$$= 0.5 \cdot 5.6 \cdot 3 \qquad \text{Write } \frac{1}{2} \text{ as the decimal 0.5.}$$

$$= 8.4$$

The area of the triangle is 8.4 square feet.

■ Work Practice 15

Objective E Using Decimals as Replacement Values

Practice 16

Evaluate $1.7y - 2$ for $y = 2.3$.

Example 16 Evaluate $-2x + 5$ for $x = 3.8$.

Solution: Replace x with 3.8 in the expression $-2x + 5$ and simplify.

$$-2x + 5 = -2(3.8) + 5 \qquad \text{Replace } x \text{ with 3.8.}$$

$$= -7.6 + 5 \qquad \text{Multiply.}$$

$$= -2.6 \qquad \text{Add.}$$

Answers

15. 7.35 sq m   **16.** 1.91

■ Work Practice 16

Vocabulary, Readiness & Video Check

Answer each exercise "true" or "false."

1. The number $0.\overline{5}$ means 0.555.

2. To write $\frac{9}{19}$ as a decimal, perform the division $19\overline{)9}$.

3. $(-1.2)^2$ means $(-1.2)(-1.2)$ or -1.44.

4. To simplify $8.6(4.8 - 9.6)$, we first subtract.

See Video 4.5

Martin-Gay Interactive Videos Watch the section lecture video and answer the following questions.

Objective A 5. In ▱ Example 2, why is the bar placed over just the 6?

Objective B 6. In ▱ Example 3, why do we write the fraction as a decimal rather than the decimal as a fraction?

Objective C 7. In ▱ Example 4, besides meaning division, what other purpose does the fraction bar serve?

Objective D 8. What formula is used to solve ▱ Example 5? What is the final answer?

Objective E 9. In ▱ Example 6, once all replacement values are put into the variable expression, what is the resulting expression to evaluate?

4.5 Exercise Set MyMathLab®

Objective A *Write each number as a decimal. See Examples 1 through 7.*

1. $\dfrac{1}{5}$

2. $\dfrac{1}{20}$

3. $\dfrac{17}{25}$

4. $\dfrac{13}{25}$

5. $\dfrac{3}{4}$

6. $\dfrac{1}{8}$

7. $-\dfrac{2}{25}$

8. $-\dfrac{3}{25}$

9. $\dfrac{9}{4}$

10. $\dfrac{8}{5}$

11. $\dfrac{11}{12}$

12. $\dfrac{5}{12}$

13. $\dfrac{17}{40}$

14. $\dfrac{19}{25}$

15. $\dfrac{9}{20}$

16. $\dfrac{31}{40}$

17. $-\dfrac{1}{3}$

18. $-\dfrac{7}{9}$

19. $\dfrac{7}{16}$

20. $\dfrac{9}{16}$

21. $\dfrac{7}{11}$

22. $\dfrac{9}{11}$

23. $5\dfrac{17}{20}$

24. $4\dfrac{7}{8}$

25. $\dfrac{78}{125}$

26. $\dfrac{159}{375}$

Round each number as indicated. See Example 4.

27. Round your answer to Exercise 17 to the nearest hundredth.

28. Round your answer to Exercise 18 to the nearest hundredth.

29. Round your answer to Exercise 19 to the nearest hundredth.

30. Round your answer to Exercise 20 to the nearest hundredth.

31. Round your answer to Exercise 21 to the nearest tenth.

32. Round your answer to Exercise 22 to the nearest tenth.

Write each fraction as a decimal. If necessary, round to the nearest hundredth. See Examples 1 through 7.

33. Of the U.S. mountains that are over 14,000 feet in elevation, $\dfrac{56}{91}$ are located in Colorado. (*Source:* U.S. Geological Survey)

34. About $\dfrac{21}{50}$ of all blood donors have type A blood. (*Source:* American Red Cross Biomedical Services)

35. About $\frac{43}{50}$ of Americans are Internet users. (*Source:* Digitalcenter)

36. About $\frac{14}{25}$ of Americans use the Internet through a wireless device. (*Source:* Digitalcenter)

37. When first launched, the Hubble Space Telescope's primary mirror was out of shape on the edges by $\frac{1}{50}$ of a human hair. This very small defect made it difficult to focus faint objects being viewed. Because the HST was in low Earth orbit, it was serviced by a shuttle and the defect was corrected.

38. The two mirrors currently in use in the Hubble Space Telescope were ground so that they do not deviate from a perfect curve by more than $\frac{1}{800,000}$ of an inch. Do not round this number.

Objective B   *Insert $<$, $>$, or $=$ to form a true statement. See Examples 8 and 9.*

39. 0.562   0.569

40. 0.983   0.988

41. 0.215 $\frac{43}{200}$

42. $\frac{29}{40}$ 0.725

43. -0.0932 -0.0923

44. -0.00563 -0.00536

45. $0.\overline{6}$ $\frac{5}{6}$

46. $0.\overline{1}$ $\frac{2}{17}$

47. $\frac{51}{91}$ $0.56\overline{4}$

48. $0.58\overline{3}$ $\frac{6}{11}$

49. $\frac{4}{7}$ 0.14

50. $\frac{5}{9}$ 0.557

51. 1.38 $\frac{18}{13}$

52. 0.372 $\frac{22}{59}$

53. 7.123 $\frac{456}{64}$

54. 12.713 $\frac{89}{7}$

Write the numbers in order from smallest to largest. See Example 10.

55. $0.34, 0.35, 0.32$

56. $0.47, 0.42, 0.40$

57. $0.49, 0.491, 0.498$

58. $0.72, 0.727, 0.728$

59. $5.23, \frac{42}{8}, 5.34$

60. $7.56, \frac{67}{9}, 7.562$

61. $\frac{5}{8}, 0.612, 0.649$

62. $\frac{5}{6}, 0.821, 0.849$

Objective C   *Simplify each expression. See Examples 11 through 14.*

63. $(0.3)^2 + 0.5$

64. $(-2.5)(3) - 4.7$

65. $\dfrac{1 + 0.8}{-0.6}$

66. $(-0.05)^2 + 3.13$

67. $(-2.3)^2(0.3 + 0.7)$

68. $(8.2)(100) - (8.2)(10)$

69. $(5.6 - 2.3)(2.4 + 0.4)$

70. $\dfrac{0.222 - 2.13}{12}$

71. $\dfrac{(4.5)^2}{100}$

72. $0.9(5.6 - 6.5)$

73. $\dfrac{7 + 0.74}{-6}$

74. $(1.5)^2 + 0.5$

Find the value of each expression. Give the result as a decimal. See Examples 11 through 14.

75. $\frac{1}{5} - 2(7.8)$

76. $\frac{3}{4} - (9.6)(5)$

77. $\frac{1}{4}(-9.6 - 5.2)$

78. $\frac{3}{8}(4.7 - 5.9)$

Objective D *Find the area of each triangle or rectangle. See Example 15.*

△ **79.**

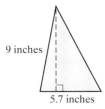

9 inches

5.7 inches

△ **80.**

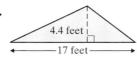

4.4 feet

17 feet

▶ **81.**

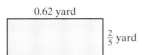

0.62 yard

$\frac{2}{5}$ yard

△ **82.**

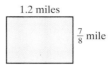

1.2 miles

$\frac{7}{8}$ mile

Objective E *Evaluate each expression for $x = 6$, $y = 0.3$, and $z = -2.4$. See Example 16.*

83. z^2 **84.** y^2 **85.** $x - y$ **86.** $x - z$ ▶ **87.** $4y - z$ **88.** $\dfrac{x}{y} + 2z$

Review

Simplify. See Sections 1.9 and 3.3.

89. $6^2 \cdot 2$ **90.** $4 \cdot 3^4$ **91.** $\left(\dfrac{2}{5}\right)\left(\dfrac{5}{2}\right)^2$ **92.** $\left(\dfrac{2}{3}\right)^2\left(\dfrac{3}{2}\right)^3$

Concept Extensions

Without calculating, describe each number as < 1, $= 1$, or > 1. See the Concept Check in this section.

93. 1.0 **94.** 1.0000 **95.** 1.00001 **96.** $\dfrac{101}{99}$ **97.** $\dfrac{99}{100}$ **98.** $\dfrac{99}{99}$

In 2013, there were 10,840 commercial radio stations in the United States. The most popular formats are shown in the graph along with their counts. Use this graph to answer Exercises 99 through 102.

99. Write as a decimal the fraction of radio stations with a classic hits music format. Round to the nearest thousandth.

100. Write as a decimal the fraction of radio stations with a Spanish format. Round to the nearest hundredth.

101. Estimate, by rounding each number in the graph to the nearest hundred, the total number of stations with the top six formats in 2013.

102. Use your estimate from Exercise 101 to write the fraction of radio stations accounted for by the top six formats as a decimal. Round to the nearest hundredth.

Top Commercial Radio Station Formats in 2013

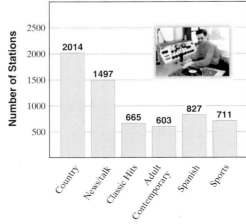

Format (Total stations: 10,840)

103. Describe two ways to write fractions as decimals.

104. Describe two ways to write mixed numbers as decimals.

Objectives

A Find the Square Root of a Number. ▷

B Approximate Square Roots. ▷

C Use the Pythagorean Theorem. ▷

Objective A Finding Square Roots ▷

The square of a number is the number times itself. For example:

The square of 5 is 25 because 5^2 or $5 \cdot 5 = 25$.
The square of 4 is 16 because 4^2 or $4 \cdot 4 = 16$.
The square of 10 is 100 because 10^2 or $10 \cdot 10 = 100$.

Recall from Chapter 1 that the reverse process of squaring is finding a **square root.** For example:

A square root of 16 is 4 because $4^2 = 16$.
A square root of 25 is 5 because $5^2 = 25$.
A square root of 100 is 10 because $10^2 = 100$.

We use the symbol $\sqrt{\ }$, called a **radical sign,** to name square roots. For example:

$\sqrt{16} = 4$ because $4^2 = 16$
$\sqrt{25} = 5$ because $5^2 = 25$

Square Root of a Number

A square root of a number a is a number b whose square is a. We use the radical sign $\sqrt{\ }$ to name square roots. In symbols,

$$\sqrt{a} = b \quad \text{if} \quad b^2 = a$$

Also,

$$\sqrt{0} = 0$$

Practice 1

Find each square root.
a. $\sqrt{100}$ **b.** $\sqrt{64}$
c. $\sqrt{121}$ **d.** $\sqrt{0}$

Example 1 Find each square root.

a. $\sqrt{49}$ **b.** $\sqrt{1}$ **c.** $\sqrt{81}$

Solution:

a. $\sqrt{49} = 7$ because $7^2 = 49$
b. $\sqrt{1} = 1$ because $1^2 = 1$
c. $\sqrt{81} = 9$ because $9^2 = 81$

■ Work Practice 1

Practice 2

Find: $\sqrt{\dfrac{1}{4}}$

Example 2 Find: $\sqrt{\dfrac{1}{36}}$

Solution: $\sqrt{\dfrac{1}{36}} = \dfrac{1}{6}$ because $\dfrac{1}{6} \cdot \dfrac{1}{6} = \dfrac{1}{36}$

■ Work Practice 2

Practice 3

Find: $\sqrt{\dfrac{9}{16}}$

Example 3 Find: $\sqrt{\dfrac{4}{25}}$

Solution: $\sqrt{\dfrac{4}{25}} = \dfrac{2}{5}$ because $\dfrac{2}{5} \cdot \dfrac{2}{5} = \dfrac{4}{25}$

■ Work Practice 3

Answers

1. a. 10 **b.** 8 **c.** 11 **d.** 0
2. $\dfrac{1}{2}$ **3.** $\dfrac{3}{4}$

Objective B Approximating Square Roots

Thus far, we have found square roots of perfect squares. Numbers like $\frac{1}{4}$, 36, $\frac{4}{25}$, and 1 are called **perfect squares** because their square root is a whole number or a fraction. A square root such as $\sqrt{5}$ cannot be written as a whole number or a fraction since 5 is not a perfect square.

Although $\sqrt{5}$ cannot be written as a whole number or a fraction, it can be approximated by estimating, by using a table (as in the appendix), or by using a calculator.

Example 4 Use Appendix A.6 or a calculator to approximate each square root to the nearest thousandth.

a. $\sqrt{43} \approx 6.557$ is approximately
b. $\sqrt{80} \approx 8.944$

■ Work Practice 4

Practice 4

Use Appendix A.6 or a calculator to approximate each square root to the nearest thousandth.
a. $\sqrt{21}$ **b.** $\sqrt{52}$

Helpful Hint

$\sqrt{80}$, above, is *approximately* 8.944. This means that if we multiply 8.944 by 8.944, the product is *close* to 80.

$$8.944 \times 8.944 \approx 79.995$$

It is possible to approximate a square root to the nearest whole number without the use of a calculator or table. To do so, study the number line below and look for patterns.

Above the number line, notice that as the numbers under the radical signs increase, their values, and thus their placement on the number line, increase also.

Example 5 Without a calculator or table:

a. Determine which two whole numbers $\sqrt{78}$ is between.
b. Use part a to approximate $\sqrt{78}$ to the nearest whole.

Solution:

a. Review perfect squares and recall that $\sqrt{64} = 8$ and $\sqrt{81} = 9$. Since 78 is between 64 and 81, $\sqrt{78}$ is between $\sqrt{64}$ (or 8) and $\sqrt{81}$ (or 9).

Thus, $\sqrt{78}$ is between 8 and 9.
b. Since 78 is closer to 81, then (as our number line shows) $\sqrt{78}$ is closer to $\sqrt{81}$, or 9.

■ Work Practice 5

Practice 5

Without a calculator or table, approximate $\sqrt{62}$ to the nearest whole.

Objective C Using the Pythagorean Theorem

One important application of square roots has to do with right triangles. Recall that a **right triangle** is a triangle in which one of the angles is a right angle, or measures 90° (degrees). The **hypotenuse** of a right triangle is the side opposite the right angle.

Answers
4. a. 4.583 **b.** 7.211
5. 8

The **legs** of a right triangle are the other two sides. These are shown in the following figure. The right angle in the triangle is indicated by the small square drawn in that angle.

The following theorem is true for all right triangles:

Pythagorean Theorem

In any right triangle,

$$(\text{leg})^2 + (\text{other leg})^2 = (\text{hypotenuse})^2$$

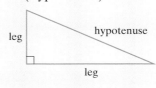

Using the Pythagorean theorem, we can use one of the following formulas to find an unknown length of a right triangle:

Finding an Unknown Length of a Right Triangle

$$\text{hypotenuse} = \sqrt{(\text{leg})^2 + (\text{other leg})^2}$$

or

$$\text{leg} = \sqrt{(\text{hypotenuse})^2 - (\text{other leg})^2}$$

Practice 6

Find the length of the hypotenuse of the given right triangle.

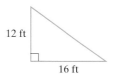

12 ft

16 ft

Example 6   Find the length of the hypotenuse of the given right triangle.

Solution: Since we are finding the hypotenuse, we use the formula

$$\text{hypotenuse} = \sqrt{(\text{leg})^2 + (\text{other leg})^2}$$

Putting the known values into the formula, we have

$$\text{hypotenuse} = \sqrt{(6)^2 + (8)^2} \quad \text{The legs are 6 feet and 8 feet.}$$
$$= \sqrt{36 + 64}$$
$$= \sqrt{100}$$
$$= 10$$

The hypotenuse is 10 feet long.

6 ft

8 ft

■ Work Practice 6

Practice 7

Approximate the length of the hypotenuse of the given right triangle. Round to the nearest whole unit.

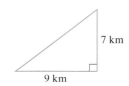

7 km

9 km

Answers

6. 20 ft   **7.** 11 km

Example 7   Approximate the length of the hypotenuse of the given right triangle. Round the length to the nearest whole unit.

Solution:

$$\text{hypotenuse} = \sqrt{(\text{leg})^2 + (\text{other leg})^2}$$
$$= \sqrt{(17)^2 + (10)^2} \quad \text{The legs are 10 meters and 17 meters.}$$
$$= \sqrt{289 + 100}$$
$$= \sqrt{389}$$
$$\approx 20 \quad \text{From Appendix A.6 or a calculator}$$

10 m

17 m

The hypotenuse is exactly $\sqrt{389}$ meters, which is approximately 20 meters.

■ Work Practice 7

Example 8   Find the length of the leg in the given right triangle. Give the exact length and a two-decimal-place approximation.

5 in.

7 in.

Solution: Notice that the hypotenuse measures 7 inches and the length of one leg measures 5 inches. Since we are looking for the length of the other leg, we use the formula

$$\text{leg} = \sqrt{(\text{hypotenuse})^2 - (\text{other leg})^2}$$

Putting the known values into the formula, we have

$$\text{leg} = \sqrt{(7)^2 - (5)^2} \qquad \text{The hypotenuse is 7 inches, and the other leg is 5 inches.}$$
$$= \sqrt{49 - 25}$$
$$= \sqrt{24} \qquad \text{Exact answer}$$
$$\approx 4.90 \qquad \text{From Appendix A.6 or a calculator}$$

The length of the leg is exactly $\sqrt{24}$ inches, which is approximately 4.90 inches.

■ Work Practice 8

✓**Concept Check**   The following lists are the lengths of the sides of two triangles. Which set forms a right triangle? Explain.
   a. 8, 15, 17   **b.** 24, 30, 40

△ Example 9   Finding the Dimensions of a Park

An inner-city park is in the shape of a square that measures 300 feet on a side. A sidewalk is to be constructed along the diagonal of the park. Find the length of the sidewalk rounded to the nearest foot.

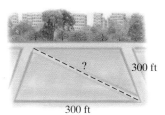

?   300 ft

300 ft

Solution: The diagonal is the hypotenuse of a right triangle, so we use the formula

$$\text{hypotenuse} = \sqrt{(\text{leg})^2 + (\text{other leg})^2}$$

Putting the known values into the formula, we have

$$\text{hypotenuse} = \sqrt{(300)^2 + (300)^2} \qquad \text{The legs are both 300 feet.}$$
$$= \sqrt{90,000 + 90,000}$$
$$= \sqrt{180,000}$$
$$\approx 424 \qquad \text{From Appendix A.6 or a calculator}$$

The length of the sidewalk is approximately 424 feet.

■ Work Practice 9

Practice 8

Find the length of the leg in the given right triangle. Give the exact length and a two-decimal-place approximation.

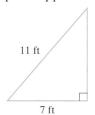

11 ft

7 ft

Practice 9

A football field is a rectangle measuring 100 yards by 53 yards. Draw a diagram and find the length of the diagonal of the football field to the nearest yard.

Answers
8. $\sqrt{72}$ ft ≈ 8.49 ft
9. 113 yd

✓**Concept Check Answer**

Set (a) forms a right triangle.

Calculator Explorations   Finding Square Roots

To simplify or approximate square roots using a calculator, locate the key marked $\boxed{\sqrt{}}$.

To simplify $\sqrt{64}$, for example, press the keys

$\boxed{64}$ $\boxed{\sqrt{}}$ or $\boxed{\sqrt{}}$ $\boxed{64}$ $\boxed{\text{ENTER}}$

The display should read $\boxed{8}$. Then

$\sqrt{64} = 8$

To *approximate* $\sqrt{10}$, press the keys

$\boxed{10}$ $\boxed{\sqrt{}}$ or $\boxed{\sqrt{}}$ $\boxed{10}$ $\boxed{\text{ENTER}}$

The display should read $\boxed{3.16227766}$. This is an *approximation* for $\sqrt{10}$. A three-decimal-place approximation is

$\sqrt{10} \approx 3.162$

Is this answer reasonable? Since 10 is between perfect squares 9 and 16, $\sqrt{10}$ is between $\sqrt{9} = 3$ and $\sqrt{16} = 4$. Our answer is reasonable since 3.162 is between 3 and 4.

Simplify.

1. $\sqrt{1024}$
2. $\sqrt{676}$

Approximate each square root. Round each answer to the nearest thousandth.

3. $\sqrt{31}$
4. $\sqrt{19}$
5. $\sqrt{97}$
6. $\sqrt{56}$

Vocabulary, Readiness & Video Check

Use the choices below to fill in each blank. Some choices may be used more than once.

squaring   Pythagorean theorem   radical   leg
hypotenuse   perfect squares   10

1. $\sqrt{100} =$ _____ because $10 \cdot 10 = 100$.

2. The _____ sign is used to denote the square root of a number.

3. The reverse process of _____ a number is finding a square root of a number.

4. The numbers 9, 1, and $\frac{1}{25}$ are called _____.

5. Label the parts of the right triangle. _____

6. The _____ can be used for right triangles.

Martin-Gay Interactive Videos   Watch the section lecture video and answer the following questions.

See Video 4.6

Objective A 7. From the lecture before ▤Example 1, explain why $\sqrt{49} = 7$. ▶

Objective B 8. In ▤Example 5, how do we know $\sqrt{15}$ is closer to 4 than to 3? ▶

Objective C 9. At the beginning of ▤Example 6, what are we reminded about regarding the Pythagorean theorem? ▶

4.6 **Exercise Set** MyMathLab®

Objective A *Find each square root. See Examples 1 through 3.*

1. $\sqrt{4}$

2. $\sqrt{9}$

3. $\sqrt{121}$

4. $\sqrt{144}$

5. $\sqrt{\dfrac{1}{81}}$

6. $\sqrt{\dfrac{1}{64}}$

7. $\sqrt{\dfrac{16}{64}}$

8. $\sqrt{\dfrac{36}{81}}$

Objective B *Use Appendix A.6 or a calculator to approximate each square root. Round the square root to the nearest thousandth. See Examples 4 and 5.*

9. $\sqrt{3}$

10. $\sqrt{5}$

11. $\sqrt{15}$

12. $\sqrt{17}$

13. $\sqrt{47}$

14. $\sqrt{85}$

15. $\sqrt{26}$

16. $\sqrt{35}$

Determine what two whole numbers each square root is between without using a calculator or table. Then use a calculator or Appendix A.6 to check. See Example 5.

17. $\sqrt{38}$

18. $\sqrt{27}$

19. $\sqrt{101}$

20. $\sqrt{85}$

Objectives A B Mixed Practice *Find each square root. If necessary, round the square root to the nearest thousandth. See Examples 1 through 5.*

21. $\sqrt{256}$

22. $\sqrt{625}$

23. $\sqrt{92}$

24. $\sqrt{18}$

25. $\sqrt{\dfrac{49}{144}}$

26. $\sqrt{\dfrac{121}{169}}$

27. $\sqrt{71}$

28. $\sqrt{62}$

Objective C *Find the unknown length in each right triangle. If necessary, approximate the length to the nearest thousandth. See Examples 6 through 8.*

29.

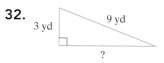

5 in. ? 12 in.

30.

? 15 ft 36 ft

31.

10 cm 12 cm ?

32.

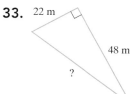

3 yd 9 yd ?

33.

22 m 48 m ?

34.

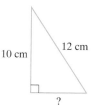

34 mi 70 mi ?

35.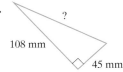

36.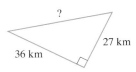

Sketch each right triangle and find the length of the side not given. If necessary, approximate the length to the nearest thousandth. (Each length is in units.) See Examples 6 through 8.

37. leg = 3, leg = 4

38. leg = 9, leg = 12

39. leg = 5, hypotenuse = 13

40. leg = 6, hypotenuse = 10

41. leg = 10, leg = 14

42. leg = 2, leg = 16

43. leg = 35, leg = 28

44. leg = 30, leg = 15

45. leg = 30, leg = 30

46. leg = 21, leg = 21

▶ **47.** hypotenuse = 2, leg = 1

48. hypotenuse = 9, leg = 8

49. leg = 7.5, leg = 4

50. leg = 12, leg = 22.5

Solve. See Example 9.

51. A standard city block is a square with each side measuring 100 yards. Find the length of the diagonal of a city block to the nearest hundredth yard.

52. A section of land is a square with each side measuring 1 mile. Find the length of the diagonal of the section of land to the nearest thousandth mile.

53. Find the height of the tree. Round the height to one decimal place.

54. Find the height of the antenna. Round the height to one decimal place.

55. The playing field for football is a rectangle that is 300 feet long by 160 feet wide. Find, to the nearest foot, the length of a straight-line run that started at one corner and went diagonally to end at the opposite corner.

56. A soccer field is in the shape of a rectangle and its dimensions depend on the age of the players. The dimensions of the soccer field below are the minimum dimensions for international play. Find the length of the diagonal of this rectangle. Round the answer to the nearest tenth of a yard.

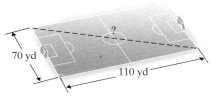

Review

Write each fraction in simplest form. See Section 3.2.

57. $\dfrac{10}{12}$ **58.** $\dfrac{10}{15}$ **59.** $\dfrac{24}{60}$

60. $\dfrac{35}{75}$ **61.** $\dfrac{30}{72}$ **62.** $\dfrac{18}{30}$

Concept Extensions

Use the results of Exercises 17–20 and approximate each square root to the nearest whole number without using a calculator or table. Then use a calculator or Appendix A.6 to check. See Example 5.

63. $\sqrt{38}$ **64.** $\sqrt{27}$ **65.** $\sqrt{101}$ **66.** $\sqrt{85}$

Solve.

67. Without using a calculator, explain how you know that $\sqrt{105}$ is *not* approximately 9.875.

68. Without using a calculator, explain how you know that $\sqrt{27}$ is *not* approximately 3.296.

Does the set form the lengths of the sides of a right triangle? See the Concept Check in this section.

69. 25, 60, 65   **70.** 20, 45, 50

71. Find the exact length of x. Then give a two-decimal-place approximation.

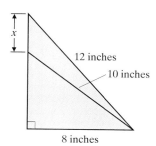

Chapter 4 Group Activity

Maintaining a Checking Account

(Sections 4.1, 4.2, 4.3, 4.4)

This activity may be completed by working in groups or individually.

A checking account is a convenient way of handling money and paying bills. To open a checking account, the bank or savings and loan association requires a customer to make a deposit. Then the customer receives a checkbook that contains checks, deposit slips, and a register for recording checks written and deposits made. It is important to record all payments and deposits that affect the account. It is also important to keep the checkbook balance current by subtracting checks written and adding deposits made.

About once a month, checking customers receive a statement from the bank listing all activity that the account has had in the last month. The statement lists a beginning balance, all checks and deposits, any service charges made against the account, and an ending balance. Because it may take several days for checks that a customer has written to clear the banking system, the check register may list checks that do not appear on the monthly bank statement. These checks are called **outstanding checks**. Deposits that are recorded in the check register but do not appear on the statement are called **deposits in transit**. Because of these differences, it is important to balance, or reconcile, the checkbook against the monthly statement. The steps for doing so are listed below.

Balancing or Reconciling a Checkbook

Step 1: Place a check mark in the checkbook register next to each check and deposit listed on the monthly bank statement. Any entries in the register without a check mark are outstanding checks or deposits in transit.

Step 2: Find the ending checkbook register balance and add to it any outstanding checks and any interest paid on the account.

Step 3: From the total in Step 2, subtract any deposits in transit and any service charges.

Step 4: Compare the amount found in Step 3 with the ending balance listed on the bank statement. If they are the same, the checkbook balances with the bank statement. Be sure to update the check register with service charges and interest.

Step 5: If the checkbook does not balance, recheck the balancing process. Next, make sure that the running checkbook register balance was calculated correctly. Finally, compare the checkbook register with the statement to make sure that each check was recorded for the correct amount.

For the checkbook register and monthly bank statement given:

a. *update the checkbook register*   **b.** *list the outstanding checks and deposits in transit and the totals of these*

c. *balance the checkbook—be sure to update the register with any interest or service fees*

#	Date	Description	Payment	✓	Deposit	Balance
		Checkbook Register				425.86
114	4/1	Market Basket	30.27			
115	4/3	May's Texaco	8.50			
	4/4	Cash at ATM	50.00			
116	4/6	UNO Bookstore	121.38			
	4/7	Deposit			100.00	
117	4/9	MasterCard	84.16			
118	4/10	Blockbuster	6.12			
119	4/12	Kroger	18.72			
120	4/14	Parking sticker	18.50			
	4/15	Direct deposit			294.36	
121	4/20	Rent	395.00			
122	4/25	Student fees	20.00			
	4/28	Deposit			75.00	

First National Bank Monthly Statement 4/30

BEGINNING BALANCE:		425.86
Date	Number	Amount
CHECKS AND ATM WITHDRAWALS		
4/3	114	30.27
4/4	ATM	50.00
4/11	117	84.16
4/13	115	8.50
4/15	119	18.72
4/22	121	395.00
DEPOSITS		
4/7		100.00
4/15	Direct deposit	294.36
SERVICE CHARGES		
Low balance fee		7.50
INTEREST		
Credited 4/30		1.15
ENDING BALANCE:		227.22

Chapter 4 Vocabulary Check

Fill in each blank with one of the words listed below.

vertically	decimal	and	right triangle	hypotenuse	legs
sum	denominator	numerator	square root	standard form	

1. Like fractional notation, _____ notation is used to denote a part of a whole.

2. To write fractions as decimals, divide the _____ by the _____.

3. To add or subtract decimals, write the decimals so that the decimal points line up _____.

4. When writing decimals in words, write "_____" for the decimal point.

5. When multiplying decimals, the decimal point in the product is placed so that the number of decimal places in the product is equal to the _____ of the number of decimal places in the factors.

6. The _____, $\sqrt{}$, of a positive number a is the positive number b whose square is a.

7. A _____ is a triangle with a right angle. The side opposite the right angle is called the _____, and the other two sides are called _____.

8. When 2 million is written as 2,000,000, we say it is written in _____.

> **Helpful Hint**
> ● Are you preparing for your test? Don't forget to take the Chapter 4 Test on page 355. Then check your answers at the back of the text and use the Chapter Test Prep Videos to see the fully worked-out solutions to any of the exercises you want to review.

 4 Chapter Highlights

Definitions and Concepts	Examples

Section 4.1 Introduction to Decimals

Place-Value Chart

hundreds	tens	ones	. decimal point	tenths	hundredths	thousandths	ten-thousandths	hundred-thousandths
		4		2	6	5		
100	10	1		$\frac{1}{10}$	$\frac{1}{100}$	$\frac{1}{1000}$	$\frac{1}{10,000}$	$\frac{1}{100,000}$

4.265 means

$$4 \cdot 1 + 2 \cdot \frac{1}{10} + 6 \cdot \frac{1}{100} + 5 \cdot \frac{1}{1000}$$

or

$$4 + \frac{2}{10} + \frac{6}{100} + \frac{5}{1000}$$

(continued)

Definitions and Concepts	Examples

Section 4.1   Introduction to Decimals (*continued*)

Writing (or Reading) a Decimal in Words

Step 1: Write the whole number part in words.

Step 2: Write "and" for the decimal point.

Step 3: Write the decimal part in words as though it were a whole number, followed by the place value of the last digit.

Write 3.08 in words.

Three and eight hundredths

A decimal written in words can be written in standard form by reversing the above procedure.

Write "negative four and twenty-one thousandths" in standard form.

$$-4.021$$

To Round a Decimal to a Place Value to the Right of the Decimal Point

Step 1: Locate the digit to the right of the given place value.

Step 2: If this digit is 5 or greater, add 1 to the digit in the given place value and delete all digits to its right. If this digit is less than 5, delete all digits to the right of the given place value.

Round 86.1256 to the nearest hundredth.

Step 1: 86.12 5 6
hundredths place
digit to the right

Step 2: Since the digit to the right is 5 or greater, we add 1 to the digit in the hundredths place and delete all digits to its right.

86.1256 rounded to the nearest hundredth is 86.13.

Section 4.2   Adding and Subtracting Decimals

To Add or Subtract Decimals

Step 1: Write the decimals so that the decimal points line up vertically.

Step 2: Add or subtract as with whole numbers.

Step 3: Place the decimal point in the sum or difference so that it lines up vertically with the decimal points in the problem.

Add: 4.6 + 0.28   Subtract: 2.8 − 1.04

```
   4.60              2.8 0
 + 0.28            − 1.0 4
 ------            -------
   4.88              1.7 6
```

Section 4.3   Multiplying Decimals and Circumference of a Circle

To Multiply Decimals

Step 1: Multiply the decimals as though they are whole numbers.

Step 2: The decimal point in the product is placed so that the number of decimal places in the product is equal to the *sum* of the number of decimal places in the factors.

Multiply: 1.48 × 5.9

```
   1.48    ← 2 decimal places
 × 5.9     ← 1 decimal place
 ------
  1332
 7400
 ------
 8.732     ← 3 decimal places
```

Definitions and Concepts	Examples

Section 4.3   Multiplying Decimals and Circumference of a Circle *(continued)*

The **circumference** of a circle is the distance around the circle.

$$C = \pi d \text{ or } C = 2\pi r$$

where $\pi \approx 3.14$ or $\pi \approx \dfrac{22}{7}$.

or

Find the exact circumference of a circle with radius 5 miles and an approximation by using 3.14 for π.

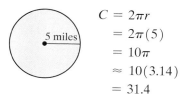

$$\begin{aligned} C &= 2\pi r \\ &= 2\pi(5) \\ &= 10\pi \\ &\approx 10(3.14) \\ &= 31.4 \end{aligned}$$

The circumference is exactly 10π miles and approximately 31.4 miles.

Section 4.4   Dividing Decimals

To Divide Decimals

Step 1: If the divisor is not a whole number, move the decimal point in the divisor to the right until the divisor is a whole number.

Step 2: Move the decimal point in the dividend to the right the *same number of places* as the decimal point was moved in Step 1.

Step 3: Divide. The decimal point in the quotient is directly over the moved decimal point in the dividend.

Divide: $1.118 \div 2.6$

$$\begin{array}{r} 0.43 \\ 2.6\overline{)1.118} \\ -104 \\ \hline 78 \\ -78 \\ \hline 0 \end{array}$$

Section 4.5   Fractions, Decimals, and Order of Operations

To **write fractions as decimals,** divide the numerator by the denominator.

Write $\dfrac{3}{8}$ as a decimal.

$$\begin{array}{r} 0.375 \\ 8\overline{)3.000} \\ -2\,4 \\ \hline 60 \\ -56 \\ \hline 40 \\ -40 \\ \hline 0 \end{array}$$

Order of Operations

1. Perform all operations within parentheses ( ), brackets [ ], or grouping symbols such as fraction bars or square roots.
2. Evaluate any expressions with exponents.
3. Multiply or divide in order from left to right.
4. Add or subtract in order from left to right.

Simplify.

$$\begin{aligned} -1.9(12.8 - 4.1) &= -1.9(8.7) \quad \text{Subtract.} \\ &= -16.53 \quad \text{Multiply.} \end{aligned}$$

Definitions and Concepts	Examples

Section 4.6   Square Roots and the Pythagorean Theorem

Square Root of a Number

A **square root** of a number a is a number b whose square is a. We use the radical sign $\sqrt{}$ to indicate square roots.

$\sqrt{9} = 3$, $\sqrt{100} = 10$, $\sqrt{1} = 1$

Pythagorean Theorem

$$(\text{leg})^2 + (\text{other leg})^2 = (\text{hypotenuse})^2$$

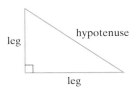

To Find an Unknown Length of a Right Triangle

$$\text{hypotenuse} = \sqrt{(\text{leg})^2 + (\text{other leg})^2}$$

$$\text{leg} = \sqrt{(\text{hypotenuse})^2 - (\text{other leg})^2}$$

Find the hypotenuse of the given triangle.

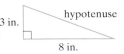

$\text{hypotenuse} = \sqrt{(\text{leg})^2 + (\text{other leg})^2}$

$\phantom{\text{hypotenuse}} = \sqrt{(3)^2 + (8)^2}$ The legs are 3 and 8 inches.

$\phantom{\text{hypotenuse}} = \sqrt{9 + 64}$

$\phantom{\text{hypotenuse}} = \sqrt{73}$ inches

$\phantom{\text{hypotenuse}} \approx 8.5$ inches

Chapter 4   Review

(4.1) *Determine the place value of the number 4 in each decimal.*

1. 23.45

2. 0.000345

Write each decimal in words.

3. −23.45

4. 0.00345

5. 109.23

6. 200.000032

Write each decimal in standard form.

7. Two and seven hundredths

8. Negative five hundred three and one hundred two thousandths

9. Sixteen thousand twenty-five and fourteen ten-thousandths

10. Fourteen and eleven thousandths

Write each decimal as a fraction or a mixed number.

11. 0.16

12. 0.55

13. −12.023

14. 25.25

Insert <, >, or = between each pair of numbers to make a true statement.

15. 0.49   0.43

16. 0.973   0.9730

17. −402.00032   −402.000032

18. −0.230505   −0.23505

Round each decimal to the given place value.

19. 0.623, nearest tenth

20. 0.9384, nearest hundredth

21. −42.895, nearest hundredth

22. −16.34925, nearest thousandth

(4.2) *Add.*

23. 2.4 + 7.1

24. 3.9 + 1.2

25. −6.4 + (−0.88)

26. −19.02 + 6.98

27. 200.49 + 16.82 + 103.002

28. 0.00236 + 100.45 + 48.29

Subtract.

29. 4.9 − 3.2

30. 5.23 − 2.74

31. −892.1 − 432.4

32. 0.064 − 10.2

33. 100 − 34.98

34. 200 − 0.00198

Solve.

35. Find the total distance between Grove City and Jerome.

36. Evaluate $x - y$ for $x = 1.2$ and $y = 6.9$.

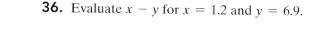
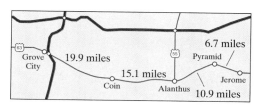

△ **37.** Find the perimeter.

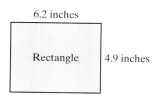

△ **38.** Find the perimeter.

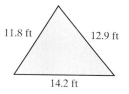

(4.3) *Multiply.*

39. 7.2 × 10

40. 9.345 × 1000

41. −34.02 × 2.3

42. −839.02 × (−87.3)

Write each number in standard notation. (See Section 4.1 and Section 4.3.)

43. Saturn is a distance of about 887 million miles from the Sun.

44. The tail of a comet can be over 600 thousand miles long.

Find the exact circumference of each circle. Then use the approximation 3.14 for π and approximate the circumference.

△ **45.**

7 meters

△ **46.**

20 inches

(4.4) *Divide. Round the quotient to the nearest thousandth if necessary.*

47. $3\overline{)0.2631}$

48. $20\overline{)316.5}$

49. $-21 \div (-0.3)$

50. $-0.0063 \div 0.03$

51. $0.34\overline{)2.74}$

52. $19.8\overline{)601.92}$

53. $\dfrac{23.65}{1000}$

54. $\dfrac{-93}{10}$

55. There are approximately 3.28 feet in 1 meter. Find how many meters are in 24 feet to the nearest tenth of a meter.

⟵——1 meter——⟶
⟵—— ≈3.28 feet ——⟶

56. George Strait pays $69.71 per month to pay back a loan of $3136.95. In how many months will the loan be paid off?

(4.5) *Write each fraction or mixed number as a decimal. Round to the nearest thousandth if necessary.*

57. $\dfrac{4}{5}$

58. $-\dfrac{12}{13}$

59. $2\dfrac{1}{3}$

60. $\dfrac{13}{60}$

Insert $<, >,$ or $=$ to make a true statement.

61. 0.392 ___ 0.39200

62. -0.0231 ___ -0.0221

63. $\dfrac{4}{7}$ ___ 0.625

64. 0.293 ___ $\dfrac{5}{17}$

Write the numbers in order from smallest to largest.

65. $0.837, 0.839, 0.832$

66. $0.685, 0.626, \dfrac{5}{8}$

67. $\dfrac{3}{7}, 0.42, 0.43$

68. $\dfrac{18}{11}, 1.63, \dfrac{19}{12}$

Simplify each expression.

69. $-7.6 \times 1.9 + 2.5$

70. $(-2.3)^2 - 1.4$

71. $0.0726 \div 10 \times 1000$

72. $0.6(2 - 0.65)$

73. $\dfrac{(1.5)^2 + 0.5}{0.05}$

74. $\dfrac{5 + 2.74}{-0.06}$

Find each area.

△ **75.**

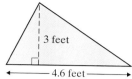

3 feet

4.6 feet

△ **76.**

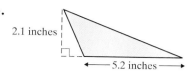

2.1 inches

5.2 inches

(4.6) *Simplify.*

77. $\sqrt{64}$

78. $\sqrt{144}$

79. $\sqrt{\dfrac{4}{25}}$

80. $\sqrt{\dfrac{1}{100}}$

Find the unknown length of each given right triangle. If necessary, round to the nearest tenth.

81. leg = 12, leg = 5

82. leg = 20, leg = 21

83. leg = 9, hypotenuse = 14

84. leg = 124, hypotenuse = 155

85. A baseball diamond is in the shape of a square and has sides of length 90 feet. Find the distance across the diamond from third base to first base, to the nearest tenth of a foot.

90 ft

?

86. Find the height of the building rounded to the nearest tenth.

126 ft

?

90 ft

Mixed Review

87. Write 200.0032 in words.

88. Write negative sixteen thousand twenty-five and fourteen thousandths in standard form.

89. Write 0.00231 as a fraction or a mixed number.

90. Write the numbers $\dfrac{6}{7}, \dfrac{8}{9}$, 0.75 in order from smallest to largest.

Write each fraction as a decimal.

91. $-\dfrac{7}{100}$

92. $\dfrac{9}{80}$

Insert $<, >,$ *or* $=$ *to make a true statement.*

93. -402.000032 _____ -402.00032

94. $\dfrac{6}{11}$ _____ 0.55

Round each decimal to the given place value.

95. 42.895, nearest hundredth

96. 16.34925, nearest thousandth

Round each money amount to the nearest dollar.

97. $123.46

98. $3645.52

Add or subtract as indicated.

99. 3.2 − 4.9

100. 5.23 − 2.74

101. 200.49 − 16.82 − 103.002

102. −0.00236 + (−100.45) + (−48.29)

Multiply or divide as indicated. Round to the nearest thousandth, if necessary.

103. $\begin{array}{r} 2.54 \\ \times\ 3.2 \\ \hline \end{array}$

104. (−3.45)(2.1)

105. $0.005\overline{)24.5}$

106. $2.3\overline{)54.98}$

Solve.

107. Tomaso is going to fertilize his lawn, a rectangle that measures 77.3 feet by 115.9 feet. Approximate the area of the lawn by rounding each measurement to the nearest ten feet.

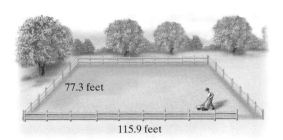
77.3 feet
115.9 feet

108. Estimate the cost of the items to see whether the groceries can be purchased with a $5 bill.

$1.89
100% WHOLE WHEAT
$1.07
3 cans for $0.99

Simplify each expression.

109. $\dfrac{(3.2)^2}{100}$

110. (2.6 + 1.4)(4.5 − 3.6)

Simplify.

111. $\sqrt{1}$

112. $\sqrt{36}$

113. $\sqrt{\dfrac{16}{81}}$

114. $\sqrt{\dfrac{1}{121}}$

Find the unknown length of each given right triangle. If necessary, round to the nearest tenth.

115. leg = 66, leg = 56

116. leg = 12, hypotenuse = 24

117. leg = 17, hypotenuse = 51

118. leg = 10, leg = 17

Write each decimal as indicated.

Answers

1. 45.092, in words

2. Three thousand and fifty-nine thousandths, in standard form

Perform each indicated operation. Round the result to the nearest thousandth if necessary.

3. $2.893 + 4.21 + 10.492$ **4.** $-47.92 - 3.28$ **5.** $9.83 - 30.25$

6. 10.2×4.01 **7.** $(-0.00843) \div (-0.23)$

Round each decimal to the indicated place value.

8. 34.8923, nearest tenth

9. 0.8623, nearest thousandth

Insert $<$, $>$, or $=$ between each pair of numbers to form a true statement.

10. 25.0909   25.9090

11. $\dfrac{4}{9}$ 0.445

Write each decimal as a fraction or a mixed number.

12. 0.345

13. -24.73

Write each fraction as a decimal. If necessary, round to the nearest thousandth.

14. $-\dfrac{13}{26}$

15. $\dfrac{16}{17}$

Simplify.

16. $(-0.6)^2 + 1.57$ **17.** $\dfrac{0.23 + 1.63}{-0.3}$ **18.** Subtract 8.6 from 20.

1. _____

2. _____

3. _____

4. _____

5. _____

6. _____

7. _____

8. _____

9. _____

10. _____

11. _____

12. _____

13. _____

14. _____

15. _____

16. _____

17. _____

18. _____

19. _____

20. _____

21. _____

22. _____

23. _____

24. _____

25. a. _____

b. _____

26. _____

Find each square root and simplify. Round to the nearest thousandth if necessary.

19. $\sqrt{49}$

20. $\sqrt{157}$

21. $\sqrt{\dfrac{64}{100}}$

Solve.

22. At its farthest, Pluto is 4583 million miles from the Sun. Write this number using standard notation.

△ **23.** Find the area.

1.1 miles

4.2 miles

△ **24.** Find the exact circumference of the circle. Then use the approximation 3.14 for π and approximate the circumference.

9 miles

25. Vivian Thomas is going to put insecticide on her lawn to control grubworms. The lawn is a rectangle that measures 123.8 feet by 80 feet. The amount of insecticide required is 0.02 ounce per square foot.

  a. Find the area of her lawn.

  b. Find how much insecticide Vivian needs to purchase.

26. Find the total distance from Bayette to Center City.

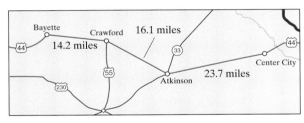

Write each number in words.

1. 85

2. 107

3. 126

4. 5026

5. Add: 23 + 136

6. Find the perimeter.

3 in.　　7 in.

9 in.

7. Subtract: 543 − 29. Check by adding.

8. Divide: 3268 ÷ 27

9. Round 278,362 to the nearest thousand.

10. Write the prime factorization of 30.

11. Multiply: 236 × 86

12. Multiply: 236 × 86 × 0

13. Find each quotient. Check by multiplying.
   a. $1\overline{)7}$
   b. 12 ÷ 1
   c. $\dfrac{6}{6}$
   d. 9 ÷ 9
   e. $\dfrac{20}{1}$
   f. $18\overline{)18}$

14. Find the average of 25, 17, 19, and 39.

15. The Hudson River in New York State is 306 miles long. The Snake River, in the northwestern United States, is 732 miles longer than the Hudson River. How long is the Snake River? (*Source:* U.S. Department of the Interior)

16. Evaluate: $\sqrt{121}$

Answers

1. _____

2. _____

3. _____

4. _____

5. _____

6. _____

7. _____

8. _____

9. _____

10. _____

11. _____

12. _____

13. a. _____

b. _____

c. _____

d. _____

e. _____

f. _____

14. _____

15. _____

16. _____

17. _____

18. _____

19. _____

20. _____

21. _____

22. _____

23. a. _____

  b. _____

  c. _____

24. a. _____

  b. _____

  c. _____

25. _____

26. _____

27. _____

28. _____

29. _____

30. _____

31. _____

32. _____

33. _____

34. _____

35. _____

36. _____

Evaluate.

17. 9^2 **18.** 5^3 **19.** 3^4 **20.** 10^3

21. Evaluate $\dfrac{x - 5y}{y}$ for $x = 21$ and $y = 3$.   **22.** Evaluate $\dfrac{2a + 4}{c}$ for $a = 7$ and $c = 3$.

23. Find the opposite of each number.
  a. 11   **b.** -2 **c.** 0

24. Find the opposite of each number.
  a. -7 **b.** 4   **c.** -1

25. Add: $-2 + (-21)$ **26.** Add: $-7 + (-15)$

Find the value of each expression.

27. $5 \cdot 6^2$ **28.** $4 \cdot 2^3$ **29.** -7^2

30. $(-2)^5$ **31.** $(-5)^2$ **32.** -3^2

Write the shaded part as an improper fraction and a mixed number.

33. 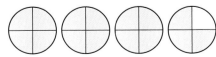   **34.**

35. 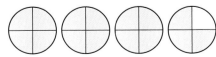   **36.**

37. Write the prime factorization of 252.

38. Find the difference of 87 and 25.

39. Write $-\dfrac{72}{26}$ in simplest form.

40. Write $9\dfrac{7}{8}$ as an improper fraction.

41. Determine whether $\dfrac{16}{40}$ and $\dfrac{10}{25}$ are equivalent.

42. Insert $<$ or $>$ to form a true statement. $\dfrac{4}{7}$ $\dfrac{5}{9}$

Multiply.

43. $\dfrac{2}{3} \cdot \dfrac{5}{11}$

44. $2\dfrac{5}{8} \cdot \dfrac{4}{7}$

45. $\dfrac{1}{4} \cdot \dfrac{1}{2}$

46. $7 \cdot 5\dfrac{2}{7}$

47. Add: $763.7651 + 22.001 + 43.89$

48. Add: $89.27 + 14.361 + 127.2318$

49. Multiply: 23.6×0.78

50. Multiply: 43.8×0.645

37. _____

38. _____

39. _____

40. _____

41. _____

42. _____

43. _____

44. _____

45. _____

46. _____

47. _____

48. _____

49. _____

50. _____

5 Ratio, Proportion, and Measurement

Having studied fractions in Chapter 3, we are ready to explore the useful notions of ratio and proportion. Ratio is another name for quotient and can be written in fraction form. A proportion is an equation with two equal ratios. In the second half of this chapter, we study the important U.S. and metric systems of measurement.

One of many 3-D cameras

Note: To date, there continues to be much research and controversy over 3-D films. The main controversy is focused on inferior conversions of 2-D-filmed movies to 3-D. The main research is focused on watching 3-D movies with adequate light and glasses or without glasses at all.

How Popular Are 3-D Films Now?

A 3-D (three-dimensional) film is a film that enhances the illusion of depth perception. Believe it or not, 3-D films have existed in some form since 1890, but because of high cost and lack of a standardized format, these films are only now starting to be widely shown and produced. The graph shows the trends in the releases of both 2-D and 3-D films in recent years.

In Section 5.1, Exercise 25, we calculate the ratio of 3-D films to total films.

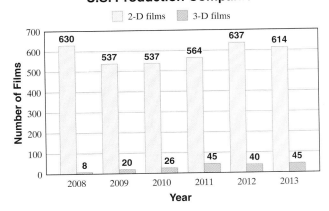

Films Released by U.S. Production Companies

Source: Motion Picture Association of America

360

Objective A Writing Ratios as Fractions

A **ratio** is the quotient of two quantities. A ratio, in fact, is no different from a fraction, except that a ratio is sometimes written using notation other than fractional notation. For example, the ratio of 1 to 2 can be written as

$$1 \text{ to } 2 \quad \text{or} \quad \frac{1}{2} \quad \text{or} \quad 1:2$$

fractional notation   colon notation

These ratios are all read as, "the ratio of 1 to 2."

✓**Concept Check** How should each ratio be read aloud?

a. $\frac{8}{5}$ **b.** $\frac{5}{8}$

In this section, we write ratios using fractional notation. If the fraction happens to be an improper fraction, do not write the fraction as a mixed number. Why? The mixed number form is not a ratio or quotient of two quantities.

Writing a Ratio as a Fraction

The order of the quantities is important when writing ratios. To write a ratio as a fraction, write the *first number* of the ratio as the *numerator* of the fraction and the *second number* as the *denominator*.

Helpful Hint

The ratio of 6 to 11 is $\frac{6}{11}$, *not* $\frac{11}{6}$.

Example 1 Write the ratio of 12 to 17 using fractional notation.

Solution: The ratio is $\frac{12}{17}$.

Helpful Hint Don't forget that order is important when writing ratios. The ratio $\frac{17}{12}$ is *not* the same as the ratio $\frac{12}{17}$.

Work Practice 1

To simplify a ratio, we just write the fraction in simplest form. Common factors as well as common units can be divided out.

Example 2 Write the ratio of $15 to $10 as a fraction in simplest form.

Solution:

$$\frac{\$15}{\$10} = \frac{15}{10} = \frac{3 \cdot \cancel{5}}{2 \cdot \cancel{5}} = \frac{3}{2}$$

Work Practice 2

Objectives

A Write Ratios as Fractions.

B Write Rates as Fractions.

C Find Unit Rates.

D Find Unit Prices.

Practice 1

Write the ratio of 20 to 23 using fractional notation.

Practice 2

Write the ratio of $8 to $6 as a fraction in simplest form.

Answers

1. $\frac{20}{23}$ **2.** $\frac{4}{3}$

✓**Concept Check Answers**

a. "eight to five"   **b.** "five to eight"

Helpful Hint

The ratio answer to Example 2 is $\frac{3}{2}$. Although $\frac{3}{2} = 1\frac{1}{2}$, a ratio is a quotient of *two* quantities. For that reason, ratios are not written as mixed numbers.

If a ratio contains decimal numbers or mixed numbers, we simplify by writing the ratio as a ratio of whole numbers.

Practice 3

Write the ratio of 3.9 to 8.8 as a fraction in simplest form.

Example 3 Write the ratio of 2.6 to 3.1 as a fraction in simplest form.

Solution: The ratio in fraction form is

$$\frac{2.6}{3.1}$$

Now let's clear the ratio of decimals.

$$\frac{2.6}{3.1} = \frac{2.6}{3.1} \cdot 1 = \frac{2.6}{3.1} \cdot \frac{10}{10} = \frac{2.6 \cdot 10}{3.1 \cdot 10} = \frac{26}{31} \quad \text{Simplest form}$$

■ Work Practice 3

Practice 4

Write the ratio of $2\frac{2}{3}$ to $1\frac{13}{15}$ as a fraction in simplest form.

Example 4 Write the ratio of $1\frac{1}{5}$ to $2\frac{7}{10}$ as a fraction in simplest form.

Solution: The ratio in fraction form is $\dfrac{1\frac{1}{5}}{2\frac{7}{10}}$.

To simplify, remember that the fraction bar means division.

$$\frac{1\frac{1}{5}}{2\frac{7}{10}} = 1\frac{1}{5} \div 2\frac{7}{10} = \frac{6}{5} \div \frac{27}{10} = \frac{6}{5} \cdot \frac{10}{27} = \frac{6 \cdot 10}{5 \cdot 27} = \frac{2 \cdot 3 \cdot 2 \cdot 5}{5 \cdot 3 \cdot 3 \cdot 3} = \frac{4}{9} \quad \text{Simplest form}$$

■ Work Practice 4

Practice 5

Use the circle graph for Example 5 to write the ratio of work miles to total miles as a fraction in simplest form.

Example 5 Writing a Ratio from a Circle Graph

The circle graph at the right shows the part of a car's total mileage that falls into a particular category. Write the ratio of family business miles to total miles as a fraction in simplest form.

Solution:

$$\frac{\text{family business miles}}{\text{total miles}} = \frac{3000 \text{ miles}}{15{,}000 \text{ miles}}$$

$$= \frac{3000}{15{,}000}$$

$$= \frac{3000}{5 \cdot 3000}$$

$$= \frac{1}{5}$$

Work   4800 miles
Medical   150 miles
Vacation/other   900 miles
Visit friends   1800 miles
Shopping   1800 miles
School/church   600 miles
Social/recreational   1950 miles
Family business   3000 miles

Total yearly mileage: 15,000

Sources: The American Automobile Manufacturers Association and The National Automobile Dealers Association.

■ Work Practice 5

Answers

3. $\frac{39}{88}$ **4.** $\frac{10}{7}$ **5.** $\frac{8}{25}$

Example 6 Given the rectangle shown:

a. Find the ratio of its width to its length.
b. Find the ratio of its length to its perimeter.

8 feet

5 feet

Solution:

a. The ratio of its width to its length is

$$\frac{\text{width}}{\text{length}} = \frac{5 \text{ feet}}{8 \text{ feet}} = \frac{5}{8}$$

b. Recall that the perimeter of the rectangle is the distance around the rectangle:
$8 + 5 + 8 + 5 = 26$ feet. The ratio of its length to its perimeter is

$$\frac{\text{length}}{\text{perimeter}} = \frac{8 \text{ feet}}{26 \text{ feet}} = \frac{8}{26} = \frac{\overset{1}{2} \cdot 2 \cdot 2}{\underset{1}{2} \cdot 13} = \frac{4}{13}$$

■ Work Practice 6

✓**Concept Check** Explain why the answer $\frac{8}{5}$ would be incorrect for part a of Example 6.

Practice 6

Given the triangle shown:

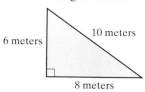

10 meters

6 meters

8 meters

a. Find the ratio of the length of the shortest side to the length of the longest side.
b. Find the ratio of the length of the longest side to the perimeter of the triangle.

Objective B Writing Rates as Fractions

A special type of ratio is a rate. **Rates** are used to compare *different* kinds of quantities. For example, suppose that a recreational runner can run 3 miles in 33 minutes. If we write this rate as a fraction, we have

$$\frac{3 \text{ miles}}{33 \text{ minutes}} = \frac{1 \text{ mile}}{11 \text{ minutes}} \quad \text{In simplest form}$$

Helpful Hint

When comparing quantities with different units, write the units as part of the comparison. They do not divide out.

Same Units: $\dfrac{3 \text{ inches}}{12 \text{ inches}} = \dfrac{1}{4}$

Different Units: $\dfrac{2 \text{ miles}}{20 \text{ minutes}} = \dfrac{1 \text{ mile}}{10 \text{ minutes}}$ Units are still written.

Examples Write each rate as a fraction in simplest form.

7. $2160 for 12 weeks is $\dfrac{2160 \text{ dollars}}{12 \text{ weeks}} = \dfrac{180 \text{ dollars}}{1 \text{ week}}$

8. 360 miles on 16 gallons of gasoline is $\dfrac{360 \text{ miles}}{16 \text{ gallons}} = \dfrac{45 \text{ miles}}{2 \text{ gallons}}$

■ Work Practice 7–8

Practice 7–8

Write each rate as a fraction in simplest form.
7. $1680 for 8 weeks
8. 236 miles on 12 gallons of gasoline

Answers

6. a. $\dfrac{3}{5}$ b. $\dfrac{5}{12}$ 7. $\dfrac{\$210}{1 \text{ wk}}$ 8. $\dfrac{59 \text{ mi}}{3 \text{ gal}}$

✓**Concept Check Answer**

$\dfrac{8}{5}$ would be the ratio of the rectangle's length to its width.

✓**Concept Check** True or false? $\dfrac{16 \text{ gallons}}{4 \text{ gallons}}$ is a rate. Explain.

Objective C Finding Unit Rates

A **unit rate** is a rate with a denominator of 1. A familiar example of a unit rate is 55 mph, read as "55 **miles per hour**." This means 55 miles per 1 hour or

$$\dfrac{55 \text{ miles}}{1 \text{ hour}}$$ Denominator of 1

> **Helpful Hint** In this context, the word "per" translates to division.

Writing a Rate as a Unit Rate

To write a rate as a unit rate, divide the numerator of the rate by the denominator.

Practice 9

Write as a unit rate: 3200 feet every 8 seconds

Example 9 Write as a unit rate: $31,500 every 7 months

Solution:

$$\dfrac{31,500 \text{ dollars}}{7 \text{ months}} \qquad \begin{array}{r} 4\,500 \\ 7\overline{)31,500} \end{array}$$

The unit rate is

$$\dfrac{4500 \text{ dollars}}{1 \text{ month}} \text{ or 4500 dollars/month}$$ Read as, "4500 dollars per month."

■ Work Practice 9

Practice 10

Write as a unit rate: 78 bushels of fruit from 12 trees

Example 10 Write as a unit rate: 337.5 miles every 15 gallons of gas

Solution:

$$\dfrac{337.5 \text{ miles}}{15 \text{ gallons}} \qquad \begin{array}{r} 22.5 \\ 15\overline{)337.5} \end{array}$$

The unit rate is

$$\dfrac{22.5 \text{ miles}}{1 \text{ gallon}} \text{ or 22.5 miles/gallon}$$ Read as, "22.5 miles per gallon."

■ Work Practice 10

Answers

9. $\dfrac{400 \text{ ft}}{1 \text{ sec}}$ or 400 ft/sec

10. $\dfrac{6.5 \text{ bushels}}{1 \text{ tree}}$ or 6.5 bushels/tree

✓**Concept Check Answer**

false; a rate compares different kinds of quantities

Objective D Finding Unit Prices

Rates are used extensively in sports, business, medicine, and science. One of the most common uses of rates is in consumer economics. When a unit rate is "money per item," it is also called a **unit price.**

$$\text{unit price} = \dfrac{\text{price}}{\text{number of units}}$$

Example 11 Finding Unit Price

A store charges $3.36 for a 16-ounce jar of picante sauce. What is the unit price in dollars per ounce?

Solution:

$$\frac{\text{unit}}{\text{price}} = \frac{\text{price}}{\text{number of units}} = \frac{\$3.36}{16\ \text{ounces}} = \frac{\$0.21}{1\ \text{ounce}} \text{ or } \$0.21 \text{ per ounce}$$

Work Practice 11

Example 12 Finding the Best Buy

Approximate each unit price to decide which is the better buy: 4 bars of soap for $0.99 or 5 bars of soap for $1.19.

Solution:

$$\frac{\text{unit}}{\text{price}} = \frac{\text{price}}{\text{no. of units}} = \frac{\$0.99}{4\ \text{bars}} \approx \$0.25 \text{ per bar of soap} \qquad \begin{array}{c} 0.247 \approx 0.25 \\ 4\overline{)0.990} \end{array} \text{("is approximately")}$$

$$\frac{\text{unit}}{\text{price}} = \frac{\text{price}}{\text{no. of units}} = \frac{\$1.19}{5\ \text{bars}} \approx \$0.24 \text{ per bar of soap} \qquad \begin{array}{c} 0.238 \approx 0.24 \\ 5\overline{)1.190} \end{array}$$

Thus, the 5-bar package is the better buy.

Work Practice 12

Practice 11

An automobile rental agency charges $170 for 5 days for a certain model car. What is the unit price in dollars per day?

Practice 12

Approximate each unit price to decide which is the better buy for a bag of nacho chips: 11 ounces for $2.32 or 16 ounces for $3.59.

Answers

11. $34 per day   **12.** 11-oz bag

Vocabulary, Readiness & Video Check

Use the choices below to fill in each blank. Not all choices will be used.

rate	division	unit price	unit
numerator	different	denominator	ratio

1. A rate with a denominator of 1 is called a _____ rate.

2. When a rate is written as money per item, a unit rate is called a _____ .

3. The word *per* translates to _____ .

4. Rates are used to compare _____ types of quantities.

5. To write a rate as a unit rate, divide the _____ of the rate by the _____ .

6. The quotient of two quantities is called a _____ .

Answer each statement true or false.

7. The ratio $\frac{7}{5}$ means the same as the ratio $\frac{5}{7}$. _____

8. The ratio $\frac{9}{10}$ is in simplest form. _____

9. The ratio $\frac{7.2}{8.1}$ is in simplest form. _____

10. The ratio $\frac{10 \text{ feet}}{30 \text{ feet}}$ is in simplest form. _____

11. The ratio $30 : 41$ equals $\frac{30}{41}$ in fractional notation. _____

12. The ratio 2 to 5 equals $\frac{5}{2}$ in fractional notation. _____

Martin-Gay Interactive Videos   Watch the section lecture video and answer the following questions.

See Video 5.1

Objective A **13.** Based on the lecture before Example 1, what three notations can we use for a ratio? For your answer, use the ratio example given in the lecture.

Objective B **14.** Why can't we divide out the units in Example 6?

Objective C **15.** Why did we divide the first quantity of the rate in Example 8 by the second quantity?

Objective D **16.** From Example 9, unit prices can be especially helpful when?

5.1 Exercise Set MyMathLab®

Objective A *Write each ratio as a ratio of whole numbers using fractional notation. Write the fraction in simplest form. See Examples 1 through 4.*

1. 16 to 24

2. 25 to 150

3. 7.7 to 10

4. 8.1 to 10

5. 4.63 to 8.21

6. 9.61 to 7.62

7. 9 inches to 12 inches

8. 14 centimeters to 20 centimeters

9. $32 to $100

10. $46 to $102

11. 24 days to 14 days

12. 80 miles to 120 miles

13. $3\frac{1}{2}$ to $12\frac{1}{4}$

14. $3\frac{1}{3}$ to $4\frac{1}{6}$

15. $7\frac{3}{5}$ hours to $1\frac{9}{10}$ hours

16. $25\frac{1}{2}$ days to $2\frac{5}{6}$ days

Write the ratio described in each exercise as a fraction in simplest form. See Examples 5 and 6.

17.

Average Weight of Mature Whales	
Blue Whale	**Fin Whale**
145 tons	50 tons

Use the table to find the ratio of the weight of an average mature fin whale to the weight of an average mature blue whale.

18.

Countries with Small Land Areas	
Tuvalu	**San Marino**
10 sq mi	24 sq mi

(*Source: World Almanac*)

Use the table to find the ratio of the land area of Tuvalu to the land area of San Marino.

△ **19.** Find the ratio of the width of a regulation size basketball court to its perimeter.

△ **20.** Find the ratio of the width to the perimeter shown of the swimming pool.

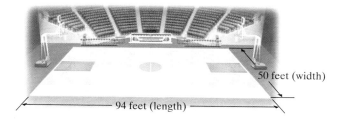

50 feet (width)
94 feet (length)

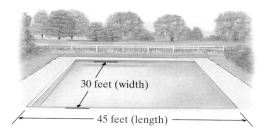

30 feet (width)
45 feet (length)

At the Hidalgo County School Board meeting one night, there were 125 women and 100 men present.

21. Find the ratio of women to men.

22. Find the ratio of men to the total number of people present.

△ **23.** Find the ratio of the longest side to the perimeter of the right-triangular-shaped billboard.

△ **24.** Find the ratio of the base to the perimeter of the triangular mainsail.

8 feet
15 feet
17 feet

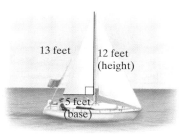

13 feet
12 feet (height)
5 feet (base)

In 2013, 659 films by U.S. production companies were released. Use this information for Exercises 25 and 26.

25. In 2013, 45 digital 3-D films were released by U.S. production companies. Find the ratio of digital films to total films for 2013. (*Source:* Motion Picture Association of America)

26. In 2013, 545 independent films were released by U.S. production companies. Find the ratio of independent films to total films for 2013. (*Source:* Motion Picture Association of America)

27. Of the U.S. mountains that are over 14,000 feet in elevation, 57 are located in Colorado and 19 are located in Alaska. Find the ratio of the number of mountains over 14,000 feet found in Alaska to the number of mountains over 14,000 feet found in Colorado. (*Source:* U.S. Geological Survey)

28. Citizens of the United States eat an average of 25 pints of ice cream per year. Residents of the New England states eat an average of 39 pints of ice cream per year. Find the ratio of the amount of ice cream eaten by New Englanders to the amount eaten by the average U.S. citizen. (*Source:* International Dairy Foods Association)

Blood contains three types of cells: red blood cells, white blood cells, and platelets. For approximately every 600 red blood cells in healthy humans, there are 40 platelets and 1 white blood cell. Use this information for Exercises 29 and 30. (Source: American Red Cross Biomedical Services)

29. Write the ratio of red blood cells to platelet cells.

30. Write the ratio of white blood cells to red blood cells.

Objective B *Write each rate as a fraction in simplest form. See Examples 7 and 8.*

31. 5 shrubs every 15 feet

32. 14 lab tables for 28 students

33. 15 returns for 100 sales

34. 150 graduate students for 8 advisors

35. 8 phone lines for 36 employees

36. 6 laser printers for 28 computers

37. 18 gallons of pesticide for 4 acres of crops

38. 4 inches of rain in 18 hours

Objective C *Write each rate as a unit rate. See Examples 9 and 10.*

39. 375 riders in 5 subway cars

40. 275 miles in 11 hours

41. A hummingbird moves its wings at a rate of 5400 wingbeats a minute. Write this rate in wingbeats per second.

42. A bat moves its wings at a rate of 1200 wingbeats a minute. Write this rate in wingbeats per second.

43. $1,000,000 lottery winnings paid over 20 years

44. 400,000 library books for 8000 students

45. The state of Delaware has 638,736 registered voters for two senators. (*Source:* Delaware.gov)

46. The 2020 projected population of Louisiana is approximately 4,758,720 residents for 64 parishes. (*Note:* Louisiana is the only U.S. state with parishes instead of counties.) (*Source:* Louisiana.gov)

47. 12,000 good assembly-line products to 40 defective products

48. 5,000,000 lottery tickets for 4 lottery winners

49. For fiscal year 2013, the National Zoo in Washington, D.C., requested an annual budget of roughly $24,300,000 for its 400 different species. (*Source:* Smithsonian Institution)

50. On average in 2013, it cost each passenger about $750 to travel 5000 miles internationally by plane. (*Source:* Airlines for America)

51. On average, it cost $1,680,000 to build 12 Habitat for Humanity houses in the Philadelphia area in 2012. (*Source:* Habitat for Humanity Philadelphia)

52. In 2013, the number of movie admissions (tickets) sold was about 1340 million for 228 million moviegoers in the U.S. and Canada. Find the number of tickets sold per moviegoer. Round to the nearest whole. (*Source:* MPAA)

53. Charlie Catlett can assemble 250 computer boards in an 8-hour shift while Suellen Catlett can assemble 402 computer boards in a 12-hour shift.
   a. Find the unit rate of Charlie.

   b. Find the unit rate of Suellen.

   c. Who can assemble computer boards faster, Charlie or Suellen?

54. Jerry Stein laid 713 bricks in 46 minutes while his associate, Bobby Burns, laid 396 bricks in 30 minutes.
   a. Find the unit rate of Jerry.

   b. Find the unit rate of Bobby.

   c. Who is the faster bricklayer?

For Exercises 55 and 56, round the rates to the nearest tenth.

55. One student drove 400 miles in his car on 14.5 gallons of gasoline. His sister drove 270 miles in her truck on 9.25 gallons of gasoline.

   a. Find the unit rate of the car.

   b. Find the unit rate of the truck.

   c. Which vehicle gets better gas mileage?

56. Charlotte Leal is a grocery scanner who can scan an average of 100 items in 3.5 minutes while her cousin Leo can scan 148 items in 5.5 minutes.

   a. Find the unit rate of Charlotte.

   b. Find the unit rate of Leo.

   c. Who is the faster scanner?

Objective D *Find each unit price. See Example 11.*

57. $57.50 for 5 DVDs

58. $0.87 for 3 apples

59. $1.19 for 7 bananas

60. $73.50 for 6 lawn chairs

Find each unit price and decide which is the better buy. Round to three decimal places. Assume that we are comparing different sizes of the same brand. See Examples 11 and 12.

61. Crackers:
$1.19 for 8 ounces
$1.59 for 12 ounces

62. Pickles:
$1.89 for 32 ounces
$0.89 for 18 ounces

63. Frozen orange juice:
$1.69 for 16 ounces
$0.69 for 6 ounces

64. Eggs:
$0.69 for a dozen
$2.10 for a flat $\left(2\frac{1}{2}\text{dozen}\right)$

65. Soy sauce:
12 ounces for $2.29
8 ounces for $1.49

66. Shampoo:
20 ounces for $1.89
32 ounces for $3.19

67. Napkins:
100 for $0.59
180 for $0.93

68. Crackers:
20 ounces for $2.39
8 ounces for $0.99

Review

Divide. See Section 4.4.

69. $9\overline{)20.7}$

70. $7\overline{)60.2}$

71. $3.7\overline{)0.555}$

72. $4.6\overline{)1.15}$

Concept Extensions

Solve.

73. Is the ratio $\frac{11}{15}$ the same as the ratio $\frac{15}{11}$? Explain your answer.

74. Explain why the ratio $\frac{40}{17}$ is incorrect for Exercise 23.

Fill in the table to calculate miles per gallon.

	Beginning Odometer Reading	Ending Odometer Reading	Miles Driven	Gallons of Gas Used	Miles per Gallon (round to the nearest tenth)
75.	29,286	29,543		13.4	
76.	16,543	16,895		15.8	
77.	79,895	80,242		16.1	
78.	31,623	32,056		11.9	

For Exercises 79 and 80, find each unit rate.

79. The longest stairway is the service stairway for the Niesenbahn Cable railway near Spiez, Switzerland. It has 11,674 steps and rises to a height of 7759 feet. Find the unit rate of steps per foot rounded to the nearest tenth of a step. (*Source: Guinness World Records*)

80. In the United States, the total number of students enrolled in public schools is 49,800,000. There are about 98,800 public schools. Write a unit rate in students per school. Round to the nearest whole. (*Source:* National Center for Education Statistics)

81. In your own words, define the phrase "unit rate."

82. In your own words, define the phrase "unit price."

83. Should the rate $\dfrac{3 \text{ lights}}{2 \text{ feet}}$ be written as $\dfrac{3}{2}$? Explain why or why not.

84. Find an item in the grocery store and calculate its unit price.

Decide whether each value is a ratio written as a fraction in simplest form. If not, write it as a fraction in simplest form.

85. $\dfrac{7.1}{4.3}$

86. $\dfrac{1 \text{ foot}}{30 \text{ inches}}$

87. $4\dfrac{1}{2}$

88. $\dfrac{12 \text{ inches}}{2 \text{ feet}}$

Solve.

89. A grocer will refuse a shipment of tomatoes if the ratio of bruised tomatoes to the total batch is at least 1 to 10. A sample is found to contain 3 bruised tomatoes and 33 good tomatoes. Determine whether the shipment should be refused.

90. A panty hose manufacturing machine will be repaired if the ratio of defective panty hose to good panty hose is at least 1 to 20. A quality control engineer found 10 defective panty hose in a batch of 200. Determine whether the machine should be repaired.

91. In 2013, 12 states had primary laws prohibiting all drivers from using handheld cell phones while driving. These laws allow law enforcement officers to ticket a driver for using a handheld cell phone, even if no other traffic offense has occurred. (*Source:* Governors Highway Safety Association)

   a. Find the ratio of states with primary handheld cell phone laws to total U.S. states.

   b. Find the number of states with no primary law prohibiting handheld cell phone use while driving.

   c. Find the ratio of states with primary handheld cell phone laws to states without such laws.

Objectives

A Write Sentences as Proportions.

B Determine Whether Proportions Are True.

C Find an Unknown Number in a Proportion.

Objective A Writing Proportions

A **proportion** is a statement that two ratios or rates are equal. For example,

$$\frac{5}{6} = \frac{10}{12}$$

is a proportion. We can read this as, "5 is to 6 as 10 is to 12."

> **Example 1** Write each sentence as a proportion.
>
> **a.** 12 diamonds is to 15 rubies as 4 diamonds is to 5 rubies.
> **b.** 5 hits is to 9 at bats as 20 hits is to 36 at bats.
>
> **Solution:**
>
> **a.** diamonds $\rightarrow$ $\dfrac{12}{15} = \dfrac{4}{5}$ $\leftarrow$ diamonds
>   rubies $\rightarrow$ $\leftarrow$ rubies
> **b.**   hits $\rightarrow$ $\dfrac{5}{9} = \dfrac{20}{36}$ $\leftarrow$ hits
>   at bats $\rightarrow$ $\leftarrow$ at bats
>
> ■ Work Practice 1

Practice 1

Write each sentence as a proportion.

a. 24 right is to 6 wrong as 4 right is to 1 wrong.
b. 32 Cubs fans is to 18 Mets fans as 16 Cubs fans is to 9 Mets fans.

> **Helpful Hint**
>
> Notice in the above examples of proportions that the numerators contain the same units and the denominators contain the same units. In this text, proportions will be written so that this is the case.

Objective B Determining Whether Proportions Are True

Like other mathematical statements, a proportion may be either true or false. A proportion is true if its ratios are equal. Since ratios are fractions, one way to determine whether a proportion is true is to write both fractions in simplest form and compare them.

Another way is to compare cross products as we did in Section 3.2.

> **Using Cross Products to Determine Whether Proportions Are True or False**
>
>   ┌──── Cross products ────┐
> $a \cdot d$ $b \cdot c$
> $$\frac{a}{b} = \frac{c}{d}$$
>
> If cross products are *equal*, the proportion is *true*.
> If cross products are *not equal*, the proportion is *false*.

Answers

1. a. $\dfrac{24}{6} = \dfrac{4}{1}$ **b.** $\dfrac{32}{18} = \dfrac{16}{9}$

Example 2 Is $\dfrac{2}{3} = \dfrac{4}{6}$ a true proportion?

Solution:

Cross products

$2 \cdot 6$ $3 \cdot 4$

$$\dfrac{2}{3} = \dfrac{4}{6}$$

$2 \cdot 6 \overset{?}{=} 3 \cdot 4$ Are cross products equal?

$12 = 12$ Equal, so proportion is true.

Since the cross products are equal, the proportion is true.

■ Work Practice 2

Practice 2

Is $\dfrac{3}{6} = \dfrac{4}{8}$ a true proportion?

Example 3 Is $\dfrac{4.1}{7} = \dfrac{2.9}{5}$ a true proportion?

Solution:

Cross products

$4.1 \cdot 5$ $7 \cdot 2.9$

$$\dfrac{4.1}{7} = \dfrac{2.9}{5}$$

$4.1 \cdot 5 \overset{?}{=} 7 \cdot 2.9$ Are cross products equal?

$20.5 \neq 20.3$ Not equal, so proportion is false.

Since the cross products are not equal, $\dfrac{4.1}{7} \neq \dfrac{2.9}{5}$. The proportion is false.

■ Work Practice 3

Practice 3

Is $\dfrac{3.6}{6} = \dfrac{5.4}{8}$ a true proportion?

Example 4 Is $\dfrac{1\frac{1}{6}}{10\frac{1}{2}} = \dfrac{\frac{1}{2}}{4\frac{1}{2}}$ a true proportion?

Solution:

$$\dfrac{1\frac{1}{6}}{10\frac{1}{2}} = \dfrac{\frac{1}{2}}{4\frac{1}{2}}$$

$1\frac{1}{6} \cdot 4\frac{1}{2} \overset{?}{=} 10\frac{1}{2} \cdot \frac{1}{2}$ Are cross products equal?

$\dfrac{7}{6} \cdot \dfrac{9}{2} \overset{?}{=} \dfrac{21}{2} \cdot \dfrac{1}{2}$ Write mixed numbers as improper fractions.

$\dfrac{21}{4} = \dfrac{21}{4}$ Equal, so proportion is true.

Since the cross products are equal, the proportion is true.

■ Work Practice 4

Practice 4

Is $\dfrac{4\frac{1}{5}}{2\frac{1}{3}} = \dfrac{3\frac{3}{10}}{1\frac{5}{6}}$ a true proportion?

Answers

2. yes   **3.** no   **4.** yes

✓**Concept Check** Think about cross products and write the true proportion $\frac{5}{8} = \frac{10}{16}$ in two other ways so that each result is also a true proportion.

(*Note:* There are no units attached in this proportion.)

Objective C Finding Unknown Numbers in Proportions ▶

When one number of a proportion is unknown, we can use cross products to find the unknown number. For example, to find the unknown number n in the proportion $\frac{n}{30} = \frac{2}{3}$, we first find the cross products.

$$n \cdot 3 \qquad\qquad 30 \cdot 2 \quad \text{Find the cross products.}$$

$$\frac{n}{30} = \frac{2}{3}$$

If the proportion is true, then the cross products are equal.

$n \cdot 3 = 30 \cdot 2$ Set the cross products equal to each other.

$n \cdot 3 = 60$ Write $2 \cdot 30$ as 60.

To find the unknown number n, we ask ourselves, "What number times 3 is 60?" The number is 20 and can be found by dividing 60 by 3.

$$n = \frac{60}{3} \quad \text{Divide 60 by the number multiplied by } n.$$

$n = 20$ Simplify.

Thus, the unknown number is 20.

Check: To *check*, let's replace n with this value, 20, and verify that a true proportion results.

$$\frac{20}{30} \stackrel{?}{=} \frac{2}{3} \quad \leftarrow \text{Replace } n \text{ with 20.}$$

$$\frac{20}{30} \stackrel{?}{=} \frac{2}{3}$$

$$3 \cdot 20 \stackrel{?}{=} 2 \cdot 30$$

$$60 = 60 \quad \text{Cross products are equal.}$$

Finding an Unknown Value n in a Proportion

Step 1: Set the cross products equal to each other.

Step 2: Divide the number not multiplied by n by the number multiplied by n.

✓**Concept Check Answer**

possible answers: $\frac{8}{5} = \frac{16}{10}$ and $\frac{5}{10} = \frac{8}{16}$

Example 5 Find the unknown number n.

$$\frac{7}{n} = \frac{6}{5}$$

Solution:

Step 1:

$$\frac{7}{n} = \frac{6}{5}$$

$7 \cdot 5 = n \cdot 6$ Set the cross products equal to each other.

$35 = n \cdot 6$ Multiply.

Step 2:

$$\frac{35}{6} = n$$ Divide 35 by 6, the number multiplied by n.

$$5\frac{5}{6} = n$$

Check: Check to see that $5\frac{5}{6}$ is the unknown number.

■ Work Practice 5

Practice 5

Find the unknown number n.

$$\frac{8}{n} = \frac{5}{9}$$

Example 6 Find the value of the unknown number n.

$$\frac{51}{-34} = \frac{3}{n}$$

Solution:

Step 1:

$$\frac{51}{-34} = \frac{3}{n}$$

$51 \cdot n = -34 \cdot 3$ Set cross products equal.

$51 \cdot n = -102$ Multiply.

Step 2:

$$n = \frac{-102}{51}$$ Divide −102 by 51, the number multiplied by n.

$n = -2$ Simplify.

Check: $\dfrac{51}{-34} \overset{?}{=} \dfrac{3}{-2}$ Replace n with its value, −2.

$$\frac{51}{-34} \overset{?}{=} \frac{3}{-2}$$

$51 \cdot -2 \overset{?}{=} -34 \cdot 3$

$-102 = -102$ Cross products are equal, so the proportion is true.

Since the proportion is true, the unknown number, n, is −2.

■ Work Practice 6

Practice 6

Find the value of the unknown number n.

$$\frac{15}{-2} = \frac{60}{n}$$

Answers

5. $n = 14\frac{2}{5}$ **6.** $n = -8$

Practice 7

Find the unknown number n.

$$\frac{n}{6} = \frac{0.7}{1.2}$$

Example 7   Find the unknown number n.

$$\frac{n}{3} = \frac{0.8}{1.5}$$

Solution:

Step 1:

$$\frac{n}{3} = \frac{0.8}{1.5}$$

$n \cdot 1.5 = 3 \cdot 0.8$ Set the cross products equal to each other.

$n \cdot 1.5 = 2.4$ Multiply.

Step 2:

$n = \dfrac{2.4}{1.5}$ Divide 2.4 by 1.5, the number multiplied by n.

$n = 1.6$ Simplify.

Check: Check to see that 1.6 is the unknown number.

■ Work Practice 7

Practice 8

Find the unknown number n.

$$\frac{n}{4\frac{1}{3}} = \frac{4\frac{1}{2}}{1\frac{3}{4}}$$

Example 8   Find the unknown number n.

$$\frac{1\frac{2}{3}}{3\frac{1}{4}} = \frac{n}{2\frac{3}{5}}$$

Solution:

Step 1:

$$\frac{1\frac{2}{3}}{3\frac{1}{4}} = \frac{n}{2\frac{3}{5}}$$

$1\frac{2}{3} \cdot 2\frac{3}{5} = 3\frac{1}{4} \cdot n$ Set the cross products equal to each other.

$\dfrac{13}{3} = 3\frac{1}{4} \cdot n$ Multiply. $1\frac{2}{3} \cdot 2\frac{3}{5} = \dfrac{5}{3} \cdot \dfrac{13}{5} = \dfrac{\cancel{5} \cdot 13}{3 \cdot \cancel{5}} = \dfrac{13}{3}$

$\dfrac{13}{3} = \dfrac{13}{4} \cdot n$ Write $3\frac{1}{4}$ as $\dfrac{13}{4}$.

Step 2:

$\dfrac{13}{3} \div \dfrac{13}{4} = n$ Divide $\dfrac{13}{3}$ by $\dfrac{13}{4}$, the number multiplied by n.

or

$n = \dfrac{13}{3} \cdot \dfrac{4}{13} = \dfrac{4}{3}$ or $1\frac{1}{3}$ Divide by multiplying by the reciprocal.

Check: Check to see that $1\frac{1}{3}$ is the unknown number.

■ Work Practice 8

Answers

7. $n = 3.5$ **8.** $n = 11\frac{1}{7}$

Vocabulary, Readiness & Video Check

Use the words and phrases below to fill in each blank.

ratio   cross products   true

false   proportion

1. $\frac{4.2}{8.4} = \frac{1}{2}$ is called a _____ while $\frac{7}{8}$ is called a(n) _____.

2. In $\frac{a}{b} = \frac{c}{d}$, $a \cdot d$ and $b \cdot c$ are called _____.

3. In a proportion, if cross products are equal, the proportion is _____.

4. In a proportion, if cross products are not equal, the proportion is _____.

Use cross products and mentally determine whether each proportion is true or false.

5. $\frac{2}{1} = \frac{6}{3}$ **6.** $\frac{3}{1} = \frac{15}{5}$ **7.** $\frac{1}{2} = \frac{3}{5}$ **8.** $\frac{2}{11} = \frac{1}{5}$ **9.** $\frac{2}{3} = \frac{40}{60}$ **10.** $\frac{3}{4} = \frac{6}{8}$

Martin-Gay Interactive Videos   Watch the section lecture video and answer the following questions.

Objective A **11.** From Example 1, what does "as" translate to in a proportion statement?

Objective B **12.** In Example 2, what are the cross products of the proportion? Is the proportion true or false?

Objective C **13.** As briefly mentioned in Example 4, what's another word for the unknown value *n*?

See Video 5.2

5.2 **Exercise Set** MyMathLab®

Objective A *Write each sentence as a proportion. See Example 1.*

1. 10 diamonds is to 6 opals as 5 diamonds is to 3 opals.

2. 8 books is to 6 courses as 4 books is to 3 courses.

3. 3 printers is to 12 computers as 1 printer is to 4 computers.

4. 4 hit songs is to 16 releases as 1 hit song is to 4 releases.

5. 6 eagles is to 58 sparrows as 3 eagles is to 29 sparrows.

6. 12 errors is to 8 pages as 1.5 errors is to 1 page.

7. $2\frac{1}{4}$ cups of flour is to 24 cookies as $6\frac{3}{4}$ cups of flour is to 72 cookies.

8. $1\frac{1}{2}$ cups milk is to 10 bagels as $\frac{3}{4}$ cup milk is to 5 bagels.

9. 22 vanilla wafers is to 1 cup of cookie crumbs as 55 vanilla wafers is to 2.5 cups of cookie crumbs. (*Source:* Based on data from *Family Circle* magazine)

10. 1 cup of instant rice is to 1.5 cups cooked rice as 1.5 cups of instant rice is to 2.25 cups of cooked rice. (*Source:* Based on data from *Family Circle* magazine)

Objective B *Determine whether each proportion is a true proportion. See Examples 2 through 4.*

11. $\dfrac{15}{9} = \dfrac{5}{3}$

12. $\dfrac{8}{6} = \dfrac{20}{15}$

13. $\dfrac{8}{6} = \dfrac{9}{7}$

14. $\dfrac{7}{12} = \dfrac{4}{7}$

15. $\dfrac{9}{36} = \dfrac{2}{8}$

16. $\dfrac{8}{24} = \dfrac{3}{9}$

17. $\dfrac{5}{8} = \dfrac{625}{1000}$

18. $\dfrac{30}{50} = \dfrac{600}{1000}$

19. $\dfrac{0.8}{0.3} = \dfrac{0.2}{0.6}$

20. $\dfrac{0.7}{0.4} = \dfrac{0.3}{0.1}$

21. $\dfrac{8}{10} = \dfrac{5.6}{0.7}$

22. $\dfrac{4.2}{8.4} = \dfrac{5}{10}$

23. $\dfrac{\frac{3}{4}}{\frac{4}{3}} = \dfrac{\frac{1}{2}}{\frac{8}{9}}$

24. $\dfrac{\frac{2}{5}}{\frac{2}{7}} = \dfrac{\frac{1}{10}}{\frac{1}{3}}$

25. $\dfrac{2\frac{2}{5}}{\frac{2}{3}} = \dfrac{1\frac{1}{9}}{\frac{1}{4}}$

26. $\dfrac{5\frac{5}{8}}{\frac{5}{3}} = \dfrac{4\frac{1}{2}}{1\frac{1}{5}}$

27. $\dfrac{\frac{4}{5}}{\frac{6}{6}} = \dfrac{\frac{6}{5}}{9}$

28. $\dfrac{\frac{6}{7}}{3} = \dfrac{\frac{10}{7}}{5}$

Objectives A B Mixed Practice—Translating *Write each sentence as a proportion. Then determine whether the proportion is a true proportion. See Examples 1 through 4.*

29. Eight is to twelve as four is to six.

30. Six is to eight as nine is to twelve.

31. Five is to two as thirteen is to five.

32. Four is to three as seven is to five.

33. One and eight tenths is to two as four and five tenths is to five.

34. Fifteen hundredths is to three as thirty-five hundredths is to seven.

35. Two thirds is to one fifth as two fifths is to one ninth.

36. Ten elevenths is to three fourths as one fourth is to one half.

Objective C *For each proportion, find the unknown number n. See Examples 5 through 8.*

37. $\dfrac{n}{5} = \dfrac{6}{10}$

38. $\dfrac{n}{3} = \dfrac{12}{9}$

39. $\dfrac{-18}{54} = \dfrac{3}{n}$

40. $\dfrac{-25}{100} = \dfrac{7}{n}$

41. $\dfrac{n}{8} = \dfrac{50}{100}$

42. $\dfrac{n}{21} = \dfrac{12}{18}$

43. $\dfrac{8}{15} = \dfrac{n}{6}$

44. $\dfrac{12}{10} = \dfrac{n}{16}$

45. $\dfrac{24}{n} = \dfrac{60}{96}$

46. $\dfrac{26}{n} = \dfrac{28}{49}$

47. $\dfrac{3.5}{12.5} = \dfrac{7}{n}$

48. $\dfrac{0.2}{0.7} = \dfrac{8}{n}$

49. $\dfrac{0.05}{12} = \dfrac{n}{0.6}$

50. $\dfrac{7.8}{13} = \dfrac{n}{2.6}$

▶ **51.** $\dfrac{8}{\frac{1}{3}} = \dfrac{24}{n}$

52. $\dfrac{12}{\frac{3}{4}} = \dfrac{48}{n}$

53. $\dfrac{\frac{1}{3}}{\frac{3}{8}} = \dfrac{\frac{2}{5}}{n}$

54. $\dfrac{\frac{7}{9}}{\frac{8}{27}} = \dfrac{\frac{1}{4}}{n}$

55. $\dfrac{12}{n} = \dfrac{\frac{2}{3}}{\frac{6}{9}}$

56. $\dfrac{24}{n} = \dfrac{\frac{8}{15}}{\frac{5}{9}}$

▶ **57.** $\dfrac{n}{1\frac{1}{5}} = \dfrac{4\frac{1}{6}}{6\frac{2}{3}}$

58. $\dfrac{n}{3\frac{1}{8}} = \dfrac{7\frac{3}{5}}{2\frac{3}{8}}$

59. $\dfrac{25}{n} = \dfrac{3}{\frac{7}{30}}$

60. $\dfrac{9}{n} = \dfrac{5}{\frac{11}{15}}$

Review

Insert < or > to form a true statement. See Sections 3.7 and 4.1.

61. 8.01   8.1

62. 7.26   7.026

63. $2\frac{1}{2}$ $2\frac{1}{3}$

64. $9\frac{1}{5}$ $9\frac{1}{4}$

65. $5\frac{1}{3}$ $6\frac{2}{3}$

66. $1\frac{1}{2}$ $2\frac{1}{2}$

Concept Extensions

Think about cross products and write each proportion in two other ways so that each result is also a true proportion. See the Concept Check in this section.

67. $\dfrac{9}{15} = \dfrac{3}{5}$

68. $\dfrac{1}{4} = \dfrac{5}{20}$

69. $\dfrac{6}{18} = \dfrac{1}{3}$

70. $\dfrac{2}{7} = \dfrac{4}{14}$

Solve.

71. If the proportion $\dfrac{a}{b} = \dfrac{c}{d}$ is a true proportion, write two other true proportions using the same letters.

72. Write a true proportion.

73. Explain the difference between a ratio and a proportion.

74. Explain how to find the unknown number in a proportion such as $\dfrac{n}{18} = \dfrac{12}{8}$.

For each proportion, find the unknown number n. For Exercises 75 through 80, round your answer to the given place value.

75. $\dfrac{3.2}{0.3} = \dfrac{n}{1.4}$

Round to the nearest tenth.

76. $\dfrac{1.8}{n} = \dfrac{2.5}{8.4}$

Round to the nearest tenth.

77. $\dfrac{n}{5.2} = \dfrac{0.08}{6}$

Round to the nearest hundredth.

78. $\dfrac{4.25}{6.03} = \dfrac{5}{n}$

Round to the nearest hundredth.

79. $\dfrac{43}{17} = \dfrac{8}{n}$

Round to the nearest thousandth.

80. $\dfrac{n}{12} = \dfrac{18}{7}$

Round to the nearest thousandth.

81. $\dfrac{n}{7} = \dfrac{0}{8}$

82. $\dfrac{0}{2} = \dfrac{n}{3.5}$

83. $\dfrac{n}{1150} = \dfrac{588}{483}$

84. $\dfrac{585}{n} = \dfrac{117}{474}$

85. $\dfrac{222}{1515} = \dfrac{37}{n}$

86. $\dfrac{1425}{1062} = \dfrac{n}{177}$

5.3   Proportions and Problem Solving

Objective

A Solve Problems by Writing Proportions.

Objective A   Solving Problems by Writing Proportions

Writing proportions is a powerful tool for solving problems in almost every field, including business, chemistry, biology, health sciences, and engineering, as well as in daily life. Given a specified ratio (or rate) of two quantities, a proportion can be used to determine an unknown quantity.

In this section, we use the same problem-solving steps that we have used earlier in this text.

Practice 1

On an architect's blueprint, 1 inch corresponds to 4 feet. How long is a wall represented by a $4\frac{1}{4}$-inch line on the blueprint?

Example 1   Determining Distances from a Map

On a chamber of commerce map of Abita Springs, 5 miles corresponds to 2 inches. How many miles correspond to 7 inches?

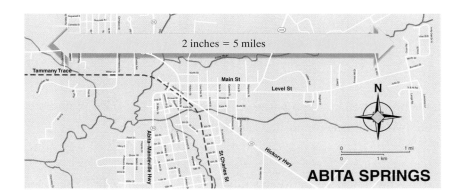

Answer

1. 17 ft

Solution:

1. UNDERSTAND. Read and reread the problem. You may want to draw a diagram.

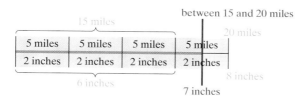

   From the diagram we can see that a reasonable solution should be between 15 and 20 miles.

2. TRANSLATE. We will let n represent our unknown number. Since 5 miles corresponds to 2 inches as n miles corresponds to 7 inches, we have the proportion

   $$\begin{array}{ll} \text{miles} \quad \rightarrow \\ \text{inches} \quad \rightarrow \end{array} \quad \frac{5}{2} = \frac{n}{7} \quad \begin{array}{l} \leftarrow \quad \text{miles} \\ \leftarrow \quad \text{inches} \end{array}$$

3. SOLVE: In earlier sections, we estimated to obtain a reasonable answer. Notice we did this in Step 1 above.

   $$\frac{5}{2} = \frac{n}{7}$$

   $$5 \cdot 7 = 2 \cdot n \qquad \text{Set the cross products equal to each other.}$$
   $$35 = 2 \cdot n \qquad \text{Multiply.}$$
   $$\frac{35}{2} = n \qquad \text{Divide 35 by 2, the number multiplied by } n.$$
   $$n = 17\frac{1}{2} \text{ or } 17.5 \quad \text{Simplify.}$$

4. INTERPRET. *Check* your work. This result is reasonable since it is between 15 and 20 miles. *State* your conclusion: 7 inches corresponds to 17.5 miles.

▪ Work Practice 1

Helpful Hint

We can also solve Example 1 by writing the proportion

$$\frac{2 \text{ inches}}{5 \text{ miles}} = \frac{7 \text{ inches}}{n \text{ miles}}$$

Although other proportions may be used to solve Example 1, we will solve by writing proportions so that the numerators have the same unit measures and the denominators have the same unit measures.

Example 2 Finding Medicine Dosage

The standard dose of an antibiotic is 4 cc (cubic centimeters) for every 25 pounds (lb) of body weight. At this rate, find the standard dose for a 140-lb woman.

Solution:

1. UNDERSTAND. Read and reread the problem. You may want to draw a diagram to estimate a reasonable solution.

(Continued on next page)

Practice 2

An auto mechanic recommends that 3 ounces of isopropyl alcohol be mixed with a tankful of gas (14 gallons) to increase the octane of the gasoline for better engine performance. At this rate, how many gallons of gas can be treated with a 16-ounce bottle of alcohol?

Answer

2. $74\frac{2}{3}$ or $74.\overline{6}$ gal

140–pound woman

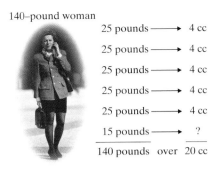

25 pounds $\longrightarrow$ 4 cc

25 pounds $\longrightarrow$ 4 cc

25 pounds $\longrightarrow$ 4 cc

25 pounds $\longrightarrow$ 4 cc

25 pounds $\longrightarrow$ 4 cc

15 pounds $\longrightarrow$?

140 pounds over 20 cc

From the diagram, we can see that a reasonable solution is a little over 20 cc.

2. TRANSLATE. We will let n represent the unknown number. From the problem, we know that 4 cc is to 25 pounds as n cc is to 140 pounds, or

cubic centimeters $\longrightarrow$ $\dfrac{4}{25} = \dfrac{n}{140}$ $\longleftarrow$ cubic centimeters

pounds $\longrightarrow$ $\phantom{\dfrac{4}{25} = \dfrac{n}{140}}$ $\longleftarrow$ pounds

3. SOLVE:

$$\frac{4}{25} = \frac{n}{140}$$

$4 \cdot 140 = 25 \cdot n$ Set the cross products equal to each other.

$560 = 25 \cdot n$ Multiply.

$\dfrac{560}{25} = n$ Divide 560 by 25, the number multiplied by n.

$n = 22\dfrac{2}{5}$ or 22.4   Simplify.

4. INTERPRET. *Check* your work. This result is reasonable since it is a little over 20 cc. *State* your conclusion: The standard dose for a 140-lb woman is 22.4 cc.

■ Work Practice 2

Practice 3

If a gallon of paint covers 400 square feet, how many gallons are needed to paint a retaining wall that is 260 feet long and 4 feet high? Round the answer up to the nearest whole gallon.

△ **Example 3**   Calculating Supplies Needed to Fertilize a Lawn

A 50-pound bag of fertilizer covers 2400 square feet of lawn. How many bags of fertilizer are needed to cover a town square containing 15,360 square feet of lawn? Round the answer up to the nearest whole bag.

Answer

3. 3 gal

Solution:

1. UNDERSTAND. Read and reread the problem. Draw a picture.

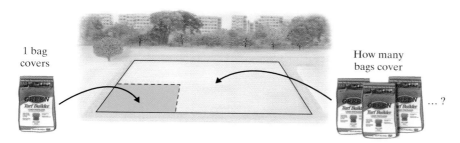

Since one bag covers 2400 square feet, let's see how many 2400s there are in 15,360. We will estimate. The number 15,360 rounded to the nearest thousand is 15,000 and 2400 rounded to the nearest thousand is 2000. Then

$$15{,}000 \div 2000 = 7\frac{1}{2} \text{ or } 7.5.$$

2. TRANSLATE. We'll let n represent the unknown number. From the problem, we know that 1 bag is to 2400 square feet as n bags is to 15,360 square feet.

$$\begin{array}{cc} \text{bags} \quad \rightarrow \\ \text{square feet} \quad \rightarrow \end{array} \frac{1}{2400} = \frac{n}{15{,}360} \begin{array}{cc} \leftarrow \quad \text{bags} \\ \leftarrow \quad \text{square feet} \end{array}$$

3. SOLVE:

$$\frac{1}{2400} = \frac{n}{15{,}360}$$

$$1 \cdot 15{,}360 = 2400 \cdot n \quad \text{Set the cross products equal to each other.}$$

$$15{,}360 = 2400 \cdot n \quad \text{Multiply.}$$

$$\frac{15{,}360}{2400} = n \quad \text{Divide 15,360 by 2400, the number multiplied by } n.$$

$$n = 6.4 \quad \text{Simplify.}$$

4. INTERPRET. *Check* that replacing n with 6.4 makes the proportion true. Is the answer reasonable? Yes, because it's close to $7\frac{1}{2}$ or 7.5. Because we must buy whole bags of fertilizer, 7 bags are needed. *State* your conclusion: To cover 15,360 square feet of lawn, 7 bags are needed.

■ Work Practice 3

✔**Concept Check** You are told that 12 ounces of ground coffee will brew enough coffee to serve 20 people. How could you estimate how much ground coffee will be needed to serve 95 people?

✔**Concept Check Answer**

Find how much will be needed for 100 people (20×5) by multiplying 12 ounces by 5, which is 60 ounces.

Vocabulary, Readiness & Video Check

Martin-Gay Interactive Videos *Watch the section lecture video and answer the following question.*

Objective A **1.** In Example 2, interpret the meaning of the answer 102.9. ▶

See Video 5.3

5.3 Exercise Set MyMathLab®

Objective A *Solve. For Exercises 1 and 2, the solutions have been started for you. See Examples 1 through 3.*

An NBA basketball player averages 45 baskets for every 100 attempts.

1. If he attempted 800 field goals, how many field goals did he make?

Start the solution:

  1. UNDERSTAND the problem. Reread it as many times as needed. Let's let
  n = how many field goals he made

  2. TRANSLATE into an equation.
  baskets (field goals) → $\dfrac{45}{100}$ attempts = $\dfrac{n}{800}$ ← baskets (field goals) ← attempts

  3. SOLVE the equation. Set cross products equal to each other and solve.

$$\dfrac{45}{100} \diagdown\!\!\!\!\times\!\!\!\!\diagup \dfrac{n}{800}$$

  After SOLVING, then
  4. INTERPRET.

2. If he made 225 baskets, how many did he attempt?

Start the solution:

  1. UNDERSTAND the problem. Reread it as many times as needed. Let's let
  n = how many baskets attempted

  2. TRANSLATE into an equation.
  baskets → $\dfrac{45}{100} = \dfrac{225}{n}$ ← baskets
  attempts → $\phantom{\dfrac{45}{100}}$ ← attempts

  3. SOLVE the equation. Set cross products equal to each other and solve.

$$\dfrac{45}{100} \diagdown\!\!\!\!\times\!\!\!\!\diagup \dfrac{225}{n}$$

  After SOLVING, then
  4. INTERPRET.

It takes a word processor 30 minutes to word process and spell check 4 pages.

3. Find how long it takes her to word process and spell check 22 pages.

4. Find how many pages she can word process and spell check in 4.5 hours.

University Law School accepts 2 out of every 7 applicants.

5. If the school accepted 180 students, find how many applications they received.

6. If the school accepted 150 students, find how many applications they received.

On an architect's blueprint, 1 inch corresponds to 8 feet.

7. Find the length of a wall represented by a line $2\frac{7}{8}$ inches long on the blueprint.

8. Find the length of a wall represented by a line $5\frac{1}{4}$ inches long on the blueprint.

A human-factors expert recommends that there be at least 9 square feet of floor space in a college classroom for every student in the class.

△ **9.** Find the minimum floor space that 30 students require.

△ **10.** Due to a lack of space, a university converts a 21-by-15-foot conference room into a classroom. Find the maximum number of students the room can accommodate.

A Honda Civic Hybrid car averages 627 miles on a 12.3-gallon tank of gas.

11. Manuel Lopez is planning a 1250-mile vacation trip in his Honda Civic Hybrid. Find how many gallons of gas he can expect to burn. Round to the nearest gallon.

12. Ramona Hatch has enough money to put 6.9 gallons of gas in her Honda Civic Hybrid. She is planning on driving home from college for the weekend. If her home is 290 miles away, should she make it home before she runs out of gas?

The scale on an Italian map states that 1 centimeter corresponds to 30 kilometers.

13. Find how far apart Milan and Rome are if their corresponding points on the map are 15 centimeters apart.

14. On the map, a small Italian village is located 0.4 centimeter from the Mediterranean Sea. Find the actual distance.

A bag of Scotts fertilizer covers 3000 square feet of lawn.

△ **15.** Find how many bags of fertilizer should be purchased to cover a rectangular lawn 260 feet by 180 feet.

△ **16.** Find how many bags of fertilizer should be purchased to cover a square lawn measuring 160 feet on each side.

A Cubs baseball player gets 3 hits every 8 times at bat.

17. If this Cubs player comes up to bat 40 times in the World Series, find how many hits he would be expected to get.

18. At this rate, if he got 12 hits, find how many times he batted.

A survey reveals that 2 out of 3 people prefer Coke to Pepsi.

19. In a room of 40 people, how many people are likely to prefer Coke? Round the answer to the nearest person.

20. In a college class of 36 students, find how many students are likely to prefer Pepsi.

A self-tanning lotion advertises that a 3-oz bottle will provide four applications.

21. Jen Haddad found a great deal on a 14-oz bottle of the self-tanning lotion she had been using. Based on the advertising claims, how many applications of the self-tanner should Jen expect? Round down to the smaller whole number.

22. The Community College thespians need fake tans for a play they are doing. If the play has a cast of 35, how many ounces of self-tanning lotion should the cast purchase? Round up to the next whole number of ounces.

The school's computer lab goes through 5 reams of printer paper every 3 weeks.

▶ 23. Find out how long a case of printer paper is likely to last (a case of paper holds 8 reams of paper). Round to the nearest week.

24. How many cases of printer paper should be purchased to last the entire semester of 15 weeks? Round up to the next case.

A recipe for pancakes calls for 2 cups flour and $1\frac{1}{2}$ cups milk to make a serving for four people.

25. Ming has plenty of flour, but only 4 cups milk. How many servings can he make?

26. The swim team has a weekly breakfast after early practice. How much flour will it take to make pancakes for 18 swimmers?

Solve.

27. In the Seattle Space Needle, the elevators whisk you to the revolving restaurant at a speed of 800 feet in 60 seconds. If the revolving restaurant is 500 feet up, how long will it take you to reach the restaurant by elevator? (*Source:* Seattle Space Needle)

28. A 16-oz grande Tazo Black Iced Tea at Starbucks has 80 calories. How many calories are there in a 24-oz venti Tazo Black Iced Tea? (*Source:* Starbucks Coffee Company)

29. Mosquitos are annoying insects. To eliminate mosquito larvae, a certain granular substance can be applied to standing water in a ratio of 1 tsp per 25 sq ft of standing water.

   a. At this rate, find how many teaspoons of granules must be used for 450 square feet.

   b. If 3 tsp = 1 tbsp, how many tablespoons of granules must be used?

30. Another type of mosquito control is liquid, where 3 oz of pesticide is mixed with 100 oz of water. This mixture is sprayed on roadsides to control mosquito breeding grounds hidden by tall grass.

   a. If one mixture of water with this pesticide can treat 150 feet of roadway, how many ounces of pesticide are needed to treat one mile? (*Hint:* 1 mile = 5280 feet)

   b. If 8 liquid ounces equals one cup, write your answer to part a in cups. Round to the nearest cup.

31. The daily supply of oxygen for one person is provided by 625 square feet of lawn. A total of 3750 square feet of lawn would provide the daily supply of oxygen for how many people? (*Source:* Professional Lawn Care Association of America)

32. In 2014, approximately $22 billion of the $60 billion Americans spent on their pets was spent on pet food. Petsmart had $6,920,000,000 in net sales that year. How much of Petsmart's net sales would you expect to have been spent on pet food? Round to the nearest thousand. (*Source:* American Pet Products Manufacturers Association and Petsmart)

33. A student would like to estimate the height of the Statue of Liberty in New York City's harbor. The length of the Statue of Liberty's right arm is 42 feet. The student's right arm is 2 feet long and her height is $5\frac{1}{3}$ feet. Use this information to estimate the height of the Statue of Liberty. How close is your estimate to the statue's actual height of 111 feet, 1 inch from heel to top of head? (*Source:* National Park Service)

34. The length of the Statue of Liberty's index finger is 8 feet while the height to the top of the head is about 111 feet. Suppose your measurements are proportionally the same as this statue and your height is 5 feet.

   a. Use this information to find the proposed length of your index finger. Give an exact measurement and then a decimal rounded to the nearest hundredth.

   b. Measure your index finger and write it as a decimal in feet rounded to the nearest hundredth. How close is the length of your index finger to the answer to part a? Explain why.

42 feet

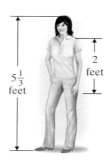

$5\frac{1}{3}$ feet

2 feet

35. There are 72 milligrams of cholesterol in a 3.5-ounce serving of lobster. How much cholesterol is in 5 ounces of lobster? Round to the nearest tenth of a milligram. (*Source:* The National Institutes of Health)

36. There are 76 milligrams of cholesterol in a 3-ounce serving of skinless chicken. How much cholesterol is in 8 ounces of chicken? (*Source:* USDA)

37. Trump World Tower in New York City is 881 feet tall and contains 72 stories. The Empire State Building contains 102 stories. If the Empire State Building has the same number of feet per floor as the Trump World Tower, approximate its height rounded to the nearest foot. (*Source:* skyscrapers.com)

38. In 2013, approximately 96 million of the 136 million U.S. employees worked in service industries. In a town of 6800 workers, how many would be expected to work in service-industry jobs? (*Source:* U.S. Bureau of Labor Statistics)

39. Medication is prescribed in 7 out of every 10 hospital emergency room visits that involve an injury. If a large urban hospital had 620 emergency room visits involving an injury in the past month, how many of these visits would you expect included a prescription for medication? (*Source:* National Center for Health Statistics)

40. One pound of firmly packed brown sugar yields $2\frac{1}{4}$ cups. How many pounds of brown sugar will be required in a recipe that calls for 6 cups of firmly packed brown sugar? (*Source:* Based on data from *Family Circle* magazine)

41. In 2013, three out of every ten wireless smartphones sold in the world were Samsung smartphones. Approximately 282 million wireless smartphones were sold in the world in 2013. How many of them were Samsung smartphones? Round to the nearest million. (*Source:* Gartner)

42. In 2013, approximately one out of every four autos sold in the U.S. was a crossover utility vehicle (CUV). If an auto dealership sold 4500 vehicles in total that year, how many of these sales would you expect involved a CUV? (*Source:* Alliance of Automobile Manufacturers)

When making homemade ice cream in a hand-cranked freezer, the tub containing the ice cream mix is surrounded by a brine (water/salt) solution. To freeze the ice cream mix rapidly so that smooth and creamy ice cream results, the brine solution should combine crushed ice and rock salt in a ratio of 5 to 1. Use this information for Exercises 43 and 44. (Source: White Mountain Freezers, The Rival Company)

43. A small ice cream freezer requires 12 cups of crushed ice. How much rock salt should be mixed with the ice to create the necessary brine solution?

44. A large ice cream freezer requires $18\frac{3}{4}$ cups of crushed ice. How much rock salt will be needed to create the necessary brine solution?

45. The gas/oil ratio for a certain chainsaw is 50 to 1.
   a. How much oil (in gallons) should be mixed with 5 gallons of gasoline?
   b. If 1 gallon equals 128 fluid ounces, write the answer to part a in fluid ounces. Round to the nearest whole ounce.

46. The gas/oil ratio for a certain tractor mower is 20 to 1.
   a. How much oil (in gallons) should be mixed with 10 gallons of gas?
   b. If 1 gallon equals 4 quarts, write the answer to part a in quarts.

47. The adult daily dosage for a certain medicine is 150 mg (milligrams) of medicine for every 20 pounds of body weight.
   a. At this rate, find the daily dose for a man who weighs 275 pounds.
   b. If the man is to receive 500 mg of this medicine every 8 hours, is he receiving the proper dosage?

48. The adult daily dosage for a certain medicine is 80 mg (milligrams) for every 25 pounds of body weight.
   a. At this rate, find the daily dose for a woman who weighs 190 pounds.
   b. If she is to receive this medicine every 6 hours, find the amount to be given every 6 hours.

Review

Find the prime factorization of each number. See Section 3.2.

49. 15   **50.** 21   **51.** 20   **52.** 24   **53.** 200   **54.** 300   **55.** 32   **56.** 81

Concept Extensions

As we have seen, proportions are often used in medicine dosage calculations. The exercises below have to do with liquid drug preparations, where the weight of the drug is contained in a volume of solution. The descriptions of mg and ml below will help. We will study metric units further in Sections 5.4 through 5.7.

mg means milligram (A paper clip weighs about a gram. A milligram is about the weight of $\frac{1}{1000}$ of a paper clip.)

ml means milliliter (A liter is about a quart. A milliliter is about the amount of liquid in $\frac{1}{1000}$ of a quart.)

One way to solve the applications below is to set up the proportion $\dfrac{mg}{ml} = \dfrac{mg}{ml}$.

A solution strength of 15 mg of medicine in 1 ml of solution is available.

57. If a patient needs 12 mg of medicine, how many ml do you administer?

58. If a patient needs 33 mg of medicine, how many ml do you administer?

A solution strength of 8 mg of medicine in 1 ml of solution is available.

59. If a patient needs 10 mg of medicine, how many ml do you administer?

60. If a patient needs 6 mg of medicine, how many ml do you administer?

Estimate the following. See the Concept Check in this section.

61. It takes 1.5 cups of milk to make 11 muffins. Estimate the amount of milk needed to make 8 dozen muffins. Explain your calculation.

62. A favorite chocolate chip recipe calls for $2\frac{1}{2}$ cups of flour to make 2 dozen cookies. Estimate the amount of flour needed to make 50 cookies. Explain your calculation.

A board such as the one pictured below will balance if the following proportion is true:

$$\frac{\text{first weight}}{\text{second distance}} = \frac{\text{second weight}}{\text{first distance}}$$

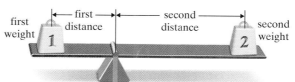

Use this proportion to solve Exercises 63 and 64.

63. Find the distance *n* that will allow the board to balance.

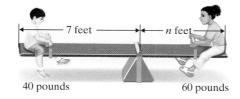

64. Find the length *n* needed to lift the weight below.

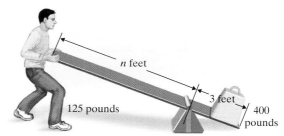

65. Describe a situation in which writing a proportion might solve a problem related to driving a car.

Ratio and Proportion

Answers

1. _____

2. _____

3. _____

4. _____

5. _____

6. _____

7. _____

8. _____

9. _____

10. _____

11. _____

12. _____

13. a. _____

b. _____

14. _____

Write each ratio as a ratio of whole numbers using fractional notation. Write the fraction in simplest form.

1. 18 to 20

2. 36 to 100

3. 8.6 to 10

4. 1.6 to 4.6

5. $8.65 to $6.95

6. 7.2 ounces to 8.4 ounces

7. $3\frac{1}{2}$ to 13

8. $1\frac{2}{3}$ to $2\frac{3}{4}$

9. 8 inches to 12 inches

10. 3 hours to 24 hours

Find the ratio described in each problem.

11. During the 2012–13 academic year, a full college professor in Columbia University's doctoral program earned $212.3 thousand. By contrast, a Columbia doctoral associate professor earned only $132.4 thousand. Find the ratio of full professor salary to associate professor salary at the doctoral level at Columbia University. (*Source:* American Association of University Professors)

12. The New York Yankees are a dynastic powerhouse. They won 27 out of the 108 Major League Baseball World Series played through 2012. (*Source:* Major League Baseball)

13. The circle graph below shows how the top 25 movies of 2013 were rated. Use this graph to answer the questions.

a. How many top 25 movies were rated R?

b. Find the ratio of top 25 PG-rated movies to PG-13-rated movies for 2013.

14. Find the ratio of the width to the length of the sign below.

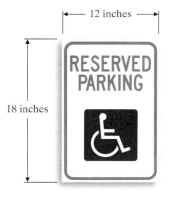

Top 25 Movies of 2013

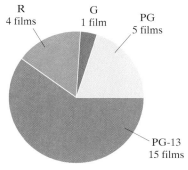

R
4 films

G
1 film

PG
5 films

PG-13
15 films

Source: MPAA

Write each rate as a fraction in simplest form.

15. 5 offices for every 20 graduate assistants

16. 6 lights every 15 feet

17. 64 computers for every 100 households

18. 45 students for every 10 computers

Write each rate as a unit rate.

19. 165 miles in 3 hours

20. 560 feet in 4 seconds

21. 115 miles every gallons

22. 112 teachers for 7 computers

Write each unit price, rounded to the nearest hundredth, and decide which is the better buy.

23. Microwave popcorn:
3 packs for $2.39
8 packs for $5.99

24. AA batteries:
4 for $3.69
10 for $9.89

Determine whether each proportion is true.

25. $\dfrac{7}{4} = \dfrac{5}{3}$

26. $\dfrac{8.2}{2} = \dfrac{16.4}{4}$

Find the unknown number n in each proportion.

27. $\dfrac{5}{3} = \dfrac{40}{n}$

28. $\dfrac{n}{10} = \dfrac{13}{4}$

29. $\dfrac{6}{11} = \dfrac{n}{5}$

30. $\dfrac{21}{n} = \dfrac{\frac{7}{2}}{3}$

15. _____

16. _____

17. _____

18. _____

19. _____

20. _____

21. _____

22. _____

23. _____

24. _____

25. _____

26. _____

27. _____

28. _____

29. _____

30. _____

5.4 Length: U.S. and Metric Systems of Measurement

Objectives

A Define U.S. Units of Length and Convert from One Unit to Another.

B Use Mixed U.S. Units of Length.

C Perform Arithmetic Operations on U.S. Units of Length.

D Define Metric Units of Length and Convert from One Unit to Another.

E Perform Arithmetic Operations on Metric Units of Length.

Objective A Defining and Converting U.S. System Units of Length

In the United States, two systems of measurement are commonly used. They are the **United States (U.S.), or English, measurement system** and the **metric system.** The U.S. measurement system is familiar to most Americans. Units such as feet, miles, ounces, and gallons are used. However, the metric system is also commonly used in fields such as medicine, sports, international marketing, and certain physical sciences. We are accustomed to buying 2-liter bottles of soft drinks, watching televised coverage of the 100-meter dash at the Olympic Games, or taking a 200-milligram dose of pain reliever.

The U.S. system of measurement uses the **inch, foot, yard, and mile** to measure **length.** The following is a summary of equivalencies between units of length:

U.S. Units of Length

$$12 \text{ inches (in.)} = 1 \text{ foot (ft)}$$
$$3 \text{ feet} = 1 \text{ yard (yd)}$$
$$36 \text{ inches} = 1 \text{ yard}$$
$$5280 \text{ feet} = 1 \text{ mile (mi)}$$

To convert from one unit of length to another, we will use **unit fractions.** We define a unit fraction to be a fraction that is equivalent to 1. Examples of unit fractions are as follows:

Unit Fractions

$$\frac{12 \text{ in.}}{1 \text{ ft}} = 1 \text{ or } \frac{1 \text{ ft}}{12 \text{ in.}} = 1 \text{ (since 12 in. = 1 ft)}$$

$$\frac{3 \text{ ft}}{1 \text{ yd}} = 1 \text{ or } \frac{1 \text{ yd}}{3 \text{ ft}} = 1 \text{ (since 3 ft = 1 yd)}$$

$$\frac{5280 \text{ ft}}{1 \text{ mi}} = 1 \text{ or } \frac{1 \text{ mi}}{5280 \text{ ft}} = 1 \text{ (since 5280 = 1 mi)}$$

Remember that multiplying a number / 1 does not change the value of the number.

Practice 1

Convert 6 feet to inches.

Example 1 Convert 8 feet to inches

Solution: We multiply 8 feet by a unit fraction that uses the equality 12 inches = 1 foot. The unit fraction would be in the form $\dfrac{\text{units to convert to}}{\text{original units}}$ or, in this case, $\dfrac{12 \text{ inches}}{1 \text{ foot}}$. We do this so that like units will divide out to 1, as shown.

$$8 \text{ ft} = \frac{8 \text{ ft}}{1} \cdot 1$$

$$= \frac{8 \text{ ft}}{1} \cdot \frac{12 \text{ in.}}{1 \text{ ft}} \quad \text{Multiply 1 in the form of } \frac{12 \text{ in.}}{1 \text{ ft}}.$$

$$= 8 \cdot 12 \text{ in.}$$

$$= 96 \text{ in.} \quad \text{Mul}$$

Answer

1. 72 in.

Thus, 8 ft = 96 in., as shown in the diagram:

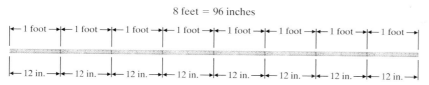

8 feet = 96 inches

■ Work Practice 1

Example 2   Convert 7 feet to yards.

Solution: We multiply by a unit fraction that compares 1 yard to 3 feet.

$$7 \text{ ft} = \frac{7 \text{ ft}}{1} \cdot 1$$

$$= \frac{7 \text{ ft}}{1} \cdot \frac{1 \text{ yd}}{3 \text{ ft}} \quad \leftarrow \text{Units to convert to}$$
$$\quad\quad\quad\quad\quad \leftarrow \text{Original units}$$

$$= \frac{7}{3} \text{ yd}$$

$$= 2\frac{1}{3} \text{ yd} \quad\quad \text{Divide.}$$

Thus, 7 ft = $2\frac{1}{3}$ yd, as shown in the diagram.

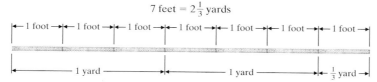

7 feet = $2\frac{1}{3}$ yards

■ Work Practice 2

Practice 2

Convert 8 yards to feet.

Helpful Hint   When converting from one unit to another, select a unit fraction with the properties below:

$$\frac{\text{units you are converting to}}{\text{original units}}$$

By using this unit fraction, the original units will divide out, as wanted.

Example 3   Finding the Length of a Pelican's Bill

The Australian pelican has the longest bill, measuring from 13 to 18.5 inches long. The pelican in the photo has a 15-inch bill. Convert 15 inches to feet, using decimals in your final answer.

Solution:

$$15 \text{ in.} = \frac{15 \text{ in.}}{1} \cdot \frac{1 \text{ ft}}{12 \text{ in.}} \quad \leftarrow \text{Units to convert to}$$
$$\quad\quad\quad\quad\quad\quad\quad \leftarrow \text{Original units}$$

$$= \frac{15}{12} \text{ ft}$$

$$= \frac{5}{4} \text{ ft} \quad\quad \text{Simplify } \frac{15}{12}.$$

$$= 1.25 \text{ ft} \quad\quad \text{Divide.}$$

Thus, 15 in. = 1.25 ft, as shown in the diagram.

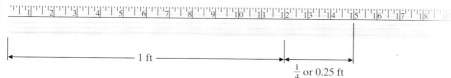

15 inches = 1.25 ft

1 ft $\frac{1}{4}$ or 0.25 ft

■ Work Practice 3

Practice 3

Suppose the pelican's bill (in the photo) measures 18 inches. Convert 18 inches to feet, using decimals.

Answers
2. 24 ft **3.** 1.5 ft

Objective B Using Mixed U.S. System Units of Length

Sometimes it is more meaningful to express a measurement of length with mixed units such as 1 ft and 5 in. We usually condense this and write 1 ft 5 in.

In Example 2, we found that 7 feet is the same as $2\frac{1}{3}$ yards. The measurement can also be written as a mixture of yards and feet. That is,

7 ft = _____ yd _____ ft

Because 3 ft = 1 yd, we divide 3 into 7 to see how many whole yards are in 7 feet. The quotient is the number of yards, and the remainder is the number of feet.

```
   2 yd 1 ft
3)7
  -6
   1
```

Thus, 7 ft = 2 yd 1 ft, as seen in the diagram:

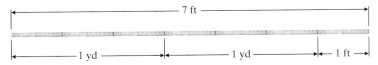

Practice 4

Convert: 68 in. = _____ ft _____ in.

Example 4 Convert: 134 in. = _____ ft _____ in.

Solution: Because 12 in. = 1 ft, we divide 12 into 134. The quotient is the number of feet. The remainder is the number of inches. To see why we divide 12 into 134, notice that

$$134 \text{ in.} = \frac{134 \text{ in.}}{1} \cdot \frac{1 \text{ ft}}{12 \text{ in.}} = \frac{134}{12} \text{ ft}$$

```
    11 ft 2 in
12)134
  - 12
    14
  - 12
     2
```

Thus, 134 in. = 11 ft 2 in.

Work Practice 4

Practice 5

Convert 5 yards 2 feet to feet.

Example 5 Convert 3 feet 7 inches to inches.

Solution: First, we convert 3 feet to inches. Then we add 7 inches.

$$3 \text{ ft} = \frac{3 \text{ ft}}{1} \cdot \frac{12 \text{ in.}}{1 \text{ ft}} = 36 \text{ in.}$$

Then

$$3 \text{ ft } 7 \text{ in.} = 36 \text{ in.} + 7 \text{ in.} = 43 \text{ in.}$$

Work Practice 5

Answers
4. 5 ft 8 in.   **5.** 17 ft

Objective C Performing Operations on U.S. System Units of Length ▶

Finding sums or differences of measurements often involves converting units, as shown in the next example. Just remember that, as usual, only like units can be added or subtracted.

Example 6 Add 3 ft 2 in. and 5 ft 11 in.

Solution: To add, we line up the similar units.

$$
\begin{array}{r}
3 \text{ ft } 2 \text{ in.}\\
+\ 5 \text{ ft } 11 \text{ in.}\\
\hline
8 \text{ ft } 13 \text{ in.}
\end{array}
$$

Since 13 inches is the same as 1 ft 1 in., we have

$$8 \text{ ft } 13 \text{ in.} = 8 \text{ ft } + 1 \text{ ft } 1 \text{ in.}$$
$$= 9 \text{ ft } 1 \text{ in.}$$

■ Work Practice 6

Practice 6

Add 4 ft 8 in. to 8 ft 11 in.

✔**Concept Check** How could you estimate the following sum?

$$
\begin{array}{r}
7 \text{ yd } 4 \text{ in.}\\
+\ 3 \text{ yd } 27 \text{ in.}
\end{array}
$$

Example 7 Multiply 8 ft 9 in. by 3.

Solution: By the distributive property, we multiply 8 ft by 3 and 9 in. by 3.

$$
\begin{array}{r}
8 \text{ ft } 9 \text{ in.}\\
\times\qquad 3\\
\hline
24 \text{ ft } 27 \text{ in.}
\end{array}
$$

Since 27 in. is the same as 2 ft 3 in., we simplify the product as

$$24 \text{ ft } 27 \text{ in.} = 24 \text{ ft } + 2 \text{ ft } 3 \text{ in.}$$
$$= 26 \text{ ft } 3 \text{ in.}$$

■ Work Practice 7

Practice 7

Multiply 4 ft 7 in. by 4.

We divide in a similar manner as above.

Example 8 Finding the Length of a Piece of Rope

A rope of length 6 yd 1 ft has 2 yd 2 ft cut from one end. Find the length of the remaining rope.

Solution: Subtract 2 yd 2 ft from 6 yd 1 ft.

$$
\begin{array}{rl}
\text{beginning length} \rightarrow & 6 \text{ yd } 1 \text{ ft}\\
-\quad \text{amount cut} \rightarrow & -2 \text{ yd } 2 \text{ ft}\\
\hline
\text{remaining length} &
\end{array}
$$

We cannot subtract 2 ft from 1 ft, so we borrow 1 yd from the 6 yd. One yard is converted to 3 ft and combined with the 1 ft already there.

(Continued on next page)

Practice 8

A carpenter cuts 1 ft 9 in. from a board of length 5 ft 8 in. Find the length of the remaining board.

Answers

6. 13 ft 7 in. **7.** 18 ft 4 in. **8.** 3 ft 11 in.

✔**Concept Check Answer**

round each to the nearest yard:
7 yd + 4 yd = 11 yd

Borrow 1 yd = 3 ft

5 yd + ⟨1 yd⟩ ⟨3 ft⟩

$$
\begin{array}{rcl}
\cancel{6 \text{ yd}}\ 1\text{ ft} & = & 5\text{ yd }4\text{ ft} \\
-2\text{ yd }2\text{ ft} & = & -2\text{ yd }2\text{ ft} \\
\hline
 & & 3\text{ yd }2\text{ ft}
\end{array}
$$

The remaining rope is 3 yd 2 ft long.

■ Work Practice 8

Objective D Defining and Converting Metric System Units of Length

The basic unit of length in the metric system is the **meter.** A meter is slightly longer than a yard. It is approximately 39.37 inches long. Recall that a yard is 36 inches long.

1 yard = 36 inches

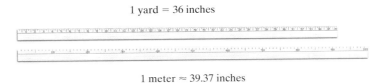

1 meter ≈ 39.37 inches

All units of length in the metric system are based on the meter. The following is a summary of the prefixes used in the metric system. Also shown are equivalencies between units of length. Like the decimal system, the metric system uses powers of 10 to define units.

Metric Units of Length
1 kilometer (km) = 1000 meters (m)
1 hectometer (hm) = 100 m
1 dekameter (dam) = 10 m
1 meter (m) = 1 m
1 decimeter (dm) = 1/10 m or 0.1 m
1 centimeter (cm) = 1/100 m or 0.01 m
1 millimeter (mm) = 1/1000 m or 0.001 m

The figure below will help you with decimeters, centimeters, and millimeters.

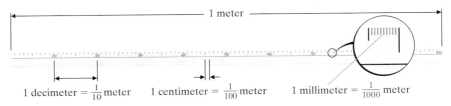

1 decimeter = $\frac{1}{10}$ meter   1 centimeter = $\frac{1}{100}$ meter   1 millimeter = $\frac{1}{1000}$ meter

Helpful Hint

Study the figure above for other equivalencies between metric units of length.

10 decimeters = 1 meter	10 millimeters = 1 centimeter
100 centimeters = 1 meter	10 centimeters = 1 decimeter
1000 millimeters = 1 meter	

These same prefixes are used in the metric system for mass and capacity. The most commonly used measurements of length in the metric system are the **meter, millimeter, centimeter,** and **kilometer.**

✓**Concept Check** Is this statement reasonable? "The screen of a home television set has a 30-meter diagonal." Why or why not?

Being comfortable with the metric units of length means gaining a "feeling" for metric lengths, just as you have a "feeling" for the lengths of an inch, a foot, and a mile. To help you accomplish this, study the following examples:

A millimeter is about the thickness of a large paper clip.

A centimeter is about the width of a large paper clip.

A meter is slightly longer than a yard.

A kilometer is about two-thirds of a mile.

The width of this book is approximately 21.5 centimeters.

The distance between New York City and Philadelphia is about 160 kilometers.

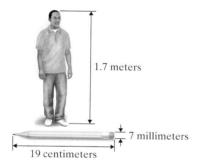

1.7 meters

19 centimeters

7 millimeters

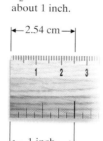

$2\frac{1}{2}$ centimeters is about 1 inch.

As with the U.S. system of measurement, unit fractions may be used to convert from one unit of length to another. For example, let's convert 1200 meters to kilometers. To do so, we will multiply by 1 in the form of the unit fraction

$$\frac{1 \text{ km}}{1000 \text{ m}} \quad \begin{array}{l} \leftarrow \text{Units to convert to} \\ \leftarrow \text{Original units} \end{array}$$

Unit fraction

$$1200 \text{ m} = \frac{1200 \text{ m}}{1} \cdot 1 = \frac{1200 \text{ m}}{1} \cdot \frac{1 \text{ km}}{1000 \text{ m}} = \frac{1200 \text{ km}}{1000} = 1.2 \text{ km}$$

The metric system does, however, have a distinct advantage over the U.S. system of measurement: the ease of converting from one unit of length to another. Since all units of length are powers of 10 of the meter, converting from one unit of length to another is as simple as moving the decimal point. Listing units of length in order from largest to smallest helps to keep track of how many places to move the decimal point when converting.

Let's again convert 1200 meters to kilometers. This time, to convert from meters to kilometers, we move along the chart shown, 3 units to the left, from meters to kilometers. This means that we move the decimal point 3 places to the left.

km   hm   dam   **m**   dm   cm   mm

3 units to the left

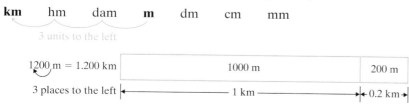

1200 m = 1.200 km

3 places to the left

1000 m

200 m

1 km

0.2 km

Thus, 1200 m = 1.2 km, as shown in the diagram.

✓**Concept Check Answer**
no; answers may vary

Practice 9

Convert 2.5 m to millimeters.

Example 9   Convert 2.3 m to centimeters.

Solution: First we will convert by using a unit fraction.

$$2.3 \text{ m} = \frac{2.3 \text{ m}}{1} \cdot \frac{\overset{\text{Unit fraction}}{100 \text{ cm}}}{1 \text{ m}} = 230 \text{ cm}$$

Now we will convert by listing the units of length in order from left to right and moving from meters to centimeters.

| km | hm | dam | m | dm | cm | mm |

2 units to the right

2.30 m = 230. cm

2 places to the right

With either method, we get 230 cm.

▣ Work Practice 9

Practice 10

Convert 3500 m to kilometers.

Example 10   Convert 450,000 mm to meters.

Solution: We list the units of length in order from left to right and move from millimeters to meters.

| km | hm | dam | m | dm | cm | mm |

3 units to the left

450,000 mm = 450.000 m   or   450 m

▣ Work Practice 10

✓**Concept Check** What is wrong with the following conversion of 150 cm to meters?

150.00 cm = 15,000 m

Objective E Performing Operations on Metric System Units of Length ▶

To add, subtract, multiply, or divide with metric measurements of length, we write all numbers using the same unit of length and then add, subtract, multiply, or divide as with decimals.

Practice 11

Subtract 640 m from 2.1 km.

Example 11   Subtract 430 m from 1.3 km.

Solution: First we convert both measurements to kilometers or both to meters.

| 430 m = 0.43 km | or | 1.3 km = 1300 m |

$$\begin{array}{r} 1.30 \text{ km} \\ -0.43 \text{ km} \\ \hline 0.87 \text{ km} \end{array} \qquad \begin{array}{r} 1300 \text{ m} \\ -430 \text{ m} \\ \hline 870 \text{ m} \end{array}$$

The difference is 0.87 km or 870 m.

▣ Work Practice 11

Answers

9. 2500 mm   **10.** 3.5 km
11. 1.46 km or 1460 m

✓**Concept Check Answer**

decimal point should be moved two places to the left: 1.5 m

Example 12 Multiply 5.7 mm by 4.

Solution: Here we simply multiply the two numbers. Note that the unit of measurement remains the same.

$$\begin{array}{r} 5.7 \text{ mm} \\ \times\quad 4 \\ \hline 22.8 \text{ mm} \end{array}$$

■ Work Practice 12

Practice 12

Multiply 18.3 hm by 5.

Example 13 Finding a Person's Height

Fritz Martinson was 1.2 meters tall on his last birthday. Since then, he has grown 14 centimeters. Find his current height in meters.

Solution:

original height	$\rightarrow$	1.20 m
+ height grown	$\rightarrow$	+ 0.14 m (Since 14 cm = 0.14 m)
current height		1.34 m

Fritz is now 1.34 meters tall.

■ Work Practice 13

Practice 13

A child was 55 centimeters at birth. Her adult height was 1.72 meters. Find how much she grew from birth to adult height.

Example 14 Finding a Crocodile's Length

A newly hatched Nile crocodile averages 26 centimeters in length. This type of crocodile normally grows 4.74 meters to reach its adult length. What is the adult length of this type of crocodile?

Solution:

original length	$\rightarrow$	0.26 m (Since 26 cm = 0.26 m)
+ length grown	$\rightarrow$	+ 4.74 m
adult length		5.00 m

The adult length is 5 meters.

■ Work Practice 14

Practice 14

Doris Blackwell is knitting a scarf that is currently 0.8 meter long. If she knits an additional 45 centimeters, how long will the scarf be?

Vocabulary, Readiness & Video Check

Use the choices below to fill in each blank. Some choices may be used more than once.

inches   yard   unit fraction

feet   meter

1. The basic unit of length in the metric system is the _____ .

2. The expression $\dfrac{1\ \text{foot}}{12\ \text{inches}}$ is an example of a(n) _____ .

3. A meter is slightly longer than a(n) _____ .

4. One foot equals 12 _____ .

5. One yard equals 3 _____ .

6. One yard equals 36 _____ .

7. One mile equals 5280 _____ .

8. One foot equals $\dfrac{1}{3}$ _____ .

Martin-Gay Interactive Videos

See Video 5.4

Watch the section lecture video and answer the following questions.

Objective A **9.** In ▣ Example 3, what units are used in the denominator of the unit fraction and why was this decided?

Objective B **10.** In ▣ Example 4, how is a mixed unit similar to a mixed number? Use examples in your answer.

Objective C **11.** In ▣ Example 5, why is the original sum of the addition problem not the final answer? Reference the sum in your answer.

Objective D **12.** Based on the lecture before ▣ Example 6, explain how to convert metric units to other metric units.

Objective E **13.** What two answers did we get for ▣ Example 8? Explain why both answers are correct.

5.4 Exercise Set MyMathLab®

Objective A *Convert each measurement as indicated. See Examples 1 through 3.*

1. 60 in. to feet

2. 84 in. to feet

3. 12 yd to feet

4. 18 yd to feet

5. 42,240 ft to miles

6. 36,960 ft to miles

7. $8\dfrac{1}{2}$ ft to inches

8. $12\dfrac{1}{2}$ ft to inches

9. 10 ft to yards

10. 25 ft to yards

11. 6.4 mi to feet

12. 3.8 mi to feet

13. 162 in. to yd (Write answer as a decimal.)

14. 7216 yd to mi (Write answer as a decimal.)

15. 3 in. to ft (Write answer as a decimal.)

16. 129 in. to ft (Write answer as a decimal.)

Objective B *Convert each measurement as indicated. See Examples 4 and 5.*

17. 40 ft = _____ yd _____ ft

18. 100 ft = _____ yd _____ ft

19. 85 in. = _____ ft _____ in.

20. 47 in. = _____ ft _____ in.

21. 10,000 ft = _____ mi _____ ft

22. 25,000 ft = _____ mi _____ ft

23. 5 ft 2 in. = _____ in.

24. 4 ft 11 in. = _____ in.

25. 8 yd 2 ft = _____ ft

26. 4 yd 1 ft = _____ ft

27. 2 yd 1 ft = _____ in.

28. 1 yd 2 ft = _____ in.

Objective C *Perform each indicated operation. Simplify the result if possible. See Examples 6 through 8.*

29. 3 ft 10 in. + 7 ft 4 in.

30. 12 ft 7 in. + 9 ft 11 in.

31. 12 yd 2 ft + 9 yd 2 ft

32. 16 yd 2 ft + 8 yd 2 ft

33. 22 ft 8 in. − 16 ft 3 in.

34. 15 ft 5 in. − 8 ft 2 in.

35. 18 ft 3 in. − 10 ft 9 in.

36. 14 ft 8 in. − 3 ft 11 in.

37. 28 ft 8 in. ÷ 2

38. 34 ft 6 in. ÷ 2

39. 16 yd 2 ft × 5

40. 15 yd 1 ft × 8

Objective D *Convert as indicated. See Examples 9 and 10.*

41. 60 m to centimeters

42. 46 m to centimeters

43. 40 mm to centimeters

44. 14 mm to centimeters

45. 500 m to kilometers

46. 400 m to kilometers

47. 1700 mm to meters

48. 6400 mm to meters

49. 1500 cm to meters

50. 6400 cm to meters

51. 0.42 km to centimeters

52. 0.95 km to centimeters

53. 7 km to meters

54. 5 km to meters

55. 8.3 cm to millimeters

56. 4.6 cm to millimeters

57. 20.1 mm to decimeters

58. 140.2 mm to decimeters

59. 0.04 m to millimeters

60. 0.2 m to millimeters

Objective **E** *Perform each indicated operation. Remember to insert units when writing your answers. See Examples 11 through 14.*

61. 8.6 m + 0.34 m

62. 14.1 cm + 3.96 cm

63. 2.9 m + 40 mm

64. 30 cm + 8.9 m

65. 24.8 mm − 1.19 cm

66. 45.3 m − 2.16 dam

67. 15 km − 2360 m

68. 14 cm − 15 mm

69. 18.3 m × 3

70. 14.1 m × 4

71. 6.2 km ÷ 4

72. 9.6 m ÷ 5

Objectives **A** **C** **D** **E** **Mixed Practice** *Solve. Remember to insert units when writing your answers. For Exercises 73 through 82, complete the charts. See Examples 1 through 14.*

		Yards	Feet	Inches
73.	Chrysler Building in New York City		1046	
74.	4-story building			792
75.	Python length		35	
76.	Ostrich height			108

		Meters	Millimeters	Kilometers	Centimeters
77.	Length of elephant	5			
78.	Height of grizzly bear	3			
79.	Tennis ball diameter				6.5
80.	Golf ball diameter				4.6
81.	Distance from London to Paris			342	
82.	Distance from Houston to Dallas			396	

A massive structure of Crazy Horse is currently being carved into the Black Hills about 8 miles from Mt. Rushmore. Use the dimensions in the photo for Exercises 83 and 84.

641 feet

563 feet

83. The total width of the Crazy Horse carving is 641 feet. Convert this width to
   a. yards
   b. inches

84. The total height of the Crazy Horse carving is 563 feet. Convert this height to
   a. yards
   b. inches

85. The National Zoo maintains a small patch of bamboo, which it grows as a food supply for its pandas. Two weeks ago, the bamboo was 6 ft 10 in. tall. Since then, the bamboo has grown 3 ft 8 in. How tall is the bamboo now?

86. While exploring in the Mariana Trench, a submarine probe was lowered to a point 1 mile 1400 feet below the ocean's surface. Later it was lowered an additional 1 mile 4000 feet below this point. How far was the probe below the surface of the Pacific?

87. At its deepest point, the Grand Canyon of the Colorado River in Arizona is about 6000 ft. The Grand Canyon of the Yellowstone River, which is in Yellowstone National Park in Wyoming, is at most 900 feet deep. How much deeper is the Grand Canyon of the Colorado River than the Grand Canyon of the Yellowstone River? (*Source:* National Park Service)

88. The Black Canyon of the Gunnison River is only 1150 ft wide at its narrowest point. At its narrowest, the Grand Canyon of the Yellowstone is $\frac{1}{2}$ mile wide. Find the difference in width between the Grand Canyon of the Yellowstone and the Black Canyon of the Gunnison. (*Note:* Notice that the dimensions are different.) (*Source:* National Park Service)

89. The tallest man in the world is recorded as Robert Pershing Wadlow of Alton, Illinois. Born in 1918, he measured 8 ft 11 in. at his tallest. The shortest man in the world is Chandra Bahadur Dangi of Nepal, who measures 21.5 in. How many times taller than Chandra was Robert? Round to one decimal place. (*Source: Guinness World Records*)

90. A 3.4-m rope is attached to a 5.8-m rope. However, when the ropes are tied, 8 cm of length is lost to form the knot. What is the length of the tied ropes?

91. The length of one of the Statue of Liberty's hands is 16 ft 5 in. One of the statue's eyes is 2 ft 6 in. across. How much longer is a hand than the width of an eye? (*Source:* National Park Service)

92. The width of the Statue of Liberty's head from ear to ear is 10 ft. The height of the statue's head from chin to cranium is 17 ft 3 in. How much taller is the statue's head than its width? (*Source:* National Park Service)

93. The ice on a pond is 5.33 cm thick. For safe skating, the owner of the pond insists that it be 80 mm thick. How much thicker must the ice be before skating is allowed?

94. The sediment on the bottom of the Towamencin Creek is normally 14 cm thick, but the recent flood washed away 22 mm of sediment. How thick is it now?

95. The Amana Corporation stacks up its microwave ovens in a distribution warehouse. Each stack is 1 ft 9 in. wide. How far from the wall would 9 of these stacks extend?

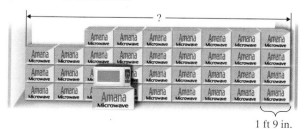

1 ft 9 in.

96. The highway commission is installing concrete sound barriers along a highway. Each barrier is 1 yd 2 ft long. Find the total length of 25 barriers placed end to end.

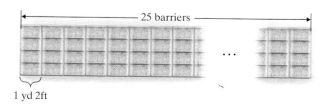

1 yd 2ft

97. A carpenter needs to cut a board into thirds. If the board is 9 ft 3 in. long originally, how long will each cut piece be?

9 feet 3 inches

98. A wall is erected exactly halfway between two buildings that are 192 ft 8 in. apart. If the wall is 8 in. wide, how far is it from the wall to either of the buildings?

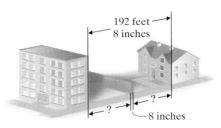

192 feet 8 inches

? ? 8 inches

99. An art class is learning how to make kites. The two sticks used for each kite have lengths of 1 m and 65 cm. What total length of wood must be ordered for the sticks if 25 kites are to be built?

100. The total pages of a hardbound economics text are 3.1 cm thick. The front and back covers are each 2 mm thick. How high would a stack of 10 of these texts be?

101. A logging firm needs to cut a 67-m-long redwood log into 20 equal pieces before loading it onto a truck for shipment. How long will each piece be?

102. An 18.3-m-tall flagpole is mounted on a 65-cm-high pedestal. How far is the top of the flagpole from the ground?

103. A 2.15-m-long sash cord has become frayed at both ends. To correct this, 1 cm is trimmed from each end. How long is the remaining cord?

104. A 112.5-foot-tall dead pine tree is removed by starting at the top and cutting off 9-foot-long sections. How many whole sections are removed?

105. The longest truck in the world is operated by Gould Transport in Australia, and is the 182-ft Road Train. How many *yards* long are 2 of these trucks? (*Source: Guinness World Records*)

106. Three hundred fifty thousand people daily see the large Coca-Cola sign in the Tokyo Ginza shopping district. It is in the shape of a rectangle whose length is 31 yards and whose width is 49 feet. Find the area of the sign in square feet. (*Source:* Coca-Cola Company) (*Hint:* Recall that the area of a rectangle is the product length times width.)

107. A floor tile is 22.86 cm wide. How many tiles in a row are needed to cross a room 3.429 m wide?

△ **108.** A standard postcard is 1.6 times longer than it is wide. If it is 9.9 cm wide, what is its length?

Review

Write each decimal as a fraction and each fraction as a decimal. See Sections 4.1 and 4.5.

109. 0.21   **110.** 0.86   **111.** $\dfrac{13}{100}$ **112.** $\dfrac{47}{100}$ **113.** $\dfrac{1}{4}$ **114.** $\dfrac{3}{20}$

Concept Extensions

Determine whether the measurement in each statement is reasonable. See the second Concept Check in this section.

115. The width of a twin-size bed is 20 meters.

116. A window measures 1 meter by 0.5 meter.

117. A drinking glass is made of glass 2 millimeters thick.

118. A paper clip is 4 kilometers long.

119. The distance across the Colorado River is 50 kilometers.

120. A model's hair is 30 centimeters long.

Estimate each sum or difference. See the first Concept Check in this section.

121.   5 yd 2 in.
     $+$ 7 yd 30 in.

122.   45 ft 1 in.
     $-$ 10 ft 11 in.

Solve.

123. Using a unit other than the foot, write a length that is equivalent to 4 feet. (*Hint:* There are many possibilities.)

124. Using a unit other than the meter, write a length that is equivalent to 7 meters. (*Hint:* There are many possibilities.)

125. To convert from meters to centimeters, the decimal point is moved two places to the right. Explain how this relates to the fact that the prefix *centi* means $\dfrac{1}{100}$.

126. Explain why conversions in the metric system are easier to make than conversions in the U.S. system of measurement.

127. An advertisement sign outside Fenway Park in Boston measures 18.3 m by 18.3 m. What is the area of this sign?

5.5 Weight and Mass: U.S. and Metric Systems of Measurement

Objective A Defining and Converting U.S. System Units of Weight

Whenever we talk about how heavy an object is, we are concerned with the object's **weight.** We discuss weight when we refer to a 12-ounce box of Rice Krispies, a 15-pound tabby cat, and a barge hauling 24 tons of garbage.

Objectives

A Define U.S. Units of Weight and Convert from One Unit to Another.

B Perform Arithmetic Operations on U.S. Units of Weight.

C Define Metric Units of Mass and Convert from One Unit to Another.

D Perform Arithmetic Operations on Metric Units of Mass.

12 ounces

15 pounds   24 tons of garbage

The most common units of weight in the U.S. measurement system are the **ounce,** the **pound,** and the **ton.** The following is a summary of equivalencies between units of weight:

U.S. Units of Weight

16 ounces (oz) = 1 pound (lb)

2000 pounds = 1 ton

Unit Fractions

$$\frac{16 \text{ oz}}{1 \text{ lb}} = \frac{1 \text{ lb}}{16 \text{ oz}} = 1$$

$$\frac{2000 \text{ lb}}{1 \text{ ton}} = \frac{1 \text{ ton}}{2000 \text{ lb}} = 1$$

✓**Concept Check** If you were describing the weight of a fully loaded semitrailer, which type of unit would you use: ounce, pound, or ton? Why?

Unit fractions, which equal 1, are used to convert between units of weight in the U.S. system. When converting using unit fractions, recall that the numerator of a unit fraction should contain the units we are converting to and the denominator should contain the original units.

Practice 1

Convert 6500 pounds to tons.

Example 1 Convert 9000 pounds to tons.

Solution: We multiply 9000 lb by a unit fraction that uses the equality

2000 pounds = 1 ton

Remember, the unit fraction should be $\dfrac{\text{units to convert to}}{\text{original units}}$ or $\dfrac{1 \text{ ton}}{2000 \text{ lb}}$.

$$9000 \text{ lb} = \frac{9000 \text{ lb}}{1} \cdot 1 = \frac{9000 \text{ lb}}{1} \cdot \frac{1 \text{ ton}}{2000 \text{ lb}} = \frac{9000 \text{ tons}}{2000} = \frac{9}{2} \text{ tons or } 4\frac{1}{2} \text{ tons}$$

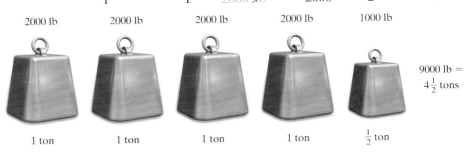

| 2000 lb | 2000 lb | 2000 lb | 2000 lb | 1000 lb |
| 1 ton | 1 ton | 1 ton | 1 ton | $\frac{1}{2}$ ton |

$9000 \text{ lb} = 4\frac{1}{2} \text{ tons}$

■ Work Practice 1

Practice 2

Convert 72 ounces to pounds.

Example 2 Convert 3 pounds to ounces.

Solution: We multiply by the unit fraction $\dfrac{16 \text{ oz}}{1 \text{ lb}}$ to convert from pounds to ounces.

$$3 \text{ lb} = \frac{3 \text{ lb}}{1} \cdot 1 = \frac{3 \text{ lb}}{1} \cdot \frac{16 \text{ oz}}{1 \text{ lb}} = 3 \cdot 16 \text{ oz} = 48 \text{ oz}$$

| 1 pound | 1 pound | 1 pound |

$3 \text{ lb} = 48 \text{ oz}$

| 16 ounces | 16 ounces | 16 ounces |

■ Work Practice 2

Answers

1. $3\frac{1}{4}$ tons   **2.** $4\frac{1}{2}$ lb

✓**Concept Check Answer**

ton

As with length, it is sometimes useful to simplify a measurement of weight by writing it in terms of mixed units.

Example 3 Convert: 33 ounces = _____ lb _____ oz

Solution: Because 16 oz = 1 lb, divide 16 into 33 to see how many pounds are in 33 ounces. The quotient is the number of pounds, and the remainder is the number of ounces. To see why we divide 16 into 33, notice that

$$33 \text{ oz} = 33 \text{ oz} \cdot \frac{1 \text{ lb}}{16 \text{ oz}} = \frac{33}{16} \text{ lb}$$

$$\begin{array}{r} 2 \text{ lb 1 oz} \\ 16\overline{)33} \\ -32 \\ \hline 1 \end{array}$$

Thus, 33 ounces is the same as 2 lb 1 oz.

16 ounces   16 ounces   1 ounce

33 oz = 2 lb 1 oz

1 pound   1 pound   1 ounce

■ Work Practice 3

Practice 3

Convert:
47 ounces = _____ lb _____ oz

Objective B Performing Operations on U.S. System Units of Weight ▶

Performing arithmetic operations on units of weight works the same way as performing arithmetic operations on units of length.

Example 4 Subtract 3 tons 1350 lb from 8 tons 1000 lb.

Solution: To subtract, we line up similar units.

$$\begin{array}{r} 8 \text{ tons 1000 lb} \\ -3 \text{ tons 1350 lb} \\ \hline \end{array}$$

Since we cannot subtract 1350 lb from 1000 lb, we borrow 1 ton from the 8 tons. To do so, we write 1 ton as 2000 lb and combine it with the 1000 lb.

7 tons + (1 ton) 2000 lb

$$\begin{array}{rcl} 8 \text{ tons 1000 lb} & = & 7 \text{ tons 3000 lb} \\ -3 \text{ tons 1350 lb} & = & -3 \text{ tons 1350 lb} \\ \hline & & 4 \text{ tons 1650 lb} \end{array}$$

To check, see that the sum of 4 tons 1650 lb and 3 tons 1350 lb is 8 tons 1000 lb.

■ Work Practice 4

Practice 4

Subtract 5 tons 1200 lb from 8 tons 100 lb.

Answers

3. 2 lb 15 oz   **4.** 2 tons 900 lb

Practice 5

Divide 5 lb 8 oz by 4.

Example 5   Divide 9 lb 6 oz by 2.

Solution: We divide each of the units by 2.

$$
\begin{array}{r}
\quad 4\text{ lb}\quad 11\text{ oz} \\
\hline
2)\ \ 9\text{ lb}\quad 6\text{ oz} \\
\underline{-8}\quad\quad\quad \\
1\text{ lb} = \underline{16\text{ oz}} \\
22\text{ oz}
\end{array}
$$

Divide 2 into 22 oz to get 11 oz.

To check, multiply 4 pounds 11 ounces by 2. The result is 9 pounds 6 ounces.

■ Work Practice 5

Practice 6

A 5-lb 14-oz batch of cookies is packed into a 6-oz container before it is mailed. Find the total weight.

Example 6   Finding the Weight of a Child

Bryan weighed 8 lb 8 oz at birth. By the time he was 1 year old, he had gained 11 lb 14 oz. Find his weight at age 1 year.

Solution:

birth weight	→	8 lb 8 oz
+ weight gained	→	+ 11 lb 14 oz
total weight	→	19 lb 22 oz

Since 22 oz equals 1 lb 6 oz,

$$19\text{ lb } 22\text{ oz} = 19\text{ lb} + 1\text{ lb } 6\text{ oz}$$
$$= 20\text{ lb } 6\text{ oz}$$

Bryan weighed 20 lb 6 oz on his first birthday.

■ Work Practice 6

Objective C Defining and Converting Metric System Units of Mass

In scientific and technical areas, a careful distinction is made between **weight** and **mass. Weight** is really a measure of the pull of gravity. The farther from Earth an object gets, the less it weighs. However, **mass** is a measure of the amount of substance in the object and does not change. Astronauts orbiting Earth weigh much less than they weigh on Earth, but they have the same mass in orbit as they do on Earth. Here on Earth, weight and mass are the same, so either term may be used.

The basic unit of mass in the metric system is the **gram.** It is defined as the mass of water contained in a cube 1 centimeter (cm) on each side.

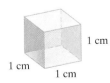

The following examples may help you get a feeling for metric masses:

A tablet contains 200 milligrams of ibuprofen.

A large paper clip weighs approximately 1 gram.

A box of crackers weighs 453 grams.

A kilogram is slightly over 2 pounds. An adult woman may weigh 60 kilograms.

The prefixes for units of mass in the metric system are the same as for units of length, as shown in the following table:

Metric Units of Mass
1 **kilo**gram (kg) = 1000 grams (g)
1 **hecto**gram (hg) = 100 g
1 **deka**gram (dag) = 10 g
1 gram (g) = 1 g
1 **deci**gram (dg) = 1/10 g or 0.1 g
1 **centi**gram (cg) = 1/100 g or 0.01 g
1 **milli**gram (mg) = 1/1000 g or 0.001 g

✓**Concept Check** True or false? A decigram is larger than a dekagram. Explain.

The **milligram,** the **gram,** and the **kilogram** are the three most commonly used units of mass in the metric system.

As with lengths, all units of mass are powers of 10 of the gram, so converting from one unit of mass to another only involves moving the decimal point. To convert from one unit of mass to another in the metric system, list the units of mass in order from largest to smallest.

Let's convert 4300 milligrams to grams. To convert from milligrams to grams, we move along the list 3 units to the left.

kg   hg   dag   **g**   dg   cg   **mg**

3 units to the left

This means that we move the decimal point 3 places to the left to convert from milligrams to grams.

4300 mg = 4.3 g

Don't forget that the same conversion can be done with unit fractions.

$$4300 \text{ mg} = \frac{4300 \text{ mg}}{1} \cdot 1 = \frac{4300 \text{ mg}}{1} \cdot \frac{0.001 \text{ g}}{1 \text{ mg}}$$

$$= 4300 \cdot 0.001 \text{ g}$$

$$= 4.3 \text{ g} \quad \text{To multiply by 0.001, move the decimal point 3 places to the left.}$$

To see that this is reasonable, study the diagram:

1000 mg   1000 mg   1000 mg   1000 mg   300 mg

4300 mg = 4.3 g

1 g   1 g   1 g   1 g   0.3 g

Thus, 4300 mg = 4.3 g.

✓**Concept Check Answer**
false

Practice 7

Convert 3.41 g to milligrams.

Example 7   Convert 3.2 kg to grams.

Solution: First we convert by using a unit fraction.

$$3.2 \text{ kg} = 3.2 \text{ kg} \cdot 1 = 3.2 \ \cancel{\text{kg}} \cdot \overset{\text{Unit fraction}}{\dfrac{1000 \text{ g}}{1 \ \cancel{\text{kg}}}} = 3200 \text{ g}$$

Now let's list the units of mass in order from left to right and move from kilograms to grams.

kg   hg   dag   g   dg   cg   mg

3 units to the right

3.200 kg = 3200. g

3 places to the right

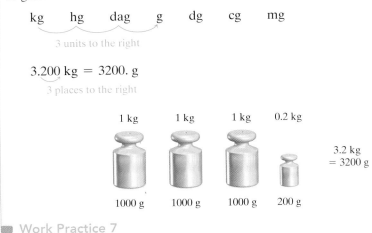

1 kg   1 kg   1 kg   0.2 kg

3.2 kg = 3200 g

1000 g   1000 g   1000 g   200 g

■ Work Practice 7

Practice 8

Convert 56.2 cg to grams.

Example 8   Convert 2.35 cg to grams.

Solution: We list the units of mass in a chart and move from centigrams to grams.

kg   hg   dag   g   dg   cg   mg

2 units to the left

02.35 cg = 0.0235 g

2 places to the left

■ Work Practice 8

Objective D Performing Operations on Metric System Units of Mass ▶

Arithmetic operations can be performed with metric units of mass just as we performed operations with metric units of length. We convert each number to the same unit of mass and add, subtract, multiply, or divide as with decimals.

Practice 9

Subtract 3.1 dg from 2.5 g.

Example 9   Subtract 5.4 dg from 1.6 g.

Solution: We convert both numbers to decigrams or to grams before subtracting.

5.4 dg = 0.54 g   or   1.6 g = 16 dg

$$\begin{array}{r} 1.60 \text{ g} \\ -\,0.54 \text{ g} \\ \hline 1.06 \text{ g} \end{array} \qquad\qquad \begin{array}{r} 16.0 \text{ dg} \\ -\,5.4 \text{ dg} \\ \hline 10.6 \text{ dg} \end{array}$$

The difference is 1.06 g or 10.6 dg.

■ Work Practice 9

Answers

7. 3410 mg   **8.** 0.562 g

9. 2.19 g or 21.9 dg

Example 10 Calculating Allowable Weight in an Elevator

An elevator has a weight limit of 1400 kg. A sign posted in the elevator indicates that the maximum capacity of the elevator is 17 persons. What is the average allowable weight for each passenger, rounded to the nearest kilogram?

Solution: To solve, notice that the total weight of 1400 kilograms ÷ 17 = average weight.

$$
\begin{array}{r}
82.3 \text{ kg} \approx 82 \text{ kg} \\
17)\overline{1400.0 \text{ kg}} \\
-136 \\
\overline{40} \\
-34 \\
\overline{6\,0} \\
-5\,1 \\
\overline{9}
\end{array}
$$

Each passenger can weigh an average of 82 kg. (Recall that a kilogram is slightly over 2 pounds, so 82 kilograms is over 164 pounds.)

📖 Work Practice 10

Practice 10

Twenty-four bags of cement weigh a total of 550 kg. Find the average weight of 1 bag, rounded to the nearest kilogram.

Answer

10. 23 kg

Vocabulary, Readiness & Video Check

Use the choices below to fill in each blank.

    mass   weight   gram

1. _____ is a measure of the amount of substance in an object. This measure does not change.

2. _____ is the measure of the pull of gravity.

3. The basic unit of mass in the metric system is the _____ .

Fill in these blanks with the correct number. Choices for these blanks are not shown in the list of terms above.

4. One pound equals _____ ounces.

5. One ton equals _____ pounds.

Convert without pencil or paper.

6. 3 tons to pounds   **7.** 32 ounces to pounds   **8.** 3 pounds to ounces   **9.** 4000 pounds to tons   **10.** 1 ton to pounds

Martin-Gay Interactive Videos

See Video 5.5 🍎

Watch the section lecture video and answer the following questions.

Objective A **11.** In ▦ Example 2, what units are used in the numerator of the unit fraction and why was this decided? ⬤

Objective B **12.** In ▦ Example 4, explain the first step taken to solve the problem. ⬤

Objective C **13.** In ▦ Example 5, how many places is the decimal point moved and in what direction? What is the final conversion? ⬤

Objective D **14.** What is the answer to ▦ Example 7 in decigrams? ⬤

5.5 Exercise Set MyMathLab®

Objective A *Convert as indicated. See Examples 1 through 3.*

1. 2 pounds to ounces

2. 5 pounds to ounces

3. 5 tons to pounds

4. 7 tons to pounds

5. 18,000 pounds to tons

6. 28,000 pounds to tons

7. 60 ounces to pounds

8. 90 ounces to pounds

9. 3500 pounds to tons

10. 11,000 pounds to tons

11. 12.75 pounds to ounces

12. 9.5 pounds to ounces

13. 4.9 tons to pounds

14. 8.3 tons to pounds

15. $4\frac{3}{4}$ pounds to ounces

16. $9\frac{1}{8}$ pounds to ounces

17. 2950 pounds to the nearest tenth of a ton

18. 51 ounces to the nearest tenth of a pound

19. $\frac{4}{5}$ oz to pounds

20. $\frac{1}{4}$ oz to pounds

21. $5\frac{3}{4}$ lb to ounces

22. $2\frac{1}{4}$ lb to ounces

23. 10 lb 1 oz to ounces

24. 7 lb 6 oz to ounces

25. 89 oz = _____ lb _____ oz

26. 100 oz = _____ lb _____ oz

Objective B *Perform each indicated operation. See Examples 4 through 6.*

27. 34 lb 12 oz + 18 lb 14 oz

28. 6 lb 10 oz + 10 lb 8 oz

29. 3 tons 1820 lb + 4 tons 930 lb

30. 1 ton 1140 lb + 5 tons 1200 lb

31. 5 tons 1050 lb − 2 tons 875 lb

32. 4 tons 850 lb − 1 ton 260 lb

33. 12 lb 4 oz − 3 lb 9 oz

34. 45 lb 6 oz − 26 lb 10 oz

35. 5 lb 3 oz × 6

36. 2 lb 5 oz × 5

37. 6 tons 1500 lb ÷ 5

38. 5 tons 400 lb ÷ 4

Objective C *Convert as indicated. See Examples 7 and 8.*

39. 500 g to kilograms

40. 820 g to kilograms

41. 4 g to milligrams

42. 9 g to milligrams

43. 25 kg to grams

44. 18 kg to grams

45. 48 mg to grams

46. 112 mg to grams

47. 6.3 g to kilograms

48. 4.9 g to kilograms

49. 15.14 g to milligrams

50. 16.23 g to milligrams

51. 6.25 kg to grams

52. 3.16 kg to grams

53. 35 hg to centigrams

54. 4.26 cg to dekagrams

Objective D *Perform each indicated operation. Remember to insert units when writing your answers. See Examples 9 and 10.*

55. 3.8 mg + 9.7 mg

56. 41.6 g + 9.8 g

57. 205 mg + 5.61 g

58. 2.1 g + 153 mg

59. 9 g − 7150 mg

60. 6.13 g − 418 mg

61. 1.61 kg − 250 g

62. 4 kg − 2410 g

63. 5.2 kg × 2.6

64. 4.8 kg × 9.3

65. 17 kg ÷ 8

66. 8.25 g ÷ 6

Objectives A B C D Mixed Practice *Solve. Remember to insert units when writing your answers. For Exercises 67 through 74, complete the charts. See Examples 1 through 10.*

	Object	Tons	Pounds	Ounces
67.	Statue of Liberty–weight of copper sheeting	100		
68.	Statue of Liberty–weight of steel	125		
69.	A 12-inch cube of osmium (heaviest metal)		1345	
70.	A 12-inch cube of lithium (lightest metal)		32	

	Object	Grams	Kilograms	Milligrams	Centigrams
71.	Capsule of amoxicillin (antibiotic)			500	
72.	Tablet of Topamax (epilepsy and migraine uses)			25	
73.	A six-year-old boy		21		
74.	A golf ball	45			

75. A can of 7-Up weighs 336 grams. Find the weight in kilograms of 24 cans.

76. Guy Green normally weighs 73 kg, but he lost 2800 grams after being sick with the flu. Find Guy's new weight.

77. Sudafed is a decongestant that comes in two strengths. Regular strength contains 60 mg of medication. Extra strength contains 0.09 g of medication. How much extra medication is in the extra-strength tablet?

78. A small can of Planters sunflower seeds weighs 177 g. If each can contains 6 servings, find the weight of one serving.

79. Doris Johnson has two open containers of Uncle Ben's rice. If she combines 1 lb 10 oz from one container with 3 lb 14 oz from the other container, how much total rice does she have?

80. Dru Mizel maintains the records of the amount of coal delivered to his department in the steel mill. In January, 3 tons 1500 lb were delivered. In February, 2 tons 1200 lb were delivered. Find the total amount delivered in these two months.

81. Carla Hamtini was amazed when she grew a 28-lb 10-oz zucchini in her garden, but later she learned that the heaviest zucchini ever grown weighed 64 lb 8 oz in Llanharry, Wales, by B. Lavery in 1990. How far below the record weight was Carla's zucchini? (*Source: Guinness World Records*)

82. The heaviest baby born in good health weighed an incredible 22 lb 8 oz. He was born in Italy in September 1955. How much heavier is this than a 7-lb 12-oz baby? (*Source: Guinness World Records*)

83. The smallest baby born in good health weighed only 8.6 ounces, less than a can of soda. She was born in Chicago in December 2004. How much lighter was she than an average baby, who weighs about 7 lb 8 ounces?

84. A large bottle of Hire's Root Beer weighs 1900 grams. If a carton contains 6 large bottles of root beer, find the weight in kilograms of 5 cartons.

85. Three milligrams of preservatives are added to a 0.5-kg box of dried fruit. How many milligrams of preservatives are in 3 cartons of dried fruit if each carton contains 16 boxes?

86. One box of Swiss Miss Cocoa Mix weighs 0.385 kg, but 39 grams of this weight is the packaging. Find the actual weight of the cocoa in 8 boxes.

87. A carton of 12 boxes of Quaker Oats Oatmeal weighs 6.432 kg. Each box includes 26 grams of packaging material. What is the actual weight of the oatmeal in the carton?

88. The supermarket prepares hamburger in 85-gram market packages. When Leo Gonzalas gets home, he divides the package in half before refrigerating the meat. How much will each package weigh?

89. The Shop 'n Bag supermarket chain ships hamburger meat by placing 10 packages of hamburger in a box, with each package weighing 3 lb 4 oz. How much will 4 boxes of hamburger weigh?

90. The Quaker Oats Company ships its 1-lb 2-oz boxes of oatmeal in cartons containing 12 boxes of oatmeal. How much will 3 such cartons weigh?

91. A carton of Del Monte Pineapple weighs 55 lb 4 oz, but 2 lb 8 oz of this weight is due to packaging. Find the actual weight of the pineapple in 4 cartons.

92. The Hormel Corporation ships cartons of canned ham weighing 43 lb 2 oz each. Of this weight, 3 lb 4 oz is due to packaging. Find the actual weight of the ham found in 3 cartons.

Review

Write each fraction as a decimal. See Section 4.5.

93. $\dfrac{4}{25}$ **94.** $\dfrac{3}{5}$ **95.** $\dfrac{7}{8}$ **96.** $\dfrac{3}{16}$

Concept Extensions

Determine whether the measurement in each statement is reasonable.

97. The doctor prescribed a pill containing 2 kg of medication.

98. A full-grown cat weighs approximately 15 g.

99. A bag of flour weighs 4.5 kg.

100. A staple weighs 15 mg.

101. A professor weighs less than 150 g.

102. A car weighs 2000 mg.

Solve.

103. Use a unit other than centigram and write a mass that is equivalent to 25 centigrams. (*Hint:* There are many possibilities.)

104. Use a unit other than pound and write a weight that is equivalent to 4000 pounds. (*Hint:* There are many possibilities.)

True or false? See the second Concept Check in this section.

105. A kilogram is larger than a gram.

106. A decigram is larger than a milligram.

107. Why is the decimal point moved to the right when grams are converted to milligrams?

108. To change 8 pounds to ounces, multiply by 16. Why is this the correct procedure?

5.6 Capacity: U.S. and Metric Systems of Measurement

Objective A Defining and Converting U.S. System Units of Capacity

Units of **capacity** are generally used to measure liquids. The number of gallons of gasoline needed to fill a gas tank in a car, the number of cups of water needed in a bread recipe, and the number of quarts of milk sold each day at a supermarket are all examples of using units of capacity. The following summary shows equivalencies between units of capacity:

U.S. Units of Capacity

$$8 \text{ fluid ounces (fl oz)} = 1 \text{ cup (c)}$$
$$2 \text{ cups} = 1 \text{ pint (pt)}$$
$$2 \text{ pints} = 1 \text{ quart (qt)}$$
$$4 \text{ quarts} = 1 \text{ gallon (gal)}$$

Objectives

A Define U.S. Units of Capacity and Convert from One Unit to Another.

B Perform Arithmetic Operations on U.S. Units of Capacity.

C Define Metric Units of Capacity and Convert from One Unit to Another.

D Perform Arithmetic Operations on Metric Units of Capacity.

Just as with units of length and weight, we can form unit fractions to convert between different units of capacity. For instance,

$$\frac{2\text{ c}}{1\text{ pt}} = \frac{1\text{ pt}}{2\text{ c}} = 1 \quad \text{and} \quad \frac{2\text{ pt}}{1\text{ qt}} = \frac{1\text{ qt}}{2\text{ pt}} = 1$$

Practice 1

Convert 43 pints to quarts.

Example 1 Convert 9 quarts to gallons.

Solution: We multiply by the unit fraction $\frac{1\text{ gal}}{4\text{ qt}}$.

$$9\text{ qt} = \frac{9\text{ qt}}{1} \cdot 1$$

$$= \frac{9\text{ qt}}{1} \cdot \frac{1\text{ gal}}{4\text{ qt}}$$

$$= \frac{9\text{ gal}}{4}$$

$$= 2\frac{1}{4}\text{ gal}$$

Thus, 9 quarts is the same as $2\frac{1}{4}$ gallons, as shown in the diagram:

Work Practice 1

Practice 2

Convert 26 quarts to cups.

Example 2 Convert 14 cups to quarts.

Solution: Our equivalency table contains no direct conversion from cups to quarts. However, from this table we know that

$$1\text{ qt} = 2\text{ pt} = \frac{2\text{ pt}}{1} \cdot 1 = \frac{2\text{ pt}}{1} \cdot \frac{2\text{ c}}{1\text{ pt}} = 4\text{ c}$$

so 1 qt = 4 c. Now we have the unit fraction $\frac{1\text{ qt}}{4\text{ c}}$. Thus,

$$14\text{ c} = \frac{14\text{ c}}{1} \cdot 1 = \frac{14\text{ c}}{1} \cdot \frac{1\text{ qt}}{4\text{ c}} = \frac{14\text{ qt}}{4} = \frac{7}{2}\text{ qt} \quad \text{or} \quad 3\frac{1}{2}\text{ qt}$$

14 cups $= 3\frac{1}{2}$ qt

Work Practice 2

Answers

1. $21\frac{1}{2}$ qt   2. 104 c

✓ **Concept Check Answer**

less than 50; answers may vary

✓ **Concept Check** If 50 cups is converted to quarts, will the equivalent number of quarts be less than or greater than 50? Explain.

Objective B Performing Operations on U.S. System Units of Capacity ▶

As is true of units of length and weight, units of capacity can be added, subtracted, multiplied, and divided.

Example 3 Subtract 3 qt from 4 gal 2 qt.

Solution: To subtract, we line up similar units.

$$
\begin{array}{r}
4 \text{ gal } 2 \text{ qt} \\
- \quad\quad 3 \text{ qt} \\
\hline
\end{array}
$$

We cannot subtract 3 qt from 2 qt. We need to borrow 1 gallon from the 4 gallons, convert it to 4 quarts, and then combine it with the 2 quarts.

3 gal + (1 gal) 4 qt

$$
\begin{array}{rcl}
4 \text{ gal } 2 \text{ qt} & = & 3 \text{ gal } 6 \text{ qt} \\
- \quad\quad 3 \text{ qt} & = & - \quad\quad 3 \text{ qt} \\
\hline
& & 3 \text{ gal } 3 \text{ qt}
\end{array}
$$

To check, see that the sum of 3 gal 3 qt and 3 qt is 4 gal 2 qt.

▸ Work Practice 3

Practice 3

Subtract 2 qt from 1 gal 1 qt.

Example 4 Divide 3 gal 2 qt by 2.

Solution: We divide each unit of capacity by 2.

$$
\begin{array}{r}
1 \text{ gal} \quad\quad 3 \text{ qt} \\
2\overline{)3 \text{ gal} \quad\quad 2 \text{ qt}} \\
-2 \quad\quad\quad\quad\quad\quad \\
\hline
1 \text{ gal} = 4 \text{ qt} \\
\quad\quad\quad 6 \text{ qt}
\end{array}
$$

Convert 1 gallon to 4 qt and add to 2 qt before continuing.
6 qt ÷ 2 = 3 qt

▸ Work Practice 4

Practice 4

Divide 7 gal 2 qt by 3.

Example 5 Finding the Amount of Water in an Aquarium

An aquarium contains 6 gal 3 qt of water. If 2 gal 2 qt of water is added, what is the total amount of water in the aquarium?

Solution:

$$
\begin{array}{rcr}
\text{beginning water} & \rightarrow & 6 \text{ gal } 3 \text{ qt} \\
+ \quad \text{water added} & \rightarrow & + 2 \text{ gal } 2 \text{ qt} \\
\hline
\text{total water} & \rightarrow & 8 \text{ gal } 5 \text{ qt}
\end{array}
$$

Since 5 qt = 1 gal 1 qt, we have

8 gal   5 qt

$$
\begin{array}{l}
= 8 \text{ gal} + 1 \text{ gal } 1 \text{ qt} \\
= 9 \text{ gal } 1 \text{ qt}
\end{array}
$$

The total amount of water is 9 gal 1 qt.

▸ Work Practice 5

Practice 5

A large oil drum contains 15 gal 3 qt of oil. How much will be in the drum if an additional 4 gal 3 qt of oil is poured into it?

Answers

3. 3 qt   **4.** 2 gal 2 qt   **5.** 20 gal 2 qt

Objective C Defining and Converting Metric System Units of Capacity

10 cm
10 cm
10 cm

Thus far, we know that the basic unit of length in the metric system is the meter and that the basic unit of mass in the metric system is the gram. What is the basic unit of capacity? The **liter.** By definition, a **liter** is the capacity or volume of a cube measuring 10 centimeters on each side.

The following examples may help you get a feeling for metric capacities:

One liter of liquid is slightly more than one quart.
Many soft drinks are packaged in 2-liter bottles.

The metric system was designed to be a consistent system. Once again, the prefixes for metric units of capacity are the same as for metric units of length and mass, as summarized in the following table:

1 liter   1 quart

Metric Units of Capacity
1 **kilo**liter (kl) = 1000 liters (L)
1 **hecto**liter (hl) = 100 L
1 **deka**liter (dal) = 10 L
1 liter (L) = 1 L
1 **deci**liter (dl) = 1/10 L or 0.1 L
1 **centi**liter (cl) = 1/100 L or 0.01 L
1 **milli**liter (ml) = 1/1000 L or 0.001 L

2 liters

The **milliliter** and the **liter** are the two most commonly used metric units of capacity.

Converting from one unit of capacity to another involves multiplying by powers of 10 or moving the decimal point to the left or to the right. Listing units of capacity in order from largest to smallest helps to keep track of how many places to move the decimal point when converting.

Let's convert 2.6 liters to milliliters. To convert from liters to milliliters, we move along the chart 3 units to the right.

kl   hl   dal   **L**   dl   cl   **ml**

3 units to the right

This means that we move the decimal point 3 places to the right to convert from liters to milliliters.

2.600 L = 2600. ml

This same conversion can be done with unit fractions.

$$2.6 \text{ L} = \frac{2.6 \text{ L}}{1} \cdot 1$$

$$= \frac{2.6 \text{ L}}{1} \cdot \frac{1000 \text{ ml}}{1 \text{ L}}$$

$$= 2.6 \cdot 1000 \text{ ml}$$

$$= 2600 \text{ ml} \quad \text{To multiply by 1000, move the decimal point 3 places to the right.}$$

To visualize the result, study the diagram below:

2.6 L

1000 ml   1000 ml   600 ml = 2600 ml

Thus, 2.6 L = 2600 ml.

Example 6 Convert 3210 ml to liters.

Solution: Let's use the unit fraction method first.

$$3210 \text{ ml} = \frac{3210 \text{ ml}}{1} \cdot 1 = 3210 \text{ ml} \cdot \frac{1 \text{ L}}{1000 \text{ ml}} = 3.21 \text{ L}$$

Now let's list the unit measures in order from left to right and move from milliliters to liters.

kl   hl   dal   **L**   dl   cl   ml

3 units to the left

3210 ml = 3.210 L, the same results as before and

3 places to the left   shown below in the diagram.

1000 ml   1000 ml   1000 ml

210 ml

3210 ml

1 L   1 L   1 L   0.210 L = 3.210 L

■ Work Practice 6

Example 7 Convert 0.185 dl to milliliters.

Solution: We list the unit measures in order from left to right and move from deciliters to milliliters.

kl   hl   dal   L   dl   cl   ml

2 units to the right

0.185 dl = 18.5 ml

2 places to the right

■ Work Practice 7

Objective D Performing Operations on Metric System Units of Capacity ▶

As was true for length and weight, arithmetic operations involving metric units of capacity can also be performed. Make sure that the metric units of capacity are the same before adding or subtracting.

Practice 6
Convert 2100 ml to liters.

Practice 7
Convert 2.13 dal to liters.

Answers
6. 2.1 L **7.** 21.3 L

Practice 8

Add 1250 ml to 2.9 L.

Example 8 Add 2400 ml to 8.9 L.

Solution: We must convert both to liters or both to milliliters before adding the capacities together.

$$2400 \text{ ml} = 2.4 \text{ L} \quad \text{or} \quad 8.9 \text{ L} = 8900 \text{ ml}$$

$$
\begin{array}{r}
2.4 \text{ L} \\
+8.9 \text{ L} \\
\hline
11.3 \text{ L}
\end{array}
\qquad
\begin{array}{r}
2400 \text{ ml} \\
+8900 \text{ ml} \\
\hline
11{,}300 \text{ ml}
\end{array}
$$

The total is 11.3 L or 11,300 ml. They both represent the same capacity.

■ Work Practice 8

✓**Concept Check** How could you estimate the following operation? Subtract 950 ml from 7.5 L.

Practice 9

If 28.6 L of water can be pumped every minute, how much water can be pumped in 85 minutes?

Example 9 Finding the Amount of Medication a Person Receives

A patient hooked up to an IV unit in the hospital is to receive 12.5 ml of medication every hour. How much medication does the patient receive in 3.5 hours?

Solution: We multiply 12.5 ml by 3.5.

$$
\begin{array}{rl}
\text{medication per hour} \quad \rightarrow & 12.5 \text{ ml} \\
\times \quad \text{hours} \quad \rightarrow & \times\ 3.5 \\
\hline
\text{total medication} & 625 \\
& 3750 \\
\hline
& 43.75 \text{ ml}
\end{array}
$$

The patient receives 43.75 ml of medication.

■ Work Practice 9

Answers

8. 4150 ml or 4.15 L   **9.** 2431 L

✓**Concept Check Answer**
950 ml = 0.95 L; round 0.95 to 1;
7.5 − 1 = 6.5 L

Vocabulary, Readiness & Video Check

Use the choices below to fill in each blank. Some choices may be used more than once.

cups	pints	liter
quarts	fluid ounces	capacity

1. Units of _____ are generally used to measure liquids.

2. The basic unit of capacity in the metric system is the _____.

3. One cup equals 8 _____.

4. One quart equals 2 _____.

5. One pint equals 2 _____.

6. One quart equals 4 _____.

7. One gallon equals 4 _____.

Convert as indicated without pencil or paper or calculator.

8. 2 c to pints

9. 4 c to pints

10. 4 qt to gallons

11. 8 qt to gallons

12. 2 pt to quarts

13. 6 pt to quarts

14. 8 fl oz to cups

15. 24 fl oz to cups

16. 3 pt to cups

Martin-Gay Interactive Videos *Watch the section lecture video and answer the following questions.*

See Video 5.6

Objective A 17. Complete this statement based on ▦ Example 1: When using a unit fraction, we are not changing the _____; we are changing the _____. ◯

Objective B 18. In ▦ Example 4, explain the first step taken to solve the problem. ◯

Objective C 19. In ▦ Example 5, how many places is the decimal point moved and in what direction? What is the final conversion? ◯

Objective D 20. What is the answer to ▦ Example 7 in dekaliters? ◯

5.6 **Exercise Set** MyMathLab®

Objective A *Convert each measurement as indicated. See Examples 1 and 2.*

1. 32 fluid ounces to cups

2. 16 quarts to gallons

3. 8 quarts to pints

4. 9 pints to quarts

5. 14 quarts to gallons

6. 11 cups to pints

7. 80 fluid ounces to pints

8. 18 pints to gallons

9. 2 quarts to cups

10. 3 pints to fluid ounces

11. 120 fluid ounces to quarts

12. 20 cups to gallons

13. 42 cups to quarts

14. 7 quarts to cups

15. $4\frac{1}{2}$ pints to cups

16. $6\frac{1}{2}$ gallons to quarts

17. 5 gal 3 qt to quarts

18. 4 gal 1 qt to quarts

19. $\frac{1}{2}$ cup to pints

20. $\frac{1}{2}$ pint to quarts

21. 58 qt = _____ gal _____ qt

22. 70 qt = _____ gal _____ qt

23. 39 pt = _____ gal _____ qt _____ pt

24. 29 pt = _____ gal _____ qt _____ pt

25. $2\frac{3}{4}$ gallons to pints

26. $3\frac{1}{4}$ quarts to cups

Objective B *Perform each indicated operation. See Examples 3 through 5.*

27. 5 gal 3 qt + 7 gal 3 qt

28. 2 gal 2 qt + 9 gal 3 qt

29. 1 c 5 fl oz + 2 c 7 fl oz

30. 2 c 3 fl oz + 2 c 6 fl oz

31. 3 gal − 1 gal 3 qt

32. 2 pt − 1 pt 1 c

33. 3 gal 1 qt − 1 qt 1 pt

34. 3 qt 1 c − 1 c 4 fl oz

35. 8 gal 2 qt × 2

36. 6 gal 1 pt × 2

37. 9 gal 2 qt ÷ 2

38. 5 gal 6 fl oz ÷ 2

Objective C *Convert as indicated. See Examples 6 and 7.*

39. 5 L to milliliters

40. 8 L to milliliters

41. 0.16 L to kiloliters

42. 0.127 L to kiloliters

43. 5600 ml to liters

44. 1500 ml to liters

45. 3.2 L to centiliters

46. 1.7 L to centiliters

47. 410 L to kiloliters

48. 250 L to kiloliters

49. 64 ml to liters

50. 39 ml to liters

51. 0.16 kl to liters

52. 0.48 kl to liters

53. 3.6 L to milliliters

54. 1.9 L to milliliters

Objective D *Perform each indicated operation. Remember to insert units when writing your answers. See Examples 8 and 9.*

55. 3.4 L + 15.9 L

56. 18.5 L + 4.6 L

57. 2700 ml + 1.8 L

58. 4.6 L + 1600 ml

59. 8.6 L − 190 ml

60. 4.8 L − 283 ml

61. 17,500 ml − 0.9 L

62. 6850 ml − 0.3 L

63. 480 ml × 8

64. 290 ml × 6

65. 81.2 L ÷ 0.5

66. 5.4 L ÷ 3.6

Objectives A B C D Mixed Practice *Solve. Remember to insert units when writing your answers. For Exercises 67 through 70, complete the chart. See Examples 1 through 9.*

	Capacity	Cups	Gallons	Quarts	Pints
67.	An average-size bath of water		21		
68.	A dairy cow's daily milk yield				38
69.	Your kidneys filter about this amount of blood every minute	4			
70.	The amount of water needed in a punch recipe	2			

71. Mike Schaferkotter drank 410 ml of Mountain Dew from a 2-liter bottle. How much Mountain Dew remains in the bottle?

72. The Werners' Volvo has a 54.5-L gas tank. Only 38 dekaliters of gasoline still remain in the tank. How much is needed to fill it?

73. Margie Phitts added 354 ml of Prestone dry gas to the 18.6 L of gasoline in her car's tank. Find the total amount of gasoline in the tank.

74. Chris Peckaitis wishes to share a 2-L bottle of Coca-Cola equally with 7 of his friends. How much will each person get?

75. A garden tool engine requires a 30-to-1 gas-to-oil mixture. This means that $\frac{1}{30}$ of a gallon of oil should be mixed with 1 gallon of gas. Convert $\frac{1}{30}$ gallon to fluid ounces. Round to the nearest tenth.

76. Henning's Supermarket sells homemade soup in 1 qt 1 pt containers. How much soup is contained in three such containers?

77. Can 5 pt 1 c of fruit punch and 2 pt 1 c of ginger ale be poured into a 1-gal container without it overflowing?

78. Three cups of prepared Jell-O are poured into 6 dessert dishes. How many fluid ounces of Jell-O are in each dish?

79. Stanley Fisher paid $30 to fill his car with 44.3 liters of gasoline. Find the price per liter of gasoline to the nearest thousandth of a dollar.

80. A student carelessly misread the scale on a cylinder in the chemistry lab and added 40 cl of water to a mixture instead of 40 ml. Find the excess amount of water.

Review

Write each fraction in simplest form. See Section 3.2.

81. $\frac{20}{25}$ **82.** $\frac{75}{100}$ **83.** $\frac{27}{45}$ **84.** $\frac{56}{60}$ **85.** $\frac{72}{80}$ **86.** $\frac{18}{20}$

Concept Extensions

Determine whether the measurement in each statement is reasonable.

87. Clair took a dose of 2 L of cough medicine to cure her cough.

88. John drank 250 ml of milk for lunch.

89. Jeannie likes to relax in a tub filled with 3000 ml of hot water.

90. Sarah pumped 20 L of gasoline into her car yesterday.

Solve. See the Concept Checks in this section.

91. If 70 pints are converted to gallons, will the equivalent number of gallons be less than or greater than 70? Explain why.

92. If 30 gallons are converted to quarts, will the equivalent number of quarts be less than or greater than 30? Explain why.

93. Explain how to estimate the following operation: Add 986 ml to 6.9 L.

94. Explain how to borrow in order to subtract 1 gal 2 qt from 3 gal 1 qt.

95. Find the number of fluid ounces in 1 gallon.

96. Find the number of fluid ounces in 1.5 gallons.

A cubic centimeter (cc) is the amount of space that a volume of 1 ml occupies. Because of this, we will say that 1 cc = 1 ml.

A common syringe is one with a capacity of 3 cc. Use the diagram and give the measurement indicated by each arrow.

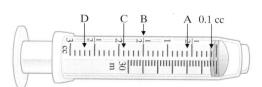

97. B   **98.** A   **99.** D   **100.** C

In order to measure small dosages, such as for insulin, u-100 syringes are used. For these syringes, 1 cc has been divided into 100 equal units (u). Use the diagram and give the measurement indicated by each arrow in units (u) and then in cubic centimeters. Use 100 u = 1 cc.

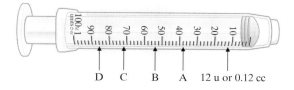

101. B   **102.** A

103. D   **104.** C

Conversions Between the U.S. and Metric Systems

Objective

A Convert Between the U.S. and Metric Systems.

Length

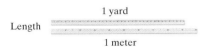

1 yard

1 meter

Capacity

1 quart   1 liter

Weight (Mass)

1 pound   1 kilogram

Objective A Converting Between the U.S. and Metric Systems

The metric system probably had its beginnings in France in the 1600s, but it was the Metric Act of 1866 that made the use of this system legal (but not mandatory) in the United States. Other laws have followed that allow for a slow, but deliberate, transfer to the modernized metric system. In April 2001, for example, the U.S. Stock Exchanges completed their change to decimal trading instead of fractions. By the end of 2009, all products sold in Europe (with some exceptions) were required to have only metric units on their labels. (*Source:* U.S. Metric Association and National Institute of Standards and Technology)

You may be surprised at the number of everyday items we use that are already manufactured in metric units. We easily recognize 1-L and 2-L soda bottles, but what about the following?

Pencil leads (0.5 mm or 0.7 mm)

Camera film (35 mm)

Sporting events (5-km or 10-km races)

Medicines (500-mg capsules)

Labels on retail goods (dual-labeled since 1994)

Since the United States has not completely converted to the metric system, we need to practice converting from one system to the other. Below is a table of mostly approximate conversions.

Length:		Capacity:		Weight (mass):	
Metric	U.S. System	Metric	U.S. System	Metric	U.S. System
1 m ≈ 1.09 yd		1 L ≈ 1.06 qt		1 kg ≈ 2.20 lb	
1 m ≈ 3.28 ft		1 L ≈ 0.26 gal		1 g ≈ 0.04 oz	
1 km ≈ 0.62 mi		3.79 L ≈ 1 gal		0.45 kg ≈ 1 lb	
2.54 cm = 1 in.		0.95 L ≈ 1 qt		28.35 g ≈ 1 oz	
0.30 m ≈ 1 ft		29.57 ml ≈ 1 fl oz			
1.61 km ≈ 1 mi					

There are many ways to perform these metric-to-U.S. conversions. We will do so by using unit fractions.

Example 1 Compact Discs

Standard-sized compact discs are 12 centimeters in diameter. Convert this length to inches. Round the result to two decimal places. (*Source:* usByte.com)

Solution: From our length conversion table, we know that 2.54 cm = 1 in. This fact gives us two unit fractions: $\frac{2.54 \text{ cm}}{1 \text{ in.}}$ and $\frac{1 \text{ in.}}{2.54 \text{ cm}}$. We use the unit fraction with cm in the denominator so that these units divide out.

$$12 \text{ cm} = \frac{12 \text{ cm}}{1} \cdot 1 = \frac{12 \text{ cm}}{1} \cdot \overbrace{\frac{1 \text{ in.}}{2.54 \text{ cm}}}^{\text{Unit fraction}} \quad \begin{matrix} \leftarrow \text{Units to convert to} \\ \leftarrow \text{Original units} \end{matrix}$$

$$= \frac{12 \text{ in.}}{2.54}$$

$$\approx 4.72 \text{ in.} \quad \text{Divide.}$$

1.5 cm

12 cm

Practice 1

The center hole of a standard-sized compact disc is 1.5 centimeters in diameter. Convert this length to inches. Round the result to 2 decimal places.

Answer
1. 0.59 in.

Thus, the diameter of a standard compact disc is exactly 12 cm or approximately 4.72 inches. For a dimension this size, you can use a ruler to check. Another method is to approximate. Our result, 4.72 in., is close to 5 inches. Since 1 in. is about 2.5 cm, then 5 in. is about $5(2.5 \text{ cm}) = 12.5 \text{ cm}$, which is close to 12 cm.

■ Work Practice 1

Example 2 Liver

The liver is your largest internal organ. It weighs about 3.5 pounds in a grown man. Convert this weight to kilograms. Round to the nearest tenth. (*Source: Some Body!* by Dr. Pete Rowan)

Unit fraction

Solution: $3.5 \text{ lb} \approx \dfrac{3.5 \text{ lb}}{1} \cdot \dfrac{0.45 \text{ kg}}{1 \text{ lb}} = 3.5(0.45 \text{ kg}) \approx 1.6 \text{ kg}$

Thus 3.5 pounds are approximately 1.6 kilograms. From the table of conversions, we know that $1 \text{ kg} \approx 2.2 \text{ lb}$. So that means $0.5 \text{ kg} \approx 1.1 \text{ lb}$ and after adding, we have $1.5 \text{ kg} \approx 3.3 \text{ lb}$. Our result is reasonable.

■ Work Practice 2

Practice 2

A full-grown human heart weighs about 8 ounces. Convert this weight to grams. If necessary, round your result to the nearest tenth of a gram.

Example 3 Postage Stamp

Australia converted to the metric system in 1973. In that year, four postage stamps were issued to publicize this conversion. One such stamp is shown. Let's check the mathematics on the stamp by converting 7 fluid ounces to milliliters. Round to the nearest hundred.

Solution: $7 \text{ fl oz} \approx \dfrac{7 \text{ fl oz}}{1} \cdot \dfrac{29.57 \text{ ml}}{1 \text{ fl oz}} = 7(29.57 \text{ ml}) = 206.99 \text{ ml}$

Unit fraction

Rounded to the nearest hundred, $7 \text{ fl oz} \approx 200 \text{ ml}$.

■ Work Practice 3

Practice 3

Convert 237 ml to fluid ounces. Round to the nearest whole fluid ounce.

Answers
2. 226.8 g   **3.** 8 fl oz

Vocabulary, Readiness & Video Check

Martin-Gay Interactive Videos

Watch the section lecture video and answer the following questions.

Objective A **1.** Write two conversions that may be used to solve ▣ Example 2.

2. Why isn't 0.1125 kg the final answer to ▣ Example 3?

See Video 5.7

5.7 Exercise Set MyMathLab®

Note: Because approximations are used, your answers may vary slightly from the answers given in the back of the book.

Objective A *Convert as indicated. If necessary, round answers to two decimal places. See Examples 1 through 3.*

1. 756 milliliters to fluid ounces

2. 18 liters to quarts

▶ **3.** 86 inches to centimeters

4. 86 miles to kilometers

5. 1000 grams to ounces

6. 100 kilograms to pounds

7. 93 kilometers to miles

8. 9.8 meters to feet

▶ **9.** 14.5 liters to gallons

10. 150 milliliters to fluid ounces

11. 30 pounds to kilograms

12. 15 ounces to grams

Fill in the chart. Give exact answers or round to one decimal place. See Examples 1 through 3.

		Meters	Yards	Centimeters	Feet	Inches
13.	The height of a woman				5	
14.	Statue of Liberty length of nose	1.37				
15.	Leaning Tower of Pisa		60			
16.	Blue whale		36			

Solve. If necessary, round answers to two decimal places. See Examples 1 through 3.

17. The balance beam for female gymnasts is 10 centimeters wide. Convert this width to inches.

18. In men's gymnastics, the rings are 250 centimeters from the floor. Convert this height to inches and then to feet.

19. In many states, the maximum speed limit for recreational vehicles is 50 miles per hour. Convert this to kilometers per hour.

20. In some states, the speed limit is 70 miles per hour. Convert this to kilometers per hour.

21. Ibuprofen comes in 200-milligram tablets. Convert this to ounces. (Round your answer to this exercise to 3 decimal places.)

22. Vitamin C tablets come in 500-milligram caplets. Convert this to ounces.

23. A stone is a unit in the British customary system. Use the conversion 14 pounds = 1 stone to check the equivalencies on this 1973 Australian stamp. Is 100 kilograms approximately 15 stone 10 pounds?

24. Convert 5 feet 11 inches to centimeters and check the conversion on this 1973 Australian stamp. Is it correct?

25. The Monarch butterfly migrates annually between the northern United States and central Mexico. The trip is about 4500 km long. Convert this to miles.

26. There is a species of African termite that builds nests up to 18 ft high. Convert this to meters.

27. A $3\frac{1}{2}$-inch diskette is not really $3\frac{1}{2}$ inches. To find its actual width, convert this measurement to centimeters and then to millimeters. Round the result to the nearest ten.

28. The average two-year-old is 84 centimeters tall. Convert this to feet and inches.

29. For an average adult, the weight of the right lung is greater than the weight of the left lung. If the right lung weighs 1.5 pounds and the left lung weighs 1.25 pounds, find the difference in grams. (*Source: Some Body!*)

30. The skin of an average adult weighs 9 pounds and is the heaviest organ. Find the weight in grams. (*Source: Some Body!*)

31. A fast sneeze has been clocked at about 167 kilometers per hour. Convert this to miles per hour. Round to the nearest whole.

32. A Boeing 747 has a cruising speed of about 980 kilometers per hour. Convert this to miles per hour. Round to the nearest whole.

33. The General Sherman giant sequoia tree has a diameter of about 8 meters at its base. Convert this to feet. (*Source: Fantastic Book of Comparisons*)

34. The largest crater on the near side of the moon is Billy Crater. It has a diameter of 303 kilometers. Convert this to miles. (*Source: Fantastic Book of Comparisons*)

35. The total length of the track on a CD is about 4.5 kilometers. Convert this to miles. Round to the nearest whole mile.

36. The distance between Mackinaw City, Michigan, and Cheyenne, Wyoming, is 2079 kilometers. Convert this to miles. Round to the nearest whole mile.

37. A doctor orders a dosage of 5 ml of medicine every 4 hours for 1 week. How many fluid ounces of medicine should be purchased? Round up to the next whole fluid ounce.

38. A doctor orders a dosage of 12 ml of medicine every 6 hours for 10 days. How many fluid ounces of medicine should be purchased? Round up to the next whole fluid ounce.

Without actually converting, choose the most reasonable answer.

39. This math book has a height of about _____.
  a. 28 mm   **b.** 28 cm
  c. 28 m   **d.** 28 km

40. A mile is _____ a kilometer.
  a. shorter than   **b.** longer than
  c. the same length as

41. A liter has _____ capacity than a quart.
  a. less   **b.** greater
  c. the same

42. A foot is _____ a meter.
   a. shorter than   **b.** longer than
   c. the same length as

43. A kilogram weighs _____ a pound.
   a. the same as   **b.** less than
   c. more than

44. A football field is 100 yards, which is about _____.
   a. 9 m   **b.** 90 m
   c. 900 m   **d.** 9000 m

45. An $8\frac{1}{2}$-ounce glass of water has a capacity of about _____.
   a. 250 L   **b.** 25 L
   c. 2.5 L   **d.** 250 ml

46. A 5-gallon gasoline can has a capacity of about _____.
   a. 19 L   **b.** 1.9 L
   c. 19 ml   **d.** 1.9 ml

47. The weight of an average man is about _____.
   a. 700 kg   **b.** 7 kg
   c. 0.7 kg   **d.** 70 kg

48. The weight of a pill is about _____.
   a. 200 kg   **b.** 20 kg
   c. 2 kg   **d.** 200 mg

Review

Perform the indicated operations. See Section 1.9.

49. $6 \cdot 4 + 5 \div 1$

50. $10 \div 2 + 9(8)$

51. $\dfrac{10 + 8}{10 - 8}$

52. $\dfrac{14 + 1}{5(3)}$

53. $3 + 5(19 - 17) - 8$

54. $1 + 4(19 - 9) + 5$

55. $3[(1 + 5) \cdot (8 - 6)]$

56. $5[(18 - 8) - 9]$

Concept Extensions

Body surface area (BSA) is often used to calculate dosages for some drugs. BSA is calculated in square meters using a person's weight and height.

$$\text{BSA} = \sqrt{\frac{(\text{weight in kg}) \times (\text{height in cm})}{3600}}$$

For Exercises 57 through 62, calculate the BSA for each person. Round to the nearest hundredth. You will need to use the square root key on your calculator.

57. An adult whose height is 182 cm and weight is 90 kg

58. An adult whose height is 157 cm and weight is 63 kg

59. A child whose height is 40 in. and weight is 50 kg (*Hint:* Don't forget to first convert inches to centimeters.)

60. A child whose height is 26 in. and weight is 13 kg

61. An adult whose height is 60 in. and weight is 150 lb

62. An adult whose height is 69 in. and weight is 172 lb

Solve.

63. Suppose the adult from Exercise 57 is to receive a drug that has a recommended dosage range of 10–12 mg per sq meter. Find the dosage range for the adult.

64. Suppose the child from Exercise 60 is to receive a drug that has a recommended dosage of 30 mg per sq meter. Find the dosage for the child.

65. A handball court is a rectangle that measures 20 meters by 40 meters. Find its area in square meters and square feet.

66. A backpack measures 16 inches by 13 inches by 5 inches. Find the volume of a box with these dimensions. Find the volume in cubic inches and cubic centimeters. Round the cubic centimeters to the nearest whole cubic centimeter.

Chapter 5 Group Activity

Consumer Price Index

Sections 5.1–5.3

Do you remember when the regular price of a candy bar was 5¢, 10¢, or 25¢? It is certainly difficult to find a candy bar for that price these days. The reason is inflation: the tendency for the price of a given product to increase over time. Businesses and government agencies use the Consumer Price Index (CPI) to track inflation. The CPI measures the change in prices of basic consumer goods and services over time.

The CPI is very useful for comparing the prices of fixed items in various years. For instance, suppose an insurance company customer submits a claim for the theft of a fishing boat purchased in 1975. Because the customer's policy includes replacement cost coverage, the insurance company must calculate how much it would cost to replace the boat at the time of the theft. (Let's assume the theft took place in August 2013.) The customer has a receipt for the boat showing that it cost $598 in 1975. The insurance company can use the following proportion to calculate the replacement cost:

$$\frac{\text{price in earlier year}}{\text{price in later year}} = \frac{\text{CPI value in earlier year}}{\text{CPI value in later year}}$$

The CPI value is 53.8 for 1975. In August 2013, the CPI value was 233.877. The insurance company would then use the following proportion for this situation. (We will let n represent the unknown price in August 2013.)

$$\frac{\text{price in 1975}}{\text{price in 2013}} = \frac{\text{CPI value in 1975}}{\text{CPI value in Aug. 2013}}$$

$$\frac{598}{n} = \frac{53.8}{233.877}$$

$$53.8 \cdot n = 598(233.877)$$

$$\frac{53.8 \cdot n}{53.8} = \frac{598(233.877)}{53.8}$$

$$n \approx 2600$$

The replacement cost of the fishing boat at August 2013 prices is $2600.

Critical Thinking

1. What trends do you see in the CPI values in the table? Do you think these trends make sense? Explain.

2. A piece of jewelry cost $800 in 1985. What was its 2012 replacement value?

3. In 2000, the cost of a loaf of bread was about $1.89. What would an equivalent loaf of bread have cost in 1950?

4. Suppose a couple purchased a house for $12,000 in 1940. At what price could they have been expected to sell the house in 1990?

5. An original Ford Model T cost about $850 in 1915. What is the equivalent cost of a Model T in 2010 dollars?

Consumer Price Index	
Year	**CPI**
1915	10.1
1920	20.0
1925	17.5
1930	16.7
1935	13.7
1940	14.0
1945	18.0
1950	24.1
1955	26.8
1960	29.6
1965	31.5
1970	38.8
1975	53.8
1980	82.4
1985	107.6
1990	130.7
1995	152.4
2000	172.2
2005	195.3
2010	218.1
2011	224.9
2012	229.6
2013	233.0

(*Source:* Bureau of Labor Statistics, U.S. Department of Labor)

Chapter 5 Vocabulary Check

Fill in each blank with one of the words or phrases listed below.

not equal   equal   cross products   rate   mass   unit fractions   unit rate

ratio   unit price   proportion   meter   liter   weight

1. A _____ is the quotient of two numbers. It can be written as a fraction, using a colon, or using the word *to*.

2. $\dfrac{x}{2} = \dfrac{7}{16}$ is an example of a _____.

3. A _____ is a rate with a denominator of 1.

4. A _____ is a "money per item" unit rate.

5. A _____ is used to compare different kinds of quantities.

6. In the proportion $\dfrac{x}{2} = \dfrac{7}{16}$, $x \cdot 16$ and $2 \cdot 7$ are called _____.

7. If cross products are _____, the proportion is true.

8. If cross products are _____, the proportion is false.

9. _____ is a measure of the pull of gravity.

10. _____ is a measure of the amount of substance in an object. This measure does not change.

11. The basic unit of length in the metric system is the _____.

12. To convert from one unit of length to another, _____ may be used.

13. The _____ is the basic unit of capacity in the metric system.

> **Helpful Hint**
> ▶ Are you preparing for your test? Don't forget to take the Chapter 5 Test on page 439. Then check your answers at the back of the text and use the Chapter Test Prep Videos to see the fully worked-out solutions to any of the exercises you want to review.

5 Chapter Highlights

Definitions and Concepts	Examples
Section 5.1 Ratios	
A **ratio** is the quotient of two quantities.	The ratio of 3 to 4 can be written as $$\dfrac{3}{4} \quad \text{or} \quad 3 : 4$$ ↑ fraction notation ↑ colon notation
Rates are used to compare different kinds of quantities.	Write the rate 12 spikes every 8 inches as a fraction in simplest form. $$\dfrac{12 \text{ spikes}}{8 \text{ inches}} = \dfrac{3 \text{ spikes}}{2 \text{ inches}}$$
A **unit rate** is a rate with a denominator of 1.	Write as a unit rate: 117 miles on 5 gallons of gas $$\dfrac{117 \text{ miles}}{5 \text{ gallons}} = \dfrac{23.4 \text{ miles}}{1 \text{ gallon}} \quad \begin{array}{l} \text{or 23.4 miles per gallon} \\ \text{or 23.4 miles/gallon} \end{array}$$
A **unit price** is a "money per item" unit rate.	Write as a unit price: $5.88 for 42 ounces of detergent $$\dfrac{\$5.88}{42 \text{ ounces}} = \dfrac{\$0.14}{1 \text{ ounce}} = \$0.14 \text{ per ounce}$$

Definitions and Concepts	Examples

Section 5.2   Proportions

A **proportion** is a statement that two ratios or rates are equal.	$\dfrac{1}{2} = \dfrac{4}{8}$ is a proportion.

Using Cross Products to Determine Whether Proportions Are True or False

Is $\dfrac{6}{10} = \dfrac{9}{15}$ a true proportion?

$$\overset{\overbrace{\qquad\text{Cross products}\qquad}}{a \cdot d \qquad\qquad\qquad b \cdot c}$$
$$\frac{a}{b} = \frac{c}{d}$$

$$\overset{\overbrace{\qquad\text{Cross products}\qquad}}{6 \cdot 15 \qquad\qquad\qquad 10 \cdot 9}$$
$$\frac{6}{10} = \frac{9}{15}$$

If cross products are equal, the proportion is true.
If $ad = bc$, then the proportion is true.
If cross products are not equal, the proportion is false.
If $ad \neq bc$, then the proportion is false.

$6 \cdot 15 \overset{?}{=} 10 \cdot 9$ Are cross products equal?
$90 = 90$

Since cross products are equal, the proportion is a true proportion.

Finding an Unknown Value n in a Proportion

Find n: $\dfrac{n}{7} = \dfrac{5}{8}$

Step 1: Set the cross products equal to each other.

Step 1:

$$\frac{n}{7} = \frac{5}{8}$$

$n \cdot 8 = 7 \cdot 5$ Set the cross products equal to each other.
$n \cdot 8 = 35$ Multiply.

Step 2: Divide the number not multiplied by n by the number multiplied by n.

Step 2:

$n = \dfrac{35}{8}$ Divide 35 by 8, the number multiplied by n.

$n = 4\dfrac{3}{8}$

Section 5.3   Proportions and Problem Solving

Given a specified ratio (or rate) of two quantities, a proportion can be used to determine an unknown quantity.	On a map, 50 miles corresponds to 3 inches. How many miles correspond to 10 inches?

1. UNDERSTAND. Read and reread the problem.

2. TRANSLATE. We let n represent the unknown number. We are given that 50 miles is to 3 inches as n miles is to 10 inches.

$$\text{miles} \rightarrow \frac{50}{3} = \frac{n}{10} \leftarrow \text{miles}$$
$$\text{inches} \rightarrow \qquad\qquad \leftarrow \text{inches}$$

(continued)

Definitions and Concepts	Examples

Section 5.3   Proportions and Problem Solving (*Continued*)

3. SOLVE:

$$\frac{50}{3} = \frac{n}{10}$$

$50 \cdot 10 = 3 \cdot n$ Set the cross products equal to each other.

$500 = 3 \cdot n$ Multiply.

$\dfrac{500}{3} = \dfrac{3n}{3}$ Divide 500 by 3, the number multiplied by n

$n = 166\dfrac{2}{3}$

4. INTERPRET. *Check* your work. *State* your conclusion:
On the map, $166\dfrac{2}{3}$ miles corresponds to 10 inches.

Section 5.4   Length: U.S. and Metric Systems of Measurement

To convert from one unit of length to another, multiply by a **unit fraction** in the form

$$\frac{\text{units to convert to}}{\text{original units}}.$$

Length: U.S. System of Measurement

$$12 \text{ inches (in.)} = 1 \text{ foot (ft)}$$
$$3 \text{ feet} = 1 \text{ yard (yd)}$$
$$5280 \text{ feet} = 1 \text{ mile (mi)}$$

The basic unit of length in the metric system is the **meter.** A meter is slightly longer than a yard.

Length: Metric System of Measurement

Metric Units of Length
1 **kilo**meter (km) = 1000 meters (m)
1 **hecto**meter (hm) = 100 m
1 **deka**meter (dam) = 10 m
1 meter (m) = 1 m
1 **deci**meter (dm) = 1/10 m or 0.1 m
1 **centi**meter (cm) = 1/100 m or 0.01 m
1 **milli**meter (mm) = 1/1000 m or 0.001 m

$$\frac{12 \text{ inches}}{1 \text{ foot}}, \frac{1 \text{ foot}}{12 \text{ inches}}, \frac{3 \text{ feet}}{1 \text{ yard}}$$

Convert 6 feet to inches.

$$6 \text{ ft} = \frac{6 \text{ ft}}{1} \cdot 1$$

$$= \frac{6 \text{ ft}}{1} \cdot \frac{12 \text{ in.}}{1 \text{ ft}} \begin{array}{l} \leftarrow \text{ units to convert to} \\ \leftarrow \text{ original units} \end{array}$$

$$= 6 \cdot 12 \text{ in.}$$

$$= 72 \text{ in.}$$

Convert 3650 centimeters to meters.

$$3650 \text{ cm} = 3650 \text{ cm} \cdot 1$$

$$= \frac{3650 \text{ cm}}{1} \cdot \frac{0.01 \text{ m}}{1 \text{ cm}} = 36.5 \text{ m}$$

or

km   hm   dam   m   dm   cm   mm

2 units to the left

$$3650 \text{ cm} = 36.5 \text{ m}$$

2 places to the left

Definitions and Concepts	Examples

Section 5.5   Weight and Mass: U.S. and Metric Systems of Measurement

Weight is really a measure of the pull of gravity. **Mass** is a measure of the amount of substance in an object and does not change.

Convert 5 pounds to ounces.

$$5 \text{ lb} = 5 \text{ lb} \cdot 1 = \frac{5 \text{ lb}}{1} \cdot \frac{16 \text{ oz}}{1 \text{ lb}} = 80 \text{ oz}$$

Weight: U.S. System of Measurement

   16 ounces (oz) = 1 pound (lb)

   2000 pounds = 1 ton

A **gram** is the basic unit of mass in the metric system. It is the mass of water contained in a cube 1 centimeter on each side. A paper clip weighs about 1 gram.

Convert 260 grams to kilograms.

$$260 \text{ g} = \frac{260 \text{ g}}{1} \cdot 1 = \frac{260 \text{ g}}{1} \cdot \frac{1 \text{ kg}}{1000 \text{ g}} = 0.26 \text{ kg}$$

or

Mass: Metric System of Measurement

Metric Units of Mass
1 kilogram (kg) = 1000 grams (g)
1 hectogram (hg) = 100 g
1 dekagram (dag) = 10 g
1 gram (g) = 1 g
1 decigram (dg) = 1/10 g or 0.1 g
1 centigram (cg) = 1/100 g or 0.01 g
1 milligram (mg) = 1/1000 g or 0.001 g

kg hg dag g dg cg mg

3 units to the left

260 g = 0.260 kg

3 places to the left

Section 5.6   Capacity: U.S. and Metric Systems of Measurement

Capacity: U.S. System of Measurement

  8 fluid ounces (fl oz) = 1 cup (c)

          2 cups = 1 pint (pt)

        2 pints = 1 quart (qt)

     4 quarts = 1 gallon (gal)

Convert 5 pints to gallons.

$$1 \text{ gal} = 4 \text{ qt} = 8 \text{ pt}$$

$$5 \text{ pt} = 5 \text{ pt} \cdot 1 = \frac{5 \text{ pt}}{1} \cdot \frac{1 \text{ gal}}{8 \text{ pt}} = \frac{5}{8} \text{ gal}$$

The **liter** is the basic unit of capacity in the metric system. It is the capacity or volume of a cube measuring 10 centimeters on each side. A liter of liquid is slightly more than 1 quart.

Convert 1.5 liters to milliliters.

$$1.5 \text{ L} = \frac{1.5 \text{ L}}{1} \cdot 1 = \frac{1.5 \text{ L}}{1} \cdot \frac{1000 \text{ ml}}{1 \text{ L}} = 1500 \text{ ml}$$

or

Capacity: Metric System of Measurement

Metric Units of Capacity
1 kiloliter (kl) = 1000 liters (L)
1 hectoliter (hl) = 100 L
1 dekaliter (dal) = 10 L
1 liter (L) = 1 L
1 deciliter (dl) = 1/10 L or 0.1 L
1 centiliter (cl) = 1/100 L or 0.01 L
1 milliliter (ml) = 1/1000 L or 0.001 L

kl hl dal L dl cl ml

3 units to the right

1.500 L = 1500 ml

3 places to the right

Definitions and Concepts	Examples
Section 5.7 **Conversions Between the U.S. and Metric Systems**	
To convert between systems, use approximate unit fractions from Section 5.7.	Convert 7 feet to meters. $7 \text{ ft} \approx \dfrac{7 \cancel{\text{ft}}}{1} \cdot \dfrac{0.30 \text{ m}}{1 \cancel{\text{ft}}} = 2.1 \text{ m}$ Convert 8 liters to quarts. $8 \text{ L} \approx \dfrac{8 \cancel{\text{L}}}{1} \cdot \dfrac{1.06 \text{ qt}}{1 \cancel{\text{L}}} = 8.48 \text{ qt}$ Convert 363 grams to ounces. $363 \text{ g} \approx \dfrac{363 \cancel{\text{g}}}{1} \cdot \dfrac{0.04 \text{ oz}}{1 \cancel{\text{g}}} = 14.52 \text{ oz}$

Chapter 5 Review

(5.1) *Write each ratio as a fraction in simplest form.*

1. 23 to 37

2. 14 to 51

3. 6000 people to 4800 people

4. $121 to $143

5. 3.5 centimeters to 7.5 centimeters

6. 4.25 yards to 8.75 yards

7. $2\dfrac{1}{4}$ to $4\dfrac{3}{8}$

8. $3\dfrac{1}{2}$ to $2\dfrac{7}{10}$

The circle graph below shows how the top 25 movies of 2013 were rated. Use this graph to answer the questions.

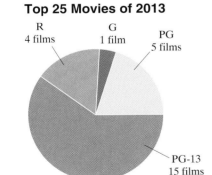

Top 25 Movies of 2013

R 4 films   G 1 film   PG 5 films   PG-13 15 films

Source: MPAA

9. a. How many top 25 movies were rated G?
   b. Find the ratio of top 25 G-rated movies to total movies for that year.

10. a. How many top 25 movies were rated PG-13?
   b. Find the ratio of top 25 PG-13-rated movies to total movies for that year.

Write each rate as a fraction in simplest form.

11. 15 word processing pages printed in 6 minutes

12. 8 computers assembled in 6 hours

Write each rate as a unit rate.

13. 468 miles in 9 hours

14. 180 feet in 12 seconds

15. $27.84 for 4 CDs

16. 8 gallons of pesticide for 6 acres of crops

Find each unit price and decide which is the better buy. Round to three decimal places. Assume that we are comparing different sizes of the same brand.

17. Taco sauce: 8 ounces for $0.99 or 12 ounces for $1.69

18. Peanut butter: 18 ounces for $1.49 or 28 ounces for $2.39

(5.2) *Write each sentence as a proportion.*

19. 16 sandwiches is to 8 players as 2 sandwiches is to 1 player.

20. 12 tires is to 3 cars as 4 tires is to 1 car.

Determine whether each proportion is true.

21. $\dfrac{21}{8} = \dfrac{14}{6}$

22. $\dfrac{3}{5} = \dfrac{60}{100}$

23. $\dfrac{3.75}{3} = \dfrac{7.5}{6}$

24. $\dfrac{3.1}{6.2} = \dfrac{0.8}{0.16}$

Find the unknown number n in each proportion.

25. $\dfrac{n}{6} = \dfrac{-15}{18}$

26. $\dfrac{n}{-9} = \dfrac{5}{3}$

27. $\dfrac{9}{2} = \dfrac{n}{\frac{3}{2}}$

28. $\dfrac{6}{\frac{5}{2}} = \dfrac{n}{3}$

29. $\dfrac{0.4}{n} = \dfrac{2}{4.7}$

30. $\dfrac{7.2}{n} = \dfrac{6}{0.3}$

31. $\dfrac{n}{4\frac{1}{2}} = \dfrac{2\frac{1}{10}}{8\frac{2}{5}}$

32. $\dfrac{n}{4\frac{2}{7}} = \dfrac{3\frac{1}{9}}{9\frac{1}{3}}$

(5.3) *Solve.*

The ratio of a quarterback's completed passes to attempted passes is 3 to 7.

33. If he attempted 32 passes, find how many passes he completed. Round to the nearest whole pass.

34. If he completed 15 passes, find how many passes he attempted.

One bag of pesticide covers 4000 square feet of garden.

△ **35.** Find how many bags of pesticide should be purchased to cover a rectangular garden that is 180 feet by 175 feet.

△ **36.** Find how many bags of pesticide should be purchased to cover a square garden that is 250 feet on each side.

On an architect's blueprint, 1 inch = 12 feet.

37. Find the length of a wall represented by a $3\frac{3}{8}$-inch line on the blueprint.

38. If an exterior wall is 99 feet long, find how long the blueprint measurement should be.

(5.4) *Convert.*

39. 108 in. to feet

40. $\frac{1}{2}$ yd to inches

41. 52 ft = _____ yd _____ ft

42. 46 in. = _____ ft _____ in.

43. 42 m to centimeters

44. 2.31 m to kilometers

Perform each indicated operation.

45. 4 yd 2 ft + 16 yd 2 ft

46. 12 ft 1 in. − 4 ft 8 in.

47. 8 cm + 15 mm

48. 19.6 km ÷ 8

Solve.

49. The trip from Philadelphia to Washington, D.C., is 217 km each way. Four friends agree to share the driving equally. How far must each drive on this round-trip vacation?

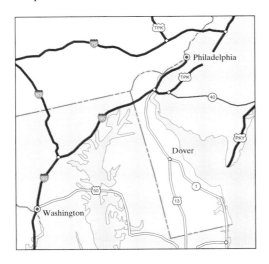

△ **50.** The college has ordered that NO SMOKING signs be placed above the doorway of each classroom. Each sign is 0.8 m long and 30 cm wide. Find the area of each sign. (*Hint:* Recall that the area of a rectangle = width · length.)

0.8 meter

30 centimeters

(5.5) *Convert.*

51. 66 oz to pounds

52. 2.3 tons to pounds

53. 52 oz = _____ lb _____ oz

54. 10,300 lb = _____ tons _____ lb

55. 27 mg to grams

56. 40 kg to grams

Perform each indicated operation.

57. 6 lb 5 oz − 2 lb 12 oz

58. 8 lb 6 oz × 4

Solve.

59. Donshay Berry ordered 1 lb 12 oz of soft-center candies and 2 lb 8 oz of chewy-center candies for his party. Find the total weight of the candy ordered.

60. Four local townships jointly purchase 38 tons 300 lb of cinders to spread on their roads during an ice storm. Determine the weight of the cinders each township receives if they share the purchase equally.

(5.6) *Convert.*

61. 16 pints to quarts

62. 40 fluid ounces to cups

63. 3 qt 1 pt to pints

64. 18 quarts to cups

65. 9 pt = _____ qt _____ pt

66. 15 qt = _____ gal _____ qt

67. 3.8 L to milliliters

68. 4.2 ml to deciliters

Perform each indicated operation.

69. 1 qt 1 pt + 3 qt 1 pt

70. 0.946 L − 210 ml

Solve.

71. Each bottle of Kiwi liquid shoe polish holds 85 ml of the polish. Find the number of liters of shoe polish contained in 8 boxes if each box contains 16 bottles.

72. Ivan Miller wants to pour three separate containers of saline solution into a single vat with a capacity of 10 liters. Will 6 liters of solution in the first container combined with 1300 milliliters in the second container and 2.6 liters in the third container fit into the vat?

(5.7) *Note: Because approximations are used in this section, your answers may vary slightly from the answers given in the back of the book.*

Convert as indicated. If necessary, round to two decimal places.

73. 7 meters to feet

74. 11.5 yards to meters

75. 17.5 liters to gallons

76. 7.8 liters to quarts

77. 15 ounces to grams

78. 23 pounds to kilograms

79. A 100-meter dash is being held today. How many yards is this?

80. If a person weighs 82 kilograms, how many pounds is this?

81. How many quarts are contained in a 3-liter bottle of cola?

82. A compact disc is 1.2 mm thick. Find the height (in inches) of 50 discs.

Mixed Review

Write each ratio as a fraction in simplest form.

83. 15 to 25

84. 6 pints to 48 pints

Write each rate as a fraction in simplest form.

85. 2 teachers for 18 students

86. 6 nurses for 24 patients

Write each rate as a unit rate.

87. 136 miles in 4 hours

88. 12 gallons of milk from 6 cows

Find each unit price and decide which is the better buy. Round to three decimal places. Assume that we are comparing different sizes of the same brand.

89. cold medicine:
$4.94 for 4 oz,
$9.98 for 8 oz

90. juice:
12 oz for $0.65,
64 oz for $2.98

Write each sentence as a proportion.

91. 2 cups of cookie dough is to 30 cookies as 4 cups of cookie dough is to 60 cookies.

92. 5 nickels is to 3 dollars as 20 nickels is to 12 dollars.

Find the unknown number n in each proportion.

93. $\dfrac{3}{n} = \dfrac{15}{8}$

94. $\dfrac{5}{4} = \dfrac{n}{20}$

Convert the following.

95. 2.5 mi to feet

96. 129 in. to feet

97. 8200 lb = _____ tons
_____ lb

98. 5 m to centimeters

99. 1400 mg to grams

100. 286 mm to kilometers

Perform the indicated operations and simplify.

101. 9.3 km − 183 m

102. 6 gal 1 qt + 2 gal 1 qt

Write each ratio or rate as a fraction in simplest form.

Answers

1. $75 to $10

2. 9 inches of rain in 30 days

3. 8.6 to 10

4. $5\frac{7}{8}$ to $9\frac{3}{4}$

5. The world's largest yacht, the *Azzam*, measures in at 591 feet long. A Boeing 787-8 Dreamliner measures 186 feet long. Find the ratio of the length of the *Azzam* to the length of a 787-8. (*Source:* Superyachts.com, Boeing)

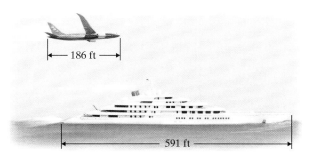

186 ft

591 ft

Find each unit rate.

6. 650 kilometers in 8 hours

7. 140 students for 5 teachers

8. QRIO (Quest for Curiosity) is the world's first bipedal robot capable of running (moving with both legs off the ground at the same time) at a rate of 108 inches each 12 seconds. (*Source: Guinness World Records*)

Find each unit price and compare them to decide which is the better buy.

9. Steak sauce:
8 ounces for $1.19
12 ounces for $1.89

Determine whether the proportion is true.

10. $\dfrac{28}{16} = \dfrac{14}{8}$

Find the unknown number n in each proportion.

11. $\dfrac{n}{3} = \dfrac{15}{9}$

12. $\dfrac{8}{n} = \dfrac{11}{6}$

13. $\dfrac{-1.5}{5} = \dfrac{2.4}{n}$

14. $\dfrac{n}{2\frac{5}{8}} = \dfrac{1\frac{1}{6}}{3\frac{1}{2}}$

Solve.

15. On an architect's drawing, 2 inches corresponds to 9 feet. Find the length of a home represented by a line that is 11 inches long.

16. If a car can be driven 80 miles in 3 hours, how long will it take to travel 100 miles?

1. _____

2. _____

3. _____

4. _____

5. _____

6. _____

7. _____

8. _____

9. _____

10. _____

11. _____

12. _____

13. _____

14. _____

15. _____

16. _____

17. _____

18. _____

19. _____

20. _____

21. _____

22. _____

23. _____

24. _____

25. _____

26. _____

27. _____

28. _____

29. _____

30. _____

31. _____

32. _____

33. _____

17. The standard dose of medicine for a dog is 10 grams for every 15 pounds of body weight. What is the standard dose for a dog that weighs 80 pounds?

Convert.

18. 280 in. = _____ ft _____ in.

19. $2\frac{1}{2}$ gal to quarts

20. 30 oz to pounds

21. 40 mg to grams

22. 3.6 cm to millimeters

23. 0.83 L to milliliters

Perform each indicated operation.

24. 8 lb 6 oz − 4 lb 9 oz   **25.** 5 gal 2 qt ÷ 2   **26.** 1.8 km + 456 m

Solve.

27. The sugar maples in front of Bette MacMillan's house are 8.4 meters tall. Because they interfere with the phone lines, the telephone company plans to remove the top third of the trees. How tall will the maples be after they are shortened?

28. A total of 15 gal 1 qt of oil has been removed from a 20-gallon drum. How much oil still remains in the container?

29. The engineer in charge of bridge construction said that the span of a certain bridge would be 88 m. But the actual construction required it to be 340 cm longer. Find the span of the bridge, in meters.

30. If 2 ft 9 in. of material is used to manufacture one scarf, how much material is needed for 6 scarves?

31. The Vietnam Veterans Memorial, inscribed with the names of 58,226 deceased and missing U.S. soldiers from the Vietnam War, is located on the National Mall in Washington, D.C. This memorial is formed from two straight sections of wall that meet at an angle at the center of the monument. Each wall is 246 ft 9 in. long. What is the total length of the Vietnam Veterans Memorial's wall? (*Source:* National Park Service)

32. Each panel making up the wall of the Vietnam Veterans Memorial is 101.6 cm wide. There are a total of 148 panels making up the wall. What is the total length of the wall in meters? (*Source:* National Park Service)

33. A 5-kilometer race is being held today. How many miles is this?

1. Subtract. Check each answer by adding.
   a. $12 - 9$
   b. $22 - 7$
   c. $35 - 35$
   d. $70 - 0$

2. Multiply:
   a. $20 \cdot 0$
   b. $20 \cdot 1$
   c. $0 \cdot 20$
   d. $1 \cdot 20$

3. Round 248,982 to the nearest hundred.

4. Round 248,982 to the nearest thousand.

5. Multiply:
   a. $\begin{array}{r} 25 \\ \times\ 8 \\ \hline \end{array}$ b. $\begin{array}{r} 246 \\ \times\ 5 \\ \hline \end{array}$

6. Divide: $10,468 \div 28$

7. The director of a learning lab at a local community college is working on next year's budget. Thirty-three new DVD players are needed at a cost of $187 each. What is the total cost of these DVD players?

8. A study is being conducted for erecting soundproof walls along the interstate of a metropolitan area. The following feet of walls are part of the proposal. Find their total: 4800 feet, 3270 feet, 2761 feet, 5760 feet.

9. Write the prime factorization of 80.

10. Find $\sqrt{64}$.

11. Write $\dfrac{12}{20}$ in simplest form.

12. Find $9^2 \cdot \sqrt{9}$.

Multiply.

13. $-\dfrac{1}{4} \cdot \dfrac{1}{2}$

14. $3\dfrac{3}{8} \cdot 4\dfrac{5}{9}$

15. $\left(-\dfrac{6}{13}\right)\left(-\dfrac{26}{30}\right)$

16. $\dfrac{2}{11} \cdot \dfrac{5}{8} \cdot \dfrac{22}{27}$

Perform the indicated operation and simplify.

17. $\dfrac{2}{7} + \dfrac{3}{7}$

18. $\dfrac{26}{30} - \dfrac{7}{30}$

19. $\dfrac{7}{8} + \dfrac{6}{8} + \dfrac{3}{8}$

20. $\dfrac{7}{10} - \dfrac{3}{10} + \dfrac{4}{10}$

21. Find the LCD of $\dfrac{3}{7}$ and $\dfrac{5}{14}$.

22. Add: $\dfrac{17}{25} + \dfrac{3}{10}$

Answers

1. a. _____
   b. _____
   c. _____
   d. _____
2. a. _____ b. _____
   c. _____ d. _____
3. _____
4. _____
5. a. _____
   b. _____
6. _____
7. _____
8. _____
9. _____
10. _____
11. _____
12. _____
13. _____
14. _____
15. _____
16. _____
17. _____
18. _____
19. _____
20. _____
21. _____
22. _____

441

23. _____

24. _____

25. _____

26. _____

27. _____

28. _____

29. _____

30. _____

31. _____

32. _____

33. _____

34. _____

35. _____

36. _____

37. _____

38. _____

39. _____

40. _____

41. _____

42. _____

43. _____

44. _____

45. _____

46. _____

47. _____

48. _____

49. _____

50. _____

23. Write an equivalent fraction with the indicated denominator. $\dfrac{3}{4} = \dfrac{}{20}$

24. Determine whether these fractions are equivalent.
$$\dfrac{10}{55} \quad \dfrac{6}{33}$$

25. Subtract: $\dfrac{2}{3} - \dfrac{10}{11}$

26. Subtract: $17\dfrac{5}{24} - 9\dfrac{5}{9}$

27. A flight from Tucson to Phoenix, Arizona, requires $\dfrac{5}{12}$ of an hour. If the plane has been flying $\dfrac{1}{4}$ of an hour, find how much time remains before landing.

Arizona

Phoenix

$\frac{5}{12}$ hour

Tucson

28. Simplify: $80 \div 8 \cdot 2 + 7$

29. Add: $2\dfrac{1}{3} + 5\dfrac{3}{8}$

30. Find the average of $\dfrac{3}{5}, \dfrac{4}{9},$ and $\dfrac{11}{15}$.

31. Insert $<$ or $>$ to form a true statement.
$$\dfrac{3}{4} \qquad \dfrac{9}{11}$$

32. Multiply: $28{,}000 \times 500$

33. Write the decimal -5.82 in words.

34. Write "seventy-five thousandths" in standard form.

35. Round 736.2359 to the nearest tenth.

36. Round 736.2359 to the nearest thousandth.

37. Add: $23.85 + 1.604$

38. Subtract: $700 - 18.76$

39. Multiply: 0.0531×16

40. Write $\dfrac{3}{8}$ as a decimal.

41. Divide: $-5.98 \div 115$

42. Write 7.9 as an improper fraction.

43. Simplify: $-0.5(8.6 - 1.2)$

44. Find the unknown number n.
$$\dfrac{n}{4} = \dfrac{12}{16}$$

45. Write the numbers in order from smallest to largest.
$$\dfrac{9}{20}, \dfrac{4}{9}, 0.456$$

46. Write the rate as a unit rate. 700 meters in 5 seconds

Write each ratio as a fraction in simplest form.

47. The ratio of $15 to $10

48. The ratio of 7 to 21

49. The ratio of 2.6 to 3.1

50. The ratio of 900 to 9000

Percent

Cell Phone–Only Households in the United States Continue to Increase

The number of U.S. households using landline telephone service continues to decrease while the number of households using cellular telephone service only continues to increase. Based on a survey conducted in 2013, the National Center for Health Statistics (NCHS) reported that 49.5% of all U.S. households have both landline and cellular telephone service. In Section 6.1, Exercises 85 and 86, we learn the percent of all U.S. households using cell phones only and landlines only, and we convert these percents to decimals.

This chapter is devoted to percent, a concept used virtually every day in ordinary and business life. Understanding percent and using it efficiently depend on understanding ratios because a percent is a ratio whose denominator is 100. We present techniques to write percents as fractions and as decimals and then solve problems related to sales tax, commission, discounts, interest, and other real-life situations that use percents.

Cell Phone–Only Households by State

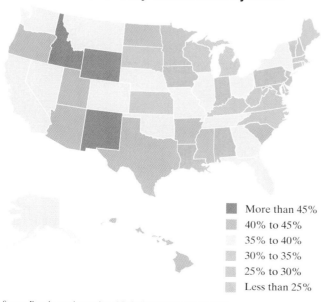

More than 45%

40% to 45%

35% to 40%

30% to 35%

25% to 30%

Less than 25%

Source: Based on estimates from Marketing Systems Group, 2013

Objectives

A Understand Percent.

B Write Percents as Decimals or Fractions.

C Write Decimals or Fractions as Percents.

D Convert Percents, Decimals, and Fractions.

Objective A Understanding Percent

The word **percent** comes from the Latin phrase *per centum*, which means **"per 100."** For example, 53% (percent) means 53 per 100. In the square below, 53 of the 100 squares are shaded. Thus, 53% of the figure is shaded.

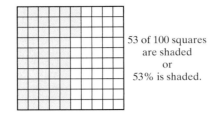

53 of 100 squares
are shaded
or
53% is shaded.

Since 53% means 53 per 100, 53% is the ratio of 53 to 100, or $\dfrac{53}{100}$.

$$53\% = \dfrac{53}{100}$$

Also,

$$7\% = \dfrac{7}{100} \quad \text{7 parts per 100 parts}$$

$$73\% = \dfrac{73}{100} \quad \text{73 parts per 100 parts}$$

$$109\% = \dfrac{109}{100} \quad \text{109 parts per 100 parts}$$

Percent

Percent means **per one hundred.** The "%" symbol is used to denote percent.

Percent is used in a variety of everyday situations. For example,

- 87.0% of the U.S. population uses the Internet. (PEW)
- The store is having a 25%-off sale.
- 78% of our pages have printed.
- The enrollment in community colleges is predicted to increase 1.3% each year.
- Let's add a 15% tip to our bill.

Practice 1

Of 100 students in a club, 23 are freshmen. What percent of the students are freshmen?

Example 1 Since 2006, white has been the world's most popular color for cars. For 2014 model cars, 25 out of every 100 were painted white. What percent of model-year 2014 cars were white? (*Source:* PPG Industries)

Solution: Since 25 out of 100 cars were painted white, the fraction is $\dfrac{25}{100}$. Then

$$\dfrac{25}{100} = 25\%$$

■ Work Practice 1

Answer
1. 23%

Example 2 46 out of every 100 college students live at home. What percent of students live at home? (*Source:* Independent Insurance Agents of America)

Solution:

$$\frac{46}{100} = 46\%$$

Work Practice 2

Practice 2

29 out of 100 executives are in their forties. What percent of executives are in their forties?

Objective B Writing Percents as Decimals or Fractions ▶

Since percent means "per hundred," we have that

$$1\% = \frac{1}{100} = 0.01$$

In other words, the percent symbol means "per hundred" or, equivalently, "$\frac{1}{100}$" or "0.01." Thus

Write 87% as a fraction: $87\% = 87 \times \frac{1}{100} = \frac{87}{100}$

or

Write 87% as a decimal: $87\% = 87 \times (0.01) = 0.87$

Results are the same.

Of course, we know that the end results are the same, that is,

$$\frac{87}{100} = 0.87$$

The above gives us two options for converting percents. We can replace the percent symbol, %, by $\frac{1}{100}$ or 0.01 and then multiply.

For consistency, when we

- convert from a percent to a *decimal*, we will drop the % symbol and multiply by 0.01
- convert from a percent to a *fraction*, we will drop the % symbol and multiply by $\frac{1}{100}$

Let's practice writing percents as decimals and then writing percents as fractions.

Writing a Percent as a Decimal

Replace the percent symbol with its decimal equivalent, 0.01; then multiply.

$$43\% = 43(0.01) = 0.43$$

Helpful Hint

If it helps, think of writing a percent as a decimal by

Percent → | Remove the % symbol and move the decimal point 2 places to the left. | → Decimal

Answer

2. 29%

Practice 3–7

Write each percent as a decimal.

3. 89%   **4.** 2.7%

5. 150%   **6.** 0.69%

7. 800%

Examples   Write each percent as a decimal.

3. $23\% = 23(0.01) = 0.23$ Replace the percent symbol with 0.01. Then multiply.

4. $4.6\% = 4.6(0.01) = 0.046$ Replace the percent symbol with 0.01. Then multiply.

5. $190\% = 190(0.01) = 1.90 \text{ or } 1.9$

6. $0.74\% = 0.74(0.01) = 0.0074$

7. $100\% = 100(0.01) = 1.00 \text{ or } 1$

> **Helpful Hint**   We just learned that 100% = 1.

■ Work Practice 3–7

✓ **Concept Check**   Why is it incorrect to write the percent 0.033% as 3.3 in decimal form?

Now let's write percents as fractions.

Writing a Percent as a Fraction

Replace the percent symbol with its fraction equivalent, $\frac{1}{100}$; then multiply. Don't forget to simplify the fraction if possible.

$$43\% = 43 \cdot \frac{1}{100} = \frac{43}{100}$$

Practice 8–12

Write each percent as a fraction or mixed number in simplest form.

8. 25%   **9.** 2.3%

10. 225%   **11.** $66\frac{2}{3}\%$

12. 8%

Examples   Write each percent as a fraction or mixed number in simplest form.

8. $40\% = 40 \cdot \frac{1}{100} = \frac{40}{100} = \frac{2 \cdot \overset{1}{\cancel{20}}}{5 \cdot \underset{1}{\cancel{20}}} = \frac{2}{5}$

9. $1.9\% = 1.9 \cdot \frac{1}{100} = \frac{1.9}{100}$. We don't want the numerator of the fraction to contain a decimal, so we multiply by 1 in the form of $\frac{10}{10}$.

$$= \frac{1.9}{100} \cdot \frac{10}{10} = \frac{1.9 \cdot 10}{100 \cdot 10} = \frac{19}{1000}$$

10. $125\% = 125 \cdot \frac{1}{100} = \frac{125}{100} = \frac{5 \cdot \overset{1}{\cancel{25}}}{4 \cdot \underset{1}{\cancel{25}}} = \frac{5}{4} \text{ or } 1\frac{1}{4}$

11. $33\frac{1}{3}\% = 33\frac{1}{3} \cdot \frac{1}{100} = \frac{100}{3} \cdot \frac{1}{100} = \frac{\overset{1}{\cancel{100}} \cdot 1}{3 \cdot \underset{1}{\cancel{100}}} = \frac{1}{3}$

    └ Write as an improper fraction. ┘

> **Helpful Hint**   Just as we did in Example 7, we confirm that 100% = 1.

12. $100\% = 100 \cdot \frac{1}{100} = \frac{100}{100} = 1$

■ Work Practice 8–12

Answers

3. 0.89   **4.** 0.027   **5.** 1.5   **6.** 0.0069

7. 8.00 or 8   **8.** $\frac{1}{4}$ **9.** $\frac{23}{1000}$

10. $\frac{9}{4}$ or $2\frac{1}{4}$ **11.** $\frac{2}{3}$ **12.** $\frac{2}{25}$

✓ **Concept Check Answer**

To write a percent as a decimal, the decimal point should be moved two places to the left, not to the right. So the correct answer is 0.00033.

Objective C Writing Decimals or Fractions
as Percents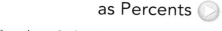

To write a decimal as a percent, we use the result of Example 7 or 12 on the previous page. In these examples, we found that $1 = 100\%$.

Write 0.38 as a percent: $0.38 = 0.38(1) = 0.38(100\%) = 38.\%$

Write $\frac{1}{4}$ as a percent: $\frac{1}{4} = \frac{1}{4}(1) = \frac{1}{4} \cdot 100\% = \frac{100}{4}\% = 25\%$

First, let's practice writing decimals as percents.

Writing a Decimal as a Percent

Multiply by 1 in the form of 100%.

$0.27 = 0.27(100\%) = 27.\%$

Helpful Hint

If it helps, think of writing a decimal as a percent by reversing the steps in the Helpful Hint on page 445.

Percent ← | Move the decimal point 2 places to the right and attach a % symbol. | ← Decimal

Examples   Write each decimal as a percent.

13. $0.65 = 0.65(100\%) = 65.\%$ or 65% Multiply by 100%.

14. $1.25 = 1.25(100\%) = 125.\%$ or 125%

15. $0.012 = 0.012(100\%) = 001.2\%$ or 1.2%

16. $0.6 = 0.6(100\%) = 060.\%$ or 60%

Helpful Hint   A zero was inserted as a placeholder.

■ Work Practice 13–16

Practice 13–16

Write each decimal as a percent.
13. 0.19   **14.** 1.75
15. 0.044   **16.** 0.7

✓**Concept Check** Why is it incorrect to write the decimal 0.0345 as 34.5% in percent form?

Now let's write fractions as percents.

Writing a Fraction as a Percent

Multiply by 1 in the form of 100%.

$\frac{1}{8} = \frac{1}{8} \cdot 100\% = \frac{1}{8} \cdot \frac{100}{1}\% = \frac{100}{8}\% = 12\frac{1}{2}\%$ or 12.5%

Answers
13. 19%   **14.** 175%   **15.** 4.4%
16. 70%

✓**Concept Check Answer**

To change a decimal to a percent, multiply by 100%, or move the decimal point *only* two places to the right. So the correct answer is 3.45%.

Helpful Hint

From Examples 7 and 12, we know that

$$100\% = 1$$

Recall that when we multiply a number by 1, we are not changing the value of that number. This means that when we multiply a number by 100%, we are not changing its value but rather writing the number as an equivalent percent.

Practice 17–19

Write each fraction or mixed number as a percent.

17. $\frac{1}{2}$ **18.** $\frac{7}{40}$ **19.** $2\frac{1}{4}$

Examples   Write each fraction or mixed number as a percent.

17. $\frac{9}{20} = \frac{9}{20} \cdot 100\% = \frac{9}{20} \cdot \frac{100}{1}\% = \frac{900}{20}\% = 45\%$

18. $\frac{2}{3} = \frac{2}{3} \cdot 100\% = \frac{2}{3} \cdot \frac{100}{1}\% = \frac{200}{3}\% = 66\frac{2}{3}\%$

19. $1\frac{1}{2} = \frac{3}{2} \cdot 100\% = \frac{3}{2} \cdot \frac{100}{1}\% = \frac{300}{2}\% = 150\%$

Helpful Hint $\frac{200}{3} = 66.\overline{6}.$ Thus, another way to write $\frac{200}{3}\%$ is $66.\overline{6}\%$.

Work Practice 17–19

✓**Concept Check** Which digit in the percent 76.4582% represents

a. A tenth percent?   **b.** A thousandth percent?
c. A hundredth percent?   **d.** A whole percent?

Practice 20

Write $\frac{3}{17}$ as a percent. Round to the nearest hundredth percent.

Example 20   Write $\frac{1}{12}$ as a percent. Round to the nearest hundredth percent.

Solution:

$$\frac{1}{12} = \frac{1}{12} \cdot 100\% = \frac{1}{12} \cdot \frac{100}{1}\% = \frac{100}{12}\% \approx 8.33\%$$

"approximately"

$$\begin{array}{r} 8.333 \approx 8.33 \\ 12\overline{)100.000} \\ -96 \\ \hline 4\,0 \\ -3\,6 \\ \hline 40 \\ -36 \\ \hline 40 \\ -36 \\ \hline 4 \end{array}$$

Thus, $\frac{1}{12}$ is approximately 8.33%.

Work Practice 20

Answers

17. 50%   **18.** $17\frac{1}{2}\%$ **19.** 225%

20. 17.65%

✓**Concept Check Answers**

a. 4   **b.** 8   **c.** 5   **d.** 6

Objective D Converting Percents, Decimals, and Fractions ▶

Let's summarize what we have learned so far about percents, decimals, and fractions:

Summary of Converting Percents, Decimals, and Fractions

- *To write a percent as a decimal,* replace the % symbol with its decimal equivalent, 0.01; then multiply.
- *To write a percent as a fraction,* replace the % symbol with its fraction equivalent, $\frac{1}{100}$; then multiply.
- *To write a decimal or fraction as a percent,* multiply by 100%.

If we let p represent a number, below we summarize using symbols.

Write a percent as a decimal:	Write a percent as a fraction:	Write a number as a percent:
$p\% = p(0.01)$	$p\% = p \cdot \dfrac{1}{100}$	$p = p \cdot 100\%$

Example 21 37.8% of automobile thefts in the continental United States occur in the South, the greatest percent. Write this percent as a decimal and as a fraction. (*Source:* Federal Bureau of Investigation's (FBI) Uniform Crime Report)

Solution:

As a decimal: $37.8\% = 37.8(0.01) = 0.378$

As a fraction: $37.8\% = 37.8 \cdot \dfrac{1}{100} = \dfrac{37.8}{100} = \dfrac{37.8}{100} \cdot \dfrac{10}{10} = \dfrac{378}{1000} = \dfrac{2 \cdot 189}{2 \cdot 500} = \dfrac{189}{500}$

Thus, 37.8% written as a decimal is 0.378 and written as a fraction is $\dfrac{189}{500}$.

■ Work Practice 21

Practice 21

A family decides to spend no more than 22.5% of its monthly income on rent. Write 22.5% as a decimal and as a fraction.

Example 22 An advertisement for a stereo system reads "$\frac{1}{4}$ off." What percent off is this?

Solution: Write $\frac{1}{4}$ as a percent.

$\dfrac{1}{4} = \dfrac{1}{4} \cdot 100\% = \dfrac{1}{4} \cdot \dfrac{100}{1}\% = \dfrac{100}{4}\% = 25\%$

Thus, "$\frac{1}{4}$ off" is the same as "25% off."

■ Work Practice 22

Practice 22

Provincetown's budget for waste disposal increased by $1\frac{1}{4}$ times over the budget from last year. What percent increase is this?

Note: It is helpful to know a few basic percent conversions. Appendix A.4 contains a handy reference of percent, decimal, and fraction equivalencies.

Also, Appendix A.5 shows how to find common percents of a number.

Answers

21. $0.225, \dfrac{9}{40}$ **22.** 125%

Vocabulary, Readiness & Video Check

Use the choices below to fill in each blank. Some choices may be used more than once.

$\dfrac{1}{100}$ 0.01   100%   percent

1. _____ means "per hundred."
2. _____ = 1.
3. The % symbol is read as _____.
4. To write a decimal or a fraction as a *percent*, multiply by 1 in the form of _____.
5. To write a percent as a *decimal*, drop the % symbol and multiply by _____.
6. To write a percent as a *fraction*, drop the % symbol and multiply by _____.

Write each fraction as a percent.

7. $\dfrac{13}{100}$ 8. $\dfrac{92}{100}$ 9. $\dfrac{87}{100}$ 10. $\dfrac{71}{100}$ 11. $\dfrac{1}{100}$ 12. $\dfrac{2}{100}$

Martin-Gay Interactive Videos   *Watch the section lecture video and answer the following questions.*

Objective A 13. From the lecture before ▤ Example 1, what is the most important thing to remember about percent?

Objective B 14. In ▤ Example 7, since the % symbol is replaced with $\dfrac{1}{100}$, why doesn't the final answer have a denominator of 100?

Objective C 15. From the lecture before ▤ Example 14, how is writing a fraction as a percent similar to writing a decimal as a percent?

Objective D 16. From ▤ Example 17, what is the main difference between writing a percent as an equivalent decimal and writing a percent as an equivalent fraction?

See Video 6.1

6.1 Exercise Set MyMathLab®

Objective A *Solve. See Examples 1 and 2.*

1. In a survey of 100 college students, 96 use the Internet. What percent use the Internet?

2. A basketball player makes 81 out of 100 attempted free throws. What percent of free throws are made?

3. Michigan leads the United States in tart cherry production, producing 75 out of every 100 tart cherries each year. (*Source:* Cherry Marketing Institute)
   a. What percent of tart cherries are produced in Michigan?
   b. What percent of tart cherries are *not* produced in Michigan?

4. The United States is the world's second-largest producer of apples. Twenty-five out of every 100 apples harvested in the United States are exported (shipped to other countries). (*Source:* U.S. Apple Association)
   a. What percent of U.S.-grown apples are exported?
   b. What percent of U.S.-grown apples are *not* exported?

One hundred adults were asked to name their favorite sport, and the results are shown in the circle graph.

5. What sport was preferred by most adults? What percent preferred this sport?

6. What sport was preferred by the least number of adults? What percent preferred this sport?

7. What percent of adults preferred football or soccer?

8. What percent of adults preferred basketball or baseball?

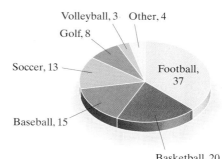

Volleyball, 3   Other, 4
Golf, 8
Soccer, 13
Football, 37
Baseball, 15
Basketball, 20

Objective B   *Write each percent as a decimal. See Examples 3 through 7.*

9. 41%   **10.** 64%   **11.** 6%   **12.** 9%

13. 100%   **14.** 136%   **15.** 61.3%   **16.** 52.7%

17. 2.8%   **18.** 1.7%   **19.** 0.6%   **20.** 0.9%

21. 300%   **22.** 700%   **23.** 32.58%   **24.** 72.18%

Write each percent as a fraction or mixed number in simplest form. See Examples 8 through 12.

25. 12%   **26.** 24%   **27.** 4%   **28.** 2%   **29.** 4.5%

30. 7.5%   **31.** 175%   **32.** 250%   **33.** 6.25%   **34.** 3.75%

35. $10\frac{1}{3}\%$ **36.** $7\frac{3}{4}\%$ **37.** $22\frac{3}{8}\%$ **38.** $15\frac{5}{8}\%$

Objective C   *Write each decimal as a percent. See Examples 13 through 16.*

39. 0.003   **40.** 0.006   **41.** 0.22   **42.** 0.45   **43.** 5.3

44. 1.6   **45.** 0.056   **46.** 0.027   **47.** 0.3328   **48.** 0.1115

49. 3   **50.** 5   **51.** 0.7   **52.** 0.8

Write each fraction or mixed number as a percent. See Examples 17 through 19.

53. $\frac{7}{10}$ **54.** $\frac{3}{10}$ **55.** $\frac{2}{5}$ **56.** $\frac{4}{5}$ **57.** $\frac{17}{50}$

58. $\frac{47}{50}$ **59.** $\frac{3}{8}$ **60.** $\frac{5}{16}$ **61.** $\frac{7}{9}$ **62.** $\frac{1}{3}$

63. $2\frac{1}{2}$ **64.** $2\frac{1}{5}$ **65.** $1\frac{9}{10}$ **66.** $2\frac{7}{10}$

Write each fraction as a percent. Round to the nearest hundredth percent. See Example 20.

67. $\dfrac{7}{11}$ **68.** $\dfrac{5}{12}$ ▶ **69.** $\dfrac{4}{15}$ **70.** $\dfrac{10}{11}$

Objective D *Complete each table. See Examples 21 and 22.*

71.

Percent	Decimal	Fraction
35%		
		$\dfrac{1}{5}$
	0.5	
70%		
		$\dfrac{3}{8}$

72.

Percent	Decimal	Fraction
50%		
		$\dfrac{2}{5}$
	0.25	
12.5%		
		$\dfrac{5}{8}$
		$\dfrac{7}{50}$

73.

Percent	Decimal	Fraction
40%		
	0.235	
		$\dfrac{4}{5}$
$33\dfrac{1}{3}\%$		
		$\dfrac{7}{8}$
7.5%		

74.

Percent	Decimal	Fraction
	0.525	
		$\dfrac{3}{4}$
$66\dfrac{2}{3}\%$		
		$\dfrac{5}{6}$
100%		

75.

Percent	Decimal	Fraction
200%		
	2.8	
705%		
		$4\dfrac{27}{50}$

76.

Percent	Decimal	Fraction
800%		
	3.2	
608%		
		$9\dfrac{13}{50}$

Solve. See Examples 21 and 22.

77. A 2012 poll revealed that 67% of American adults are in favor of keeping the penny in circulation. Write this percent as a decimal and a fraction. (*Source:* Americans for Common Cents)

78. China was responsible for 42% of the world's smelter production of aluminum in 2012. Write this percent as a decimal and a fraction. (*Source:* U.S. Geological Survey)

79. In 2013, 55.4% of all practicing veterinarians were female. Write this percent as a decimal and a fraction. (*Source:* American Veterinary Medical Association)

80. The U.S. penny is 97.5% zinc. Write this percent as a decimal and a fraction. (*Source:* Americans for Common Cents)

81. The average American wastes $\frac{9}{50}$ of all grain products brought into the home. Write this fraction as a percent. (*Source:* Natural Resources Defense Council)

82. Canada produces $\frac{1}{4}$ of the uranium produced in the world. Write this fraction as a percent. (*Source:* World Nuclear Association)

Canada

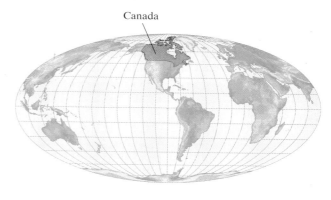

Write each percent as a decimal. See Examples 3 through 7.

83. People take aspirin for a variety of reasons. The most common use of aspirin is to prevent heart disease, accounting for 38% of all aspirin use. (*Source:* Bayer Market Research)

84. Exports to Europe accounted for 17.7% of all of Japan's motor vehicle exports. (*Source:* Japan Automobile Manufacturers Association)

85. In 2013, 39.4% of households in the United States had no landline telephones, just cell phones. (*Source:* National Center for Health Statistics)

86. In 2013, 8.5% of households in the United States had no cell phones, just landline telephones. (*Source:* National Center for Health Statistics)

In Exercises 87 through 92, write the percent from the circle graph as a decimal and a fraction.

World Population by Continent

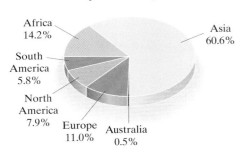

Africa 14.2%
Asia 60.6%
South America 5.8%
North America 7.9%
Europe 11.0%
Australia 0.5%

87. Australia: 0.5%

88. Europe: 11%

89. Africa: 14.2%

90. Asia: 60.6%

91. North America: 7.9%

92. South America: 5.8%

Review

Find the value of n. See Section 5.2.

93. $3 \cdot n = 45$

94. $2 \cdot n = 16$

95. $6 \cdot n = 72$

96. $5 \cdot n = 35$

Concept Extensions

97. Write 0.7682 as a percent rounded to the nearest percent.

98. Write 0.2371 as a percent rounded to the nearest percent.

99. Write 1.07835 as a percent rounded to the nearest tenth of a percent.

100. Write 1.25348 as a percent rounded to the nearest tenth of a percent.

Solve. See the Concept Checks in this section.

101. Given the percent 52.8647%, round as indicated.
 a. Round to a tenth of a percent.
 b. Round to a hundredth of a percent.

102. Given the percent 0.5269%, round as indicated.
 a. Round to a tenth of a percent.
 b. Round to a hundredth of a percent.

103. Which of the following are correct?
 a. 6.5% = 0.65
 b. 7.8% = 0.078
 c. 120% = 0.12
 d. 0.35% = 0.0035

104. Which of the following are correct?
 a. 0.231 = 23.1%
 b. 5.12 = 0.0512%
 c. 3.2 = 320%
 d. 0.0175 = 0.175%

Recall that 1 = 100%. This means that 1 whole is 100%. Use this for Exercises 105 and 106. (Source: Some Body! by Dr. Pete Rowan)

105. The four blood types are A, B, O, and AB. (Each blood type can also be further classified as Rh-positive or Rh-negative depending upon whether your blood contains protein or not.) Given the percent blood types for the United States below, calculate the percent of the U.S. population with AB blood type.

45%   40%   11%   ?%

106. The components of bone are all listed in the categories below. Find the missing percent.
 1. Minerals—45%
 2. Living tissue—30%
 3. Water—20%
 4. Other—?

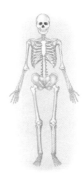

The bar graph shows the predicted fastest-growing occupations. Use the graph for Exercises 107 through 110. (Source: Bureau of Labor Statistics)

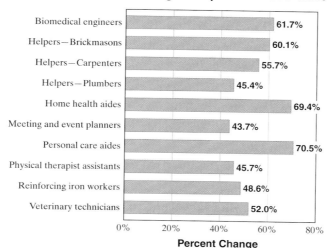

Fastest-Growing Occupations 2010–2020

Source: Bureau of Labor Statistics

107. What occupation is predicted to be the fastest growing?

108. What occupation is predicted to be the second fastest growing?

109. Write the percent change for biomedical engineers as a decimal.

110. Write the percent change for meeting and event planners as a decimal.

What percent of the figure is shaded?

111.

112.

113.

114.

Fill in the blanks.

115. A fraction written as a percent is greater than 100% when the numerator is _____ than the denominator. (greater/less)

116. A decimal written as a percent is less than 100% when the decimal is _____ than 1. (greater/less)

117. In your own words, explain how to write a percent as a fraction.

118. In your own words, explain how to write a fraction as a decimal.

Write each fraction as a decimal and then write each decimal as a percent. Round the decimal to three decimal places (nearest thousandth) and the percent to the nearest tenth of a percent.

119. $\dfrac{21}{79}$

120. $\dfrac{56}{102}$

121. $\dfrac{850}{736}$

122. $\dfrac{506}{248}$

Solving Percent Problems Using Equations

Objectives

A Write Percent Problems as Equations.

B Solve Percent Problems.

Note: Sections 6.2 and 6.3 introduce two methods for solving percent problems. It is not necessary that you study both sections. You may want to check with your instructor for further advice.

Throughout this text, we have written mathematical statements such as $3 + 10 = 13$, or area = length · width. These statements are called "equations." An **equation** is a mathematical statement that contains an equal sign. To solve percent problems in this section, we translate the problems into such mathematical statements, or equations.

Objective A Writing Percent Problems as Equations

Recognizing key words in a percent problem is helpful in writing the problem as an equation. Three key words in the statement of a percent problem and their meanings are as follows:

> **of** means **multiplication** (·)
> **is** means **equals** (=)
> **what** (or some equivalent) means **the unknown number**

In our examples, we let the letter n stand for the unknown number.

> **Helpful Hint**
>
> Any letter of the alphabet can be used to represent the unknown number. In this section, we mostly use the letter n.

Practice 1

Translate: 6 is what percent of 24?

Example 1 Translate to an equation.

5 is what percent of 20?

Solution: 5   is   what percent of   20?
$$5 = n \cdot 20$$

■ Work Practice 1

> **Helpful Hint**
>
> Remember that an equation is simply a mathematical statement that contains an equal sign (=).
>
> $$5 = n \cdot 20$$
> ↑
> equal sign

Practice 2

Translate: 1.8 is 20% of what number?

Example 2 Translate to an equation.

1.2 is 30% of what number?

Solution: 1.2   is   30%   of   what number?
$$1.2 = 30\% \cdot n$$

■ Work Practice 2

Answers
1. $6 = n \cdot 24$ **2.** $1.8 = 20\% \cdot n$

Example 3 Translate to an equation.

What number is 25% of 0.008?

Solution: What number is 25% of 0.008?

$$n = 25\% \cdot 0.008$$

■ Work Practice 3

Examples Translate each of the following to an equation:

4. 38% of 200 is what number?

$$38\% \cdot 200 = n$$

5. 40% of what number is 80?

$$40\% \cdot n = 80$$

6. What percent of 85 is 34?

$$n \cdot 85 = 34$$

■ Work Practice 4–6

✓**Concept Check** In the equation $2 \cdot n = 10$, what step should be taken to solve the equation for n?

Objective B Solving Percent Problems ▶

You may have noticed by now that each percent problem has contained three numbers—in our examples, two are known and one is unknown. Each of these numbers is given a special name.

15%   of   60   is   9

15% · 60 = 9
percent   base   amount

We call this equation the **percent equation.**

Percent Equation

percent · base = amount

Helpful Hint

Notice that the percent equation given above is a true statement. To see this, simplify the left side as shown:

$$15\% \cdot 60 = 9$$
$$0.15 \cdot 60 = 9 \quad \text{Write 15\% as 0.15.}$$
$$9 = 9 \quad \text{Multiply.}$$

The statement $9 = 9$ is true.

Practice 3

Translate: What number is 40% of 3.6?

Practice 4–6

Translate each to an equation.
4. 42% of 50 is what number?
5. 15% of what number is 9?
6. What percent of 150 is 90?

Answers
3. $n = 40\% \cdot 3.6$ **4.** $42\% \cdot 50 = n$
5. $15\% \cdot n = 9$ **6.** $n \cdot 150 = 90$

✓**Concept Check Answer**

If $2 \cdot n = 10$, then $n = \dfrac{10}{2}$, or $n = 5$.

After a percent problem has been written as a percent equation, we can use the equation to find the unknown number. This is called **solving** the equation.

Practice 7

What number is 20% of 85?

Example 7 Solving a Percent Equation for the Amount

What number is 35% of 40?

Solution:

$$n = 35\% \cdot 40 \quad \text{Translate to an equation.}$$
$$n = 0.35 \cdot 40 \quad \text{Write 35\% as 0.35.}$$
$$n = 14 \quad \text{Multiply } 0.35 \cdot 40 = 14.$$

Thus, 14 is 35% of 40.

Is this reasonable? To see, round 35% to 40%. Then 40% of 40 or 0.40(40) is 16. Our result is reasonable since 16 is close to 14.

■ Work Practice 7

Helpful Hint

When solving a percent equation, write the percent as a decimal (or fraction).

Practice 8

90% of 150 is what number?

Example 8 Solving a Percent Equation for the Amount

85% of 300 is what number?

Solution:

$$85\% \cdot 300 = n \quad \text{Translate to an equation.}$$
$$0.85 \cdot 300 = n \quad \text{Write 85\% as 0.85.}$$
$$255 = n \quad \text{Multiply } 0.85 \cdot 300 = 255.$$

Thus, 85% of 300 is 255.

Is this result reasonable? To see, round 85% to 90%. Then 90% of 300 or 0.90(300) = 270, which is close to 255.

■ Work Practice 8

Practice 9

15% of what number is 1.2?

Example 9 Solving a Percent Equation for the Base

12% of what number is 0.6?

Solution:

$$12\% \cdot n = 0.6 \quad \text{Translate to an equation.}$$
$$0.12 \cdot n = 0.6 \quad \text{Write 12\% as 0.12.}$$

Recall from Section 5.2 that if "0.12 times some number is 0.6," then the number is 0.6 divided by 0.12.

$$n = \frac{0.6}{0.12} \quad \text{Divide 0.6 by 0.12, the number multiplied by } n.$$
$$n = 5$$

Thus, 12% of 5 is 0.6.

Is this reasonable? To see, round 12% to 10%. Then 10% of 5 or 0.10(5) = 0.5, which is close to 0.6.

■ Work Practice 9

Answers

7. 17   **8.** 135   **9.** 8

Example 10 Solving a Percent Equation for the Base

$$13 \quad \text{is} \quad 6\frac{1}{2}\% \quad \text{of} \quad \text{what number?}$$

27 is $4\frac{1}{2}\%$ of what number?

Solution: $13 = 6\frac{1}{2}\% \cdot n$ Translate to an equation.

$13 = 0.065 \cdot n$ $6\frac{1}{2}\% = 6.5\% = 0.065.$

$\dfrac{13}{0.065} = n$ Divide 13 by 0.065, the number multiplied by n.

$200 = n$

Thus, 13 is $6\frac{1}{2}\%$ of 200.

Check to see if this result is reasonable.

Work Practice 10

Example 11 Solving a Percent Equation for the Percent

$$\text{What percent} \quad \text{of} \quad 12 \quad \text{is} \quad 9?$$

What percent of 80 is 8?

Solution: $n \cdot 12 = 9$ Translate to an equation.

$n = \dfrac{9}{12}$ Divide 9 by 12, the number multiplied by n.

$n = 0.75$

Next, since we are looking for percent, we write 0.75 as a percent.

$n = 75\%$

So, 75% of 12 is 9. To check, see that $75\% \cdot 12 = 9$.

Work Practice 11

Helpful Hint

If your unknown in the percent equation is the percent, don't forget to convert your answer to a percent.

Example 12 Solving a Percent Equation for the Percent

$$78 \quad \text{is} \quad \text{what percent} \quad \text{of} \quad 65?$$

35 is what percent of 25?

Solution: $78 = n \cdot 65$ Translate to an equation.

$\dfrac{78}{65} = n$ Divide 78 by 65, the number multiplied by n.

$1.2 = n$

$120\% = n$ Write 1.2 as a percent.

So, 78 is 120% of 65. Check this result.

Work Practice 12

Answers
10. 600   **11.** 10%   **12.** 140%

✓ **Concept Check** Consider these problems.

1. 75% of 50 =
   a. 50   **b.** a number greater than 50   **c.** a number less than 50

2. 40% of a number is 10. Is the number
   a. 10?   **b.** less than 10?   **c.** greater than 10?

3. 800 is 120% of what number? Is the number
   a. 800?   **b.** less than 800?   **c.** greater than 800?

Helpful Hint

Use the following to see if your answers are reasonable.

$$(100\%) \text{ of a number } = \text{ the number}$$

$$\left(\begin{array}{c} \text{a percent} \\ \text{greater than} \\ 100\% \end{array}\right) \text{ of a number } = \begin{array}{c} \text{a number greater} \\ \text{than the original number} \end{array}$$

$$\left(\begin{array}{c} \text{a percent} \\ \text{less than } 100\% \end{array}\right) \text{ of a number } = \begin{array}{c} \text{a number less} \\ \text{than the original number} \end{array}$$

✓ **Concept Check Answers**
1. c   **2.** c   **3.** b

Vocabulary, Readiness & Video Check

Use the choices below to fill in each blank.

percent	amount	of	less
base	the number	is	greater

1. The word _____ translates to " =".

2. The word _____ usually translates to "multiplication."

3. In the statement "10% of 90 is 9," the number 9 is called the _____, 90 is called the _____, and 10 is called the _____.

4. 100% of a number = _____.

5. Any "percent greater than 100%" of "a number" = "a number _____ than the original number."

6. Any "percent less than 100%" of "a number" = "a number _____ than the original number."

Identify the percent, the base, and the amount in each equation. Recall that percent · base = amount.

7. 42% · 50 = 21

8. 30% · 65 = 19.5

9. 107.5 = 125% · 86

10. 99 = 110% · 90

Martin-Gay Interactive Videos   *Watch the section lecture video and answer the following questions.*

Objective A **11.** From the lecture before Example 1, what are the key words and their translations that we need to remember?

Objective B **12.** What is the difference between the translated equation in Example 5 and those in Examples 4 and 6?

See Video 6.2

6.2 **Exercise Set** MyMathLab®

Objective A **Translating** *Translate each to an equation. Do not solve. See Examples 1 through 6.*

1. 18% of 81 is what number?

2. 36% of 72 is what number?

3. 20% of what number is 105?

4. 40% of what number is 6?

5. 0.6 is 40% of what number?

6. 0.7 is 20% of what number?

7. What percent of 80 is 3.8?

8. 9.2 is what percent of 92?

9. What number is 9% of 43?

10. What number is 25% of 55?

11. What percent of 250 is 150?

12. What percent of 375 is 300?

Objective B *Solve. See Examples 7 and 8.*

13. 10% of 35 is what number?

14. 25% of 68 is what number?

15. What number is 14% of 205?

16. What number is 18% of 425?

Solve. See Examples 9 and 10.

17. 1.2 is 12% of what number?

18. 0.22 is 44% of what number?

19. $8\frac{1}{2}$% of what number is 51?

20. $4\frac{1}{2}$% of what number is 45?

Solve. See Examples 11 and 12.

21. What percent of 80 is 88?

22. What percent of 40 is 60?

23. 17 is what percent of 50?

24. 48 is what percent of 50?

Objectives **A B** Mixed Practice *Solve. See Examples 1 through 12.*

25. 0.1 is 10% of what number?

26. 0.5 is 5% of what number?

27. 150% of 430 is what number?

28. 300% of 56 is what number?

29. 82.5 is $16\frac{1}{2}$% of what number?

30. 7.2 is $6\frac{1}{4}$% of what number?

31. 2.58 is what percent of 50?

32. 2.64 is what percent of 25?

33. What number is 42% of 60?

34. What number is 36% of 80?

35. What percent of 184 is 64.4?

36. What percent of 120 is 76.8?

37. 120% of what number is 42?

38. 160% of what number is 40?

39. 2.4% of 26 is what number?

40. 4.8% of 32 is what number?

41. What percent of 600 is 3?

42. What percent of 500 is 2?

43. 6.67 is 4.6% of what number?

44. 9.75 is 7.5% of what number?

45. 1575 is what percent of 2500?

46. 2520 is what percent of 3500?

47. 2 is what percent of 50?

48. 2 is what percent of 40?

Review

Find the value of n in each proportion. See Section 5.2.

49. $\dfrac{27}{n} = \dfrac{9}{10}$

50. $\dfrac{35}{n} = \dfrac{7}{5}$

51. $\dfrac{n}{5} = \dfrac{8}{11}$

52. $\dfrac{n}{3} = \dfrac{6}{13}$

Write each phrase as a proportion.

53. 17 is to 12 as n is to 20

54. 20 is to 25 as n is to 10

55. 8 is to 9 as 14 is to n

56. 5 is to 6 as 15 is to n

Concept Extensions

For each equation in Exercises 57–60, determine the next step taken to find the value of n. See the first Concept Check in this section.

57. $5 \cdot n = 32$

    a. $n = 5 \cdot 32$ **b.** $n = \dfrac{5}{32}$ **c.** $n = \dfrac{32}{5}$ **d.** none of these

58. $n = 0.7 \cdot 12$

    a. $n = 8.4$ **b.** $n = \dfrac{12}{0.7}$ **c.** $n = \dfrac{0.7}{12}$ **d.** none of these

59. $0.06 = n \cdot 7$

    a. $n = 0.06 \cdot 7$ **b.** $n = \dfrac{0.06}{7}$ **c.** $n = \dfrac{7}{0.06}$ **d.** none of these

60. $0.01 = n \cdot 8$

    a. $n = 0.01 \cdot 8$ **b.** $n = \dfrac{8}{0.01}$ **c.** $n = \dfrac{0.01}{8}$ **d.** none of these

61. Write a word statement for the equation $20\% \cdot n = 18.6$. Use the phrase "some number" for "n."

62. Write a word statement for the equation $n = 33\frac{1}{3}\% \cdot 24$. Use the phrase "some number" for "n."

For each exercise, determine whether the percent, n, is (a) 100%, (b) greater than 100%, or (c) less than 100%. See the second Concept Check in this section.

63. $n\%$ of 20 is 30

64. $n\%$ of 98 is 98

65. $n\%$ of 120 is 85

66. $n\%$ of 35 is 50

For each exercise, determine whether the number, n, is (a) equal to 45, (b) greater than 45, or (c) less than 45.

67. 55% of 45 is n

68. 230% of 45 is n

69. 100% of 45 is n

70. 30% of n is 45

71. 100% of n is 45

72. 180% of n is 45

Solve.

73. In your own words, explain how to solve a percent equation.

74. Write a percent problem that uses the percent 50%.

75. 1.5% of 45,775 is what number?

76. What percent of 75,528 is 27,945.36?

77. 22,113 is 180% of what number?

6.3 Solving Percent Problems Using Proportions

There is more than one method that can be used to solve percent problems. (See the note at the beginning of Section 6.2.) In the last section, we used the percent equation. In this section, we will use proportions.

Objectives

A Write Percent Problems as Proportions.

B Solve Percent Problems.

Objective A Writing Percent Problems as Proportions

To understand the proportion method, recall that 70% means the ratio of 70 to 100, or $\frac{70}{100}$.

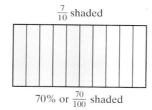

$\frac{7}{10}$ shaded

70% or $\frac{70}{100}$ shaded

$$70\% = \frac{70}{100} = \frac{7}{10}$$

Since the ratio $\frac{70}{100}$ is equal to the ratio $\frac{7}{10}$, we have the proportion

$$\frac{7}{10} = \frac{70}{100}$$

We call this proportion the "percent proportion." In general, we can name the parts of this proportion as follows:

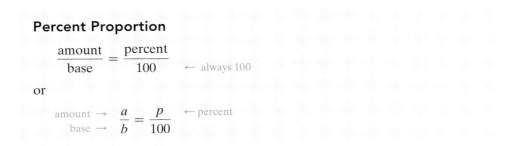

Percent Proportion

$$\frac{\text{amount}}{\text{base}} = \frac{\text{percent}}{100} \qquad \leftarrow \text{always 100}$$

or

$$\begin{array}{l}\text{amount} \rightarrow \\ \text{base} \rightarrow\end{array} \frac{a}{b} = \frac{p}{100} \qquad \leftarrow \text{percent}$$

When we translate percent problems to proportions, the **percent,** p, can be identified by looking for the symbol % or the word *percent*. The **base,** b, usually follows the word *of*. The **amount,** a, is the part compared to the whole.

Helpful Hint

Part of Proportion	How It's Identified
Percent	% or percent
Base	Appears after *of*
Amount	Part compared to whole

Practice 1

Translate to a proportion.
15% of what number is 55?

Example 1 Translate to a proportion.

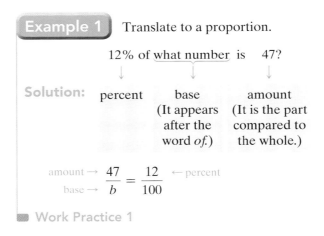

$$12\% \text{ of } \underline{\text{what number}} \text{ is } 47?$$

Solution: percent   base   amount
                      (It appears   (It is the part
                       after the   compared to
                       word *of.*)   the whole.)

$$\begin{array}{l}\text{amount} \rightarrow \\ \text{base} \rightarrow\end{array} \frac{47}{b} = \frac{12}{100} \qquad \leftarrow \text{percent}$$

■ Work Practice 1

Practice 2

Translate to a proportion.
35 is what percent of 70?

Example 2 Translate to a proportion.

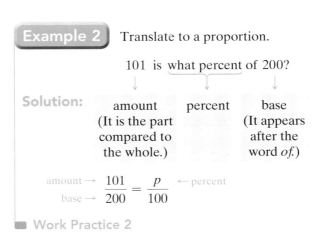

$$101 \text{ is } \underline{\text{what percent}} \text{ of } 200?$$

Solution: amount   percent   base
                      (It is the part   (It appears
                       compared to   after the
                       the whole.)   word *of.*)

$$\begin{array}{l}\text{amount} \rightarrow \\ \text{base} \rightarrow\end{array} \frac{101}{200} = \frac{p}{100} \qquad \leftarrow \text{percent}$$

■ Work Practice 2

Answers

1. $\frac{55}{b} = \frac{15}{100}$ **2.** $\frac{35}{70} = \frac{p}{100}$

Example 3 Translate to a proportion.

What number is 90% of 45?

Solution:

amount — It is the part compared to the whole.

percent

base — It appears after the word *of.*

amount → $\dfrac{a}{45} = \dfrac{90}{100}$ ← percent
base →

Work Practice 3

Practice 3

Translate to a proportion.
What number is 25% of 68?

Example 4 Translate to a proportion.

238 is 40% of what number?

Solution: amount percent base

$$\dfrac{238}{b} = \dfrac{40}{100}$$

Work Practice 4

Practice 4

Translate to a proportion.
520 is 65% of what number?

Example 5 Translate to a proportion.

What percent of 30 is 75?

Solution: percent base amount

$$\dfrac{75}{30} = \dfrac{p}{100}$$

Work Practice 5

Practice 5

Translate to a proportion.
What percent of 50 is 65?

Example 6 Translate to a proportion.

45% of 105 is what number?

Solution: percent base amount

$$\dfrac{a}{105} = \dfrac{45}{100}$$

Work Practice 6

Practice 6

Translate to a proportion.
36% of 80 is what number?

Objective B Solving Percent Problems

The proportions that we have written in this section contain three values that can change: the percent, the base, and the amount. If any two of these values are known, we can find the third (the unknown value). To do this, we write a percent proportion and find the unknown value as we did in Section 5.2.

Answers

3. $\dfrac{a}{68} = \dfrac{25}{100}$ **4.** $\dfrac{520}{b} = \dfrac{65}{100}$

5. $\dfrac{65}{50} = \dfrac{p}{100}$ **6.** $\dfrac{a}{80} = \dfrac{36}{100}$

Practice 7

What number is 8% of 120?

Example 7   Solving a Percent Proportion for the Amount

What number   is   30%   of   9?

Solution: amount   percent   base

$$\frac{a}{9} = \frac{30}{100}$$

To solve, we set cross products equal to each other.

$$\frac{a}{9} = \frac{30}{100}$$

$a \cdot 100 = 9 \cdot 30$ Set cross products equal.

$a \cdot 100 = 270$ Multiply.

Recall from Section 5.2 that if "some number times 100 is 270," then the number is 270 divided by 100.

$a = \dfrac{270}{100}$ Divide 270 by 100, the number multiplied by a.

$a = 2.7$ Simplify.

Thus, 2.7 is 30% of 9.

■ Work Practice 7

✓**Concept Check**   Consider the statement: "78 is what percent of 350?"
Which part of the percent proportion is unknown?
a. the amount   **b.** the base   **c.** the percent

Consider another statement: "14 is 10% of some number."
Which part of the percent proportion is unknown?
a. the amount   **b.** the base   **c.** the percent

Practice 8

75% of what number is 60?

Example 8   Solving a Percent Proportion for the Base

150%   of   what number   is   30?

Solution: percent   base   amount

$$\frac{30}{b} = \frac{150}{100}$$ Write the proportion.

$$\frac{30}{b} = \frac{3}{2}$$ Write $\dfrac{150}{100}$ as $\dfrac{3}{2}$.

$30 \cdot 2 = b \cdot 3$ Set cross products equal.

$60 = b \cdot 3$ Multiply.

$\dfrac{60}{3} = b$ Divide 60 by 3, the number multiplied by b.

$20 = b$ Simplify.

Thus, 150% of 20 is 30.

■ Work Practice 8

Answers
7. 9.6   **8.** 80

✓**Concept Check Answers**
c, b

✓**Concept Check** When solving a percent problem by using a proportion, describe how you can check the result.

Example 9 Solving a Percent Proportion for the Base

20.8 is 40% of what number?

$\downarrow$ $\downarrow$ $\downarrow$

Solution: amount percent   base

$$\frac{20.8}{b} = \frac{40}{100} \quad \text{or} \quad \frac{20.8}{b} = \frac{2}{5} \qquad \text{Write the proportion and simplify } \frac{40}{100}.$$

$$20.8 \cdot 5 = b \cdot 2 \qquad \text{Set cross products equal.}$$

$$104 = b \cdot 2 \qquad \text{Multiply.}$$

$$\frac{104}{2} = b \qquad \text{Divide 104 by 2, the number multiplied by } b.$$

$$52 = b \qquad \text{Simplify.}$$

So, 20.8 is 40% of 52.

■ Work Practice 9

Example 10 Solving a Percent Proportion for the Percent

What percent of 50 is 8?

$\downarrow$ $\downarrow$ $\downarrow$

Solution: percent   base   amount

$$\frac{8}{50} = \frac{p}{100} \quad \text{or} \quad \frac{4}{25} = \frac{p}{100} \qquad \text{Write the proportion and simplify } \frac{8}{50}.$$

$$4 \cdot 100 = 25 \cdot p \qquad \text{Set cross products equal.}$$

$$400 = 25 \cdot p \qquad \text{Multiply.}$$

$$\frac{400}{25} = p \qquad \text{Divide 400 by 25, the number multiplied by } p.$$

$$16 = p \qquad \text{Simplify.}$$

So, 16% of 50 is 8.

■ Work Practice 10

Helpful Hint

Recall from our percent proportion that this number already is a percent. Just keep the number as is and attach a % symbol.

Example 11 Solving a Percent Proportion for the Percent

504 is what percent of 360?

$\downarrow$ $\downarrow$ $\downarrow$

Solution: amount   percent   base

$$\frac{504}{360} = \frac{p}{100}$$

(Continued on next page)

Practice 9

15.2 is 5% of what number?

Practice 10

What percent of 40 is 6?

Practice 11

336 is what percent of 160?

Answers

9. 304   **10.** 15%   **11.** 210%

✓**Concept Check Answer**

by putting the result into the proportion and checking that the proportion is true

Let's choose not to simplify the ratio $\dfrac{504}{360}$.

$$504 \cdot 100 = 360 \cdot p \quad \text{Set cross products equal.}$$
$$50{,}400 = 360 \cdot p \quad \text{Multiply.}$$
$$\dfrac{50{,}400}{360} = p \quad \text{Divide 50,400 by 360, the number multiplied by } p.$$
$$140 = p \quad \text{Simplify.}$$

Notice that by choosing not to simplify $\dfrac{504}{360}$, we had larger numbers in our equation.

Either way, we find that 504 is 140% of 360.

▨ Work Practice 11

You may have noticed the following while working the examples.

Helpful Hint

Use the following to see whether your answers are reasonable.

$$100\% \text{ of a number} = \text{the number}$$

$$\left(\begin{array}{c}\text{a percent}\\ \text{greater than}\\ 100\%\end{array}\right) \text{of a number} = \begin{array}{c}\text{a number larger}\\ \text{than the original number}\end{array}$$

$$\left(\begin{array}{c}\text{a percent}\\ \text{less than } 100\%\end{array}\right) \text{of a number} = \begin{array}{c}\text{a number less}\\ \text{than the original number}\end{array}$$

Vocabulary, Readiness & Video Check

Use the choices below to fill in each blank. These choices will be used more than once.

amount   base   percent

1. When translating the statement "20% of 15 is 3" to a proportion, the number 3 is called the _____, 15 is the _____, and 20 is the _____.

2. In the question "50% of what number is 28?", which part of the percent proportion is unknown? _____

3. In the question "What number is 25% of 200?", which part of the percent proportion is unknown? _____

4. In the question "38 is what percent of 380?", which part of the percent proportion is unknown? _____

Identify the amount, the base, and the percent in each equation. Recall that $\dfrac{\text{amount}}{\text{base}} = \dfrac{\text{percent}}{100}$.

5. $\dfrac{12.6}{42} = \dfrac{30}{100}$

6. $\dfrac{201}{300} = \dfrac{67}{100}$

7. $\dfrac{20}{100} = \dfrac{102}{510}$

8. $\dfrac{40}{100} = \dfrac{248}{620}$

Martin-Gay Interactive Videos   Watch the section lecture video and answer the following questions.

Objective A   **9.** In ▣ Example 1, how did we identify what part of the percent proportion 45 is?

Objective B   **10.** From ▣ Examples 4–6, what number is *always* part of the cross product equation of a percent proportion? ◯

See Video 6.3 ●

6.3   **Exercise Set** MyMathLab® ◯

Objective A Translating *Translate each to a proportion. Do not solve. See Examples 1 through 6.*

1. 98% of 45 is what number?

2. 92% of 30 is what number?

3. What number is 4% of 150?

4. What number is 7% of 175?

5. 14.3 is 26% of what number?

6. 1.2 is 47% of what number?

7. 35% of what number is 84?

8. 85% of what number is 520?

9. What percent of 400 is 70?

10. What percent of 900 is 216?

11. 8.2 is what percent of 82?

12. 9.6 is what percent of 96?

Objective B *Solve. See Example 7.*

13. 40% of 65 is what number?

14. 25% of 84 is what number?

15. What number is 18% of 105?

16. What number is 60% of 29?

Solve. See Examples 8 and 9.

17. 15% of what number is 90?

18. 55% of what number is 55?

19. 7.8 is 78% of what number?

20. 1.1 is 44% of what number?

Solve. See Examples 10 and 11.

21. What percent of 35 is 42?

22. What percent of 98 is 147?

23. 14 is what percent of 50?

24. 24 is what percent of 50?

Objectives A B Mixed Practice *Solve. See Examples 1 through 11.*

25. 3.7 is 10% of what number?

26. 7.4 is 5% of what number?

27. 2.4% of 70 is what number?

28. 2.5% of 90 is what number?

29. 160 is 16% of what number?

30. 30 is 6% of what number?

31. 394.8 is what percent of 188?

32. 550.4 is what percent of 172?

33. What number is 89% of 62?

34. What number is 53% of 130?

35. What percent of 6 is 2.7?

36. What percent of 5 is 1.6?

37. 140% of what number is 105?

38. 170% of what number is 221?

39. 1.8% of 48 is what number?

40. 7.8% of 24 is what number?

41. What percent of 800 is 4?

42. What percent of 500 is 3?

43. 3.5 is 2.5% of what number?

44. 9.18 is 6.8% of what number?

45. 20% of 48 is what number?

46. 75% of 14 is what number?

47. 2486 is what percent of 2200?

48. 9310 is what percent of 3800?

Review

Add or subtract the fractions. See Sections 3.4, 3.5, and 3.7.

49. $\dfrac{11}{16} + \dfrac{3}{16}$

50. $\dfrac{5}{8} - \dfrac{7}{12}$

51. $3\dfrac{1}{2} - \dfrac{11}{30}$

52. $2\dfrac{2}{3} + 4\dfrac{1}{2}$

Add or subtract the decimals. See Section 4.2.

53. $\begin{array}{r} 0.41 \\ +\,0.29 \\ \hline \end{array}$

54. $\begin{array}{r} 10.78 \\ 4.3 \\ +\;\,0.21 \\ \hline \end{array}$

55. $\begin{array}{r} 2.38 \\ -\,0.19 \\ \hline \end{array}$

56. $\begin{array}{r} 16.37 \\ -\;\,2.61 \\ \hline \end{array}$

Concept Extensions

57. Write a word statement for the proportion $\dfrac{x}{28} = \dfrac{25}{100}$. Use the phrase "what number" for "*x*."

58. Write a percent statement that translates to $\dfrac{16}{80} = \dfrac{20}{100}$.

Suppose you have finished solving four percent problems using proportions that you set up correctly. Check each answer to see if each makes the proportion a true proportion. If any proportion is not true, solve it to find the correct solution. See the Concept Checks in this section.

59. $\dfrac{a}{64} = \dfrac{25}{100}$

Is the amount equal to 17?

60. $\dfrac{520}{b} = \dfrac{65}{100}$

Is the base equal to 800?

61. $\dfrac{p}{100} = \dfrac{13}{52}$

Is the percent equal to 25 (25%)?

62. $\dfrac{36}{12} = \dfrac{p}{100}$

Is the percent equal to 50 (50%)?

Solve.

63. In your own words, describe how to identify the percent, the base, and the amount in a percent problem.

64. In your own words, explain how to use a proportion to solve a percent problem.

Solve. Round to the nearest tenth, if necessary.

65. What number is 22.3% of 53,862?

66. What percent of 110,736 is 88,542?

67. 8652 is 119% of what number?

Percent and Percent Problems

Answers

1. _____

2. _____

3. _____

4. _____

5. _____

6. _____

7. _____

8. _____

9. _____

10. _____

11. _____

12. _____

13. _____

14. _____

15. _____

16. _____

17. _____

18. _____

19. _____

20. _____

21. _____

22. _____

23. _____

24. _____

Write each number as a percent.

1. 0.12

2. 0.68

3. $\dfrac{1}{8}$

4. $\dfrac{5}{2}$

5. 5.2

6. 8

7. $\dfrac{3}{50}$

8. $\dfrac{11}{25}$

9. $7\dfrac{1}{2}$

10. $3\dfrac{1}{4}$

11. 0.03

12. 0.05

Write each percent as a decimal.

13. 65%

14. 31%

15. 8%

16. 7%

17. 142%

18. 400%

19. 2.9%

20. 6.6%

Write each percent as a decimal and as a fraction or mixed number in simplest form. (If necessary when writing as a decimal, round to the nearest thousandth.)

21. 3%

22. 5%

23. 5.25%

24. 12.75%

25. 38%   **26.** 45%   **27.** $12\frac{1}{3}\%$ **28.** $16\frac{2}{3}\%$

Solve each percent problem.

29. 12% of 70 is what number?   **30.** 36 is 36% of what number?

31. 212.5 is 85% of what number?   **32.** 66 is what percent of 55?

33. 23.8 is what percent of 85?   **34.** 38% of 200 is what number?

35. What number is 25% of 44?   **36.** What percent of 99 is 128.7?

37. What percent of 250 is 215?   **38.** What number is 45% of 84?

39. 42% of what number is 63?   **40.** 95% of what number is 58.9?

25. _____

26. _____

27. _____

28. _____

29. _____

30. _____

31. _____

32. _____

33. _____

34. _____

35. _____

36. _____

37. _____

38. _____

39. _____

40. _____

Objectives

A Solve Applications Involving Percent.

B Find Percent of Increase and Percent of Decrease.

Objective A Solving Applications Involving Percent

Percent is used in a variety of everyday situations. The next four examples show just a few ways that percent occurs in real-life settings. (Each of these examples shows two ways of solving these problems. If you studied Section 6.2 only, see *Method 1*. If you studied Section 6.3 only, see *Method 2*.)

The first example has to do with the Appalachian Trail, a hiking trail conceived by a forester in 1921 and diagrammed to the right.

Mount Katahdin, Maine

The Appalachian Trail

Springer Mountain, Georgia

Practice 1

If the total mileage of the Appalachian Trail is 2174, use the circle graph to determine the number of miles in the state of Virginia.

Appalachian Trail Mileage by State Percent

Georgia 4%
Maine 13%
North Carolina 4%
Tennessee 14%
New Hampshire 7%
Vermont 7%
Virginia 25%
Massachusetts 4%
Connecticut 2%
New York 4%
West Virginia 0.2%
New Jersey 3%
Pennsylvania 11%
Maryland 2%

Total miles: 2174
Note: The sum of the percents is 100.2% because of rounding.
Source: purebound.com

Example 1 The circle graph in the margin shows the Appalachian Trail mileage by state. If the total mileage of the trail is 2174, use the circle graph to determine the number of miles in the state of New York. Round to the nearest whole mile.

Solution: *Method 1.* First, we state the problem in words.

In words: What number   is   4%   of   2174?

Translate: n $=$ 4% $\cdot$ 2174

To solve for n, we find $4\% \cdot 2174$.

$n = 0.04 \cdot 2174$ Write 4% as a decimal.

$n = 86.96$ Multiply.

$n \approx 87$ Round to the nearest whole.

Rounded to the nearest whole mile, we have that approximately 87 miles of the Appalachian Trail are in New York state.

Method 2. State the problem in words; then translate.

In words: What number   is   4%   of   2174?

                amount   percent   base

Translate: amount → $\dfrac{a}{2174} = \dfrac{4}{100}$ ← percent
             base →

Next, we solve for a.

$a \cdot 100 = 2174 \cdot 4$ Set cross products equal.

$a \cdot 100 = 8696$ Multiply.

$\dfrac{a \cdot 100}{100} = \dfrac{8696}{100}$ Divide both sides by 100.

$a = 86.96$ Simplify.

$a \approx 87$ Round to the nearest whole.

Rounded to the nearest whole mile, we have that approximately 87 miles of the Appalachian Trail are in New York state.

Work Practice 1

Answer

1. 543.5 mi

Example 2 Finding Percent of Nursing Job Openings Due to Retirements

There is a worldwide shortage of nurses. It is expected that there will be 1,207,500 job openings for nurses in 2020. About 495,500 of these job openings will be to replace retiring nurses. What percent of job openings for nurses in 2020 will be due to retirements? Round to the nearest whole percent. (*Source:* Bureau of Labor Statistics)

Solution: *Method 1.* First, we state the problem in words.

In words: 495,500 is <u>what percent</u> of 1,207,500?

Translate: 495,500 = n · 1,207,500

Next, solve for n.

$$\frac{495,500}{1,207,500} = n \qquad \text{Divide 495,500 by 1,207,500, the number multiplied by } n.$$

$$0.41 \approx n \qquad \text{Divide and round to the nearest hundredth.}$$

$$41 \approx n \qquad \text{Write as a percent.}$$

In 2020, about 41% of job openings for nurses will be due to retirements.

Method 2.

In words: 495,500 is what percent of 1,207,500?

 amount percent base

Translate: amount → $\dfrac{495,500}{1,207,500} = \dfrac{p}{100}$ ← percent
base →

Next, solve for p.

$$495,500 \cdot 100 = 1,207,500 \cdot p \qquad \text{Set cross products equal.}$$

$$49,550,000 = 1,207,500 \cdot p \qquad \text{Multiply.}$$

$$\frac{49,550,000}{1,207,500} = p \qquad \text{Divide 49,550,000 by 1,207,500, the number multiplied by } p.$$

$$41 \approx p$$

In 2020, about 41% of job openings for nurses will be due to retirements.

■ Work Practice 2

Practice 2

In Florida, about 34,000 new nurses were recently needed and hired. If there are now 130,000 nurses, what percent of new nurses were needed in Florida? Round to the nearest whole percent. (*Source:* St. Petersburg Times and The Registered Nurse Population)

Example 3 Finding the Base Number of Absences

Mr. Buccaran, the principal at Slidell High School, counted 31 freshmen absent during a particular day. If this is 4% of the total number of freshmen, how many freshmen are there at Slidell High School?

Solution: *Method 1.* First we state the problem in words; then we translate.

In words: 31 is 4% of <u>what number</u>?

Translate: 31 = 4% · n

(Continued on next page)

Practice 3

The freshmen class of 775 students is 31% of all students at Euclid University. How many students go to Euclid University?

Answers
2. 26% **3.** 2500

Next, we solve for n.

$$31 = 0.04 \cdot n \qquad \text{Write 4\% as a decimal.}$$

$$\frac{31}{0.04} = n \qquad \text{Divide 31 by 0.04, the number multiplied by } n.$$

$$775 = n \qquad \text{Simplify.}$$

There are 775 freshmen at Slidell High School.

Method 2.   First we state the problem in words; then we translate.

In words: 31   is   4%   of   what number?

amount   percent   base

Translate: amount → $\dfrac{31}{b} = \dfrac{4}{100}$ ← percent
          base →

Next, we solve for b.

$$31 \cdot 100 = b \cdot 4 \qquad \text{Set cross products equal.}$$

$$3100 = b \cdot 4 \qquad \text{Multiply.}$$

$$\frac{3100}{4} = b \qquad \text{Divide 3100 by 4, the number multiplied by } b.$$

$$775 = b \qquad \text{Simplify.}$$

There are 775 freshmen at Slidell High School.

■ Work Practice 3

Practice 4

From 2010 to 2013, the number of full-time workers in the United States increased by approximately 5%. In 2010, the number of full-time workers was 110 million. (*Source:* U.S. Bureau of Labor Statistics)

a. Find the increase in the number of full-time workers from 2010 to 2013.

b. Find the total number of full-time workers in 2013.

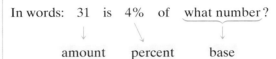

Example 4   Finding the Base Increase in Part-Time Workers

From 2003 to 2013, the number of part-time workers in the United States increased by 12%. In 2003, there were 24.5 million part-time workers. (*Source:* U.S. Bureau of Labor Statistics)

a. Find the increase in the number of part-time workers from 2003 to 2013.

b. Find the number of part-time workers in 2013.

Solution: *Method 1.*   First we find the increase in the number of part-time workers.

In words: What number   is   12%   of   24.5?

Translate: n =   12%   ·   24.5

Next, we solve for n.

$n = 0.12 \cdot 24.5$ Write 12% as a decimal.

$n = 2.94$ Multiply.

a. The increase in the number of part-time workers was 2.94 million.

b. This means that the number of part-time workers in 2013 was

$$\begin{array}{ccc} \text{Number of} & \text{Number of} & \text{Increase} \\ \text{part-time workers} = & \text{part-time workers} + & \text{in number of} \\ \text{in 2013} & \text{in 2003} & \text{part-time workers} \end{array}$$

$$= 24.5 \text{ million} + 2.94 \text{ million}$$

$$= 27.44 \text{ million}$$

Method 2.   First we find the increase in the number of part-time workers.

In words: What number   is   12%   of   24.5?

amount   percent   base

Translate: amount → $\dfrac{a}{24.5} = \dfrac{12}{100}$ ← percent   base →

Next, we solve for a.

$a \cdot 100 = 24.5 \cdot 12$ Set cross products equal.

$a \cdot 100 = 294$ Multiply.

$\dfrac{a \cdot 100}{100} = \dfrac{294}{100}$ Divide both sides by 100.

$a = 2.94$ Simplify.

a. The increase in the number of part-time workers was 2.94 million.

b. This means that the number of part-time workers in 2013 was

$$\begin{array}{ccc} \text{Number of} & \text{Number of} & \text{Increase} \\ \text{part-time workers} = & \text{part-time workers} + & \text{in number of} \\ \text{in 2013} & \text{in 2003} & \text{part-time workers} \end{array}$$

$$= 24.5 \text{ million} + 2.94 \text{ million}$$

$$= 27.44 \text{ million}$$

■ Work Practice 4

Objective B Finding Percent of Increase and Percent of Decrease

We often use percents to show how much an amount has increased or decreased.

Suppose that the population of a town is 10,000 people and then it increases by 2000 people. The **percent of increase** is

amount of increase → $\dfrac{2000}{10,000} = 0.2 = 20\%$ ← original amount

In general, we have the following.

Percent of Increase

$$\text{percent of increase} = \frac{\text{amount of increase}}{\text{original amount}}$$

Then write the quotient as a percent.

Practice 5

The number of people attending the local play, *Peter Pan,* increased from 285 on Friday to 333 on Saturday. Find the percent of increase in attendance. Round to the nearest tenth percent.

Helpful Hint   Make sure that this number is the original number and not the new number.

Example 5   Finding Percent of Increase

The number of applications for a mathematics scholarship at Yale increased from 34 to 45 in one year. What is the percent of increase? Round to the nearest whole percent.

Solution: First we find the amount of increase by subtracting the original number of applicants from the new number of applicants.

$$\text{amount of increase} = 45 - 34 = 11$$

The amount of increase is 11 applicants. To find the percent of increase,

$$\text{percent of increase} = \frac{\text{amount of increase}}{\text{original amount}} = \frac{11}{34} \approx 0.32 = 32\%$$

The number of applications increased by about 32%.

■ Work Practice 5

✓**Concept Check** A student is calculating the percent of increase in enrollment from 180 students one year to 200 students the next year. Explain what is wrong with the following calculations:

$$\text{Amount of increase} = 200 - 180 = 20$$

$$\text{Percent of increase} = \frac{20}{200} = 0.1 = 10\%$$

Suppose that your income was $300 a week and then it decreased by $30. The **percent of decrease** is

$$\begin{aligned}\text{amount of decrease} &\rightarrow \\ \text{original amount} &\rightarrow\end{aligned} \frac{\$30}{\$300} = 0.1 = 10\%$$

Percent of Decrease

$$\text{percent of decrease} = \frac{\text{amount of decrease}}{\text{original amount}}$$

Then write the quotient as a percent.

Practice 6

A town's population of 20,200 in 1995 decreased to 18,483 in 2005. What was the percent of decrease?

Example 6   Finding Percent of Decrease

In response to a decrease in sales, a company with 1500 employees reduces the number of employees to 1230. What is the percent of decrease?

Solution: First we find the amount of decrease by subtracting 1230 from 1500.

$$\text{amount of decrease} = 1500 - 1230 = 270$$

The amount of decrease is 270. To find the percent of decrease,

$$\text{percent of decrease} = \frac{\text{amount of decrease}}{\text{original amount}} = \frac{270}{1500} = 0.18 = 18\%$$

The number of employees decreased by 18%.

■ Work Practice 6

Answers

5. 16.8%   **6.** 8.5%

✓**Concept Check Answers**

To find the percent of increase, you have to divide the amount of increase (20) by the original amount (180); 10% decrease

✓**Concept Check** An ice cream stand sold 6000 ice cream cones last summer. This year the same stand sold 5400 cones. Was there a 10% increase, a 10% decrease, or neither? Explain.

Vocabulary, Readiness & Video Check

Martin-Gay Interactive Videos   *Watch the section lecture video and answer the following questions.*

Objective A 1. How do we interpret the answer 175,000 in 🎞 Example 1?

Objective B 2. In 🎞 Example 3, what does the improper fraction tell us?

See Video 6.4

6.4 Exercise Set MyMathLab®

Objective A *Solve. For Exercises 1 and 2, the solutions have been started for you. See Examples 1 through 4. If necessary, round percents to the nearest tenth and all other answers to the nearest whole.*

1. An inspector found 24 defective bolts during an inspection. If this is 1.5% of the total number of bolts inspected, how many bolts were inspected?

Start the solution:

1. UNDERSTAND the problem. Reread it as many times as needed.

Go to *Method 1* or *Method 2*.

Method 1.

2. TRANSLATE into an equation. (Fill in the boxes.)

$$24 \quad \text{is} \quad 1.5\% \quad \text{of} \quad \underline{\text{what number?}}$$
$$\downarrow \quad \downarrow \quad \downarrow \quad \downarrow \quad \downarrow$$
$$24 \quad \square \quad 1.5\% \quad \square \quad n$$

3. SOLVE for n. (See Example 3, Method 1, for help.)

4. INTERPRET. The total number of bolts inspected was _____.

Method 2.

2. TRANSLATE into a proportion. (Fill in the blanks with "amount" or "base.")

$$24 \quad \text{is} \quad 1.5\% \quad \text{of} \quad \underline{\text{what number?}}$$
$$\downarrow \qquad \searrow \qquad \downarrow$$
$$\underline{\quad\quad} \qquad \text{percent} \qquad \underline{\quad\quad}$$

$$\text{amount} \rightarrow \frac{\underline{\quad\quad}}{\underline{\quad\quad}} = \frac{1.5}{100} \leftarrow \text{percent}$$
$$\text{base} \rightarrow$$

3. SOLVE the proportion. (See Example 3, Method 2, for help.)

4. INTERPRET. The total number of bolts inspected was _____.

2. A day care worker found 28 children absent one day during an epidemic of chicken pox. If this was 35% of the total number of children attending the day care center, how many children attend this day care center?

Start the solution:

1. UNDERSTAND the problem. Reread it as many times as needed.

Go to *Method 1* or *Method 2*.

Method 1.

2. TRANSLATE into an equation. (Fill in the boxes.)

$$28 \quad \text{is} \quad 35\% \quad \text{of} \quad \underline{\text{what number?}}$$
$$\downarrow \quad \downarrow \quad \downarrow \quad \downarrow \quad \downarrow$$
$$28 \quad \square \quad 35\% \quad \square \quad n$$

3. SOLVE for n. (See Example 3, Method 1, for help.)

4. INTERPRET. The total number of children attending the day care center is _____.

Method 2.

2. TRANSLATE into a proportion. (Fill in the blanks with "amount" or "base.")

$$28 \quad \text{is} \quad 35\% \quad \text{of} \quad \underline{\text{what number?}}$$
$$\downarrow \qquad \searrow \qquad \downarrow$$
$$\underline{\quad\quad} \qquad \text{percent} \qquad \underline{\quad\quad}$$

$$\text{amount} \rightarrow \frac{\underline{\quad\quad}}{\underline{\quad\quad}} = \frac{35}{100} \leftarrow \text{percent}$$
$$\text{base} \rightarrow$$

3. SOLVE the proportion. (See Example 3, Method 2, for help.)

4. INTERPRET. The total number of children attending the day care center is _____.

3. The Total Gym® provides weight resistance through adjustments of incline. The minimum weight resistance is 4% of the weight of the person using the Total Gym. Find the minimum weight resistance possible for a 220-pound man. (*Source:* Total Gym)

4. The maximum weight resistance for the Total Gym is 60% of the weight of the person using it. Find the maximum weight resistance possible for a 220-pound man. (See Exercise 3 if needed.)

5. A student's cost for last semester at her community college was $2700. She spent $378 of that on books. What percent of last semester's college costs was spent on books?

6. Pierre Sampeau belongs to his local food cooperative, where he receives a percentage of what he spends each year as a dividend. He spent $3850 last year at the food cooperative store and received a dividend of $154. What percent of his total spending at the food cooperative did he receive as a dividend?

7. The United States' motion picture and television industry is made up of over 108,000 businesses. About 85% of these are small businesses with fewer than 10 employees. How many motion picture and television industry businesses have fewer than 10 employees? (*Source:* Motion Picture Association of America)

8. In 2013, 11% of the population of the United States and Canada were considered frequent moviegoers, purchasing movie tickets at least once per month. If the combined population of the United States and Canada was 351,300,000 in 2013, how many people were considered frequent moviegoers? (*Source:* Motion Picture Association of America)

9. The average wedding in the United States cost $28,400 in 2013. The average cost of a wedding reception venue was approximately $12,780. Determine the percent of an average wedding budget that is devoted to the reception venue. (*Source:* Wedding Channel/TheKnot.com)

10. Of the 64,700 veterinarians in private practice in the United States in 2013, approximately 34,291 were female. Determine the percent of female veterinarians in private practice in the United States. (*Source:* American Veterinary Medical Association)

11. A furniture company currently produces 6200 chairs per month. If production decreases by 8%, find the decrease and the new number of chairs produced each month.

12. The enrollment at a local college decreased by 5% over last year's enrollment of 7640. Find the decrease in enrollment and the current enrollment.

13. From 2010 to 2020, the number of people employed as physician assistants in the United States is expected to increase by 30%. The number of people employed as physician assistants in 2010 was 83,600. Find the predicted number of physician assistants in 2020. (*Source:* Bureau of Labor Statistics)

14. From 2007 to 2013, the number of U.S. veterinarians increased by 19%. The number of U.S. veterinarians in 2007 was about 83,700. Find the number of U.S. veterinarians in 2013. (*Source:* American Veterinary Medical Association)

Let's look at the populations of two states, Michigan and Louisiana. Their locations are shown on the partial U.S. map below. Round each answer to the nearest thousand. (Source: U.S. Census Bureau)

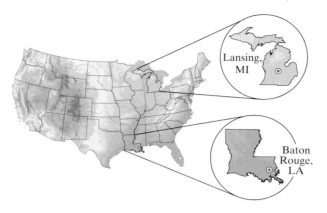

Lansing, MI

Baton Rouge, LA

15. In 2000, the population of Michigan was approximately 9940 thousand. If the population decreased by 0.4% between 2000 and 2013, find the population of Michigan in 2013.

16. In 2000, the population of Louisiana was approximately 4470 thousand. If the population increased about 3% between 2000 and 2013, find the population of Louisiana in 2013.

A popular extreme sport is snowboarding. Ski trails are marked with difficulty levels of easy ●, intermediate ■, difficult ◆, expert ◆◆, and other variations. Use this information for Exercises 17 and 18. Round each percent to the nearest whole. See Example 2.

17. At Keystone ski area in Colorado, 38 of the 131 total ski runs are rated intermediate. What percent of the runs are intermediate? (*Source:* Vail Resorts Management Company)

18. At Telluride ski area in Colorado, 29 of the 127 total ski trails are rated easy. What percent of the trails are easy? (*Source:* Telluride Ski & Golf Resort)

For each food described, find the percent of total calories from fat. If necessary, round to the nearest tenth percent. See Example 2.

19. Ranch dressing serving size of 2 tablespoons

	Calories
Total	40
From fat	20

20. Unsweetened cocoa powder serving size of 1 tablespoon

	Calories
Total	20
From fat	5

21.

Nutrition Facts
Serving Size 1 pouch (20g)
Servings Per Container 6

Amount Per Serving

| Calories | 80 |
| Calories from Fat | 10 |

	% Daily Value*
Total Fat 1g	**2%**
Sodium 45mg	**2%**
Total Carbohydrate 17g	**6%**
Sugars 9g	
Protein 0g	

| Vitamin C | 25% |

Not a significant source of saturated fat, cholesterol, dietary fiber, vitamin A, calcium and iron.

*Percent Daily Values are based on a 2,000 calorie diet.

Artificial Fruit Snacks

22.

Nutrition Facts
Serving Size ¼ cup (33g)
Servings Per Container About 9

Amount Per Serving

| Calories 190 | Calories from Fat 130 |

	% Daily Value
Total Fat 16g	**24%**
Saturated Fat 3g	**16%**
Cholesterol 0mg	**0%**
Sodium 135mg	**6%**
Total Carbohydrate 9g	**3%**
Dietary Fiber 1g	**5%**
Sugars 2g	
Protein 5g	

| Vitamin A 0% • Vitamin C 0% |
| Calcium 0% • Iron 8% |

Peanut Mixture

23.

Nutrition Facts

Serving Size 18 crackers (29g)
Servings Per Container About 9

Amount Per Serving

Calories 120 Calories from Fat 35

	% Daily Value*
Total Fat 4g	**6%**
Saturated Fat 0.5g	**3%**
Polyunsaturated Fat 0g	
Monounsaturated Fat 1.5g	
Cholesterol 0mg	**0%**
Sodium 220mg	**9%**
Total Carbohydrate 21g	**7%**
Dietary Fiber 2g	**7%**
Sugars 3g	
Protein 2g	

Vitamin A 0% • Vitamin C 0%

Calcium 2% • Iron 4%
Phosphorus 10%

Snack Crackers

24.

Nutrition Facts

Serving Size 28 crackers (31g)
Servings Per Container About 6

Amount Per Serving

Calories 130 Calories from Fat 35

	% Daily Value*
Total Fat 4g	**6%**
Saturated Fat 2g	**10%**
Polyunsaturated Fat 1g	
Monounsaturated Fat 1g	
Cholesterol 0mg	**0%**
Sodium 470mg	**20%**
Total Carbohydrate 23g	**8%**
Dietary Fiber 1g	**4%**
Sugars 4g	
Protein 2g	

Vitamin A 0% • Vitamin C 0%

Calcium 0% • Iron 2%

Snack Crackers

Solve. If necessary, round money amounts to the nearest cent and all other amounts to the nearest tenth. See Examples 1 through 4.

25. A family paid $26,250 as a down payment for a home. If this represents 15% of the price of the home, find the price of the home.

26. A banker learned that $842.40 is withheld from his monthly check for taxes and insurance. If this represents 18% of his total pay, find the total pay.

27. An owner of a repair service company estimates that for every 40 hours a repairperson is on the job, he can bill for only 78% of the hours. The remaining hours, the repairperson is idle or driving to or from a job. Determine the number of hours per 40-hour week the owner can bill for a repairperson.

28. A manufacturer of electronic components expects 1.04% of its products to be defective. Determine the number of defective components expected in a batch of 28,350 components. Round to the nearest whole component.

29. A car manufacturer announced that next year the price of a certain model of car will increase by 4.5%. This year the price is $19,286. Find the increase in price and the new price.

30. A union contract calls for a 6.5% salary increase for all employees. Determine the increase and the new salary that a worker currently making $58,500 under this contract can expect.

A popular extreme sport is artificial wall climbing. The photo shown is an artificial climbing wall. Exercises 31 and 32 are about the Footsloggers Climbing Tower in Boone, North Carolina.

31. A climber is resting at a height of 24 feet while on the Footsloggers Climbing Tower. If this is 60% of the tower's total height, find the height of the tower.

32. A group plans to climb the Footsloggers Climbing Tower at the group rate, once they save enough money. Thus far, $175 has been saved. If this is 70% of the total amount needed for the group, find the total price.

Solve.

33. Tuition for an Ohio resident at the Columbus campus of Ohio State University was $8406 in 2009. The tuition increased by 14.4% during the period from 2009 to 2013. Find the increase and the tuition for the 2013–2014 school year. Round the increase to the nearest whole dollar. (*Source:* The Ohio State University)

34. The population of Americans aged 65 and older was 43 million in 2012. That population is projected to increase by 69% by 2030. Find the increase and the projected 2030 population. (*Source:* Bureau of the Census)

35. From 2010–2011 to 2020–2021, the number of associate degrees awarded is projected to increase by 14.5%. If the number of associate degrees awarded in 2010–2011 was 888,000, find the increase and the projected number of associate degrees awarded in the 2020–2021 school year. (*Source:* National Center for Education Statistics)

36. From 2010–2011 to 2020–2021, the number of bachelor degrees awarded is projected to increase by 16%. If the number of bachelor degrees awarded in 2010–2011 was 1,703,000, find the increase and the projected number of bachelor degrees awarded in the 2020–2021 school year. (*Source:* National Center for Education Statistics)

Objective B *Find the amount of increase and the percent of increase. See Example 5.*

	Original Amount	New Amount	Amount of Increase	Percent of Increase
37.	50	80		
38.	8	12		
39.	65	117		
40.	68	170		

Find the amount of decrease and the percent of decrease. See Example 6.

	Original Amount	New Amount	Amount of Decrease	Percent of Decrease
41.	8	6		
42.	25	20		
43.	160	40		
44.	200	162		

Solve. Round percents to the nearest tenth, if necessary. See Examples 5 and 6.

45. There are 150 calories in a cup of whole milk and only 84 in a cup of skim milk. In switching to skim milk, find the percent of decrease in the number of calories per cup.

46. In reaction to a slow economy, the number of employees at a soup company decreased from 530 to 477. What was the percent of decrease in the number of employees?

47. The number of cable TV systems recently decreased from 10,845 to 10,700. Find the percent of decrease.

48. Before taking a typing course, Geoffry Landers could type 32 words per minute. By the end of the course, he was able to type 76 words per minute. Find the percent of increase.

49. In 1940, the average size of a privately owned farm in the United States was 174 acres. In a recent year, the average size of a privately owned farm in the United States had increased to 421 acres. Find the percent of increase. (*Source:* National Agricultural Statistics Service)

50. In 2006, the average size of a privately owned farm in the United States was 443 acres. In 2012, the average size of a privately owned farm in the United States had decreased to 421 acres. Find the percent of decrease. (*Source:* National Agricultural Statistics Service)

51. Total U.S. music industry revenues fell from $7.8 billion in 2009 to $7.0 billion in 2013. Find the percent of decrease. (*Source:* Recording Industry Association of America)

52. In 2008, the number of milk cow operations in the United States was 67,000. By 2012, this number had decreased to 58,000. What was the percent of decrease? (*Source:* National Agricultural Statistics Service)

53. In 2004, 21 million people in the U.S. subscribed to high-speed Internet service via their cable provider. By 2012, this number had increased to 50.3 million. What was the percent of increase? (*Source:* SNL Kagan)

54. In 2010, there were 1038 thousand high school teachers employed in the United States. This number is expected to increase to 1110 thousand teachers in 2020. What is the percent of increase? (*Source:* Bureau of Labor Statistics)

55. Between 2000 and July 2013, the number of indoor cinema sites in the United States decreased from 6550 to 5317. Find the percent of decrease. (*Source:* National Association of Theater Owners)

56. In 2003, the average price of a cinema ticket was $6.03. By 2013, this price had increased to $8.13. What was the percent of increase? (*Source:* National Association of Theatre Owners)

57. In 2010, there were 64,400 dieticians employed in the United States. This number is expected to increase to 77,100 dieticians in 2020. What is the percent of increase? (*Source:* Bureau of Labor Statistics)

58. In 2010, there were 242,900 coaches and scouts employed in the United States. This number is expected to increase to 314,300 in 2020. What is the percent of increase? (*Source:* Bureau of Labor Statistics)

59. The number of cell phone sites in the United States was 195,613 in 2006. By the beginning of 2014, the number of cell sites had increased to 304,360. Find the percent of increase. (*Source:* CTIA—The Wireless Association)

60. The population of Japan is expected to decrease from 127,050 thousand in 2014 to 100,600 thousand in 2050. Find the percent of decrease. (*Source:* Department of Population Dynamics Research)

Review

Perform each indicated operation. See Sections 4.2 and 4.3.

61. $\begin{array}{r} 0.12 \\ \times\ 38 \\ \hline \end{array}$

62. $\begin{array}{r} 42 \\ \times\ 0.7 \\ \hline \end{array}$

63. $9.20 + 1.98$

64. $46 + 7.89$

65. $78 - 19.46$

66. $64.80 - 10.72$

Concept Extensions

67. If a number is increased by 100%, how does the increased number compare with the original number? Explain your answer.

68. In your own words, explain what is wrong with the following statement: "Last year we had 80 students attend. This year we have a 50% increase or a total of 160 students attending."

Explain what errors were made by each student when solving percent of increase or decrease problems and then correct the errors. See the Concept Checks in this section.

The population of a certain rural town was 150 in 1980, 180 in 1990, and 150 in 2000.

69. Find the percent of increase in population from 1980 to 1990.

Miranda's solution: Percent of increase $= \dfrac{30}{180} = 0.1\overline{6} \approx 16.7\%$

70. Find the percent of decrease in population from 1990 to 2000.

Jeremy's solution: Percent of decrease $= \dfrac{30}{150} = 0.20 = 20\%$

71. The percent of increase from 1980 to 1990 is the same as the percent of decrease from 1990 to 2000. True or false?

Chris's answer: True because they had the same amount of increase as the amount of decrease.

72. Refer to Exercises 49 and 50. They are similar except that one asks us to find the percent of increase in the size of U.S. privately owned farms and one asks us to find the percent of decrease. In your own words, explain how these can both be correct.

6.5 Percent and Problem Solving: Sales Tax, Commission, and Discount

Objective A Calculating Sales Tax and Total Price

Percents are frequently used in the retail trade. For example, most states charge a tax on certain items when purchased. This tax is called a **sales tax,** and retail stores collect it for the state. Sales tax is almost always stated as a percent of the purchase price.

A 9% sales tax rate on a purchase of a $10 calculator gives a sales tax of

sales tax $= 9\%$ of $\$10 = 0.09 \cdot \$10.00 = \$0.90$

Objectives

A Calculate Sales Tax and Total Price.

B Calculate Commissions.

C Calculate Discount and Sale Price.

The total price to the customer would be

$$\underbrace{\text{purchase price}}_{\downarrow} \quad \underbrace{\text{plus}}_{\downarrow} \quad \underbrace{\text{sales tax}}_{\downarrow}$$

$$\$10.00 \qquad + \qquad \$0.90 = \$10.90$$

This example suggests the following equations:

Sales Tax and Total Price

$$\text{sales tax} = \text{tax rate} \cdot \text{purchase price}$$
$$\text{total price} = \text{purchase price} + \text{sales tax}$$

In this section we round dollar amounts to the nearest cent.

Practice 1

If the sales tax rate is 8.5%, what is the sales tax and the total amount due on a $59.90 Goodgrip tire? (Round the sales tax to the nearest cent.)

Example 1 Finding Sales Tax and Purchase Price

Find the sales tax and the total price on the purchase of an $85.50 atlas in a city where the sales tax rate is 7.5%.

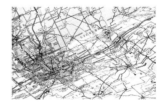

Solution: The purchase price is $85.50 and the tax rate is 7.5%.

$$\text{sales tax} = \text{tax rate} \cdot \text{purchase price}$$
$$\downarrow \qquad \downarrow \qquad \nearrow$$
$$\text{sales tax} = 7.5\% \cdot \$85.50$$
$$= 0.075 \cdot \$85.50 \quad \text{\small Write 7.5\% as a decimal.}$$
$$\approx \$6.41 \quad \text{\small Round to the nearest cent.}$$

Thus, the sales tax is $6.41. Next, find the total price.

$$\text{total price} = \text{purchase price} + \text{sales tax}$$
$$\downarrow \qquad \downarrow \qquad \swarrow$$
$$\text{total price} = \$85.50 + \$6.41$$
$$= \$91.91$$

The sales tax on $85.50 is $6.41, and the total price is $91.91.

■ Work Practice 1

✓**Concept Check** The purchase price of a textbook is $50 and sales tax is 10%. If you are told by the cashier that the total price is $75, how can you tell that a mistake has been made?

Answer

1. tax: $5.09; total: $64.99

✓ **Concept Check Answer**

Since $10\% = \dfrac{1}{10}$, the sales tax is $\dfrac{\$50}{10} = \$5$. The total price should have been $55.

Example 2 Finding a Sales Tax Rate

The sales tax on a $406 Sony flat-screen, digital, 27-inch television is $34.51. Find the sales tax rate.

Solution: Let *r* represent the unknown sales tax rate. Then

sales tax = tax rate · purchase price

$$\$34.51 = r \cdot \$406$$

$$\frac{34.51}{406} = \frac{r \cdot 406}{406} \quad \text{Divide both sides by 406.}$$

$$0.085 = r \quad \text{Simplify.}$$

$$8.5\% = r \quad \text{Write 0.085 as a percent.}$$

The sales tax rate is 8.5%.

■ Work Practice 2

Practice 2

The sales tax on an $18,500 automobile is $1665. Find the sales tax rate.

Objective B Calculating Commissions

A **wage** is payment for performing work. Hourly wage, commissions, and salary are some of the ways wages can be paid. Many people who work in sales are paid a commission. An employee who is paid a **commission** is paid a percent of his or her total sales.

Commission

commission = commission rate · sales

Example 3 Finding the Amount of Commission

Sherry Souter, a real estate broker for Wealth Investments, sold a house for $214,000 last week. If her commission rate is 1.5% of the selling price of the home, find the amount of her commission.

Solution:

commission	=	commission rate	·	sales	
commission	=	1.5%	·	$214,000	
	=	0.015	·	$214,000	Write 1.5% as 0.015.
	=	$3210			Multiply.

Her commission on the house is $3210.

■ Work Practice 3

Practice 3

A sales representative for Office Product Copiers sold $47,632 worth of copier equipment and supplies last month. What is his commission for the month if he is paid a commission rate of 6.6% of his total sales for the month?

Answers
2. 9% **3.** $3143.71

Practice 4

A salesperson earns $645 for selling $4300 worth of appliances. Find the commission rate.

Example 4   Finding a Commission Rate

A salesperson earned $1560 for selling $13,000 worth of electronics equipment. Find the commission rate.

Solution: Let r stand for the unknown commission rate. Then

$$\text{commission} = \text{commission rate} \cdot \text{sales}$$

$1560	=	r	$\cdot$	$13,000

$$\dfrac{1560}{13,000} = r \quad \text{Divide 1560 by 13,000, the number multiplied by } r.$$

$$0.12 = r \quad \text{Simplify.}$$

$$12\% = r \quad \text{Write 0.12 as a percent.}$$

The commission rate is 12%.

■ Work Practice 4

Objective C Calculating Discount and Sale Price

Suppose that an item that normally sells for $40 is on sale for 25% off. This means that the **original price** of $40 is reduced, or **discounted,** by 25% of $40, or $10. The **discount rate** is 25%, the **amount of discount** is $10, and the **sale price** is $40 − $10, or $30. Study the diagram below to visualize these terms.

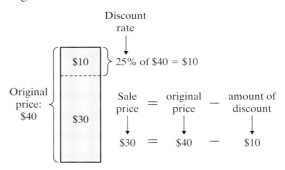

To calculate discounts and sale prices, we can use the following equations:

Discount and Sale Price

$$\text{amount of discount} = \text{discount rate} \cdot \text{original price}$$
$$\text{sale price} = \text{original price} - \text{amount of discount}$$

Practice 5

A discontinued washer and dryer combo is advertised on sale for 35% off the regular price of $700. Find the amount of discount and the sale price.

Example 5   Finding a Discount and a Sale Price

An electric rice cooker that normally sells for $65 is on sale for 25% off. What is the amount of discount and what is the sale price?

Solution: First we find the amount of discount, or simply the discount.

$$\text{amount of discount} = \text{discount rate} \cdot \text{original price}$$

amount of discount	=	25%	$\cdot$	$65

$$= 0.25 \cdot \$65 \quad \text{Write 25\% as 0.25.}$$

$$= \$16.25 \quad \text{Multiply.}$$

Answers
4. 15%   **5.** $245; $455

The discount is $16.25. Next, find the sale price.

sale price = original price − discount

$\qquad\downarrow\qquad\qquad\downarrow\qquad\qquad\downarrow$

sale price = $65   − $16.25

$\qquad\qquad\quad = \$48.75$ Subtract.

The sale price is $48.75.

■ Work Practice 5

Vocabulary, Readiness & Video Check

Use the choices below to fill in each blank. Some choices may be used more than once.

amount of discount   sale price   sales tax

commission   total price

1. _____ = tax rate · purchase price

2. _____ = purchase price + sales tax

3. _____ = commission rate · sales

4. _____ = discount rate · original price

5. _____ = original price − amount of discount

6. sale price = original price − _____

Martin-Gay Interactive Videos   *Watch the section lecture video and answer the following questions.*

Objective A **7.** In ▦ Example 1, what is our first step after translating the problem into an equation?

Objective B **8.** What is our final step in solving ▦ Example 2? ◯

Objective C **9.** In the lecture before ▦ Example 3, since both equations shown involve the "amount of discount," how can the two equations be combined into one equation? ◯

See Video 6.5 ●

6.5 Exercise Set MyMathLab® ▶

Objective A *Solve. See Examples 1 and 2.*

1. What is the sales tax on a jacket priced at $150 if the sales tax rate is 5%?

2. If the sales tax rate is 6%, find the sales tax on a microwave oven priced at $188.

3. The purchase price of a camcorder is $799. What is the total price if the sales tax rate is 7.5%?

4. A stereo system has a purchase price of $426. What is the total price if the sales tax rate is 8%?

5. A new large-screen television has a purchase price of $4790. If the sales tax on this purchase is $335.30, find the sales tax rate.

6. The sales tax on the purchase of a $6800 used car is $374. Find the sales tax rate.

7. The sales tax on a table saw is $10.20.

   a. What is the purchase price of the table saw (before tax) if the sales tax rate is 8.5%? (*Hint:* Use the sales tax equation and insert the replacement values.)

   b. Find the total price of the table saw.

8. The sales tax on a one-half-carat diamond ring is $76.

   a. Find the purchase price of the ring (before tax) if the sales tax rate is 9.5%. (See the hint for Exercise 7a.)

   b. Find the total price of the ring.

9. A gold and diamond bracelet sells for $1800. Find the sales tax and the total price if the sales tax rate is 6.5%.

10. The purchase price of a personal computer is $1890. If the sales tax rate is 8%, what is the sales tax and the total price?

11. The sales tax on the purchase of a futon is $24.25. If the tax rate is 5%, find the purchase price of the futon.

12. The sales tax on the purchase of a TV-DVD combination is $32.85. If the tax rate is 9%, find the purchase price of the TV-DVD.

13. The sales tax is $98.70 on a stereo sound system purchase of $1645. Find the sales tax rate.

14. The sales tax is $103.50 on a necklace purchase of $1150. Find the sales tax rate.

15. A cell phone costs $210, a battery recharger costs $15, and batteries cost $5. What is the sales tax and total price for purchasing these items if the sales tax rate is 7%?

16. Ms. Warner bought a blouse for $35, a skirt for $55, and a blazer for $95. Find the sales tax and the total price she paid, given a sales tax rate of 6.5%.

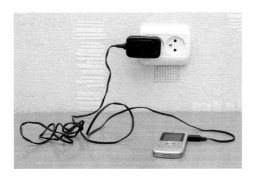

Objective B *Solve. See Examples 3 and 4.*

17. A sales representative for a large furniture warehouse is paid a commission rate of 4%. Find her commission if she sold $1,329,401 worth of furniture last year.

18. Rosie Davis-Smith is a beauty consultant for a home cosmetic business. She is paid a commission rate of 12.8%. Find her commission if she sold $1638 in cosmetics last month.

19. A salesperson earned a commission of $1380.40 for selling $9860 worth of paper products. Find the commission rate.

20. A salesperson earned a commission of $3575 for selling $32,500 worth of books to various bookstores. Find the commission rate.

21. How much commission will Jack Pruet make on the sale of a $325,900 house if he receives 1.5% of the selling price?

22. Frankie Lopez sold $9638 of jewelry this week. Find her commission for the week if she receives a commission rate of 5.6%.

23. A real estate agent earned a commission of $5565 for selling a house. If his rate is 3%, find the selling price of the house. (*Hint:* Use the commission equation and insert the replacement values.)

24. A salesperson earned $1750 for selling fertilizer. If her commission rate is 7%, find the selling price of the fertilizer. (See the hint for Exercise 23.)

Objective C *Find the amount of discount and the sale price. See Example 5.*

	Original Price	Discount Rate	Amount of Discount	Sale Price
25.	$89	10%		
26.	$74	20%		
27.	$196.50	50%		
28.	$110.60	40%		
29.	$410	35%		
30.	$370	25%		
31.	$21,700	15%		
32.	$17,800	12%		

33. A $300 fax machine is on sale for 15% off. Find the amount of discount and the sale price.

34. A $4295 designer dress is on sale for 30% off. Find the amount of discount and the sale price.

Objectives A B **Mixed Practice** *Complete each table.*

	Purchase Price	Tax Rate	Sales Tax	Total Price
35.	$305	9%		
36.	$243	8%		
37.	$56	5.5%		
38.	$65	8.4%		

	Sale	Commission Rate	Commission
39.	$235,800	3%	
40.	$195,450	5%	
41.	$17,900		$1432
42.	$25,600		$2304

Review

Multiply. See Sections 4.3 and 4.5.

43. $2000 \cdot \dfrac{3}{10} \cdot 2$

44. $500 \cdot \dfrac{2}{25} \cdot 3$

45. $400 \cdot \dfrac{3}{100} \cdot 11$

46. $1000 \cdot \dfrac{1}{20} \cdot 5$

47. $600 \cdot 0.04 \cdot \dfrac{2}{3}$

48. $6000 \cdot 0.06 \cdot \dfrac{3}{4}$

Concept Extensions

Solve. See the Concept Check in this section.

49. Your purchase price is $68 and the sales tax rate is 9.5%. Round each amount and use the rounded amounts to estimate the total price. Choose the best estimate.
 a. $105   **b.** $58   **c.** $93   **d.** $77

50. Your purchase price is $200 and the tax rate is 10%. Choose the best estimate of the total price.
 a. $190   **b.** $210   **c.** $220   **d.** $300

Tipping

One very useful application of percent is mentally calculating a tip. Recall that to find 10% of a number, simply move the decimal point one place to the left. To find 20% of a number, just double 10% of the number. To find 15% of a number, find 10% and then add to that number half of the 10% amount. Mentally fill in the chart below. To do so, start by rounding the bill amount to the nearest dollar.

| | **Tipping Chart** | | | |
	Bill Amount	**10%**	**15%**	**20%**
51.	$40.21			
52.	$15.89			
53.	$72.17			
54.	$9.33			

55. Suppose that the original price of a shirt is $50. Which is better, a 60% discount or a discount of 30% followed by a discount of 35% of the reduced price? Explain your answer.

56. Which is better, a 30% discount followed by an additional 25% off or a 20% discount followed by an additional 40% off? To see, suppose an item costs $100 and calculate each discounted price. Explain your answer.

57. A diamond necklace sells for $24,966. If the tax rate is 7.5%, find the total price.

58. A house recently sold for $562,560. The commission rate on the sale is 5.5%. If the real estate agent is to receive 60% of the commission, find the amount received by the agent.

6.6 **Percent and Problem Solving: Interest**

Objectives

A Calculate Simple Interest.

B Calculate Compound Interest.

C Calculate Monthly Payments.

Objective A Calculating Simple Interest

Interest is money charged for using other people's money. When you borrow money, you pay interest. When you loan or invest money, you earn interest. The money borrowed, loaned, or invested is called the **principal amount,** or simply **principal.** Interest is normally stated in terms of a percent of the principal for a given period of time. The **interest rate** is the percent used in computing the interest. Unless stated otherwise, *the rate is understood to be per year.* When the interest is computed on the original principal, it is called **simple interest.** Simple interest is calculated using the following equation:

Simple Interest

Simple Interest = Principal · Rate · Time
$$I = P \cdot R \cdot T$$

where the rate is understood to be per year and time is in years.

Example 1 Finding Simple Interest

Find the simple interest after 2 years on $500 at an interest rate of 12%.

Solution: In this example, $P = \$500$, $R = 12\%$, and $T = 2$ years. Replace the variables in the formula $I = PRT$ with values.

$$I = P \cdot R \cdot T$$
$$I = \$500 \cdot 12\% \cdot 2 \quad \text{Let } P = \$500, R = 12\%, \text{ and } T = 2.}$$
$$= \$500 \cdot (0.12) \cdot 2 \quad \text{Write 12\% as a decimal.}$$
$$= \$120 \quad \text{Multiply.}$$

The simple interest is $120.

Work Practice 1

Practice 1

Find the simple interest after 5 years on $875 at an interest rate of 7%.

If time is not given in years, we need to convert the given time to years.

Example 2 Finding Simple Interest

Ivan Borski borrowed $2400 at 10% simple interest for 8 months to buy a used Toyota Corolla. Find the simple interest he paid.

Solution: Since there are 12 months in a year, we first find what part of a year 8 months is.

$$8 \text{ months} = \frac{8}{12} \text{ year} = \frac{2}{3} \text{ year}$$

Now we find the simple interest.

simple interest	=	principal	·	rate	·	time
↓		↓		↓		↓
simple interest	=	$2400	·	10%	·	$\frac{2}{3}$
	=	$2400	·	0.10	·	$\frac{2}{3}$
	=	$160				

The interest on Ivan's loan is $160.

Work Practice 2

Practice 2

A student borrowed $1500 for 9 months on her credit card at a simple interest rate of 20%. How much interest did she pay?

✓**Concept Check** Suppose in Example 2 you had obtained an answer of $16,000. How would you know that you had made a mistake in this problem?

When money is borrowed, the borrower pays the original amount borrowed, or the principal, as well as the interest. When money is invested, the investor receives the original amount invested, or the principal, as well as the interest. In either case, the **total amount** is the sum of the principal and the interest.

Finding the Total Amount of a Loan or Investment

total amount (paid or received) = principal + interest

Answers
1. $306.25   **2.** $225

✓**Concept Check Answer**
$16,000 is too much interest.

Practice 3

If $2100 is borrowed at a simple interest rate of 13% for 6 months, find the total amount paid.

Example 3   Finding the Total Amount of an Investment

An accountant invested $2000 at a simple interest rate of 10% for 2 years. What total amount of money will she have from her investment in 2 years?

Solution: First we find her interest.

$$I = P \cdot R \cdot T$$
$$= \$2000 \cdot (0.10) \cdot 2 \quad \text{Let } P = \$2000, R = 10\% \text{ or } 0.10, \text{ and } T = 2.$$
$$= \$400$$

The interest is $400.

Next, we add the interest to the principal.

$$\text{total amount} = \text{principal} + \text{interest}$$
$$\downarrow \qquad\qquad \downarrow \qquad\qquad \downarrow$$
$$\text{total amount} = \$2000 + \$400$$
$$= \$2400$$

After 2 years, she will have a total amount of $2400.

■ Work Practice 3

✓**Concept Check** Which investment would earn more interest: an amount of money invested at 8% interest for 2 years, or the same amount of money invested at 8% for 3 years? Explain.

Objective B   Calculating Compound Interest

Recall that simple interest depends on the original principal only. Another type of interest is compound interest. **Compound interest** is computed not only on the principal, but also on the interest already earned in previous compounding periods. Compound interest is used more often than simple interest.

Let's see how compound interest differs from simple interest. Suppose that $2000 is invested at 7% interest **compounded annually** for 3 years. This means that interest is added to the principal at the end of each year and that next year's interest is computed on this new amount. In this section, we round dollar amounts to the nearest cent.

	Amount at Beginning of Year	Principal	·	Rate	·	Time	= Interest	Amount at End of Year
1st year	$2000	$2000	·	0.07	·	1	= $140	$2000 + 140 = $2140
2nd year	$2140	$2140	·	0.07	·	1	= $149.80	$2140 + 149.80 = $2289.80
3rd year	$2289.80	$2289.80	·	0.07	·	1	= $160.29	$2289.80 + 160.29 = $2450.09

The compound interest earned can be found by

$$\text{total amount} - \text{original principal} = \text{compound interest}$$
$$\downarrow \qquad\qquad \downarrow \qquad\qquad \downarrow$$
$$\$2450.09 - \$2000 = \$450.09$$

The simple interest earned would have been

$$\text{principal} \cdot \text{rate} \cdot \text{time} = \text{interest}$$
$$\downarrow \qquad \downarrow \qquad \downarrow \qquad \downarrow$$
$$\$2000 \cdot 0.07 \cdot 3 = \$420$$

Answer

3. $2236.50

✓**Concept Check Answer**

8% for 3 years. Since the interest rate is the same, the longer you keep the money invested, the more interest you earn.

Since compound interest earns "interest on interest," compound interest earns more than simple interest.

Computing compound interest using the method just shown can be tedious. We can use a calculator and the compound interest formula below to compute compound interest more quickly.

Compound Interest Formula

The total amount A in an account is given by

$$A = P\left(1 + \frac{r}{n}\right)^{n \cdot t}$$

where P is the principal, r is the interest rate written as a decimal, t is the length of time in years, and n is the number of times compounded per year.

Example 4 $1800 is invested at 2% interest compounded annually. Find the total amount after 3 years.

Solution: "Compounded annually" means 1 time a year, so $n = 1$. Also, $P = \$1800$, $r = 2\% = 0.02$, and $t = 3$ years.

$$A = P\left(1 + \frac{r}{n}\right)^{n \cdot t}$$

$$= 1800\left(1 + \frac{0.02}{1}\right)^{1 \cdot 3}$$

$$= 1800(1.02)^3$$

$$\approx 1910.17 \qquad \text{Round to 2 decimal places.}$$

> **Helpful Hint** Remember order of operations. **First** evaluate $(1.02)^3$; then multiply by 1800.

The total amount at the end of 3 years is about $1910.17.

Work Practice 4

Example 5 Finding Total Amount Received from an Investment

$4000 is invested at 5.3% compounded quarterly for 10 years. Find the total amount at the end of 10 years.

Solution: "Compounded quarterly" means 4 times a year, so $n = 4$. Also, $P = \$4000$, $r = 5.3\% = 0.053$, and $t = 10$ years.

$$A = P\left(1 + \frac{r}{n}\right)^{n \cdot t}$$

$$= 4000\left(1 + \frac{0.053}{4}\right)^{4 \cdot 10}$$

$$= 4000(1.01325)^{40}$$

$$\approx 6772.12$$

The total amount after 10 years is about $6772.12.

Work Practice 5

Note: Part of the compound interest formula, $\left(1 + \dfrac{r}{n}\right)^{n \cdot t}$, is called the **compound interest factor.** Appendix A.7 contains a table of various calculated compound interest factors. Another way to calculate the total amount, A, in the compound interest formula is to multiply the principal, P, by the appropriate compound interest factor found in Appendix A.7.

Practice 4

$3000 is invested at 4% interest compounded annually. Find the total amount after 6 years.

Practice 5

$5500 is invested at $6\frac{1}{4}\%$ compounded *daily* for 5 years. Find the total amount at the end of 5 years. (Use 1 year = 365 days.)

Answers

4. $3795.96   **5.** $7517.41

The Calculator Explorations box below shows how compound interest factors are calculated.

Objective C Calculating a Monthly Payment

We conclude this section with a method to find the monthly payment on a loan.

Finding the Monthly Payment on a Loan

$$\text{monthly payment} = \frac{\text{principal} + \text{interest}}{\text{total number of payments}}$$

Practice 6

Find the monthly payment on a $3000 3-year loan if the interest on the loan is $1123.58.

Answer

6. $114.54

Example 6 Finding a Monthly Payment

Find the monthly payment on a $2000 loan for 2 years. The interest on the 2-year loan is $435.88.

Solution: First we determine the total number of monthly payments. The loan is for 2 years. Since there are 12 months per year, the number of payments is $2 \cdot 12$, or 24. Now we calculate the monthly payment.

$$\text{monthly payment} = \frac{\text{principal} + \text{interest}}{\text{total number of payments}}$$

$$\text{monthly payment} = \frac{\$2000 + \$435.88}{24}$$

$$\approx \$101.50.$$

The monthly payment is about $101.50.

■ Work Practice 6

 Calculator Explorations Compound Interest Factor

A compound interest factor may be found by using your calculator and evaluating the formula

$$\textbf{compound interest factor} = \left(1 + \frac{r}{n}\right)^{n \cdot t}$$

where r is the interest rate, t is the time in years, and n is the number of times compounded per year. For example, the compound interest factor for 10 years at 8% compounded semiannually is about 2.19112. Let's find this factor by evaluating the compound interest factor formula when $r = 8\%$ or 0.08, $t = 10$, and $n = 2$ (compounded semiannually means 2 times per year). Thus,

$$\text{compound interest factor} = \left(1 + \frac{0.08}{2}\right)^{2 \cdot 10}$$

$$\text{or } \left(1 + \frac{0.08}{2}\right)^{20}$$

To evaluate, press the keys

(1 + 0.08 ÷ 2) y^x or ^ 20 and then = or ENTER . The display will read 2.1911231 . Rounded to 5 decimal places, this is 2.19112.

Find the compound interest factors. Use the table in Appendix A.7 to check your answers. For Exercises 1–4, round to 5 decimal places. For Exercises 5 and 6, round to 2 decimal places.

1. 5 years, 9%, compounded quarterly
2. 15 years, 14%, compounded daily
3. 20 years, 11%, compounded annually
4. 1 year, 7%, compounded semiannually
5. Find the total amount after 4 years when $500 is invested at 6% compounded quarterly. (Multiply the appropriate compound interest factor by $500.)
6. Find the total amount for 19 years when $2500 is invested at 5% compounded daily.

Vocabulary, Readiness & Video Check

Use the choices below to fill in each blank. Choices may be used more than once.

    total amount   simple   principal amount   compound

1. To calculate _____ interest, use $I = P \cdot R \cdot T$.

2. To calculate _____ interest, use $A = P\left(1 + \dfrac{r}{n}\right)^{n \cdot t}$.

3. _____ interest is computed not only on the original principal, but also on interest already earned in previous compounding periods.

4. When interest is computed on the original principal only, it is called _____ interest.

5. _____ (paid or received) = principal + interest.

6. The _____ is the money borrowed, loaned, or invested.

Martin-Gay Interactive Videos Watch the section lecture video and answer the following questions.

See Video 6.6

Objective A **7.** Complete this statement based on the lecture before ⊞ Example 1: Simple interest is charged on the _____ only.

Objective B **8.** In ⊞ Example 2, how often is the interest compounded and what number does this translate to in the formula?

Objective C **9.** In ⊞ Example 3, how was the denominator of 48 determined?

6.6 Exercise Set MyMathLab®

Objective A *Find the simple interest. See Examples 1 and 2.*

	Principal	Rate	Time
1.	$200	8%	2 years
3.	$160	11.5%	4 years
5.	$5000	10%	$1\frac{1}{2}$ years
7.	$375	18%	6 months
9.	$2500	16%	21 months

	Principal	Rate	Time
2.	$800	9%	3 years
4.	$950	12.5%	5 years
6.	$1500	14%	$2\frac{1}{4}$ years
8.	$775	15%	8 months
10.	$1000	10%	18 months

Solve. See Examples 1 through 3.

11. A company borrows $162,500 for 5 years at a simple interest rate of 12.5%. Find the interest paid on the loan and the total amount paid back.

12. $265,000 is borrowed to buy a house. If the simple interest rate on the 30-year loan is 8.25%, find the interest paid on the loan and the total amount paid back.

13. A money market fund advertises a simple interest rate of 9%. Find the total amount received on an investment of $5000 for 15 months.

14. The Real Service Company takes out a 270-day (9-month) short-term, simple interest loan of $4500 to finance the purchase of some new equipment. If the interest rate is 14%, find the total amount that the company pays back.

15. Marsha borrows $8500 and agrees to pay it back in 4 years. If the simple interest rate is 17%, find the total amount she pays back.

16. An 18-year-old is given a high school graduation gift of $2000. If this money is invested at 8% simple interest for 5 years, find the total amount.

Objective B *Find the total amount in each compound interest account. See Examples 4 and 5.*

17. $6150 is compounded semiannually at a rate of 14% for 15 years.

18. $2060 is compounded annually at a rate of 15% for 10 years.

19. $1560 is compounded daily at a rate of 8% for 5 years.

20. $1450 is compounded quarterly at a rate of 10% for 15 years.

21. $10,000 is compounded semiannually at a rate of 9% for 20 years.

22. $3500 is compounded daily at a rate of 8% for 10 years.

23. $2675 is compounded annually at a rate of 9% for 1 year.

24. $6375 is compounded semiannually at a rate of 10% for 1 year.

25. $2000 is compounded annually at a rate of 8% for 5 years.

26. $2000 is compounded semiannually at a rate of 8% for 5 years.

27. $2000 is compounded quarterly at a rate of 8% for 5 years.

28. $2000 is compounded daily at a rate of 8% for 5 years.

Objective C *Solve. See Example 6.*

29. A college student borrows $1500 for 6 months to pay for a semester of school. If the interest is $61.88, find the monthly payment.

30. Jim Tillman borrows $1800 for 9 months. If the interest is $148.90, find his monthly payment.

31. $20,000 is borrowed for 4 years. If the interest on the loan is $10,588.70, find the monthly payment.

32. $105,000 is borrowed for 15 years. If the interest on the loan is $181,125, find the monthly payment.

Review

Find the perimeter of each figure. See Section 1.3.

33.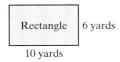
Rectangle | 6 yards
10 yards

34.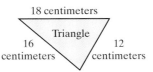
18 centimeters
Triangle
16 centimeters   12 centimeters

35.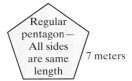
Regular pentagon— All sides are same length / 7 meters

36.
Square | 21 miles

Concept Extensions

37. Explain how to look up a compound interest factor in the compound interest table.

38. Explain how to find the amount of interest in a compounded account.

39. Compare the following accounts: Account 1: $1000 is invested for 10 years at a simple interest rate of 6%. Account 2: $1000 is compounded semiannually at a rate of 6% for 10 years. Discuss how the interest is computed for each account. Determine which account earns more interest. Why?

Chapter 6 Group Activity

Fastest-Growing Occupations

According to U.S. Bureau of Labor Statistics projections, the careers listed below are the top ten fastest-growing jobs ranked by expected percent of increase through the year 2020. (*Source:* Bureau of Labor Statistics)

	Occupation	Employment in 2010	Percent Change	Expected Employment in 2020
1	Personal care aides	861,000	70.5%	
2	Home health aides	1,017,700	69.4%	
3	Biomedical engineers	15,700	61.7%	
4	Helpers—masons	29,400	60.1%	
5	Helpers—carpenters	46,500	55.7%	
6	Veterinary technicians	80,200	52.0%	
7	Reinforcing iron/rebar workers	19,100	48.6%	
8	Physical therapist assistants	67,400	45.7%	
9	Helpers—plumbers	57,900	45.4%	
10	Meeting/event planners	71,600	43.7%	

What do most of these fast-growing occupations have in common? They require knowledge of math! For some careers, such as home health aides, event planners, and biomedical engineers, the ways math is used on the job may be obvious. For other occupations, the use of math may not be quite as apparent. However, tasks common to many jobs—filling in a time sheet, writing up an expense or mileage report, planning a budget, figuring a bill, ordering supplies, and even making a work schedule—all require math.

This activity may be completed by working in groups or individually.

1. List the top five occupations by order of employment figures for 2010.

2. Using the 2010 employment figures and the percent increase from 2010 to 2020, find the expected 2020 employment figure for each occupation listed in the table. Round to the nearest thousand.

3. List the top five occupations by order of employment figures for 2020. Did the order change at all from 2010? Explain.

Chapter 6 Vocabulary Check

Fill in each blank with one of the words or phrases listed below. Some choices may be used more than once.

percent	sales tax	is	0.01	$\dfrac{1}{100}$	amount of discount	percent of decrease	total price
base	of	amount	100%	compound interest	percent of increase	sale price	commission

1. In a mathematical statement, _____ usually means "multiplication."

2. In a mathematical statement, _____ means "equals."

3. _____ means "per hundred."

4. _____ is computed not only on the principal, but also on interest already earned in previous compounding periods.

5. In the percent proportion, $\dfrac{\underline{\hspace{2cm}}}{\underline{\hspace{3cm}}} = \dfrac{\text{percent}}{100}$.

6. To write a decimal or fraction as a percent, multiply by $\underline{\hspace{3cm}}$.

7. The decimal equivalent of the % symbol is $\underline{\hspace{3cm}}$.

8. The fraction equivalent of the % symbol is $\underline{\hspace{3cm}}$.

9. The percent equation is $\underline{\hspace{2.5cm}}$ · percent $= \underline{\hspace{3cm}}$.

10. $\underline{\hspace{3cm}} = \dfrac{\text{amount of decrease}}{\text{original amount}}$.

11. $\underline{\hspace{3cm}} = \dfrac{\text{amount of increase}}{\text{original amount}}$.

12. $\underline{\hspace{2.5cm}} = $ tax rate · purchase price.

13. $\underline{\hspace{2.5cm}} = $ purchase price $+$ sales tax.

14. $\underline{\hspace{2.5cm}} = $ commission rate · sales.

15. $\underline{\hspace{2.5cm}} = $ discount rate · original price.

16. $\underline{\hspace{2.5cm}} = $ original price $-$ amount of discount.

> **Helpful Hint**
>
> ▶ Are you preparing for your test? Don't forget to take the Chapter 6 Test on page 507. Then check your answers at the back of the text and use the Chapter Test Prep Videos to see the fully worked-out solutions to any of the exercises you want to review.

 Chapter Highlights

Definitions and Concepts	Examples
Section 6.1 Percents, Decimals, and Fractions	
Percent means "per hundred." The % symbol denotes percent.	$51\% = \dfrac{51}{100}$ 51 per 100 $7\% = \dfrac{7}{100}$ 7 per 100
To write a percent as a decimal, replace the % symbol with its decimal equivalent, 0.01, and multiply. **To write a decimal as a percent,** multiply by 100%.	$32\% = 32(0.01) = 0.32$ $0.08 = 0.08(100\%) = 08.\% = 8\%$
To write a percent as a fraction, replace the % symbol with its fraction equivalent, $\dfrac{1}{100}$, and multiply. **To write a fraction as a percent,** multiply by 100%.	$25\% = \dfrac{25}{100} = \dfrac{\overset{1}{\cancel{25}}}{4 \cdot \cancel{25}} = \dfrac{1}{4}$ $\dfrac{1}{6} = \dfrac{1}{6} \cdot 100\% = \dfrac{1}{6} \cdot \dfrac{100}{1}\% = \dfrac{100}{6}\% = 16\dfrac{2}{3}\%$

Definitions and Concepts	Examples

Section 6.2   Solving Percent Problems Using Equations

Three key words in the statement of a percent problem are

of, which means **multiplication** ($\cdot$)

is, which means **equals** ($=$)

what (or some equivalent word or phrase), which stands for **the unknown number**

Solve:

$$
\begin{array}{ccccc}
6 & \text{is} & 12\% & \text{of} & \text{what number?} \\
\downarrow & \downarrow & \downarrow & \downarrow & \downarrow \\
6 & = & 12\% & \cdot & n
\end{array}
$$

$6 = 0.12 \cdot n$ Write 12% as a decimal.

$\dfrac{6}{0.12} = n$ Divide 6 by 0.12, the number multiplied by n.

$50 = n$

Thus, 6 is 12% of 50.

Section 6.3   Solving Percent Problems Using Proportions

Percent Proportion

$$\frac{\text{amount}}{\text{base}} = \frac{\text{percent}}{100} \quad \leftarrow \text{always 100}$$

or

$$\text{amount} \rightarrow \frac{a}{b} = \frac{p}{100} \leftarrow \text{percent}$$
$$\text{base} \rightarrow$$

Solve:

$$
\begin{array}{ccc}
20.4 \text{ is} & \text{what percent} & \text{of } 85? \\
\searrow & \downarrow & \searrow \\
\text{amount} & \text{percent} & \text{base}
\end{array}
$$

$$\text{amount} \rightarrow \frac{20.4}{85} = \frac{p}{100} \leftarrow \text{percent}$$
$$\text{base} \rightarrow$$

$20.4 \cdot 100 = 85 \cdot p$ Set cross products equal.

$2040 = 85 \cdot p$ Multiply.

$\dfrac{2040}{85} = p$ Divide 2040 by 85, the number multiplied by p.

$24 = p$ Simplify.

Thus, 20.4 is 24% of 85.

Section 6.4   Applications of Percent

Percent of Increase

$$\text{percent of increase} = \frac{\text{amount of increase}}{\text{original amount}}$$

Percent of Decrease

$$\text{percent of decrease} = \frac{\text{amount of decrease}}{\text{original amount}}$$

A town's population of 16,480 decreased to 13,870 over a 12-year period. Find the percent of decrease. Round to the nearest whole percent.

$$\text{amount of decrease} = 16{,}480 - 13{,}870$$
$$= 2610$$

$$\text{percent of decrease} = \frac{\text{amount of decrease}}{\text{original amount}}$$

$$= \frac{2610}{16{,}480} \approx 0.16$$

$$= 16\%$$

The town's population decreased by 16%.

Section 6.5   Percent and Problem Solving: Sales Tax, Commission, and Discount

Sales Tax and Total Price

$$\text{sales tax} = \text{sales tax rate} \cdot \text{purchase price}$$
$$\text{total price} = \text{purchase price} + \text{sales tax}$$

Find the sales tax and the total price on a purchase of $42 if the sales tax rate is 9%.

$$
\begin{array}{ccccc}
\text{sales tax} & = & \text{sales tax rate} & \cdot & \text{purchase price} \\
\downarrow & & \downarrow & & \downarrow \\
\text{sales tax} & = & 9\% & \cdot & \$42
\end{array}
$$

$$= 0.09 \cdot \$42$$

$$= \$3.78$$

(continued)

Definitions and Concepts	Examples

Section 6.5   Percent and Problem Solving: Sales Tax, Commission, and Discount (*continued*)

The total price is

$$\begin{array}{ccccc} \text{total price} & = & \text{purchase price} & + & \text{sales tax} \\ \downarrow & & \downarrow & & \downarrow \\ \text{total price} & = & \$42 & + & \$3.78 \\ & = & \$45.78 \end{array}$$

Commission

commission = commission rate · total sales

A salesperson earns a commission of 3%. Find the commission from sales of $12,500 worth of appliances.

$$\begin{array}{ccccc} \text{commission} & = & \text{commission rate} & \cdot & \text{sales} \\ \downarrow & & \downarrow & & \downarrow \\ \text{commission} & = & 3\% & \cdot \$12,500 \\ & = & 0.03 \cdot \$12,500 \\ & = & \$375 \end{array}$$

Discount and Sale Price

amount of discount = discount rate · original price

sale price = original price − amount of discount

A suit is priced at $320 and is on sale today for 25% off. What is the sale price?

$$\begin{array}{ccccc} \text{amount of discount} & = & \text{discount rate} & \cdot & \text{original price} \\ \downarrow & & \downarrow & & \downarrow \\ \text{amount of discount} & = & 25\% & \cdot & \$320 \\ & = & 0.25 \cdot \$320 \\ & = & \$80 \end{array}$$

$$\begin{array}{ccccc} \text{sale price} & = & \text{original price} & - & \text{amount of discount} \\ \downarrow & & \downarrow & & \downarrow \\ \text{sale price} & = & \$320 & - & \$80 \\ & = & \$240 \end{array}$$

The sale price is $240.

Section 6.6   Percent and Problem Solving: Interest

Simple Interest

interest = principal · rate · time

where the rate is understood to be per year.

Find the simple interest after 3 years on $800 at an interest rate of 5%.

$$\begin{array}{ccccccc} \text{interest} & = & \text{principal} & \cdot & \text{rate} & \cdot & \text{time} \\ \downarrow & & \downarrow & & \downarrow & & \downarrow \\ \text{interest} & = & \$800 & \cdot & 5\% & \cdot & 3 \\ & = & \$800 \cdot 0.05 \cdot 3 & & \text{\small Write 5\% as 0.05.} \\ & = & \$120 & & \text{\small Multiply.} \end{array}$$

The interest is $120.

Compound interest is computed not only on the principal, but also on interest already earned in previous compounding periods. (See Appendix A.7 for various compound interest factors.)

$$A = P\left(1 + \frac{r}{n}\right)^{n \cdot t}$$

where n is the number of times compounded per year.

$800 is invested at 5% compounded quarterly for 10 years. Find the total amount at the end of 10 years.

$$A = \$800\left(1 + \frac{0.05}{4}\right)^{4 \cdot 10}$$
$$= \$800(1.0125)^{40}$$
$$\approx \$1314.90$$

(6.1) *Solve.*

1. In a survey of 100 adults, 37 preferred pepperoni on their pizzas. What percent preferred pepperoni?

2. A basketball player made 77 out of 100 attempted free throws. What percent of free throws were made?

Write each percent as a decimal.

3. 83%

4. 75%

5. 73.5%

6. 1.5%

7. 125%

8. 145%

9. 0.5%

10. 0.7%

11. 200%

12. 400%

13. 26.25%

14. 85.34%

Write each decimal as a percent.

15. 2.6

16. 1.02

17. 0.35

18. 0.055

19. 0.725

20. 0.252

21. 0.076

22. 0.085

23. 0.71

24. 0.65

25. 4

26. 9

Write each percent as a fraction or mixed number in simplest form.

27. 1%

28. 10%

29. 25%

30. 8.5%

31. 10.2%

32. $16\frac{2}{3}\%$

33. $33\frac{1}{3}\%$

34. 110%

Write each fraction or mixed number as a percent.

35. $\frac{1}{5}$

36. $\frac{7}{10}$

37. $\frac{5}{6}$

38. $\frac{3}{5}$

39. $1\frac{1}{4}$

40. $1\frac{2}{3}$

41. $\frac{1}{16}$

42. $\frac{5}{8}$

(6.2) *Translate each to an equation and solve.*

43. 1250 is 1.25% of what number?

44. What number is $33\frac{1}{3}$% of 24,000?

45. 124.2 is what percent of 540?

46. 22.9 is 20% of what number?

47. What number is 40% of 7500?

48. 693 is what percent of 462?

(6.3) *Translate each to a proportion and solve.*

49. 104.5 is 25% of what number?

50. 16.5 is 5.5% of what number?

51. What number is 36% of 180?

52. 63 is what percent of 35?

53. 93.5 is what percent of 85?

54. What number is 33% of 500?

(6.4) *Solve.*

55. In a survey of 2000 people, it was found that 1320 have a microwave oven. Find the percent of people who own microwaves.

56. Of the 12,360 freshmen entering County College, 2000 are enrolled in basic college mathematics. Find the percent of entering freshmen who are enrolled in basic college mathematics. Round to the nearest whole percent.

57. The number of violent crimes in a city decreased from 675 to 534. Find the percent of decrease. Round to the nearest tenth of a percent.

58. The current charge for dumping waste in a local landfill is $16 per cubic foot. To cover new environmental costs, the charge will increase to $33 per cubic foot. Find the percent of increase.

59. This year the fund drive for a charity collected $215,000. Next year, a 4% decrease is expected. Find how much is expected to be collected in next year's drive.

60. A local union negotiated a new contract that increases the hourly pay 15% over last year's pay. The old hourly rate was $11.50. Find the new hourly rate rounded to the nearest cent.

(6.5) *Solve.*

61. If the sales tax rate is 5.5%, what is the total amount charged for a $250 coat?

62. Find the sales tax paid on a $25.50 purchase if the sales tax rate is 4.5%.

63. Russ James is a sales representative for a chemical company and is paid a commission rate of 5% on all sales. Find his commission if he sold $100,000 worth of chemicals last month.

64. Carol Sell is a sales clerk in a clothing store. She receives a commission of 7.5% on all sales. Find her commission for the week if her sales for the week were $4005. Round to the nearest cent.

65. A $3000 mink coat is on sale for 30% off. Find the discount and the sale price.

66. A $90 calculator is on sale for 10% off. Find the discount and the sale price.

(6.6) *Solve.*

67. Find the simple interest due on $4000 loaned for 4 months at 12% interest.

68. Find the simple interest due on $6500 loaned for 3 months at 20%.

69. Find the total amount in an account if $5500 is compounded annually at 12% for 15 years.

70. Find the total amount in an account if $6000 is compounded semiannually at 11% for 10 years.

71. Find the compound interest earned if $100 is compounded quarterly at 12% for 5 years.

72. Find the compound interest earned if $1000 is compounded quarterly at 18% for 20 years.

Mixed Review

Write each percent as a decimal.

73. 3.8%

74. 24.5%

75. 0.9%

Write each decimal as a percent.

76. 0.54

77. 95.2

78. 0.3

Write each percent as a fraction or mixed number in simplest form.

79. 47%

80. $6\frac{2}{5}\%$

81. 5.6%

Write each fraction or mixed number as a percent.

82. $\frac{3}{8}$

83. $\frac{2}{13}$

84. $\frac{6}{5}$

Translate each into an equation and solve.

85. 43 is 16% of what number?

86. 27.5 is what percent of 25?

87. What number is 36% of 1968?

88. 67 is what percent of 50?

Translate each into a proportion and solve.

89. 75 is what percent of 25?

90. What number is 16% of 240?

91. 28 is 5% of what number?

92. 52 is what percent of 16?

Solve.

93. The total number of cans in a soft drink machine is 300. If 78 soft drinks have been sold, find the percent of soft drink cans that have been sold.

94. A home valued at $96,950 last year has lost 7% of its value this year. Find the loss in value.

95. A dinette set sells for $568.00. If the sales tax rate is 8.75%, find the purchase price of the dinette set.

96. The original price of a video game is $23.00. It is on sale for 15% off. What is the amount of the discount?

97. A candy salesman makes a commission of $1.60 from each case of candy he sells. If a case of candy costs $12.80, what is his rate of commission?

98. Find the total amount due on a 6-month loan of $1400 at a simple interest rate of 13%.

99. $8800 is invested at 8% interest compounded quarterly. Find the total amount after 9 years.

100. Find the total amount due on a loan of $5500 for 9 years at 12.5% simple interest.

Write each percent as a decimal.

1. 85%

2. 500%

3. 0.8%

Write each decimal as a percent.

4. 0.056

5. 6.1

6. 0.39

Write each percent as a fraction or mixed number in simplest form.

7. 120%

8. 38.5%

9. 0.2%

Write each fraction or mixed number as a percent.

10. $\frac{11}{20}$

11. $\frac{3}{8}$

12. $1\frac{5}{9}$

Solve.

13. What number is 42% of 80?

14. 0.6% of what number is 7.5?

15. 567 is what percent of 756?

Answers

1. _____

2. _____

3. _____

4. _____

5. _____

6. _____

7. _____

8. _____

9. _____

10. _____

11. _____

12. _____

13. _____

14. _____

15. _____

16. _____

Solve. Round all dollar amounts to the nearest cent.

16. An alloy is 12% copper. How much copper is contained in 320 pounds of this alloy?

17. _____

17. A farmer in Nebraska estimates that 20% of his potential crop, or $11,350, has been lost to a hard freeze. Find the total value of his potential crop.

18. _____

18. If the local sales tax rate is 1.25%, find the total amount charged for a stereo system priced at $354.

19. _____

19. A town's population increased from 25,200 to 26,460. Find the percent of increase.

20. _____

20. A $120 framed picture is on sale for 15% off. Find the discount and the sale price.

21. _____

21. Randy Nguyen is paid a commission rate of 4% on all sales. Find Randy's commission if his sales were $9875.

22. _____

22. A sales tax of $1.53 is added to an item's price of $152.99. Find the sales tax rate. Round to the nearest whole percent.

23. _____

23. Find the simple interest earned on $2000 saved for $3\frac{1}{2}$ years at an interest rate of 9.25%.

24. _____

24. $1365 is compounded annually at 8%. Find the total amount in the account after 5 years.

25. _____

25. A couple borrowed $400 from a bank at 13.5% simple interest for 6 months for car repairs. Find the total amount due the bank at the end of the 6-month period.

1. How many cases can be filled with 9900 cans of jalapeños if each case holds 48 cans? How many cans will be left over? Will there be enough cases to fill an order for 200 cases?

2. Multiply: 409×76

3. Write each fraction as a mixed number or a whole number.

    a. $\dfrac{30}{7}$ **b.** $\dfrac{16}{15}$ **c.** $\dfrac{84}{6}$

4. Write each mixed number as an improper fraction.

    a. $2\dfrac{5}{7}$ **b.** $10\dfrac{1}{10}$ **c.** $5\dfrac{3}{8}$

5. Write $-\dfrac{10}{27}$ in simplest form.

6. Find the average of 28, 34, and 70.

7. Multiply and simplify: $\dfrac{23}{32} \cdot \dfrac{4}{7}$

8. Round 76,498 to the nearest ten.

9. Divide and simplify: $\dfrac{2}{5} \div \dfrac{1}{2}$

10. Write the shaded part of the figure as an improper fraction and as a mixed number.

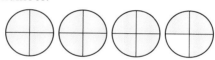

△ **11.** Find the perimeter of the rectangle.

$\frac{2}{15}$ inch

$\frac{4}{15}$ inch

12. Find $2 \cdot 5^2$.

13. Find the LCD of $\dfrac{11}{12}$ and $\dfrac{7}{20}$.

14. Subtract $\dfrac{7}{9}$ from $\dfrac{10}{9}$.

15. Add: $\dfrac{2}{5} + \dfrac{4}{15}$

16. Find $\dfrac{2}{3}$ of 510.

17. Subtract: $7\dfrac{3}{14} - 3\dfrac{6}{7}$

18. Simplify: $9 \cdot \sqrt{25} - 6 \cdot \sqrt{4}$

Perform each indicated operation.

19. $\dfrac{3}{4} \div 5$

20. $20\dfrac{4}{5} + 12\dfrac{7}{8}$

21. $\left(-\dfrac{1}{4}\right)^2$

22. $1\dfrac{7}{8} \cdot 3\dfrac{2}{5}$

Answers

1. _____

2. _____

3. a. _____

   b. _____

   c. _____

4. a. _____

   b. _____

   c. _____

5. _____

6. _____

7. _____

8. _____

9. _____

10. _____

11. _____

12. _____

13. _____

14. _____

15. _____

16. _____

17. _____

18. _____

19. _____

20. _____

21. _____

22. _____

Write each fraction as a decimal.

23. _____

24. _____

23. $-\dfrac{5}{8}$ **24.** $\dfrac{9}{100}$

25. _____

26. _____

25. $\dfrac{22}{7}$ (Round to the nearest hundredth.)   **26.** $\dfrac{48}{10,000}$

27. _____

28. _____

27. A high school teacher's taxable income is \$41,567.72. The tax tables in the teacher's state use amounts rounded to the nearest dollar. Round the teacher's income to the nearest whole dollar.

28. Subtract: $38 - 10.06$

29. _____

30. _____

29. Add: $763.7651 + 22.001 + 43.89$ **30.** 12.483×100

31. _____

32. _____

31. Multiply: 23.6×0.78 **32.** 76.3×1000

33. _____

34. _____

Divide.

33. $\dfrac{786.1}{1000}$ **34.** $0.5\overline{)0.638}$

35. _____

36. _____

35. $\dfrac{-0.12}{10}$ **36.** $0.23\overline{)11.6495}$

37. _____

38. _____

37. Simplify: $723.6 \div 1000 \times 10$ **38.** Simplify: $\dfrac{3.19 - 0.707}{13}$

39. _____

40. _____

39. Write $\dfrac{1}{4}$ as a decimal.   **40.** Write $\dfrac{5}{9}$ as a decimal. Give the exact answer and a three-decimal-place approximation.

41. _____

42. _____

41. Is $\dfrac{4.1}{7} = \dfrac{2.9}{5}$ a true proportion?   **42.** Find each unit rate and decide on the better buy.
  \$0.93 for 18 flour tortillas
  \$1.40 for 24 flour tortillas

43. _____

44. a. _____

b. _____

43. On a chamber of commerce map of Abita Springs, 5 miles corresponds to 2 inches. How many miles correspond to 7 inches?

44. Write each percent as a decimal.
  a. 7%   **b.** 200%   **c.** 0.5%

c. _____

45. _____

46. _____

45. Translate to an equation: What number is 25% of 0.008?   **46.** Write $\dfrac{3}{8}$ as a percent.

Statistics and Probability

Can We Experience Weightlessness?

Today, space travel, while still not commonplace, is well within everyone's comprehension. From landing on the Moon, to probes to Mars and beyond, the Hubble Space Telescope, and the International Space Station, there are many exciting explorations in space. There are even space tourists who pay big money to travel in space.

If space travel is not for you but you'd like to experience weightlessness, there are commercial companies that offer rides on Boeing 727s with flight patterns similar to that shown below to simulate zero gravity.

In Section 7.1, Example 2, we explore lunar and planetary exploration since 1957.

Flight Pattern to Simulate Weightlessness

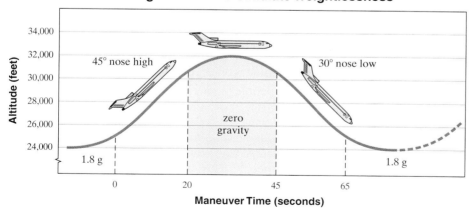
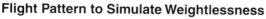

Check Your Progress

Vocabulary Check

Chapter Highlights

Chapter Review

Chapter Test

Cumulative Review

We often need to make decisions based on known statistics or the probability of an event occurring. For example, we decide whether or not to bring an umbrella to work based on the probability of rain. We choose an investment based on its mean, or average, return. We can predict which football team will win based on the trend in its previous wins and losses. This chapter reviews presenting data in a usable form on a graph and the basic ideas of statistics and probability.

Objectives

A Read Pictographs.

B Read and Construct Bar Graphs.

C Read and Construct Histograms.

D Read Line Graphs.

Often data are presented visually in a graph. In this section, we practice reading several kinds of graphs, including pictographs, bar graphs, histograms, and line graphs.

Objective A Reading Pictographs

A **pictograph** such as the one below is a graph in which pictures or symbols are used. This type of graph contains a key that explains the meaning of the symbol used. An advantage of using a pictograph to display information is that comparisons can easily be made. A disadvantage of using a pictograph is that it is often hard to tell what fractional part of a symbol is shown. For example, in the pictograph below, Arabic shows a part of a symbol, but it's hard to read with any accuracy what fractional part of the symbol is shown.

Practice 1

Use the pictograph shown in Example 1 to answer the following questions:

a. Approximate the number of people who primarily speak Spanish.

b. Approximate how many more people primarily speak Spanish than Arabic.

Example 1 Calculating Languages Spoken

The following pictograph shows the top eight most-spoken (primary) languages. Use this pictograph to answer the questions.

Top 8 Most-Spoken (Primary) Languages

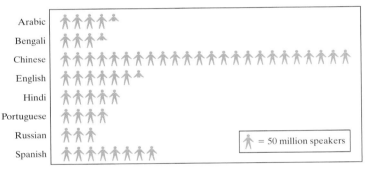

Source: www.ethnologue.com

a. Approximate the number of people who primarily speak Hindi.

b. Approximate how many more people primarily speak English than Hindi.

Solution:

a. Hindi corresponds to 5 symbols, and each symbol represents 50 million speakers. This means that the number of people who primarily speak Hindi is approximately $5 \cdot (50 \text{ million})$ or 250 million people. This can also be written as 250,000,000 people.

b. English shows $1\frac{1}{2}$ more symbols than Hindi. This means that $1\frac{1}{2} \cdot (50 \text{ million})$ or 75 million or 75,000,000 more people primarily speak English than Hindi.

■ Work Practice 1

Example 2 Calculating Solar System Exploration

The following pictograph shows the approximate number of solar system exploration missions by various countries or space consortia from 1957 to the present day. Use this pictograph to answer the questions.

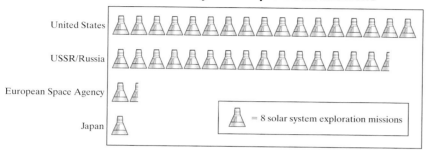

Solar System Exploration Missions

a. Approximate the number of solar system exploration missions undertaken by the United States.

b. Approximate how many more solar system exploration missions have been undertaken by the United States than by the USSR/Russia.

Solution:

a. The United States corresponds to 17 symbols, and each symbol represents 8 solar system exploration missions. This means that the United States has undertaken approximately $17 \cdot 8 = 136$ missions for solar system exploration.

b. The USSR/Russia shows $15\frac{1}{2}$ symbols, or $1\frac{1}{2}$ fewer than the United States. This means that the United States has undertaken $1\frac{1}{2} \cdot 8 = 12$ more solar system exploration missions than the USSR/Russia.

Work Practice 2

Practice 2

Use the pictograph shown in Example 2 to answer the following questions:

a. Approximate the number of solar system exploration missions undertaken by the European Space Agency.

b. Approximate the total number of solar system exploration missions undertaken by the European Space Agency and Japan.

Objective B Reading and Constructing Bar Graphs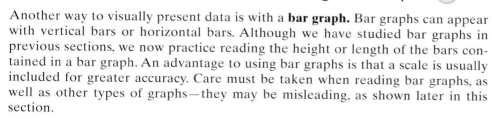

Another way to visually present data is with a **bar graph.** Bar graphs can appear with vertical bars or horizontal bars. Although we have studied bar graphs in previous sections, we now practice reading the height or length of the bars contained in a bar graph. An advantage to using bar graphs is that a scale is usually included for greater accuracy. Care must be taken when reading bar graphs, as well as other types of graphs—they may be misleading, as shown later in this section.

Practice 3

Use the bar graph in Example 3 to answer the following questions:

a. Approximate the number of endangered species that are amphibians.

b. Which category shows the fewest endangered species?

Example 3 Finding the Number of Endangered Species

The following bar graph shows the number of endangered species in the United States in 2013. Use this graph to answer the questions.

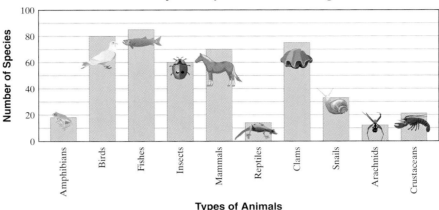

How Many U.S. Species Are Endangered?

Source: U.S. Fish and Wildlife Service

a. Approximate the number of endangered species that are clams.

b. Which category has the most endangered species?

Solution:

a. To approximate the number of endangered species that are clams, we go to the top of the bar that represents clams. From the top of this bar, we move horizontally to the left until the scale is reached. We read the height of the bar on the scale as approximately 75. There are approximately 75 clam species that are endangered, as shown.

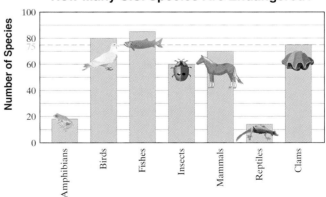

How Many U.S. Species Are Endangered?

Source: U.S. Fish and Wildlife Service

b. The most endangered species is represented by the tallest (longest) bar. The tallest bar corresponds to fishes.

■ Work Practice 3

Answers
3. a. 18   **b.** arachnids

Next, we practice constructing a bar graph.

Example 4 Draw a vertical bar graph using the information in the table below, which gives the caffeine content of selected foods.

Average Caffeine Content of Selected Foods

Food	Milligrams	Food	Milligrams
Brewed coffee (percolator, 8 ounces)	124	Instant coffee (8 ounces)	104
Brewed decaffeinated coffee (8 ounces)	3	Brewed tea (U.S. brands, 8 ounces)	64
Coca-Cola Classic (8 ounces)	31	Mr. Pibb (8 ounces)	27
Dark chocolate (semisweet, $1\frac{1}{2}$ ounces)	30	Milk chocolate (8 ounces)	9

(*Sources:* International Food Information Council and the Coca-Cola Company)

Solution: We draw and label a vertical line and a horizontal line as shown below on the left. These lines are also called axes. We place the different food categories along the horizontal axis. Along the vertical axis, we place a scale.

There are many choices of scales that would be appropriate. Notice that the milligrams range from a low of 3 to a high of 124. From this information, we use a scale that starts at 0 and then shows multiples of 20 so that the scale is not too cluttered. The scale stops at 140, the smallest multiple of 20 that will allow all milligrams to be graphed. It may also be helpful to draw horizontal lines along the scale markings to help draw the vertical bars at the correct heights. The finished bar graph is shown below on the right.

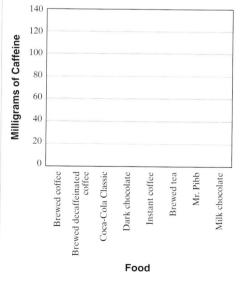

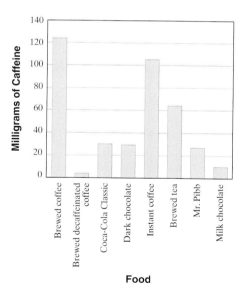

Work Practice 4

As mentioned previously, graphs can be misleading. Both graphs on the next page show the same information, but with different scales. Special care should be taken when forming conclusions from the appearance of a graph.

Practice 4

Draw a vertical bar graph using the information in the table about selected states' electoral votes for President in the 2012, 2016, and 2020 presidential elections.

Total Electoral Votes by Selected States

State	Electoral Votes
Texas	38
California	55
Florida	29
Nebraska	5
Indiana	11
Georgia	16

(*Source:* U.S. Electoral College)

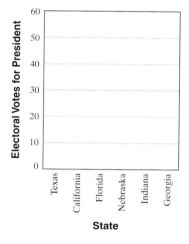

Answer

4.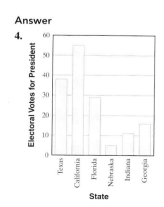

Notice the ⚡ symbol on each vertical scale on the graphs below. This symbol alerts us that numbers are missing from that scale.

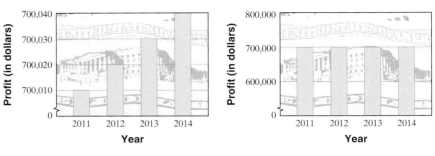

Are profits shown in the graphs above greatly increasing, or are they remaining about the same?

Objective C Reading and Constructing Histograms

Suppose that the test scores of 36 students are summarized in the table below:

Student Scores	Frequency (Number of Students)
40–49	1
50–59	3
60–69	2
70–79	10
80–89	12
90–99	8

The results in the table can be displayed in a histogram. A **histogram** is a special bar graph. The width of each bar represents a range of numbers called a **class interval.** The height of each bar corresponds to how many times a number in the class interval occurs and is called the **class frequency.** The bars in a histogram lie side by side with no space between them.

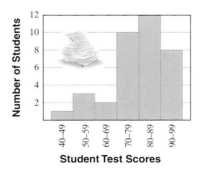

Student Test Scores

Practice 5

Use the histogram on the right to determine how many students scored 80–89 on the test.

Example 5 Reading a Histogram on Student Test Scores

Use the preceding histogram to determine how many students scored 50–59 on the test.

Solution: We find the bar representing 50–59. The height of this bar is 3, which means 3 students scored 50–59 on the test.

▧ Work Practice 5

Answer

5. 12

Example 6 Reading a Histogram on Student Test Scores

Use the preceding histogram to determine how many students scored 80 or above on the test.

Solution: We see that two different bars fit this description. There are 12 students who scored 80–89 and 8 students who scored 90–99. The sum of these two categories is 12 + 8 or 20 students. Thus, 20 students scored 80 or above on the test.

■ Work Practice 6

Now we will look at a way to construct histograms.

The daily high temperatures for 1 month in New Orleans, Louisiana, are recorded in the following list:

85°	90°	95°	89°	88°	94°
87°	90°	95°	92°	95°	94°
82°	92°	96°	91°	94°	92°
89°	89°	90°	93°	95°	91°
88°	90°	88°	86°	93°	89°

The data in this list have not been organized and can be hard to interpret. One way to organize the data is to place them in a **frequency distribution table.** We will do this in Example 7.

Example 7 Completing a Frequency Distribution on Temperature

Complete the frequency distribution table for the preceding temperature data.

Solution: Go through the data and place a tally mark in the second column of the table next to the class interval. Then count the tally marks and write each total in the third column of the table.

Class Intervals (Temperatures)	Tally	Class Frequency (Number of Days)
82°–84°			1
85°–87°	|||	3
88°–90°	⫼⫼ ⫼⫼|	11
91°–93°	⫼⫼ ||	7
94°–96°	⫼⫼ |||	8

■ Work Practice 7

Example 8 Constructing a Histogram

Construct a histogram from the frequency distribution table in Example 7.

Solution:

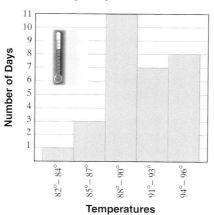

■ Work Practice 8

Practice 6

Use the histogram above Example 5 to determine how many students scored less than 80 on the test.

Practice 7

Complete the frequency distribution table for the data below. Each number represents a credit card owner's unpaid balance for one month.

0	53	89	125
265	161	37	76
62	201	136	42

Class Intervals (Credit Card Balances)	Tally	Class Frequency (Number of Months)
$0–$49	____	____
$50–$99	____	____
$100–$149	____	____
$150–$199	____	____
$200–$249	____	____
$250–$299	____	____

Practice 8

Construct a histogram from the frequency distribution table above.

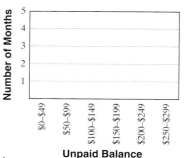

Answers

6. 16

7. table in class interval order:

Tally	Class Frequency (Number of Months)	Tally	Class Frequency (Number of Months)
				3			1
					4			1
			2			1

8.

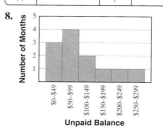

✓**Concept Check** Which of the following sets of data is better suited to representation by a histogram? Explain.

Set 1		Set 2	
Grade on Final	# of Students	Section Number	Avg. Grade on Final
51–60	12	150	78
61–70	18	151	83
71–80	29	152	87
81–90	23	153	73
91–100	25		

Objective D Reading Line Graphs

Another common way to display information with a graph is by using a **line graph.** An advantage of a line graph is that it can be used to visualize relationships between two quantities. A line graph can also be very useful in showing a change over time.

Practice 9

Use the temperature graph in Example 9 to answer the following questions:

a. During what month is the average daily temperature the lowest?

b. During what month is the average daily temperature 25°F?

c. During what months is the average daily temperature greater than 70°F?

Example 9 Reading Temperatures from a Line Graph

The following line graph shows the average daily temperature for each month in Omaha, Nebraska. Use this graph to answer the questions below.

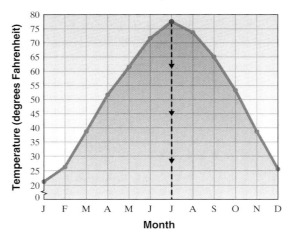

Source: National Climatic Data Center

a. During what month is the average daily temperature the highest?

b. During what month, from July through December, is the average daily temperature 65°F?

c. During what months is the average daily temperature less than 30°F?

Solution:

a. The month with the highest temperature corresponds to the highest point. This is the red point shown on the graph above. We follow this highest point downward to the horizontal month scale and see that this point corresponds to July.

Answers

9. a. January   b. December
c. June, July, and August

✓**Concept Check Answer**

Set 1; the grades are arranged in ranges of scores.

b. The months July through December correspond to the right side of the graph. We find the 65°F mark on the vertical temperature scale and move to the right until a point on the right side of the graph is reached. From that point, we move downward to the horizontal month scale and read the corresponding month. During the month of September, the average daily temperature is 65°F.

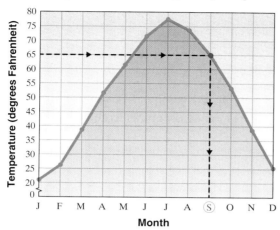

Source: National Climatic Data Center

c. To see what months the temperature is less than 30°F, we find what months correspond to points that fall below the 30°F mark on the vertical scale. These months are January, February, and December.

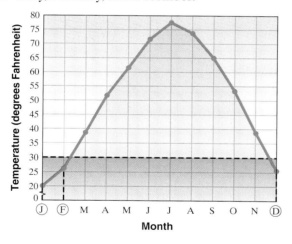

Source: National Climatic Data Center

■■ Work Practice 9

Vocabulary, Readiness & Video Check

Fill in each blank with one of the choices below.

pictograph   bar   class frequency

histogram   line   class interval

1. A(n) _____ graph presents data using vertical or horizontal bars.

2. A(n) _____ is a graph in which pictures or symbols are used to visually present data.

3. A(n) _____ graph displays information with a line that connects data points.

4. A(n) _____ is a special bar graph in which the width of each bar represents a(n) _____ and the height of each bar represents the _____ .

Martin-Gay Interactive Videos

See Video 7.1

Watch the section lecture video and answer the following questions.

Objective A 5. From the pictograph in Example 1, how would you approximate the number of wildfires for any given year? 🔘

Objective B 6. From 🔲 Example 5, what is one advantage of displaying data in a bar graph? 🔘

Objective C 7. Complete this statement based on the lecture before 🔲 Example 6: A histogram is a special kind of _____. 🔘

Objective D 8. From the line graph in 🔲 Examples 10–13, during which year(s) were total points scored greater than 40? 🔘

7.1 **Exercise Set** MyMathLab® 🔘

Objective A *The following pictograph shows the number of acres devoted to wheat production in selected states in 2014. Use this graph to answer Exercises 1 through 8. See Examples 1 and 2. (Source: National Agricultural Statistics Service)*

1. Which of the states shown planted the greatest quantity of acreage in wheat?

2. Which of the states shown planted the least amount of wheat acreage?

3. Approximate the number of acres of wheat planted in Oklahoma.

4. Approximate the number of acres of wheat planted in Kansas.

5. Which state planted about 2,500,000 acres of wheat?

6. Which state planted about 6,000,000 acres of wheat?

7. How many more acres of wheat were planted in Kansas than in Montana?

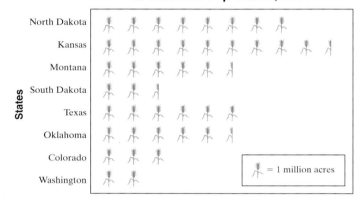

Wheat Acreage in Selected Top States, 2014

States: North Dakota, Kansas, Montana, South Dakota, Texas, Oklahoma, Colorado, Washington

🌾 = 1 million acres

8. How many more acres of wheat were planted in Oklahoma than in Washington?

The following pictograph shows the annual number of wildfires in the United States between 2006 and 2013. Use this graph to answer Exercises 9 through 16. See Examples 1 and 2. (Source: National Interagency Fire Center)

9. Approximate the number of wildfires in 2012.

10. Approximately how many wildfires were there in 2006?

11. Which year, of the years shown, had the most wildfires?

12. In what years were the number of wildfires greater than 72,000?

Wildfires in the United States

Years: 2006, 2007, 2008, 2009, 2010, 2011, 2012, 2013

🔥 = 12,000 fires

13. What was the amount of increase in wildfires from 2010 to 2011?

14. What was the amount of decrease in wildfires from 2006 to 2012?

15. What was the average annual number of wildfires from 2010 to 2012? (*Hint:* How do you calculate the average?)

16. Give an explanation for the large number of wildfires in 2006.

Objective B *The National Weather Service has exacting definitions for hurricanes; they are tropical storms with winds in excess of 74 mph. The following bar graph shows the number of hurricanes, by month, that have made landfall on the mainland United States between 1851 and 2013. Use this graph to answer Exercises 17 through 22. See Example 3. (Source: NOAA: Hurricane Research Division)*

17. In which month did the most hurricanes make landfall in the United States?

18. In which month did the fewest hurricanes make landfall in the United States?

19. Approximate the number of hurricanes that made landfall in the United States during the month of August.

20. Approximate the number of hurricanes that made landfall in the United States in September.

21. In 2008, two hurricanes made landfall during the month of August. What fraction of all the 76 hurricanes that made landfall during August is this?

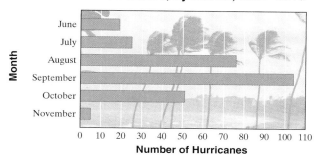

Hurricanes Making Landfall in the United States, by Month, 1851–2013

22. In 2012, only one hurricane, Hurricane Sandy, made landfall in the United States during the month of October. If there have been 51 hurricanes to make landfall in the month of October since 1851, approximately what percent of these occurred in 2012?

The following horizontal bar graph shows the recent population of the world's largest cities (including their suburbs). Use this graph to answer Exercises 23 through 28. See Example 3. (Source: CityPopulation)

23. Name the city with the largest population, and estimate its population.

24. Name the city whose population is between 26 million and 27 million, and estimate its population.

25. Name the city on this list with the smallest population, and estimate its population.

26. Name the two cities that have approximately the same population.

27. How much larger (in terms of population) is Seoul, South Korea, than Mexico City, Mexico?

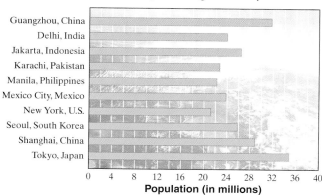

World's Largest Cities (including suburbs)

28. How much larger (in terms of population) is Guangzhou, China, than Karachi, Pakistan?

Use the information given to draw a vertical bar graph. Clearly label the bars. See Example 4.

29.

Fiber Content of Selected Foods

Food	Grams of Total Fiber
Kidney beans $\left(\frac{1}{2} \text{ c}\right)$	4.5
Oatmeal, cooked $\left(\frac{3}{4} \text{ c}\right)$	3.0
Peanut butter, chunky (2 tbsp)	1.5
Popcorn (1 c)	1.0
Potato, baked, with skin (1 med)	4.0
Whole wheat bread (1 slice)	2.5

(*Sources:* American Dietetic Association and National Center for Nutrition and Dietetics.)

30.

U.S. Annual Food Sales

Year	Sales in Billions of Dollars
2007	1079
2008	1117
2009	1086
2010	1139
2011	1274
2012	1345

(*Source:* U.S. Department of Agriculture)

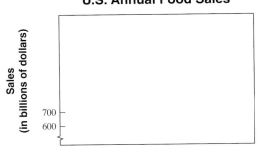

31.

Best-Selling Albums of All Time (U.S. sales)

Album	Estimated Units Sold (in millions)
Led Zeppelin: *Led Zeppelin IV* (1971)	23
Eagles: *Their Greatest Hits* (1976)	29
Pink Floyd: *The Wall* (1979)	23
Michael Jackson: *Thriller* (1982)	29
Billy Joel: *Greatest Hits: Volumes I & II* (1985)	23

(*Source:* Recording Industry Association of America)

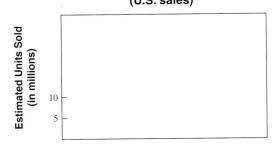

32.

Selected Worldwide Commercial Space Launches

Source	Total Commercial Space Launches 1990–2013
United States	162
Europe	157
Russia	153
China	23
Sea Launch[*]	40

[*]Sea Launch is an international venture involving 4 countries that uses its own launch facility outside national borders.

(*Source:* Bureau of Transportation Statistics)

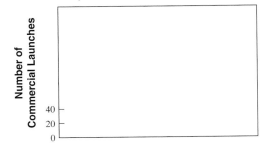

Objective C *The following histogram shows the number of miles that each adult, from a survey of 100 adults, drives per week. Use this histogram to answer Exercises 33 through 42. See Examples 5 and 6.*

33. How many adults drive 100–149 miles per week?

34. How many adults drive 200–249 miles per week?

▶ 35. How many adults drive fewer than 150 miles per week?

36. How many adults drive 200 miles or more per week?

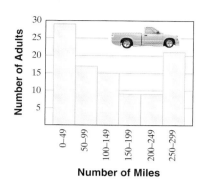

Number of Miles

37. How many adults drive 100–199 miles per week?

38. How many adults drive 150–249 miles per week?

▶ 39. How many more adults drive 250–299 miles per week than 200–249 miles per week?

40. How many more adults drive 0–49 miles per week than 50–99 miles per week?

41. What is the ratio of adults who drive 150–199 miles per week to the total number of adults surveyed?

42. What is the ratio of adults who drive 50–99 miles per week to the total number of adults surveyed?

The following histogram shows the projected population (in millions), by age groups, for the United States for the year 2020. Use this histogram to answer Exercises 43 through 50. For Exercises 45 through 48, estimate to the nearest whole million. See Examples 5 and 6.

43. What age range will be the largest population group in 2020?

44. What age range will be the smallest population group in 2020?

45. How large is the population of 20- to 44-year-olds expected to be in 2020?

46. How large is the population of 45- to 64-year-olds expected to be in 2020?

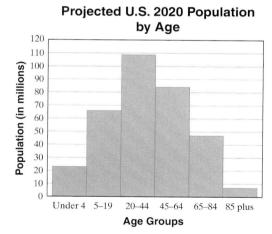

Projected U.S. 2020 Population by Age

Age Groups

47. How large is the population of those less than 4 years old expected to be in 2020?

48. How large is the population of 5- to 19-year-olds expected to be in 2020?

49. Which bar represents the age range you expect to be in during 2020?

50. How many more 20- to 44-year-olds are there expected to be than 45- to 64-year-olds in 2020?

The following list shows the golf scores for an amateur golfer. Use this list to complete the frequency distribution table to the right. See Example 7.

78	84	91	93	97
97	95	85	95	96
101	89	92	89	100

Class Intervals (Scores)	Tally	Class Frequency (Number of Games)
▶ **51.** 70–79		
▶ **52.** 80–89		
▶ **53.** 90–99		
▶ **54.** 100–109		

Twenty-five people in a survey were asked to give their current checking account balances. Use the balances shown in the following list to complete the frequency distribution table to the right. See Example 7.

$53	$105	$162	$443	$109
$468	$47	$259	$316	$228
$207	$357	$15	$301	$75
$86	$77	$512	$219	$100
$192	$288	$352	$166	$292

Class Intervals (Account Balances)	Tally	Class Frequency (Number of People)
55. $0–$99		
56. $100–$199		
57. $200–$299		
58. $300–$399		
59. $400–$499		
60. $500–$599		

▶ **61.** Use the frequency distribution table from Exercises 51 through 54 to construct a histogram. See Example 8.

Golf Scores

62. Use the frequency distribution table from Exercises 55 through 60 to construct a histogram. See Example 8.

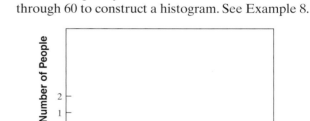

Account Balances

Objective D *The following line graph shows the total points scored by both teams in the NFL Super Bowl from 2006 through 2014. Use this graph to answer Exercises 63 through 70. See Example 9. (Source: superbowlhistory.net)*

▶ **63.** Find the total points scored in the Super Bowl in 2009.

64. Find the total points scored in the Super Bowl in 2013.

▶ **65.** During which year(s) shown were the total points scored in the Super Bowl greater than 50?

66. During which year(s) shown was the total score in the Super Bowl the highest?

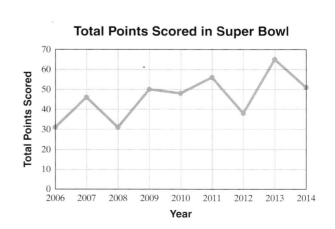

Total Points Scored in Super Bowl

67. During which year(s) shown was the total score in the Super Bowl the lowest?

68. Between 2011 and 2012, did the total score in the Super Bowl increase or decrease?

69. During which year(s) was the total score in the Super Bowl less than 40?

70. Between 2008 and 2009, did the total score in the Super Bowl increase or decrease?

Review

Find each percent. See Section 6.2 or 6.3.

71. 30% of 12

72. 45% of 120

73. 10% of 62

74. 95% of 50

Write each fraction as a percent. See Section 6.1.

75. $\frac{1}{4}$

76. $\frac{2}{5}$

77. $\frac{17}{50}$

78. $\frac{9}{10}$

Concept Extensions

The following double line graph shows temperature highs and lows for a week. Use this graph to answer Exercises 79 through 84.

79. What was the high temperature reading on Thursday?

80. What was the low temperature reading on Thursday?

81. What day was the temperature the lowest? What was this low temperature?

82. What day of the week was the temperature the highest? What was this high temperature?

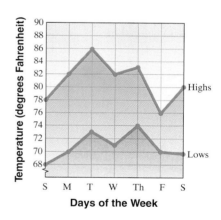

83. On what day of the week was the difference between the high temperature and the low temperature the greatest? What was this difference in temperature?

84. On what day of the week was the difference between the high temperature and the low temperature the least? What was this difference in temperature?

85. True or false? With a bar graph, the width of the bar is just as important as the height of the bar. Explain your answer.

86. Kansas plants about 24% of the wheat acreage in the United States. About how many acres of wheat are planted in the United States, according to the pictograph for Exercises 1 through 8? Round to the nearest million acres.

Objectives

A Read Circle Graphs.

B Draw Circle Graphs.

Objective **A** Reading Circle Graphs

In Exercise Set 6.1, the following **circle graph** was shown. This particular graph shows the favorite sport for 100 adults.

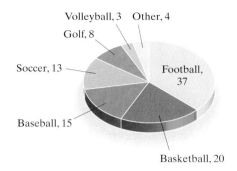

Each sector of the graph (shaped like a piece of pie) shows a category and the relative size of the category. In other words, the most popular sport is football, and it is represented by the largest sector.

Practice 1

Find the ratio of adults preferring golf to total adults. Write the ratio as a fraction in simplest form.

Example 1 Find the ratio of adults preferring basketball to total adults. Write the ratio as a fraction in simplest form.

Solution: The ratio is

$$\frac{\text{people preferring basketball}}{\text{total adults}} = \frac{20}{100} = \frac{1}{5}$$

■ Work Practice 1

A circle graph is often used to show percents in different categories, with the whole circle representing 100%.

Practice 2

Using the circle graph shown in Example 2, determine the percent of visitors to the United States that came from Europe, Asia, and South America.

Example 2 Using a Circle Graph

The following graph shows the percent of visitors to the United States in 2012 from various regions. Using the circle graph shown, determine the percent of visitors who came to the United States from Mexico and Canada.

Solution: To find this percent, we add the percents corresponding to Mexico and Canada. The percent of visitors to the United States that came from Mexico and Canada is

$$21.3\% + 34.1\% = 55.4\%$$

■ Work Practice 2

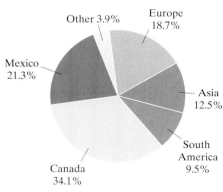

Visitors to U.S. by Region

Source: Office of Travel and Tourism Industries

Answers

1. $\frac{2}{25}$ **2.** 40.7%

Helpful Hint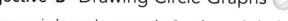

Since a circle graph represents a whole, the percents should add to 100% or 1. Notice this is true for Example 2.

Example 3 Finding Percent of Population

The U.S. Department of Commerce recorded 67 million international visitors to the United States in 2012. Use the circle graph from Example 2 and estimate the number of tourists that might have come from Europe.

Solution: We use the percent equation.

$$\text{amount} = \text{percent} \cdot \text{base}$$

$$\text{amount} = 0.187 \cdot 67{,}000{,}000$$

$$= 0.187(67{,}000{,}000)$$

$$= 12{,}529{,}000$$

Thus, 12,529,000 tourists might have come from Europe in 2012.

■ Work Practice 3

Practice 3

Use the information in Example 3 and the circle graph from Example 2 to estimate the number of tourists that might have come from Mexico in 2012.

✓**Concept Check** Can the following data be represented by a circle graph? Why or why not?

Responses to the Question, "In Which Activities Are You Involved?"	
Intramural sports	60%
On-campus job	42%
Fraternity/sorority	27%
Academic clubs	21%
Music programs	14%

Objective B Drawing Circle Graphs ▶

To draw a circle graph, we use the fact that a whole circle contains 360° (degrees).

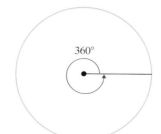

Answer

3. 14,271,000 tourists from Mexico

✓**Concept Check Answer**

no; the percents add up to more than 100%

Practice 4

Use the data shown to draw a circle graph.

Freshmen	30%
Sophomores	27%
Juniors	25%
Seniors	18%

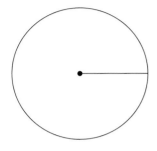

Example 4   Drawing a Circle Graph for U.S. Armed Forces Personnel

The following table shows the percent of U.S. armed forces personnel that were in each branch of service in 2011. (*Source:* U.S. Department of Defense)

Branch of Service	Percent
Army	38
Navy	23
Marine Corps	14
Air Force	22
Coast Guard	3

Draw a circle graph showing this data.

Solution: First we find the number of degrees in each sector representing each branch of service. Remember that the whole circle contains 360°. (We will round degrees to the nearest whole.)

Sector	Degrees in Each Sector
Army	$38\% \times 360° = 0.38 \times 360° = 136.8° \approx 137°$
Navy	$23\% \times 360° = 0.23 \times 360° = 82.8° \approx 83°$
Marine Corps	$14\% \times 360° = 0.14 \times 360° = 50.4° \approx 50°$
Air Force	$22\% \times 360° = 0.22 \times 360° = 79.2° \approx 79°$
Coast Guard	$3\% \times 360° = 0.03 \times 360° = 10.8° \approx 11°$

Helpful Hint

Check your calculations by finding the sum of the degrees.

$$137° + 83° + 50° + 79° + 11° = 360°$$

The sum should be 360°. (It may vary only slightly because of rounding.)

Next we draw a circle and mark its center. Then we draw a line from the center of the circle to the circle itself.

To construct the sectors, we will use a **protractor.** A protractor measures the number of degrees in an angle. We place the hole in the protractor over the center of the circle. Then we adjust the protractor so that 0° on the protractor is aligned with the line that we drew.

It makes no difference which sector we draw first. To construct the "Army" sector, we find 137° on the protractor and mark our circle. Then we remove the protractor and use this mark to draw a second line from the center to the circle itself.

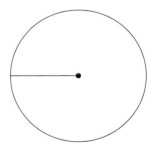

Answer

4.

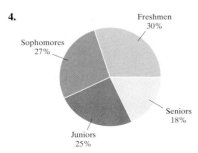

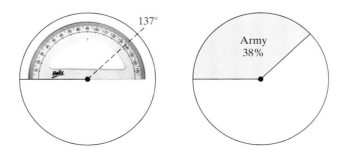

To construct the "Navy" sector, we follow the same procedure, except that we line up 0° with the second line we drew and mark the protractor at 83°.

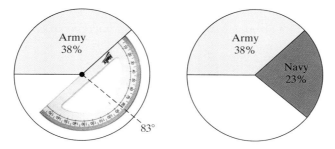

We continue in this manner until the circle graph is complete.

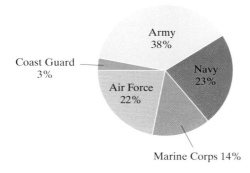

■ Work Practice 4

✓**Concept Check**   True or false? The larger a sector in a circle graph, the larger the percent of the total it represents. Explain your answer.

✓**Concept Check Answer**
true

Vocabulary, Readiness & Video Check

Use the choices below to fill in each blank.

    sector   circle   100   360

1. In a _____ graph, each section (shaped like a piece of pie) shows a category and the relative size of the category.

2. A circle graph contains pie-shaped sections, each called a _____ .

3. The number of degrees in a whole circle is _____ .

4. If a circle graph has percent labels, the percents should add up to _____ .

Martin-Gay Interactive Videos   *Watch the section lecture video and answer the following questions.*

Objective A   **5.** From ▦ Example 3, the circle graph shows percents in different categories. What does the whole circle graph represent? ◯

Objective B   **6.** From ▦ Example 6, when looking at the sector degree measures of a circle graph, the whole circle graph corresponds to what degree measure? ◯

See Video 7.2 ◯

7.2 Exercise Set MyMathLab®

Objective A *The following circle graph is a result of surveying 700 college students. They were asked where they live while attending college. Use this graph to answer Exercises 1 through 6. Write all ratios as fractions in simplest form. See Example 1.*

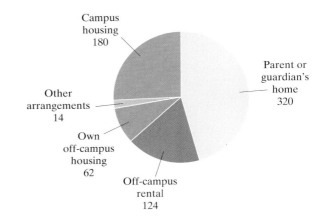

1. Where do most of these college students live?

2. Besides the category "Other arrangements," where do the fewest of these college students live?

3. Find the ratio of students living in campus housing to total students.

4. Find the ratio of students living in off-campus rentals to total students.

5. Find the ratio of students living in campus housing to students living in a parent or guardian's home.

6. Find the ratio of students living in off-campus rentals to students living in a parent or guardian's home.

The following circle graph shows the percent of the land area of the continents on Earth. Use this graph for Exercises 7 through 14. See Example 2.

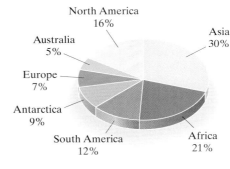

Source: National Geographic Society

7. Which continent is the largest?

8. Which continent is the smallest?

9. What percent of the land on Earth is accounted for by Asia and Europe together?

10. What percent of the land on Earth is accounted for by North and South America?

The total amount of land from the continents is approximately 57,000,000 square miles. Use the graph to find the area of the continents given in Exercises 11 through 14. See Example 3.

11. Asia

12. South America

13. Australia

14. Europe

The following circle graph shows the percent of the types of books available at Midway Memorial Library. Use this graph for Exercises 15 through 24. See Example 2.

15. What percent of books are classified as some type of fiction?

16. What percent of books are nonfiction or reference?

17. What is the second-largest category of books?

18. What is the third-largest category of books?

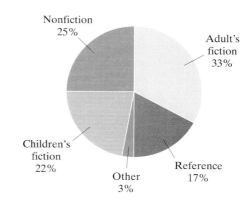

If this library has 125,600 books, find how many books are in each category given in Exercises 19 through 24. See Example 3.

19. Nonfiction

20. Reference

21. Children's fiction

22. Adult's fiction

23. Reference or other

24. Nonfiction or other

Objective B *Fill in the tables. Round to the nearest degree. Then draw a circle graph to represent the information given in each table. (Remember: The total of "Degrees in Sector" column should equal 360° or very close to 360° because of rounding.) See Example 4.*

25.

Types of Apples Grown in Washington State		
Type of Apple	**Percent**	**Degrees in Sector**
Red Delicious	37%	
Golden Delicious	13%	
Fuji	14%	
Gala	15%	
Granny Smith	12%	
Other varieties	6%	
Braeburn	3%	
(*Source:* U.S. Apple Association)		

26.

Color Distribution of M&M's Milk Chocolate		
Color	**Percent**	**Degrees in Sector**
Blue	22.1%	
Orange	16.7%	
Green	16.7%	
Red	16.7%	
Brown	16.7%	
Yellow	11.1%	
(*Source:* M&M Mars)		

27.

Distribution of Large Dams by Continent		
Continent	**Percent**	**Degrees in Sector**
Europe	19%	
North America	32%	
South America	3%	
Asia	39%	
Africa	5%	
Australia	2%	
(*Source:* International Commission on Large Dams)		

28.

Number of Times the "Are We There Yet?" Question Is Asked of Parents During Road Trips		
	Percent	**Degrees in Sector**
Never	20%	
Once	11%	
2–5 times	36%	
6–10 times	14%	
More than 10 times	19%	
(*Source:* KRC Research for Goodyear Tire & Rubber Co.)		

Review

Write the prime factorization of each number. See Section 3.2.

29. 20

30. 25

31. 40

32. 16

33. 85

34. 105

Concept Extensions

The following circle graph shows the relative sizes of the great oceans. Use this graph for Exercises 35 through 40.

35. Without calculating, determine which ocean is the largest. How can you answer this question by looking at the circle graph?

36. Without calculating, determine which ocean is the smallest. How can you answer this question by looking at the circle graph?

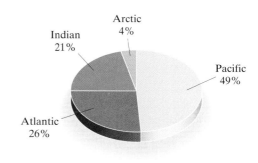

Source: *Philip's World Atlas*

These oceans together make up 264,489,800 square kilometers of the Earth's surface. Find the square kilometers for each ocean.

37. Pacific Ocean

38. Atlantic Ocean

39. Indian Ocean

40. Arctic Ocean

The following circle graph summarizes the results of a survey of 2800 Internet users who make purchases online. Use this graph for Exercises 41 through 46. Round to the nearest whole.

41. How many of the survey respondents said that they spend $0–$15 online each month?

42. How many of the survey repondents said that they spend $15–$175 online each month?

43. How many of the survey respondents said that they spend $0 to $175 online each month?

44. How many of the survey respondents said that they spend $15 to over $175 online each month?

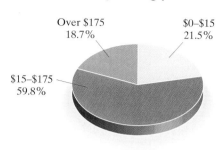

Online Spending per Month

Over $175
18.7%

$0–$15
21.5%

$15–$175
59.8%

Source: UCLA Center for Communication Policy

45. Find the ratio of *number* of respondents who spend $0–$15 online to *number* of respondents who spend $15–$175 online. Write the ratio as a fraction with integers in the numerator and denominator.

46. Find the ratio of *percent* of respondents who spend $0–$15 online to *percent* of those who spend $15–$175. Write the ratio as a fraction with integers in the numerator and denominator.

Solve. See the Concept Checks in this section.

47. Can the data below be represented by a circle graph? Why or why not?

Responses to the Question, "What Classes Are You Taking?"	
Math	80%
English	72%
History	37%
Biology	21%
Chemistry	14%

48. True or false? The smaller a sector in a circle graph, the smaller the percent of the total it represents. Explain why.

Answers

Reading Graphs

The following pictograph shows the six occupations with the largest estimated numerical increase in employment in the United States between 2010 and 2020. Use this graph to answer Exercises 1 through 4.

1. _____

2. _____

3. _____

4. _____

5. _____

6. _____

7. _____

8. _____

Jobs with Projected Highest Numerical Increase: 2010–2020

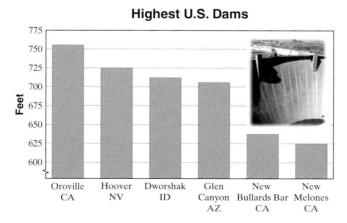

Source: Bureau of Labor Statistics

1. Approximate the increase in the number of registered nurses from 2010 to 2020.

2. Approximate the increase in the number of general office clerks from 2010 to 2020.

3. Which occupations are expected to show approximately the same increase in numbers of employees between the years shown?

4. Which of the listed occupations is expected to show the least increase in number of employees between the years shown?

The following bar graph shows the highest U.S. dams. Use this graph to answer Exercises 5 through 8.

Highest U.S. Dams

Source: Committee on Register of Dams

5. Name the U.S. dam with the greatest height and estimate its height.

6. Name the U.S. dam whose height is between 625 and 650 feet and estimate its height.

7. Estimate how much higher the Hoover Dam is than the Glen Canyon Dam.

8. How many U.S. dams have heights over 700 feet?

The following line graph shows the daily high temperatures for 1 week in Annapolis, Maryland. Use this graph to answer Exercises 9 through 12.

9. Name the day(s) of the week with the highest temperature and give that high temperature.

10. Name the day(s) of the week with the lowest temperature and give that low temperature.

11. On what days of the week was the temperature less than 90° Fahrenheit?

12. On what days of the week was the temperature greater than 90° Fahrenheit?

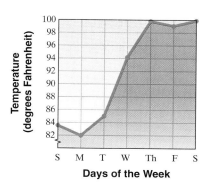

Days of the Week

The following circle graph shows the types of milk beverage consumed in the United States. Use this graph for Exercises 13 through 16.

If a store in Kerrville, Texas, sells 200 quart containers of milk per week, estimate how many quart containers are sold in each category below.

Types of Milk Beverage Consumed

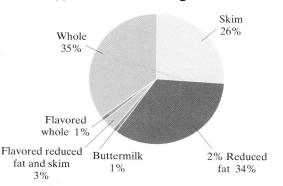

Source: U.S. Department of Agriculture

13. Whole milk

14. Skim milk

15. Buttermilk

16. Flavored reduced fat and skim milk

The following list shows weekly quiz scores for a student in basic college mathematics. Use this list to complete the frequency distribution table.

50	80	71	83	86
67	89	93	88	97
	53	90		
75	80	78	93	99

	Class Intervals (Scores)	Tally	Class Frequency (Number of Quizzes)
17.	50–59		
18.	60–69		
19.	70–79		
20.	80–89		
21.	90–99		

22. Use the table from Exercises 17 through 21 to construct a histogram.

Quiz Scores

9. _____

10. _____

11. _____

12. _____

13. _____

14. _____

15. _____

16. _____

17. _____

18. _____

19. _____

20. _____

21. _____

22. _____

Objectives

A Find the Mean of a List of Numbers.

B Find the Median of a List of Numbers.

C Find the Mode of a List of Numbers.

Objective A Finding the Mean

Sometimes we want to summarize data by displaying them in a graph, but sometimes it is also desirable to be able to describe a set of data, or a set of numbers, by a single "middle" number. Three such **measures of central tendency** are the **mean,** the **median,** and the **mode.**

The most common measure of central tendency is the mean (sometimes called the "arithmetic mean" or the "average"). Recall that we first introduced finding the average of a list of numbers in Section 1.7.

The **mean (average)** of a set of number items is the sum of the items divided by the number of items.

$$\text{mean} = \frac{\text{sum of items}}{\text{number of items}}$$

Practice 1

Find the mean of the following test scores: 87, 75, 96, 91, and 78.

Example 1   Finding the Mean Time in an Experiment

Seven students in a psychology class conducted an experiment on mazes. Each student was given a pencil and asked to successfully complete the same maze. The timed results are below:

Student	Ann	Thanh	Carlos	Jesse	Melinda	Ramzi	Dayni
Time (Seconds)	13.2	11.8	10.7	16.2	15.9	13.8	18.5

a. Who completed the maze in the shortest time? Who completed the maze in the longest time?

b. Find the mean time.

c. How many students took longer than the mean time? How many students took shorter than the mean time?

Solution:

a. Carlos completed the maze in 10.7 seconds, the shortest time. Dayni completed the maze in 18.5 seconds, the longest time.

b. To find the mean (or average), we find the sum of the items and divide by 7, the number of items.

$$\text{mean} = \frac{13.2 + 11.8 + 10.7 + 16.2 + 15.9 + 13.8 + 18.5}{7}$$

$$= \frac{100.1}{7} = 14.3$$

c. Three students, Jesse, Melinda, and Dayni, had times longer than the mean time. Four students, Ann, Thanh, Carlos, and Ramzi, had times shorter than the mean time.

 Work Practice 1

✓**Concept Check** Estimate the mean of the following set of data:

5, 10, 10, 10, 10, 15

Often in college, the calculation of a **grade point average** (GPA) is a **weighted mean** and is calculated as shown in Example 2.

Answer
1. 85.4

✓**Concept Check Answer**
10

Example 2 Calculating Grade Point Average (GPA)

The following grades were earned by a student during one semester. Find the student's grade point average.

Course	Grade	Credit Hours
College mathematics	A	3
Biology	B	3
English	A	3
PE	C	1
Social studies	D	2

Solution: To calculate the grade point average, we need to know the point values for the different possible grades. The point values of grades commonly used in colleges and universities are given below:

A: 4, B: 3, C: 2, D: 1, F: 0

Now, to find the grade point average, we multiply the number of credit hours for each course by the point value of each grade. The grade point average is the sum of these products divided by the sum of the credit hours.

Course	Grade	Point Value of Grade	Credit Hours	Point Value × Credit Hours
College mathematics	A	4	3	12
Biology	B	3	3	9
English	A	4	3	12
PE	C	2	1	2
Social studies	D	1	2	2
		Totals:	12	37

$$\text{grade point average} = \frac{37}{12} \approx 3.08 \text{ rounded to two decimal places}$$

The student earned a grade point average of 3.08.

■ Work Practice 2

Objective B Finding the Median

You may have noticed that a very low number or a very high number can affect the mean of a list of numbers. Because of this, you may sometimes want to use another measure of central tendency. A second measure of central tendency is called the **median.** The median of a list of numbers is not affected by a low or high number in the list.

The **median** of a set of numbers in numerical order is the middle number. If the number of items is odd, the median is the middle number. If the number of items is even, the median is the mean of the two middle numbers.

Example 3 Find the median of the following list of numbers:

25, 54, 56, 57, 60, 71, 98

Solution: Because this list is in numerical order, the median is the middle number, 57.

■ Work Practice 3

Practice 2

Find the grade point average if the following grades were earned in one semester.

Grade	Credit Hours
A	2
B	4
C	5
D	2
A	2

Practice 3

Find the median of the list of numbers: 5, 11, 14, 23, 24, 35, 38, 41, 43

Answers

2. 2.67 **3.** 24

Practice 4

Find the median of the list of scores:

36, 91, 78, 65, 95, 95, 88, 71

Example 4   Find the median of the following list of scores: 67, 91, 75, 86, 55, 91

Solution: First we list the scores in numerical order and then find the middle number.

55, 67, 75, 86, 91, 91

Since there is an even number of scores, there are two middle numbers, 75 and 86. The median is the mean of the two middle numbers.

$$\text{median} = \frac{75 + 86}{2} = 80.5$$

The median is 80.5.

■ Work Practice 4

> **Helpful Hint**   Don't forget to write the numbers in order from smallest to largest before finding the median.

Objective C Finding the Mode

The last common measure of central tendency is called the **mode.**

> The **mode** of a set of numbers is the number that occurs most often. (It is possible for a set of numbers to have more than one mode or to have no mode.)

Practice 5

Find the mode of the list of numbers:

14, 10, 10, 13, 15, 15, 15, 17, 18, 18, 20

Example 5   Find the mode of the list of numbers:

11, 14, 14, 16, 31, 56, 65, 77, 77, 78, 79

Solution: There are two numbers that occur the most often. They are 14 and 77. This list of numbers has two modes, 14 and 77.

■ Work Practice 5

Practice 6

Find the median and the mode of the list of numbers:

26, 31, 15, 15, 26, 30, 16, 18, 15, 35

Example 6   Find the median and the mode of the following set of numbers. These numbers were high temperatures for 14 consecutive days in a city in Montana.

76, 80, 85, 86, 89, 87, 82, 77, 76, 79, 82, 89, 89, 92

Solution: First we write the numbers in numerical order.

76, 76, 77, 79, 80, 82, 82, 85, 86, 87, 89, 89, 89, 92

Since there is an even number of items, the median is the mean of the two middle numbers, 82 and 85.

$$\text{median} = \frac{82 + 85}{2} = 83.5$$

The mode is 89, since 89 occurs most often.

■ Work Practice 6

Answers

4. 83   **5.** 15   **6.** median: 22; mode: 15

✓ **Concept Check Answer**

false; a set of numbers may have no mode

✓ **Concept Check**   True or false? Every set of numbers *must* have a mean, median, and mode. Explain your answer.

Helpful Hint

Don't forget that it is possible for a list of numbers to have no mode. For example, the list

2, 4, 5, 6, 8, 9

has no mode. There is no number or numbers that occur more often than the others.

Vocabulary, Readiness & Video Check

Use the choices below to fill in each blank. Some choices may be used more than once.

mean	mode	grade point average
median	average	

1. Another word for "mean" is _____ .

2. The number that occurs most often in a set of numbers is called the _____ .

3. The _____ of a set of number items is $\dfrac{\text{sum of items}}{\text{number of items}}$.

4. The _____ of a set of numbers is the middle number. If the number of numbers is even, it is the _____ of the two middle numbers.

5. An example of a weighted mean is a calculation of _____ .

Martin-Gay Interactive Videos   Watch the section lecture video and answer the following questions.

See Video 7.3

Objective A 6. Why is the ≈ symbol used in Example 1?

Objective B 7. From Example 3, what is always the first step when finding the median of a set of data numbers?

Objective C 8. From Example 4, why do you think it is helpful to have data numbers in numerical order when finding the mode?

7.3 Exercise Set MyMathLab®

Objectives A B C Mixed Practice *For each set of numbers, find the mean, median, and mode. If necessary, round the mean to one decimal place. See Examples 1 and 3 through 6.*

1. 15, 23, 24, 18, 25

2. 45, 36, 28, 46, 52

3. 7.6, 8.2, 8.2, 9.6, 5.7, 9.1

4. 4.9, 7.1, 6.8, 6.8, 5.3, 4.9

5. 0.5, 0.2, 0.2, 0.6, 0.3, 1.3, 0.8, 0.1, 0.5

6. 0.6, 0.6, 0.8, 0.4, 0.5, 0.3, 0.7, 0.8, 0.1

7. 231, 543, 601, 293, 588, 109, 334, 268

8. 451, 356, 478, 776, 892, 500, 467, 780

The ten tallest buildings in the world, completed as of 2013, are listed in the following table. Use this table to answer Exercises 9 through 14. If necessary, round results to one decimal place. See Examples 1 and 3 through 6.

9. Find the mean height of the five tallest buildings.

10. Find the median height of the five tallest buildings.

11. Find the median height of the eight tallest buildings.

12. Find the mean height of the eight tallest buildings.

Building	Height (in feet)
Burj Khalifa	2717
Makkah Royal Clock Tower Hotel	1972
Taipei 101	1667
Shanghai World Financial Center	1614
International Commerce Centre	1588
Petronas Tower 1	1483
Petronas Tower 2	1483
Zifeng Tower	1476
Willis Tower	1451
KK100	1449

(*Source:* Council on Tall Buildings and Urban Habitat)

13. Given the building heights, explain how you know, without calculating, that the answer to Exercise 10 is greater than the answer to Exercise 11.

14. Given the building heights, explain how you know, without calculating, that the answer to Exercise 12 is less than the answer to Exercise 9.

For Exercises 15 through 18, the grades are given for a student for a particular semester. Find the grade point average. If necessary, round the grade point average to the nearest hundredth. See Example 2.

15.

Grade	Credit Hours
B	3
C	3
A	4
C	4

16.

Grade	Credit Hours
D	1
F	1
C	4
B	5

17.

Grade	Credit Hours
A	3
A	3
A	4
B	3
C	1

18.

Grade	Credit Hours
B	2
B	2
C	3
A	3
B	3

For Exercises 19 through 27, find the mean, median, and mode, as requested. See Examples 1 and 3 through 6. During an experiment, the following times (in seconds) were recorded:

7.8, 6.9, 7.5, 4.7, 6.9, 7.0.

19. Find the mean.

20. Find the median.

21. Find the mode.

In a mathematics class, the following test scores were recorded for a student: 93, 85, 89, 79, 88, 92.

22. Find the mean. Round to the nearest hundredth.

23. Find the median.

24. Find the mode.

The following pulse rates were recorded for a group of 15 students:

78, 80, 66, 68, 71, 64, 82, 71, 70, 65, 70, 75, 77, 86, 72.

25. Find the mean.

26. Find the median.

27. Find the mode.

28. How many pulse rates were higher than the mean?

29. How many pulse rates were lower than the mean?

Review

Write each fraction in simplest form. See Section 3.2.

30. $\dfrac{12}{20}$

31. $\dfrac{6}{18}$

32. $\dfrac{4}{36}$

33. $\dfrac{18}{30}$

34. $\dfrac{35}{100}$

35. $\dfrac{55}{75}$

Concept Extensions

Find the missing numbers in each set of numbers.

36. 16, 18, _____, _____, _____. The mode is 21. The median is 20.

37. _____, _____, _____, 40, _____. The mode is 35. The median is 37. The mean is 38.

Solve.

38. Write a list of numbers for which you feel the median would be a better measure of central tendency than the mean.

39. Without making any computations, decide whether the median of the following list of numbers will be a whole number. Explain your reasoning.

36, 77, 29, 58, 43

7.4 Counting and Introduction to Probability

Objective A Using a Tree Diagram

In our daily conversations, we often talk about the likelihood or **probability** of a given result occurring. For example:

The *chance* of thundershowers is 70 percent.

What are the *odds* that the New Orleans Saints will go to the Super Bowl?

What is the *probability* that you will finish cleaning your room today?

Each of these chance happenings—thundershowers, the New Orleans Saints playing in the Super Bowl, and cleaning your room today—is called an **experiment.** The possible results of an experiment are called **outcomes.** For example, flipping a coin is an experiment, and the possible outcomes are heads (H) or tails (T).

One way to picture the outcomes of an experiment is to draw a **tree diagram.** Each outcome is shown on a separate branch. For example, the outcomes of flipping a coin are

Objectives

A Use a Tree Diagram to Count Outcomes.

B Find the Probability of an Event.

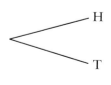

Heads

Tails

Practice 1

Draw a tree diagram for tossing a coin three times. Then use the diagram to find the number of possible outcomes.

Practice 2

Draw a tree diagram for an experiment consisting of tossing a coin and then rolling a die. Then use the diagram to find the number of possible outcomes.

Example 1   Draw a tree diagram for tossing a coin twice. Then use the diagram to find the number of possible outcomes.

Solution:

First Coin Toss	Second Coin Toss	Outcomes
H	H	H, H
	T	H, T
T	H	T, H
	T	T, T

There are 4 possible outcomes when tossing a coin twice.

■ Work Practice 1

Example 2   Draw a tree diagram for an experiment consisting of rolling a die and then tossing a coin. Then use the diagram to find the number of possible outcomes.

Die

Solution: Recall that a die has six sides and that each side represents a number, 1 through 6.

Roll a Die	Toss a Coin	Outcomes
1	H	1, H
	T	1, T
2	H	2, H
	T	2, T
3	H	3, H
	T	3, T
4	H	4, H
	T	4, T
5	H	5, H
	T	5, T
6	H	6, H
	T	6, T

There are 12 possible outcomes for rolling a die and then tossing a coin.

■ Work Practice 2

Any number of outcomes considered together is called an **event.** For example, when tossing a coin twice, H, H is an event. The event is tossing heads first and tossing heads second. Another event would be tossing tails first and then heads (T, H), and so on.

Answers

1.

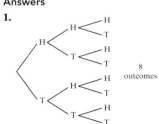

8 outcomes

2.

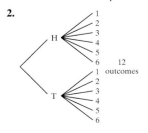

12 outcomes

Objective B Finding the Probability of an Event

As we mentioned earlier, the **probability of an event is a measure of the chance or likelihood of it occurring.** For example, if a coin is tossed, what is the probability that heads occurs? Since one of two equally likely possible outcomes is heads, the probability is $\frac{1}{2}$.

The Probability of an Event

$$\text{probability of an event} = \frac{\text{number of ways that the event can occur}}{\text{number of possible outcomes}}$$

Note from the definition of probability that the probability of an event is always between 0 and 1, inclusive (i.e., including 0 and 1). A probability of 0 means that an event won't occur, and a probability of 1 means that an event is certain to occur.

Example 3 If a coin is tossed twice, find the probability of tossing heads on the first toss and then heads again on the second toss (H, H).

Solution: 1 way the event can occur

$$\underbrace{\text{H, T,　H, H,　T, H,　T, T}}_{\text{4 possible outcomes}}$$

$$\text{probability} = \frac{1}{4} \quad \begin{array}{l}\text{Number of ways the event can occur} \\ \text{Number of possible outcomes}\end{array}$$

The probability of tossing heads and then heads is $\frac{1}{4}$.

■ Work Practice 3

Practice 3
If a coin is tossed three times, find the probability of tossing tails, then heads, and then tails (T, H, T).

Example 4 If a die is rolled one time, find the probability of rolling a 3 or a 4.

Solution: Recall that there are 6 possible outcomes when rolling a die.

2 ways that the event can occur

$$\text{possible outcomes:}\quad \underbrace{1,\ \ 2,\ \ 3,\ \ 4,\ \ 5,\ \ 6}_{\text{6 possible outcomes}}$$

$$\text{probability of a 3 or a 4} = \frac{2}{6} \quad \begin{array}{l}\text{Number of ways the event can occur} \\ \text{Number of possible outcomes}\end{array}$$

$$= \frac{1}{3} \quad \text{Simplest form}$$

■ Work Practice 4

Practice 4
If a die is rolled one time, find the probability of rolling a 2 or a 5.

Answers
3. $\frac{1}{8}$　4. $\frac{1}{3}$

✔**Concept Check** Suppose you have calculated a probability of $\frac{11}{9}$. How do you know that you have made an error in your calculation?

✔**Concept Check Answer**
The number of ways an event can occur can't be larger than the number of possible outcomes.

Practice 5

Use the diagram and information in Example 5 and find the probability of choosing a blue marble from the box.

Example 5 Find the probability of choosing a red marble from a box containing 1 red, 1 yellow, and 2 blue marbles.

Solution: 1 way that event can occur

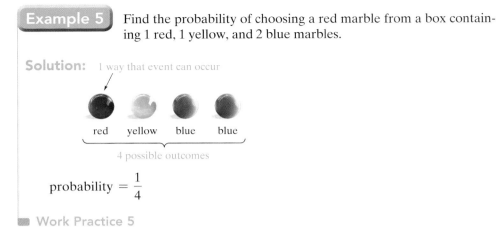

red   yellow   blue   blue

4 possible outcomes

$$\text{probability} = \frac{1}{4}$$

Work Practice 5

Answer

5. $\frac{1}{2}$

Vocabulary, Readiness & Video Check

Use the choices below to fill in each blank. Choices may be used more than once.

0   probability   tree diagram

1   outcome

1. A possible result of an experiment is called a(n) _____ .

2. A(n) _____ shows each outcome of an experiment as a separate branch.

3. The _____ of an event is a measure of the likelihood of it occurring.

4. _____ is calculated by the number of ways that the event can occur divided by the number of possible outcomes.

5. A probability of _____ means that an event won't occur.

6. A probability of _____ means that an event is certain to occur.

Martin-Gay Interactive Videos

See Video 7.4

Watch the section lecture video and answer the following questions.

Objective A **7.** In ▤ Example 1, how was the possible number of outcomes to the experiment determined from the tree diagram? ◉

Objective B **8.** In ▤ Example 2, what is the probability of getting a 7? Explain this result. ◉

7.4 **Exercise Set** MyMathLab®

Objective A *Draw a tree diagram for each experiment. Then use the diagram to find the number of possible outcomes. See Examples 1 and 2.*

1. Choosing a letter in the word MATH and then a number (1, 2, or 3)

2. Choosing a number (1 or 2) and then a vowel (a, e, i, o, u)

Use the following spinners to draw a tree diagram and find the number of outcomes for Exercises 3–10.

Spinner A

Spinner B

3. Spinning Spinner A once

4. Spinning Spinner B once

5. Spinning Spinner B twice

6. Spinning Spinner A twice

7. Spinning Spinner A and then Spinner B

8. Spinning Spinner B and then Spinner A

9. Tossing a coin and then spinning Spinner B

10. Spinning Spinner A and then tossing a coin

Objective B *If a single die is tossed once, find the probability of each event. See Examples 3 through 5.*

▶ **11.** A 5

12. A 9

13. A 1 or a 6

14. A 2 or a 3

▶ **15.** An even number

16. An odd number

17. A number greater than 2

18. A number less than 6

Suppose the spinner shown is spun once. Find the probability of each event. See Examples 3 through 5.

▶ **19.** The result of the spin is 2.

20. The result of the spin is 3.

21. The result of the spin is 1, 2, or 3.

22. The result of the spin is not 3.

23. The result of the spin is an odd number.

24. The result of the spin is an even number.

If a single choice is made from the bag of marbles shown, find the probability of each event. See Examples 3 through 5.

25. A red marble is chosen.

26. A blue marble is chosen.

27. A yellow marble is chosen.

28. A green marble is chosen.

29. A green or red marble is chosen.

30. A blue or yellow marble is chosen.

A new drug is being tested that is supposed to lower blood pressure. This drug was given to 200 people and the results are shown below. See Examples 3 through 5.

Lower Blood Pressure 152	Higher Blood Pressure 38	Blood Pressure Not Changed 10

31. If a person is testing this drug, what is the probability that his or her blood pressure will be higher?

32. If a person is testing this drug, what is the probability that his or her blood pressure will be lower?

33. If a person is testing this drug, what is the probability that his or her blood pressure will not change?

34. What is the sum of the answers to Exercises 31, 32, and 33? In your own words, explain why.

Review

Perform each indicated operation. See Sections 3.3 and 3.5.

35. $\dfrac{1}{2} + \dfrac{1}{3}$

36. $\dfrac{7}{10} - \dfrac{2}{5}$

37. $\dfrac{1}{2} \cdot \dfrac{1}{3}$

38. $\dfrac{7}{10} \div \dfrac{2}{5}$

39. $5 \div \dfrac{3}{4}$

40. $\dfrac{3}{5} \cdot 10$

Concept Extensions

Recall that a deck of cards contains 52 cards. These cards consist of four suits (hearts, spades, clubs, and diamonds) of each of the following: 2, 3, 4, 5, 6, 7, 8, 9, 10, jack, queen, king, and ace. If a card is chosen from a deck of cards, find the probability of each event.

41. The king of hearts

42. The 10 of spades

43. A king

44. A 10

45. A heart

46. A club

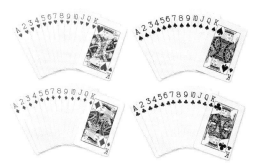

47. A card in black ink

48. A queen or ace

Two dice are tossed. Find the probability of each sum of the dice. (Hint: Draw a tree diagram of the possibilities of two tosses of a die, and then find the sum of the numbers on each branch.)

49. A sum of 6

50. A sum of 10

51. A sum of 13

52. A sum of 2

Solve. See the Concept Check in this section.

53. In your own words, explain why the probability of an event cannot be greater than 1.

54. In your own words, explain when the probability of an event is 0.

Chapter 7 Group Activity

Sections 7.1, 7.3

This activity may be completed by working in groups or individually.

How often have you read an article in a newspaper or in a magazine that included results from a survey or poll? Surveys seem to have become very popular ways of getting feedback on anything from a political candidate, to a new product, to services offered by a health club. In this activity, you will conduct a survey and analyze the results.

1. Conduct a survey of 30 students in one of your classes. Ask each student to report his or her age.

2. Classify each age according to the following categories: under 20, 20 to 24, 25 to 29, 30 to 39, 40 to 49, and 50 or over. Tally the number of your survey respondents that fall into each category. Make a bar graph of your results. What does this graph tell you about the ages of your survey respondents?

3. Find the average age of your survey respondents.

4. Find the median age of your survey respondents.

5. Find the mode of the ages of your survey respondents.

6. Compare the mean, median, and mode of your age data. Are these measures similar? Which is largest? Which is smallest? If there is a noticeable difference between any of these measures, can you explain why?

Chapter 7 Vocabulary Check

Fill in each blank with one of the words or phrases listed below.

outcomes	bar	experiment	mean	tree diagram
pictograph	line	class interval	median	probability
histogram	circle	class frequency	mode	

1. A(n) _____ graph presents data using vertical or horizontal bars.

2. The _____ of a set of number items is $\dfrac{\text{sum of items}}{\text{number of items}}$.

3. The possible results of an experiment are the _____.

4. A(n) _____ is a graph in which pictures or symbols are used to visually present data.

5. The _____ of a set of numbers is the number that occurs most often.

6. A(n) _____ graph displays information with a line that connects data points.

7. The _____ of an ordered set of numbers is the middle number.

8. A(n) _____ is one way to picture and count outcomes.

9. A(n) _____ is an activity being considered, such as tossing a coin or rolling a die.

10. In a(n) _____ graph, each section (shaped like a piece of pie) shows a category and the relative size of the category.

11. The _____ of an event is $\dfrac{\text{number of ways that the event can occur}}{\text{number of possible outcomes}}$.

12. A(n) _____ is a special bar graph in which the width of each bar represents a(n) _____ and the height of each bar represents the _____.

Helpful Hint

▶ Are you preparing for your test? Don't forget to take the Chapter 7 Test on page 556. Then check your answers at the back of your text and use the Chapter Test Prep Videos to see the fully worked-out solutions to any of the exercises you want to review.

7 Chapter Highlights

Definitions and Concepts	Examples

Section 7.1   Reading Pictographs, Bar Graphs, Histograms, and Line Graphs

A **pictograph** is a graph in which pictures or symbols are used to visually present data.

A **line graph** displays information with a line that connects data points.

A **bar graph** presents data using vertical or horizontal bars.

The bar graph on the right shows the number of acres of wheat harvested in 2014 for leading states.

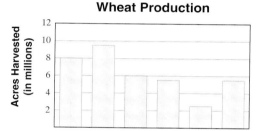

Wheat Production

Source: National Agricultural Statistics Service

1. Approximately how many acres of wheat were harvested in Kansas?

9,500,000 acres

2. About how many more acres of wheat were harvested in North Dakota than South Dakota?

$$\begin{array}{r} 8 \text{ million} \\ -2.5 \text{ million} \\ \hline 5.5 \text{ million} \end{array} \quad \text{or } 5{,}500{,}000 \text{ acres}$$

A **histogram** is a special bar graph in which the width of each bar represents a **class interval** and the height of each bar represents the **class frequency.** The histogram on the right shows student quiz scores.

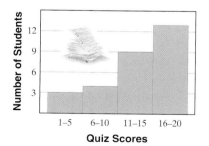

1. How many students received a score of 6–10?

4 students

2. How many students received a score of 11–20?

$9 + 13 = 22$ students

Definitions and Concepts	Examples

Section 7.2   Reading Circle Graphs

In a **circle graph,** each section (shaped like a piece of pie) shows a category and the relative size of the category.

The circle graph on the right classifies tornadoes by wind speed.

Tornado Wind Speeds

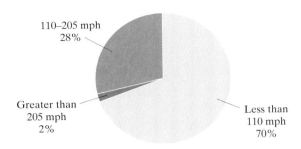

110–205 mph
28%

Greater than
205 mph
2%

Less than
110 mph
70%

Source: National Oceanic and Atmospheric Administration

1. What percent of tornadoes have wind speeds of 110 mph or greater?

   28% + 2% = 30%

2. If there were 1235 tornadoes in the United States in 1995, how many of these might we expect to have had wind speeds less than 110 mph? Find 70% of 1235.

   $70\%(1235) = 0.70(1235) = 864.5 \approx 865$

   Around 865 tornadoes would be expected to have had wind speeds of less than 110 mph.

Section 7.3   Mean, Median, and Mode

The **mean** (or **average**) of a set of number items is

$$\text{mean} = \frac{\text{sum of items}}{\text{number of items}}$$

Find the mean, median, and mode of the following set of numbers: 33, 35, 35, 43, 68, 68

$$\text{mean} = \frac{33 + 35 + 35 + 43 + 68 + 68}{6} = 47$$

The **median** of a set of numbers in numerical order is the middle number. If the number of items is even, the median is the mean of the two middle numbers.

The median is the mean of the two middle numbers, 35 and 43

$$\text{median} = \frac{35 + 43}{2} = 39$$

The **mode** of a set of numbers is the number that occurs most often. (A set of numbers may have no mode or more than one mode.)

There are two modes because there are two numbers that occur twice:

35 and 68

Section 7.4   Counting and Introduction to Probability

An **experiment** is an activity being considered, such as tossing a coin or rolling a die. The possible results of an experiment are the **outcomes**. A **tree diagram** is one way to picture and count outcomes.

Draw a tree diagram for tossing a coin and then choosing a number from 1 to 4.

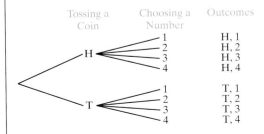

Tossing a Coin	Choosing a Number	Outcomes
H	1	H, 1
	2	H, 2
	3	H, 3
	4	H, 4
T	1	T, 1
	2	T, 2
	3	T, 3
	4	T, 4

Definitions and Concepts	Examples

Section 7.4 Counting and Introduction to Probability (*continued*)

Any number of outcomes considered together is called an **event**. The **probability** of an event is a measure of the chance or likelihood of it occurring. $$\text{probability of an event} = \frac{\text{number of ways that the event can occur}}{\text{number of possible outcomes}}$$	Find the probability of tossing a coin twice and tails occurring each time. 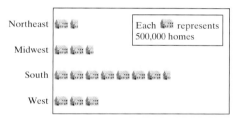 1 way the event can occur HH, HT, TH, TT 4 possible outcomes $$\text{probability} = \frac{1}{4}$$

Chapter 7 Review

(7.1) *The following pictograph shows the number of new homes constructed from August 2012 to August 2013, by region. Use this graph to answer Exercises 1 through 6.*

New Home Construction

Northeast

Each 🏠 represents 500,000 homes

Midwest

South

West

Source: U.S. Census Bureau

1. How many new homes were constructed in the Midwest during the given year?

2. How many new homes were constructed in the West during the given year?

3. Which region had the most new homes constructed?

4. Which region had the fewest new homes constructed?

5. Which region(s) had 3,000,000 or more new homes constructed?

6. Which region(s) had fewer than 3,000,000 new homes constructed?

The following bar graph shows the percent of persons age 25 or over who completed four or more years of college. Use this graph to answer Exercises 7 through 10.

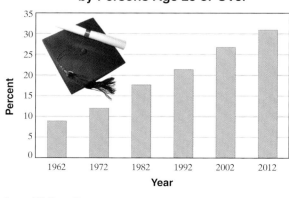

Four or More Years of College by Persons Age 25 or Over

Source: U.S. Census Bureau

7. Approximate the percent of persons who had completed four or more years of college in 1972.

8. What year shown had the greatest percent of persons completing four or more years of college?

9. What years shown had 20% or more of persons completing four or more years of college?

10. Describe any patterns you notice in this graph.

The following line graph shows the total number of Olympic medals awarded during the Summer Olympics between 1992 and 2012. Use this graph to answer Exercises 11 through 16.

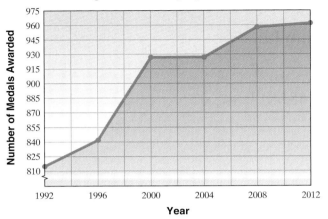

Number of Medals Awarded During Summer Olympics: 1992–2012

Source: International Olympic Committee

11. Approximate the number of medals awarded during the Summer Olympics of 2012.

12. Approximate the number of medals awarded during the Summer Olympics of 2000.

13. Between which two Summer Olympics did the number of medals awarded not change?

14. Between which two Summer Olympics did the number of medals awarded change the most?

15. How many more medals were awarded at the Summer Olympics of 1996 than at the Summer Olympics of 1992?

16. How many more medals were awarded at the Summer Olympics of 2012 than at the Summer Olympics of 1996?

The following histogram shows the hours worked per week by the employees of Southern Star Furniture. Use this histogram to answer Exercises 17 through 20.

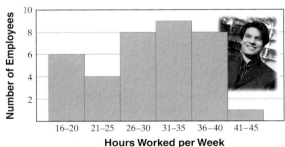

Southern Star Furniture

17. How many employees work 41–45 hours per week?

18. How many employees work 21–25 hours per week?

19. How many employees work 30 hours or less per week?

20. How many employees work 36 hours or more per week?

Following is a list of monthly record high temperatures for New Orleans, Louisiana. Use this list to complete the frequency distribution table below.

83	96	101	92
85	100	92	102
89	101	87	84

	Class Intervals (Temperatures)	Tally	Class Frequency (Number of Months)
21.	80°–89°		
22.	90°–99°		
23.	100°–109°		

24. Use the table from Exercises 21–23 to draw a histogram.

Temperatures

(7.2) *The following circle graph shows a family's $4000 monthly budget. Use this graph to answer Exercises 25 through 30. Write all ratios as fractions in simplest form.*

25. What is the largest budget item?

26. What is the smallest budget item?

27. How much money is budgeted for the mortgage payment and utilities?

28. How much money is budgeted for savings and contributions?

29. Find the ratio of the mortage payment to the total monthly budget.

30. Find the ratio of food to the total monthly budget.

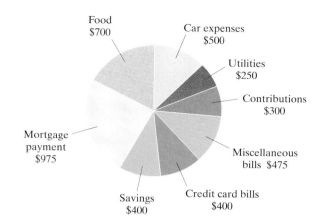

Food $700
Car expenses $500
Utilities $250
Contributions $300
Miscellaneous bills $475
Credit card bills $400
Savings $400
Mortgage payment $975

In 2013, there were 65 buildings over 1000 feet tall in the world. The following circle graph shows the percent of buildings over 1000 feet tall in the world by region in 2013. Use this graph to determine the number of tall buildings on each continent in Exercises 31 through 34. Round each answer to the nearest whole.

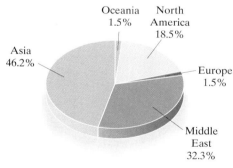

Oceania 1.5%
North America 18.5%
Asia 46.2%
Europe 1.5%
Middle East 32.3%

Source: Council on Tall Buildings and Urban Habitat

31. How many tall buildings were located in Asia?

32. How many tall buildings were located in North America?

33. How many tall buildings were located in the Middle East?

34. How many tall buildings were located in Europe?

(7.3) *Find the mean, median, and any mode(s) for each list of numbers. If necessary, round to the nearest tenth.*

35. 13, 23, 33, 14, 6

36. 45, 86, 21, 60, 86, 64, 45

37. 14,000, 20,000, 12,000, 20,000, 36,000, 45,000

38. 560, 620, 123, 400, 410, 300, 400, 780, 430, 450

For Exercises 39 and 40, the grades are given for a student for a particular semester. Find each grade point average. If necessary, round the grade point average to the nearest hundredth.

39.

Grade	Credit Hours
A	3
A	3
C	2
B	3
C	1

40.

Grade	Credit Hours
B	3
B	4
C	2
D	2
B	3

(7.4) *Draw a tree diagram for each experiment. Then use the diagram to determine the number of outcomes.*

Spinner 1   Spinner 2

41. Tossing a coin and then spinning Spinner 1

42. Spinning Spinner 2 and then tossing a coin

43. Spinning Spinner 1 twice

44. Spinning Spinner 2 twice

45. Spinning Spinner 1 and then Spinner 2

Find the probability of each event.

46. Rolling a 4 on a die

47. Rolling a 3 on a die

48. Spinning a 4 on the spinner

49. Spinning a 3 on the spinner

50. Spinning either a 1, 3, or 5 on the spinner

51. Spinning either a 2 or a 4 on the spinner

52. Rolling an even number on a die

53. Rolling a number greater than 3 on a die

Mixed Review

Find the mean, median, and any mode(s) for each list of numbers. If needed, round answers to two decimal places.

54. 73, 82, 95, 68, 54

55. 25, 27, 32, 98, 62

56. 750, 500, 427, 322, 500, 225

57. 952, 327, 566, 814, 327, 729

Given a bag containing 2 red marbles, 2 blue marbles, 3 yellow marbles, and 1 green marble, find the following:

58. The probability of choosing a blue marble from the bag

59. The probability of choosing a yellow marble from the bag

60. The probability of choosing a red marble from the bag

61. The probability of choosing a green marble from the bag

Answers

The following pictograph shows the money collected each week from a wrapping paper fundraiser. Use this graph to answer Exercises 1 through 3.

Weekly Wrapping Paper Sales

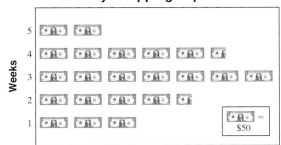

1. _____

1. How much money was collected during the second week?

2. During which week was the most money collected? How much money was collected during that week?

3. What was the total money collected for the fundraiser?

2. _____

The bar graph shows the normal monthly precipitation in centimeters for Chicago, Illinois. Use this graph to answer Exercises 4 through 6.

Chicago Precipitation

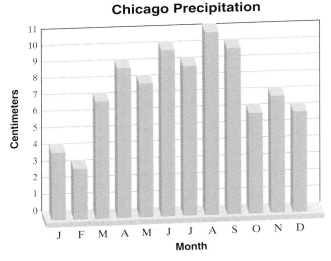

3. _____

Source: U.S. National Oceanic and Atmospheric Administration, *Climatography of the United States*, No. 81

4. _____

4. During which month(s) does Chicago normally have more than 9 centimeters of precipitation?

5. _____

5. During which month does Chicago normally have the least amount of precipitation? How much precipitation occurs during that month?

6. During which month(s) does 7 centimeters of precipitation normally occur?

6. _____

7. Use the information in the table to draw a bar graph. Clearly label each bar.

Most Common Blood Types	
Blood Type	% of Population with This Blood Type
O+	38%
A+	34%
B+	9%
O−	7%
A−	6%
AB+	3%
B−	2%
AB−	1%

Most Common Blood Types by Percent in the Population

7. _____

8. _____

The following line graph shows the average annual inflation rate in the United States for the years 2003–2013. Use this graph to answer Exercises 8 through 10.

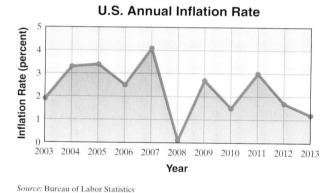

U.S. Annual Inflation Rate

Source: Bureau of Labor Statistics

8. Approximate the annual inflation rate in 2011.

9. During which of the years shown was the inflation rate greater than 3%?

10. During which sets of years was the inflation rate increasing?

9. _____

10. _____

The result of a survey of 200 people is shown in the following circle graph. Each person was asked to tell his or her favorite type of music. Use this graph to answer Exercises 11 and 12.

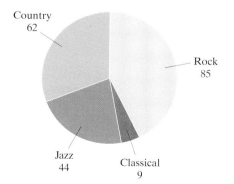

11. Find the ratio of those who prefer rock music to the total number surveyed.

12. Find the ratio of those who prefer country music to those who prefer jazz.

11. _____

12. _____

13. _____

The following graph (from earlier in this chapter) shows the percent of visitors to the United States in 2012 by various regions. During 2012, there were approximately 66,700,000 foreign visitors to the United States. Use the graph to find how many people visited the United States in 2012 from the world regions given in Exercises 13 and 14. (Source: Office of Travel & Tourism Industries)

Visitors to U.S. by Region

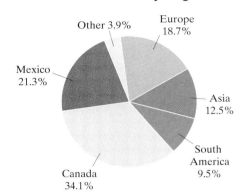

14. _____

13. Canada   **14.** Asia

A professor measures the heights of the students in her class. The results are shown in the following histogram. Use this histogram to answer Exercises 15 and 16.

Student Heights

15. _____

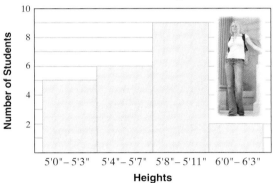

15. How many students are 5′8″–5′11″ tall?

16. How many students are 5′7″ or shorter?

16. _____

17. The history test scores of 25 students are shown below. Use these scores to complete the frequency distribution table.

70	86	81	65	92
43	72	85	69	97
82	51	75	50	68
88	83	85	77	99
77	63	59	84	90

Class Intervals (Scores)	Tally	Class Frequency (Number of Students)
40–49		
50–59		
60–69		
70–79		
80–89		
90–99		

17. _____

18. Use the results of Exercise 17 to draw a histogram.

Scores

Find the mean, median, and mode of each list of numbers.

19. 26, 32, 42, 43, 49

20. 8, 10, 16, 16, 14, 12, 12, 13

Find the grade point average. If necessary, round to the nearest hundredth.

21.

Grade	Credit Hours
A	3
B	3
C	3
B	4
A	1

22. Draw a tree diagram for the experiment of spinning the spinner twice. State the number of outcomes.

23. Draw a tree diagram for the experiment of tossing a coin twice. State the number of outcomes.

Suppose that the numbers 1 to 10 are each written on a scrap of paper and placed in a bag. You then select one number from the bag.

24. What is the probability of choosing a 6 from the bag?

25. What is the probability of choosing a 3 or a 4 from the bag?

Answers

1. Write 106,052,447 in words.

2. Write 276,004 in words.

△ **3.** Find the perimeter of the polygon shown.

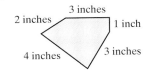

4. Find the perimeter of the rectangle shown.

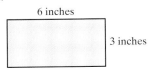

1. _____

2. _____

3. _____

4. _____

5. _____

6. _____

7. _____

8. _____

9. _____

10. _____

11. _____

12. _____

13. _____

14. _____

15. _____

16. _____

17. a. _____

b. _____

c. _____

18. a. _____

b. _____

19. _____

20. _____

21. _____

22. _____

5. Subtract: $900 - 174$. Check by adding.

6. Subtract: $17{,}801 - 8216$. Check by adding.

7. Round 248,982 to the nearest hundred.

8. Round 844,497 to the nearest thousand.

9. Multiply: 25×8

10. Multiply: 395×74

11. Divide and check: $1872 \div 9$

12. Divide and check: $3956 \div 46$

13. Simplify: $2 \cdot 4 - 3 \div 3$

14. Simplify: $8 \cdot 4 + 9 \div 3$

15. Evaluate $x^2 + z - 3$ for $x = 5$ and $z = 4$.

16. Evaluate $2a^2 + 5 - c$ for $a = 2$ and $c = 3$.

17. Insert $<$ or $>$ between each pair of numbers to make a true statement.
   a. -7 7
   b. 0 -4
   c. -9 -11

18. Insert $<$ or $>$ between each pair of numbers to make a true statement.
   a. -14 0
   b. $-(-7)$ -8

19. Add using a number line: $5 + (-2)$

20. Add using a number line: $-3 + (-4)$

Add.

21. $-5 + (-10)$

22. $3 + (-7)$

23. $2 + 6$

24. $21 + 15 + (-19)$

Subtract.

25. $-4 - 10$

26. $-2 - 3$

27. $6 - (-5)$

28. $19 - (-10)$

29. $-11 - (-7)$

30. $-16 - (-13)$

Divide.

31. $\dfrac{-12}{6}$

32. $\dfrac{-30}{-5}$

33. $-20 \div (-4)$

34. $26 \div (-2)$

35. $\dfrac{48}{-3}$

36. $\dfrac{-120}{12}$

37. Add: $2\dfrac{4}{5} + 5 + 1\dfrac{1}{2}$

38. Multiply: $5\dfrac{1}{3} \cdot 2\dfrac{1}{8}$

Write each rate as a fraction in simplest form.

39. $2160 for 12 weeks

40. 340 miles every 5 hours

41. Convert 7 feet to yards.

42. Convert 2.5 tons to pounds.

43. Convert 2.35 cg to grams.

44. Convert 106 cm to millimeters.

45. Find: $\sqrt{\dfrac{1}{36}}$

46. Find: $\sqrt{\dfrac{1}{25}}$

47. Find the mode of the list of numbers: $11, 14, 14, 16, 31, 56, 65, 77, 77, 78, 79$

48. Find the median of the numbers in Exercise 47.

49. If a coin is tossed twice, find the probability of tossing heads on the first toss and then heads again on the second toss (H, H).

50. A bag contains 3 red marbles and 2 blue marbles. Find the probability of choosing a red marble.

23. _____

24. _____

25. _____

26. _____

27. _____

28. _____

29. _____

30. _____

31. _____

32. _____

33. _____

34. _____

35. _____

36. _____

37. _____

38. _____

39. _____

40. _____

41. _____

42. _____

43. _____

44. _____

45. _____

46. _____

47. _____

48. _____

49. _____

50. _____

Introduction to Algebra

In this chapter we make the transition from arithmetic to algebra. In algebra, letters are used to stand for unknown quantities. Using variables is a very powerful tool for solving problems that cannot be solved with arithmetic alone. This chapter introduces variables, algebraic expressions, and solving variable equations.

The Top 4 Social Networking Sites Have Changed!

A social networking site or service consists of a profile or representation of each user, his or her social links, and a variety of other services, including a way for users to interact over the Internet. Although Google is currently the most popular Internet site, with an estimated 1,000,000,000 unique monthly visitors, social networking giant Facebook is the third most popular Internet site and tops among social networking sites. Some social networking sites concentrate on entertainment, some on career-minded professionals, and some, like Twitter, let members send out short, 140-character messages called "tweets."

In Section 8.5, Exercises 65 and 66, we explore numbers of personal computers in use in select countries.

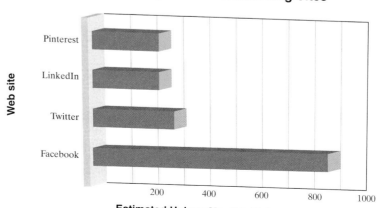

Top 4 Social Networking Sites

Source: www.ebizmba.com, September 2014

8.1 Variable Expressions

Objective A Evaluating Algebraic Expressions

Recall from Section 2.1 that a combination of numbers, letters (variables), and operation symbols is called an **algebraic expression** or simply an **expression.** For example,

$$3 + x, \quad 5 \cdot y, \quad \text{and} \quad 2 \cdot z - 1 + x$$

are expressions.

If two variables or a number and a variable are next to each other, with no operation sign between them, the indicated operation is multiplication. For example,

$$2x \quad \text{means} \quad 2 \cdot x$$

and

$$xy \text{ or } x(y) \quad \text{means} \quad x \cdot y$$

Also, the meaning of an exponent remains the same when the base is a variable. For example,

$$x^2 = \underbrace{x \cdot x}_{\text{2 factors of } x} \quad \text{and} \quad y^5 = \underbrace{y \cdot y \cdot y \cdot y \cdot y}_{\text{5 factors of } y}$$

Throughout this text, we have practiced replacing variables in an expression by numbers and then finding the value of the expression. Remember that this is called **evaluating the expression.** Let's review this process. When finding the value of an expression, don't forget to follow the order of operations.

Objectives

A Evaluate Algebraic Expressions for Given Replacement Values for the Variables.

B Use Properties of Numbers to Combine Like Terms.

C Use Properties of Numbers to Multiply Expressions.

D Simplify Expressions by Multiplying and Then Combining Like Terms.

E Find the Perimeter and Area of Figures.

Example 1 Evaluate: $2x + y$ when $x = 8$ and $y = -7$

Solution: Replace x with 8 and y with -7 in $2x + y$.

$$\begin{aligned}
2x + y &= 2 \cdot 8 + (-7) \quad \text{Replace } x \text{ with 8 and } y \text{ with } -7. \\
&= 16 + (-7) \quad \text{Multiply first because of the order of operations.} \\
&= 9 \quad \text{Add.}
\end{aligned}$$

■ Work Practice 1

Practice 1

Evaluate: $5x - y$ when $x = 2$ and $y = -3$

Example 2 Evaluate: $\dfrac{3m - 2n}{-2q}$ when $m = 8, n = 4,$ and $q = 1$

Solution:

$$\begin{aligned}
\frac{3m - 2n}{-2q} &= \frac{3 \cdot 8 - 2 \cdot 4}{-2 \cdot 1} \quad \text{Replace } m \text{ with 8, } n \text{ with 4, and } q \text{ with 1.} \\
&= \frac{24 - 8}{-2} \quad \text{Multiply.} \\
&= \frac{16}{-2} \quad \text{Subtract in the numerator.} \\
&= -8 \quad \text{Divide.}
\end{aligned}$$

■ Work Practice 2

Practice 2

Evaluate: $\dfrac{5r - 2s}{-3q}$ when $r = 3, s = 3,$ and $q = 1$

Answers

1. 13 **2.** −3

Practice 3

Evaluate: $13 - (3a + 8)$ when $a = -2$

Example 3 Evaluate: $8 - (6a - 5)$ when $a = -3$

Solution:

$$
\begin{aligned}
8 - (6a - 5) &= 8 - (6 \cdot (-3) - 5) && \text{Replace } a \text{ with } -3. \\
&= 8 - (-18 - 5) && \text{Multiply.} \\
&= 8 - (-23) && \text{Simplify inside the parentheses.} \\
&= 8 + 23 \\
&= 31 && \text{Add.}
\end{aligned}
$$

■ Work Practice 3

Practice 4

Evaluate: $a^2 - 0.7b$ when $a = 6$ and $b = -2$

Example 4 Evaluate: $x^3 - 1.1y$ when $x = 4$ and $y = -1$

Solution:

$$
\begin{aligned}
x^3 - 1.1y &= 4^3 - 1.1(-1) && \text{Replace } x \text{ with 4 and } y \text{ with } -1. \\
&= 64 - 1.1(-1) && \text{Evaluate } 4^3. \\
&= 64 - (-1.1) && \text{Multiply.} \\
&= 64 + 1.1 \\
&= 65.1 && \text{Add.}
\end{aligned}
$$

■ Work Practice 4

Objective B Combining Like Terms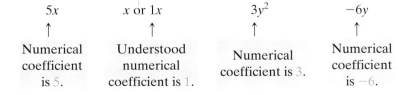

The addends of an algebraic expression are called the **terms** of the expression.

$x + 3$

└───┴──── 2 terms

$3y^2 + (-6y) + 4$

└────┴────┴──── 3 terms

A term that is only a number has a special name. It is called a **constant term,** or simply a **constant.** A term that contains a variable is called a **variable term.**

x	$+$	3	$3y^2 + (-6y) +$		4
↑		↑	↑	↑	↑
variable term		constant term	variable terms		constant term

Helpful Hint

Recall that $1 \cdot$ any number = that number. This means that

$1 \cdot x = x$ or that $1x = x$.

Thus x can always be replaced by $1x$ or $1 \cdot x$.

The number factor of a variable term is called the **numerical coefficient.** A numerical coefficient of 1 is usually not written.

$5x$	x or $1x$	$3y^2$	$-6y$
↑	↑	↑	↑
Numerical coefficient is 5.	Understood numerical coefficient is 1.	Numerical coefficient is 3.	Numerical coefficient is -6.

Answers

3. 11 **4.** 37.4

Terms with the same variable factors, except that they may have different numerical coefficients, are called **like terms.**

Like Terms	Unlike Terms
$3x, \dfrac{1}{2}x$	$5x, x^2$
$-6y, 2y, y$	$7x, 7y$

✓**Concept Check** True or false? The terms $-7xy$ and $-7yx$ are like terms. Explain.

A sum or difference of like terms can be simplified using the **distributive property.** Recall from Section 1.6 that the distributive property says that multiplication distributes over addition (and subtraction). Using variables, we can write the distributive property as follows:

$$(a + b)c = ac + bc$$

If we write the right side of the equation first and then the left side, we have the following:

Distributive Property

If a, b, and c are numbers, then

$$ac + bc = (a + b)c$$

Also,

$$ac - bc = (a - b)c$$

The distributive property guarantees that, no matter what number x is, $7x + 5x$ (for example) has the same value as $(7 + 5)x$, or $12x$. We then have that

$$7x + 5x = (7 + 5)x = 12x$$

This is an example of **combining like terms.** An algebraic expression is **simplified** when all like terms have been combined.

Example 5   Simplify each expression by combining like terms.

a. $3x + 2x$ **b.** $y - 7y$

Solution: We add or subtract like terms.

a. $3x + 2x = (3 + 2)x$
$$= 5x$$

     Understood 1

b. $y - 7y = 1y - 7y$
$$= (1 - 7)y$$
$$= -6y$$

■ Work Practice 5

The commutative and associative properties of addition and multiplication can also help us simplify expressions. We presented these properties in Sections 1.3 and 1.6 and state them again using variables.

Properties of Addition and Multiplication

If a, b, and c are numbers, then

$$a + b = b + a \quad \text{Commutative property of addition}$$
$$a \cdot b = b \cdot a \quad \text{Commutative property of multiplication}$$

That is, the **order** of adding or multiplying two numbers can be changed without changing their sum or product.

$$(a + b) + c = a + (b + c) \quad \text{Associative property of addition}$$
$$(a \cdot b) \cdot c = a \cdot (b \cdot c) \quad \text{Associative property of multiplication}$$

That is, the **grouping** of numbers in addition or multiplication can be changed without changing their sum or product.

Helpful Hint

- Examples of these properties are

$$2 + 3 = 3 + 2 \quad \text{Commutative property of addition}$$
$$7 \cdot 9 = 9 \cdot 7 \quad \text{Commutative property of multiplication}$$
$$(1 + 8) + 10 = 1 + (8 + 10) \quad \text{Associative property of addition}$$
$$(4 \cdot 2) \cdot 3 = 4 \cdot (2 \cdot 3) \quad \text{Associative property of multiplication}$$

- These properties are not true for subtraction or division.

Practice 6

Simplify: $8m + 5 + m - 4$

Example 6   Simplify: $2y - 6 + 4y + 8$

Solution: We begin by writing subtraction as the addition of opposites.

$$
\begin{aligned}
2y - 6 + 4y + 8 &= 2y + (-6) + 4y + 8 \\
&= 2y + 4y + (-6) + 8 \quad \text{Apply the commutative property of addition.} \\
&= (2 + 4)y + (-6) + 8 \quad \text{Apply the distributive property.} \\
&= 6y + 2 \quad \text{Simplify.}
\end{aligned}
$$

■ Work Practice 6

Practice 7–10

Simplify each expression by combining like terms.
7. $7y + 11y - 8$
8. $2y - 6 + y + 7y$
9. $3.7x + 5 - 4.2x + 15$
10. $-9y + 2 - 4y - 8x + 12 - x$

Examples   Simplify each expression by combining like terms.

7. $6x + 2x - 5 = 8x - 5$
8. $4x + 3 - 5x + 2x = 4x - 5x + 2x + 3$
$$= 1x + 3 \quad \text{or} \quad x + 3$$
9. $1.2y + 10 - 5.7y - 9 = 1.2y - 5.7y + 10 - 9$
$$= -4.5y + 1$$
10. $2x - 5 + 3y + 4x - 10y + 11 = 6x - 7y + 6$

■ Work Practice 7–10

Objective C Multiplying Expressions

We can also use properties of numbers to multiply expressions such as $3(2x)$. By the associative property of multiplication, we can write the product $3(2x)$ as $(3 \cdot 2)x$, which simplifies to $6x$.

Examples Multiply.

11. $5(3y) = (5 \cdot 3)y$ Apply the associative property of multiplication.

$\qquad = 15y$ Multiply.

12. $-2(4x) = (-2 \cdot 4)x$ Apply the associative property of multiplication.

$\qquad = -8x$ Multiply.

◼ Work Practice 11–12

Practice 11–12

Multiply.
11. $7(8a)$
12. $-5(9x)$

We can use the distributive property to combine like terms, which we have done, and also to multiply expressions such as $2(3 + x)$. By the distributive property, we have that

$2(3 + x) = 2 \cdot 3 + 2 \cdot x$ Apply the distributive property.

$\qquad = 6 + 2x$ Multiply.

Example 13 Use the distributive property to multiply: $6(x + 4)$

Solution: By the distributive property,

$6(x + 4) = 6 \cdot x + 6 \cdot 4$ Apply the distributive property.

$\qquad = 6x + 24$ Multiply.

◼ Work Practice 13

Practice 13

Use the distributive property to multiply: $7(y + 2)$

✓**Concept Check** What's wrong with the following?

$8(a - b) = 8a - b$

Example 14 Multiply: $-3(5a + 2)$

Solution: By the distributive property,

$-3(5a + 2) = -3(5a) + (-3)(2)$ Apply the distributive property.

$\qquad = (-3 \cdot 5)a + (-6)$ Apply the associative property. Also, write $(-3)(2)$ as -6.

$\qquad = -15a - 6$ Multiply.

◼ Work Practice 14

Practice 14

Multiply: $4(7a - 5)$

Objective D Simplifying Expressions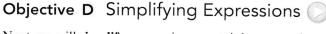

Next, we will **simplify** expressions containing parentheses by first using the distributive property to multiply and then **combining** any like terms.

Example 15 Simplify: $2(3 + 7x) - 15$

Solution: First we use the distributive property to remove parentheses.

$2(3 + 7x) - 15 = 2(3) + 2(7x) - 15$ Apply the distributive property.

$\qquad = 6 + 14x - 15$ Multiply.

$\qquad = 14x + (-9)$ or $14x - 9$ Combine like terms.

◼ Work Practice 15

Practice 15

Simplify: $5(2y - 3) - 8$

Helpful Hint 2 is *not* distributed to the -15 since it is not within the parentheses.

Answers

11. $56a$ **12.** $-45x$ **13.** $7y + 14$
14. $28a - 20$ **15.** $10y - 23$

✓**Concept Check Answer**

did not distribute the 8 to the b (or $-b$)

Practice 16

Simplify:
$-7(x - 1) + 5(2x + 3)$

Example 16 Simplify: $-2(x - 5) + 4(2x + 2)$

Solution: First we use the distributive property to remove parentheses.

$$-2(x - 5) + 4(2x + 2) = -2(x) - (-2)(5) + 4(2x) + 4(2) \quad \text{Apply the distributive property.}$$

$$= -2x + 10 + 8x + 8 \qquad \text{Multiply.}$$

$$= 6x + 18 \qquad \text{Combine like terms.}$$

■ Work Practice 16

Objective E Finding Perimeter and Area

Practice 17

Find the perimeter of the square.

2x centimeters

Example 17 Find the perimeter of the triangle.

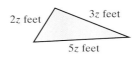

2z feet   3z feet

5z feet

Solution: Recall that the perimeter of a figure is the distance around the figure. To find the perimeter, then, we find the sum of the lengths of the sides. We use the letter P to represent perimeter.

$$P = 2z + 3z + 5z$$

$$= 10z$$

The perimeter is $10z$ feet.

Helpful Hint   Don't forget to insert proper units.

■ Work Practice 17

Practice 18

Find the area of the rectangular garden.

$(12y + 9)$ yards   3 yards

Example 18 Finding the Area of a Basketball Court

Find the area of this YMCA basketball court.

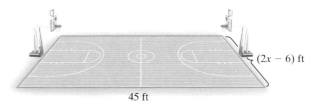

$(2x - 6)$ ft

45 ft

Solution: Recall how to find the area of a rectangle. **A**rea = **L**ength · **W**idth, or if A represents area, l represents length, and w represents width, we have $A = l \cdot w$.

$$A = l \cdot w$$

$$= 45(2x - 6) \quad \text{Let } l = 45 \text{ and } w = (2x - 6).$$

$$= 90x - 270 \quad \text{Multiply.}$$

The area is $(90x - 270)$ *square* feet.

■ Work Practice 18

Answers

16. $3x + 22$ **17.** $8x$ cm
18. $(36y + 27)$ sq yd

Helpful Hint

Don't forget…

Area:	Perimeter:
• surface enclosed	• distance around
• measured in square units	• measured in units

Vocabulary, Readiness & Video Check

Use the choices below to fill in each blank. Some choices may be used more than once.

numerical coefficient	combine like terms	like	term	variable	associative
constant	expression	unlike	distributive	commutative	

1. $14y^2 + 2x - 23$ is called a(n) _____ while $14y^2$, $2x$, and -23 are each called a(n) _____.
2. To multiply $3(-7x + 1)$, we use the _____ property.
3. To simplify an expression like $y + 7y$, we _____.
4. By the _____ properties, the *order* of adding or multiplying two numbers can be changed without changing their sum or product.
5. The term $5x$ is called a(n) _____ term while the term 7 is called a(n) _____ term.
6. The term z has an understood _____ of 1.
7. By the _____ properties, the *grouping* of numbers in addition or multiplication can be changed without changing their sum or product.
8. The terms $-x$ and $5x$ are _____ terms.
9. For the term $-3x^2y$, -3 is called the _____.
10. The terms $5x$ and $5y$ are _____ terms.

Identify each pair of terms as like terms or unlike terms.

11. $5x$ and $5y$
12. $-3a$ and $-3b$
13. x and $-2x$
14. $7y$ and y
15. $-5n$ and $6n^2$
16. $4m^2$ and $2m$
17. $8b$ and $-6b$
18. $12a$ and $-11a$

Martin-Gay Interactive Videos   *Watch the section lecture video and answer the following questions.*

See Video 8.1

Objective A 19. Complete this statement based on the lecture before Example 1: When a letter and a variable are next to each other, the operation is an understood _____.

Objective B 20. From Example 3, what is the numerical coefficient of the term $-x$?

Objective C 21. In Example 6, what two properties are used to multiply the expressions and in what order?

Objective D 22. In Example 7, why can't we add 3 to 6?

Objective E 23. In Example 8, what operation is used to find P? What operation is used to find A? What do P and A stand for?

8.1 Exercise Set MyMathLab®

Objective A *Evaluate each expression when $x = -2$, $y = 5$, and $z = -3$. See Examples 1 through 4.*

1. $3 + 2z$

2. $7 + 3z$

3. $-y - z$

4. $-y - x$

5. $z - x + y$

6. $x + y - z$

7. $3x - z$

8. $y + 5z$

9. $8 - (5y - 7)$

10. $5 + (2x - 1)$

11. $y^3 - 4x$

12. $y^2 - 2z$

13. $\dfrac{6xy}{4}$

14. $\dfrac{8yz}{15}$

15. $\dfrac{2y - 2}{x}$

16. $\dfrac{6 + 3x}{z}$

17. $\dfrac{x + 2y}{2z}$

18. $\dfrac{2z - y}{3x}$

19. $\dfrac{5x}{y} - 10$

20. $7 - \dfrac{3y}{z}$

21. $\dfrac{xz}{y} + \dfrac{3}{10}$

22. $\dfrac{x}{yz} + \dfrac{31}{30}$

23. $|x| - |y| - 7.6$

24. $|z| - |y| - 12.7$

Objective B *Simplify each expression by combining like terms. See Examples 5 through 10.*

25. $3x + 5x$

26. $8y + 3y$

27. $5n - 9n$

28. $7z - 10z$

29. $4c + c - 7c$

30. $5b - 8b - b$

31. $5x - 7x + x - 3x$

32. $8y + y - 2y - y$

33. $4a + 3a + 6a - 8$

34. $5b - 4b + b - 15$

35. $1.7x + 3.4 - 2.6x + 7.8$

36. $-8.6y + 1.3 - 2.9y - 14.7$

37. $3x + 7 - x - 14$

38. $9x - 6 + x - 10$

39. $4x + 5y + 2 - y - 9x - 7$

40. $a + 4b + 3 - 7a - 5b - 10$

41. $\dfrac{5}{6} - \dfrac{7}{12}x - \dfrac{1}{3} - \dfrac{3}{10}x$

42. $-\dfrac{2}{5} + \dfrac{4}{9}y - \dfrac{4}{15} + \dfrac{1}{6}y$

43. $-5m - 2.3m + 11 + 2.5m - 15.1$ **44.** $-13n - 4.8n + 13 + 6.9n - 13.6$

Objective C *Multiply. See Examples 11 through 14.*

45. $6(5x)$

46. $4(4x)$

47. $-2(11y)$

48. $-3(21z)$

49. $-0.6(7a)$

50. $-0.4(9a)$

51. $\dfrac{2}{3}(-6a)$

52. $\dfrac{3}{4}(-8a)$

53. $2(y + 2)$

54. $3(x + 1)$

55. $5(3a - 8)$ **56.** $4(5y - 6)$ **57.** $-4(3x + 7)$ **58.** $-8(8y + 10)$

59. $1.2(5x - 0.1)$ **60.** $3.1(7x - 0.3)$ **61.** $\frac{1}{2}(-8x - 3)$ **62.** $\frac{1}{5}(-20x - 7)$

Objective D *Simplify each expression. Use the distributive property to remove parentheses first. See Examples 15 and 16.*

63. $2(x + 4) - 17$ **64.** $5(6 + y) - 2$ **65.** $4(6n - 5) + 3n$

66. $3(5 - 2b) - 4b$ **67.** $3 + 6(w + 2) + w$ **68.** $8z + 5(6 + z) + 20$

69. $-2(3x + 1) - 5(x - 2)$ **70.** $-3(5x - 2) - 2(3x + 1)$

Objective E *Find the perimeter of each figure. See Example 17.*

71.
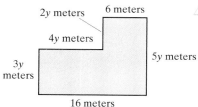

72.
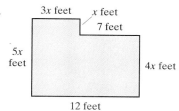

73.
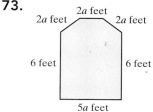

74.
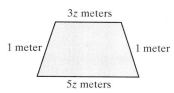

75.
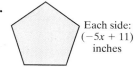

76.
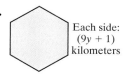

Find the area of each rectangle. See Example 18.

77.
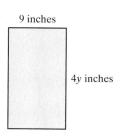

78.
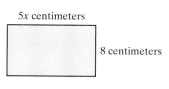

79.
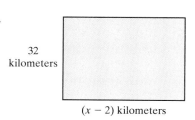

80.
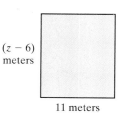

81.

82.
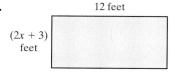

Objectives **A** **B** **C** **D** **E**   **Mixed Practice** *Solve. See Examples 1 through 18.*

83. Find the area of a regulation NCAA basketball court that is 94 feet long and 50 feet wide.

84. Find the area of a rectangular movie screen that is 50 feet long and 40 feet high.

85. A decorator wishes to put a wallpaper border around a rectangular room that measures 14 feet by 18 feet. Find the room's perimeter.

86. How much fencing will a rancher need for a rectangular cattle lot that measures 80 feet by 120 feet?

87. Find the perimeter of a triangular garden that measures 5 feet by x feet by $(2x + 1)$ feet.

88. Find the perimeter of a triangular picture frame that measures x inches by x inches by $(x - 14)$ inches.

89. How much interest will $3000 in a passbook savings account earn in 2 years at Money Bank, which pays 6% simple interest? Use $I = PRT$.

90. How much interest will a $12,000 certificate of deposit earn in 1 year at a rate of 8% simple interest? Use $I = PRT$.

91. Find the area of a circular braided rug with a radius of 5 feet. Use $A = \pi r^2$ and $\pi \approx 3.14$.

92. Mario's Pizza sells one 16″ cheese pizza or two 10″ cheese pizzas for $9.99. Which deal gives you more pizza? Use $A = \pi r^2$ with $\pi \approx 3.14$.

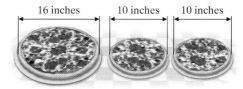

16 inches   10 inches   10 inches

93. Convert Paris, France's, low temperature of $-5°C$ to Fahrenheit. Use $F = \dfrac{9}{5}C + 32$.

94. Convert Nome, Alaska's, 18°F high temperature to Celsius. Use $C = \dfrac{5}{9}(F - 32)$.

95. Find the volume of a box that measures 12 inches by 6 inches by 4 inches. Use $V = lwh$.

96. How many cubic meters does a space shuttle cargo compartment have if its dimensions are 8 meters long by 4 meters wide by 3 meters high? Use $V = lwh$.

97. Find the volume. Use $V = lwh$.

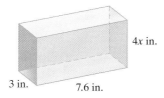

4x in.

3 in.   7.6 in.

98. Find the volume. Use $V = lwh$.

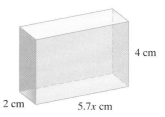

4 cm

2 cm   5.7x cm

Review

Perform each indicated operation. See Sections 2.3 and 2.4.

99. $-13 + 10$

100. $-15 + 23$

101. $-4 - (-12)$

102. $-7 - (-4)$

103. $-4 + 4$

104. $8 + (-8)$

Concept Extensions

If the expression on the left side of the equal sign is equivalent to the right, write "correct." If not, write "incorrect" and then write an expression that is equivalent to the left side. See the second Concept Check in this section.

105. $5(3x - 2) \overset{?}{=} 15x - 2$

106. $2(xy) \overset{?}{=} 2x \cdot 2y$

107. $7x - (x + 2) \overset{?}{=} 7x - x + 2$

108. $4(y - 3) + 11 \overset{?}{=} 4y - 3 + 11$

For Exercises 109 through 112, review the commutative, associative, and distributive properties. Then identify which property allows us to write the equivalent expression on the right side of the equal sign.

109. $6(2x - 3) + 5 = 12x - 18 + 5$

110. $9 + 7x + (-2) = 7x + 9 + (-2)$

111. $-7 + (4 + y) = (-7 + 4) + y$

112. $(x + y) + 11 = 11 + (x + y)$

113. If x is a whole number, which expression is the largest: $2x, 5x,$ or $\frac{1}{3}x$? Explain your answer.

114. If x is a whole number, which expression is the smallest: $2x, 5x,$ or $\frac{1}{3}x$? Explain your answer.

Find the area of each figure.

△ **115.**

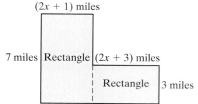

(2x + 1) miles
7 miles | Rectangle | (2x + 3) miles
Rectangle | 3 miles

△ **116.**

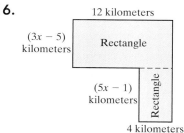

12 kilometers
(3x − 5) kilometers | Rectangle
(5x − 1) kilometers | Rectangle
4 kilometers

To appraise the value of a large tree in landscaping, the trunk area is used. The trunk area A is calculated with the formula $A = 0.7854d^2$, where d is the diameter of the tree.

117. *Tane Mahuta* is the name of New Zealand's largest known living Kauri tree. Its trunk has a diameter of 173 inches. Use the formula to find the trunk area of this tree. Round your result to the nearest tenth. (*Source:* New Zealand Parks)

118. *Lost Monarch*, which was discovered only in 1998, is a coast redwood tree in northern California. It is located among other giant redwoods in "The Grove of Titans," in Jedediah Smith Redwoods State Park. This mammoth tree has a confirmed diameter of 312 inches. Use the formula to find the trunk area of this tree. Round your result to the nearest tenth. (*Source:* Sierra Club)

Simplify.

119. $9684q - 686 - 4860q + 12,960$

120. $76(268x + 592) - 2960$

Objectives

A Determine Whether a Given Number Is a Solution of an Equation.

B Use the Addition Property of Equality to Solve Equations.

As we mentioned earlier, in this book we have frequently written statements like $7 + 4 = 11$ or Area = length · width. Each of these statements is called an **equation.** An equation is of the form

expression = expression

An equation can be labeled as

equal sign
↓
$$\underset{\text{left side}}{\underline{x + 7}} \; = \; \underset{\text{right side}}{10}$$

It is very important to know the difference between an **expression** and an **equation.** An equation contains an equal sign and an expression does not.

	Equations	Expressions	
equal signs	$7x = 6x + 4$	$7x - 6x + 4$	no equal signs
	$3(3y - 5) = 10y$	$y - 1 + 11y - 21$	

Objective A Determining Whether a Number Is a Solution

When an equation contains a variable, finding which values of the variable make the equation a true statement is called **solving** the equation for the variable. A **solution** of an equation is a value for the variable that makes the equation a true statement. For example, 2 is a solution of the equation $x + 5 = 7$ since replacing x with 2 results in the *true* statement $2 + 5 = 7$. Similarly, 3 is not a solution of $x + 5 = 7$ since replacing x with 3 results in the *false* statement $3 + 5 = 7$.

Practice 1

Determine whether 4 is a solution of the equation $3(y - 6) = 6$.

Example 1 Determine whether 6 is a solution of the equation $4(x - 3) = 12$.

Solution: We replace x with 6 in the equation.

$$4(x - 3) = 12$$
$$\downarrow$$
$$4(6 - 3) \overset{?}{=} 12 \quad \text{Replace } x \text{ with 6.}$$
$$4(3) \overset{?}{=} 12$$
$$12 \overset{?}{=} 12 \quad \text{True}$$

Since $12 = 12$ is a true statement, 6 *is* a solution of the equation.

■ Work Practice 1

Practice 2

Determine whether -2 is a solution of the equation $-4x - 3 = 5$.

Example 2 Determine whether -1 is a solution of the equation $3y + 1 = 3$.

Solution:

$$3y + 1 = 3$$
$$3(-1) + 1 \overset{?}{=} 3$$
$$-3 + 1 \overset{?}{=} 3$$
$$-2 \overset{?}{=} 3 \quad \text{False}$$

Since $-2 = 3$ is false, -1 is *not* a solution of the equation.

■ Work Practice 2

Answers

1. no   **2.** yes

Objective B Using the Addition Property to Solve Equations

To solve an equation, we use properties of equality to write simpler equations, all equivalent to the original equation, until the final equation has the form

x = **number**   or   **number** = x

Equivalent equations have the same solution, so the word "number" above represents the solution of the original equation. The first property of equality to help us write simpler, equivalent equations is the **addition property of equality.**

Addition Property of Equality

Let a, b, and c represent numbers. Then

$a = b$	Also, $a = b$
and $a + c = b + c$	and $a - c = b - c$
are equivalent equations.	are equivalent equations.

In other words, the **same number** may be **added to or subtracted from both sides** of an equation without changing the solution of the equation.

A good way to visualize a true equation is to picture a balanced scale. Since it is balanced, each side of the scale weighs the same amount. Similarly, in a true equation the expressions on each side have the same value. Picturing our balanced scale, if we add the same weight to each side, the scale remains balanced.

Example 3 Solve the equation for x: $x - 2 = 1$

Solution: To solve the equation for x, we need to rewrite the equation in the form x = number. In other words, our goal is to get x alone on one side of the equation. To do so, we add 2 to both sides of the equation.

$$x - 2 = 1$$
$$x - 2 + 2 = 1 + 2 \quad \text{Add 2 to both sides of the equation.}$$
$$x + 0 = 3 \quad \text{Replace } -2 + 2 \text{ with 0.}$$
$$x = 3 \quad \text{Simplify by replacing } x + 0 \text{ with } x.$$

Check: To check, we replace x with 3 in the *original* equation.

$$x - 2 = 1 \quad \text{Original equation}$$
$$3 - 2 \stackrel{?}{=} 1 \quad \text{Replace } x \text{ with 3.}$$
$$1 \stackrel{?}{=} 1 \quad \text{True}$$

Since $1 = 1$ is a true statement, 3 is the solution of the equation.

■ Work Practice 3

Practice 3

Solve the equation for y:
$y - 5 = -3$

Helpful Hint

Note that it is always a good idea to check the solution in the *original* equation to see that it makes the equation a true statement.

Answer

3. 2

Let's visualize how we used the addition property of equality to solve the equation in Example 3. Picture the original equation, $x - 2 = 1$, as a balanced scale. The left side of the equation has the same value as the right side.

If the same weight is added to each side of a scale, the scale remains balanced. Likewise, if the same number is added to each side of an equation, the left side continues to have the same value as the right side.

Practice 4

Solve: $-1 = z + 9$

Example 4   Solve: $-8 = x + 1$

Solution: To get x alone on one side of the equation, we subtract 1 from both sides of the equation.

$$-8 = x + 1$$
$$-8 - 1 = x + 1 - 1 \quad \text{Subtract 1 from both sides.}$$
$$-9 = x + 0 \quad \text{Replace } 1 - 1 \text{ with } 0.$$
$$-9 = x \quad \text{Simplify.}$$

Check:

$$-8 = x + 1$$
$$-8 \stackrel{?}{=} -9 + 1 \quad \text{Replace } x \text{ with } -9.$$
$$-8 \stackrel{?}{=} -8 \quad \text{True}$$

The solution is -9.

■ Work Practice 4

> **Helpful Hint**   Remember that we can get the variable alone on either side of the equation. For example, the equations $x = 2$ and $2 = x$ both have the solution of 2.

Practice 5

Solve: $x - 2.6 = -1.8 - 5.9$

Example 5   Solve: $y - 1.2 = -3.2 - 6.6$

Solution: First we simplify the right side of the equation.

$$y - 1.2 = -3.2 - 6.6$$
$$y - 1.2 = -9.8$$

Next, we get y alone on the left side by adding 1.2 to both sides of the equation.

$$y - 1.2 + 1.2 = -9.8 + 1.2 \quad \text{Add 1.2 to both sides.}$$
$$y = -8.6 \quad \text{Simplify.}$$

Check to see that -8.6 is the solution.

■ Work Practice 5

Answers

4. -10 **5.** -5.1

✓ **Concept Check Answer**

Subtract 2.1 from both sides.

✓ **Concept Check**   What number should be added to or subtracted from both sides of the equation in order to solve the equation $-3.75 = y + 2.1$?

Example 6 Solve: $5x + 2 - 4x = 7 - 9$

Solution: First we simplify each side of the equation separately.

$$5x + 2 - 4x = 7 - 9$$
$$\underbrace{5x - 4x} + 2 = \underbrace{7 - 9}$$
$$1x + 2 = -2$$

To get x alone on the left side, we subtract 2 from both sides.

$$1x + 2 - 2 = -2 - 2$$
$$1x = -4 \text{ or } x = -4$$

Check to verify that -4 is the solution.

■ Work Practice 6

Practice 6
Solve:
$-6y + 1 + 7y = 6 - 11$

Example 7 Solve: $\dfrac{7}{8} = y - \dfrac{1}{2}$

Solution: We use the addition property of equality to add $\dfrac{1}{2}$ to both sides.

$$\dfrac{7}{8} = y - \dfrac{1}{2}$$

$$\dfrac{7}{8} + \dfrac{1}{2} = y - \dfrac{1}{2} + \dfrac{1}{2} \qquad \text{Add } \tfrac{1}{2} \text{ to both sides.}$$

$$\dfrac{7}{8} + \dfrac{4}{8} = y \qquad \text{Simplify.}$$

$$\dfrac{11}{8} = y \qquad \text{Simplify.}$$

Check to see that $\dfrac{11}{8}$ is the solution. (Although $\dfrac{11}{8} = 1\dfrac{3}{8}$, we will leave solutions as improper fractions.)

■ Work Practice 7

Practice 7
Solve: $\dfrac{2}{3} = x - \dfrac{4}{9}$

Example 8 Solve: $3(3x - 5) = 10x$

Solution: First we multiply on the left side to remove the parentheses.

$$3(3x - 5) = 10x$$
$$3 \cdot 3x - 3 \cdot 5 = 10x \qquad \text{Use the distributive property.}$$
$$9x - 15 = 10x$$

Now we subtract $9x$ from both sides.

$$9x - 15 - 9x = 10x - 9x \qquad \text{Subtract } 9x \text{ from both sides.}$$
$$-15 = 1x \quad \text{or} \quad x = -15 \qquad \text{Simplify.}$$

■ Work Practice 8

Practice 8
Solve: $13x = 4(3x - 1)$

Recall that the addition property of equality allows us to add the same number to or subtract the same number from both sides of an equation. Let's see how adding the same number to both sides of an equation also allows us to subtract the same number from both sides. To do so, let's add $(-c)$ to both sides of $a = b$. Then we have

$$a + (-c) = b + (-c)$$

which is the same as $a - c = b - c$.

Answers

6. -6 **7.** $\dfrac{10}{9}$ **8.** -4

Vocabulary, Readiness & Video Check

Use the choices below to fill in each blank.

equation	addition	simplifying
solving	equivalent	expression

1. The equations $x + 6 = 10$ and $x + 6 - 6 = 10 - 6$ are called _____ equations.
2. The difference between an equation and an expression is that a(n) _____ contains an equal sign, while a(n) _____ does not.
3. The process of writing $-3x + 10x$ as $7x$ is called _____ the expression.
4. For the equation $x - 1 = -21$, the process of finding that -20 is the solution is called _____ the equation.
5. By the _____ property of equality, $x = -2$ and $x + 7 = -2 + 7$ are equivalent equations.

Martin-Gay Interactive Videos   *Watch the section lecture video and answer the following questions.*

Objective A 6. From the lecture before ▥ Example 1, what does an equation have that an expression does not? ▷

Objective B 7. In the lecture before ▥ Example 2, what does the addition property of equality mean in words? ▷

8. When solving ▥ Example 7, what must be done before applying the addition property of equality? ▷

See Video 8.2 🍎

8.2 Exercise Set MyMathLab® ▷

Objective A *Decide whether the given number is a solution of the given equation. See Examples 1 and 2.*

1. Is 10 a solution of $x - 8 = 2$?

2. Is 9 a solution of $y - 2 = 7$?

▷ 3. Is -5 a solution of $x + 12 = 17$?

4. Is -7 a solution of $a + 23 = -16$?

5. Is -8 a solution of $-9f = 64 - f$?

6. Is -6 a solution of $-3k = 12 - k$?

7. Is 3 a solution of $5(c - 5) = 10$?

8. Is 1 a solution of $2(b - 3) = 10$?

Objective B *Solve. Check each solution. See Examples 3 through 7.*

▷ 9. $a + 5 = 23$

10. $f + 4 = -6$

11. $d - 9 = -17$

12. $s - 7 = 15$

▷ 13. $7 = y - 2$

14. $1 = y + 7$

15. $-12 = x + 4$

16. $-10 = z - 15$

17. $x + \dfrac{1}{2} = \dfrac{7}{2}$

18. $x + \dfrac{1}{3} = \dfrac{4}{3}$

19. $y - \dfrac{3}{4} = -\dfrac{5}{8}$

20. $y - \dfrac{5}{6} = -\dfrac{11}{12}$

21. $x - 3 = -1 + 4$

22. $y - 8 = -5 - 1$

23. $-7 + 10 = m - 5$

24. $1 - 8 = n + 2$

25. $x - 0.6 = 4.7$

26. $y - 1.2 = 7.5$

27. $-2 - 3 = -4 + x$

28. $7 - (-10) = x - 5$

29. $y + 2.3 = -9.2 - 8.6$

30. $x + 4.7 = -7.5 - 3.4$

31. $-8x + 4 + 9x = -1 + 7$

32. $3x - 2x + 5 = 5 - 2$

33. $5 + (-12) = 5x - 7 - 4x$

34. $11 + (-15) = 6x - 4 - 5x$

35. $7x + 14 - 6x = -4 + (-10)$

36. $-10x + 11x + 5 = -9 + (-5)$

Solve. First multiply to remove parentheses. See Example 8.

37. $2(5x - 3) = 11x$

38. $6(3x + 1) = 19x$

39. $3y = 2(y + 12)$

40. $17x = 4(4x - 6)$

41. $21y = 5(4y - 6)$

42. $28z = 9(3z - 2)$

43. $-3(-4 - 2z) = 7z$

44. $-2(-1 - 3y) = 7y$

Review

Perform each indicated operation. See Section 3.3.

45. $\dfrac{-7}{-7}$

46. $\dfrac{4.2}{4.2}$

47. $\dfrac{1}{3} \cdot 3$

48. $\dfrac{1}{5} \cdot 5$

49. $-\dfrac{2}{3} \cdot -\dfrac{3}{2}$

50. $-\dfrac{7}{2} \cdot -\dfrac{2}{7}$

Concept Extensions

What number should be added to or subtracted from both sides of each equation in order to solve the equation? See the Concept Check in this section.

51. $\dfrac{2}{3} + x = \dfrac{1}{12}$

52. $12.5 = -3.75 + x$

53. $-\dfrac{1}{7} = -\dfrac{4}{5} + x$

54. $9.1 = 5.9 + x$

55. In your own words, explain what is meant by the phrase "a number is a solution of an equation."

56. In your own words, explain how to check a possible solution of an equation.

Solve.

57. $x - 76{,}862 = 86{,}102$

58. $-968 + 432 = 86y - 508 - 85y$

A football team's total offense T is found by adding the total passing yardage P to the total rushing yardage R: $T = P + R$.

59. During the 2013 regular football season, the Pittsburgh Steelers' total offense was 5400 yards. The Steelers' passing yardage for the season was 4017 yards. How many yards did the Steelers gain by rushing during the season? (*Source:* National Football League)

60. During the 2013 regular football season, the New Orleans Saints' total offense was 6391 yards. The Saints' rushing yardage for the season was 4918 yards. How many yards did the Saints gain by passing during the season? (*Source:* National Football League)

In accounting, a company's annual net income I can be computed using the relation $I = R - E$, where R is the company's total revenues for the year and E is the company's total expenses for the year.

61. At the end of fiscal year 2013, Kohl's had a net income of $986,000,000. During the year, Kohl's had total expenses of $18,293,000,000. What were Kohl's total revenues for the year? (*Source:* Kohl's Corporation)

62. At the end of fiscal year 2013, Target had a net income of $2,999,000,000. During the year, Target had total expenses of $70,302,000,000. What were Target's total revenues for the year? (*Source:* Target Corporation)

8.3 Solving Equations: The Multiplication Property

Objective

A Use the Multiplication Property to Solve Equations.

Objective A Using the Multiplication Property to Solve Equations

Although the addition property of equality is a powerful tool for helping us solve equations, it cannot help us solve all types of equations. For example, it cannot help us solve an equation such as $2x = 6$. To solve this equation, we use a second property of equality called the **multiplication property of equality.**

Multiplication Property of Equality

Let a, b, and c represent numbers and let $c \neq 0$. Then

$$a = b$$

and $a \cdot c = b \cdot c$

are equivalent equations.

Also, $a = b$

and $\dfrac{a}{c} = \dfrac{b}{c}$

are equivalent equations.

In other words, both sides of an equation may be multiplied or divided by the same nonzero number without changing the solution of the equation.

Picturing again our balanced scale, if we multiply or divide the weight on each side by the same nonzero number, the scale (or equation) remains balanced.

To solve $2x = 6$ for x, we use the multiplication property of equality to divide both sides of the equation by 2, and simplify as follows:

$$2x = 6$$
$$\frac{2x}{2} = \frac{6}{2} \quad \text{Divide both sides by 2.}$$
$$\frac{2}{2} \cdot x = 3$$
$$1 \cdot x = 3$$
$$x = 3$$

Example 1   Solve: $-5x = 15$

Solution: To get x by itself, we divide both sides by -5.

$$-5x = 15 \quad \text{Original equation}$$
$$\frac{-5x}{-5} = \frac{15}{-5} \quad \text{Divide both sides by } -5.$$
$$\frac{-5}{-5} \cdot x = \frac{15}{-5}$$
$$1 \cdot x = -3 \quad \text{Simplify.}$$
$$x = -3$$

Check: To check, we replace x with -3 in the original equation.

$$-5x = 15 \quad \text{Original equation}$$
$$-5(-3) \overset{?}{=} 15 \quad \text{Let } x = -3.$$
$$15 \overset{?}{=} 15 \quad \text{True}$$

The solution is -3.

■ Work Practice 1

Practice 1

Solve: $-3y = 18$

Example 2   Solve: $-8 = 2y$

Solution: To get y alone, we divide both sides of the equation by 2.

$$-8 = 2y$$
$$\frac{-8}{2} = \frac{2y}{2} \quad \text{Divide both sides by 2.}$$
$$-4 = 1 \cdot y \quad \text{or} \quad y = -4$$

Check to see that -4 is the solution.

■ Work Practice 2

Practice 2

Solve: $-16 = 8x$

Answers

1. -6 **2.** -2

Practice 3

Solve: $-0.3y = -27$

Example 3 Solve: $-1.2x = -36$

Solution: We divide both sides of the equation by the numerical coefficient of x, which is -1.2.

$$-1.2x = -36$$

$$\frac{-1.2x}{-1.2} = \frac{-36}{-1.2}$$

$$1 \cdot x = 30$$

$$x = 30$$

Check to see that 30 is the solution.

■ Work Practice 3

Practice 4

Solve: $\frac{5}{7}b = 25$

Example 4 Solve: $\frac{3}{5}a = 9$

Solution: Recall that the product of a number and its reciprocal is 1. To get a alone, then, we multiply both sides by $\frac{5}{3}$, the reciprocal of $\frac{3}{5}$.

$$\frac{3}{5}a = 9$$

$$\frac{5}{3} \cdot \frac{3}{5}a = \frac{5}{3} \cdot 9 \qquad \text{Multiply both sides by } \frac{5}{3}.$$

$$1 \cdot a = \frac{5 \cdot \overset{3}{\cancel{9}}}{\cancel{3} \cdot 1} \qquad \text{Multiply.}$$

$$a = 15 \qquad \text{Simplify.}$$

Check: To check, we replace a with 15 in the original equation.

$$\frac{3}{5}a = 9 \qquad \text{Original equation}$$

$$\frac{3}{5} \cdot 15 \overset{?}{=} 9 \qquad \text{Replace } a \text{ with 15.}$$

$$\frac{3}{\cancel{5}} \cdot \frac{\overset{3}{\cancel{15}}}{1} \overset{?}{=} 9 \qquad \text{Multiply.}$$

$$9 \overset{?}{=} 9 \qquad \text{True}$$

Since $9 = 9$ is true, 15 is the solution of $\frac{3}{5}a = 9$.

■ Work Practice 4

Copyright © 2016 Pearson Education, Inc.

Answers

3. 90   **4.** 35

Example 5 Solve: $-\dfrac{1}{4}x = \dfrac{1}{8}$

Solution: We multiply both sides of the equation by $-\dfrac{4}{1}$, the reciprocal of $-\dfrac{1}{4}$.

$$-\dfrac{1}{4}x = \dfrac{1}{8}$$

$$-\dfrac{4}{1} \cdot -\dfrac{1}{4}x = -\dfrac{4}{1} \cdot \dfrac{1}{8} \quad \text{Multiply both sides by } -\dfrac{4}{1}.$$

$$1 \cdot x = -\dfrac{\overset{1}{\cancel{4}} \cdot 1}{1 \cdot \underset{2}{\cancel{8}}} \quad \text{Multiply.}$$

$$x = -\dfrac{1}{2} \quad \text{Simplify.}$$

Check to see that $-\dfrac{1}{2}$ is the solution.

■ Work Practice 5

Practice 5

Solve: $-\dfrac{7}{10}x = \dfrac{2}{5}$

✓**Concept Check** Which operation is appropriate for solving each of the following equations, addition or division?

a. $6 = -4x$ **b.** $6 = x - 4$

We often need to simplify one or both sides of an equation before applying the properties of equality to get the variable alone.

Example 6 Solve: $3y - 7y = 12$

Solution: First we combine like terms.

$$3y - 7y = 12$$

$$-4y = 12 \quad \text{Combine like terms.}$$

$$\dfrac{-4y}{-4} = \dfrac{12}{-4} \quad \text{Divide both sides by } -4.$$

$$y = -3 \quad \text{Simplify.}$$

Check: We replace y with -3.

$$3y - 7y = 12$$

$$3(-3) - 7(-3) \overset{?}{=} 12$$

$$-9 + 21 \overset{?}{=} 12$$

$$12 \overset{?}{=} 12 \quad \text{True}$$

The solution is -3.

■ Work Practice 6

Practice 6

Solve: $2m - 4m = 10$

Answers

5. $-\dfrac{4}{7}$ **6.** -5

✓**Concept Check Answers**

a. division   **b.** addition

Practice 7

Solve: $-3a + 2a = -8 + 6$

Example 7 Solve: $-2z + z = 11 - 5$

Solution: We simplify both sides of the equation first.

$$-2z + z = 11 - 5$$
$$-1z = 6 \qquad \text{Combine like terms.}$$
$$\frac{-1z}{-1} = \frac{6}{-1} \qquad \text{Divide both sides by } -1.$$
$$z = -6 \qquad \text{Simplify.}$$

Check to see that -6 is the solution.

◼ Work Practice 7

Practice 8

Solve: $-8 + 6 = \dfrac{a}{3}$

Example 8 Solve: $\dfrac{z}{-4} = 11 - 5$

Solution: Simplify the right side of the equation first.

$$\frac{z}{-4} = 11 - 5$$
$$\frac{z}{-4} = 6$$

Next, to get z alone, multiply both sides by -4.

$$-4 \cdot \frac{z}{-4} = -4 \cdot 6 \qquad \text{Multiply both sides by } -4.$$
$$\frac{-4}{-4} \cdot z = -4 \cdot 6$$
$$1z = -24 \quad \text{or} \quad z = -24$$

Check to see that -24 is the solution.

◼ Work Practice 8

Answers

7. 2 **8.** −6

Vocabulary, Readiness & Video Check

Use the choices below to fill in each blank.

equation	multiplication	simplifying
solving	equivalent	expression

1. The equations $-3x = 51$ and $\dfrac{-3x}{-3} = \dfrac{51}{-3}$ are called _____ equations.

2. The difference between an equation and an expression is that a(n) _____ contains an equal sign, while a(n) _____ does not.

3. The process of writing $-3x + x$ as $2x$ is called _____ the expression.

4. For the equation $-5x = -20$, the process of finding that 4 is the solution is called _____ the equation.

5. By the _____ property of equality, $y = 8$ and $3 \cdot y = 3 \cdot 8$ are equivalent equations.

Martin-Gay Interactive Videos   *Watch the section lecture video and answer the following questions.*

Objective A 6. In solving ▯ Example 1, how is the multiplication property of equality used?

7. When solving ▯ Example 5, what was done before applying the multiplication property of equality?

See Video 8.3

8.3 Exercise Set MyMathLab®

Objective A *Solve. See Examples 1 through 5.*

1. $5x = 20$

2. $6y = 48$

3. $-3z = 12$

4. $-2x = 26$

5. $0.4y = -12$

6. $0.8x = -8$

7. $2z = -34$

8. $7y = -21$

9. $-0.3x = -15$

10. $-0.4z = -16$

11. $10 = \dfrac{2}{5}x$

12. $27 = \dfrac{3}{7}x$

13. $\dfrac{1}{6}y = -5$

14. $\dfrac{1}{8}y = -3$

15. $\dfrac{5}{6}x = \dfrac{5}{18}$

16. $\dfrac{4}{7}y = \dfrac{8}{21}$

17. $-\dfrac{2}{9}z = \dfrac{4}{27}$

18. $-\dfrac{3}{4}v = \dfrac{9}{14}$

Solve. First combine any like terms on each side of the equation. See Examples 6 and 7.

19. $2w - 12w = 40$

20. $-8y + y = 35$

21. $16 = 10t - 8t$

22. $100 = 15y - 5y$

23. $2z = 1.2 + 1.4$

24. $3x = 1.1 + 0.7$

25. $4 - 10 = -3z$

26. $12 - 20 = -4x$

Mixed Practice *Solve. See Examples 1 through 8.*

27. $-7x = 0$

28. $-20y = 0$

29. $0.4 = -8z$

30. $0.5 = -20x$

31. $\dfrac{8}{5}t = -\dfrac{3}{8}$

32. $\dfrac{7}{4}r = -\dfrac{2}{7}$

33. $-\dfrac{3}{5}x = -\dfrac{6}{15}$

34. $-\dfrac{6}{7}y = -\dfrac{1}{14}$

35. $-3.6 = -0.9u + 0.3u$

36. $-5.4 = -1.4y + 1.3y$

37. $5 - 5 = 2x + 7x$

38. $12 + (-12) = 7x + 8x$

39. $-42 + 20 = -2x + 13x$

40. $-4y + 9y = -20 + 15$

41. $-3x - 3x = 50 - 2$

42. $5y - 9y = -14 + (-14)$

43. $23x - 25x = 7 - 9$

44. $6x - 8x = 12 - 22$

45. $\frac{1}{4}x - \frac{5}{8}x = 20 - 47$

46. $\frac{1}{2}x - \frac{4}{5}x = 10 - 19$

47. $18 - 11 = \frac{x}{5}$

48. $14 - 9 = \frac{x}{12}$

49. $\frac{x}{-4} = 1 - (-6)$

50. $\frac{y}{-6} = 6 - (-1)$

Review

Evaluate each expression when $x = 5$. See Section 8.1.

51. $3x + 10$

52. $40x$

53. $\frac{x - 3}{2}$

54. $7x - 20$

55. $\frac{3x + 5}{x - 7}$

56. $\frac{2x - 1}{x - 8}$

The bar graph shows the number of acres burned by wildfires in the United States in recent years. Use the bar graph to answer Exercises 57 through 60. See Section 7.1.

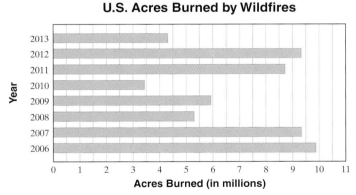

U.S. Acres Burned by Wildfires

Year / Acres Burned (in millions)

Source: National Interagency Fire Center

57. During which year shown is the number of acres burned by wildfires the greatest?

58. During which year shown is the number of acres burned by wildfires the least?

59. Use the length of the bar to estimate the number of acres burned by wildfires in 2012.

60. Describe any trends shown in this graph.

Concept Extensions

What operation is appropriate for solving each equation: addition or division? See the Concept Check in this section.

61. $12 = x - 5$ **62.** $12 = -5x$ **63.** $-7x = 21$ **64.** $-7 + x = 21$

Solve.

65. Why does the multiplication property of equality not allow us to divide both sides of an equation by zero?

66. Is the equation $-x = 6$ solved for the variable? Explain why or why not.

The equation $d = r \cdot t$ describes the relationship between distance d in miles, rate r in miles per hour, and time t in hours. If necessary, round answers to the nearest tenth.

67. The distance between New Orleans, Louisiana, and Memphis, Tennessee, by road is 390 miles. How long will it take to drive from New Orleans to Memphis if the driver maintains a speed of 60 miles per hour? (*Source: World Almanac*)

68. The distance between Boston, Massachusetts, and Milwaukee, Wisconsin, by road is 1050 miles. How long will it take to drive from Boston to Milwaukee if the driver maintains a speed of 55 miles per hour? (*Source: World Almanac*)

69. The distance between Cleveland, Ohio, and Indianapolis, Indiana, by road is 294 miles. At what speed should a driver drive if he or she would like to make the trip in 5 hours? (*Source: World Almanac*)

70. The distance between St. Louis, Missouri, and Minneapolis, Minnesota, by road is 552 miles. If it took 9 hours to drive from St. Louis to Minneapolis, what was the driver's average speed? (*Source: World Almanac*)

Solve.

71. $-0.025x = 91.2$

72. $3.6y = -1.259 - 3.277$

73. $\dfrac{y}{72} = -86 - (-1029)$

74. $\dfrac{x}{-13} = 4^6 - 5^7$

75. $\dfrac{x}{-2} = 5^2 - |-10| - (-9)$

76. $\dfrac{y}{10} = (-8)^2 - |20| + (-2)^2$

Expressions and Equations

For the table below, identify each as an expression or an equation.

Expression or Equation	
1. $7x - 5y + 14$	
2. $7x = 35 + 14$	
3. $3(x - 2) = 5(x + 1) - 17$	
4. $-9(2x + 1) - 4(x - 2) + 14$	

1. _____

2. _____

3. _____

4. _____

Fill in each blank with "simplify" or "solve."

5. To _____ an expression, we combine any like terms.

5. _____

6. _____

6. To _____ an equation, we use the properties of equality to find any value of the variable that makes the equation a true statement.

7. _____

8. _____

Evaluate each expression when $x = -1$ and $y = 3$.

9. _____

7. $y - x$ **8.** $\dfrac{8y}{4x}$ **9.** $5x + 2y$ **10.** $\dfrac{y^2 + x}{2x}$

10. _____

11. _____

Simplify each expression by combining like terms.

11. $7x + x$ **12.** $6y - 10y$

12. _____

13. _____

13. $2a + 5a - 9a - 2$ **14.** $3x - y + 4 - 5x + 4y - 11$

14. _____

15. _____

Multiply and simplify if possible.

16. _____

15. $-2(4x + 7)$ **16.** $-3(2x - 10)$

17. $5(y + 2) - 20$

18. $12x + 3(x - 6) - 13$

△ **19.** Find the area.

Rectangle | 3 meters

$(4x - 2)$ meters

△ **20.** Find the perimeter.

Square

$5y$ inches

19. _____

20. _____

21. _____

Solve and check.

21. $x + 7 = 20$

22. $-11 = x - 2$

22. _____

23. _____

23. $n - \dfrac{2}{5} = \dfrac{3}{10}$

24. $-7y = 0$

24. _____

25. _____

25. $12 = 11x - 14x$

26. $\dfrac{3}{5}x = 15$

26. _____

27. _____

27. $x - 1.2 = -4.5 + 2.3$

28. $8y + 7y = -45$

28. _____

29. _____

29. $6 - (-5) = x + 5$

30. $-0.2m = -1.6$

30. _____

31. _____

31. $-\dfrac{2}{3}n = \dfrac{6}{11}$

32. $11x = 55$

32. _____

Solving Equations Using Addition and Multiplication Properties

Practice 1

Solve: $5y - 8 = 17$

Objective A Solving Equations Using Addition and Multiplication Properties

We will now solve equations using more than one property of equality. To solve an equation such as $2x - 6 = 18$, we will first get the variable term $2x$ alone on one side of the equation.

Example 1 Solve: $2x - 6 = 18$

Solution: We start by adding 6 to both sides to get the variable term $2x$ alone.

$$2x - 6 = 18$$
$$2x - 6 + 6 = 18 + 6 \quad \text{Add 6 to both sides.}$$
$$2x = 24 \quad \text{Simplify.}$$

To finish solving, we divide both sides by 2.

$$\frac{2x}{2} = \frac{24}{2} \quad \text{Divide both sides by 2.}$$
$$1 \cdot x = 12 \quad \text{Simplify.}$$
$$x = 12$$

Check:

$$2x - 6 = 18$$
$$2(12) - 6 \stackrel{?}{=} 18 \quad \text{Replace } x \text{ with 12 and simplify.}$$
$$24 - 6 \stackrel{?}{=} 18$$
$$18 \stackrel{?}{=} 18 \quad \text{True}$$

The solution is 12.

■ **Work Practice 1**

Helpful Hint

Make sure you understand which property to use to solve an equation.

Addition
$$x + 2 = 10$$
To undo addition of 2, we subtract 2 from both sides.
$$x + 2 - 2 = 10 - 2 \quad \text{Use addition property of equality.}$$
$$x = 8$$

Check:
$$x + 2 = 10$$
$$8 + 2 \stackrel{?}{=} 10$$
$$10 \stackrel{?}{=} 10 \quad \text{True}$$

Understood multiplication
$$2x = 10$$
To undo multiplication of 2, we divide both sides by 2.
$$\frac{2x}{2} = \frac{10}{2} \quad \text{Use multiplication property of equality.}$$
$$x = 5$$

Check:
$$2x = 10$$
$$2 \cdot 5 \stackrel{?}{=} 10$$
$$10 \stackrel{?}{=} 10 \quad \text{True}$$

Answer

1. 5

590

Example 2 Solve: $17 - x + 3 = 15 - (-6)$

Solution: First we simplify each side of the equation.

$$17 - x + 3 = 15 - (-6)$$

$$20 - x = 21 \qquad \text{Combine like terms on each side of the equation.}$$

Next, we get the variable term alone on one side of the equation.

$$20 - x - 20 = 21 - 20 \qquad \text{Subtract 20 from both sides.}$$

$$-1x = 1 \qquad \text{Simplify. Recall that } -x \text{ means } -1x.$$

$$\frac{-1x}{-1} = \frac{1}{-1} \qquad \text{Divide both sides by } -1.$$

$$1 \cdot x = -1 \qquad \text{Simplify.}$$

$$x = -1$$

Check:

$$17 - x + 3 = 15 - (-6)$$

$$17 - (-1) + 3 \stackrel{?}{=} 15 - (-6) \qquad \text{Replace } x \text{ with } -1 \text{ and simplify.}$$

$$17 + 1 + 3 \stackrel{?}{=} 15 + 6$$

$$21 \stackrel{?}{=} 21 \qquad \text{True}$$

The solution is -1.

🔲 Work Practice 2

Practice 2

Solve:
$7 - y + 3 = 20 - (-25)$

Example 3 Solve: $1 = \frac{2}{3}x + 7$

Solution: Subtract 7 from both sides to get the variable term alone.

$$1 - 7 = \frac{2}{3}x + 7 - 7 \qquad \text{Subtract 7 from both sides.}$$

$$-6 = \frac{2}{3}x \qquad \text{Simplify.}$$

$$\frac{3}{2} \cdot -6 = \frac{3}{2} \cdot \frac{2}{3}x \qquad \text{Multiply both sides by } \frac{3}{2}.$$

$$\frac{3}{2} \cdot \frac{\overset{-3}{\cancel{-6}}}{1} = 1 \cdot x \qquad \text{Simplify.}$$

$$-9 = x \qquad \text{Simplify.}$$

Check to see that the solution is -9.

🔲 Work Practice 3

Practice 3

Solve: $11 = \frac{3}{4}y + 20$

Helpful Hint Don't forget that we can get the variable alone on either side of the equation.

If an equation contains variable terms on both sides, we use the addition property of equality to get all the variable terms on one side and all the constants, or numbers, on the other side.

Answers

2. -35 **3.** -12

Practice 4

Solve: $9x - 12 = x + 4$

Example 4   Solve: $3a - 6 = a + 4$

Solution:

$$3a - 6 = a + 4$$
$$3a - 6 + 6 = a + 4 + 6 \qquad \text{Add 6 to both sides.}$$
$$3a = a + 10 \qquad \text{Simplify.}$$
$$3a - a = a + 10 - a \qquad \text{Subtract } a \text{ from both sides.}$$
$$2a = 10 \qquad \text{Simplify.}$$
$$\frac{2a}{2} = \frac{10}{2} \qquad \text{Divide both sides by 2.}$$
$$a = 5 \qquad \text{Simplify.}$$

Check to see that the solution is 5.

■ Work Practice 4

Practice 5

Solve: $8x + 4.2 = 10x - 11.6$

Example 5   Solve: $7x + 3.2 = 4x - 1.6$

Solution:

$$7x + 3.2 = 4x - 1.6$$
$$7x + 3.2 - 3.2 = 4x - 1.6 - 3.2 \qquad \text{Subtract 3.2 from both sides.}$$
$$7x = 4x - 4.8 \qquad \text{Simplify.}$$
$$7x - 4x = 4x - 4.8 - 4x \qquad \text{Subtract } 4x \text{ from both sides.}$$
$$3x = -4.8 \qquad \text{Simplify.}$$
$$\frac{3x}{3} = \frac{-4.8}{3} \qquad \text{Divide both sides by 3.}$$
$$x = -1.6 \qquad \text{Simplify.}$$

Check to see that -1.6 is the solution.

■ Work Practice 5

Objective B   Solving Equations Containing Parentheses ▶

If an equation contains parentheses, we must first use the distributive property to remove them.

Practice 6

Solve: $6(a - 5) = 7a - 13$

Example 6   Solve: $7(x - 2) = 9x - 6$

Solution: First we apply the distributive property.

$$7(x - 2) = 9x - 6$$
$$7x - 14 = 9x - 6 \qquad \text{Apply the distributive property.}$$

Next, we move variable terms to one side of the equation and constants to the other side.

Answers

4. 2   **5.** 7.9   **6.** -17

$$7x - 14 - 9x = 9x - 6 - 9x \quad \text{Subtract } 9x \text{ from both sides.}$$
$$-2x - 14 = -6 \quad \text{Simplify.}$$
$$-2x - 14 + 14 = -6 + 14 \quad \text{Add 14 to both sides.}$$
$$-2x = 8 \quad \text{Simplify.}$$
$$\frac{-2x}{-2} = \frac{8}{-2} \quad \text{Divide both sides by } -2.$$
$$x = -4 \quad \text{Simplify.}$$

Check to see that -4 is the solution.

▧ Work Practice 6

You may want to use the steps shown below to solve equations.

Steps for Solving an Equation

Step 1: If parentheses are present, use the distributive property.

Step 2: Combine any like terms on each side of the equation.

Step 3: Use the addition property of equality to rewrite the equation so that variable terms are on one side of the equation and constant terms are on the other side.

Step 4: Use the multiplication property of equality to divide both sides by the numerical coefficient of the variable to solve.

Step 5: Check the solution in the *original equation*.

Example 7 Solve: $3(2x - 6) + 6 = 0$

Solution:

$$3(\overset{\frown}{2x - 6}) + 6 = 0$$

Step 1: $6x - 18 + 6 = 0 \quad$ Apply the distributive property.

Step 2: $6x - 12 = 0 \quad$ Combine like terms on the left side of the equation.

Step 3: $6x - 12 + 12 = 0 + 12 \quad$ Add 12 to both sides.

$$6x = 12 \quad \text{Simplify.}$$

Step 4: $\dfrac{6x}{6} = \dfrac{12}{6} \quad$ Divide both sides by 6.

$$x = 2 \quad \text{Simplify.}$$

Check:

Step 5: $3(2x - 6) + 6 = 0$
$$3(2 \cdot 2 - 6) + 6 \overset{?}{=} 0$$
$$3(4 - 6) + 6 \overset{?}{=} 0$$
$$3(-2) + 6 \overset{?}{=} 0$$
$$-6 + 6 \overset{?}{=} 0$$
$$0 \overset{?}{=} 0 \quad \text{True}$$

The solution is 2.

▧ Work Practice 7

Practice 7

Solve: $4(2x - 3) + 4 = 0$

Answer

7. 1

Objective C Writing Sentences as Equations

Next, we practice translating sentences into equations. Below are key words and phrases that translate to an equal sign:

Key Words or Phrases	Examples	Symbols
equals	3 equals 2 plus 1	$3 = 2 + 1$
gives	the quotient of 10 and -5 gives -2	$\dfrac{10}{-5} = -2$
is/was	17 minus 12 is 5	$17 - 12 = 5$
yields	11 plus 2 yields 13	$11 + 2 = 13$
amounts to	twice -15 amounts to -30	$2(-15) = -30$
is equal to	-24 is equal to 2 times -12	$-24 = 2(-12)$

Practice 8

Translate each sentence into an equation.

a. The difference of 110 and 80 is 30.

b. The product of 3 and the sum of -9 and 11 amounts to 6.

c. The quotient of 24 and -6 yields -4.

Example 8 Translate each sentence into an equation.

a. The product of 7 and 6 is 42.

b. Twice the sum of 3 and 5 is equal to 16.

c. The quotient of -45 and 5 yields -9.

Solution:

a. In words: The product of 7 and 6   is   42

Translate: $7 \cdot 6$ $=$ 42

b. In words: Twice   the sum of 3 and 5   is equal to   16

Translate: 2 $(3 + 5)$ $=$ 16

c. In words: The quotient of -45 and 5   yields -9

Translate: $\dfrac{-45}{5}$ $=$ -9

Answers

8. a. $110 - 80 = 30$

b. $3(-9 + 11) = 6$ **c.** $\dfrac{24}{-6} = -4$

Work Practice 8

Calculator Explorations Checking Possible Solutions

A calculator can be used to check possible solutions of equations. To do this, replace the variable by the possible solution and evaluate each side of the equation separately. For example, to see whether 7 is a solution of the equation $52x = 15x + 259$, replace x with 7 and use your calculator to evaluate each side separately.

Equation: $52x = 15x + 259$

$52 \cdot 7 \overset{?}{=} 15 \cdot 7 + 259$ Replace x with 7.

Evaluate left side: $\boxed{52}$ $\boxed{\times}$ $\boxed{7}$ and then $\boxed{=}$ or $\boxed{\text{ENTER}}$.

Display: $\boxed{364}$.

Evaluate right side: $\boxed{15}$ $\boxed{\times}$ $\boxed{7}$ $\boxed{+}$ $\boxed{259}$ and then $\boxed{=}$ or $\boxed{\text{ENTER}}$. Display: $\boxed{364}$.

Since the left side equals the right side, 7 is a solution of the equation $52x = 15x + 259$.

Use a calculator to determine whether the numbers given are solutions of each equation.

1. $76(x - 25) = -988;$ 12

2. $-47x + 862 = -783;$ 35

3. $x + 562 = 3x + 900;$ -170

4. $55(x + 10) = 75x + 910;$ -18

5. $29x - 1034 = 61x - 362;$ -21

6. $-38x + 205 = 25x + 120;$ 25

Vocabulary, Readiness & Video Check

Use the choices below to fill in each blank. Some choices may be used more than once.

addition   multiplication   combine like terms

$5(2x + 6) - 1 = 39$ $3x - 9 + x - 16$ distributive

1. An example of an expression is _____ while an example of an equation is _____.

2. To solve $\dfrac{x}{-7} = -10$, we use the _____ property of equality.

3. To solve $x - 7 = -10$, we use the _____ property of equality.

Use the order of the Steps for Solving an Equation in this section to answer Exercises 4 through 6.

4. To solve $9x - 6x = 10 + 6$, first _____.
5. To solve $5(x - 1) = 25$, first use the _____ property.
6. To solve $4x + 3 = 19$, first use the _____ property of equality.

Martin-Gay Interactive Videos   *Watch the section lecture video and answer the following questions.*

See Video 8.4

Objective A 7. In ⊞ Example 1, the number 1 is subtracted from the left side of the equation. What property tells us we must also subtract 1 from the right side? Why is it important to do the same thing to both sides? ▶

Objective B 8. From ⊞ Example 3, what is the first step when solving an equation that contains parentheses? What property do we use to perform this step? ▶

Objective C 9. What word or phrase translates to "equals" in ⊞ Example 5? In ⊞ Example 6? ▶

8.4 Exercise Set MyMathLab®

Objective A *Solve each equation. See Examples 1 through 5.*

1. $2x - 6 = 0$
2. $3y - 12 = 0$
3. $3n + 3.6 = 9.3$
4. $4z + 0.8 = 5.2$

5. $6 - n = 10$
6. $7 - y = 9$
7. $-\dfrac{2}{5}x + 19 = -21$
8. $-\dfrac{3}{7}y - 14 = 7$

9. $1.7 = 2y + 9.5$
10. $-5.1 = 3x + 2.4$
11. $2n + 8 = 0$
12. $8w + 40 = 0$

13. $3x - 7 = 4x + 5$
14. $7x - 1 = 8x + 4$
▶ 15. $10x + 15 = 6x + 3$
16. $5x - 3 = 2x - 18$

17. $9 - 3x = 14 + 2x$
18. $4 - 7m = -3m + 4$
19. $-1.4x - 2 = -1.2x + 7$

20. $5.7y + 14 = 5.4y - 10$
21. $x + 20 + 2x = -10 - 2x - 15$
22. $2x + 10 + 3x = -12 - x - 20$

23. $40 + 4y - 16 = 13y - 12 - 3y$
24. $19x - 2 - 7x = 31 + 6x - 15$

Objective B *Solve each equation. See Examples 6 and 7.*

25. $-2(y + 4) = 2$

26. $-1(y + 3) = 10$

▶ 27. $3(x - 1) - 12 = 0$

28. $2(x + 5) + 8 = 0$

29. $35 - 17 = 3(x - 2)$

30. $22 - 42 = 4(x - 1)$

31. $2(y - 3) = y - 6$

32. $3(z + 2) = 5z + 6$

33. $2t - 1 = 3(t + 7)$

34. $-4 + 3c = 4(c + 2)$

35. $3(5c + 1) - 12 = 13c + 3$

36. $4(3t + 4) - 20 = 3 + 5t$

Mixed Practice (*Sections 8.1–8.4*) *Solve each equation. See Examples 1 through 7.*

37. $-4x = 44$

38. $-3x = 51$

39. $x + 9 = 2$

40. $y - 6 = -11$

41. $8 - b = 13$

42. $7 - z = 15$

43. $3r + 4 = 19$

44. $5m + 1 = 46$

45. $2x - 1 = -7$

46. $3t - 2 = -11$

47. $7 = 4c - 1$

48. $9 = 2b - 5$

49. $9a + 29 = -7$

50. $10 + 4v = -6$

51. $0 = 4x + 4$

52. $0 = 5y + 5$

53. $11(x - 2) = 22$

54. $5(a - 4) = 20$

▶ 55. $-7c + 1 = -20$

56. $-2b + 5 = -7$

57. $3(x - 5) = -7 - 11$

58. $4(x - 2) = -20 - 4$

59. $-5 + 7k = -13 + 8k$

60. $-7 + 9d = -17 + 10d$

61. $4x + 3 = 2x + 11$

62. $6y - 8 = 3y + 7$

63. $-8(n + 2) + 17 = -6n - 5$

64. $-10(x + 1) + 2 = -x + 10$

65. $\frac{3}{8}x + 14 = \frac{5}{8}x - 2$

66. $\frac{2}{7}x - 9 = \frac{5}{7}x - 15$

67. $10 + 5(z - 2) = -4z + 1$

68. $20 + 4(w - 5) = 5 - 2w$

69. $\frac{5}{8}a = \frac{1}{8}a + \frac{3}{4}$

70. $\frac{4}{9}a = \frac{1}{9}a + \frac{5}{6}$

71. $7(6 + w) = 6(w - 2)$

72. $6(5 + c) = 5(c - 4)$

73. $3 + 2(2n - 5) = 1$

74. $5 + 4(3x - 2) = 21$

75. $2(3z - 2) - 2(5 - 2z) = 4$

76. $2(3w + 7) - 4(5 - 2w) = 6$

77. $-20 - (-50) = \frac{x}{9}$

78. $-2 - 10 = \frac{z}{10}$

79. $12 + 5t = 6(t + 2)$ **80.** $4 + 3c = 2(c + 2)$ **81.** $3(5c - 1) - 2 = 13c + 3$

82. $4(2t + 5) - 21 = 7t - 6$ **83.** $10 + 5(z - 2) = 4z + 1$ **84.** $14 + 4(w - 5) = 6 - 2w$

Objective C *Write each sentence as an equation. See Example 8.*

85. The sum of -42 and 16 is -26.

86. The difference of -30 and 10 equals -40.

87. The product of -5 and -29 gives 145.

88. The quotient of -16 and 2 yields -8.

89. Three times the difference of -14 and 2 amounts to -48.

90. The product of -2 and the sum of 3 and 12 is -30.

91. The quotient of 100 and twice 50 is equal to 1.

92. Seventeen subtracted from -12 equals -29.

Review

The following bar graph shows the number of U.S. federal individual income tax returns filed electronically during the years shown (some years are projected). Use this graph to answer Exercises 93 through 96. See Section 7.1.

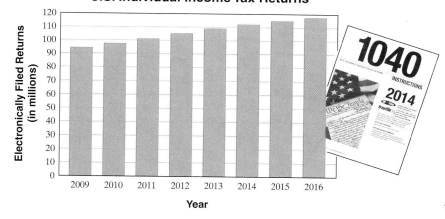

93. Approximate the number of electronically filed returns for 2010.

94. Determine the number of electronically filed returns for 2009.

95. By how much is the number of electronically filed returns projected to increase from 2013 to 2016?

Source: IRS Compliance Research Division; some years are projected

96. Describe any trends shown in this graph.

Concept Extensions

Using the Steps for Solving an Equation, choose the next operation for solving the given equation.

97. $2x - 5 = -7$
   a. Add 7 to both sides.
   b. Add 5 to both sides.
   c. Divide both sides by 2.

98. $3x + 2x = -x - 4$
   a. Add 4 to both sides.
   b. Subtract $2x$ from both sides.
   c. Add $3x$ and $2x$.

99. $-3x = -12$
   a. Divide both sides by -3.
   b. Add 12 to both sides.
   c. Add $3x$ to both sides.

100. $9 - 5x = 15$
   a. Divide both sides by -5.
   b. Subtract 15 from both sides.
   c. Subtract 9 from both sides.

A classmate shows you steps for solving an equation. The solution does not check, but the classmate is unable to find the error. For each set of steps, check the solution, find the error, and correct it.

101.
$$2(3x - 5) = 5x - 7$$
$$6x - 5 = 5x - 7$$
$$6x - 5 + 5 = 5x - 7 + 5$$
$$6x = 5x - 2$$
$$6x - 5x = 5x - 2 - 5x$$
$$x = -2$$

102.
$$37x + 1 = 9(4x - 7)$$
$$37x + 1 = 36x - 7$$
$$37x + 1 - 1 = 36x - 7 - 1$$
$$37x = 36x - 8$$
$$37x - 36x = 36x - 8 - 36x$$
$$x = -8$$

Solve.

103. $(-8)^2 + 3x = 5x + 4^3$

104. $3^2 \cdot x = (-9)^3$

105. $2^3(x + 4) = 3^2(x + 4)$

106. $x + 45^2 = 54^2$

107. A classmate tries to solve $3x = 39$ by subtracting 3 from both sides of the equation. Will this step solve the equation for x? Why or why not?

108. A classmate tries to solve $2 + x = 20$ by dividing both sides by 2. Will this step solve the equation for x? Why or why not?

The equation $C = \dfrac{5}{9}(F - 32)$ gives the relationship between Celsius temperatures C and Fahrenheit temperatures F.

109. The highest recorded temperature in Australia occurred in January 1960 at Oodnadatta, South Australia. The temperature reached 50.7°C. Use the given equation to convert this temperature to degrees Fahrenheit. (*Source:* World Weather Centre at Perth)

110. The highest recorded temperature in Africa occurred in July 1931 at Kebili, Tunisia. The temperature reached 55.0°C. Use the given equation to convert this temperature to degrees Fahrenheit. (*Source:* World Meteorological Organization)

111. The lowest recorded temperature in Australia occurred in June 1994 at Charlotte Pass, New South Wales. The temperature plummeted to −23.0°C. Use the given equation to convert this temperature to degrees Fahrenheit. (*Source:* World Weather Centre at Perth)

112. The lowest recorded temperature in North America occurred in February 1947 at Snag, Canada. The temperature plummeted to −63.0°C. Use the given equation to convert this temperature to degrees Fahrenheit. (*Source:* World Weather Centre at Perth)

Equations and Problem Solving

Objective A Writing Phrases as Algebraic Expressions

Now that we have practiced solving equations for a variable, we can extend considerably our problem-solving skills. We begin by writing phrases as algebraic expressions using the following key words and phrases as a guide:

Addition	Subtraction	Multiplication	Division	Equal Sign
sum	difference	product	quotient	equals
plus	minus	times	divided by	gives
added to	subtracted from	multiply	into	is/was
more than	less than	twice	per	yields
increased by	decreased by	of		amounts to
total	less	double		is equal to

Objectives

A Write Phrases as Algebraic Expressions.

B Write Sentences as Equations.

C Use Problem-Solving Steps to Solve Problems.

Example 1 Write each phrase as an algebraic expression. Use x to represent "a number."

a. 7 increased by a number

b. 15 decreased by a number

c. the product of 2 and a number

d. the quotient of a number and 5

e. 2 subtracted from a number

f. the sum of 9 and twice a number

Solution:

a. In words: 7   increased by   a number

Translate: 7 $+$ x

b. In words: 15   decreased by   a number

Translate: 15 $-$ x

c. In words: the product of

    2   and   a number

Translate: 2 $\cdot$ x or $2x$

d. In words: the quotient of

    a number   and   5

Translate: x $\div$ 5   or $\dfrac{x}{5}$

e. In words: 2   subtracted from   a number

Translate: x $-$ 2

Practice 1

Write each phrase as an algebraic expression. Use x to represent "a number."

a. twice a number

b. 8 increased by a number

c. 10 minus a number

d. 10 subtracted from a number

e. the quotient of 6 and a number

f. the sum of 14 and triple a number

(Continued on next page)

Answers

1. a. $2x$ **b.** $8 + x$ **c.** $10 - x$
d. $x - 10$ **e.** $6 \div x$ or $\dfrac{6}{x}$ **f.** $14 + 3x$

f. In words: the sum of

9 and twice a number

↓ ↓ ↓

Translate: 9 + 2x

■ Work Practice 1

Objective B Writing Sentences as Equations

Now that we have practiced writing phrases as algebraic expressions, let's write sentences as equations. You may want to first study the key words and phrases chart to review some key words and phrases that translate to an equal sign.

Practice 2

Write each sentence as an equation. Use x to represent "a number."

a. Five times a number is 20.

b. The sum of a number and -5 yields 14.

c. Ten subtracted from a number amounts to -23.

d. Five times the difference of a number and 7 is equal to -8.

e. The quotient of triple a number and 5 gives 1.

Example 2   Write each sentence as an equation. Use x to represent "a number."

a. Nine increased by a number is 5.

b. Twice a number equals -10.

c. A number minus 6 amounts to 168.

d. Three times the sum of a number and 5 is -30.

e. The quotient of twice a number and 8 is equal to 2.

Solution:

a. In words: Nine increased by a number is 5

↓ ↓ ↓ ↓ ↓

Translate: 9 + x = 5

b. In words: Twice a number equals -10

↓ ↓ ↓

Translate: 2x = -10

c. In words: A number minus 6 amounts to 168

↓ ↓ ↓ ↓ ↓

Translate: x − 6 = 168

d. In words: Three times the sum of a number and 5 is -30

↓ ↓ ↓ ↓

Translate: 3 (x + 5) = -30

e. In words: The quotient of

twice a number and 8 is equal to 2

↓ ↓ ↓ ↓ ↓

Translate: 2x ÷ 8 = 2

or $\dfrac{2x}{8} = 2$

■ Work Practice 2

Answers

2. a. $5x = 20$ **b.** $x + (-5) = 14$

c. $x - 10 = -23$ **d.** $5(x - 7) = -8$

e. $3x \div 5 = 1$ or $\dfrac{3x}{5} = 1$

Objective C Using Problem-Solving Steps to Solve Problems

Our main purpose for studying arithmetic and algebra is to solve problems. The same problem-solving steps that have been used throughout this text are used in this section also. Those steps are next.

Problem-Solving Steps

1. UNDERSTAND the problem. During this step, become comfortable with the problem. Some ways of doing this are as follows:
   - Read and reread the problem.
   - Construct a drawing.
   - Propose a solution and check. Pay careful attention to how you check your proposed solution. This will help when writing an equation to model the problem.
   - Choose a variable to represent an unknown. Use this variable to represent any other unknowns.
2. TRANSLATE the problem into an equation.
3. SOLVE the equation.
4. INTERPRET the results: *Check* the proposed solution in the stated problem and *state* your conclusion.

The first problem that we solve consists of finding an unknown number.

Example 3 Finding an Unknown Number

Twice a number plus 3 is the same as the number minus 6. Find the unknown number.

Solution:

1. UNDERSTAND the problem. To do so, we read and reread the problem.

   Let's propose a solution to help us understand. Suppose the unknown number is 5. Twice this number plus 3 is $2 \cdot 5 + 3$ or 13. Is this the same as the number minus 6, or $5 - 6$, or -1? Since 13 is not the same as -1, we know that 5 is not the solution. However, remember that the purpose of proposing a solution is not to guess correctly, but to better understand the problem.

   Now let's choose a variable to represent the unknown. Let's let

   x = unknown number

2. TRANSLATE the problem into an equation.

In words:	Twice a number	plus 3	is the same as	the number minus 6
	↓	↓	↓	↓
Translate:	$2x$	$+ 3$	$=$	$x - 6$

3. SOLVE the equation. To solve the equation, we first subtract x from both sides.

$$2x + 3 = x - 6$$
$$2x + 3 - x = x - 6 - x$$
$$x + 3 = -6 \qquad \text{Simplify.}$$
$$x + 3 - 3 = -6 - 3 \qquad \text{Subtract 3 from both sides.}$$
$$x = -9 \qquad \text{Simplify.}$$

(*Continued on next page*)

Practice 3

Translate "The difference of a number and 2 equals 6 added to three times the number" into an equation and solve.

Answer

3. $x - 2 = 6 + 3x$; -4

4. INTERPRET the results. First, *check* the proposed solution in the stated problem. Twice "−9" is −18 and −18 + 3 is −15. This is equal to the number minus 6, or "−9" − 6, or −15. Then *state* your conclusion: The unknown number is −9.

■ Work Practice 3

✓**Concept Check** Suppose you have solved an equation involving perimeter to find the length of a rectangular table. Explain why you would want to recheck your math if you obtain the result of −5.

Example 4 Determining Distances

The distance by road from Chicago, Illinois, to Los Angeles, California, is 1091 miles *more* than the distance from Chicago to Boston, Massachusetts. If the total of these two distances is 3017 miles, find the distance from Chicago to Boston. (*Source: World Almanac*)

Solution:

1. UNDERSTAND the problem. We read and reread the problem.

Let's propose and check a solution to help us better understand the problem. Suppose the distance from Chicago to Boston is 600 miles. Since the distance from Chicago to Los Angeles is 1091 *more* miles, then this distance is 600 + 1091 = 1691 miles. With these numbers, the total of the distances is 600 + 1691 = 2291 miles. This is less than the given total of 3017 miles, so we are incorrect. But not only do we have a better understanding of this exercise, we also know that the distance from Boston to Chicago is greater than 600 miles since this proposed solution led to a total too small. Now let's choose a variable to represent an unknown. Then we'll use this variable to represent any other unknown quantities. Let

x = distance from Chicago to Boston

Then

$x + 1091$ = distance from Chicago to Los Angeles

since that distance is 1091 more miles.

2. TRANSLATE the problem into an equation.

In words:	Chicago to Boston distance	+	Chicago to Los Angeles distance	=	total miles
	↓		↓		↓
Translate:	x	+	$x + 1091$	=	3017

3. SOLVE the equation:

$$x + x + 1091 = 3017$$
$$2x + 1091 = 3017 \qquad \text{Combine like terms.}$$
$$2x + 1091 - 1091 = 3017 - 1091 \qquad \text{Subtract 1091 from both sides.}$$
$$2x = 1926 \qquad \text{Simplify.}$$
$$\frac{2x}{2} = \frac{1926}{2} \qquad \text{Divide both sides by 2.}$$
$$x = 963 \qquad \text{Simplify.}$$

4. INTERPRET the results. First *check* the proposed solution in the stated problem. Since x represents the distance from Chicago to Boston, this is 963 miles. The distance from Chicago to Los Angeles is $x + 1091 = 963 + 1091 = 2054$ miles.

Practice 4

The distance by road from Cincinnati, Ohio, to Denver, Colorado, is 71 miles *less* than the distance from Denver to San Francisco, California. If the total of these two distances is 2399 miles, find the distance from Denver to San Francisco.

Answer
4. 1235 miles

✓**Concept Check Answer**
Length cannot be negative.

To check, notice that the total number of miles is $963 + 2054 = 3017$ miles, the given total of miles. Also, 2054 is 1091 more miles than 963, so the solution checks. Then, *state* your conclusion: The distance from Chicago to Boston is 963 miles.

■ Work Practice 4

Example 5 Calculating Separate Costs

A salesperson at an electronics store sold a computer system and software for $2100, receiving four times as much money for the computer system as for the software. Find the price of each.

Practice 5

A woman's $57,000 estate is to be divided so that her husband receives twice as much as her son. How much will each receive?

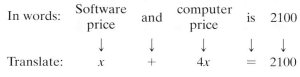

Solution:

1. UNDERSTAND the problem. We read and re-read the problem. Then we choose a variable to represent an unknown. We use this variable to represent any other unknown quantities. We let

   $x = $ the software price

   $4x = $ the computer system price

2. TRANSLATE the problem into an equation.

In words:	Software price	and	computer price	is	2100
	↓	↓	↓	↓	↓
Translate:	x	$+$	$4x$	$=$	2100

3. SOLVE the equation:

$$x + 4x = 2100$$
$$5x = 2100 \quad \text{Combine like terms.}$$
$$\frac{5x}{5} = \frac{2100}{5} \quad \text{Divide both sides by 5.}$$
$$x = 420 \quad \text{Simplify.}$$

4. INTERPRET the results. *Check* the proposed solution in the stated problem. The software sold for $420. The computer system sold for $4x = 4(\$420) = \$1680$. Since $\$420 + \$1680 = \$2100$, the total price, and $1680 is four times $420, the solution checks. *State* your conclusion: The software sold for $420, and the computer system sold for $1680.

■ Work Practice 5

Answer

5. husband: $38,000; son: $19,000

Vocabulary, Readiness & Video Check

Martin-Gay Interactive Videos

See Video 8.5 🍎

Watch the section lecture video and answer the following questions.

Objective A 1. In ⊟ Example 2, what phrase translates to subtraction? ◯

Objective B 2. In ⊟ Example 4, why does the left side of the equation translate to $-20 - x$ and not $x - (-20)$? ◯

Objective C 3. Why are parentheses used in the translation of the left side of the equation in ⊟ Example 6? ◯

4. In ⊟ Example 7, the solution to the equation is $x = 37$. Why is this not the solution to the application? ◯

8.5 Exercise Set MyMathLab®

Objective A **Translating** *Write each phrase as a variable expression. Use x to represent "a number." See Example 1.*

1. The sum of a number and five

2. Ten plus a number

3. The total of a number and eight

4. The difference of a number and five hundred

5. Twenty decreased by a number

6. A number less thirty

7. The product of 512 and a number

8. A number times twenty

9. A number divided by 2

10. The quotient of six and a number

11. The sum of seventeen, a number, and the product of five and the number

12. The difference of twice a number, and four

Objective B **Translating** *Write each sentence as an equation. Use x to represent "a number." See Example 2.*

13. A number added to -5 is -7.

14. Five subtracted from a number equals 10.

15. Three times a number yields 27.

16. The quotient of 8 and a number is -2.

17. A number subtracted from -20 amounts to 104.

18. Two added to twice a number gives -14.

Objectives A B **Mixed Practice Translating** *Write each phrase as a variable expression or each sentence as an equation. Use x to represent "a number." See Examples 1 and 2.*

19. The product of five and a number

20. The quotient of twenty and a number, decreased by three

21. A number subtracted from 11

22. Twelve subtracted from a number

23. Twice a number gives 108.

24. Five times a number is equal to -75.

25. Fifty decreased by eight times a number

26. Twenty decreased by twice a number

27. The product of 5 and the sum of -3 and a number is -20.

28. Twice the sum of -17 and a number is -14.

Objective C *Translate each to an equation. Then solve the equation. See Example 3.*

29. Three times a number, added to 9, is 33. Find the number.

30. Twice a number, subtracted from 60, is 20. Find the number.

31. The sum of 3, 4, and a number amounts to 16. Find the number.

32. The sum of 7, 9, and a number is 40. Find the number.

33. The difference of a number and 3 is equal to the quotient of 10 and 5. Find the number.

34. Eight decreased by a number equals the quotient of 15 and 5. Find the number.

35. Thirty less a number is equal to the product of 3 and the sum of the number and 6. Find the number.

36. The product of a number and 3 is twice the sum of that number and 5. Find the number.

37. 40 subtracted from five times a number is 8 more than the number. Find the number.

38. Five times the sum of a number and 2 is 11 less than the number times 8. Find the number.

39. Three times the difference of a number and 5 amounts to the quotient of 108 and 12. Find the number.

40. Seven times the difference of a number and 1 gives the quotient of 70 and 10. Find the number.

41. The product of 4 and a number is the same as 30 less twice that same number. Find the number.

42. Twice a number equals 25 less triple that same number. Find the number.

Solve. For Exercises 43 and 44, the solutions have been started for you. See Examples 4 and 5.

43. Currently, Florida has 26 fewer electoral votes for president than California. If the total number of electoral votes for these two states is 84, find the number for each state. (*Source:* U.S. Electoral College)

Start the solution:

1. UNDERSTAND the problem. Reread it as many times as needed. Let's let

   x = number of electoral votes for California
   Then
   $x - 26$ = number of electoral votes for Florida

2. TRANSLATE into an equation. (Fill in the blanks below.)

$$
\begin{array}{ccccc}
\text{votes for} & & \text{votes for} & & \\
\text{California} & + & \text{Florida} & = & 84 \\
\downarrow & & \downarrow & & \\
\rule{2cm}{0.4pt} & + & \rule{2cm}{0.4pt} & = & 84
\end{array}
$$

   Now, you finish with

3. SOLVE the equation.
4. INTERPRET the results.

44. Currently, Ohio has twice the number of electoral votes for president as South Carolina. If the total number of electoral votes for these two states is 27, find the number for each state. (*Source:* U.S. Electoral College)

Start the solution:

1. UNDERSTAND the problem. Reread it as many times as needed. Let's let

   x = number of electoral votes for South Carolina
   Then
   $2x$ = number of electoral votes for Ohio

2. TRANSLATE into an equation. (Fill in the blanks below.)

$$
\begin{array}{ccccc}
\text{votes for South} & & \text{votes for} & & \\
\text{Carolina} & + & \text{Ohio} & = & 27 \\
\downarrow & & \downarrow & & \\
\rule{2cm}{0.4pt} & + & \rule{2cm}{0.4pt} & = & 27
\end{array}
$$

   Now, you finish with

3. SOLVE the equation.
4. INTERPRET the results.

45. A falcon, when diving, can travel five times as fast as a pheasant's top speed. If the total speed for these two birds is 222 miles per hour, find the fastest speed of the falcon and the fastest speed of the pheasant. (*Source: Fantastic Book of Comparisons*)

46. Norway has had three times as many rulers as Liechtenstein. If the total number of rulers for both countries is 56, find the number of rulers for Norway and the number for Liechtenstein.

Norway   Liechtenstein

47. The largest university (by enrollment) is Indira Gandhi National Open University in India, followed by Anadolu University in Turkey. If the enrollment in the Indian university is 1.5 million more students than the Turkish university and their combined enrollment is 5.5 million students, find the enrollment for each university. (*Source:* Wikipedia.org)

48. The average life expectancy for an elephant is 24 years longer than the life expectancy for a chimpanzee. If the total of these life expectancies is 130 years, find the life expectancy of each.

49. An Xbox 360 game system and several games are sold for $560. The cost of the Xbox 360 is 3 times as much as the cost of the games. Find the cost of the Xbox 360 and the cost of the games.

50. The two top-selling video games for 2012 were *Call of Duty: Black Ops II* and *Madden NFL 13*. The retail price of *Call of Duty: Black Ops II* was $30 more than the retail price of *Madden NFL 13*. If the total of these two prices was $88, find the price of each game. (*Source:* NPD Group, Amazon.com)

51. By air, the distance from New York City to London is 2001 miles *less* than the distance from Los Angeles to Tokyo. If the total of these two distances is 8939 miles, find the distance from Los Angeles to Tokyo.

52. By air, the distance from Melbourne, Australia, to Cairo, Egypt, is 2338 miles *more* than the distance from Madrid, Spain, to Bangkok, Thailand. If the total of these distances is 15,012 miles, find the distance from Madrid to Bangkok.

53. The two NCAA stadiums with the largest capacities are Michigan Stadium (Univ. of Michigan) and Beaver Stadium (Penn State). Michigan Stadium has a capacity of 3329 more than Beaver Stadium. If the combined capacity for the two stadiums is 216,473, find the capacity for each stadium. (*Source: Wikipedia.org*)

54. In 2020, China is projected to be the country with the greatest number of visiting tourists. This number is twice the number of tourists projected for Spain. If the total number of tourists for these two countries is projected to be 210 million, find the number projected for each. (*Source: The State of the World Atlas by Dan Smith*)

55. California contains the largest state population of native Americans. This population is three times the native American population of Washington state. If the total of these two populations is 412 thousand, find the native American population in each of these two states. (*Source: U.S. Census Bureau*)

56. In 2014, the number of Kohl's stores in California exceeded the number of Kohl's stores in Texas by 43 stores. If the combined number of Kohl's stores in California and Texas was 213, find the number of Kohl's stores in each state. (*Source: Kohl's Corporation*)

57. During the 2014 Super Bowl, the Seattle Seahawks scored 35 more points than the Denver Broncos. Together, both teams scored a total of 51 points. How many points did the 2014 champion Seattle Seahawks score during this game? (*Source: NFL*)

58. During the 2014 Men's NCAA Division I basketball championship game, the Kentucky Wildcats scored 6 fewer points than the Connecticut Huskies. Together, both teams scored a total of 114 points. How many points did the 2014 champion Connecticut Huskies score during this game? (*Source: National Collegiate Athletic Association*)

59. In 2020, the shortage of nurses is projected to be 533,201 more nurses than the shortage in 2010. If the total number of nurse shortages for these two years is 1,083,631, find the nurse shortage for each year.

60. The percent shortage of nurses in 2019 is predicted to be three times the percent shortage in 2007. If the total percent shortages for these two years is 36%, find the percent shortage in 2007 and the percent shortage in 2019.

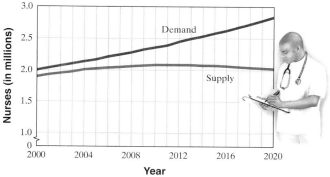

National Supply and Demand Projections for Full-Time Equivalent Registered Nurses: 2000 to 2020

Source: Bureau of Health Professions, RN Supply and Demand Projections

61. In Germany, about twice as many cars are manufactured per day than in Spain. If the total number of cars manufactured per day is 19,827, find the number manufactured in Spain and the number manufactured in Germany.

62. A Toyota Camry is traveling twice as fast as a Dodge truck. If their combined speed is 105 miles per hour, find the speed of the car and find the speed of the truck.

63. A biker sold his used mountain bike and accessories for $270. If he received five times as much money for the bike as he did for the accessories, find how much money he received for the bike.

64. A tractor and a plow attachment are worth $1200. The tractor is worth seven times as much money as the plow. Find the value of the tractor and the value of the plow.

65. In a 2014 report, the United States had the highest number of personal computers in use, followed by China. If the United States had 116 million more personal computers in use than China and the total number of computers for both countries was 506 million, find the number of personal computers in use for each country. (*Source of data:* Computer Industry Almanac Inc.)

66. Based on a 2014 report, the total number of personal computers in use for Russia and Italy was 99 million. If Russia had 9 million more personal computers than Italy, find the number of personal computers for each country. (*Source of data:* Computer Industry Almanac Inc.)

Review

Round each number to the given place value. See Section 1.5.

67. 586 to the nearest ten

68. 82 to the nearest ten

69. 1026 to the nearest hundred

70. 52,333 to the nearest thousand

71. 2986 to the nearest thousand

72. 101,552 to the nearest hundred

Concept Extensions

73. Solve Example 4 again, but this time let x be the distance from Chicago to Los Angeles. Did you get the same results? Explain why or why not.

74. Solve Exercise 43 again, but this time let x be the number of electoral votes for Florida. Did you get the same results? Explain why or why not.

In real estate, a house's selling price P is found by adding the real estate agent's commission C to the amount A that the seller of the house receives: $P = A + C$.

75. A house sold for $230,000. The owner's real estate agent received a commission of $13,800. How much did the seller receive? (*Hint:* Substitute the known values into the equation, and then solve the equation for the remaining unknown.)

76. A homeowner plans to use a real estate agent to sell his house. He hopes to sell the house for $165,000 and keep $156,750 of that. If everything goes as he has planned, how much will his real estate agent receive as a commission?

In retailing, the retail price P of an item can be computed using the equation P = C + M, where C is the wholesale cost of the item and M is the amount of markup.

77. The retail price of a computer system is $999 after a markup of $450. What is the wholesale cost of the computer system? (*Hint:* Substitute the known values into the equation, and then solve the equation for the remaining unknown.)

78. Slidell Feed and Seed sells a bag of cat food for $12. If the store paid $7 for the cat food, what is the markup on the cat food?

Chapter 8 Group Activity

Modeling Equation Solving with Addition and Subtraction

Sections 8.1–8.4

We can use positive counters ● and negative counters ● to help us model the equation-solving process. We also need to use an object that represents a variable. We use small slips of paper with the variable name written on them.

Taking a ● and ● together creates a neutral or zero pair. After a neutral pair has been formed, it can be removed from or added to an equation model without changing the overall value. We also need to remember that we can add or remove the same number of positive or negative counters from both sides of an equation without changing the overall value.

We can represent the equation $x + 5 = 2$ as follows:

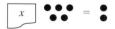

To get the variable by itself, we must remove 5 black counters from the left side of the model. To do so, we must add 5 negative counters to both sides of the model. Then we can remove neutral pairs: 5 from the left side and 2 from the right side (since there are only 2 black counters on the right side).

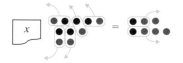

We are left with the following model, which represents the solution, $x = -3$.

$$\boxed{x} \; = \; \bullet\bullet\bullet$$

Similarly, we can represent the equation $x - 4 = -6$ as follows:

$$\boxed{x} \; \begin{matrix}\bullet\bullet\\\bullet\bullet\end{matrix} \; = \; \begin{matrix}\bullet\bullet\bullet\\\bullet\bullet\bullet\end{matrix}$$

To get the variable by itself, we must remove 4 red counters from both sides of the model

$$\boxed{x} \; \boxed{\begin{matrix}\bullet\bullet\\\bullet\bullet\end{matrix}} \; = \; \boxed{\begin{matrix}\bullet\bullet\\\bullet\bullet\end{matrix}}\bullet\bullet$$

We are left with the following model, which represents the solution, $x = -2$.

$$\boxed{x} \; = \; \begin{matrix}\bullet\\\bullet\end{matrix}$$

Use the counter model to solve each equation.

1. $x - 3 = -7$ **2.** $x - 1 = -9$

3. $x + 2 = 8$ **4.** $x + 4 = 5$

5. $x + 8 = 3$ **6.** $x - 5 = -1$

7. $x - 2 = 1$ **8.** $x - 5 = 10$

9. $x + 3 = -7$ **10.** $x + 8 = -2$

Chapter 8 Vocabulary Check

Fill in each blank with one of the words or phrases listed below.

variable	addition	constant	algebraic expression	equation
terms	simplified	multiplication	evaluating the expression	solution
like	combined	numerical coefficient	distributive	

1. An algebraic expression is _____ when all like terms have been _____.

2. Terms that are exactly the same, except that they may have different numerical coefficients, are called _____ terms.

3. A letter used to represent a number is called a(n) _____.

4. A combination of operations on variables and numbers is called a(n) _____.

5. The addends of an algebraic expression are called the _____ of the expression.

6. The number factor of a variable term is called the _____.

7. Replacing a variable in an expression by a number and then finding the value of the expression is called _____ for the variable.

8. A term that is a number only is called a(n) _____.

9. A(n) _____ is of the form expression = expression.

10. A(n) _____ of an equation is a value for the variable that makes the equation a true statement.

11. To multiply $-3(2x + 1)$, we use the _____ property.

12. By the _____ property of equality, we may multiply or divide both sides of an equation by any nonzero number without changing the solution of the equation.

13. By the _____ property of equality, the same number may be added to or subtracted from both sides of an equation without changing the solution of the equation.

> **Helpful Hint**
>
> ▶ Are you preparing for your test? Don't forget to take the Chapter 8 Test on page 618. Then check your answers at the back of the text and use the Chapter Test Prep Videos to see the fully worked-out solutions to any of the exercises you want to review.

8 Chapter Highlights

Definitions and Concepts	Examples
Section 8.1 Variable Expressions	
A letter used to represent a number is called a **variable.** A combination of numbers, letters (variables), and operation symbols is called an **algebraic expression,** or an **expression.**	x, y, z, a, b $3 + x, 7y, x^3 + y - 10$
Replacing a variable in an expression by a number and then finding the value of the expression is called **evaluating the expression.**	Evaluate: $2x + y$ when $x = 22$ and $y = 4$ $2x + y = 2 \cdot 22 + 4$ Replace x with 22 and y with 4. $ = 44 + 4$ Multiply. $ = 48$ Add.
The addends of an algebraic expression are called the **terms** of the expression.	$5x^2 + (-4x) + (-2)$ ⎣___⎦ ⎣___⎦ ⎣___⎦— 3 terms

Definitions and Concepts	Examples

Section 8.1   Introduction to Variables (*continued*)

The number factor of a variable term is called the **numerical coefficient.**

Term	**Numerical Coefficient**
$7x$	7
$-6y$	-6
x or $1x$	1

Terms that are exactly the same, except that they may have different numerical coefficients, are called **like terms.**

$$5x + 11x = (5 + 11)x = 16x$$

like terms

$$y - 6y = (1 - 6)y = -5y$$

An algebraic expression is **simplified** when all like terms have been **combined.**

Use the distributive property to multiply an algebraic expression within parentheses by a term. Once any like terms are then combined, the algebraic expression is simplified.

Simplify:
$$-4(x + 2) + 3(5x - 7)$$
$$= -4(x) + (-4)(2) + 3(5x) - 3(7)$$
$$= -4x + (-8) + 15x - (21)$$
$$= -4x + 15x + (-8) + (-21)$$
$$= 11x + (-29) \quad \text{or} \quad 11x - 29$$

Section 8.2   Solving Equations: The Addition Property

Addition Property of Equality

Let a, b, and c represent numbers. Then

$a = b$	Also, $a = b$
and $a + c = b + c$	and $a - c = b - c$
are equivalent equations.	are equivalent equations.

In other words, the same number may be added to or subtracted from both sides of an equation without changing the solution of the equation.

Solve for x:
$$x + 8 = 2 + (-1)$$
$$x + 8 = 1$$
$$x + 8 - 8 = 1 - 8 \qquad \text{Subtract 8 from both sides.}$$
$$x = -7 \qquad \text{Simplify.}$$

The solution is -7.

Section 8.3   Solving Equations: The Multiplication Property

Multiplication Property of Equality

Let a, b, and c represent numbers and let $c \neq 0$. Then

$a = b$	Also, $a = b$
and $a \cdot c = b \cdot c$	and $\dfrac{a}{c} = \dfrac{b}{c}$
are equivalent equations.	are equivalent equations.

In other words, both sides of an equation may be multiplied or divided by the same nonzero number without changing the solution of the equation.

Solve: $-7x = 42$
$$\frac{-7x}{-7} = \frac{42}{-7} \qquad \text{Divide both sides by } -7.$$
$$x = -6 \qquad \text{Simplify.}$$

Solve: $\dfrac{2}{3}x = -10$

$$\frac{3}{2} \cdot \frac{2}{3}x = \frac{3}{2} \cdot -10 \qquad \text{Multiply both sides by } \frac{3}{2}.$$

$$x = -15 \qquad \text{Simplify.}$$

Definitions and Concepts	Examples

Section 8.4   Solving Equations Using Addition and Multiplication Properties

Steps for Solving an Equation

Step 1: If parentheses are present, use the distributive property.

Step 2: Combine any like terms on each side of the equation.

Step 3: Use the addition property of equality to rewrite the equation so that variable terms are on one side of the equation and constant terms are on the other side.

Step 4: Use the multiplication property of equality to divide both sides by the numerical coefficient of the variable to solve.

Step 5: Check the solution in the *original equation*.

Solve for x: $5(3x - 1) + 15 = -5$

Step 1: $15x - 5 + 15 = -5$ Apply the distributive property.

Step 2: $15x + 10 = -5$ Combine like terms.

Step 3: $15x + 10 - 10 = -5 - 10$ Subtract 10 from both sides.

$$15x = -15$$

Step 4: $\dfrac{15x}{15} = \dfrac{-15}{15}$ Divide both sides by 15.

$$x = -1$$

Step 5: Check to see that -1 is the solution.

Section 8.5   Equations and Problem Solving

Problem-Solving Steps

1. UNDERSTAND the problem. Some ways of doing this are

   Read and reread the problem.
   Construct a drawing.
   Choose a variable to represent an unknown in the problem.

The incubation period for a golden eagle is three times the incubation period for a hummingbird. If the total of their incubation periods is 60 days, find the incubation period for each bird. (*Source: Wildlife Fact File,* International Masters Publishers)

1. UNDERSTAND the problem. Then choose a variable to represent an unknown. Let

$$x = \text{incubation period of a hummingbird}$$
$$3x = \text{incubation period of a golden eagle}$$

2. TRANSLATE the problem into an equation.

2. TRANSLATE.

Incubation of hummingbird	+	incubation of golden eagle	is	60
↓		↓	↓	↓
x	+	$3x$	=	60

3. SOLVE the equation.

3. SOLVE:

$$x + 3x = 60$$
$$4x = 60$$
$$\dfrac{4x}{4} = \dfrac{60}{4}$$
$$x = 15$$

4. INTERPRET the results. *Check* the proposed solution in the stated problem and *state* your conclusion.

4. INTERPRET the results in the stated problem. The incubation period for a hummingbird is 15 days. The incubation period for a golden eagle is $3x = 3 \cdot 15 = 45$ days.

Since 15 days $+$ 45 days $=$ 60 days and 45 is $3(15)$, the solution checks.

State your conclusion: The incubation period for a hummingbird is 15 days. The incubation period for a golden eagle is 45 days.

(8.1) *Evaluate each expression when* $x = 5$, $y = 0$, *and* $z = -2$.

1. $\dfrac{2x}{z}$

2. $4x - 3$

3. $\dfrac{x + 7}{y}$

4. $\dfrac{y}{5x}$

5. $x^3 - 2z$

6. $\dfrac{7 + x}{3z}$

△ **7.** Find the volume of a storage cube whose sides measure 2 feet. Use $V = s^3$.

2 feet

△ **8.** Find the volume of a wooden crate in the shape of a cube 4 feet on each side. Use $V = s^3$.

4 feet

9. Lamar deposited his $5000 bonus into an account paying 6% annual interest. How much interest will he earn in 6 years? Use $I = PRT$.

10. Jennifer Lewis borrowed $2000 from her grandmother and agreed to pay her 5% simple interest. How much interest will she owe after 3 years? Use $I = PRT$.

Simplify each expression by combining like terms.

11. $-6x - 9x$

12. $\dfrac{2}{3}x - \dfrac{9}{10}x$

13. $2y - 10 - 8y$

14. $8a + a - 7 - 15a$

15. $y + 3 - 9y - 1$

16. $1.7x - 3.2 + 2.9x - 8.7$

Multiply.

17. $-2(4y)$

18. $3(5y - 8)$

Simplify.

19. $7x + 3(x - 4) + x$

20. $4(x - 7) + 21$

21. $3(5a - 2) + 10(-2a + 1)$

22. $6y + 3 + 2(3y - 6)$

△ **23.** Find the area.

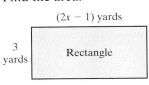

(2x − 1) yards

3 yards | Rectangle

△ **24.** Find the perimeter.

7y meters

Square

(8.2)

25. Is 4 a solution of $5(2 - x) = -10$?

26. Is 0 a solution of $6y + 2 = 23 + 4y$?

Solve.

27. $z - 5 = -7$

28. $x + 1 = 8$

29. $x + \dfrac{7}{8} = \dfrac{3}{8}$

30. $y + \dfrac{4}{11} = -\dfrac{2}{11}$

31. $n + 18 = 10 - (-2)$

32. $15 = 8x + 35 - 7x$

33. $m - 3.9 = -2.6$

34. $z - 4.6 = -2.2$

(8.3) *Solve.*

35. $-3y = -21$

36. $-8x = 72$

37. $-5n = -5$

38. $-3a = 15$

39. $\dfrac{2}{3}x = -\dfrac{8}{15}$

40. $-\dfrac{7}{8}y = 21$

41. $-1.2x = 144$

42. $-0.8y = -10.4$

43. $-5x = 100 - 120$

44. $18 - 30 = -4x$

(8.4) *Solve.*

45. $3x - 4 = 11$

46. $6y + 1 = 73$

47. $-\dfrac{5}{9}x + 23 = -12$

48. $-\dfrac{2}{3}x - 11 = \dfrac{2}{3}x - 55$

49. $6.8 + 4y = -2.2$

50. $-9.6 + 5y = -3.1$

51. $2x + 7 = 6x - 1$

52. $5x - 18 = -4x + 36$

53. $5(n - 3) = 7 + 3n$

54. $7(2 + x) = 4x - 1$

55. $2(4n - 11) + 8 = 5n + 4$

56. $3(5x - 6) + 9 = 13x + 7$

Write each sentence as an equation.

57. The difference of 20 and -8 is 28.

58. Nineteen subtracted from -2 amounts to -21.

59. The quotient of -75 and the sum of 5 and 20 is equal to -3.

60. The product of -5 and the sum of -2 and 6 yields -20.

(8.5) *Write each phrase as an algebraic expression. Use x to represent "a number."*

61. The quotient of 70 and a number

62. The difference of a number and 13

63. A number subtracted from 85

64. Eleven added to twice a number

Write each sentence as an equation. Use x to represent "a number."

65. A number increased by 8 is 40.

66. Twelve subtracted from twice a number is 10.

Solve.

67. Five times a number subtracted from 40 is the same as three times the number. Find the number.

68. The product of a number and 3 is twice the difference of that number and 8. Find the number.

69. Bamboo and Pacific Kelp, a kind of seaweed, are two fast-growing plants. Bamboo grows twice as fast as kelp. If in one day both can grow a total of 54 inches, find how many inches each plant can grow in one day.

70. In an election between the incumbent and a challenger, the incumbent received 11,206 more votes than the challenger. If a total of 18,298 votes were cast, find the number of votes for each candidate.

Bamboo

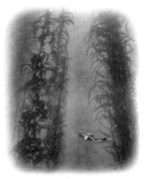
Kelp

Mixed Review

Evaluate each expression when $x = 4$, $y = -3$, and $z = 5$.

71. $18 - (9 - 5x)$

72. $\dfrac{z}{100} + \dfrac{y}{10}$

Simplify.

73. $9x - 20x$

74. $-5(7x)$

75. $12x + 5(2x - 3) - 4$

76. $-7(x + 6) - 2(x - 5)$

Write each phrase as an algebraic expression. Use x to represent "a number."

77. Seventeen less than a number

78. Three times the sum of a number and five

Write each sentence as an equation using x to represent "a number."

79. The difference of a number and 3 is the quotient of the number and 4.

80. The product of a number and 6 is equal to the sum of the number and 2.

81. Is 3 a solution of $4y + 2 - 6y = 5 + 7$?

82. Is 7 a solution of $4(z - 8) + 12 = 8$?

Solve.

83. $c - 5 = -13 + 7$

84. $7x + 5 - 6x = -20$

85. $-7x + 3x = -50 - 2$

86. $-x + 8x = -38 - 4$

87. $14 - y = -3$

88. $7 - z = 0$

89. $9x + 12 - 8x = -6 + (-4)$

90. $-17x + 14 + 20x - 2x = 5 - (-3)$

91. $\frac{4}{9}x = -\frac{1}{3}$

92. $-\frac{5}{24}x = \frac{5}{6}$

93. $2y + 6y = 24 - 8$

94. $13x - 7x = -4 - 12$

95. $\frac{2}{3}x - 12 = -4$

96. $\frac{7}{8}x + 5 = -2$

97. $-5z + 3z - 7 = 8z - 7$

98. $4x - 3 + 6x = 5x - 3$

99. Three times a number added to twelve is 27. Find the number.

100. Twice the sum of a number and four is ten. Find the number.

Answers

1. Evaluate $\dfrac{3x - 5}{2y}$ when $x = 7$ and $y = -8$.

2. Simplify $7x - 5 - 12x + 10$ by combining like terms.

3. Multiply: $-2(3y + 7)$

4. Simplify: $5(3z + 2) - z - 18$

1. _____

2. _____

△ **5.** Write a product that represents the area of the rectangle. Then multiply.

4 meters	
Rectangle	$(3x - 1)$ meters

3. _____

4. _____

5. _____

Solve.

6. _____

6. $x - 17 = -10$

7. $y + \dfrac{3}{4} = \dfrac{1}{4}$

7. _____

8. _____

8. $-4x = 48$

9. $-\dfrac{5}{8}x = -25$

9. _____

10. _____

10. $5x + 12 - 4x - 14 = 22$

11. $2 - c + 2c = 5$

11. _____

12. _____

12. $3x - 5 = -11$

13. $-4x + 7 = 15$

13. _____

618

14. $3.6 - 2x = -5.4$ **15.** $12 = 3(4 + 2y)$

16. $5x - 2 = x - 10$ **17.** $10y - 1 = 7y + 21$

18. $6 + 2(3n - 1) = 28$ **19.** $4(5x + 3) = 2(7x + 6)$

Solve.

△ **20.** A lawn is in the shape of a trapezoid with a height of 60 feet and bases of 70 feet and 130 feet. Find the area of the lawn. Use
$$A = \frac{1}{2}h(B + b).$$

△ **21.** If the height of a triangular-shaped jib sail is 12 feet and its base is 5 feet, find the area of the sail. Use
$$A = \frac{1}{2}bh.$$

22. Translate the following phrases into mathematical expressions. Use x to represent "a number."
    a. The product of a number and 17
    b. Twice a number subtracted from 20

23. The difference of three times a number and five times the same number is 4. Find the number.

24. In a championship basketball game, Paula Zimmerman scored twice as many points as Maria Kaminsky. If the total number of points made by both women was 51, find how many points Paula scored.

25. In a 10-kilometer race, there are 112 more men entered than women. Find the number of women runners if the total number of runners in the race is 600.

14. _____

15. _____

16. _____

17. _____

18. _____

19. _____

20. _____

21. _____

22. a. _____

    b. _____

23. _____

24. _____

25. _____

Answers

1. _____

2. _____

3. _____

4. _____

5. _____

6. _____

7. _____

8. _____

9. _____

10. _____

11. a. _____

b. _____

12. _____

1. Find the place value of the digit 3 in the whole number 396,418.

2. Write 2036 in words.

3. Add: 34,285 + 149,761

4. Find the average of 56, 18, and 43.

△ 5. Find the perimeter of the polygon shown.

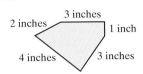

2 inches · 3 inches · 1 inch · 4 inches · 3 inches

6. Subtract 8 from 25.

7. In 2011, a total of 12,734,424 passenger vehicles were sold in the United States. In 2012, total passenger vehicle sales in the United States had increased by 1,705,636 vehicles. Find the total number of passenger vehicles sold in the United States in 2012. (*Source:* Alliance of Automobile Manufacturers)

8. Find $\sqrt{25}$.

9. Subtract 7826 − 505. Check by adding.

10. Find 8^2.

11. In the following graph, each bar represents a country and the height of each bar represents the number of endangered species identified in that country.

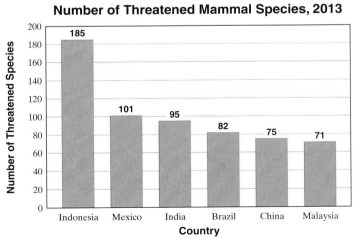

Number of Threatened Mammal Species, 2013

Source: International Union for Conservation of Nature

a. Which country shown has the greatest number of threatened mammal species?
b. Find the total number of threatened mammal species for Malaysia, China, and Indonesia.

12. Evaluate: $\left(-\dfrac{1}{2}\right)^3$

Simplify.

13. $-(-4)$ **14.** $-|20|$ **15.** $-|-5|$ **16.** $|0|$

Add.

17. $-2 + (-21)$ **18.** $-8.2 + 4.6$

19. $(-3) + 4 + (-11)$ **20.** $\dfrac{2}{5} + \left(-\dfrac{3}{10}\right)$

Subtract.

21. $8 - 15$ **22.** $4.6 - (-1.2)$

23. $-4 - (-5)$ **24.** $\dfrac{7}{10} - \dfrac{23}{24}$

Multiply.

25. $-3(-5)$ **26.** $-8(1.2)$

27. $(-1)(-2)(-3)(-4)$ **28.** $-2\dfrac{2}{9}\left(1\dfrac{4}{5}\right)$

29. Simplify: $(-3) \cdot |-5| - (-2) + 4^2$ **30.** Solve: $4x - 7.1 = 3x + 2.6$

31. Multiply: 0.0531×16 **32.** Multiply: 0.0531×1000

33. Given the rectangle shown:

8 feet

5 feet

34. Add: $\dfrac{5}{12} + \dfrac{2}{9}$

a. Find the ratio of its width to its length.
b. Find the ratio of its length to its
perimeter.

13. _____

14. _____

15. _____

16. _____

17. _____

18. _____

19. _____

20. _____

21. _____

22. _____

23. _____

24. _____

25. _____

26. _____

27. _____

28. _____

29. _____

30. _____

31. _____

32. _____

33. a. _____

b. _____

34. _____

35. _____

36. _____

37. _____

38. _____

39. _____

40. _____

41. _____

42. _____

43. _____

44. _____

45. _____

46. _____

47. _____

48. _____

49. _____

50. _____

51. _____

52. _____

35. 12% of what number is 0.6?

36. Multiply: $\dfrac{7}{8} \cdot \dfrac{2}{3}$

37. What percent of 12 is 9?

38. Divide: $1\dfrac{4}{5} \div 2\dfrac{3}{10}$

39. Convert 3 pounds to ounces.

40. Round 23,781 to the nearest thousand.

41. Add 2400 ml to 8.9 L.

42. Round 0.02351 to the nearest thousandth.

43. Is $\dfrac{1\frac{1}{6}}{10\frac{1}{2}} = \dfrac{\frac{1}{2}}{4\frac{1}{2}}$ a true proportion?

44. Is $\dfrac{7.8}{3} = \dfrac{5.2}{2}$ a true proportion?

45. The standard dose of an antibiotic is 4 cc (cubic centimeters) for every 25 pounds (lb) of body weight. At this rate, find the standard dose for a 140-lb woman.

46. On a certain map, 2 inches represents 75 miles. How many miles are represented by 7 inches?

Write each percent as a decimal.

47. 4.6%

48. 452%

Write each percent as a fraction in simplest form.

49. $33\dfrac{1}{3}\%$

50. 27%

51. Translate to an equation: Five is what percent of 20?

52. Translate to a proportion: Five is what percent of 20?

53. Find the sales tax and the total price on the purchase of an $85.50 atlas in a city where the sales tax rate is 7.5%.

54. A salesperson makes a 7% commission rate on her total sales. If her total sales are $23,000, what is her commission?

55. The following bar graph shows the number of endangered species in 2013. Use this graph to answer the questions.

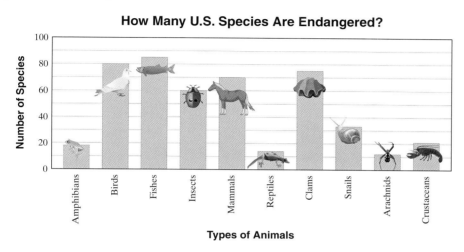

How Many U.S. Species Are Endangered?

Source: U.S. Fish and Wildlife Service

a. Approximate the number of endangered species that are clams.
b. Which category has the most endangered species?

56. Find the mean, median, and mode of 1, 7, 8, 10, 11, 11.

---

53. _____

54. _____

55. a. _____

    b. _____

56. _____

9 Geometry

The word *geometry* is formed from the Greek words *geo*, meaning Earth, and *metron*, meaning measure. Geometry literally means to measure the Earth. In this chapter we learn about various geometric figures and their properties such as perimeter, area, and volume. Knowledge of geometry can help us solve practical problems in real-life situations. For instance, knowing certain measures of a circular swimming pool allows us to calculate how much water it can hold.

A Swiss company created the Zaugg Pipe Monster (shown above) specifically for building superpipes.

What Is a Superpipe?

For winter sports, the term *superpipe* is used to describe a halfpipe built of snow that has walls 22 feet high from the flat bottom on both sides. The length of a superpipe ranges from 400 feet to 600 feet.

Halfpipes in snow were originally formed by hand tools or with heavy machinery. The current method of halfpipe cutting and grooming is by use of a Zaugg Pipe Monster, shown above. Because of the high expense of constructing and maintaining them, there are very few true superpipes. During the 2013–2014 northern hemisphere winter, only 14 superpipes existed globally.

Throughout this chapter, we work with lines, angles, circumferences, and other geometric concepts that give us an appreciation of the work that goes into constructing a halfpipe.

**1998 Olympics
Nagano, Japan**

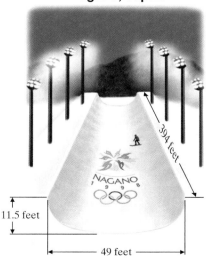

11.5 feet

394 feet

49 feet

**2014 Olympics
Sochi, Russia**

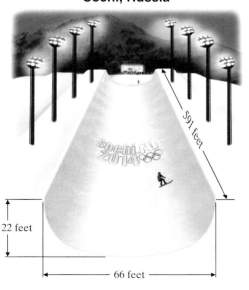

22 feet

591 feet

66 feet

Objective A Identifying Lines, Line Segments, Rays,
and Angles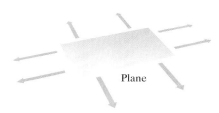

Let's begin with a review of two important concepts—space and plane.
Space extends in all directions indefinitely. Examples of objects in space are houses,
grains of salt, bushes, your *Basic College Mathematics with Early Integers* textbook,
and you.
A **plane** is a flat surface that extends indefinitely. Surfaces like a plane are a
classroom floor and a blackboard or whiteboard.

Objectives

A Identify Lines, Line Segments,
Rays, and Angles.

B Classify Angles as Acute,
Right, Obtuse, or
Straight.

C Identify Complementary and
Supplementary Angles.

D Find Measures of Angles.

Plane

The most basic concept of geometry is the idea of a point in space. A **point** has
no length, no width, and no height, but it does have location. We represent a point
by a dot, and we usually label points with capital letters.

P

Point *P*

A **line** is a set of points extending indefinitely in two directions. A line has
no width or height, but it does have length. We can name a line by any two of its
points or by a single lowercase letter. A **line segment** is a piece of a line with two
endpoints.

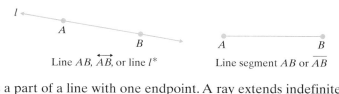

Line *AB*, $\overleftrightarrow{AB}$, or line *l**   Line segment *AB* or $\overline{AB}$

A **ray** is a part of a line with one endpoint. A ray extends indefinitely in one direction. An **angle** is made up of two rays that share the same endpoint. The common
endpoint is called the **vertex.**

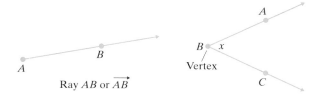

Ray *AB* or $\overrightarrow{AB}$ Vertex

The angle in the figure above can be named

$\angle ABC$ $\angle CBA$ $\angle B$ or $\angle x$

The vertex is the
middle point.

Rays *BA* and *BC* are **sides** of the angle.

---

*Although line *l* is also line *BA* or $\overleftrightarrow{BA}$, we will use only one order of points to name a line or line segment.

Helpful Hint

Naming an Angle
When there is no confusion as to what angle is being named, you may use the vertex alone.

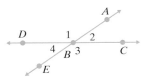

Name of ∠B is all right.
There is no confusion. ∠B means ∠1.

Name of ∠B is *not* all right.
There is confusion. Does ∠B mean
∠1, ∠2, ∠3, or ∠4?

Practice 1

Identify each figure as a line, a ray, a line segment, or an angle. Then name the figure using the given points.

a.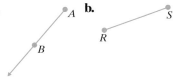

b.

c.

d.

Example 1 Identify each figure as a line, a ray, a line segment, or an angle. Then name the figure using the given points.

a.

b.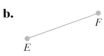

c.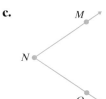

d.

Solution:

Figure (a) extends indefinitely in two directions. It is line CD or $\overleftrightarrow{CD}$.
Figure (b) has two endpoints. It is line segment EF or $\overline{EF}$.
Figure (c) has two rays with a common endpoint. It is ∠MNO, ∠ONM, or ∠N.
Figure (d) is part of a line with one endpoint. It is ray PT or $\overrightarrow{PT}$.

■ Work Practice 1

Practice 2

Use the figure in Example 2 to list other ways to name ∠z.

Example 2 List other ways to name ∠y.

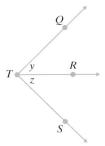

Solution: Two other ways to name ∠y are ∠QTR and ∠RTQ. We may *not* use the vertex alone to name this angle because three different angles have T as their vertex.

■ Work Practice 2

Answers

1. a. ray; ray AB or $\overrightarrow{AB}$ **b.** line segment; line segment RS or $\overline{RS}$ **c.** line; line EF or $\overleftrightarrow{EF}$ **d.** angle; ∠TVH or ∠HVT or ∠V

2. ∠RTS, ∠STR

Objective B Classifying Angles as Acute, Right, Obtuse, or Straight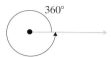

An angle can be measured in **degrees.** The symbol for degrees is a small, raised circle, °. There are 360° in a full revolution, or a full circle.

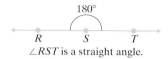

$\frac{1}{2}$ of a revolution measures $\frac{1}{2}(360°) = 180°$. An angle that measures 180° is called a **straight angle.**

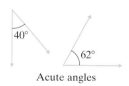

∠RST is a straight angle.

$\frac{1}{4}$ of a revolution measures $\frac{1}{4}(360°) = 90°$. An angle that measures 90° is called a **right angle.** The symbol ∟ is used to denote a right angle.

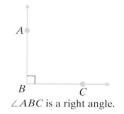

∠ABC is a right angle.

An angle whose measure is between 0° and 90° is called an **acute angle.**

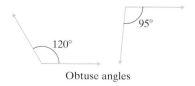

Acute angles

An angle whose measure is between 90° and 180° is called an **obtuse angle.**

Obtuse angles

Example 3 Classify each angle as acute, right, obtuse, or straight.

a.

R

b.
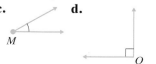
S

c.

T

d.

Q

(*Continued on next page*)

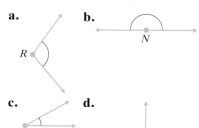

Solution:

a. $\angle R$ is a right angle, denoted by ∟. It measures 90°.
b. $\angle S$ is a straight angle. It measures 180°.
c. $\angle T$ is an acute angle. It measures between 0° and 90°.
d. $\angle Q$ is an obtuse angle. It measures between 90° and 180°.

■ Work Practice 3

Let's look at $\angle B$ below, whose measure is 62°.

There is a shorthand notation for writing the measure of this angle. To write "The measure of $\angle B$ is 62°," we can write

$$m\angle B = 62°$$

By the way, note that $\angle B$ is an acute angle because $m\angle B$ is between 0° and 90°.

Objective C Identifying Complementary and Supplementary Angles ▶

Two angles that have a sum of 90° are called **complementary angles.** We say that each angle is the **complement** of the other.

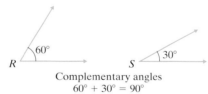

Complementary angles
60° + 30° = 90°

$\angle R$ and $\angle S$ are complementary angles because

$$m\angle R + m\angle S = 60° + 30° = 90°$$

Two angles that have a sum of 180° are called **supplementary angles.** We say that each angle is the **supplement** of the other.

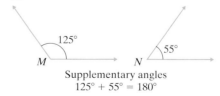

Supplementary angles
125° + 55° = 180°

$\angle M$ and $\angle N$ are supplementary angles because

$$m\angle M + m\angle N = 125° + 55° = 180°$$

Example 4 Find the complement of a 48° angle.

Solution: Two angles that have a sum of 90° are complementary. This means that the complement of an angle that measures 48° is an angle that measures $90° - 48° = 42°$.

■ Work Practice 4

Practice 4

Find the complement of a 29° angle.

Answer
4. 61°

Example 5   Find the supplement of a 107° angle.

Solution: Two angles that have a sum of 180° are supplementary. This means that the supplement of an angle that measures 107° is an angle that measures 180° − 107° = 73°.

■ Work Practice 5

Practice 5

Find the supplement of a 67° angle.

✔**Concept Check** True or false? The supplement of a 48° angle is 42°. Explain.

Objective D Finding Measures of Angles

Measures of angles can be added or subtracted to find measures of related angles.

Example 6   Find the measure of ∠x. Then classify ∠x as an acute, obtuse, or right angle.

Solution: $m\angle x = m\angle QTS - m\angle RTS$
$$= 87° - 52°$$
$$= 35°$$

Thus, the measure of ∠x (m∠x) is 35°.
    Since ∠x measures between 0° and 90°, it is an acute angle.

■ Work Practice 6

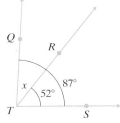

Practice 6

a. Find the measure of ∠y.

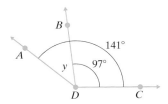

b. Find the measure of ∠x.

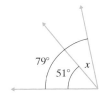

c. Classify ∠x and ∠y as acute, obtuse, or right angles.

Two lines in a plane can be either parallel or intersecting. **Parallel lines** never meet. **Intersecting lines** meet at a point. The symbol ∥ is used to indicate "is parallel to." For example, in the figure, p ∥ q.

Parallel lines   Intersecting lines

Some intersecting lines are perpendicular. Two lines are **perpendicular** if they form right angles when they intersect. The symbol ⊥ is used to denote "is perpendicular to." For example, in the figure below, m ⊥ n.

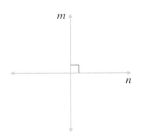

Perpendicular lines

When two lines intersect, four angles are formed. Two angles that are opposite each other are called **vertical angles.** Vertical angles have the same measure.
    Two angles that share a common side are called **adjacent angles.** Adjacent angles formed by intersecting lines are supplementary. That is, the sum of their measures is 180°.

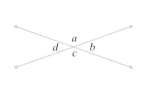

Vertical angles:
∠a and ∠c
∠d and ∠b

Adjacent angles:
∠a and ∠b
∠b and ∠c
∠c and ∠d
∠d and ∠a

Answers
5. 113° **6. a.** 44° **b.** 28° **c.** both acute

✔**Concept Check Answer**
false; the *complement* of a 48° angle is 42°; the *supplement* of a 48° angle is 132°

Here are a few real-life examples of the lines we just discussed.

Parallel lines

Vertical angles

Perpendicular lines

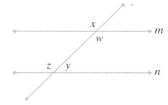

Practice 7

Find the measures of $\angle a$, $\angle b$, and $\angle c$.

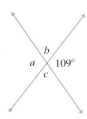

Example 7

Find the measures of $\angle x$, $\angle y$, and $\angle z$ if the measure of $\angle t$ is 42°.

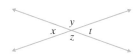

Solution: Since $\angle t$ and $\angle x$ are vertical angles, they have the same measure, so $\angle x$ measures 42°.

Since $\angle t$ and $\angle y$ are adjacent angles, their measures have a sum of 180°. So $\angle y$ measures $180° - 42° = 138°$.

Since $\angle y$ and $\angle z$ are vertical angles, they have the same measure. So $\angle z$ measures 138°.

■ Work Practice 7

A line that intersects two or more lines at different points is called a **transversal.** Line l is a transversal that intersects lines m and n. The eight angles formed have special names. Some of these names are:

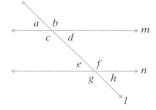

Corresponding angles: $\angle a$ and $\angle e$, $\angle c$ and $\angle g$, $\angle b$ and $\angle f$, $\angle d$ and $\angle h$

Alternate interior angles: $\angle c$ and $\angle f$, $\angle d$ and $\angle e$

When two lines cut by a transversal are *parallel,* the following statement is true:

Practice 8

Given that $m \parallel n$ and that the measure of $\angle w = 45°$, find the measures of all the angles shown.

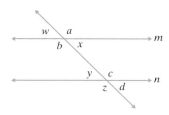

Parallel Lines Cut by a Transversal

If two parallel lines are cut by a transversal, then the measures of **corresponding angles are equal** and the measures of the **alternate interior angles are equal.**

Example 8

Given that $m \parallel n$ and that the measure of $\angle w$ is 100°, find the measures of $\angle x$, $\angle y$, and $\angle z$.

Answers

7. $m\angle a = 109°; m\angle b = 71°;$
$\quad m\angle c = 71°$

8. $m\angle x = 45°; m\angle y = 45°;$
$\quad m\angle z = 135°;$
$\quad m\angle a = 135°;$
$\quad m\angle b = 135°;$
$\quad m\angle c = 135°;$
$\quad m\angle d = 45°$

Solution:

$m\angle x = 100°$ $\angle x$ and $\angle w$ are vertical angles.

$m\angle z = 100°$ $\angle x$ and $\angle z$ are corresponding angles.

$m\angle y = 180° - 100° = 80°$ $\angle z$ and $\angle y$ are supplementary angles.

■ Work Practice 8

Vocabulary, Readiness & Video Check

Use the choices below to fill in each blank.

acute	straight	degrees	adjacent	parallel	intersecting
obtuse	space	plane	point	vertical	vertex
right	angle	ray	line	perpendicular	transversal

1. A(n) _____ is a flat surface that extends indefinitely.
2. A(n) _____ has no length, no width, and no height.
3. _____ extends in all directions indefinitely.
4. A(n) _____ is a set of points extending indefinitely in two directions.
5. A(n) _____ is part of a line with one endpoint.
6. A(n) _____ is made up of two rays that share a common endpoint. The common endpoint is called the _____ .
7. A(n) _____ angle measures 180°.
8. A(n) _____ angle measures 90°.
9. A(n) _____ angle measures between 0° and 90°.
10. A(n) _____ angle measures between 90° and 180°.
11. _____ lines never meet and _____ lines meet at a point.
12. Two intersecting lines are _____ if they form right angles when they intersect.
13. An angle can be measured in _____ .
14. A line that intersects two or more lines at different points is called a(n) _____ .
15. When two lines intersect, four angles are formed. The angles that are opposite each other are called _____ angles.
16. Two angles that share a common side are called _____ angles.

Martin-Gay Interactive Videos   Watch the section lecture video and answer the following questions.

See Video 9.1

Objective A 17. In the lecture after ▣ Example 2, what are the four ways we can name the angle shown?

Objective B 18. In the lecture before ▣ Example 3, what type of angle forms a line? What is its measure?

Objective C 19. What calculation is used to find the answer to ▣ Example 6?

Objective D 20. In the lecture before ▣ Example 7, two lines in a plane that aren't parallel must what?

9.1 Exercise Set MyMathLab®

Objective A *Identify each figure as a line, a ray, a line segment, or an angle. Then name the figure using the given points. See Examples 1 and 2.*

1.
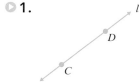

2.
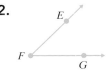

3.

4.
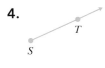

5.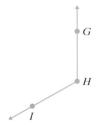

6.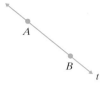

7.

8.

List two other ways to name each angle. See Example 2.

9. $\angle x$

10. $\angle w$

11. $\angle z$

12. $\angle y$

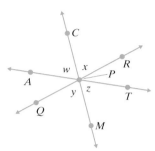

Objective B   *Classify each angle as acute, right, obtuse, or straight. See Example 3.*

13.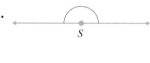

14.

15.

16.

17.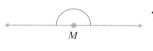

18.

19.

20.

Objective C   *Find each complementary or supplementary angle as indicated. See Examples 4 and 5.*

21. Find the complement of a 23° angle.

22. Find the complement of a 77° angle.

23. Find the supplement of a 17° angle.

24. Find the supplement of a 77° angle.

25. Find the complement of a 58° angle.

26. Find the complement of a 22° angle.

27. Find the supplement of a 150° angle.

28. Find the supplement of a 130° angle.

29. Identify the pairs of complementary angles.

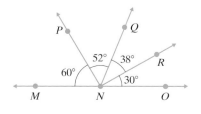

30. Identify the pairs of complementary angles.

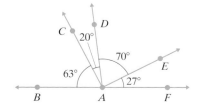

31. Identify the pairs of supplementary angles.

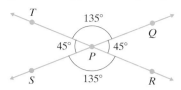

32. Identify the pairs of supplementary angles.

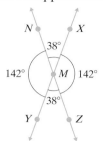

Objective D *Find the measure of ∠x in each figure. See Example 6.*

33.

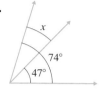

34.

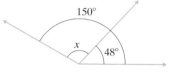

35.

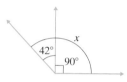

36.

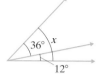

Find the measures of angles x, y, and z in each figure. See Examples 7 and 8.

37.

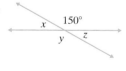

38.

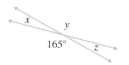

▶ **39.**

40.

41. *m ∥ n*

42. *m ∥ n*

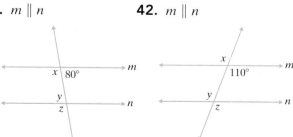

▶ **43.** *m ∥ n*

44. *m ∥ n*

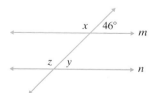

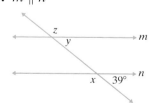

Objectives **A** **D** **Mixed Practice** *Find two other ways of naming each angle. See Example 2.*

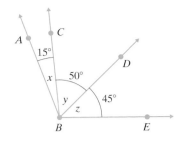

45. $\angle x$

46. $\angle y$

47. $\angle z$

48. $\angle ABE$ (just name one other way)

Find the measure of each angle in the figure above. See Example 6.

49. $\angle ABC$

50. $\angle EBD$

51. $\angle CBD$

52. $\angle CBA$

53. $\angle DBA$

54. $\angle EBC$

55. $\angle CBE$

56. $\angle ABE$

Review

Perform each indicated operation. See Sections 3.3, 3.5, and 3.6.

57. $\frac{7}{8} + \frac{1}{4}$

58. $\frac{7}{8} - \frac{1}{4}$

59. $\frac{7}{8} \cdot \frac{1}{4}$

60. $\frac{7}{8} \div \frac{1}{4}$

61. $3\frac{1}{3} - 2\frac{1}{2}$

62. $3\frac{1}{3} + 2\frac{1}{2}$

63. $3\frac{1}{3} \div 2\frac{1}{2}$

64. $3\frac{1}{3} \cdot 2\frac{1}{2}$

Concept Extensions

Solve.

65. The angle between the two walls of the Vietnam Veterans Memorial in Washington, D.C., is 125.2°. Find the supplement of this angle. (*Source:* National Park Service)

66. The faces of Khafre's Pyramid at Giza, Egypt, are inclined at an angle of 53.13°. Find the complement of this angle. (*Source:* PBS *NOVA* Online)

Answer true or false for Exercises 67 through 70. See the Concept Check in this section. If false, explain why.

67. The complement of a 100° angle is an 80° angle.

68. It is possible to find the complement of a 120° angle.

69. It is possible to find the supplement of a 120° angle.

70. The supplement of a 5° angle is a 175° angle.

71. If lines *m* and *n* are parallel, find the measures of angles *a* through *e*.

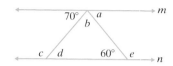

72. Below is a rectangle. List which segments, if extended, would be parallel lines.

73. Can two supplementary angles both be acute? Explain why or why not.

74. In your own words, describe how to find the complement and the supplement of a given angle.

75. Find two complementary angles with the same measure.

76. Is the figure below possible? Why or why not?

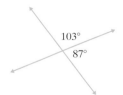

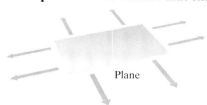

△ **9.2** **Plane Figures and Solids**

In order to prepare for the sections ahead in this chapter, we first review plane figures and solids.

Objectives

A Identify Plane Figures.

B Identify Solids.

Objective A Identifying Plane Figures

Recall from Section 9.1 that a **plane** is a flat surface that extends indefinitely.

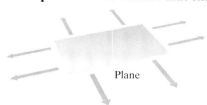

Plane

A **plane figure** is a figure that lies on a plane. Plane figures, like planes, have length and width but no thickness or depth.

A **polygon** is a closed plane figure that basically consists of three or more line segments that meet at their endpoints.

A **regular polygon** is one whose sides are all the same length and whose angles are the same measure.

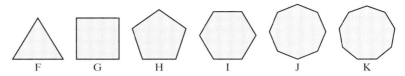

F   G   H   I   J   K

A polygon is named according to the number of its sides.

Polygons		
Number of Sides	**Name**	**Figure Examples**
3	Triangle	A, F
4	Quadrilateral	B, E, G
5	Pentagon	H
6	Hexagon	I
7	Heptagon	C
8	Octagon	J
9	Nonagon	K
10	Decagon	D

Some triangles and quadrilaterals are given special names, so let's study these polygons further. We begin with triangles.

The sum of the measures of the angles of a triangle is 180°.

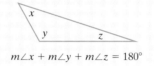

$$m\angle x + m\angle y + m\angle z = 180°$$

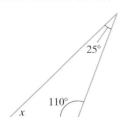

25°

110°

x

Example 1   Find the measure of $\angle a$.

a

95°

35°

Solution: Since the sum of the measures of the three angles is 180°, we have

measure of $\angle a$, or $m\angle a = 180° - 95° - 35° = 50°$

To check, see that $95° + 35° + 50° = 180°$.

■ Work Practice 1

We can classify triangles according to the lengths of their sides. (We will use tick marks to denote the sides and angles of a figure that are equal.)

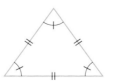      

Equilateral triangle

All three sides are the same length. Also, all three angles have the same measure.

Isosceles triangle

Two sides are the same length. Also, the angles opposite the equal sides have equal measure.

Scalene triangle

No sides are the same length. No angles have the same measure.

Answer
1. 45°

One other important type of triangle is a right triangle. A **right triangle** is a triangle with a right angle. The side opposite the right angle is called the **hypotenuse,** and the other two sides are called **legs.**

leg

hypotenuse

leg

Example 2 Find the measure of ∠b.

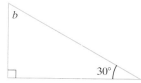

b

30°

Solution: We know that the measure of the right angle, ∟, is 90°. Since the sum of the measures of the angles is 180°, we have

measure of ∠b, or $m\angle b = 180° - 90° - 30° = 60°$

■ Work Practice 2

Practice 2

Find the measure of ∠y.

25°

y

Helpful Hint

From the previous example, can you see that in a right triangle, the sum of the other two acute angles is 90°? This is because

$$90° + 90° = 180°$$

↑   ↑   ↑

right
angle's
measure

sum of
other two
angles'
measures

sum of
angles'
measures

Now we review some special quadrilaterals. A **parallelogram** is a special quadrilateral with opposite sides parallel and equal in length.

A **rectangle** is a special **parallelogram** that has four right angles.

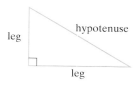

A **square** is a special **rectangle** that has all four sides equal in length.

Answer

2. 65°

A **rhombus** is a special **parallelogram** that has all four sides equal in length.

A **trapezoid** is a quadrilateral with exactly one pair of opposite sides parallel.

✓**Concept Check** True or false? All quadrilaterals are parallelograms. Explain.

In addition to triangles, quadrilaterals, and other polygons, circles are also plane figures. A **circle** is a plane figure that consists of all points that are the same fixed distance from a point c. The point c is called the **center** of the circle. The **radius** of a circle is the distance from the center of the circle to any point on the circle. The **diameter** of a circle is the distance across the circle passing through the center. Notice that the diameter is twice the radius, and the radius is half the diameter.

$$\text{diameter} \ = \ 2 \ \cdot \ \text{radius} \qquad\qquad \text{radius} \ = \ \frac{\text{diameter}}{2}$$

$$\downarrow \qquad\quad \downarrow \quad\ \downarrow \qquad\qquad\qquad\quad \downarrow \qquad\quad \downarrow$$

$$d \ = \ 2 \ \cdot \ r \qquad\qquad\qquad\quad r \ = \ \frac{d}{2}$$

Practice 3

Find the radius of the circle.

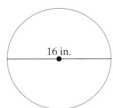

16 in.

Example 3 Find the diameter of the circle.

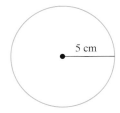

5 cm

Solution: The diameter is twice the radius.

$$d = 2 \cdot r$$
$$d = 2 \cdot 5 \text{ cm} = 10 \text{ cm}$$

The diameter is 10 centimeters.

 Work Practice 3

Objective B Identifying Solids ▶

Recall from Section 9.1 that space extends in all directions indefinitely.

A **solid** is a figure that lies in space. Solids have length, width, and height or depth.

Answer

3. 8 in.

✓**Concept Check Answer**

false

A **rectangular solid** is a solid that consists of six sides, or faces, all of which are rectangles.

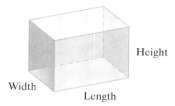

A **cube** is a rectangular solid whose six sides are squares.

A **pyramid** is shown below. The pyramids we will study have square bases and heights that are perpendicular to their base.

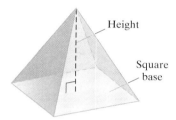

A **sphere** consists of all points in space that are the same distance from a point c. The point c is called the **center** of the sphere. The **radius** of a sphere is the distance from the center to any point on the sphere. The **diameter** of a sphere is the distance across the sphere passing through the center.

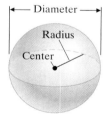

The radius and diameter of a sphere are related in the same way that the radius and diameter of a circle are related.

$$d = 2 \cdot r \quad \text{or} \quad r = \frac{d}{2}$$

Example 4 Find the radius of the sphere.

Solution: The radius is half the diameter.

$$r = \frac{d}{2}$$

$$r = \frac{36 \text{ feet}}{2} = 18 \text{ feet}$$

The radius is 18 feet.

Work Practice 4

Practice 4

Find the diameter of the sphere.

Answer

4. 14 mi

The **cylinders** we will study have bases that are in the shape of circles and heights that are perpendicular to their base.

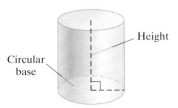

Height

Circular base

The **cones** we will study have bases that are circles and heights that are perpendicular to their base.

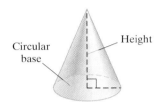

Height

Circular base

Vocabulary, Readiness & Video Check

Martin-Gay Interactive Videos

See Video 9.2

Watch the section lecture video and answer the following questions.

Objective A **1.** From the lecture after ▯ Example 2, since all angles of an equilateral triangle have the same measure, what is the measure of each angle? ▶

Objective B **2.** What solid is identified in ▯ Example 6? What two real-life examples of the solid are given? ▶

9.2 Exercise Set MyMathLab®

Objective A *Identify each polygon. See the table at the beginning of this section.*

1.

2.

3.

4.

5.

6.

7.

8.

Classify each triangle as equilateral, isosceles, or scalene. Also identify any triangles that are also right triangles. See the triangle classification after Example 1.

9.

10.

11.

12.

13.

14.

Find the measure of ∠x in each figure. See Examples 1 and 2.

15.
70°
85°
x

16.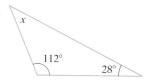
x
112°
28°

17.
95°
72°
x

18.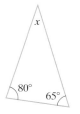
x
80°
65°

19.
x
50°

20.
x
20°

Fill in each blank.

21. Twice the radius of a circle is its _____.

22. A rectangle with all four sides equal is a(n) _____.

23. A parallelogram with four right angles is a(n) _____.

24. Half the diameter of a circle is its _____.

25. A quadrilateral with opposite sides parallel is a(n) _____.

26. A quadrilateral with exactly one pair of opposite sides parallel is a(n) _____.

27. The side opposite the right angle of a right triangle is called the _____.

28. A triangle with no equal sides is a(n) _____.

Find the unknown diameter or radius in each figure. See Example 3.

29.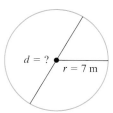
d = ?
r = 7 m

30.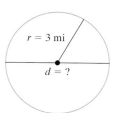
r = 3 mi
d = ?

31.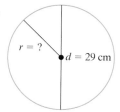
r = ?
d = 29 cm

32.
r = ?
d = 13 ft

33.
d = ? r = 20.3 cm

34.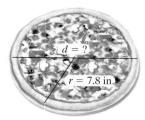
d = ?
r = 7.8 in.

35.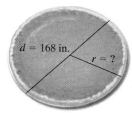
d = 168 in.
r = ?

Largest pumpkin pie
(*Source:* Circleville, Ohio, Pumpkin Festival)

36.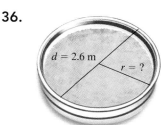
d = 2.6 m
r = ?

Largest cereal bowl
(*Source: Guinness World Records*)

Objective B *Identify each solid.*

37.

38.

39.

40.

41.

42.

Identify the basic shape of each item.

43.

44.

45.

46.

47.

48.

49.

50.

Find each unknown radius or diameter. See Example 4.

51. The radius of a sphere is 7.4 inches. Find its diameter.

52. The radius of a sphere is 5.8 meters. Find its diameter.

53. Find the radius of the sphere.

54. Find the radius of the sphere.

55. Saturn has a radius of approximately 36,184 miles. What is its diameter?

56. A sphere-shaped wasp nest found in Japan had a radius of approximately 15 inches. What was its diameter? (*Source: Guinness World Records*)

Review

Perform each indicated operation. See Sections 1.3, 1.6, 4.2, and 4.3.

57. $2(18) + 2(36)$

58. $4(87)$

59. $4(3.14)$

60. $2(7.8) + 2(9.6)$

Concept Extensions

Determine whether each statement is true or false. See the Concept Check in this section.

61. A square is also a rhombus.

62. A square is also a regular polygon.

63. A rectangle is also a parallelogram.

64. A trapezoid is also a parallelogram.

65. A pentagon is also a quadrilateral.

66. A rhombus is also a parallelogram.

67. Is an isosceles right triangle possible? If so, draw one.

68. In your own words, explain whether a square is also a rhombus.

69. The following demonstration is credited to the mathematician Pascal, who is said to have developed it as a young boy.

Cut a triangle from a piece of paper. The length of the sides and the size of the angles are unimportant. Tear the points off the triangle as shown in the top right figure.

Place the points of the triangle together, as shown in the bottom right figure. Notice that a straight line is formed. What was Pascal trying to show?

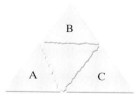

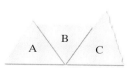

Objectives

A Use Formulas to Find Perimeters. ▶

B Use Formulas to Find Circumferences. ▶

Objective A Using Formulas to Find Perimeters ▶

Recall from Section 1.3 that the perimeter of a polygon is the distance around the polygon. This means that the perimeter of a polygon is the sum of the lengths of its sides.

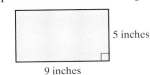 **Example 1** Find the perimeter of the rectangle below.

5 inches

9 inches

Solution:

$$\text{perimeter} = 9 \text{ inches} + 9 \text{ inches} + 5 \text{ inches} + 5 \text{ inches}$$
$$= 28 \text{ inches}$$

■ Work Practice 1

Practice 1

a. Find the perimeter of the rectangle.

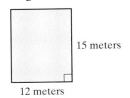

15 meters

12 meters

b. Find the perimeter of the rectangular lot shown below:

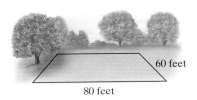

60 feet

80 feet

Notice that the perimeter of the rectangle in Example 1 can be written as

$$2 \cdot (9 \text{ inches}) + 2 \cdot (5 \text{ inches}).$$
↑ length   ↑ width

In general, we can say that the perimeter of a rectangle is always

$$2 \cdot \text{length} + 2 \cdot \text{width}$$

As we have just seen, the perimeters of some special figures such as rectangles form patterns. These patterns are given as **formulas.** The formula for the perimeter of a rectangle is shown next:

Perimeter of a Rectangle

$$\text{perimeter} = 2 \cdot \text{length} + 2 \cdot \text{width}$$

In symbols, this can be written as

$$P = 2 \cdot l + 2 \cdot w$$

length

width   width

length

Practice 2

Find the perimeter of a rectangle with a length of 22 centimeters and a width of 10 centimeters.

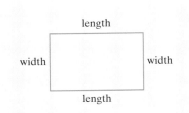 **Example 2** Find the perimeter of a rectangle with a length of 11 inches and a width of 3 inches.

11 in.

3 in.

Solution: We use the formula for perimeter and replace the letters by their known lengths.

$$P = 2 \cdot l + 2 \cdot w$$
$$= 2 \cdot 11 \text{ in.} + 2 \cdot 3 \text{ in.} \quad \text{Replace } l \text{ with 11 in. and } w \text{ with 3 in.}$$
$$= 22 \text{ in.} + 6 \text{ in.}$$
$$= 28 \text{ in.}$$

The perimeter is 28 inches.

■ Work Practice 2

Answers
1. a. 54 m **b.** 280 ft **2.** 64 cm

Recall that a square is a special rectangle with all four sides the same length. The formula for the perimeter of a square is shown next:

Perimeter of a Square

Perimeter = side + side + side + side
$$= 4 \cdot \text{side}$$

In symbols,

$$P = 4 \cdot s$$

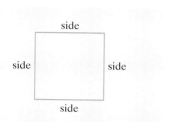

Example 3　Finding the Perimeter of a Field

How much fencing is needed to enclose a square field 50 yards on a side?

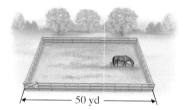

Solution: To find the amount of fencing needed, we find the distance around, or perimeter. The formula for the perimeter of a square is $P = 4 \cdot s$. We use this formula and replace s by 50 yards.

$$P = 4 \cdot s$$
$$= 4 \cdot 50 \text{ yd}$$
$$= 200 \text{ yd}$$

The amount of fencing needed is 200 yards.

■ Work Practice 3

Practice 3

Find the perimeter of a square tabletop if each side is 5 feet long.

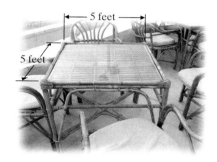

The formula for the perimeter of a triangle with sides of lengths a, b, and c is given next:

Perimeter of a Triangle

Perimeter = side a + side b + side c

In symbols,

$$P = a + b + c$$

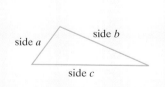

Example 4　Find the perimeter of a triangle if the sides are 3 inches, 7 inches, and 6 inches.

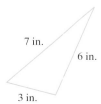

7 in.

6 in.

3 in.

Practice 4

Find the perimeter of a triangle if the sides are 5 centimeters, 10 centimeters, and 6 centimeters in length.

(Continued on next page)

Answers
3. 20 ft　**4.** 21 cm

Solution: The formula for the perimeter is $P = a + b + c$, where a, b, and c are the lengths of the sides. Thus,

$$P = a + b + c$$
$$= 3 \text{ in.} + 7 \text{ in.} + 6 \text{ in.}$$
$$= 16 \text{ in.}$$

The perimeter of the triangle is 16 inches.

■ Work Practice 4

Recall that to find the perimeter of other polygons, we find the sum of the lengths of their sides.

Practice 5

Find the perimeter of the trapezoid shown.

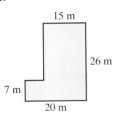

Example 5 Find the perimeter of the trapezoid shown below:

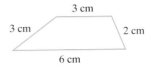

Solution: To find the perimeter, we find the sum of the lengths of its sides.

$$\text{perimeter} = 3 \text{ cm} + 2 \text{ cm} + 6 \text{ cm} + 3 \text{ cm} = 14 \text{ cm}$$

The perimeter is 14 centimeters.

■ Work Practice 5

Practice 6

Find the perimeter of the room shown.

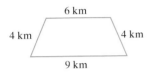

Example 6 Finding the Perimeter of a Room

Find the perimeter of the room shown below:

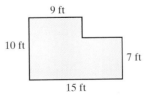

Solution: To find the perimeter of the room, we first need to find the lengths of all sides of the room.

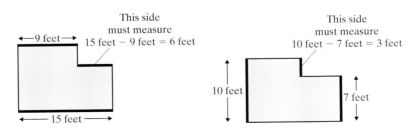

Answers

5. 23 km   **6.** 92 m

Now that we know the measures of all sides of the room, we can add the measures to find the perimeter.

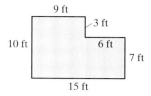

perimeter = 10 ft + 9 ft + 3 ft + 6 ft + 7 ft + 15 ft

= 50 ft

The perimeter of the room is 50 feet.

■ Work Practice 6

Example 7 Calculating the Cost of Wallpaper Border

A rectangular room measures 10 feet by 12 feet. Find the cost to hang a wallpaper border on the walls close to the ceiling if the cost of the wallpaper border is $1.09 per foot.

Solution: First we find the perimeter of the room.

$P = 2 \cdot l + 2 \cdot w$

$= 2 \cdot 12 \text{ ft} + 2 \cdot 10 \text{ ft}$ Replace l with 12 feet and w with 10 feet.

$= 24 \text{ ft} + 20 \text{ ft}$

$= 44 \text{ ft}$

The cost of the wallpaper is

cost = $1.09 \cdot 44$ ft = 47.96

The cost of the wallpaper is $47.96.

■ Work Practice 7

Practice 7

A rectangular lot measures 60 feet by 120 feet. Find the cost to install fencing around the lot if the cost of fencing is $1.90 per foot.

Objective B Using Formulas to Find Circumferences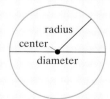

Recall from Section 4.3 that the distance around a circle is called the **circumference.** This distance depends on the radius or the diameter of the circle.

The formulas for circumference are shown next:

Circumference of a Circle

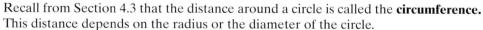

Circumference = $2 \cdot \pi \cdot$ radius   or   Circumference = $\pi \cdot$ diameter

In symbols,

$C = 2 \cdot \pi \cdot r$ or $C = \pi \cdot d$,

where $\pi \approx 3.14$ or $\pi \approx \dfrac{22}{7}$.

Answer

7. $684

To better understand circumference and π (pi), try the following experiment. Take any can and measure its circumference and its diameter.

The can in the figure above has a circumference of 23.5 centimeters and a diameter of 7.5 centimeters. Now divide the circumference by the diameter.

$$\frac{\text{circumference}}{\text{diameter}} = \frac{23.5 \text{ cm}}{7.5 \text{ cm}} \approx 3.13$$

Try this with other sizes of cylinders and circles–you should always get a number close to 3.1. The exact ratio of circumference to diameter is π. (Recall that $\pi \approx 3.14$ or $\pi \approx \frac{22}{7}$.)

Practice 8

a. An irrigation device waters a circular region with a diameter of 20 yards. Find the exact circumference of the watered region, and then use $\pi \approx 3.14$ to give an approximation.

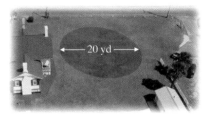

b. A manufacturer of clocks is designing a new model. To help the designer calculate the cost of materials to make the new clock, calculate the circumference of a clock with a face diameter of 12 inches. Give the exact circumference; then use $\pi \approx 3.14$ to approximate.

Answers

8. a. exactly 20π yd ≈ 62.8 yd

b. exactly 12π in. ≈ 37.68 in.

✔**Concept Check Answer**

a square with side length 5 in.

Example 8   Finding Circumference of a Spa

Mary Catherine Dooley plans to install a border of new tiling around the circumference of her circular spa. If her spa has a diameter of 14 feet, find its exact circumference. Then use the approximation 3.14 for π to approximate the circumference.

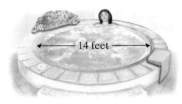

Solution: Because we are given the diameter, we use the formula $C = \pi \cdot d$.

$$C = \pi \cdot d$$
$$= \pi \cdot 14 \text{ ft} \quad \text{Replace } d \text{ with 14 feet.}$$
$$= 14\pi \text{ ft}$$

The circumference of the spa is *exactly* 14π feet. By replacing π with the *approximation* 3.14, we find that the circumference is *approximately* 14 feet $\cdot$ 3.14 = 43.96 feet.

🔲 Work Practice 8

✔**Concept Check** The distance around which figure is greater: a square with side length 5 inches or a circle with radius 3 inches?

Vocabulary, Readiness & Video Check

Use the choices below to fill in each blank.

circumference	radius	π	$\frac{22}{7}$
diameter	perimeter	3.14	

1. The _____ of a polygon is the sum of the lengths of its sides.

2. The distance around a circle is called the _____.

3. The exact ratio of circumference to diameter is _____.

4. The diameter of a circle is double its _____.

5. Both _____ and _____ are approximations for π.

6. The radius of a circle is half its _____.

Martin-Gay Interactive Videos   *Watch the section lecture video and answer the following questions.*

Objective A **7.** In Example 1, how can the perimeter be found if we forget the formula?

Objective B **8.** From the lecture before ⊞ Example 6, circumference is a special name for what?

See Video 9.3

9.3 **Exercise Set** MyMathLab®

Objective A *Find the perimeter of each figure. See Examples 1 through 6.*

1.
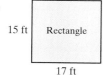
15 ft Rectangle
17 ft

2.
Rectangle | 14 m
5 m

3.
Parallelogram
25 cm
35 cm

4.
Parallelogram
3 yd
2 yd

5.
5 in.   7 in.
9 in.

6.
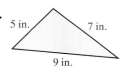
5 units   11 units
10 units

7.
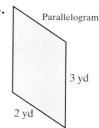
10 ft   8 ft
7 ft   8 ft
15 ft

8.
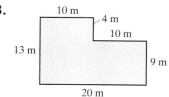
10 m   4 m
10 m
13 m
9 m
20 m

Find the perimeter of each regular polygon. (The sides of a regular polygon have the same length.)

9.  14 inches

10. 50 m

11. 31 cm

12. 15 yd

Solve. See Examples 1 through 7.

13. A polygon has sides of length 5 feet, 3 feet, 2 feet, 7 feet, and 4 feet. Find its perimeter.

14. A triangle has sides of length 8 inches, 12 inches, and 10 inches. Find its perimeter.

15. A line-marking machine lays down lime powder to mark both foul lines on a baseball field. If each foul line for this field measures 312 feet, how many feet of lime powder will be deposited?

16. A baseball diamond has 4 sides, with each side length 90 feet. If a baseball player hits a home run, how far does the player run (home plate, around the bases, and then back to home plate)?

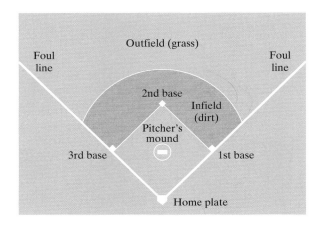

17. If a football field is 53 yards wide and 120 yards long, what is the perimeter?

18. A stop sign has eight equal sides of length 12 inches. Find its perimeter.

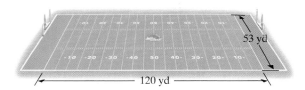

19. A metal strip is being installed around a workbench that is 8 feet long and 3 feet wide. Find how much stripping is needed for this project.

20. Find how much fencing is needed to enclose a rectangular garden 70 feet by 21 feet.

21. If the stripping in Exercise 19 costs $2.50 per foot, find the total cost of the stripping.

22. If the fencing in Exercise 20 costs $2 per foot, find the total cost of the fencing.

23. A regular octagon has a side length of 9 inches. Find its perimeter.

24. A regular pentagon has a side length of 14 meters. Find its perimeter.

25. Find the perimeter of the top of a square compact disc case if the length of one side is 7 inches.

26. Find the perimeter of a square ceramic tile with a side of length 3 inches.

27. A rectangular room measures 10 feet by 11 feet. Find the cost of installing a strip of wallpaper around the room if the wallpaper costs $0.86 per foot.

28. A rectangular house measures 85 feet by 70 feet. Find the cost of installing gutters around the house if the cost is $2.36 per foot.

Find the perimeter of each figure. See Example 6.

29.
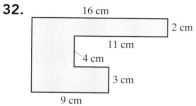
17 m
28 m
20 m
20 m

30.
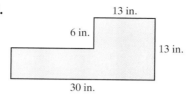
13 in.
6 in.
13 in.
30 in.

31.
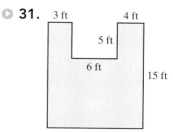
3 ft   4 ft
5 ft
6 ft
15 ft

32.
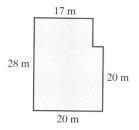
16 cm
2 cm
11 cm
4 cm
3 cm
9 cm

33.
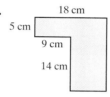
18 cm
5 cm
9 cm
14 cm

34.
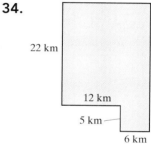
22 km
12 km
5 km
6 km

Objective B *Find the circumference of each circle. Give the exact circumference and then an approximation. Use π ≈ 3.14. See Example 8.*

35.
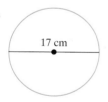
17 cm

36.
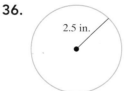
2.5 in.

37.

8 mi

38.
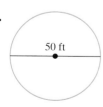
50 ft

39.
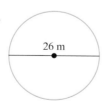
26 m

40.
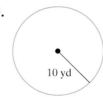
10 yd

41. Wyley Robinson just bought a trampoline for his children to use. The trampoline has a diameter of 15 feet. If Wyley wishes to buy netting to go around the outside of the trampoline, how many feet of netting does he need?

42. The largest round barn in the world is located at the Marshfield Fairgrounds in Wisconsin. The barn has a diameter of 150 ft. What is the circumference of the barn? (*Source: The Milwaukee Journal Sentinel*)

43. Meteor Crater, near Winslow, Arizona, is 4000 feet in diameter. Approximate the distance around the crater. Use 3.14 for π. (*Source: The Handy Science Answer Book*)

44. The largest pearl, the *Pearl of Lao-tze*, has a diameter of $5\frac{1}{2}$ inches. Approximate the distance around the pearl. Use $\frac{22}{7}$ for π. (*Source: The Guinness Book of World Records*)

Objectives A B Mixed Practice *Find the distance around each figure. For circles, give the exact circumference and then an approximation. Use $\pi \approx 3.14$. See Examples 1 through 8.*

45.

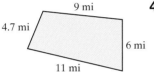
9 mi
4.7 mi
6 mi
11 mi

46.

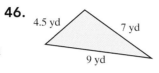
4.5 yd
7 yd
9 yd

47.

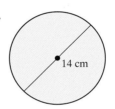
14 cm

48.

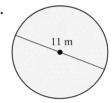
11 m

49.

Regular Pentagon
8 mm

50.

Regular Parallelogram
19 km

51.

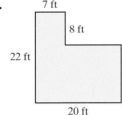
7 ft
8 ft
22 ft
20 ft

52.

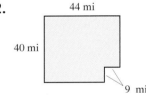
44 mi
40 mi
9 mi

Review

Simplify. See Section 1.9.

53. $5 + 6 \cdot 3$

54. $25 - 3 \cdot 7$

55. $(20 - 16) \div 4$

56. $6 \cdot (8 + 2)$

57. $72 \div (2 \cdot 6)$

58. $(72 \div 2) \cdot 6$

59. $(18 + 8) - (12 + 4)$

60. $4^1 \cdot (2^3 - 8)$

Concept Extensions

There are a number of factors that determine the dimensions of a rectangular soccer field. Use the table below to answer Exercises 61 and 62.

Soccer Field Width and Length		
Age	**Width Min–Max**	**Length Min–Max**
Under 6/7:	15–20 yards	25–30 yards
Under 8:	20–25 yards	30–40 yards
Under 9:	30–35 yards	40–50 yards
Under 10:	40–50 yards	60–70 yards
Under 11:	40–50 yards	70–80 yards
Under 12:	40–55 yards	100–105 yards
Under 13:	50–60 yards	100–110 yards
International:	70–80 yards	110–120 yards

61. a. Find the minimum length and width of a soccer field for 8-year-old children. (Carefully consider the age.)

   b. Find the perimeter of this field.

62. a. Find the maximum length and width of a soccer field for 12-year-old children.

   b. Find the perimeter of this field.

Solve. See the Concept Check in this section. Choose the figure that has the greater distance around.

63. a. A square with side length 3 inches

   b. A circle with diameter 4 inches

64. a. A circle with diameter 7 inches

   b. A square with side length 7 inches

Solve.

65. a. Find the circumference of each circle. Approximate the circumference by using 3.14 for π.

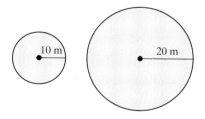

   b. If the radius of a circle is doubled, is its corresponding circumference doubled?

66. a. Find the circumference of each circle. Approximate the circumference by using 3.14 for π.

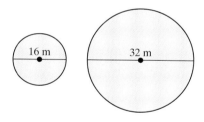

   b. If the diameter of a circle is doubled, is its corresponding circumference doubled?

67. In your own words, explain how to find the perimeter of any polygon.

68. In your own words, explain how perimeter and circumference are the same and how they are different.

Find the perimeter. Round your results to the nearest tenth.

69.

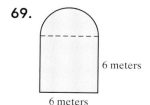

6 meters

6 meters

70.

6 meters

6 meters

71.

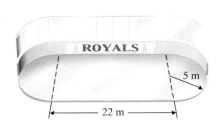

72.

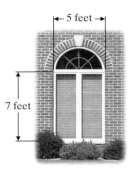

73. The perimeter of this rectangle is 31 feet. Find its width.

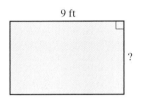

74. The perimeter of this square is 18 inches. Find the length of a side.

△ **9.4** **Area** ◯

Objective

A Find the Areas of Geometric Figures. ▶

Objective A Finding Areas of Geometric Figures ◯

Recall that area measures the amount of surface of a region. Thus far, we know how to find the area of a rectangle and a square. These formulas, as well as formulas for finding the areas of other common geometric figures, are given next:

Area Formulas of Common Geometric Figures

Geometric Figure	Area Formula
RECTANGLE 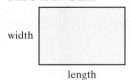	Area of a rectangle: **A**rea = **l**ength · **w**idth $A = l \cdot w$
SQUARE 	Area of a square: **A**rea = **s**ide · **s**ide $A = s \cdot s = s^2$
TRIANGLE 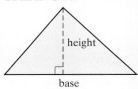	Area of a triangle: **A**rea = $\dfrac{1}{2}$ · **b**ase · **h**eight $A = \dfrac{1}{2} \cdot b \cdot h$

Area Formulas of Common Geometric Figures (*continued*)

Geometric Figure   **Area Formula**

PARALLELOGRAM

Area of a parallelogram:
Area = **base** · **height**
$$A = b \cdot h$$

TRAPEZOID

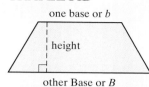

Area of a trapezoid:
$$\textbf{Area} = \frac{1}{2} \cdot (\text{one } \textbf{base} + \text{other } \textbf{Base}) \cdot \textbf{height}$$

$$A = \frac{1}{2} \cdot (b + B) \cdot h$$

Use these formulas for the following examples.

Helpful Hint

Area is always measured in square units.

Example 1   Find the area of the triangle.

Solution: $A = \frac{1}{2} \cdot b \cdot h$

$$= \frac{1}{2} \cdot 14 \text{ cm} \cdot 8 \text{ cm}$$

$$= \frac{\overset{1}{\cancel{2}} \cdot 7 \cdot 8}{\underset{1}{\cancel{2}}} \text{ sq cm}$$

$$= 56 \text{ square cm}$$

The area is 56 square centimeters.

Helpful Hint   You may see 56 sq cm, for example, written with the notation 56 cm². Both of these notations mean the same quantity.

■ Work Practice 1

Practice 1

Find the area of the triangle.

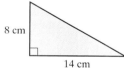

Example 2   Find the area of the parallelogram.

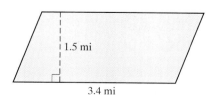

Solution: $A = b \cdot h$

$$= 3.4 \text{ miles} \cdot 1.5 \text{ miles}$$

$$= 5.1 \text{ square miles}$$

The area is 5.1 square miles.

■ Work Practice 2

Practice 2

Find the area of the square.

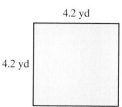

Answers
1. 25 sq in.   **2.** 17.64 sq yd

Practice 3

Find the area of the figure.

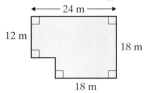

Example 3 Find the area of the figure.

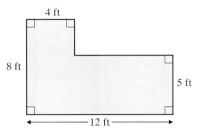

Solution: Split the figure into two rectangles. To find the area of the figure, we find the sum of the areas of the two rectangles.

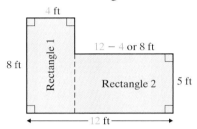

$$\text{Area of Rectangle 1} = l \cdot w$$
$$= 8 \text{ feet} \cdot 4 \text{ feet}$$
$$= 32 \text{ square feet}$$

Notice that the length of Rectangle 2 is 12 feet − 4 feet, or 8 feet.

$$\text{Area of Rectangle 2} = l \cdot w$$
$$= 8 \text{ feet} \cdot 5 \text{ feet}$$
$$= 40 \text{ square feet}$$

$$\text{Area of the Figure} = \text{Area of Rectangle 1} + \text{Area of Rectangle 2}$$
$$= 32 \text{ square feet} + 40 \text{ square feet}$$
$$= 72 \text{ square feet}$$

■ Work Practice 3

Helpful Hint

The figure in Example 3 can also be split into two rectangles as shown:

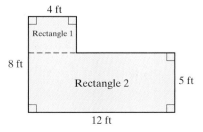

Answer

3. 396 sq m

To better understand the formula for area of a circle, try the following. Cut a circle into many pieces, as shown:

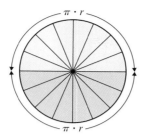

The circumference of a circle is $2 \cdot \pi \cdot r$. This means that the circumference of half a circle is half of $2 \cdot \pi \cdot r$, or $\pi \cdot r$.

Then unfold the two halves of the circle and place them together, as shown:

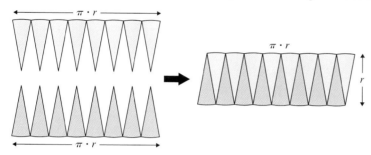

The figure on the right is almost a parallelogram with a base of $\pi \cdot r$ and a height of r. The area is

$$A = \text{base} \cdot \text{height}$$
$$= (\pi \cdot r) \cdot r$$
$$= \pi \cdot r^2$$

This is the formula for area of a circle.

Area Formula of a Circle

CIRCLE

Area of a circle:

Area $= \pi \cdot (\text{radius})^2$

$A = \pi \cdot r^2$

(A fraction approximation for π is $\dfrac{22}{7}$.)

(A decimal approximation for π is 3.14.)

Practice 4

Find the area of the given circle. Find the exact area and an approximation. Use 3.14 as an approximation for π.

Example 4 Find the area of a circle with a radius of 3 feet. Find the exact area and an approximation. Use 3.14 as an approximation for π.

Solution: We let $r = 3$ ft and use the formula.

$$A = \pi \cdot r^2$$
$$= \pi \cdot (3 \text{ ft})^2$$
$$= \pi \cdot 9 \text{ square ft, or } 9 \cdot \pi \text{ square ft}$$

(Continued on next page)

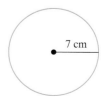

Answer

4. 49π sq cm ≈ 153.86 sq cm

To approximate this area, we substitute 3.14 for π.

$$9 \cdot \pi \text{ square feet} \approx 9 \cdot 3.14 \text{ square feet}$$
$$= 28.26 \text{ square feet}$$

The *exact* area of the circle is 9π square feet, which is *approximately* 28.26 square feet.

■ Work Practice 4

✓ **Concept Check Answer**
a square 10 in. long on each side

✓ **Concept Check** Use diagrams to decide which figure would have a larger area: a circle of diameter 10 inches or a square 10 inches long on each side.

Vocabulary, Readiness & Video Check

Martin-Gay Interactive Videos   Watch the section lecture video and answer the following question.

Objective A 1. What formula was used twice and why did we use it twice to solve ⊞ Example 3? ⊙

See Video 9.4 🍎

9.4 **Exercise Set** MyMathLab®

Objective A *Find the area of the geometric figure. If the figure is a circle, give the exact area and then use the given* ***approximation*** *for π to approximate the area. See Examples 1 through 4.*

1.

2 m | Rectangle
3.5 m

2.

2.75 ft | Rectangle
7 ft

3.

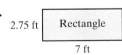
3 yd
$6\frac{1}{2}$ yd

4.

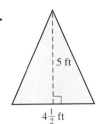
5 ft
$4\frac{1}{2}$ ft

5.

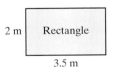
6 yd
5 yd

6.

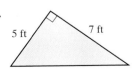
5 ft   7 ft

7. Use 3.14 for π.

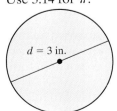
$d = 3$ in.

8. Use $\dfrac{22}{7}$ for π.

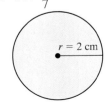
$r = 2$ cm

9.

Square | 4.2 ft

10.

Square | 2.6 m

11.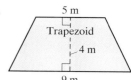

5 m
Trapezoid
4 m
9 m

12.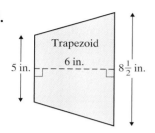

Trapezoid
5 in. 6 in. $8\frac{1}{2}$ in.

13.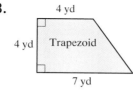

4 yd
4 yd Trapezoid
7 yd

14.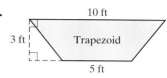

10 ft
3 ft Trapezoid
5 ft

15.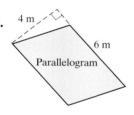

7 ft
Parallelogram
$5\frac{1}{4}$ ft

16.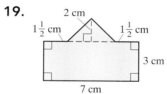

Parallelogram $4\frac{1}{4}$ cm
3 cm

17.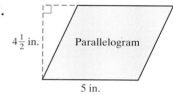

$4\frac{1}{2}$ in. Parallelogram
5 in.

18.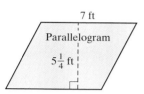

4 m
6 m
Parallelogram

19.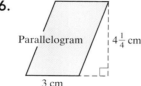

2 cm
$1\frac{1}{2}$ cm $1\frac{1}{2}$ cm
3 cm
7 cm

20.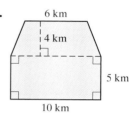

6 km
4 km
5 km
10 km

21.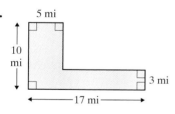

5 mi
10 mi
3 mi
17 mi

22.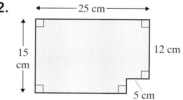

25 cm
15 cm 12 cm
5 cm

23.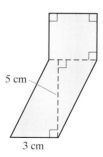

5 cm
3 cm

24.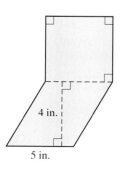

4 in.
5 in.

25. Use $\frac{22}{7}$ for π.

$r = 6$ in.

26. Use 3.14 for π.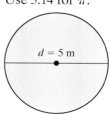

$d = 5$ m

Solve. See Examples 1 through 4.

27. A $10\frac{1}{2}$-foot by 16-foot concrete wall is to be built using concrete blocks. Find the area of the wall.

28. The floor of Terry's attic is 24 feet by 35 feet. Find how many square feet of insulation are needed to cover the attic floor.

29. The world's largest U.S. flag is the "Superflag," which measures 505 feet by 255 feet. Find its area. (*Source:* Superflag.com)

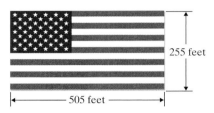

255 feet

505 feet

30. The world's largest illuminated indoor advertising sign is located in the Dubai International Airport in Dubai, UAE. It measures 28.0 meters in length by 6.2 meters in height. Find its area. (*Source:* World Record Academy)

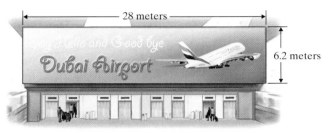

28 meters

6.2 meters

31. The face of a watch has a diameter of 2 centimeters. What is its area? Give the exact answer and then an approximation using 3.14 for π.

2 cm

32. The world's largest commercially available pizza is sold by Big Mama's & Big Papa's Pizzeria in Los Angeles, CA. This huge square pizza, called "The Giant Sicilian," measures 54 inches on each side and sells for $199.99 plus tax. Find the area of the top of the pizza. (*Source: Guinness World Records*, Big Mama's & Big Papa's Pizzeria Inc.)

54 in.

33. One side of a concrete block measures 8 inches by 16 inches. Find the area of the side in square inches. Find the area in square feet (144 sq in. = 1 sq ft).

34. A standard *double* roll of wallpaper is $6\frac{5}{6}$ feet wide and 33 feet long. Find the area of the *double* roll.

35. A picture frame measures 20 inches by $25\frac{1}{2}$ inches. Find how many square inches of glass the frame requires.

36. A mat to go under a tablecloth is made to fit a round dining table with a 4-foot diameter. Approximate how many square feet of mat there are. Use 3.14 as an approximation for π.

▶ **37.** A drapery panel measures 6 feet by 7 feet. Find how many square feet of material are needed for *four* panels.

38. A page in a book measures 27.5 centimeters by 20.5 centimeters. Find its area.

39. Find how many square feet of land are in the plot shown:

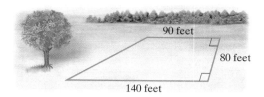

90 feet

80 feet

140 feet

40. For Gerald Gomez to determine how much grass seed he needs to buy, he must know the size of his yard. Use the drawing to determine how many square feet are in his yard.

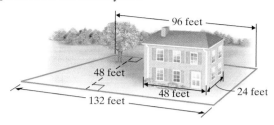

96 feet

48 feet

48 feet

24 feet

132 feet

41. The outlined part of the roof shown is in the shape of a trapezoid and needs to be shingled. The number of shingles to buy depends on the area.

a. Use the dimensions given to find the area of the outlined part of the roof to the nearest whole square foot.

36 ft

$12\frac{1}{2}$ ft 25 ft

b. Shingles are packaged in a unit called a "square." If a "square" covers 100 square feet, how many whole squares need to be purchased to shingle this part of the roof?

42. The entire side of the building shaded in the drawing is to be bricked. The number of bricks to buy depends on the area.

a. Find the area.

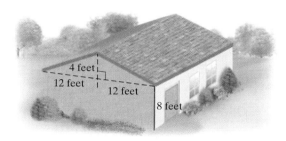

4 feet

12 feet 12 feet

8 feet

b. If the side area of each brick (including mortar room) is $\frac{1}{6}$ square foot, find the number of bricks needed to brick the end of the building.

Review

Find the perimeter or circumference of each geometric figure. See Section 9.3.

43. Give the exact circumference and an approximation. Use 3.14 for π.

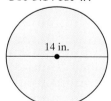

14 in.

44.

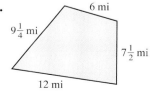

4 cm

5 cm

Rectangle

45.

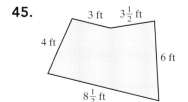

3 ft $3\frac{1}{2}$ ft

4 ft

6 ft

$8\frac{1}{2}$ ft

46.

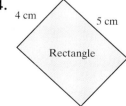

6 mi

$9\frac{1}{4}$ mi

$7\frac{1}{2}$ mi

12 mi

47.

$2\frac{1}{8}$ ft

Regular hexagon

48.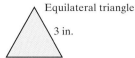

Equilateral triangle

3 in.

Concept Extensions

Given the following situations, tell whether you are more likely to be concerned with area or perimeter.

49. ordering fencing to fence a yard

50. ordering grass seed to plant in a yard

51. buying carpet to install in a room

52. buying gutters to install on a house

53. ordering paint to paint a wall

54. ordering baseboards to install in a room

55. buying a wallpaper border to go on the walls around a room

56. buying fertilizer for your yard

Solve.

57. A pizza restaurant recently advertised two specials. The first special was a 12-inch-diameter pizza for $10. The second special was two 8-inch-diameter pizzas for $9. Determine the better buy. (*Hint:* First find and compare the areas of the pizzas in the two specials. Then find a price per square inch for the pizzas in both specials.)

58. Find the approximate area of the state of Utah.

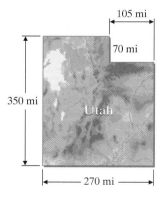

105 mi

70 mi

350 mi

Utah

270 mi

59. Find the area of a rectangle that measures 2 *feet* by 8 *inches*. Give the area in square feet and in square inches.

60. In your own words, explain why perimeter is measured in units and area is measured in square units. (*Hint:* See Section 1.6 for an introduction on the meaning of area.)

61. Find the area of the shaded region. Use the approximation 3.14 for π.

6 in.

62. Estimate the cost of a piece of carpet for a rectangular room 10 feet by 15 feet. The cost of the carpet is $31.50 per square yard.

63. The largest pumpkin pie was made for the 100th anniversary of the Circleville, Ohio, Pumpkin Festival in October 2008. The pie had a diameter of 168 inches. Find the exact area of the top of the pie and an approximation. Use $\pi \approx 3.14$. (*Source:* Circleville, Ohio, Pumpkin Festival)

64. The largest cereal bowl in the world was made by Kellogg's South Africa in July 2007. The bowl had a 2.6-meter diameter. Calculate the exact area of the circular base of the bowl and an approximation. Use $\pi \approx 3.14$. (*Source: Guinness Book of World Records*)

Find the area of each figure. If needed, use $\pi \approx 3.14$ and round results to the nearest tenth.

65. Find the skating area.

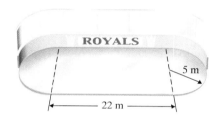

66.

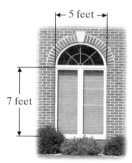

There are a number of factors that determine the dimensions of a rectangular soccer field. Use the table below to answer Exercises 67 and 68.

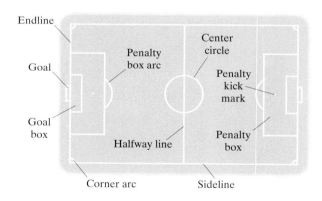

Soccer Field Width and Length		
Age	**Width Min–Max**	**Length Min–Max**
Under 6/7:	15–20 yards	25–30 yards
Under 8:	20–25 yards	30–40 yards
Under 9:	30–35 yards	40–50 yards
Under 10:	40–50 yards	60–70 yards
Under 11:	40–50 yards	70–80 yards
Under 12:	40–55 yards	100–105 yards
Under 13:	50–60 yards	100–110 yards
International:	70–80 yards	110–120 yards

67. a. Find the minimum length and width of a soccer field for 9-year-old children. (Carefully consider the age.)
b. Find the area of this field.

68. a. Find the maximum length and width of a soccer field for 11-year-old children.
b. Find the area of this field.

69. Do two rectangles with the same perimeter have the same area? To see, find the perimeter and the area of each rectangle.

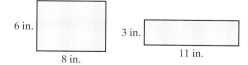

Objective

A Find the Volume and Surface Area of Solids.

Objective A Finding Volume and Surface Area of Solids

A **convex solid** is a set of points, S, not all in one plane, such that for any two points A and B in S, all points between A and B are also in S. In this section, we will find the volume and surface area of special types of solids called polyhedrons. A solid formed by the intersection of a finite number of planes is called a **polyhedron.** The box below is an example of a polyhedron.

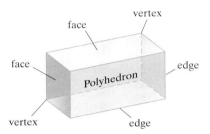

Each of the plane regions of a polyhedron is called a **face** of the polyhedron. If the intersection of two faces is a line segment, this line segment is an **edge** of the polyhedron. The intersections of the edges are the **vertices** of the polyhedron.

Volume is a measure of the space of a region. The volume of a box or can, for example, is the amount of space inside. Volume can be used to describe the amount of juice in a pitcher or the amount of concrete needed to pour a foundation for a house.

The volume of a solid is the number of **cubic units** in the solid. A cubic centimeter and a cubic inch are illustrated.

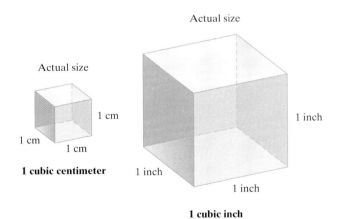

The **surface area** of a polyhedron is the sum of the areas of the faces of the polyhedron. For example, each face of the cube on the left above has an area of 1 square centimeter. Since there are 6 faces of the cube, the sum of the areas of the faces is 6 square centimeters. Surface area can be used to describe the amount of material needed to cover a solid. Surface area is measured in square units.

Formulas for finding the volumes, V, and surface areas, SA, of some common solids are given next.

Volume and Surface Area Formulas of Common Solids	
Solid	**Formulas**
Rectangular Solid	$V = lwh$ $SA = 2lh + 2wh + 2lw$ where h = height, w = width, l = length
Cube	$V = s^3$ $SA = 6s^2$ where s = side
Sphere	$V = \dfrac{4}{3}\pi r^3$ $SA = 4\pi r^2$ where r = radius
Circular Cylinder	$V = \pi r^2 h$ $SA = 2\pi rh + 2\pi r^2$ where h = height, r = radius
Cone	$V = \dfrac{1}{3}\pi r^2 h$ $SA = \pi r \sqrt{r^2 + h^2} + \pi r^2$ where h = height, r = radius
Square-Based Pyramid	$V = \dfrac{1}{3}s^2 h$ $SA = B + \dfrac{1}{2}pl$ where B = area of base, p = perimeter of base, h = height, s = side, l = slant height

Example 1 Find the volume and surface area of a rectangular box that is 12 inches long, 6 inches wide, and 3 inches high.

3 in.

6 in.   12 in.

(*Continued on next page*)

Practice 1

Find the volume and surface area of a rectangular box that is 7 feet long, 3 feet wide, and 4 feet high.

Answer

1. $V = 84$ cu ft; $SA = 122$ sq ft

Solution: Let $h = 3$ in., $l = 12$ in., and $w = 6$ in.

$$V = lwh$$
$$V = 12 \text{ inches} \cdot 6 \text{ inches} \cdot 3 \text{ inches} = 216 \text{ cubic inches}$$

The volume of the rectangular box is 216 cubic inches.

$$SA = 2lh + 2wh + 2lw$$
$$= 2(12 \text{ in.})(3 \text{ in.}) + 2(6 \text{ in.})(3 \text{ in.}) + 2(12 \text{ in.})(6 \text{ in.})$$
$$= 72 \text{ sq in.} + 36 \text{ sq in.} + 144 \text{ sq in.}$$
$$= 252 \text{ sq in.}$$

The surface area of the rectangular box is 252 square inches.

■ Work Practice 1

✓**Concept Check** Juan is calculating the volume of the following rectangular solid. Find the error in his calculation.

Volume $= l + w + h$
$$= 14 \text{ cm} + 8 \text{ cm} + 5 \text{ cm}$$
$$= 27 \text{ cu cm}$$

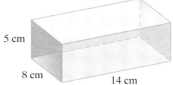

5 cm

8 cm   14 cm

Practice 2

Find the volume and surface area of a ball of radius $\frac{1}{2}$ centimeter. Give the exact volume and surface area. Then use $\frac{22}{7}$ for π and approximate the values.

Example 2 Find the volume and surface area of a ball of radius 2 inches. Give the exact volume and surface area. Then use the approximation $\frac{22}{7}$ for π.

2 in.

Solution:

$$V = \frac{4}{3}\pi r^3 \qquad \text{Formula for volume of a sphere}$$

$$V = \frac{4}{3} \cdot \pi (2 \text{ in.})^3 \qquad \text{Let } r = 2 \text{ inches.}$$

$$= \frac{32}{3}\pi \text{ cu in.} \qquad \text{Exact volume}$$

$$\approx \frac{32}{3} \cdot \frac{22}{7} \text{ cu in.} \qquad \text{Approximate } \pi \text{ with } \frac{22}{7}.$$

$$= \frac{704}{21} \text{ or } 33\frac{11}{21} \text{ cu in.} \qquad \text{Approximate volume}$$

Answer

2. $V = \frac{1}{6}\pi$ cu cm $\approx \frac{11}{21}$ cu cm;

$SA = \pi$ sq cm $\approx 3\frac{1}{7}$ sq cm

✓**Concept Check Answer**

Volume $= l \cdot w \cdot h$
$$= 14 \text{ cm} \cdot 8 \text{ cm} \cdot 5 \text{ cm}$$
$$= 560 \text{ cu cm}$$

The volume of the sphere is exactly $\frac{32}{3}\pi$ cubic inches or approximately $33\frac{11}{21}$ cubic inches.

$$SA = 4\pi r^2 \qquad \text{Formula for surface area}$$
$$SA = 4 \cdot \pi (2\text{ in.})^2 \qquad \text{Let } r = 2 \text{ inches.}$$
$$= 16\pi \text{ sq in.} \qquad \text{Exact surface area}$$
$$\approx 16 \cdot \frac{22}{7} \text{ sq in.} \qquad \text{Approximate } \pi \text{ with } \frac{22}{7}.$$
$$= \frac{352}{7} \text{ or } 50\frac{2}{7} \text{ sq in.} \qquad \text{Approximate surface area}$$

The surface area of the sphere is exactly 16π square inches or approximately $50\frac{2}{7}$ square inches.

■ Work Practice 2

Example 3 Approximate the volume of a can that has a $3\frac{1}{2}$-inch radius and a height of 6 inches. Use $\frac{22}{7}$ for π. Give the exact volume and an approximate volume.

Practice 3

Approximate the volume of a cylinder of radius 5 inches and height 9 inches. Use 3.14 for π. Give the exact answer and an approximate answer.

Solution: Using the formula for a circular cylinder, we have

$$V = \pi \cdot r^2 \cdot h \qquad 3\frac{1}{2} = \frac{7}{2}$$
$$= \pi \cdot \left(\frac{7}{2}\text{ in.}\right)^2 \cdot 6 \text{ in.}$$
$$= \pi \cdot \frac{49}{4} \text{ sq in.} \cdot 6 \text{ in.}$$
$$= \frac{\pi \cdot 49 \cdot \overset{1}{\cancel{2}} \cdot 3}{\cancel{2} \cdot 2} \text{ cu in.}$$
$$= 73\frac{1}{2}\pi \text{ cu in. or } 73.5\pi \text{ cu in.}$$

This is the exact volume. To approximate the volume, use the approximation $\frac{22}{7}$ for π.

$$V = 73\frac{1}{2}\pi \text{ or } \approx \frac{147}{2} \cdot \frac{22}{7} \text{ cu in.} \qquad \text{Replace } \pi \text{ with } \frac{22}{7}.$$
$$= \frac{21 \cdot \cancel{7} \cdot \overset{1}{\cancel{2}} \cdot 11}{\cancel{2} \cdot \cancel{7}} \text{ cu in.}$$
$$= 231 \text{ cubic in.}$$

The volume is approximately 231 cubic inches.

■ Work Practice 3

Answer

3. 225π cu in. ≈ 706.5 cu in.

Practice 4

Find the volume of a square-based pyramid that has a 3-meter side and a height of 5.1 meters.

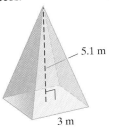

5.1 m

3 m

Example 4 Approximate the volume of a cone that has a height of 14 centimeters and a radius of 3 centimeters. Use 3.14 for π. Give the exact answer and an approximate answer.

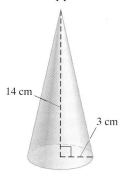

14 cm

3 cm

Slution: Using the formula for volume of a cone, we have

$$V = \frac{1}{3} \cdot \pi \cdot r^2 \cdot h$$

$$= \frac{1}{3} \cdot \pi \cdot (3 \text{ cm})^2 \cdot 14 \text{ cm} \quad \text{Replace } r \text{ with 3 cm and } h \text{ with 14 cm.}$$

$$= 42\pi \text{ cu cm}$$

Th, 42π cubic centimeters is the exact volume. To approximate the volume, use theproximation 3.14 for π.

$$' \approx 42 \cdot 3.14 \text{ cu cm} \quad \text{Replace } \pi \text{ with 3.14.}$$

$$= 131.88 \text{ cu cm}$$

Tholume is approximately 131.88 cubic centimeters.

Work Practice 4

Answer

4. 15.3 cu m

Vocabulary, Readiness & Videcheck

Use the choices below to fill in each blank. e exercises are based on Section 9.3.

units   volume   squ:
perimeter   surface area   cubi

1. The _____ of a polyhedron is the of the areas of its faces.
2. The measure of the amount of space insisolid is its _____ .
3. Volume is measured in _____ unit
4. Surface area is measured in _____ .
5. The _____ of a polygon is the sum e lengths of its sides.
6. Perimeter is measured in _____ .

Martin-Gay Interactive Videos   Watch the section lecture video and answer the following question.

See Video 9.5

Objective A 7. In ▣ Examples 2 and 3, explain the difference in the two answers found for each. ◯

9.5 Exercise Set MyMathLab®

Objective A *Find the volume and surface area of each solid. See Examples 1 through 4. For formulas containing* π, *give the exact answer and then approximate using* $\dfrac{22}{7}$ *for* π.

1.

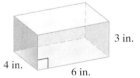

3 in.
4 in.   6 in.

2.

4 cm

4 cm   8 cm

3.

8 cm

8 cm
8 cm

4.

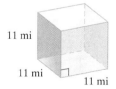

11 mi

11 mi   11 mi

5. For surface area, round the approximation to 2 decimal places.

3 yd
2 yd

6. For surface area, round the approximation to 2 decimal places.

$1\frac{3}{4}$ in.

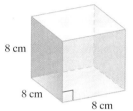

9 in.

7.

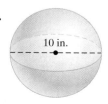

10 in.

8.

3 mi

9. Find the volume only.

2 in.

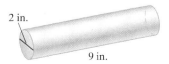

9 in.

10. Find the volume only.

10 ft

6 ft

11. Find the volume only.

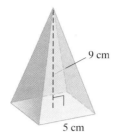

9 cm

5 cm

12. Find the volume only.

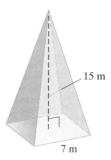

15 m

7 m

Solve. See Examples 1 through 4.

▶ **13.** Find the volume of a cube with edges of $1\frac{1}{3}$ inches.

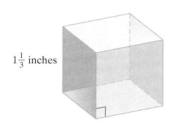

$1\frac{1}{3}$ inches

14. A water storage tank is in the shape of a cone with the pointed end down. If the radius is 14 ft and the depth of the tank is 15 ft, approximate the volume of the tank in cubic feet. Use $\frac{22}{7}$ for π.

14 ft

15 ft

15. Find the volume and surface area of a rectangular box 2 ft by 1.4 ft by 3 ft.

16. Find the volume and surface area of a box in the shape of a cube that is 5 ft on each side.

17. A paperweight is in the shape of a square-based pyramid 20 centimeters tall. If an edge of the base is 12 centimeters, find the volume of the paperweight.

18. A birdbath is made in the shape of a hemisphere (half-sphere). If its radius is 10 inches, approximate its volume. Use $\frac{22}{7}$ for π.

10 in.

19. Find the exact volume and surface area of a sphere with a radius of 7 inches.

20. A tank is in the shape of a cylinder 8 feet tall and 3 feet in radius. Find the exact volume and surface area of the tank.

21. Find the volume of a rectangular block of ice 2 feet by $2\frac{1}{2}$ feet by $1\frac{1}{2}$ feet.

22. Find the capacity (volume in cubic feet) of a rectangular ice chest with inside measurements of 3 feet by $1\frac{1}{2}$ feet by $1\frac{3}{4}$ feet.

23. Find the exact volume of a waffle ice cream cone with a 3-in. diameter and a height of 7 inches.

24. A snow globe has a diameter of 6 inches. Find its exact volume. Then approximate its volume using 3.14 for π.

25. The largest cereal bowl in the world was made by Kellogg's South Africa in July 2007. The bowl had a 2.6-m diameter and a height of 1.5 m. Calculate the volume of cereal you could put into this bowl. Use $\pi \approx 3.14$ and round to the nearest hundredth of a cubic meter. (*Source: Guinness Book of World Records*)

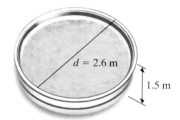

$d = 2.6$ m

1.5 m

26. Mount Fuji, in Japan, is considered the most beautiful composite volcano in the world. The mountain is in the shape of a cone whose height is about 3.5 kilometers and whose base radius is about 3 kilometers. Approximate the volume of Mt. Fuji in cubic kilometers. Use $\dfrac{22}{7}$ for π.

27. Find the volume of a pyramid with a square base 5 inches on a side and a height of $1\dfrac{3}{10}$ inches.

28. Approximate to the nearest hundredth the volume of a sphere with a radius of 2 centimeters. Use 3.14 for π.

29. In 2013, the largest free-floating soap bubble made with a wand had a diameter between 11 and 12 feet. Calculate the exact volume of a sphere with a diameter of 12 feet. (*Source: Guinness World Records*)

30. The largest inflatable beach ball was created in Poland in 2012. It has a diameter of just under 54 feet. Calculate the exact volume of a sphere with a diameter of 54 feet. (*Source: Guinness World Records*)

31. An ice cream cone with a 4-centimeter diameter and 3-centimeter depth is filled exactly level with the top of the cone. Approximate how much ice cream (in cubic centimeters) is in the cone. Use $\dfrac{22}{7}$ for π.

32. A child's toy is in the shape of a square-based pyramid 10 inches tall. If an edge of the base is 7 inches, find the volume of the toy.

The Space Cube is supposed to be the world's smallest computer, with dimensions of 2 inches by 2 inches by 2.2 inches.

33. Find the volume of the Space Cube.

34. Find the volume of an actual cube that measures 2 inches by 2 inches by 2 inches.

35. Find the volume of an actual cube that measures 2.2 inches by 2.2 inches by 2.2 inches.

36. Comment on the results of Exercises 33–35. Were you surprised when you compared volumes? Why or why not?

Review

Evaluate. See Section 1.9.

37. 5^2

38. 7^2

39. 3^2

40. 20^2

41. $1^2 + 2^2$

42. $5^2 + 3^2$

43. $4^2 + 2^2$

44. $1^2 + 6^2$

Concept Extensions

Solve.

45. The Hayden Planetarium, at the Museum of Natural History in New York City, boasts a dome that has a diameter of 20 m. The dome is a hemisphere, or half a sphere. What is the volume enclosed by the dome at the Hayden Planetarium? Use 3.14 for π and round to the nearest hundredth. (*Source:* Hayden Planetarium)

diameter

hemisphere

46. The Adler Museum in Chicago recently added a new planetarium, its StarRider Theater, which has a diameter of 55 feet. Find the volume of its hemispheric (half a sphere) dome. Use 3.14 for π and round to the nearest hundredth. (*Source:* The Adler Museum)

47. Do two rectangular solids with the same volume have the same shape? To see, find the volume of each rectangular solid.

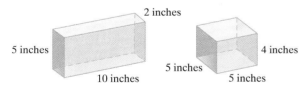

2 inches

5 inches

10 inches

4 inches

5 inches

5 inches

48. Do two rectangular solids with the same volume have the same surface area? To see, find the volume and surface area of each rectangular solid.

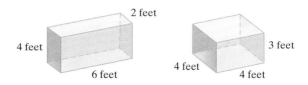

2 feet

4 feet

6 feet

3 feet

4 feet

4 feet

49. Two kennels are offered at a hotel. The kennels measure

a. 2'1″ by 1'8″ by 1'7″ and
b. 1'1″ by 2' by 2'8″

What is the volume of each kennel rounded to the nearest tenth of a cubic foot? Which is larger?

50. The centerpiece of the New England Aquarium in Boston is its Giant Ocean Tank. This exhibit is a four-story cylindrical saltwater tank containing sharks, sea turtles, stingrays, and tropical fish. The radius of the tank is 16.3 feet and its height is 32 feet (assuming that a story is 8 feet). What is the volume of the Giant Ocean Tank? Use $\pi \approx 3.14$ and round to the nearest tenth of a cubic foot. (*Source:* New England Aquarium)

51. Find the volume of the figure shown. Give the exact measure and then a whole number approximation.

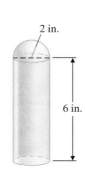

2 in.

6 in.

52. Can you compute the volume of a rectangle? Why or why not?

△Geometry Concepts

Answers

1. Find the supplement and the complement of a 27° angle.

1. _____

Find the measures of angles x, y, and z in each figure.

2. _____

2.

3. $m \parallel n$

3. _____

4. _____

4. Find the measure of ∠x.   **5.** Find the diameter.    **6.** Find the radius.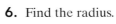

5. _____

6. _____

7. _____

For Exercises 7 through 11, find the perimeter (or circumference) and area of each figure. For the circle, give the exact circumference and area. Then use π ≈ 3.14 to approximate each. Don't forget to attach correct units.

8. _____

7. Square 5 m   **8.** 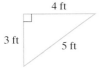 4 ft 3 ft 5 ft   **9.** 5 cm   **10.** 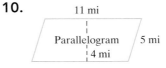 11 mi Parallelogram 5 mi 4 mi

9. _____

10. _____

11. 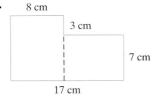 8 cm 3 cm 7 cm 17 cm

12. The smallest cathedral is in Highlandville, Missouri. The rectangular floor of the cathedral measures 14 feet by 17 feet. Find its perimeter and its area. (*Source: The Guinness Book of Records*)

11. _____

12. _____

13. _____

Find the volume of each solid. Don't forget to attach correct units. For Exercises 13 and 14, find the surface area of each solid also.

14. _____

13. A cube with edges of 4 inches each

14. A rectangular box 2 feet by 3 feet by 5.1 feet

15. _____

15. A pyramid with a square base 10 centimeters on a side and a height of 12 centimeters

16. A sphere with a diameter of 3 miles. Give the exact volume and then use $\pi \approx \dfrac{22}{7}$ to approximate.

16. _____

Objectives

A Decide Whether Two Triangles Are Congruent.

B Find the Ratio of Corresponding Sides in Similar Triangles.

C Find Unknown Lengths of Sides in Similar Triangles.

Objective A Deciding Whether Two Triangles Are Congruent

Congruent angles are angles that have the same measure. Two triangles are **congruent** when they have the same shape and the same size. In congruent triangles, the measures of corresponding angles are equal and the lengths of corresponding sides are equal. The following triangles are congruent:

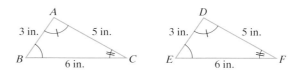

Since these triangles are congruent, the measures of their corresponding angles are equal.

Angles with equal measure: $\angle A$ and $\angle D$, $\angle B$ and $\angle E$, $\angle C$ and $\angle F$. Also, the lengths of their corresponding sides are equal.

Equal corresponding sides: $\overline{AB}$ and $\overline{DE}$, $\overline{BC}$ and $\overline{EF}$, $\overline{CA}$ and $\overline{FD}$

Any one of the following may be used to determine whether two triangles are congruent:

Congruent Triangles

Angle-Side-Angle (ASA)

If the measures of two angles of a triangle equal the measures of two angles of another triangle, and the lengths of the sides between each pair of angles are equal, the triangles are congruent.

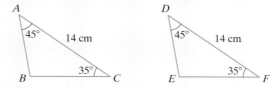

For example, these two triangles are congruent by Angle-Side-Angle.

Side-Side-Side (SSS)

If the lengths of the three sides of a triangle equal the lengths of the corresponding sides of another triangle, the triangles are congruent.

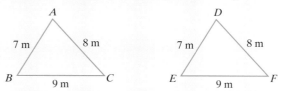

For example, these two triangles are congruent by Side-Side-Side.

Side-Angle-Side (SAS)

If the lengths of two sides of a triangle equal the lengths of corresponding sides of another triangle, and the measures of the angles between each pair of sides are equal, the triangles are congruent.

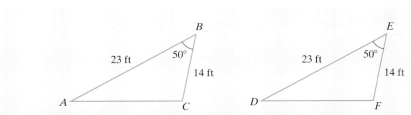

For example, these two triangles are congruent by Side-Angle-Side.

Example 1 Determine whether triangle ABC is congruent to triangle DEF.

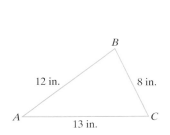
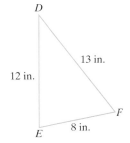

Solution: Since the lengths of all three sides of triangle ABC equal the lengths of all three sides of triangle DEF, the triangles are congruent.

📀 Work Practice 1

In Example 1, notice that as soon as we know that the two triangles are congruent, we know that all three corresponding angles are congruent.

Objective B Finding the Ratio of Corresponding Sides in Similar Triangles ▶

Two triangles are **similar** when they have the same shape but not necessarily the same size. In similar triangles, the measures of corresponding angles are equal and corresponding sides are in proportion. The following triangles are similar:

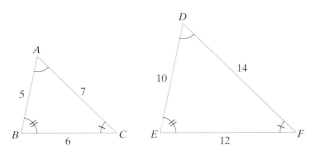

Since these triangles are similar, the measures of their corresponding angles are equal.

Angles with equal measure: $\angle A$ and $\angle D$, $\angle B$ and $\angle E$, $\angle C$ and $\angle F$. Also, the lengths of their corresponding sides are in proportion.

Sides in proportion: $\dfrac{AB}{DE} = \dfrac{BC}{EF} = \dfrac{CA}{FD}$ or, in this particular case,

$$\frac{AB}{DE} = \frac{5}{10} = \frac{1}{2}, \frac{BC}{EF} = \frac{6}{12} = \frac{1}{2}, \frac{CA}{FD} = \frac{7}{14} = \frac{1}{2}$$

The ratio of corresponding sides is $\dfrac{1}{2}$.

Practice 1

a. Determine whether triangle MNO is congruent to triangle RQS.

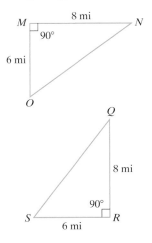

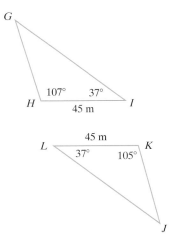

b. Determine whether triangle GHI is congruent to triangle JKL.

Answers

1. a. congruent **b.** not congruent

Practice 2

Find the ratio of corresponding sides for the similar triangles *QRS* and *XYZ*.

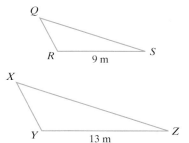

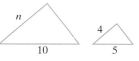

Example 2   Find the ratio of corresponding sides for the similar triangles *ABC* and *DEF*.

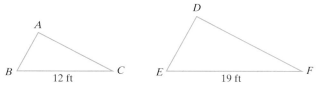

Solution: We are given the lengths of two corresponding sides. Their ratio is

$$\frac{12 \text{ feet}}{19 \text{ feet}} = \frac{12}{19}$$

■ Work Practice 2

Objective C   Finding Unknown Lengths of Sides in Similar Triangles ▶

Because the ratios of lengths of corresponding sides are equal, we can use proportions to find unknown lengths in similar triangles.

Practice 3

Given that the triangles are similar, find the missing length *n*.

a.

b.

Example 3   Given that the triangles are similar, find the missing length *n*.

Solution: Since the triangles are similar, corresponding sides are in proportion. Thus, the ratio of 2 to 3 is the same as the ratio of 10 to *n*, or

$$\frac{2}{3} = \frac{10}{n}$$

To find the unknown length *n*, we set cross products equal.

$$\frac{2}{3} = \frac{10}{n}$$

$$2 \cdot n = 3 \cdot 10 \qquad \text{Set cross products equal.}$$

$$2 \cdot n = 30 \qquad \text{Multiply.}$$

$$n = \frac{30}{2} \qquad \text{Divide 30 by 2, the number multiplied by } n.$$

$$n = 15$$

The missing length is 15 units.

■ Work Practice 3

Answers

2. $\frac{9}{13}$ **3. a.** $n = 8$ **b.** $n = \frac{10}{3}$ or $3\frac{1}{3}$

✔ Concept Check Answer

A corresponds to *O*; *B* corresponds to *N*; *C* corresponds to *M*

✔**Concept Check**   The following two triangles are similar. Which vertices of the first triangle appear to correspond to which vertices of the second triangle?

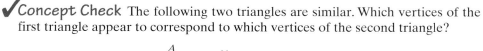

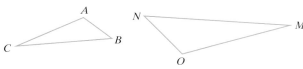

Many applications involve diagrams containing similar triangles. Surveyors, astronomers, and many other professionals continually use similar triangles in their work.

Example 4 Finding the Height of a Tree

Mel Rose is a 6-foot-tall park ranger who needs to know the height of a particular tree. He measures the shadow of the tree to be 69 feet long when his own shadow is 9 feet long. Find the height of the tree.

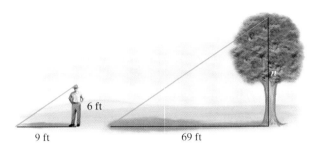

6 ft

9 ft   69 ft

Solution:

1. UNDERSTAND. Read and reread the problem. Notice that the triangle formed by the Sun's rays, Mel, and his shadow is similar to the triangle formed by the Sun's rays, the tree, and its shadow.

2. TRANSLATE. Write a proportion from the similar triangles formed.

$$\frac{\text{Mel's height}}{\text{height of tree}} \rightarrow \frac{6}{n} = \frac{9}{69} \leftarrow \frac{\text{length of Mel's shadow}}{\text{length of tree's shadow}}$$

$$\text{or } \frac{6}{n} = \frac{3}{23} \quad \text{Simplify } \tfrac{9}{69} \text{ (ratio in lowest terms).}$$

3. SOLVE for n:

$$\frac{6}{n} = \frac{3}{23}$$

$$6 \cdot 23 = n \cdot 3 \quad \text{Set cross products equal.}$$

$$138 = n \cdot 3 \quad \text{Multiply.}$$

$$\frac{138}{3} = n \quad \text{Divide 138 by 3, the number multiplied by } n.$$

$$46 = n$$

4. INTERPRET. *Check* to see that replacing n with 46 in the proportion makes the proportion true. *State* your conclusion: The height of the tree is 46 feet.

 Work Practice 4

Practice 4

Tammy Shultz, a firefighter, needs to estimate the height of a burning building. She estimates the length of her shadow to be 8 feet long and the length of the building's shadow to be 60 feet long. Find the approximate height of the building if she is 5 feet tall.

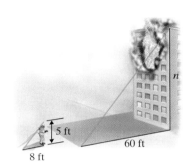

5 ft

8 ft   60 ft

Answer

4. approximately 37.5 ft

Vocabulary, Readiness & Video Check

Answer each question true or false.

1. Two triangles that have the same shape but not necessarily the same size are congruent.

2. Two triangles are congruent if they have the same shape and size.

3. Congruent triangles are also similar.

4. Similar triangles are also congruent.

5. For the two similar triangles, the ratio of corresponding sides is $\frac{5}{6}$.

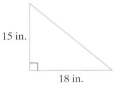

15 in.

18 in.

5 in.

6 in.

Each pair of triangles is similar. Name the congruent angles and the corresponding sides that are proportional.

6.

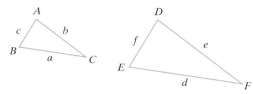

7.

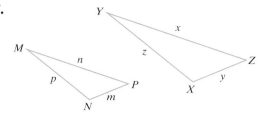

Martin-Gay Interactive Videos   Watch the section lecture video and answer the following questions.

Objective A 8. How did we decide which congruency rule to use to determine if the two triangles in ▤ Example 1 are congruent? ◉

Objective B 9. From ▤ Example 2, what does "corresponding sides are in proportion" mean? ◉

Objective C 10. In ▤ Example 3, what is another proportion named that we could have used to solve the application? ◉

See Video 9.6 🍎

9.6 Exercise Set MyMathLab® ▶

Objective A *Determine whether each pair of triangles is congruent. If congruent, state the reason why, such as SSS, SAS, or ASA. See Example 1.*

1.

5 in.   6 in.   6 in.   7 in.

7 in.   5 in.

2.

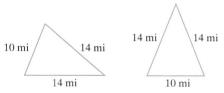

10 mi   14 mi   14 mi   14 mi

14 mi   10 mi

3. ▶

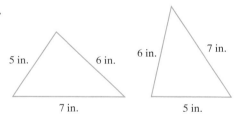

25 m   25 m

40 m   40 m

24 m

23 m

4.

21 cm   7 cm

17 cm   21 cm

7 cm   16 cm

5.

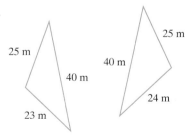

48°   48°

30 m   30 m

42°   42°

6.

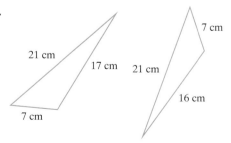

15 yd

23°   36°

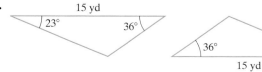

36°   23°

15 yd

7.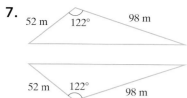

8.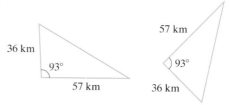

Objective B *Find each ratio of the corresponding sides of the given similar triangles. See Example 2.*

 9.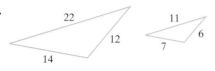

10.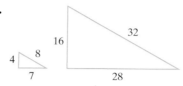

11.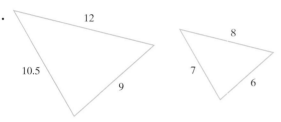

12.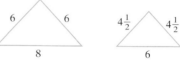

Objective C *Given that the pairs of triangles are similar, find the unknown length of the side labeled n.*
See Example 3.

13.

14.

15.

16.

17.

18.

19.

20.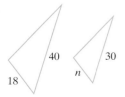

21.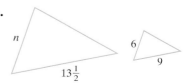

22.

23.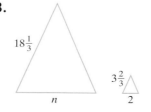

24.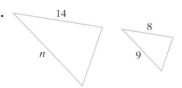

25.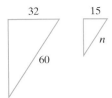

26.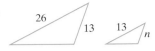

27.

28.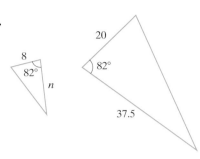

Solve. For Exercises 29 and 30, the solutions have been started for you. See Example 4.

29. Given the following diagram, approximate the height of the observation deck in the Seattle Space Needle in Seattle, Washington. (*Source:* Seattle Space Needle)

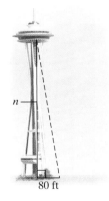 

Start the solution:

1. UNDERSTAND the problem. Reread it as many times as needed.
2. TRANSLATE into a proportion using the similar triangles formed. (Fill in the blanks.)

height of
observation deck → $\dfrac{n}{13} = \dfrac{}{}$ ← length of Space Needle shadow
height of pole → ← length of pole shadow

3. SOLVE by setting cross products equal.
4. INTERPRET.

30. A fountain in Fountain Hills, Arizona, sits in a 28-acre lake and shoots up a column of water every hour. Based on the diagram below, what is the height of the fountain?

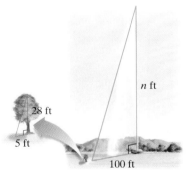

Start the solution:

1. UNDERSTAND the problem. Reread it as many times as needed.
2. TRANSLATE into a proportion using the similar triangles formed. (Fill in the blanks.)

height of tree → $\dfrac{28}{n} = \dfrac{}{}$ ← length of tree shadow
height of fountain → ← length of fountain shadow

3. SOLVE by setting cross products equal.
4. INTERPRET.

31. Given the following diagram, approximate the height of the Chase Tower in Oklahoma City, Oklahoma. Here, we use *x* to represent the unknown number. (*Source:* Council on Tall Buildings and Urban Habitat)

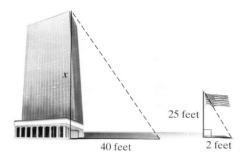

25 feet

40 feet   2 feet

32. The tallest tree standing today is a redwood located in the Humboldt Redwoods State Park near Ukiah, California. Given the following diagram, approximate its height. Here, we use *x* to represent the unknown number. (*Source: Guinness World Records*)

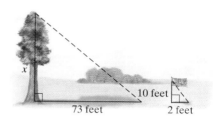

x

10 feet

73 feet   2 feet

33. Samantha Black, a 5-foot-tall park ranger, needs to know the height of a tree. She notices that when the shadow of the tree is 48 feet long, her shadow is 4 feet long. Find the height of the tree.

34. Lloyd White, a firefighter, needs to estimate the height of a burning building. He estimates the length of his shadow to be 9 feet long and the length of the building's shadow to be 75 feet long. Find the approximate height of the building if he is 6 feet tall.

35. If a 30-foot tree casts an 18-foot shadow, find the length of the shadow cast by a 24-foot tree.

36. If a 24-foot flagpole casts a 32-foot shadow, find the length of the shadow cast by a 44-foot antenna. Round to the nearest tenth.

Review

Solve. See Section 5.3.

37. For the health of his fish, the owner of Pete's Sea World uses the standard that a 20-gallon tank should house only 19 neon tetras. Find the number of neon tetras that Pete should place into a 55-gallon tank.

38. A local package express deliveryman is traveling the city expressway at 45 mph when he is forced to slow down due to traffic ahead. His truck slows at the rate of 3 mph every 5 seconds. Find his speed 8 seconds after braking.

Solve. See Section 4.6.

39. Launch Umbilical Tower 1 is the name of the gantry used for the *Apollo* launch that took Neil Armstrong and Buzz Aldrin to the moon. Find the height of the gantry to the nearest whole foot.

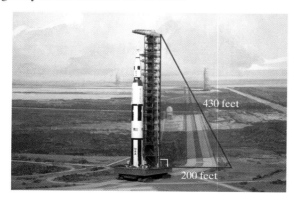

430 feet

200 feet

40. Arena polo, popular in the United States and England, is played on a field that is 100 yards long and usually 50 yards wide. Find the length, to the nearest yard, of the diagonal of this field.

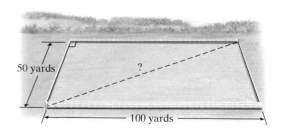

50 yards   ?

100 yards

Perform the indicated operation. See Sections 4.2 through 4.4.

41. $3.6 + 0.41$

42. $3.6 - 0.41$

43. $(0.41)(3)$

44. $0.48 \div 3$

Concept Extensions

Solve.

45. The print area on a particular page measures 7 inches by 9 inches. A printing shop is to copy the page and reduce the print area so that its length is 5 inches. What will its width be? Will the print now fit on a 3-by-5-inch index card?

46. The art sample for a banner measures $\frac{1}{3}$ foot in width by $1\frac{1}{2}$ feet in length. If the completed banner is to have a length of 9 feet, find its width.

Given that the pairs of triangles are similar, find the length of the side labeled n. Round your results to 1 decimal place.

 47.

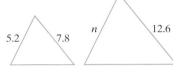

 **48.**

49. In your own words, describe any differences in similar triangles and congruent triangles.

50. Describe a situation where similar triangles would be useful for a contractor building a house.

51. A triangular park is planned and waiting to be approved by the city zoning commission. A drawing of the park shows sides of lengths 5 inches, $7\frac{1}{2}$ inches, and $10\frac{5}{8}$ inches. If the scale on the drawing is $\frac{1}{4}$ in. = 10 ft, find the actual proposed dimensions of the park.

52. John and Robyn Costello draw a triangular deck on their house plans. Robyn measures sides of the deck drawing on the plans to be 3 inches, $4\frac{1}{2}$ inches, and 6 inches. If the scale on the drawing is $\frac{1}{4}$ in. = 1 foot, find the lengths of the sides of the deck they want built.

Chapter 9 Group Activity

The Cost of Road Signs

Sections 9.1, 9.2, 9.4

There are nearly 4 million miles of streets and roads in the United States. With streets, roads, and highways comes the need for traffic control, guidance, warning, and regulation. Road signs perform many of these tasks. Just in our routine travels, we see a wide variety of road signs every day. Think how many road signs must exist on the 4 million miles of roads in the United States. Have you ever wondered how much signs like these cost?

The cost of a road sign generally depends on the type of sign. Costs for several types of signs and signposts are listed in the table. Examples of various types of signs are shown below.

Road Sign Costs	
Type of Sign	**Cost**
Regulatory, warning, marker	$15–$18 per square foot
Large guide	$20–$25 per square foot
Type of Post	**Cost**
U-channel	$125–$200 each
Square tube	$10–$15 per foot
Steel breakaway	$15–$25 per foot

The cost of a sign is based on its area. For diamond, square, or rectangular signs, the area is found by multiplying the length (in feet) times the width (in feet). Then the area is multiplied by the cost per square foot. For signs with irregular shapes, costs are generally figured *as if* the sign were a rectangle, multiplying the height and width at the tallest and widest parts of the sign.

Group Activity

Locate four different kinds of road signs on or near your campus. Measure the dimensions of each sign, including the height of the post on which it is mounted. Using the cost data given in the table, find the minimum and maximum costs of each sign, including its post. Summarize your results in a table, and include a sketch of each sign.

Regulatory   Warning   Marker   Large Guide   Posts

U-channel

Square tube

Steel breakaway

Chapter 9 Vocabulary Check

Fill in each blank with one of the words or phrases listed below.

transversal	line	congruent	hypotenuse	legs	acute
right	line segment	complementary	polygon	vertical	supplementary
right triangle	volume	obtuse	vertex	ray	angle
similar	perimeter	area	straight	adjacent	

1. A(n) _____ is a triangle with a right angle. The side opposite the right angle is called the _____, and the other two sides are called _____ .

2. A(n) _____ is a piece of a line with two endpoints.

3. Two angles that have a sum of 90° are called _____ angles.

4. A(n) _____ is a set of points extending indefinitely in two directions.

5. The _____ of a polygon is the distance around the polygon.

6. A(n) _____ is made up of two rays that share the same endpoint. The common endpoint is called the _____ .

7. _____ triangles have the same shape and the same size.

8. _____ measures the amount of surface of a region.

9. A(n) _____ is a part of a line with one endpoint. It extends indefinitely in one direction.

10. A(n) _____ is a closed plane figure that basically consists of three or more line segments that meet at their endpoints.

11. A line that intersects two or more lines at different points is called a(n) _____ .

12. A angle that measures 180° is called a(n) _____ angle.

13. The measure of the space of a solid is called its _____ .

14. When two lines intersect, four angles are formed. The angles that are opposite each other are called _____ angles.

15. When two of the four angles from Exercise 14 share a common side, they are called _____ angles.

16. An angle whose measure is between 90° and 180° is called a(n) _____ angle.

17. An angle that measures 90° is called a(n) _____ angle.

18. An angle whose measure is between 0° and 90° is called a(n) _____ angle.

19. Two angles that have a sum of 180° are called _____ angles.

20. _____ triangles have exactly the same shape but not necessarily the same size.

Helpful Hint ▶ Are you preparing for your test? Don't forget to take the Chapter 9 Test on page 694. Then check your answers at the back of the text and use the Chapter Test Prep Videos to see the fully worked-out solutions to any of the exercises you want to review.

9 Chapter Highlights

Definitions and Concepts	Examples
Section 9.1 Lines and Angles	
A **line** is a set of points extending indefinitely in two directions. A line has no width or height, but it does have length. We name a line by any two of its points.	Line AB or $\overleftrightarrow{AB}$
A **line segment** is a piece of a line with two endpoints.	Line segment AB or $\overline{AB}$
A **ray** is a part of a line with one endpoint. A ray extends indefinitely in one direction.	Ray AB or $\overrightarrow{AB}$
An **angle** is made up of two rays that share the same endpoint. The common endpoint is called the **vertex.**	Angle ABC, $\angle ABC$, $\angle CBA$, or $\angle B$
An angle that measures 180° is called a **straight angle.**	$\angle RST$ is a straight angle.

Definitions and Concepts	Examples

Section 9.1   Lines and Angles (*continued*)

An angle that measures 90° is called a **right angle.** The symbol ∟ is used to denote a right angle.

∠*ABC* is a right angle.

An angle whose measure is between 0° and 90° is called an **acute angle.**

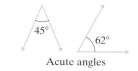

Acute angles

An angle whose measure is between 90° and 180° is called an **obtuse angle.**

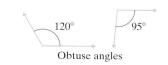

Obtuse angles

Two angles that have a sum of 90° are called complementary **angles.** We say that each angle is the **complement** of the other.

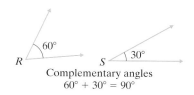

Complementary angles
60° + 30° = 90°

Two angles that have a sum of 180° are called supplementary **angles.** We say that each angle is the **supplement** of the other.

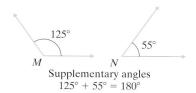

Supplementary angles
125° + 55° = 180°

When two lines intersect, four angles are formed. Two of these angles that are opposite each other are called **vertical angles.** Vertical angles have the same measure.

Two of these angles that share a common side are called **adjacent angles.** Adjacent angles formed by intersecting lines are supplementary.

Vertical angles:
∠*a* and ∠*c*
∠*d* and ∠*b*

Adjacent angles:
∠*a* and ∠*b*
∠*b* and ∠*c*
∠*c* and ∠*d*
∠*d* and ∠*a*

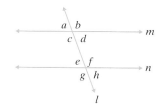

A line that intersects two or more lines at different points is called a **transversal.** Line *l* is a transversal that intersects lines *m* and *n*. The eight angles formed have special names. Some of these names are:

Corresponding angles: ∠*a* and ∠*e*, ∠*c* and ∠*g*, ∠*b* and ∠*f*, ∠*d* and ∠*h*

Alternate interior angles: ∠*c* and ∠*f*, ∠*d* and ∠*e*

Parallel Lines Cut by a Transversal

If two parallel lines are cut by a transversal, then the measures of **corresponding angles are equal** and the measures of **alternate interior angles are equal.**

Definitions and Concepts	Examples

Section 9.2   Plane Figures and Solids

The **sum of the measures** of the angles of a triangle is 180°.

Find the measure of $\angle x$.

The measure of $\angle x = 180° - 85° - 45° = 50°$

A **right triangle** is a triangle with a right angle. The side opposite the right angle is called the **hypotenuse,** and the other two sides are called **legs.**

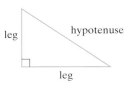

For a circle or a sphere:

$$\text{diameter} = 2 \cdot \text{radius}$$

$$d = 2 \cdot r$$

$$\text{radius} = \frac{\text{diameter}}{2}$$

$$r = \frac{d}{2}$$

Find the diameter of the circle.

$$d = 2 \cdot r$$
$$= 2 \cdot 6 \text{ feet} = 12 \text{ feet}$$

Section 9.3   Perimeter

Perimeter Formulas

Rectangle: $P = 2 \cdot l + 2 \cdot w$

Square: $P = 4 \cdot s$

Triangle: $P = a + b + c$

Circumference of a Circle: $C = 2 \cdot \pi \cdot r$ or $C = \pi \cdot d$,

where $\pi \approx 3.14$ or $\pi \approx \dfrac{22}{7}$

Find the perimeter of a rectangle with length 28 meters and width 15 meters.

$$P = 2 \cdot l + 2 \cdot w$$
$$= 2 \cdot 28 \text{ m} + 2 \cdot 15 \text{ m}$$
$$= 56 \text{ m} + 30 \text{ m}$$
$$= 86 \text{ m}$$

The perimeter is 86 meters.

Section 9.4   Area

Area Formulas

Rectangle: $A = l \cdot w$

Square: $A = s^2$

Triangle: $A = \dfrac{1}{2} \cdot b \cdot h$

Parallelogram: $A = b \cdot h$

Trapezoid: $A = \dfrac{1}{2} \cdot (b + B) \cdot h$

Circle: $A = \pi \cdot r^2$

Find the area of a square with side length 8 centimeters.

$$A = s^2$$
$$= (8 \text{ cm})^2$$
$$= 64 \text{ square centimeters}$$

The area of the square is 64 square centimeters.

Definitions and Concepts	Examples

Section 9.5 Volume and Surface Area

Volume Formulas

Rectangular Solid:

$$V = l \cdot w \cdot h$$

Cube:

$$V = s^3$$

Sphere:

$$V = \frac{4}{3} \cdot \pi \cdot r^3$$

Right Circular Cylinder:

$$V = \pi \cdot r^2 \cdot h$$

Cone:

$$V = \frac{1}{3} \cdot \pi \cdot r^2 \cdot h$$

Square-Based Pyramid:

$$V = \frac{1}{3} \cdot s^2 \cdot h$$

Surface Area Formulas

See page 665.

Find the volume of the sphere. Use $\frac{22}{7}$ for π.

4 in.

$$V = \frac{4}{3} \cdot \pi \cdot r^3$$

$$\approx \frac{4}{3} \cdot \frac{22}{7} \cdot (4 \text{ inches})^3$$

$$\approx \frac{4 \cdot 22 \cdot 64}{3 \cdot 7} \text{ cubic inches}$$

$$\approx \frac{5632}{21} \quad \text{or} \quad 268\frac{4}{21} \text{ cubic inches}$$

Section 9.6 Congruent and Similar Triangles

Congruent triangles have the same shape and the same size. Corresponding angles are equal, and corresponding sides are equal.

Similar triangles have exactly the same shape but not necessarily the same size. Corresponding angles are equal, and the ratios of the lengths of corresponding sides are equal.

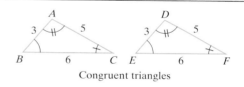

Congruent triangles

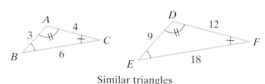

Similar triangles

$$\frac{AB}{DE} = \frac{3}{9} = \frac{1}{3}, \frac{BC}{EF} = \frac{6}{18} = \frac{1}{3},$$

$$\frac{CA}{FD} = \frac{4}{12} = \frac{1}{3}$$

Chapter 9   Review

(9.1) *Classify each angle as acute, right, obtuse, or straight.*

1.

2.

3. *C*

4.

5. Find the complement of a 25° angle.

6. Find the supplement of a 105° angle.

Find the measure of angle x in each figure.

7.

8.

9.
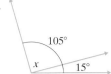

10.
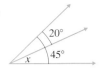

11. Identify the pairs of supplementary angles.

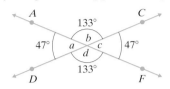

12. Identify the pairs of complementary angles.

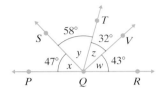

Find the measures of angles x, y, and z in each figure.

13.

14.
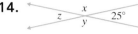

15. Given that $m \parallel n$.

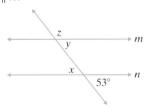

16. Given that $m \parallel n$.

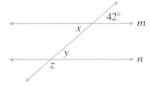

(9.2) *Find the measure of ∠x in each figure.*

17.

18.

19.

20.
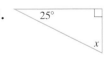

Find the unknown diameter or radius as indicated.

21.
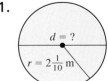

22.

23.
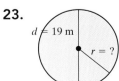

24.
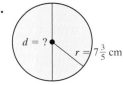

Identify each solid.

25.

26.

27.

28.

Find the unknown radius or diameter as indicated.

29. The radius of a sphere is 9 inches. Find its diameter.

30. The diameter of a sphere is 4.7 meters. Find its radius.

Identify each regular polygon.

31.

32.

Identify each triangle as equilateral, isosceles, or scalene. Also identify any triangle that is a right triangle.

33.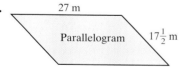

34.

(9.3) *Find the perimeter of each figure.*

35.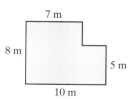

27 m

Parallelogram $17\frac{1}{2}$ m

36.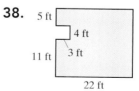

11 cm   7.6 cm

12 cm

37.

7 m

8 m   5 m

10 m

38.

5 ft

4 ft

11 ft   3 ft

22 ft

Solve.

39. Find the perimeter of a rectangular sign that measures 6 feet by 10 feet.

40. Find the perimeter of a town square that measures 110 feet on a side.

Find the circumference of each circle. Use $\pi \approx 3.14$.

41.

1.7 in.

42.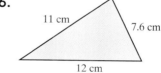

5 yd

(9.4) *Find the area of each figure. For the circles, find the exact area and then use* $\pi \approx 3.14$ *to approximate the area.*

43.

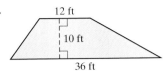

12 ft
10 ft
36 ft

44.

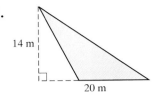

14 m
20 m

45.

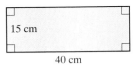

15 cm
40 cm

46.

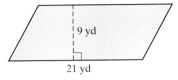

9 yd
21 yd

47.

7 ft

48.

Square   9.1 m

49.

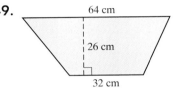

64 cm
26 cm
32 cm

50.

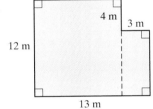

4 m   3 m
12 m
13 m

51. The amount of sealer necessary to seal a driveway depends on the area. Find the area of a rectangular driveway 36 feet by 12 feet.

52. Find how much carpet is necessary to cover the floor of the room shown.

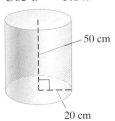

10 feet   13 feet

(9.5) *Find the volume and surface area of the solids in Exercises 53 and 54. For Exercises 55 and 56, give the exact volume and an approximation.*

53.

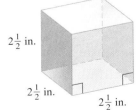

$2\frac{1}{2}$ in.
$2\frac{1}{2}$ in.
$2\frac{1}{2}$ in.

54.

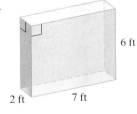

6 ft
2 ft   7 ft

55. Use $\pi \approx 3.14$.

50 cm
20 cm

56. Use $\pi \approx \frac{22}{7}$.

$\frac{1}{2}$ km

57. Find the volume of a pyramid with a square base 2 feet on a side and a height of 2 feet.

58. Approximate the volume of a tin can 8 inches high and 3.5 inches in radius. Use 3.14 for π.

59. A chest has 3 drawers. If each drawer has inside measurements of $2\frac{1}{2}$ feet by $1\frac{1}{2}$ feet by $\frac{2}{3}$ foot, find the total volume of the 3 drawers.

60. A cylindrical canister for a shop vacuum is 2 feet tall and 1 foot in *diameter*. Find its exact volume.

(9.6) *Given that the pairs of triangles are similar, find the unknown length n.*

61.

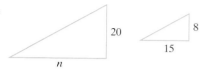

62.

63.

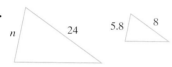

64.

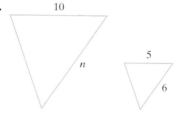

Solve.

65. A housepainter needs to estimate the height of a condominium. He estimates the length of his shadow to be 7 feet long and the length of the building's shadow to be 42 feet long. Find the approximate height of the building if the housepainter is $5\frac{1}{2}$ feet tall.

66. A toy company is making a triangular sail for a toy sailboat. The toy sail is to be the same shape as a real sailboat's sail. Use the following diagram to find the unknown lengths *x* and *y*.

Mixed Review

Find the following.

67. The supplement of a 72° angle

68. The complement of a 1° angle

Find the measure of angle x in each figure.

69.

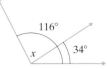

70.

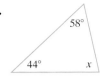

71.

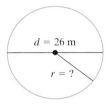

72.

$m \parallel n$

Find the unknown diameter or radius as indicated.

73.

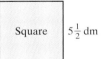

74.

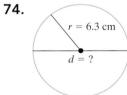

Find the perimeter of each figure.

75.

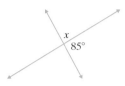

76.

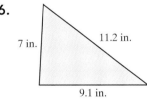

77.

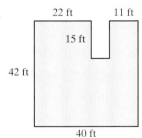

Find the area of each figure. For the circles, find the exact area and then use $\pi \approx 3.14$ to approximate the area.

78.

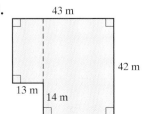

79.

Find the volume of each solid.

80. Give an approximation using $\frac{22}{7}$ for π.

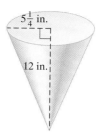

81.

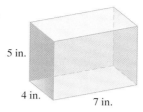

Solve.

82. Find the volume of air in a rectangular room 15 feet by 12 feet with a 7-foot ceiling.

83. A mover has two boxes left for packing. Both are cubical, one 3 feet on a side and the other 1.2 feet on a side. Find their combined volume.

Given that the pairs of triangles are similar, find the unknown length n.

84.

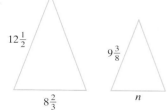

85.

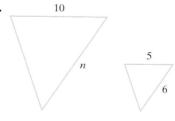

Chapter 9   Test

Step-by-step test solutions are found on the Chapter Test Prep Videos. Where available: **MyMathLab®** or **You Tube**

Answers

1. Find the complement of a 78° angle.

2. Find the supplement of a 124° angle.

3. Find the measure of ∠x.

Find the measures of x, y, and z in each figure.

4.

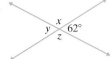

5. Given: $m \parallel n$.

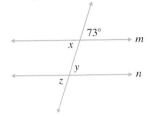

1. _____

2. _____

3. _____

4. _____

Find the unknown diameter or radius as indicated.

6.

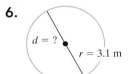

7.

5. _____

6. _____

8. Find the measure of ∠x.

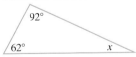

7. _____

8. _____

Find the perimeter (or circumference) and area of each figure. For the circle, give the exact value and then use $\pi \approx 3.14$ for an approximation.

9.

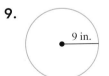

10.

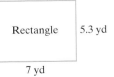

9. _____

10. _____

11.

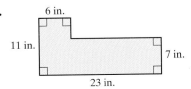

11. _____

Find the volume of each solid. For the cylinder, use $\pi \approx \dfrac{22}{7}$.

12.

5 in.

2 in.

13. Find the surface area also.

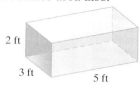

2 ft

3 ft   5 ft

Solve.

14. Find the perimeter of a square photo with a side length of 4 inches.

15. How much soil is needed to fill a rectangular hole 3 feet by 3 feet by 2 feet?

16. Find how much baseboard is needed to go around a rectangular room that measures 18 feet by 13 feet. If baseboard costs $1.87 per foot, also calculate the total cost needed for materials.

17. Vivian Thomas is going to put insecticide on her lawn to control grubworms. The lawn is a rectangle measuring 123.8 feet by 80 feet. The amount of insecticide required is 0.02 ounces per square foot. Find how much insecticide Vivian needs to purchase.

18. Given that the triangles are similar, find the missing length *n*.

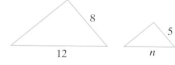

8

12

5

n

19. Tamara Watford, a surveyor, needs to estimate the height of a tower. She estimates the length of her shadow to be 4 feet long and the length of the tower's shadow to be 48 feet long. Find the approximate height of the tower if she is $5\dfrac{3}{4}$ feet tall.

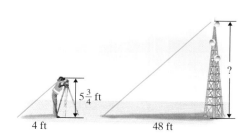

$5\dfrac{3}{4}$ ft

4 ft   48 ft

?

12. _____

13. _____

14. _____

15. _____

16. _____

17. _____

18. _____

19. _____

Answers

1. _____

2. _____

3. _____

4. _____

5. _____

6. _____

7. _____

8. _____

9. _____

10. _____

11. _____

12. _____

13. _____

14. _____

15. _____

16. _____

17. _____

18. _____

19. _____

20. _____

21. _____

22. _____

1. Write the decimal -50.82 in words.

2. Add: $\dfrac{7}{11} + \dfrac{1}{6}$

3. Round 736.2359 to the nearest tenth.

4. Round 736.2359 to the nearest hundred.

5. Add: $45 + 2.06$

6. Divide: $-3\dfrac{1}{3} \div 1\dfrac{5}{6}$

Multiply.

7. 7.68×10

8. $\dfrac{7}{11} \cdot \dfrac{1}{6}$

9. $(-76.3)(1000)$

10. $5\dfrac{1}{2} \cdot 2\dfrac{1}{11}$

11. Divide: $270.2 \div 7$. Check your answer.

12. Divide: $\dfrac{56.7}{100}$

13. Simplify: $-0.5(8.6 - 1.2)$

14. Simplify: $\dfrac{5 + 2(8 - 3)}{30 \div 6 \cdot 5}$

15. Insert $<, >,$ or $=$ to form a true statement. $\dfrac{1}{8}$ ___ 0.12

16. Insert $<, >,$ or $=$ to form a true statement. 0.75 ___ $\dfrac{13}{16}$

17. Write the ratio of 2.6 to 3.1 as a fraction in simplest form.

18. Find: $\dfrac{2}{9} + \dfrac{7}{15} - \dfrac{1}{3}$

19. Is $\dfrac{2}{3} = \dfrac{4}{6}$ a true proportion?

20. Solve for x: $\dfrac{7}{8} = \dfrac{x}{20}$

21. For 2014 model cars, 25 out of every 100 were painted white. What percent of model-year 2014 cars were white?

22. Solve for x: $4x - 7x = -30$

Write each percent as a fraction or mixed number in simplest form.

23. 1.9%   **24.** 26%   **25.** 125%   **26.** 560%

27. 85% of 300 is what number?   **28.** What percent of 16 is 2.4?

29. 20.8 is 40% of what number?   **30.** Find: $(7 - \sqrt{16})^2$

31. Mr. Buccaran, the principal at Slidell High School, counted 31 freshmen absent during a particular day. If this is 4% of the total number of freshmen, how many freshmen are there at Slidell High School?

32. Flooring tiles cost $90 for a box with 40 tiles. Each tile is 1 square foot. Find the unit price in dollars per square foot.

33. Sherry Souter, a real estate broker for Wealth Investments, sold a house for $214,000 last week. If her commission rate is 1.5% of the selling price of the home, find the amount of her commission.

34. A student can complete 7 exercises in 6 minutes. At this rate, how many exercises can be completed in 30 minutes?

35. Convert 8 feet to inches.   **36.** 100 inches = ____ yd ____ ft ____ in.

37. Convert 3.2 kilograms to grams.   **38.** Convert 70 mm to meters.

39. Subtract 3 quarts from 4 gallons 2 quarts.   **40.** Write seventy thousand, fifty-two in standard form.

41. Find the measure of ∠a.

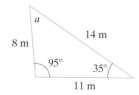

42. Find the perimeter of the triangle in Exercise 41.

43. Find the perimeter of the rectangle below:

44. Solve for x: $7(x - 2) = 9x - 6$

45. Find $\sqrt{\dfrac{4}{25}}$.   **46.** Find $\sqrt{\dfrac{9}{16}}$.

23. _____
24. _____
25. _____
26. _____
27. _____
28. _____
29. _____
30. _____
31. _____
32. _____
33. _____
34. _____
35. _____
36. _____
37. _____
38. _____
39. _____
40. _____
41. _____
42. _____
43. _____
44. _____
45. _____
46. _____

Tables

A.1 Addition Table and One Hundred Addition Facts

+	0	1	2	3	4	5	6	7	8	9
0	0	1	2	3	4	5	6	7	8	9
1	1	2	3	4	5	6	7	8	9	10
2	2	3	4	5	6	7	8	9	10	11
3	3	4	5	6	7	8	9	10	11	12
4	4	5	6	7	8	9	10	11	12	13
5	5	6	7	8	9	10	11	12	13	14
6	6	7	8	9	10	11	12	13	14	15
7	7	8	9	10	11	12	13	14	15	16
8	8	9	10	11	12	13	14	15	16	17
9	9	10	11	12	13	14	15	16	17	18

One Hundred Addition Facts

Knowledge of the basic addition facts found above is an important prerequisite for a course in basic college mathematics with early integers. Study the table above and then perform the additions. Check your answers either by comparing them with those found in the back-of-the-book answer section or by using the table. Review any facts that you missed.

1. $\begin{array}{r} 1 \\ +4 \\ \hline \end{array}$
2. $\begin{array}{r} 5 \\ +6 \\ \hline \end{array}$
3. $\begin{array}{r} 2 \\ +3 \\ \hline \end{array}$
4. $\begin{array}{r} 7 \\ +8 \\ \hline \end{array}$
5. $\begin{array}{r} 3 \\ +9 \\ \hline \end{array}$
6. $\begin{array}{r} 6 \\ +1 \\ \hline \end{array}$

7. $\begin{array}{r} 4 \\ +4 \\ \hline \end{array}$
8. $\begin{array}{r} 0 \\ +6 \\ \hline \end{array}$
9. $\begin{array}{r} 9 \\ +5 \\ \hline \end{array}$
10. $\begin{array}{r} 8 \\ +2 \\ \hline \end{array}$
11. $\begin{array}{r} 5 \\ +7 \\ \hline \end{array}$
12. $\begin{array}{r} 3 \\ +2 \\ \hline \end{array}$

13. $\begin{array}{r} 5 \\ +5 \\ \hline \end{array}$
14. $\begin{array}{r} 1 \\ +1 \\ \hline \end{array}$
15. $\begin{array}{r} 8 \\ +1 \\ \hline \end{array}$
16. $\begin{array}{r} 6 \\ +6 \\ \hline \end{array}$
17. $\begin{array}{r} 2 \\ +9 \\ \hline \end{array}$
18. $\begin{array}{r} 3 \\ +5 \\ \hline \end{array}$

19. $\begin{array}{r} 9 \\ +9 \\ \hline \end{array}$
20. $\begin{array}{r} 5 \\ +2 \\ \hline \end{array}$
21. $\begin{array}{r} 6 \\ +4 \\ \hline \end{array}$
22. $\begin{array}{r} 0 \\ +0 \\ \hline \end{array}$
23. $\begin{array}{r} 1 \\ +9 \\ \hline \end{array}$
24. $\begin{array}{r} 3 \\ +7 \\ \hline \end{array}$

25. $\begin{array}{r} 9 \\ +8 \\ \hline \end{array}$
26. $\begin{array}{r} 0 \\ +8 \\ \hline \end{array}$
27. $\begin{array}{r} 4 \\ +9 \\ \hline \end{array}$
28. $\begin{array}{r} 3 \\ +0 \\ \hline \end{array}$
29. $\begin{array}{r} 7 \\ +5 \\ \hline \end{array}$
30. $\begin{array}{r} 8 \\ +9 \\ \hline \end{array}$

31. $\begin{array}{r} 9 \\ +7 \\ \hline \end{array}$ 32. $\begin{array}{r} 2 \\ +6 \\ \hline \end{array}$ 33. $\begin{array}{r} 4 \\ +3 \\ \hline \end{array}$ 34. $\begin{array}{r} 8 \\ +5 \\ \hline \end{array}$ 35. $\begin{array}{r} 3 \\ +1 \\ \hline \end{array}$ 36. $\begin{array}{r} 0 \\ +3 \\ \hline \end{array}$

37. $\begin{array}{r} 7 \\ +1 \\ \hline \end{array}$ 38. $\begin{array}{r} 3 \\ +4 \\ \hline \end{array}$ 39. $\begin{array}{r} 8 \\ +0 \\ \hline \end{array}$ 40. $\begin{array}{r} 6 \\ +3 \\ \hline \end{array}$ 41. $\begin{array}{r} 2 \\ +4 \\ \hline \end{array}$ 42. $\begin{array}{r} 0 \\ +9 \\ \hline \end{array}$

43. $\begin{array}{r} 8 \\ +8 \\ \hline \end{array}$ 44. $\begin{array}{r} 5 \\ +3 \\ \hline \end{array}$ 45. $\begin{array}{r} 3 \\ +6 \\ \hline \end{array}$ 46. $\begin{array}{r} 6 \\ +9 \\ \hline \end{array}$ 47. $\begin{array}{r} 4 \\ +8 \\ \hline \end{array}$ 48. $\begin{array}{r} 0 \\ +1 \\ \hline \end{array}$

49. $\begin{array}{r} 2 \\ +5 \\ \hline \end{array}$ 50. $\begin{array}{r} 6 \\ +0 \\ \hline \end{array}$ 51. $\begin{array}{r} 2 \\ +0 \\ \hline \end{array}$ 52. $\begin{array}{r} 4 \\ +2 \\ \hline \end{array}$ 53. $\begin{array}{r} 8 \\ +3 \\ \hline \end{array}$ 54. $\begin{array}{r} 7 \\ +4 \\ \hline \end{array}$

55. $\begin{array}{r} 1 \\ +7 \\ \hline \end{array}$ 56. $\begin{array}{r} 4 \\ +6 \\ \hline \end{array}$ 57. $\begin{array}{r} 0 \\ +5 \\ \hline \end{array}$ 58. $\begin{array}{r} 9 \\ +1 \\ \hline \end{array}$ 59. $\begin{array}{r} 8 \\ +6 \\ \hline \end{array}$ 60. $\begin{array}{r} 5 \\ +1 \\ \hline \end{array}$

61. $\begin{array}{r} 6 \\ +7 \\ \hline \end{array}$ 62. $\begin{array}{r} 4 \\ +0 \\ \hline \end{array}$ 63. $\begin{array}{r} 1 \\ +6 \\ \hline \end{array}$ 64. $\begin{array}{r} 4 \\ +5 \\ \hline \end{array}$ 65. $\begin{array}{r} 0 \\ +7 \\ \hline \end{array}$ 66. $\begin{array}{r} 5 \\ +8 \\ \hline \end{array}$

67. $\begin{array}{r} 7 \\ +6 \\ \hline \end{array}$ 68. $\begin{array}{r} 7 \\ +0 \\ \hline \end{array}$ 69. $\begin{array}{r} 4 \\ +1 \\ \hline \end{array}$ 70. $\begin{array}{r} 5 \\ +4 \\ \hline \end{array}$ 71. $\begin{array}{r} 0 \\ +4 \\ \hline \end{array}$ 72. $\begin{array}{r} 1 \\ +2 \\ \hline \end{array}$

73. $\begin{array}{r} 7 \\ +9 \\ \hline \end{array}$ 74. $\begin{array}{r} 3 \\ +8 \\ \hline \end{array}$ 75. $\begin{array}{r} 7 \\ +7 \\ \hline \end{array}$ 76. $\begin{array}{r} 9 \\ +4 \\ \hline \end{array}$ 77. $\begin{array}{r} 1 \\ +0 \\ \hline \end{array}$ 78. $\begin{array}{r} 4 \\ +7 \\ \hline \end{array}$

79. $\begin{array}{r} 2 \\ +2 \\ \hline \end{array}$ 80. $\begin{array}{r} 1 \\ +3 \\ \hline \end{array}$ 81. $\begin{array}{r} 2 \\ +8 \\ \hline \end{array}$ 82. $\begin{array}{r} 5 \\ +9 \\ \hline \end{array}$ 83. $\begin{array}{r} 6 \\ +2 \\ \hline \end{array}$ 84. $\begin{array}{r} 9 \\ +6 \\ \hline \end{array}$

85. $\begin{array}{r} 5 \\ +0 \\ \hline \end{array}$ 86. $\begin{array}{r} 8 \\ +7 \\ \hline \end{array}$ 87. $\begin{array}{r} 7 \\ +3 \\ \hline \end{array}$ 88. $\begin{array}{r} 0 \\ +2 \\ \hline \end{array}$ 89. $\begin{array}{r} 9 \\ +2 \\ \hline \end{array}$ 90. $\begin{array}{r} 3 \\ +3 \\ \hline \end{array}$

91. $\begin{array}{r} 9 \\ +3 \\ \hline \end{array}$ 92. $\begin{array}{r} 1 \\ +5 \\ \hline \end{array}$ 93. $\begin{array}{r} 2 \\ +7 \\ \hline \end{array}$ 94. $\begin{array}{r} 6 \\ +5 \\ \hline \end{array}$ 95. $\begin{array}{r} 7 \\ +2 \\ \hline \end{array}$ 96. $\begin{array}{r} 1 \\ +8 \\ \hline \end{array}$

97. $\begin{array}{r} 6 \\ +8 \\ \hline \end{array}$ 98. $\begin{array}{r} 8 \\ +4 \\ \hline \end{array}$ 99. $\begin{array}{r} 9 \\ +0 \\ \hline \end{array}$ 100. $\begin{array}{r} 2 \\ +1 \\ \hline \end{array}$

×	0	1	2	3	4	5	6	7	8	9
0	0	0	0	0	0	0	0	0	0	0
1	0	1	2	3	4	5	6	7	8	9
2	0	2	4	6	8	10	12	14	16	18
3	0	3	6	9	12	15	18	21	24	27
4	0	4	8	12	16	20	24	28	32	36
5	0	5	10	15	20	25	30	35	40	45
6	0	6	12	18	24	30	36	42	48	54
7	0	7	14	21	28	35	42	49	56	63
8	0	8	16	24	32	40	48	56	64	72
9	0	9	18	27	36	45	54	63	72	81

One Hundred Multiplication Facts

Knowledge of the basic multiplication facts found above is an important prerequisite for a course in basic college mathematics with early integers. Study the table above and then perform the multiplications. Check your answers either by comparing them with those found in the back-of-the-book answer section or by using the table. Review any facts that you missed.

1. $\begin{array}{r} 1 \\ \times\,1 \\ \hline \end{array}$ **2.** $\begin{array}{r} 5 \\ \times\,7 \\ \hline \end{array}$ **3.** $\begin{array}{r} 7 \\ \times\,8 \\ \hline \end{array}$ **4.** $\begin{array}{r} 3 \\ \times\,3 \\ \hline \end{array}$ **5.** $\begin{array}{r} 8 \\ \times\,4 \\ \hline \end{array}$ **6.** $\begin{array}{r} 9 \\ \times\,5 \\ \hline \end{array}$

7. $\begin{array}{r} 4 \\ \times\,7 \\ \hline \end{array}$ **8.** $\begin{array}{r} 7 \\ \times\,1 \\ \hline \end{array}$ **9.** $\begin{array}{r} 2 \\ \times\,2 \\ \hline \end{array}$ **10.** $\begin{array}{r} 0 \\ \times\,5 \\ \hline \end{array}$ **11.** $\begin{array}{r} 9 \\ \times\,7 \\ \hline \end{array}$ **12.** $\begin{array}{r} 8 \\ \times\,8 \\ \hline \end{array}$

13. $\begin{array}{r} 3 \\ \times\,2 \\ \hline \end{array}$ **14.** $\begin{array}{r} 6 \\ \times\,0 \\ \hline \end{array}$ **15.** $\begin{array}{r} 5 \\ \times\,6 \\ \hline \end{array}$ **16.** $\begin{array}{r} 2 \\ \times\,5 \\ \hline \end{array}$ **17.** $\begin{array}{r} 4 \\ \times\,6 \\ \hline \end{array}$ **18.** $\begin{array}{r} 0 \\ \times\,7 \\ \hline \end{array}$

19. $\begin{array}{r} 6 \\ \times\,3 \\ \hline \end{array}$ **20.** $\begin{array}{r} 8 \\ \times\,9 \\ \hline \end{array}$ **21.** $\begin{array}{r} 5 \\ \times\,8 \\ \hline \end{array}$ **22.** $\begin{array}{r} 7 \\ \times\,2 \\ \hline \end{array}$ **23.** $\begin{array}{r} 4 \\ \times\,8 \\ \hline \end{array}$ **24.** $\begin{array}{r} 1 \\ \times\,2 \\ \hline \end{array}$

25. $\begin{array}{r} 9 \\ \times\,6 \\ \hline \end{array}$ **26.** $\begin{array}{r} 3 \\ \times\,1 \\ \hline \end{array}$ **27.** $\begin{array}{r} 8 \\ \times\,7 \\ \hline \end{array}$ **28.** $\begin{array}{r} 2 \\ \times\,8 \\ \hline \end{array}$ **29.** $\begin{array}{r} 6 \\ \times\,9 \\ \hline \end{array}$ **30.** $\begin{array}{r} 5 \\ \times\,5 \\ \hline \end{array}$

31. $\begin{array}{r} 2 \\ \times\,1 \\ \hline \end{array}$ **32.** $\begin{array}{r} 8 \\ \times\,0 \\ \hline \end{array}$ **33.** $\begin{array}{r} 4 \\ \times\,9 \\ \hline \end{array}$ **34.** $\begin{array}{r} 8 \\ \times\,3 \\ \hline \end{array}$ **35.** $\begin{array}{r} 6 \\ \times\,2 \\ \hline \end{array}$ **36.** $\begin{array}{r} 4 \\ \times\,5 \\ \hline \end{array}$

37. 9
    ×4

38. 2
    ×9

39. 3
    ×4

40. 1
    ×6

41. 8
    ×6

42. 9
    ×8

43. 1
    ×8

44. 5
    ×1

45. 9
    ×0

46. 7
    ×4

47. 9
    ×3

48. 0
    ×3

49. 3
    ×5

50. 6
    ×8

51. 5
    ×9

52. 2
    ×6

53. 1
    ×0

54. 3
    ×9

55. 9
    ×9

56. 5
    ×4

57. 0
    ×6

58. 1
    ×9

59. 5
    ×0

60. 6
    ×1

61. 9
    ×2

62. 1
    ×7

63. 1
    ×3

64. 7
    ×3

65. 6
    ×6

66. 4
    ×0

67. 7
    ×9

68. 4
    ×3

69. 7
    ×5

70. 2
    ×0

71. 6
    ×7

72. 0
    ×8

73. 8
    ×5

74. 2
    ×4

75. 0
    ×1

76. 3
    ×8

77. 9
    ×1

78. 7
    ×0

79. 5
    ×3

80. 4
    ×4

81. 1
    ×5

82. 6
    ×5

83. 3
    ×0

84. 1
    ×4

85. 3
    ×7

86. 4
    ×2

87. 0
    ×2

88. 7
    ×7

89. 8
    ×2

90. 6
    ×4

91. 0
    ×0

92. 2
    ×7

93. 4
    ×1

94. 0
    ×4

95. 2
    ×3

96. 8
    ×1

97. 3
    ×6

98. 5
    ×2

99. 0
    ×9

100. 7
    ×6

Plane Figures Have Length and Width but No Thickness or Depth		
Name	Description	Figure
Polygon	Union of three or more coplanar line segments that intersect with each other only at each endpoint, with each endpoint shared by two segments.	
Triangle	Polygon with three sides (sum of measures of three angles is 180°).	
Scalene Triangle	Triangle with no sides of equal length.	
Isosceles Triangle	Triangle with two sides of equal length.	
Equilateral Triangle	Triangle with all sides of equal length.	
Right Triangle	Triangle that contains a right angle.	
Quadrilateral	Polygon with four sides (sum of measures of four angles is 360°).	
Trapezoid	Quadrilateral with exactly one pair of opposite sides parallel.	
Isosceles Trapezoid	Trapezoid with legs of equal length.	
Parallelogram	Quadrilateral with both pairs of opposite sides parallel.	
Rhombus	Parallelogram with all sides of equal length.	
Rectangle	Parallelogram with four right angles.	

(*Continued*)

Plane Figures Have Length and Width but No Thickness or Depth (*continued*)

Name	Description	Figure
Square	Rectangle with all sides of equal length.	
Circle	All points in a plane the same distance from a fixed point called the **center.**	radius center diameter

Solid Figures Have Length, Width, and Height or Depth

Name	Description	Figure
Rectangular Solid	A solid with six sides, all of which are rectangles.	
Cube	A rectangular solid whose six sides are squares.	
Sphere	All points the same distance from a fixed point called the **center.**	radius center
Right Circular Cylinder	A cylinder having two circular bases that are perpendicular to its altitude.	
Right Circular Cone	A cone with a circular base that is perpendicular to its altitude.	

Table of Percents, Decimals, and Fraction Equivalents

Percent	Decimal	Fraction
1%	0.01	$\frac{1}{100}$
5%	0.05	$\frac{1}{20}$
10%	0.1	$\frac{1}{10}$
12.5% or $12\frac{1}{2}$%	0.125	$\frac{1}{8}$
$16.\overline{6}$% or $16\frac{2}{3}$%	$0.1\overline{6}$	$\frac{1}{6}$
20%	0.2	$\frac{1}{5}$
25%	0.25	$\frac{1}{4}$
30%	0.3	$\frac{3}{10}$
$33.\overline{3}$% or $33\frac{1}{3}$%	$0.\overline{3}$	$\frac{1}{3}$
37.5% or $37\frac{1}{2}$%	0.375	$\frac{3}{8}$
40%	0.4	$\frac{2}{5}$
50%	0.5	$\frac{1}{2}$
60%	0.6	$\frac{3}{5}$
62.5% or $62\frac{1}{2}$%	0.625	$\frac{5}{8}$
$66.\overline{6}$% or $66\frac{2}{3}$%	$0.\overline{6}$	$\frac{2}{3}$
70%	0.7	$\frac{7}{10}$
75%	0.75	$\frac{3}{4}$
80%	0.8	$\frac{4}{5}$
$83.\overline{3}$% or $83\frac{1}{3}$%	$0.8\overline{3}$	$\frac{5}{6}$
87.5% or $87\frac{1}{2}$%	0.875	$\frac{7}{8}$
90%	0.9	$\frac{9}{10}$
100%	1.0	1
110%	1.1	$1\frac{1}{10}$
125%	1.25	$1\frac{1}{4}$
$133.\overline{3}$% or $133\frac{1}{3}$%	$1.\overline{3}$	$1\frac{1}{3}$
150%	1.5	$1\frac{1}{2}$
$166.\overline{6}$% or $166\frac{2}{3}$%	$1.\overline{6}$	$1\frac{2}{3}$
175%	1.75	$1\frac{3}{4}$
200%	2.0	2

Common Percent Equivalences*	Shortcut Method for Finding Percent	Example
$1\% = 0.01 \left(\text{or } \frac{1}{100}\right)$	To find 1% of a number, multiply by 0.01. To do so, move the decimal point two places to the left.	1% of 210 is 2.10 or 2.1. 1% of 1500 is 15. 1% of 8.6 is 0.086.
$10\% = 0.1 \left(\text{or } \frac{1}{10}\right)$	To find 10% of a number, multiply by 0.1, or move the decimal point of the number one place to the left.	10% of 140 is 14. 10% of 30 is 3. 10% of 17.6 is 1.76.
$25\% = \frac{1}{4}$	To find 25% of a number, find $\frac{1}{4}$ of the number, or divide the number by 4.	25% of 20 is $\frac{20}{4}$ or 5. 25% of 8 is 2. 25% of 10 is $\frac{10}{4}$ or $2\frac{1}{2}$.
$50\% = \frac{1}{2}$	To find 50% of a number, find $\frac{1}{2}$ of the number, or divide the number by 2.	50% of 64 is $\frac{64}{2}$ or 32. 50% of 1000 is 500. 50% of 9 is $\frac{9}{2}$ or $4\frac{1}{2}$.
$100\% = 1$	To find 100% of a number, multiply the number by 1. In other words, 100% of a number is the number.	100% of 98 is 98. 100% of 1407 is 1407. 100% of 18.4 is 18.4.
$200\% = 2$	To find 200% of a number, multiply the number by 2.	200% of 31 is $31 \cdot 2$ or 62. 200% of 750 is 1500. 200% of 6.5 is 13.

*See Appendix A.4.

n	n^2	$\sqrt{n}$	n	n^2	$\sqrt{n}$
1	1	1.000	51	2601	7.141
2	4	1.414	52	2704	7.211
3	9	1.732	53	2809	7.280
4	16	2.000	54	2916	7.348
5	25	2.236	55	3025	7.416
6	36	2.449	56	3136	7.483
7	49	2.646	57	3249	7.550
8	64	2.828	58	3364	7.616
9	81	3.000	59	3481	7.681
10	100	3.162	60	3600	7.746
11	121	3.317	61	3721	7.810
12	144	3.464	62	3844	7.874
13	169	3.606	63	3969	7.937
14	196	3.742	64	4096	8.000
15	225	3.873	65	4225	8.062
16	256	4.000	66	4356	8.124
17	289	4.123	67	4489	8.185
18	324	4.243	68	4624	8.246
19	361	4.359	69	4761	8.307
20	400	4.472	70	4900	8.367
21	441	4.583	71	5041	8.426
22	484	4.690	72	5184	8.485
23	529	4.796	73	5329	8.544
24	576	4.899	74	5476	8.602
25	625	5.000	75	5625	8.660
26	676	5.099	76	5776	8.718
27	729	5.196	77	5929	8.775
28	784	5.292	78	6084	8.832
29	841	5.385	79	6241	8.888
30	900	5.477	80	6400	8.944
31	961	5.568	81	6561	9.000
32	1024	5.657	82	6724	9.055
33	1089	5.745	83	6889	9.110
34	1156	5.831	84	7056	9.165
35	1225	5.916	85	7225	9.220
36	1296	6.000	86	7396	9.274
37	1369	6.083	87	7569	9.327
38	1444	6.164	88	7744	9.381
39	1521	6.245	89	7921	9.434
40	1600	6.325	90	8100	9.487
41	1681	6.403	91	8281	9.539
42	1764	6.481	92	8464	9.592
43	1849	6.557	93	8649	9.644
44	1936	6.633	94	8836	9.695
45	2025	6.708	95	9025	9.747
46	2116	6.782	96	9216	9.798
47	2209	6.856	97	9409	9.849
48	2304	6.928	98	9604	9.899
49	2401	7.000	99	9801	9.950
50	2500	7.071	100	10,000	10.000

Compound Interest Table

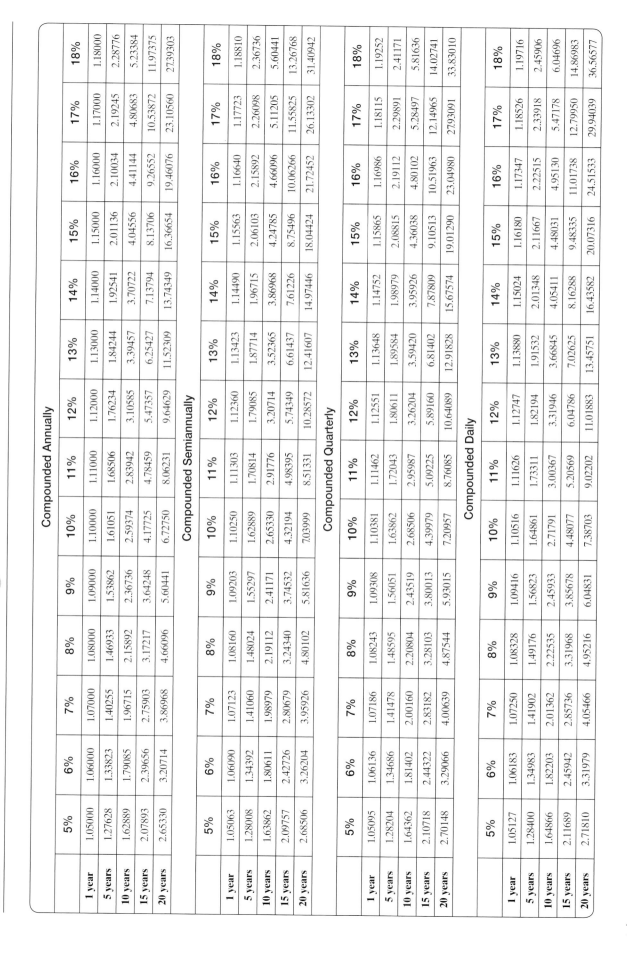

Compounded Annually

	5%	6%	7%	8%	9%	10%	11%	12%	13%	14%	15%	16%	17%	18%
1 year	1.05000	1.06000	1.07000	1.08000	1.09000	1.10000	1.11000	1.12000	1.13000	1.14000	1.15000	1.16000	1.17000	1.18000
5 years	1.27628	1.33823	1.40255	1.46933	1.53862	1.61051	1.68506	1.76234	1.84244	1.92541	2.01136	2.10034	2.19245	2.28776
10 years	1.62889	1.79085	1.96715	2.15892	2.36736	2.59374	2.83942	3.10585	3.39457	3.70722	4.04556	4.41144	4.80683	5.23384
15 years	2.07893	2.39656	2.75903	3.17217	3.64248	4.17725	4.78459	5.47357	6.25427	7.13794	8.13706	9.26552	10.53872	11.97375
20 years	2.65330	3.20714	3.86968	4.66096	5.60441	6.72750	8.06231	9.64629	11.52309	13.74349	16.36654	19.46076	23.10560	27.39303

Compounded Semiannually

	5%	6%	7%	8%	9%	10%	11%	12%	13%	14%	15%	16%	17%	18%
1 year	1.05063	1.06090	1.07123	1.08160	1.09203	1.10250	1.11303	1.12360	1.13423	1.14490	1.15563	1.16640	1.17723	1.18810
5 years	1.28008	1.34392	1.41060	1.48024	1.55297	1.62889	1.70814	1.79085	1.87714	1.96715	2.06103	2.15892	2.26098	2.36736
10 years	1.63862	1.80611	1.98979	2.19112	2.41171	2.65330	2.91776	3.20714	3.52365	3.86968	4.24785	4.66096	5.11205	5.60441
15 years	2.09757	2.42726	2.80679	3.24340	3.74532	4.32194	4.98395	5.74349	6.61437	7.61226	8.75496	10.06266	11.55825	13.26768
20 years	2.68506	3.26204	3.95926	4.80102	5.81636	7.03999	8.51331	10.28572	12.41607	14.97446	18.04424	21.72452	26.13302	31.40942

Compounded Quarterly

	5%	6%	7%	8%	9%	10%	11%	12%	13%	14%	15%	16%	17%	18%
1 year	1.05095	1.06136	1.07186	1.08243	1.09308	1.10381	1.11462	1.12551	1.13648	1.14752	1.15865	1.16986	1.18115	1.19252
5 years	1.28204	1.34686	1.41478	1.48595	1.56051	1.63862	1.72043	1.80611	1.89584	1.98979	2.08815	2.19112	2.29891	2.41171
10 years	1.64362	1.81402	2.00160	2.20804	2.43519	2.68506	2.95987	3.26204	3.59420	3.95926	4.36038	4.80102	5.28497	5.81636
15 years	2.10718	2.44322	2.83182	3.28103	3.80013	4.39979	5.09225	5.89160	6.81402	7.87809	9.10513	10.51963	12.14965	14.02741
20 years	2.70148	3.29066	4.00639	4.87544	5.93015	7.20957	8.76085	10.64089	12.91828	15.67574	19.01290	23.04980	27.93091	33.83010

Compounded Daily

	5%	6%	7%	8%	9%	10%	11%	12%	13%	14%	15%	16%	17%	18%
1 year	1.05127	1.06183	1.07250	1.08328	1.09416	1.10516	1.11626	1.12747	1.13880	1.15024	1.16180	1.17347	1.18526	1.19716
5 years	1.28400	1.34983	1.41902	1.49176	1.56823	1.64861	1.73311	1.82194	1.91532	2.01348	2.11667	2.22515	2.33918	2.45906
10 years	1.64866	1.82203	2.01362	2.22535	2.45933	2.71791	3.00367	3.31946	3.66845	4.05411	4.48031	4.95130	5.47178	6.04696
15 years	2.11689	2.45942	2.85736	3.31968	3.85678	4.48077	5.20569	6.04786	7.02625	8.16288	9.48335	11.01738	12.79950	14.86983
20 years	2.71810	3.31979	4.05466	4.95216	6.04831	7.38703	9.02202	11.01883	13.45751	16.43582	20.07316	24.51533	29.94039	36.56577

Exponents and Polynomials

Adding and Subtracting Polynomials

Objectives

A Add Polynomials.

B Subtract Polynomials.

C Evaluate Polynomials at Given Replacement Values.

Before we add and subtract polynomials, let's first review some definitions presented in Section 8.1. Recall that the *addends* of an algebraic expression are the *terms* of the expression.

Expression

$$3x + 5 \qquad\qquad 7y^2 + (-6y) + 4$$

$3x + 5$ — 2 terms $7y^2 + (-6y) + 4$ — 3 terms

Also, recall that *like terms* can be added or subtracted by using the distributive property. For example,

$$7x + 3x = (7 + 3)x = 10x$$

Objective A Adding Polynomials

Some terms are also **monomials.** A term is a monomial if the term contains only whole number exponents and no variable in the denominator.

Monomials	**Not Monomials**	
$3x^2$	$\dfrac{2}{y}$	Variable in denominator
$-\dfrac{1}{2}a^2bc^3$	$-2x^{-5}$	Not a whole number exponent
7		

A monomial or a sum and/or difference of monomials is called a **polynomial.**

Polynomial

A **polynomial** is a monomial or a sum and/or difference of monomials.

Examples of Polynomials

$$5x^3 - 6x^2 + 2x + 10, \quad -1.2y^3 + 0.7y, \quad z, \quad \frac{1}{3}r - \frac{1}{2}, \quad 0$$

Some polynomials are given special names depending on their number of terms.

Types of Polynomials

A **monomial** is a polynomial with exactly one term.
A **binomial** is a polynomial with exactly two terms.
A **trinomial** is a polynomial with exactly three terms.

Below are examples of monomials, binomials, and trinomials. Each of these examples is also a polynomial.

Polynomials			
Monomials	**Binomials**	**Trinomials**	**More than Three Terms**
z	$x + 2$	$x^2 - 2x + 1$	$5x^3 - 6x^2 + 2x - 10$
4	$\dfrac{1}{3}r - \dfrac{1}{2}$	$y^5 + 3y^2 - 1.7$	$t^7 - t^5 + t^3 - t + 1$
$0.2x^2$	$-1.2y^3 + 0.7y$	$-a^3 + 2a^2 - 5a$	$z^8 - z^4 + 3z^2 - 2z$

↑ 1 term   ↑ 2 terms   ↑ 3 terms

To add polynomials, we use the commutative and associative properties to rearrange and group like terms. Then, we combine like terms.

Adding Polynomials

To add polynomials, combine like terms.

Example 1 Add: $(3x - 1) + (-6x + 2)$

Solution:

$$(3x - 1) + (-6x + 2) = (3x - 6x) + (-1 + 2) \quad \text{Group like terms.}$$
$$= (-3x) + (1) \quad \text{Combine like terms.}$$
$$= -3x + 1$$

■ Work Practice 1

Practice 1

Add: $(2y + 7) + (9y - 14)$

Example 2 Add: $(9y^2 - 6y) + (7y^2 + 10y + 2)$

Solution:

$$(9y^2 - 6y) + (7y^2 + 10y + 2) = 9y^2 + 7y^2 - 6y + 10y + 2 \quad \text{Group like terms.}$$
$$= 16y^2 + 4y + 2$$

■ Work Practice 2

Practice 2

Add:
$(5x^2 + 4x - 3) + (x^2 - 6x)$

Example 3 Find the sum of $(-y^2 + 2y + 1.7)$ and $(12y^2 - 6y - 3.6)$.

Solution: Recall that "sum" means addition.

$$(-y^2 + 2y + 1.7) + (12y^2 - 6y - 3.6)$$
$$= \underbrace{-y^2 + 12y^2} + \underbrace{2y - 6y} + \underbrace{1.7 - 3.6} \quad \text{Group like terms.}$$
$$= 11y^2 - 4y - 1.9 \quad \text{Combine like terms.}$$

■ Work Practice 3

Practice 3

Find the sum of
$(7z^2 - 4.2z + 11)$ and
$(-9z^2 - 1.9z + 4)$.

Polynomials can also be added vertically. To do this, line up like terms underneath one another. Let's vertically add the polynomials in Example 3.

Answers
1. $11y - 7$ **2.** $6x^2 - 2x - 3$
3. $-2z^2 - 6.1z + 15$

Practice 4

Add the polynomials in Practice 3 vertically.

Example 4   Find the sum of $(-y^2 + 2y + 1.7)$ and $(12y^2 - 6y - 3.6)$. Use a vertical format.

Solution: Line up like terms underneath one another.

$$\begin{array}{r} -y^2 + 2y + 1.7 \\ +12y^2 - 6y - 3.6 \\ \hline 11y^2 - 4y - 1.9 \end{array}$$

■ Work Practice 4

Notice that we are finding the same sum in Example 4 as we found in Example 3. Of course, the results are the same.

Objective B Subtracting Polynomials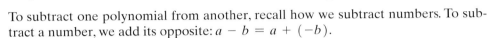

To subtract one polynomial from another, recall how we subtract numbers. To subtract a number, we add its opposite: $a - b = a + (-b)$.

For example,

$$7 - 10 = 7 + (-10)$$
$$= -3$$

To subtract a polynomial, we also add its opposite. Just as the opposite of 3 is -3, the opposite of $(2x^2 - 5x + 1)$ is $-(2x^2 - 5x + 1)$. Let's practice simplifying the opposite of a polynomial.

Practice 5

Simplify: $-(7y^2 + 4y - 6)$

Example 5   Simplify: $-(2x^2 - 5x + 1)$

Solution: Rewrite $-(2x^2 - 5x + 1)$ as $-1(2x^2 - 5x + 1)$ and use the distributive property.

$$\begin{aligned} -(2x^2 - 5x + 1) &= -1(2x^2 - 5x + 1) \\ &= -1(2x^2) + (-1)(-5x) + (-1)(1) \\ &= -2x^2 + 5x - 1 \end{aligned}$$

■ Work Practice 5

Notice the result of Example 5.

$$-(2x^2 - 5x + 1) = -2x^2 + 5x - 1$$

This means that **the opposite of a polynomial can be found by changing the signs of the terms of the polynomial.** This leads to the following.

Subtracting Polynomials

To subtract polynomials, change the signs of the terms of the polynomial being subtracted, and then add.

Practice 6

Subtract:
$(3b - 2) - (7b + 23)$

Example 6   Subtract: $(5a + 7) - (2a - 10)$

Solution:

$$\begin{aligned} (5a + 7) - (2a - 10) &= (5a + 7) + (-2a + 10) && \text{Add the opposite of } 2a - 10. \\ &= 5a - 2a + 7 + 10 && \text{Group like terms.} \\ &= 3a + 17 \end{aligned}$$

Answers
4. $-2z^2 - 6.1z + 15$
5. $-7y^2 - 4y + 6$
6. $-4b - 25$

■ Work Practice 6

Example 7 Subtract: $(8x^2 - 4x + 1) - (10x^2 + 4)$

Solution:

$$(8x^2 - 4x + 1) - (10x^2 + 4) = (8x^2 - 4x + 1) + (-10x^2 - 4) \quad \text{Add the opposite of } 10x^2 + 4.$$

$$= 8x^2 - 10x^2 - 4x + 1 - 4 \quad \text{Group like terms.}$$

$$= -2x^2 - 4x - 3$$

■ Work Practice 7

Practice 7

Subtract:

$(11x^2 + 7x + 2) - (15x^2 + 4x)$

Example 8 Subtract $(-6z^2 - 2z + 13)$ from $(4z^2 - 20z)$.

Solution: Be careful when arranging the polynomials in this example.

$$(4z^2 - 20z) - (-6z^2 - 2z + 13) = (4z^2 - 20z) + (6z^2 + 2z - 13)$$

$$= 4z^2 + 6z^2 - 20z + 2z - 13 \quad \text{Group like terms.}$$

$$= 10z^2 - 18z - 13$$

■ Work Practice 8

Practice 8

Subtract $(3x^2 - 12x)$ from $(-4x^2 + 20x + 17)$.

✓**Concept Check** Find and explain the error in the following subtraction.

$$(3x^2 + 4) - (x^2 - 3x)$$
$$= (3x^2 + 4) + (-x^2 - 3x)$$
$$= 3x^2 - x^2 - 3x + 4$$
$$= 2x^2 - 3x + 4$$

Just as with adding polynomials, we can subtract polynomials using a vertical format. Let's subtract the polynomials in Example 8 using a vertical format.

Example 9 Subtract $(-6z^2 - 2z + 13)$ from $(4z^2 - 20z)$. Use a vertical format.

Solution: Line up like terms underneath one another.

$$\begin{array}{r} 4z^2 - 20z \\ -(-6z^2 - 2z + 13) \\ \end{array} \qquad \begin{array}{r} 4z^2 - 20z \\ +6z^2 + 2z - 13 \\ \hline 10z^2 - 18z - 13 \end{array}$$

■ Work Practice 9

Practice 9

Subtract $(3x^2 - 12x)$ from $(-4x^2 + 20x + 17)$. Use a vertical format.

Objective C Evaluating Polynomials ▶

Polynomials have different values depending on the replacement values for the variables.

Example 10 Find the value of the polynomial $3t^3 - 2t + 5$ when $t = 1$.

Solution: Replace t with 1 and simplify.

$$3t^3 - 2t + 5 = 3(1)^3 - 2(1) + 5 \quad \text{Let } t = 1.$$
$$= 3(1) - 2 + 5 \quad (1)^3 = 1.$$
$$= 3 - 2 + 5$$
$$= 6$$

The value of $3t^3 - 2t + 5$ when $t = 1$ is 6.

■ Work Practice 10

Practice 10

Find the value of the polynomial $2y^3 + y^2 - 6$ when $y = 3$.

Answers

7. $-4x^2 + 3x + 2$ **8.** $-7x^2 + 32x + 17$
9. $-7x^2 + 32x + 17$ **10.** 57

✓**Concept Check Answer**

$$(3x^2 + 4) - (x^2 - 3x)$$
$$= (3x^2 + 4) + (-x^2 + 3x)$$
$$= 3x^2 - x^2 + 3x + 4$$
$$= 2x^2 + 3x + 4$$

Many real-world applications are modeled by polynomials.

An object is dropped from the top of a 530-foot cliff. Its height in feet at time t seconds is given by the polynomial $-16t^2 + 530$. Find the height of the object when $t = 1$ second and when $t = 4$ seconds.

Example 11 Finding the Height of an Object

An object is dropped from the top of an 800-foot-tall building. Its height at time t seconds is given by the polynomial $-16t^2 + 800$. Find the height of the object when $t = 1$ second and when $t = 3$ seconds.

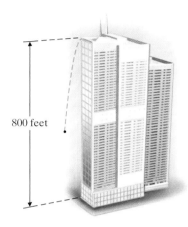

800 feet

Solution: To find each height, we evaluate the polynomial when $t = 1$ and when $t = 3$.

$$-16t^2 + 800 = -16(1)^2 + 800$$
$$= -16 + 800$$
$$= 784$$

The height of the object at 1 second is 784 feet.

$$-16t^2 + 800 = -16(3)^2 + 800$$
$$= -16(9) + 800$$
$$= -144 + 800$$
$$= 656$$

Helpful Hint

Don't forget to insert units, if appropriate.

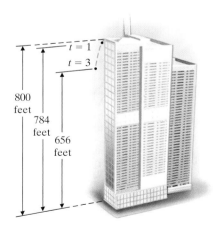

800 feet
784 feet
656 feet
$t = 1$
$t = 3$

The height of the object at 3 seconds is 656 feet.

▣ Work Practice 11

Answer
11. 514 feet; 274 feet

B.1 **Exercise Set** MyMathLab®

Objective A *Add the polynomials. See Examples 1 through 4.*

1. $(2x + 3) + (-7x - 27)$

2. $(9y - 16) + (-43y + 16)$

3. $(-4z^2 - 6z + 1) + (-5z^2 + 4z + 5)$

4. $(17a^2 - 6a + 3) + (16a^2 - 6a - 10)$

5. $(12y - 20) + (9y^2 + 13y - 20)$

6. $(5x^2 - 6) + (-3x^2 + 17x - 2)$

7. $(4.3a^4 + 5) + (-8.6a^4 - 2a^2 + 4)$

8. $(-12.7z^3 - 14z) + (-8.9z^3 + 12z + 2)$

Objective B *Subtract the polynomials. See Examples 5 through 9.*

9. $(5a - 6) - (a + 2)$

10. $(12b + 7) - (-b - 5)$

11. $(3x^2 - 2x + 1) - (5x^2 - 6x)$

12. $(-9z^2 + 6z + 2) - (3z^2 + 1)$

13. $(10y^2 - 7) - (20y^3 - 2y^2 - 3)$

14. $(11x^3 + 15x - 9) - (-x^3 + 10x^2 - 9)$

15. Subtract $(3x - 4)$ from $(2x + 12)$.

16. Subtract $(6a + 1)$ from $(-7a + 7)$.

17. Subtract $(5y^2 + 4y - 6)$ from $(13y^2 - 6y - 14)$.

18. Subtract $(16x^2 - x + 1)$ from $(12x^2 - 3x - 12)$.

Objectives A B **Mixed Practice** *Perform each indicated operation. See Examples 1 through 9.*

19. $(25x - 5) + (-20x - 7)$

20. $(14x + 2) + (-7x - 1)$

21. $(4y + 4) - (3y + 8)$

22. $(6z - 3) - (8z + 5)$

23. $(9x^2 - 6) + (-5x^2 + x - 10)$

24. $(12a^2 - 4a - 4) + (-5a - 5)$

25. $(10x + 4.5) + (-x - 8.6)$

26. $(20x - 0.8) + (x + 1.2)$

27. $(12a - 5) - (-3a + 2)$

28. $(8t + 9) - (-2t + 6)$

29. $(21y - 4.6) - (36y - 8.2)$

30. $(8.6x + 4) - (9.7x - 93)$

31. $(18t^2 - 4t + 2) - (-t^2 + 7t - 1)$

32. $(35x^2 + x - 5) - (17x^2 - x + 5)$

33. $(b^3 - 2b^2 + 10b + 11) + (b^2 - 3b - 12)$

34. $(-2z^3 + 5z^2 - 13z + 6) + (3z^2 - 7z - 6)$

35. Add $(6x^2 - 7)$ and $(-11x^2 - 11x + 20)$.

36. Add $(-2x^2 + 3x)$ and $(9x^2 - x + 14)$.

37. Subtract $\left(3z - \dfrac{3}{7}\right)$ from $\left(3z + \dfrac{6}{7}\right)$.

38. Subtract $\left(8y^2 - \dfrac{7}{10}y\right)$ from $\left(-5y^2 + \dfrac{3}{10}y\right)$.

Objective C *Find the value of each polynomial when $x = 2$. See Examples 10 and 11.*

39. $-3x + 7$

40. $-5x - 7$

41. $x^2 - 6x + 3$

42. $5x^2 + 4x - 100$

43. $\dfrac{3x^2}{2} - 14$

44. $\dfrac{7x^3}{14} - x + 5$

Find the value of each polynomial when $x = 5$. See Examples 10 and 11.

45. $2x + 10$

46. $-5x - 6$

47. x^2

48. x^3

49. $2x^2 + 4x - 20$

50. $4x^2 - 5x + 10$

Solve. See Example 11.

The distance in feet traveled by a free-falling object in t seconds is given by the polynomial

$$16t^2$$

Use this polynomial for Exercises 51 and 52.

51. Find the distance traveled by an object that falls for 6 seconds.

52. It takes 8 seconds for a hard hat to fall from the top of a building. How high is the building?

Office Supplies, Inc. manufactures office products. They determine that the total cost for manufacturing x file cabinets is given by the polynomial

$$3000 + 20x$$

Use this polynomial for Exercises 53 and 54.

53. Find the total cost to manufacture 10 file cabinets.

54. Find the total cost to manufacture 100 file cabinets.

An object is dropped from the deck of the Royal Gorge Bridge, which stretches across Royal Gorge at a height of 1053 feet above the Arkansas River. The height of the object above the river after t seconds is given by the polynomial

$$1053 - 16t^2$$

Use this polynomial for Exercises 55 and 56. (Source: Royal Gorge Bridge Co.)

55. How far above the river is an object that has been falling for 3 seconds?

56. How far above the river is an object that has been falling for 7 seconds?

Concept Extensions

Find the perimeter of each figure.

△ **57.**
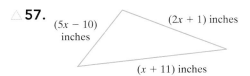
$(5x - 10)$ inches, $(2x + 1)$ inches, $(x + 11)$ inches

△ **58.**
$(x^2 - 6)$ meters, $(x + 1)$ meters, $(3x - 10)$ meters, $(5x^2 + 2x)$ meters

Given the lengths in the figure below, we find the unknown length by subtracting. Use the information to find the unknown lengths in Exercises 59 and 60.

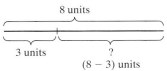

8 units
3 units
?
$(8 - 3)$ units

59.
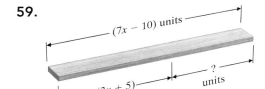
$(7x - 10)$ units, $(3x + 5)$ units, ? units

60.
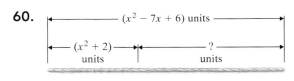
$(x^2 - 7x + 6)$ units, $(x^2 + 2)$ units, ? units

Fill in the blanks.

61. $(3x^2 + \underline{\quad} x - \underline{\quad}) + (\underline{\quad} x^2 - 6x + 2) = 5x^2 + 14x - 4$

62. $(\underline{\quad} y^2 + 4y - 3) + (8y^2 - \underline{\quad} y + \underline{\quad}) = 9y^2 + 2y + 7$

63. Find the value of $7a^4 - 6a^2 + 2a - 1$ when $a = 1.2$.

64. Find the value of $3b^3 + 4b^2 - 100$ when $b = -2.5$.

65. For Exercises 55 and 56, the polynomial $1053 - 16t^2$ was used to give the height of an object above the river after t seconds. Find the height when $t = 8$ seconds and $t = 9$ seconds. Explain what happened and why.

Objectives

A Use the Product Rule for Exponents.

B Use the Power Rule for Exponents.

C Use the Power of a Product Rule for Exponents.

Objective A Using the Product Rule

Recall from Section 8.1 that an exponent has the same meaning whether the base is a number or a variable. For example,

$$5^3 = \underbrace{5 \cdot 5 \cdot 5}_{\text{3 factors of 5}} \quad \text{and} \quad x^3 = \underbrace{x \cdot x \cdot x}_{\text{3 factors of } x}$$

We can use this definition of an exponent to discover properties that will help us to simplify products and powers of exponential expressions.

For example, let's use the definition of an exponent to find the product of x^3 and x^4.

$$x^3 \cdot x^4 = (x \cdot x \cdot x)(x \cdot x \cdot x \cdot x)$$
$$= \underbrace{x \cdot x \cdot x \cdot x \cdot x \cdot x \cdot x}_{\text{7 factors of } x}$$
$$= x^7$$

Notice that the result is the same if we add the exponents.

$$x^3 \cdot x^4 = x^{3+4} = x^7$$

This suggests the following product rule or property for exponents.

Product Property for Exponents

If m and n are positive integers and a is a real number, then

$$a^m \cdot a^n = a^{m+n}$$

In other words, to multiply two exponential expressions with the same base, keep the base and add the exponents.

Practice 1

Multiply: $z^4 \cdot z^8$

Example 1 Multiply: $y^7 \cdot y^2$

Solution:

$$y^7 \cdot y^2 = y^{7+2} \quad \text{Use the product property for exponents.}$$
$$= y^9 \quad \text{Simplify.}$$

■ Work Practice 1

Practice 2

Multiply: $7y^5 \cdot 4y^9$

Example 2 Multiply: $3x^5 \cdot 6x^3$

Solution:

$$3x^5 \cdot 6x^3 = (3 \cdot 6)(x^5 \cdot x^3) \quad \text{Apply the commutative and associative properties.}$$
$$= 18x^{5+3} \quad \text{Use the product property for exponents.}$$
$$= 18x^8 \quad \text{Simplify.}$$

■ Work Practice 2

Answers

1. z^{12} **2.** $28y^{14}$

Example 3   Multiply: $(-2a^4b^{10})(9a^5b^3)$

Solution: Use properties of multiplication to group numbers and like variables together.

$$\begin{aligned}(-2a^4b^{10})(9a^5b^3) &= (-2 \cdot 9)(a^4 \cdot a^5)(b^{10} \cdot b^3) \\ &= -18a^{4+5}b^{10+3} \\ &= -18a^9b^{13}\end{aligned}$$

■ Work Practice 3

Example 4   Multiply: $2x^3 \cdot 3x \cdot 5x^6$

Solution: First notice the factor $3x$. Since there is one factor of x in $3x$, it can also be written as $3x^1$.

$$\begin{aligned}2x^3 \cdot 3x^1 \cdot 5x^6 &= (2 \cdot 3 \cdot 5)(x^3 \cdot x^1 \cdot x^6) \\ &= 30x^{10}\end{aligned}$$

■ Work Practice 4

Practice 3

Multiply: $(-7r^6s^2)(-3r^2s^5)$

Practice 4

Multiply: $9y^4 \cdot 3y^2 \cdot y$. (Recall that $y = y^1$.)

Helpful Hint   Don't forget that if an exponent is not written, it is assumed to be 1.

Helpful Hint

These examples will remind you of the difference between adding and multiplying terms.

Addition

$5x^3 + 3x^3 = (5 + 3)x^3 = 8x^3$

$7x + 4x^2 = 7x + 4x^2$

Multiplication

$(5x^3)(3x^3) = 5 \cdot 3 \cdot x^3 \cdot x^3 = 15x^{3+3} = 15x^6$

$(7x)(4x^2) = 7 \cdot 4 \cdot x \cdot x^2 = 28x^{1+2} = 28x^3$

Objective B   Using the Power Rule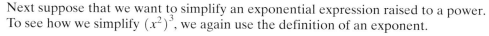

Next suppose that we want to simplify an exponential expression raised to a power. To see how we simplify $(x^2)^3$, we again use the definition of an exponent.

$$(x^2)^3 = \underbrace{(x^2) \cdot (x^2) \cdot (x^2)}_{\text{3 factors of } x^2} \quad \text{Apply the definition of an exponent.}$$

$$= x^{2+2+2} \qquad\qquad\qquad \text{Use the product property for exponents.}$$

$$= x^6 \qquad\qquad\qquad\quad \text{Simplify.}$$

Notice the result is exactly the same if we multiply the exponents.

$$(x^2)^3 = x^{2 \cdot 3} = x^6$$

This suggests the following power rule or property for exponents.

Power Property for Exponents

If m and n are positive integers and a is a real number, then

$$(a^m)^n = a^{m \cdot n}$$

Answers

3. $21r^8s^7$ **4.** $27y^7$

In other words, to raise a power to a power, keep the base and multiply the exponents.

Helpful Hint

Take a moment to make sure that you understand when to apply the product rule and when to apply the power rule.

Product Property → Add Exponents	Power Property → Multiply Exponents
$x^5 \cdot x^7 = x^{5+7} = x^{12}$	$(x^5)^7 = x^{5 \cdot 7} = x^{35}$
$y^6 \cdot y^2 = y^{6+2} = y^8$	$(y^6)^2 = y^{6 \cdot 2} = y^{12}$

Practice 5

Simplify: $(z^3)^{10}$

Example 5 Simplify: $(y^8)^2$

Solution:

$$(y^8)^2 = y^{8 \cdot 2} \quad \text{Use the power property.}$$
$$= y^{16}$$

■ Work Practice 5

Practice 6

Simplify: $(z^4)^5 \cdot (z^3)^7$

Example 6 Simplify: $(a^3)^4 \cdot (a^2)^9$

Solution:

$$(a^3)^4 \cdot (a^2)^9 = a^{12} \cdot a^{18} \quad \text{Use the power property.}$$
$$= a^{12+18} \quad \text{Use the product property.}$$
$$= a^{30} \quad \text{Simplify.}$$

■ Work Practice 6

Objective C Using the Power of a Product Rule

Next, let's simplify the power of a product.

$$(xy)^3 = xy \cdot xy \cdot xy \quad \text{Apply the definition of an exponent.}$$
$$= (x \cdot x \cdot x)(y \cdot y \cdot y) \quad \text{Group like bases.}$$
$$= x^3 y^3 \quad \text{Simplify.}$$

Notice that the power of a product can be written as the product of powers. This leads to the following power of a product rule or property.

Power of a Product Property for Exponents

If n is a positive integer and a and b are real numbers, then

$$(ab)^n = a^n b^n$$

In other words, to raise a product to a power, raise each factor to the power.

Answers

5. z^{30} **6.** z^{41}

✓**Concept Check** Which property is needed to simplify $(x^6)^3$? Explain.

a. Product property for exponents

b. Power property for exponents

c. Power of a product property for exponents

Example 7 Simplify: $(5t)^3$

Solution:

$(5t)^3 = 5^3 t^3$ Apply the power of a product property.

$\qquad = 125t^3$ Write 5^3 as 125.

■ Work Practice 7

Practice 7

Simplify: $(3b)^4$

Example 8 Simplify: $(2a^5b^3)^3$

Solution:

$(2a^5b^3)^3 = 2^3(a^5)^3(b^3)^3$ Apply the power of a product property.

$\qquad = 8a^{15}b^9$ Apply the power property.

■ Work Practice 8

Practice 8

Simplify: $(4x^2y^6)^3$

Example 9 Simplify: $(3y^4z^2)^4(2y^3z^5)^5$

Solution:

$(3y^4z^2)^4(2y^3z^5)^5 = 3^4(y^4)^4(z^2)^4 \cdot 2^5(y^3)^5(z^5)^5$ Apply the power of a product property.

$\qquad = 81y^{16}z^8 \cdot 32y^{15}z^{25}$ Apply the power property.

$\qquad = (81 \cdot 32)(y^{16} \cdot y^{15})(z^8 \cdot z^{25})$ Group like bases.

$\qquad = 2592y^{31}z^{33}$ Apply the product property.

■ Work Practice 9

Practice 9

Simplify: $(2x^2y^4)^4(3x^6y^9)^2$

Answers

7. $81b^4$ **8.** $64x^6y^{18}$ **9.** $144x^{20}y^{34}$

✓**Concept Check Answer**

b

B.2 **Exercise Set** MyMathLab®

Objective A *Multiply. See Examples 1 through 4.*

1. $x^5 \cdot x^9$

2. $y^4 \cdot y^7$

3. $a^6 \cdot a$

4. $b \cdot b^8$

5. $3z^3 \cdot 5z^2$

6. $8r^2 \cdot 2r^{15}$

7. $-4x \cdot 10x$

8. $-9y \cdot 3y$

9. $(-5x^2y^3)(-5x^4y)$

10. $(-2xy^4)(-6x^3y^7)$

11. $(7ab)(4a^4b^5)$

12. $(3a^3b^6)(12a^2b^9)$

13. $2x \cdot 3x \cdot 7x$

14. $4y \cdot 3y \cdot 5y$

15. $a \cdot 4a^{11} \cdot 3a^5$

16. $b \cdot 7b^{10} \cdot 5b^8$

Objectives A B C Mixed Practice *Simplify. See Examples 1 through 9.*

17. $(x^5)^3$

18. $(y^4)^7$

19. $(z^2)^{10}$

20. $(a^6)^9$

21. $(b^7)^6(b^2)^{10}$

22. $(x^2)^9 \cdot (x^5)^3$

23. $(3a)^4$

24. $(2y)^5$

25. $(a^{11}b^8)^3$

26. $(x^7y^4)^8$

27. $(11x^3y^6)^2$

28. $(9a^4b^3)^2$

29. $(-3y)(2y^7)^3$

30. $(-2x)(5x^2)^4$

31. $(4xy)^3(2x^3y^5)^2$

32. $(2xy)^4(3x^4y^3)^3$

Concept Extensions

Find the area of each figure.

△ **33.**

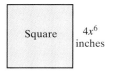

Square $4x^6$ inches

△ **34.** $9y^2$ centimeters

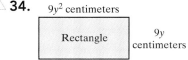

Rectangle $9y$ centimeters

△ **35.**

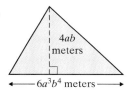

$4ab$ meters

$6a^3b^4$ meters

△ **36.**

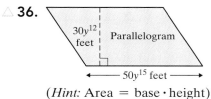

$30y^{12}$ feet   Parallelogram

$50y^{15}$ feet

(*Hint:* Area = base · height)

Multiply and simplify.

▦ **37.** $(14a^7b^6)^3(9a^6b^3)^4$

▦ **38.** $(5x^{14}y^6)^7(3x^{20}y^{19})^5$

39. $(8.1x^{10})^5$

40. $(4.6a^{14})^4$

41. $(a^{20}b^{10}c^5)^5(a^9b^{12})^3$

42. $(x^{90}y^{72})^3$

✎ **43.** In your own words, explain why $x^2 \cdot x^3 = x^5$ and $(x^2)^3 = x^6$.

Objective A Multiplying a Monomial and a Polynomial

Recall that a polynomial that consists of one term is called a **monomial.** For example, $5x$ is a monomial. To multiply a monomial and any polynomial, we use the distributive property

$$a(b + c) = a \cdot b + a \cdot c$$

and apply properties of exponents.

Example 1 Multiply: $5x(3x^2 + 2)$

Solution:

$$5x(3x^2 + 2) = 5x \cdot 3x^2 + 5x \cdot 2 \quad \text{Apply the distributive property.}$$
$$= 15x^3 + 10x$$

■ Work Practice 1

Example 2 Multiply: $2z(4z^2 + 6z - 9)$

Solution:

$$2z(4z^2 + 6z - 9) = 2z \cdot 4z^2 + 2z \cdot 6z + 2z(-9)$$
$$= 8z^3 + 12z^2 - 18z$$

■ Work Practice 2

To visualize multiplication by a monomial, let's look at two ways we can represent the area of the same rectangle.

The width of the rectangle is x and its length is $x + 3$. One way to calculate the area of the rectangle is

area = width · length
$$= x(x + 3)$$

Another way to calculate the area of the rectangle is to find the sum of the areas of the smaller figures.

area = $x^2 + 3x$

Since the areas must be equal, we have that

$$x(x + 3) = x^2 + 3x \quad \text{As expected by the distributive property}$$

Objectives

A Multiply a Monomial and Any Polynomial.

B Multiply Two Binomials.

C Square a Binomial.

D Use the FOIL Order to Multiply Binomials.

E Multiply Any Two Polynomials.

Practice 1
Multiply: $3y(7y^2 + 5)$

Practice 2
Multiply: $5r(8r^2 - r + 11)$

Answers
1. $21y^3 + 15y$ **2.** $40r^3 - 5r^2 + 55r$

Objective B Multiplying Binomials

Recall from Appendix B.1 that a polynomial that consists of exactly two terms is called a **binomial.** To multiply two binomials, we use a version of the distributive property:

$$(b + c)a = b \cdot a + c \cdot a$$

Practice 3

Multiply: $(b + 7)(b + 5)$

Example 3 Multiply: $(x + 2)(x + 3)$

Solution:

$$
\begin{aligned}
(x + 2)(x + 3) &= x(x + 3) + 2(x + 3) && \text{Apply the distributive property.} \\
&= x \cdot x + x \cdot 3 + 2 \cdot x + 2 \cdot 3 && \text{Apply the distributive property.} \\
&= x^2 + 3x + 2x + 6 && \text{Multiply.} \\
&= x^2 + 5x + 6 && \text{Combine like terms.}
\end{aligned}
$$

■ Work Practice 3

Practice 4

Multiply: $(5x - 1)(5x + 4)$

Example 4 Multiply: $(4y + 9)(3y - 2)$

Solution:

$$
\begin{aligned}
(4y + 9)(3y - 2) &= 4y(3y - 2) + 9(3y - 2) && \text{Apply the distributive property.} \\
&= 4y \cdot 3y + 4y(-2) + 9 \cdot 3y + 9(-2) && \text{Apply the distributive property.} \\
&= 12y^2 - 8y + 27y - 18 && \text{Multiply.} \\
&= 12y^2 + 19y - 18 && \text{Combine like terms}
\end{aligned}
$$

■ Work Practice 4

Objective C Squaring a Binomial

Raising a binomial to the power of 2 is also called squaring a binomial. To square a binomial, we use the definition of an exponent, and then multiply.

Practice 5

Multiply: $(6y - 1)^2$

Example 5 Multiply: $(2x + 1)^2$

Solution:

$$
\begin{aligned}
(2x + 1)^2 &= (2x + 1)(2x + 1) && \text{Apply the definition of an exponent.} \\
&= 2x(2x + 1) + 1(2x + 1) && \text{Apply the distributive property.} \\
&= 2x \cdot 2x + 2x \cdot 1 + 1 \cdot 2x + 1 \cdot 1 && \text{Apply the distributive property.} \\
&= 4x^2 + 2x + 2x + 1 && \text{Multiply.} \\
&= 4x^2 + 4x + 1 && \text{Combine like terms.}
\end{aligned}
$$

■ Work Practice 5

Objective D Using the FOIL Order to Multiply Binomials

Recall from Example 3 that

$$
\begin{aligned}
(x + 2)(x + 3) &= x \cdot x + x \cdot 3 + 2 \cdot x + 2 \cdot 3 \\
&= x^2 + 5x + 6
\end{aligned}
$$

One way to remember these products—$x \cdot x$, $x \cdot 3$, $2 \cdot x$, and $2 \cdot 3$—is to use a special order for multiplying binomials called the FOIL order. Of course, the product is the same no matter what order or method you choose to use.

FOIL stands for the products of the First terms, Outer terms, Inner terms, and then Last terms. For example,

$$(x + 2)(x + 3) = x \cdot x + x \cdot 3 + 2 \cdot x + 2 \cdot 3 = x^2 + 3x + 2x + 6$$
$$= x^2 + 5x + 6$$

Helpful Hint   The product is the same no matter what order or method you choose to use.

Examples   Use the FOIL order to multiply.

6. $(3x - 6)(2x + 1) = 3x \cdot 2x + 3x \cdot 1 + (-6)(2x) + (-6)(1)$
$$= 6x^2 + 3x - 12x - 6 \quad \text{Multiply.}$$
$$= 6x^2 - 9x - 6 \quad \text{Combine like terms.}$$

7. $(3x - 5)^2 = (3x - 5)(3x - 5)$

$$\qquad\qquad = 3x \cdot 3x + 3x(-5) + (-5)(3x) + (-5)(-5)$$
$$= 9x^2 - 15x - 15x + 25 \quad \text{Multiply.}$$
$$= 9x^2 - 30x + 25 \quad \text{Combine like terms.}$$

▩ Work Practice 6–7

Practice 6–7

Use the FOIL order to multiply.

6. $(10x - 7)(x + 3)$

7. $(3x + 2)^2$

Helpful Hint

Remember that the FOIL order can only be used to multiply **two binomials.**

Objective E   Multiplying Polynomials

Recall from Appendix B.1 that a polynomial that consists of exactly three terms is called a **trinomial.** Next, we multiply a binomial by a trinomial.

Example 8   Multiply: $(3a + 2)(a^2 - 6a + 3)$

Solution: Use the distributive property to multiply $3a$ by the trinomial $(a^2 - 6a + 3)$ and then 2 by the trinomial.

$$(3a + 2)(a^2 - 6a + 3) = 3a(a^2 - 6a + 3) + 2(a^2 - 6a + 3) \quad \text{Apply the distributive property.}$$
$$= 3a \cdot a^2 + 3a(-6a) + 3a \cdot 3 + \quad \text{Apply the distributive property.}$$
$$\quad 2 \cdot a^2 + 2(-6a) + 2 \cdot 3$$
$$= 3a^3 - 18a^2 + 9a + 2a^2 - 12a + 6 \quad \text{Multiply.}$$
$$= 3a^3 - 16a^2 - 3a + 6 \quad \text{Combine like terms.}$$

▩ Work Practice 8

Practice 8

Multiply:
$(2x + 5)(x^2 + 4x - 1)$

Answers

6. $10x^2 + 23x - 21$ **7.** $9x^2 + 12x + 4$

8. $2x^3 + 13x^2 + 18x - 5$

In general, we have the following.

To Multiply Two Polynomials

Multiply each term of the first polynomial by each term of the second polynomial, and then combine like terms.

✓ **Concept Check** True or false? When a trinomial is multiplied by a trinomial, the result will have at most nine terms. Explain.

A convenient method of multiplying polynomials is to use a vertical format similar to multiplying real numbers.

Practice 9

Multiply $(x^2 + 4x - 1)$ and $(2x + 5)$ vertically.

Example 9 Find the product of $(a^2 - 6a + 3)$ and $(3a + 2)$ vertically.

Solution:

$$
\begin{array}{r}
a^2 - 6a + 3 \\
\times \qquad 3a + 2 \\
\hline
2a^2 - 12a + 6 \\
3a^3 - 18a^2 + 9a \\
\hline
3a^3 - 16a^2 - 3a + 6
\end{array}
$$

Multiply $a^2 - 6a + 3$ by 2.

Multiply $a^2 - 6a + 3$ by $3a$. Line up like terms.

Combine like terms.

Notice that this example is the same as Example 8 and of course the products are the same.

■ Work Practice 9

Answer

9. $2x^3 + 13x^2 + 18x - 5$

✓ **Concept Check Answer**

True

B.3 **Exercise Set** MyMathLab®

Objective A *Multiply. See Examples 1 and 2.*

1. $3x(9x^2 - 3)$

2. $4y(10y^3 + 2y)$

3. $-5a(4a^2 - 6a + 1)$

4. $-2b(3b^2 - 2b + 5)$

5. $7x^2(6x^2 - 5x + 7)$

6. $6z^2(-3z^2 - z + 4)$

Objectives B C D **Mixed Practice** *Multiply. See Examples 3 through 7.*

7. $(x + 3)(x + 10)$

8. $(y + 5)(y + 9)$

9. $(2x - 6)(x + 4)$

10. $(7z + 1)(z - 6)$

11. $(6a + 4)^2$

12. $(8b - 3)^2$

Objective E *Multiply. See Examples 8 and 9.*

13. $(a + 6)(a^2 - 6a + 3)$

14. $(y + 4)(y^2 + 8y - 2)$

15. $(4x - 5)(2x^2 + 3x - 10)$

16. $(9z - 2)(2z^2 + z + 1)$

17. $(x^3 + 2x + x^2)(3x + 1 + x^2)$

18. $(y^2 - 2y + 5)(y^3 + 2 + y)$

Objectives A B C D E **Mixed Practice** *Multiply.*

19. $10r(-3r + 2)$

20. $5x(4x^2 + 5)$

21. $-2y^2(3y + y^2 - 6)$

22. $3z^3(4z^4 - 2z + z^3)$

23. $(x + 2)(x + 12)$

24. $(6s + 1)(3s - 1)$

25. $(2a + 3)(2a - 3)$

26. $(y + 7)(y - 7)$

27. $(x + 5)^2$

28. $(x + 3)^2$

29. $\left(b + \dfrac{3}{5}\right)\left(b + \dfrac{4}{5}\right)$

30. $\left(a - \dfrac{7}{10}\right)\left(a + \dfrac{3}{10}\right)$

31. $(6x + 1)(x^2 + 4x + 1)$

32. $(9y - 1)(y^2 + 3y - 5)$

33. $(7x + 5)^2$

34. $(5x + 9)^2$

35. $(2x - 1)^2$

36. $(4a - 3)^2$

37. $(2x^2 - 3)(4x^3 + 2x - 3)$

38. $(3y^2 + 2)(5y^2 - y + 2)$

39. $(x^3 + x^2 + x)(x^2 + x + 1)$

40. $(a^4 + a^2 + 1)(a^4 + a^2 - 1)$

41. $(2z^2 - z + 1)(5z^2 + z - 2)$

42. $(2b^2 - 4b + 3)(b^2 - b + 2)$

Concept Extensions

Find the area of each figure.

△ **43.**

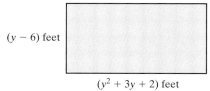

$(y - 6)$ feet

$(y^2 + 3y + 2)$ feet

△ **44.**

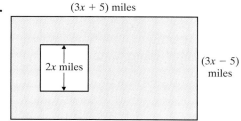

Square $(2x + 11)$ centimeters

Find the area of the shaded figure. To do so, subtract the area of the smaller square from the area of the larger geometric figure.

△ **45.**

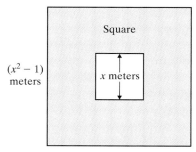

Square

$(x^2 - 1)$ meters

x meters

△ **46.**

$(3x + 5)$ miles

$2x$ miles

$(3x - 5)$ miles

✎ **47.** Suppose that a classmate asked you why $(2x + 1)^2$ is not $4x^2 + 1$. Write down your response to this classmate.

Inductive and Deductive Reasoning

Logic and logical reasoning have applications in many fields, including science, law, psychology, and mathematics. For example, computers must have logic built into their circuits in order to process information correctly. We begin our study of logic by examining inductive reasoning.

Objective A Using Inductive Reasoning

Inductive Reasoning

This is the process of forming a general conclusion based on observing a number of specific examples or outcomes.

Specific Observations $\xrightarrow{\text{lead to a}}$ general conclusion

---

Example 1 Find the next number in the sequence, or listing, of numbers.

$1, 5, 25, 125$

Solution: Each number after the first is obtained by multiplying the previous number by 5. If we assume that this pattern continues, the next number is $125 \times 5 = 625$.

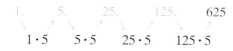

■ Work Practice 1

Example 2 Find the next two numbers in the given sequence.

$1, 1, 2, 3, 5, 8$

Solution: Each number after the first two numbers is obtained by adding the two previous numbers in the list. Notice that $1 + 1 = 2, 1 + 2 = 3, 2 + 3 = 5$, and so on. If this pattern is to continue, the next number in the sequence is $5 + 8 = 13$, and the next number is $8 + 13 = 21$.

■ Work Practice 2

The sequence described in Example 2 is called the *Fibonacci sequence*. There are many examples of this sequence found in nature. This sequence also has many applications in science, business, economics, operations research, archeology, fine arts, architecture, and poetry.

Practice 3

Find the next letter in each sequence.

a. S, M, T, W, T

b. A, F, D, I, G

Example 3 Find the next letter in each sequence.

a. J, F, M, A, M

b. A, E, D, H, G

Solution:

a. Each letter is the first letter of some of the months of the year, in order: January, February, March, April, May. The next month is June, so the letter is J.

b. Let's look for a pattern by corresponding each letter of the alphabet to a number, according to order.

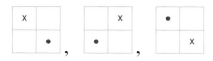

From this pattern, the next letter is

G   K
7 + 4 = 11

■ Work Practice 3

Practice 4

Use inductive reasoning to find the next shape in the sequence.

Example 4 Use inductive reasoning to find the next shape in the sequence.

Solution: Notice that the "*x*" rotates clockwise in each square. Also, there is a dot always diagonally across from each *x*. Thus, we might reason that the next shape is the following.

■ Work Practice 4

Next, let's study deductive reasoning. We begin with a definition.

Objective B Using Deductive Reasoning ▶

Deductive Reasoning

This is the process of forming a specific conclusion based on accepted assumptions.

lead to a
General Assumptions → specific conclusion

In short, with inductive reasoning, we reason from observed specific examples to a general conclusion; with deductive reasoning, we reason logically from general statements or assumptions to a specific conclusion.

Answers

3. a. F (each letter is the first letter of the days of the week) **b.** L

4.

Diagrams called *Venn diagrams* can help us reason deductively. Let's discuss the diagram below, which has to do with pet ownership.

Pet Ownership

Since 8 people are in both regions, these 8 people have both cats and dogs as pets. See if you understand each answer below.

How many people have cats? $12 + 8 = 20$

How many people have cats and no dogs? 12

How many people have dogs? $8 + 7 = 15$

How many people have dogs and no cats? 7

Example 5   The results of a survey of 50 people are as follows.

    27 people like red apples.
    25 people like green apples.
    20 people like both red and green apples.

How many people like neither red nor green apples?

Solution: We draw a Venn diagram to organize the information in the survey. This survey concerns red apples and green apples, which indicates that our diagram will consist of two circles, one representing people who like red apples and one representing people who like green apples. Since there are 20 people who like both red and green apples, we draw two overlapping circles and place 20 in the overlapping section.

A total of 27 people like red apples. This means that $27 - 20 = 7$ is the number of people who like red apples only. Since a total of 25 people like green apples, $25 - 20 = 5$ people like green apples only.

Like red apples 7 (20) 5 Like green apples

This means that $7 + 20 + 5 = 32$ people like red or green apples. Since 50 people were polled, $50 - 32 = 18$ people like neither red nor green apples.

▦ Work Practice 5

Practice 5

The results of a survey of 30 people are as follows.

    20 people like potato chips.
    17 people like tortilla chips.
    13 people like both potato and tortilla chips.

How many people like neither potato nor tortilla chips?

Answer

5. 6 people

Sometimes a grid may be useful in organizing information.

Practice 6

Persons A, B, C, and D sit in the front row of their mathematics class. Use the statements below to determine their order.

1. Person D is sitting between persons A and B.
2. Person C is sitting next to person B only.
3. Person C is not on the far left.

Example 6 Max, Jim, Michael, and Dong Ming have careers as an accountant, a mathematician, a computer programmer, and a manager, not necessarily in the given order. Use the statements below to determine which person is the mathematician.

1. Jim and Max went to lunch with the mathematician.
2. The accountant and the computer programmer taught Jim in college.
3. Max and Dong Ming went to Florida with the computer programmer.

Solution: Draw a grid to organize the information. Then record the information given in each statement. From statement 1, we know that both Jim and Max are not the mathematician. To indicate this on the grid, place X's in the appropriate places. (See below on the left.)

	math.	comp. prog.	acct.	manager
Jim	X			
Max	X			
Michael				
Dong Ming				

	math.	comp. prog.	acct.	manager
Jim	X	X	X	
Max	X			
Michael				
Dong Ming				

From statement 2, we know Jim is not the accountant and he is not the computer programmer, so place an X in the grid corresponding to Jim/accountant and an X in the grid corresponding to Jim/computer programmer. (See above on the right.)

At this point, notice that Jim must be the manager. Place a check mark in the grid corresponding to Jim/manager. Since Jim is the manager, no one else is and we can place X's in the rest of the manager column to indicate that no one else is the manager. (See below on the left.)

	math.	comp. prog.	acct.	manager
Jim	X	X	X	✓
Max	X			X
Michael				X
Dong Ming				X

	math.	comp. prog.	acct.	manager
Jim	X	X	X	✓
Max	X	X		X
Michael				X
Dong Ming		X		X

From statement 3, we know Max and Dong Ming are not the computer programmer. Record this information in the grid. (See above on the right.) Notice that Max must be the accountant. Place a check mark in the grid under Max/accountant and mark the rest of that column with X's. (See below on the left.) **Now Dong Ming must be the mathematician** and Michael must be the computer programmer. (See below on the right.)

	math.	comp. prog.	acct.	manager
Jim	X	X	X	✓
Max	X	X	✓	X
Michael			X	X
Dong Ming		X	X	X

	math.	comp. prog.	acct.	manager
Jim	X	X	X	✓
Max	X	X	✓	X
Michael	X	✓	X	X
Dong Ming	✓	X	X	X

■ Work Practice 6

Answer

6. A, D, B, C

Exercise Set MyMathLab

Objectives A B   *Determine whether each is an example of inductive or deductive reasoning.*

1. My coat is red. My neighbor's coat is red. Therefore, all coats are red.

2. All typewriters type the letter *b*. I have a typewriter. Therefore, my typewriter will type the letter *b*.

3. Rabbits do not lay eggs. Therefore, my pet rabbit will not lay an egg.

4. The last two times Ken flew on a commercial jet, his luggage was lost. Ken reasons that the next time he flies on a commercial jet, his luggage will again be lost.

5. A scientist holds a piece of salt over a burning candle and notices that it burns with a yellow flame. She does this again with another piece of salt and notices that it also burns with a yellow flame. She therefore reasons that all salt burns with a yellow flame.

6. Sherlock Holmes knows that the murderer was the butler, the maid, or the cook. The night of the murder, the cook catered a party in a nearby town and has plenty of witnesses who saw him. The butler was in the hospital with pneumonia. Therefore, Detective Holmes reasons that the maid did it.

Use inductive reasoning to determine the next number, letter, or figure in each sequence. See Examples 1–4.

7. 1, 3, 9, 27

8. 2, 4, 6, 8

9. A, Z, B, Y

10. A, C, E, G

11. 1, 4, 9, 16, 25

12. 2, 4, 8, 16

13. O, T, T, F, F, S, S

14. S, S, M, T, W, T

15. 9, 99, 999, 9999

16. 1, 10, 100, 1000

17. 2, 7, 4, 8, 6

18. 3, 1, 4, 2, 5

19. 9, 12, 20, 33, 51

20. 4, 13, 29, 52, 82

21. 1, 11, 38, 84, 151

22. 2, 12, 24, 41, 66

23.

24.

25.

26.

27.

28.

29.

30.

31.

32.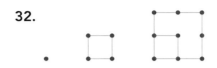

33.

34.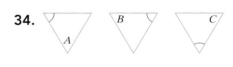

35. Give an example occurring outside the classroom where you used inductive reasoning.

36. Give an example occurring outside the classroom where you used deductive reasoning.

Draw a Venn diagram illustrating each of the following statements. Draw a region representing birds and let a point represent Wendy. See Example 5.

37. Wendy is a bird.

38. Wendy is not a bird.

Draw a Venn diagram illustrating each of the following statements. See Example 5.

39. No bicycles are tricycles.

40. All whales are mammals.

41. Some roosters crow.

42. No trees have wheels.

43. All politicians make promises.

44. Some dryers are heated by electricity.

Use diagrams and deductive reasoning to solve each problem. See Example 6.

45. Four men in a wheelchair race finished 1st, 2nd, 3rd, and 4th.

> Mark beat Bob.
> Joey finished between Mark and Bob.
> Sal beat Mark.

Who finished 3rd in the race?

46. Four women in a marathon finished 1st, 2nd, 3rd, and 4th.

> Wynonna beat Mary.
> Sally beat Wynonna.
> June did not beat Mary.

Who finished 1st in the race?

47. 75 people in a health club were surveyed and the following information was gathered.

> 55 people drink water after exercising.
> 40 people drink Gatorade after exercising.
> 35 people drink water and Gatorade after exercising.

How many people drink neither water nor Gatorade after exercising?

48. In a poll of 200 people, the following information was gathered.

> 80 people listen to country music.
> 100 people listen to rock-n-roll music.
> 50 people listen to both country and rock-n-roll music.

How many people listen to country music only?

49. Aaron, Marty, Donna, and Maria have careers as a civil engineer, an electrical engineer, a mechanical engineer, and a nautical engineer, not necessarily in the given order. Given the information below, determine who is the civil engineer.

a. Aaron and Donna wrote a research paper with the electrical engineer.

b. Marty and Maria carpool with the civil engineer.

c. Aaron works on the same floor as the civil engineer and the nautical engineer.

50. Celeste, Bryan, Clay, and Eric are majoring in business, elementary education, art, and computer science but not necessarily in that order. Given the information below, determine who is majoring in art.

a. Celeste, Bryan, and the elementary education major are in a class together.

b. Bryan and the art major had lunch with the business major.

c. Clay and the elementary education major are cousins.

d. Celeste and the art major carpool together.

51. John, Leon, Alberto, and Julio sit in a theater according to the following.

> Alberto is on the far left.
> If Leon is on the far right, then John is not sitting next to him.

If Leon is on the far right, who is sitting next to him on his left?

52. Ralph, Anoa, Tumulish, and Pier sit on a bench during graduation ceremonies according to the following.

> Tumulish is sitting on the far left.
> If Anoa is sitting on the far right, then Ralph is not sitting next to her.

If Anoa is indeed sitting on the far right, list their seating order starting with the person sitting on the far left.

53. 100 people were surveyed and the results are as follows.

> 65 people said they drink Coke.
> 40 people said they drink Pepsi.
> 10 people said they drink both Coke and Pepsi.

How many people drink neither Coke nor Pepsi?

54. A poll was taken in Gotham City one day about the type of transportation used to commute to work. One hundred fifty-five people were polled.

> 70 people ride the bus.
> 95 people ride the subway.
> 30 people ride both the bus and the subway.

How many people ride neither the bus nor the subway?

Contents of Student Resources

Study Skills Builders

Bigger Picture—Study Guide Outline

Practice Final Exam

Answers to Selected Exercises

Student Resources

Study Skills Builders

Attitude and Study Tips

Study Skills Builder 1

Have You Decided to Complete This Course Successfully?

Ask yourself if one of your current goals is to complete this course successfully.

If it is not a goal of yours, ask yourself why. One common reason is fear of failure. Amazingly enough, fear of failure alone can be strong enough to keep many of us from doing our best in any endeavor.

Another common reason is that you simply haven't taken the time to think about or write down your goals for this course. To help accomplish this, answer the questions below.

Exercises

1. Write down your goal(s) for this course.

2. Now list steps you will take to make sure your goal(s) in Exercise 1 are accomplished.

3. Rate your commitment to this course with a number between 1 and 5. Use the diagram below to help.

High Commitment		Average Commitment		Not Committed at All
5	4	3	2	1

4. If you have rated your personal commitment level (from the exercise above) as a 1, 2, or 3, list the reasons why this is so. Then determine whether it is possible to increase your commitment level to a 4 or 5.

Good luck, and don't forget that a positive attitude will make a big difference.

Study Skills Builder 2

Tips for Studying for an Exam
To prepare for an exam, try the following study techniques:

- Start the study process days before your exam.
- Make sure that you are up to date on your assignments.
- If there is a topic that you are unsure of, use one of the many resources that are available to you. For example,

  See your instructor.
  View a lecture video on the topic.
  Visit a learning resource center on campus.
  Read the textbook material and examples on the topic.

- Reread your notes and carefully review the Chapter Highlights at the end of any chapter.
- Work the review exercises at the end of the chapter.
- Find a quiet place to take the Chapter Test found at the end of the chapter. Do not use any resources when taking this sample test. This way, you will have a clear indication of how prepared you are for your exam. Check your answers and use the Chapter Test Prep Videos to make sure that you correct any missed exercises.

Good luck, and keep a positive attitude.

Exercises

Let's see how you did on your last exam.

1. How many days before your last exam did you start studying for that exam?
2. Were you up to date on your assignments at that time or did you need to catch up on assignments?
3. List the most helpful text supplement (if you used one).
4. List the most helpful campus supplement (if you used one).
5. List your process for preparing for a mathematics test.
6. Was this process helpful? In other words, were you satisfied with your performance on your exam?
7. If not, what changes can you make in your process that will make it more helpful to you?

Study Skills Builder 3

What to Do the Day of an Exam
Your first exam may be soon. On the day of an exam, don't forget to try the following:

- Allow yourself plenty of time to arrive.
- Read the directions on the test carefully.
- Read each problem carefully as you take your test. Make sure that you answer the question asked.
- Watch your time and pace yourself so that you may attempt each problem on your test.
- Check your work and answers.
- ***Do not turn your test in early.*** If you have extra time, spend it double-checking your work.

Good luck!

Exercises

Answer the following questions based on your most recent mathematics exam, whenever that was.

1. How soon before class did you arrive?
2. Did you read the directions on the test carefully?
3. Did you make sure you answered the question asked for each problem on the exam?
4. Were you able to attempt each problem on your exam?
5. If your answer to Exercise 4 is no, list reasons why.
6. Did you have extra time on your exam?
7. If your answer to Exercise 6 is yes, describe how you spent that extra time.

Study Skills Builder 4

Are You Satisfied with Your Performance on a Particular Quiz or Exam?

If not, don't forget to analyze your quiz or exam and look for common errors. Were most of your errors a result of:

- *Carelessness?* Did you turn in your quiz or exam before the allotted time expired? If so, resolve to use any extra time to check your work.

- *Running out of time?* Answer the questions you are sure of first. Then attempt the questions you are unsure of, and delay checking your work until all questions have been answered.

- *Not understanding a concept?* If so, review that concept and correct your work so that you make sure you understand it before the next quiz or the final exam.

- *Test conditions?* When studying for a quiz or exam, make sure you place yourself in conditions similar to test conditions. For example, before your next quiz or exam, take a sample test without the aid of your notes or text.

(For a sample test, see your instructor or use the Chapter Test at the end of each chapter.)

Exercises

1. Have you corrected all your previous quizzes and exams?

2. List any errors you have found common to two or more of your graded papers.

3. Is one of your common errors not understanding a concept? If so, are you making sure you understand all the concepts for the next quiz or exam?

4. Is one of your common errors making careless mistakes? If so, are you now taking all the time allotted to check over your work so that you can minimize the number of careless mistakes?

5. Are you satisfied with your grades thus far on quizzes and tests?

6. If your answer to Exercise 5 is no, are there any more suggestions you can make to your instructor or yourself to help? If so, list them here and share these with your instructor.

Study Skills Builder 5

How Are You Doing?

If you haven't done so yet, take a few moments and think about how you are doing in this course. Are you working toward your goal of successfully completing this course? Is your performance on homework, quizzes, and tests satisfactory? If not, you might want to see your instructor to see if he/she has any suggestions on how you can improve your performance. Reread Section 1.1 for ideas on places to get help with your mathematics course.

Exercises

Answer the following.

1. List any textbook supplements you are using to help you through this course.

2. List any campus resources you are using to help you through this course.

3. Write a short paragraph describing how you are doing in your mathematics course.

4. If improvement is needed, list ways that you can work toward improving your situation as described in Exercise 3.

Study Skills Builder 6

Are You Preparing for Your Final Exam?

To prepare for your final exam, try the following study techniques:

- Review the material that you will be responsible for on your exam. This includes material from your textbook, your notebook, and any handouts from your instructor.
- Review any formulas that you may need to memorize.
- Check to see if your instructor or mathematics department will be conducting a final exam review.
- Check with your instructor to see whether previous final exams are available to students for review.

- Use your previously taken exams as a practice final exam. To do so, rewrite the test questions in mixed order on blank sheets of paper. This will help you prepare for exam conditions.
- If you are unsure of a few concepts, see your instructor or visit a learning lab for assistance. Also, view the video segments of any troublesome sections.
- If you need further exercises to work, try the Cumulative Reviews at the end of the chapters.
- When you feel you are prepared for your final exam, take the Practice Final Exam on page 748. Make sure you check your answers in the answer section of the text. Also, video solutions are available.

Organizing Your Work

Study Skills Builder 7

Learning New Terms

Many of the terms used in this text may be new to you. It will be helpful to make a list of new mathematical terms and symbols as you encounter them and to review them frequently. Placing these new terms (including page references) on 3×5 index cards might help you later when you're preparing for a quiz.

Exercises

1. Name one way you might place a word and its definition on a 3×5 card.

2. How do new terms stand out in this text so that they can be found?

Study Skills Builder 8

Are You Organized?

Have you ever had trouble finding a completed assignment? When it's time to study for a test, are your notes neat and organized? Have you ever had trouble reading your own mathematics handwriting? (Be honest—I have.)

When any of these things happens, it's time to get organized. Here are a few suggestions:

- Write your notes and complete your homework assignments in a notebook with pockets (spiral or ring binder).

- Take class notes in this notebook, and then follow the notes with your completed homework assignment.

- When you receive graded papers or handouts, place them in the notebook pocket so that you will not lose them.

- Mark (possibly with an exclamation point) any note(s) that seem extra important to you.

- Mark (possibly with a question mark) any notes or homework that you are having trouble with.

- See your instructor or a math tutor to help you with the concepts or exercises that you are having trouble understanding.

- If you are having trouble reading your own handwriting, *slow down* and write your mathematics work clearly!

Exercises

1. Have you been completing your assignments on time?

2. Have you been correcting any exercises you may be having difficulty with?

3. If you are having trouble with a mathematical concept or correcting any homework exercises, have you visited your instructor, a tutor, or your campus math lab?

4. Are you taking lecture notes in your mathematics course? (By the way, these notes should include worked-out examples solved by your instructor.)

5. Is your mathematics course material (handouts, graded papers, lecture notes) organized?

6. If your answer to Exercise 5 is no, take a moment and review your course material. List at least two ways that you might better organize it.

Study Skills Builder 9

Organizing a Notebook

It's never too late to get organized. If you need ideas about organizing a notebook for your mathematics course, try some of these:

- Use a spiral or ring binder notebook with pockets and use it for mathematics only.
- Start each page by writing the book's section number you are working on at the top.
- When your instructor is lecturing, take notes. *Always* include any examples your instructor works for you.
- Place your worked-out homework exercises in your notebook immediately after the lecture notes from that section. This way, a section's worth of material is together.
- Homework exercises: Attempt and check all assigned homework.
- Place graded quizzes in the pockets of your notebook or a special section of your binder.

Exercises

Check your notebook organization by answering the following questions.

1. Do you have a spiral or ring binder notebook for your mathematics course only?

2. Have you ever had to flip through several sheets of notes and work in your mathematics notebook to determine what section's work you are in?

3. Are you now writing the textbook's section number at the top of each notebook page?

4. Have you ever lost or had trouble finding a graded quiz or test?

5. Are you now placing all your graded work in a dedicated place in your notebook?

6. Are you attempting all of your homework and placing all of your work in your notebook?

7. Are you checking and correcting your homework in your notebook? If not, why not?

8. Are you writing in your notebook the examples your instructor works for you in class?

Study Skills Builder 10

How Are Your Homework Assignments Going?

It is very important in mathematics to keep up with homework. Why? Many concepts build on each other. Often your understanding of a day's concepts depends on an understanding of the previous day's material.

Remember that completing your homework assignment involves a lot more than attempting a few of the problems assigned.

To complete a homework assignment, remember these four things:

- Attempt all of it.
- Check it.
- Correct it.
- If needed, ask questions about it.

Exercises

Take a moment and review your completed homework assignments. Answer the questions below based on this review.

1. Approximate the fraction of your homework you have attempted.

2. Approximate the fraction of your homework you have checked (if possible).

3. If you are able to check your homework, have you corrected it when errors have been found?

4. When working homework, if you do not understand a concept, what do you do?

MyMathLab and MathXL

Study Skills Builder 11

Tips for Turning In Your Homework on Time

It is very important to keep up with your mathematics homework assignments. Why? Many concepts in mathematics build upon each other.

Remember these 4 tips to help ensure your work is completed on time:

- Know the assignments and due dates set by your instructor.
- Do not wait until the last minute to submit your homework.
- Set a goal to submit your homework 6–8 hours before the scheduled due date in case you have unexpected technology trouble.
- Schedule enough time to complete each assignment.

Following the tips above will also help you avoid potentially losing points for late or missed assignments.

Exercises

Take a moment to consider your work on your homework assignments to date and answer the following questions:

1. What percentage of your assignments have you turned in on time?

2. Why might it be a good idea to submit your homework 6–8 hours before the scheduled deadline?

3. If you have missed submitting any homework by the due date, list some of the reasons why this occurred.

4. What steps do you plan to take in the future to ensure your homework is submitted on time?

Study Skills Builder 12

Tips for Doing Your Homework Online

Practice is one of the main keys to success in any mathematics course. Did you know that MyMathLab/MathXL provides you with **immediate feedback** for each exercise? If you are incorrect, you are given hints to work the exercise correctly. You have **unlimited practice opportunities** and can rework any exercises you have trouble with until you master them, and can submit homework assignments unlimited times before the deadline.

Remember these success tips when doing your homework online:

- Attempt all assigned exercises.
- Write down (neatly) your step-by-step work for each exercise before entering your answer.
- Use the immediate feedback provided by the program to help you check and correct your work for each exercise.
- Rework any exercises you have trouble with until you master them.
- Work through your homework assignment as many times as necessary until you are satisfied.

Exercises

Take a moment to think about your homework assignments to date and answer the following:

1. Have you attempted all assigned exercises?

2. Of the exercises attempted, have you also written out your work before entering your answer—so that you can check it?

3. Are you familiar with how to enter answers using the MathXL player so that you avoid answer entry–type errors?

4. List some ways the immediate feedback and practice supports have helped you with your homework. If you have not used these supports, how do you plan to use them with the success tips above on your next assignment?

Study Skills Builder 13

Organizing Your Work

Have you ever used any readily available paper (such as the back of a flyer, another course assignment, Post-its, etc.) to work out homework exercises before entering the answer in MathXL? To save time, have you ever entered answers directly into MathXL without working the exercises on paper? When it's time to study, have you ever been unable to find your completed work or read and follow your own mathematics handwriting?

When any of these things happen, it's time to get organized. Here are some suggestions:

- Write your step-by-step work for each homework exercise (neatly) on lined, loose-leaf paper and keep this in a 3-ring binder.

- Refer to your step-by-step work when you receive feedback that your answer is incorrect in MathXL. Double-check against the steps and hints provided by the program and correct your work accordingly.

- Keep your written homework with your class notes for that section.

- Identify any exercises you are having trouble with and ask questions about them.

- Keep all graded quizzes and tests in this binder as well to study later.

If you follow the suggestions above, you and your instructor or tutor will be able to follow your steps and correct any mistakes. You will have a written copy of your work to refer to later to ask questions and study for tests.

Exercises

1. Why is it important that you write out your step-by-step work on homework exercises and keep a hard copy of all work submitted online?

2. If you have gotten an incorrect answer, are you able to follow your steps and find your error?

3. If you were asked today to review your previous homework assignments and first test, could you find them? If not, list some ways you might better organize your work.

Study Skills Builder 14

Getting Help with Your Homework Assignments

There are many helpful resources available to you through MathXL to help you work through any homework exercises you may have trouble with. It is important that you know what these resources are and know when and how to use them.

Let's review these features found in the homework exercises:

- **Help Me Solve This**—provides step-by-step help for the exercise you are working. You must work an additional exercise of the same type (without this help) before you can get credit for having worked it correctly.

- **View an Example**—allows you to view a correctly worked exercise similar to the one you are having trouble with. You can go back to your original exercise and work it on your own.

- **E-Book**—allows you to read examples from your text and find similar exercises.

- **Video**—your text author, Elayn Martin-Gay, works an exercise similar to the one you need help with. **Not all exercises have an accompanying video clip.

- **Ask My Instructor**—allows you to e-mail your instructor for help with an exercise.

Exercises

1. How does the "Help Me Solve This" feature work?

2. If the "View an Example" feature is used, is it necessary to work an additional problem before continuing the assignment?

3. When might be a good time to use the "Video" feature? Do all exercises have an accompanying video clip?

4. Which of the features above have you used? List those you found the most helpful to you.

5. If you haven't used the features discussed, list those you plan to try on your next homework assignment.

Study Skills Builder 15

Tips for Preparing for an Exam

Did you know that you can rework your previous homework assignments in MyMathLab and MathXL? This is a great way to prepare for tests. To do this, open a previous homework assignment and click "similar exercise." This will generate new exercises similar to the homework you have submitted. You can then rework the exercises and assignments until you feel confident that you understand them.

To prepare for an exam, follow these tips:

- Review your written work for your previous homework assignments along with your class notes.

- Identify any exercises or topics that you have questions on or have difficulty understanding.

- Rework your previous assignments in MyMathLab and MathXL until you fully understand them and can do them without help.

- Get help for any topics you feel unsure of or for which you have questions.

Exercises

1. Are your current homework assignments up to date and is your written work for them organized in a binder or notebook? If the answer is no, it's time to get organized. For tips on this, see Study Skills Builder 13—Organizing Your Work.

2. How many days in advance of an exam do you usually start studying?

3. List some ways you think that practicing previous homework assignments can help you prepare for your test.

4. List two or three resources you can use to get help for any topics you are unsure of or have questions on.

Good luck!

Study Skills Builder 16

How Well Do You Know the Resources Available to You in MyMathLab?

There are many helpful resources available to you in MyMathLab. Let's take a moment to locate and explore a few of them now. Go into your MyMathLab course, and visit the Multimedia Library, Tools for Success, and E-Book.

Let's see what you found.

Exercises

1. List the resources available to you in the Multimedia Library.

2. List the resources available to you in the Tools for Success folder.

3. Where did you find the English/Spanish Audio Glossary?

4. Can you view videos from the E-Book?

5. Did you find any resources you did not know about? If so, which ones?

6. Which resources have you used most often or found most helpful?

Additional Help Inside and Outside Your Textbook

Study Skills Builder 17

How Well Do You Know Your Textbook?

The questions below will help determine whether you are familiar with your textbook. For additional information, see Section 1.1 in this text.

Exercises

1. What does the ⊙ icon mean?

2. What does the ＼ icon mean?

3. What does the △ icon mean?

4. Where can you find a review for each chapter? What answers to this review can be found in the back of your text?

5. Each chapter contains an overview of the chapter along with examples. What is this feature called?

6. Each chapter contains a review of vocabulary. What is this feature called?

7. There are practice exercises that are contained in this text. Where are they and how can they be used?

8. This text contains a student section in the back entitled Student Resources. List the contents of this section and how they might be helpful.

9. What exercise answers are available in this text? Where are they located?

Study Skills Builder 18

Are You Familiar with Your Textbook Supplements?
Below is a review of some of the student supplements available for additional study. Check to see if you are using the ones most helpful to you.

- **Chapter Test Prep Videos.** These videos provide video clip solutions to the Chapter Test exercises in this text. You will find this extremely useful when studying for tests or exams.

- **Interactive DVD Lecture Series.** These are keyed to each section of the text. The material is presented by me, Elayn Martin-Gay, and I have placed a ⊙ by the exercises in the text that I have worked on the video.

- **The *Student Solutions Manual*.** This contains worked-out solutions to odd-numbered exercises as well as every exercise in the Integrated Reviews, Chapter Reviews, Chapter Tests, and Cumulative Reviews.

- **Pearson Tutor Center.** Mathematics questions may be phoned, faxed, or e-mailed to this center.

- **MyMathLab and MathXL.** MyMathLab is a text-specific online course. MathXL is an online homework, tutorial, and assessment system. Take a moment and determine whether these are available to you.

As usual, your instructor is your best source of information.

Exercises

Let's see how you are doing with textbook supplements.

1. Name one way the Lecture Videos can be helpful to you.

2. Name one way the Chapter Test Prep Videos can help you prepare for a chapter test.

3. List any textbook supplements that you have found useful.

4. Have you located and visited a learning resource lab on your campus?

5. List the textbook supplements that are currently housed in your campus's learning resource lab.

Study Skills Builder 19

Are You Getting All the Mathematics Help That You Need?

Remember that, in addition to your instructor, there are many places to get help with your mathematics course. For example:

- This text has an accompanying video lesson for every section. There are also worked-out solutions to every Chapter Test exercise as well as the Practice Final Exam.
- The back of the book contains answers to odd-numbered exercises.
- A *Student Solutions Manual* is available that contains worked-out solutions to odd-numbered exercises as well as solutions to every exercise in the Integrated Reviews, Chapter Reviews, Chapter Tests, and Cumulative Reviews.
- Don't forget to check with your instructor for other local resources available to you, such as a tutoring center.

Exercises

1. List items you find helpful in the text and all student supplements to this text.

2. List all the campus help that is available to you for this course.

3. List any help (besides the textbook) from Exercises 1 and 2 above that you are using.

4. List any help (besides the textbook) that you feel you should try.

5. Write a goal for yourself that includes trying everything you listed in Exercise 4 during the next week.

Bigger Picture—
Study Guide Outline

I. Operations on Sets of Numbers

    A. Whole Numbers

        1. Add or Subtract:

$$\begin{array}{r} 14 \\ +\,39 \\ \hline 53 \end{array} \qquad \begin{array}{r} 300 \\ -\,27 \\ \hline 273 \end{array}$$

        2. Multiply or Divide:

$$\begin{array}{r} 238 \\ \times\,47 \\ \hline 1666 \\ 9520 \\ \hline 11{,}186 \end{array} \qquad \begin{array}{r} 127\ \text{R2} \\ 7\overline{)891} \\ \underline{-7} \\ 19 \\ \underline{-14} \\ 51 \\ \underline{-49} \\ 2 \end{array}$$

        3. Exponent:

$$3^4 = \overbrace{3 \cdot 3 \cdot 3 \cdot 3}^{4 \text{ factors of } 3} = 81$$

        4. Square Root:

$$\sqrt{25} = 5 \text{ } because \text{ } 5 \cdot 5 = 25 \text{ and } 5 \text{ is a positive number.}$$

        5. Order of Operations:

$$\begin{aligned} 24 \div 3 \cdot 2 - (2 + 8) &= 24 \div 3 \cdot 2 - (10) && \text{Simplify within parentheses.} \\ &= 8 \cdot 2 - 10 && \text{Multiply or divide from left to right.} \\ &= 16 - 10 && \text{Multiply or divide from left to right.} \\ &= 6 && \text{Add or subtract from left to right.} \end{aligned}$$

    B. Integers

        1. Add: $-5 + (-2) = -7$ Adding like signs
Add absolute values. Attach the common sign.

            $-5 + 2 = -3$ Adding unlike signs
Subtract absolute values. Attach the sign of the number with the larger absolute value.

        2. Subtract: Add the first number to the opposite of the second number.
$7 - 10 = 7 + (-10) = -3$

        3. Multiply or Divide: Multiply or divide as usual. If the signs of the two numbers are the same, the answer is positive. If the signs of the two numbers are different, the answer is negative.
$$-5 \cdot 5 = -25, \frac{-32}{-8} = 4$$

    C. Fractions

        1. Simplify: Factor the numerator and denominator. Then divide out factors of 1 by dividing out common factors in the numerator and denominator.
$$\text{Simplify: } \frac{20}{28} = \frac{4 \cdot 5}{4 \cdot 7} = \frac{5}{7}$$

        2. Multiply: Numerator times numerator over denominator times denominator.
$$\frac{5}{9} \cdot \frac{2}{7} = \frac{10}{63}$$

3. **Divide:** First fraction times the reciprocal of the second fraction.

$$\frac{2}{11} \div \frac{3}{4} = \frac{2}{11} \cdot \frac{4}{3} = \frac{8}{33}$$

4. **Add or Subtract:** Must have same denominators. If not, find the LCD, and write each fraction as an equivalent fraction with the LCD as denominator.

$$\frac{2}{5} + \frac{1}{15} = \frac{2}{5} \cdot \frac{3}{3} + \frac{1}{15} = \frac{6}{15} + \frac{1}{15} = \frac{7}{15}$$

D. **Decimals**

1. **Add or Subtract:** Line up decimal points.

$$\begin{array}{r} 1.27 \\ +0.6 \\ \hline 1.87 \end{array}$$

2. **Multiply:**

$$\begin{array}{r} 2.56 \\ \times\, 3.2 \\ \hline 512 \\ 768 \\ \hline 8.192 \end{array}$$

2 decimal places

1 decimal place

$2 + 1 = 3$

3 decimal places

3. **Divide:** $8\overline{)5.6}$ quotient 0.7 $0.6\overline{)0.786}$ quotient 1.31

II. Solving Equations

A. **Proportions:** Set cross products equal to each other. Then solve.

$$\frac{14}{3} = \frac{2}{n} \; or \; 14 \cdot n = 3 \cdot 2 \; or \; 14 \cdot n = 6 \; or \; n = \frac{6}{14} = \frac{3}{7}$$

B. **Percent Problems**

1. **Solved by Equations:** Remember that "of" means multiplication and "is" means equals.

"12% of some number is 6" translates to

$$12\% \cdot n = 6 \; or \; 0.12 \cdot n = 6 \; or \; n = \frac{6}{0.12} \; or \; n = 50$$

2. **Solved by Proportions:** Remember that percent, p, is identified by % or "percent"; base, b, usually appears after "of"; and amount, a, is the part compared to the whole.

"12% of some number is 6" translates to

$$\frac{6}{b} = \frac{12}{100} \; or \; 6 \cdot 100 = b \cdot 12 \; or \; \frac{600}{12} = b \; or \; 50 = b$$

C. **Equations in General:** Simplify both sides of the equation by removing parentheses and adding any like terms. Then use the addition property to write variable terms on one side and constants (or numbers) on the other side. Then use the multiplication property to solve for the variable by dividing both sides of the equation by the coefficient of the variable.

Solve: $2(x - 5) = 80$

$$2x - 10 = 80 \qquad \text{Use the distributive property.}$$
$$2x - 10 + 10 = 80 + 10 \qquad \text{Add 10 to both sides.}$$
$$2x = 90 \qquad \text{Simplify.}$$
$$\frac{2x}{2} = \frac{90}{2} \qquad \text{Divide both sides by 2.}$$
$$x = 45 \qquad \text{Simplify.}$$

Practice Final Exam

1. _____

2. _____

3. _____

4. _____

5. _____

6. _____

7. _____

8. _____

9. _____

10. _____

11. _____

12. _____

13. _____

14. _____

15. _____

16. _____

17. _____

18. _____

19. _____

20. _____

21. _____

22. _____

23. _____

Simplify by performing the indicated operations.

1. $600 - 487$

2. $(2^4 - 5) \cdot 3$

3. $-\dfrac{16}{3} \div -\dfrac{3}{12}$

4. $\dfrac{11}{12} - \dfrac{3}{8} + \dfrac{5}{24}$

5. $\dfrac{64 \div 8 \cdot 2}{(\sqrt{9} - \sqrt{4})^2 + 1}$

6. 10.2×4.01

7. $\dfrac{0.23 + 1.63}{-0.3}$

8. $\begin{array}{r} 19 \\ -\ 2\frac{3}{11} \\ \hline \end{array}$

9. $3\dfrac{1}{3} \cdot 6\dfrac{3}{4}$

10. $9.83 - 30.25$

11. $\left(-\dfrac{3}{4}\right)^2 \div \left(\dfrac{2}{3} + \dfrac{5}{6}\right)$

12. $(-5)^3 - 24 \div (-3)$

13. $-7 - (-19)$

14. $\dfrac{-3(-2) + 12}{-1(-4 - 5)}$

15. $-\dfrac{2}{7} \cdot \left(6 - \dfrac{1}{6}\right)$

16. Round 52,369 to the nearest thousand.

17. Round 34.8923 to the nearest tenth.

18. Write $\dfrac{16}{17}$ as a decimal. Round to the nearest thousandth.

19. Write 85% as a decimal.

20. Write 6.1 as a percent.

21. Write $\dfrac{3}{8}$ as a percent.

22. Write 0.2% as a fraction in simplest form.

23. Find the perimeter and the area of the rectangle below.

Rectangle — $\frac{2}{3}$ foot, 1 foot

748

Write each ratio or rate as a fraction in simplest form.

24. $75 to $10

25. 9 inches of rain in 30 days

26. Find the unit rate:
650 kilometers in 8 hours

27. Find each unit price and decide which is the better buy.
Steak sauce:
8 ounces for $1.19
12 ounces for $1.89

28. Find the unknown number, *n*, in the proportion:
$$\frac{8}{n} = \frac{11}{6}$$

Solve.

29. Subtract 8.6 from 20.

30. During a 258-mile trip, a car used $10\frac{3}{4}$ gallons of gas. How many miles would we expect the car to travel on 1 gallon of gas?

31. The standard dose of medicine for a dog is 10 grams for every 15 pounds of body weight. What is the standard dose for a dog that weighs 80 pounds?

32. Twenty-nine cans of Sherwin-Williams paint cost $493. How much was each can?

33. 0.6% of what number is 7.5?

34. 567 is what percent of 756?

35. An alloy is 12% copper. How much copper is contained in 320 pounds of this alloy?

36. A $120 framed picture is on sale for 15% off. Find the discount and the sale price.

Convert.

37. 40 mg to grams

38. $2\frac{1}{2}$ gal to quarts

Solve.

39. If 2 ft 9 in. of material is used to manufacture one scarf, how much material is needed for 6 scarves?

Find the mean, median, and mode of the list of numbers.

40. 26, 32, 42, 43, 49

24. _____

25. _____

26. _____

27. _____

28. _____

29. _____

30. _____

31. _____

32. _____

33. _____

34. _____

35. _____

36. _____

37. _____

38. _____

39. _____

40. _____

41. _____

42. _____

43. _____

44. _____

45. _____

46. _____

47. _____

48. _____

49. _____

50. _____

51. _____

A professor measures the heights of the students in her class. The results are shown in the following histogram. Use this histogram to answer Exercise 41.

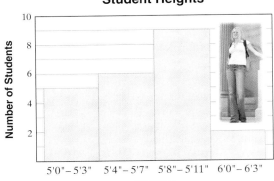

Student Heights

41. How many students are 5′7″ or shorter?

42. Evaluate: $\dfrac{3x - 5}{2y}$ when $x = 7$ and $y = -8$

43. Multiply and then simplify:
$5(3z + 2) - z - 18$

Solve.

44. $-\dfrac{5}{8}x = -25$

45. $5x + 12 - 4x - 14 = 22$

46. $3x - 5 = -11$

47. In a 10-kilometer race, there are 112 more men entered than women. Find the number of women runners if the total number of runners in the race is 600.

48. Find the complement of a 78° angle.

49. Find the measures of angles $x, y,$ and z.

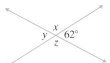

50. Find the measure of $\angle x$.

51. Given that the following triangles are similar, find the missing length n.

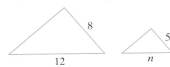

Answers to Selected Exercises

Chapter 1 The Whole Numbers

Section 1.2

Vocabulary, Readiness & Video Check   **1.** whole   **3.** words   **5.** period   **7.** hundreds   **9.** 80,000

Exercise Set 1.2   **1.** tens   **3.** thousands   **5.** hundred-thousands   **7.** millions   **9.** three hundred fifty-four   **11.** eight thousand, two hundred seventy-nine   **13.** twenty-six thousand, nine hundred ninety   **15.** two million, three hundred eighty-eight thousand   **17.** twenty-four million, three hundred fifty thousand, one hundred eighty-five   **19.** three hundred fifteen thousand, two hundred eighty-one   **21.** two thousand, seven hundred seventeen   **23.** thirty-two million, one hundred thousand   **25.** fourteen thousand, four hundred thirty-three   **27.** seventeen million   **29.** 6587   **31.** 59,800   **33.** 13,601,011   **35.** 7,000,017   **37.** 260,997   **39.** 418   **41.** 16,732   **43.** $91,071,000   **45.** 101,000   **47.** 400 + 6   **49.** 3000 + 400 + 70   **51.** 80,000 + 700 + 70 + 4   **53.** 60,000 + 6000 + 40 + 9   **55.** 30,000,000 + 9,000,000 + 600,000 + 80,000   **57.** 5532; five thousand, five hundred thirty-two   **59.** 5000 + 400 + 90 + 2   **61.** Mt. Washington   **63.** National Gallery   **65.** five million, sixty-five thousand   **67.** 4   **69.** 9861   **71.** no; one hundred five   **73.** answers may vary   **75.** 34,000,000,000,000,000   **77.** Canton

Section 1.3

Calculator Explorations   **1.** 134   **3.** 340   **5.** 2834

Vocabulary, Readiness & Video Check   **1.** number   **3.** sum; addend   **5.** grouping; associative   **7.** place; right; left   **9.** increased by

Exercise Set 1.3   **1.** 36   **3.** 292   **5.** 49   **7.** 5399   **9.** 117   **11.** 512   **13.** 209,078   **15.** 25   **17.** 62   **19.** 212   **21.** 94   **23.** 910   **25.** 8273   **27.** 11,926   **29.** 1884   **31.** 16,717   **33.** 1110   **35.** 8999   **37.** 35,901   **39.** 632,389   **41.** 42 in.   **43.** 25 ft   **45.** 24 in.   **47.** 8 yd   **49.** 29 in.   **51.** 44 m   **53.** 2093   **55.** 266   **57.** 544   **59.** 3452   **61.** 21,141 thousand   **63.** 6684 ft   **65.** 340 ft   **67.** 2425 ft   **69.** 249,849   **71.** 1,474,711 vehicles   **73.** 124 ft   **75.** 3095   **77.** California   **79.** 529 stores   **81.** New York and Pennsylvania   **83.** 6358 mi   **85.** answers may vary   **87.** answers may vary   **89.** 1,044,473,765   **91.** correct   **93.** incorrect: 530

Section 1.4

Calculator Explorations   **1.** 770   **3.** 109   **5.** 8978

Vocabulary, Readiness & Video Check   **1.** 0   **3.** minuend; subtrahend   **5.** 0   **7.** 600   **9.** We cannot take 7 from 2 in the ones place, so we borrow one ten from the tens place and move it over to the ones place to give us 10 + 2 or 12.

Exercise Set 1.4   **1.** 44   **3.** 265   **5.** 135   **7.** 2254   **9.** 5545   **11.** 600   **13.** 25   **15.** 45   **17.** 146   **19.** 288   **21.** 168   **23.** 106   **25.** 447   **27.** 5723   **29.** 504   **31.** 89   **33.** 79   **35.** 39,914   **37.** 32,711   **39.** 5041   **41.** 31,213   **43.** 4   **45.** 20   **47.** 7   **49.** 72   **51.** 88   **53.** 264 pages   **55.** 7 million sq km   **57.** 264,000 sq mi   **59.** 283,000 sq mi   **61.** 6065 ft   **63.** 28 ft   **65.** 358 mi   **67.** $619   **69.** 2630 thousand   **71.** 100 dB   **73.** 58 dB   **75.** 346   **77.** 5920 sq ft   **79.** Hartsfield-Jackson Atlanta International   **81.** 14 million   **83.** Jo; by 271 votes   **85.** 1034   **87.** 9   **89.** 8518   **91.** 22,876   **93.** minuend: 48; subtrahend: 1   **95.** minuend: 70; subtrahend: 7   **97.** incorrect: 685   **99.** correct   **101.** 5269 − 2385 = 2884   **103.** no; answers may vary   **105.** no: 1089 more pages

Section 1.5

Vocabulary, Readiness & Video Check   **1.** graph   **3.** 70; 60   **5.** 3 is in the place value we're rounding to (tens), and the digit to the right of this place value is 5 or greater, so we need to add 1 to the 3.   **7.** Each circled digit is to the right of the place value being rounded to and is used to determine whether or not we add 1 to the digit in the place value being rounded to.

Exercise Set 1.5   **1.** 420   **3.** 640   **5.** 2800   **7.** 500   **9.** 21,000   **11.** 34,000   **13.** 328,500   **15.** 36,000   **17.** 39,990   **19.** 30,000,000   **21.** 5280; 5300; 5000   **23.** 9440; 9400; 9000   **25.** 14,880; 14,900; 15,000   **27.** 28,000   **29.** 38,000 points   **31.** $98,000,000,000   **33.** $5,022,000   **35.** $129,000,000,000   **37.** 130   **39.** 80   **41.** 5700   **43.** 300   **45.** 11,400   **47.** incorrect   **49.** correct   **51.** correct   **53.** $3400   **55.** 900 mi   **57.** 6000 ft   **59.** 8,000,000,000   **61.** 3000 children   **63.** $3,067,000,000; $3,070,000,000; $3,100,000,000   **65.** $2,910,000,000; $2,910,000,000; $2,900,000,000   **67.** 5723, for example   **69. a.** 8550   **b.** 8649   **71.** answers may vary   **73.** 140 m

Section 1.6

Calculator Explorations   **1.** 3456   **3.** 15,322   **5.** 272,291

Vocabulary, Readiness & Video Check   **1.** 0   **3.** product; factor   **5.** grouping; associative   **7.** length   **9.** distributive property   **11.** Think of the problem as 50 times 9 and then attach the two zeros from 900, or think of the problem as 5 times 9 and then attach the three zeros at the end of 50 and 900. Both approaches give us 45,000.   **13.** Multiplication is also an application of addition since it is addition of the same addend.

A1

Exercise Set 1.6 **1.** 24 **3.** 0 **5.** 0 **7.** 87 **9.** $6 \cdot 3 + 6 \cdot 8$ **11.** $4 \cdot 3 + 4 \cdot 9$ **13.** $20 \cdot 14 + 20 \cdot 6$ **15.** 512 **17.** 3678 **19.** 1662
21. 6444 **23.** 1157 **25.** 24,418 **27.** 24,786 **29.** 15,600 **31.** 0 **33.** 6400 **35.** 48,126 **37.** 142,506 **39.** 2,369,826 **41.** 64,790
43. 3,949,935 **45.** 800 **47.** 11,000 **49.** 74,060 **51.** 24,000 **53.** 45,000 **55.** 3,280,000 **57.** area: 63 sq m; perimeter: 32 m
59. area: 680 sq ft; perimeter: 114 ft **61.** 240,000 **63.** 300,000 **65.** c **67.** c **69.** 880 **71.** 4200 **73.** 4480 **75.** 375 cal **77.** $3290
79. a. 20 boxes **b.** 100 boxes **c.** 2000 lb **81.** 8800 sq ft **83.** 56,000 sq ft **85.** 5828 pixels **87.** 2100 characters **89.** 1360 cal
91. $10, $60; $10, $200; $12, $36, $12, $36; total cost: $372 **93.** 1,440,000 tea bags **95.** 135 **97.** 2144 **99.** 23 **101.** 15 **103.** $4 \cdot 7$ or $7 \cdot 4$
105. a. $5 + 5 + 5$ or $3 + 3 + 3 + 3 + 3$ **b.** answers may vary **107.** 203
113. 506 windows

$$\begin{array}{r} 203 \\ \times\ 14 \\ \hline 812 \\ 2030 \\ \hline 2842 \end{array}$$

109. $\begin{array}{r} 42 \\ \times\ 93 \end{array}$ **111.** answers may vary

Section 1.7

Calculator Explorations **1.** 53 **3.** 62 **5.** 261 **7.** 0

Vocabulary, Readiness & Video Check **1.** quotient; dividend; divisor **3.** 1 **5.** undefined **7.** 0 **9.** $202 \cdot 102 + 15 = 20,619$
11. addition and division

Exercise Set 1.7 **1.** 6 **3.** 12 **5.** 0 **7.** 31 **9.** 1 **11.** 8 **13.** undefined **15.** 1 **17.** 0 **19.** 9 **21.** 29 **23.** 74 **25.** 338
27. undefined **29.** 9 **31.** 25 **33.** 68 R 3 **35.** 236 R 5 **37.** 38 R 1 **39.** 326 R 4 **41.** 13 **43.** 49 **45.** 97 R 8 **47.** 209 R 11
49. 506 **51.** 202 R 7 **53.** 54 **55.** 99 R 100 **57.** 202 R 15 **59.** 579 R 72 **61.** 17 **63.** 511 R 3 **65.** 2132 R 32 **67.** 6080
69. 23 R 2 **71.** 5 R 25 **73.** 20 R 2 **75.** 33 students **77.** 165 lb **79.** 310 yd **81.** 89 bridges **83.** 11 light poles **85.** 5 mi
87. 1760 yd **89.** 20 **91.** 387 **93.** 79 **95.** 74° **97.** 9278 **99.** 15,288 **101.** 679 **103.** undefined **105.** 9 R 12 **107.** c
109. b **111.** 84 **113.** increase; answers may vary **115.** no; answers may vary **117.** 12 ft **119.** answers may vary **121.** 5 R 1

Integrated Review **1.** 148 **2.** 6555 **3.** 1620 **4.** 562 **5.** 79 **6.** undefined **7.** 9 **8.** 1 **9.** 0 **10.** 0 **11.** 0 **12.** 3 **13.** 2433
14. 9826 **15.** 213 R 3 **16.** 79,317 **17.** 27 **18.** 9 **19.** 138 **20.** 276 **21.** 1099 R 2 **22.** 111 R 1 **23.** 663 R 6 **24.** 1076 R 60
25. 1024 **26.** 9899 **27.** 30,603 **28.** 47,500 **29.** 65 **30.** 456 **31.** 6 R 8 **32.** 53 **33.** 183 **34.** 231 **35.** 9740; 9700; 10,000
36. 1430; 1400; 1000 **37.** 20,800; 20,800; 21,000 **38.** 432,200; 432,200; 432,000 **39.** perimeter: 24 ft; area: 36 sq ft
40. perimeter: 42 in.; area: 98 sq in. **41.** 28 mi **42.** 26 m **43.** 24 **44.** 124 **45.** Lake Pontchartrain Bridge; 2175 ft **46.** 730 qt

Section 1.8

Vocabulary, Readiness & Video Check **1.** The George Washington Bridge has a length of 3500 feet.

Exercise Set 1.8 **1.** 49 **3.** 237 **5.** 42 **7.** 600 **9. a.** 400 ft **b.** 9600 sq ft **11.** $15,500 **13.** 168 hr **15.** 3500 ft **17.** 141 yr
19. 372 billion bricks **21.** 719 towns **23.** $21 **25.** 55 cal **27.** 22 hot dogs **29.** $32,842,230 **31.** 1,215,051 people **33.** 3987 mi
35. 13 paychecks **37.** $239 **39.** $1045 **41.** b will be cheaper by $3 **43.** Asia **45.** 1053 million **47.** 19 million **49.** 798 million
51. $14,754 **53.** 16,800 mg **55. a.** 3750 sq ft **b.** 375 sq ft **c.** 3375 sq ft **57.** $2 **59.** answers may vary

Section 1.9

Calculator Explorations **1.** 4096 **3.** 3125 **5.** 2048 **7.** 2526 **9.** 4295 **11.** 8

Vocabulary, Readiness & Video Check **1.** base; exponent **3.** addition **5.** division **7.** exponent; base **9.** Because $8 \cdot 8 = 64$.
11. The area of a rectangle is length $\cdot$ width. A square is a special rectangle where length $=$ width. Thus, the area of a square is
side $\cdot$ side or (side)^2.

Exercise Set 1.9 **1.** 4^3 **3.** 7^6 **5.** 12^3 **7.** $6^2 \cdot 5^3$ **9.** $9 \cdot 8^2$ **11.** $3 \cdot 2^4$ **13.** $3 \cdot 2^4 \cdot 5^5$ **15.** 64 **17.** 125 **19.** 32 **21.** 1
23. 7 **25.** 128 **27.** 256 **29.** 256 **31.** 729 **33.** 144 **35.** 100 **37.** 20 **39.** 729 **41.** 192 **43.** 162 **45.** 3 **47.** 8 **49.** 12
51. 4 **53.** 21 **55.** 7 **57.** 5 **59.** 16 **61.** 46 **63.** 8 **65.** 64 **67.** 83 **69.** 2 **71.** 48 **73.** 4 **75.** undefined **77.** 59 **79.** 52
81. 44 **83.** 12 **85.** 21 **87.** 24 **89.** 28 **91.** 3 **93.** 25 **95.** 23 **97.** 13 **99.** area: 49 sq m; perimeter: 28 m **101.** area: 529 sq mi;
perimeter: 92 mi **103.** true **105.** false **107.** $(2 + 3) \cdot 6 - 2$ **109.** $24 \div (3 \cdot 2) + 2 \cdot 5$ **111.** 1260 ft **113.** 6,384,814
115. answers may vary; $(20 - 10) \cdot 5 \div 25 + 3$

Chapter 1 Vocabulary Check **1.** whole numbers **2.** perimeter **3.** place value **4.** exponent **5.** area **6.** square root
7. digits **8.** average **9.** divisor **10.** dividend **11.** quotient **12.** factor **13.** product **14.** minuend **15.** subtrahend
16. difference **17.** addend **18.** sum

Chapter 1 Review **1.** tens **2.** ten-millions **3.** seven thousand, six hundred forty **4.** forty-six million, two hundred thousand,
one hundred twenty **5.** $3000 + 100 + 50 + 8$ **6.** $400,000,000 + 3,000,000 + 200,000 + 20,000 + 5000$ **7.** 81,900
8. 6,304,000,000 **9.** 518,512,109 **10.** 184,177,220 **11.** Asia **12.** Oceania/Australia **13.** 63 **14.** 67 **15.** 48 **16.** 77 **17.** 956
18. 840 **19.** 7950 **20.** 7250 **21.** 4211 **22.** 1967 **23.** 1326 **24.** 886 **25.** 27,346 **26.** 39,300 **27.** 8032 mi **28.** $197,699
29. 276 ft **30.** 66 km **31.** 14 **32.** 34 **33.** 65 **34.** 304 **35.** 3914 **36.** 7908 **37.** 17,897 **38.** 34,658 **39.** 238,305 **40.** 249,795
41. 397 pages **42.** $25,626 **43.** May **44.** August **45.** $110 **46.** $240 **47.** 90 **48.** 50 **49.** 470 **50.** 500 **51.** 4800 **52.** 58,000

53. 50,000,000 **54.** 800,000 **55.** 126,000,000 **56.** 99,000 **57.** 7400 **58.** 4100 **59.** 2500 mi **60.** 900,000 **61.** 1911 **62.** 1396 **63.** 1410 **64.** 2898 **65.** 800 **66.** 900 **67.** 3696 **68.** 1694 **69.** 0 **70.** 0 **71.** 16,994 **72.** 8954 **73.** 113,634 **74.** 44,763 **75.** 411,426 **76.** 636,314 **77.** 375,000 **78.** 108,000 **79.** 12,000 **80.** 35,000 **81.** 5,100,000 **82.** 7,600,000 **83.** 1150 **84.** 4920 **85.** 108 **86.** 112 **87.** 24 g **88.** $152,340 **89.** 60 sq mi **90.** 500 sq cm **91.** 3 **92.** 4 **93.** 6 **94.** 7 **95.** 5 R 2 **96.** 4 R 2 **97.** undefined **98.** 0 **99.** 1 **100.** 10 **101.** 0 **102.** undefined **103.** 33 R 2 **104.** 19 R 7 **105.** 24 R 2 **106.** 35 R 15 **107.** 506 R 10 **108.** 907 R 40 **109.** 2793 R 140 **110.** 2012 R 60 **111.** 18 R 2 **112.** 21 R 2 **113.** 458 ft **114.** 13 mi **115.** 51 **116.** 59 **117.** 27 boxes **118.** $192 **119.** $5 billion **120.** 75¢ **121.** $898 **122.** 23,150 sq ft **123.** 49 **124.** 125 **125.** 45 **126.** 400 **127.** 13 **128.** 10 **129.** 15 **130.** 7 **131.** 12 **132.** 9 **133.** 42 **134.** 33 **135.** 9 **136.** 2 **137.** 1 **138.** 0 **139.** 6 **140.** 29 **141.** 40 **142.** 72 **143.** 5 **144.** 7 **145.** 49 sq m **146.** 9 sq in. **147.** 307 **148.** 682 **149.** 2169 **150.** 2516 **151.** 901 **152.** 1411 **153.** 458 R 8 **154.** 237 R 1 **155.** 70,848 **156.** 95,832 **157.** 1644 **158.** 8481 **159.** 740 **160.** 258,000 **161.** 2000 **162.** 40,000 **163.** thirty-six thousand, nine hundred eleven **164.** one hundred fifty-four thousand, eight hundred sixty-three **165.** 70,943 **166.** 43,401 **167.** 64 **168.** 125 **169.** 12 **170.** 10 **171.** 12 **172.** 1 **173.** 2 **174.** 6 **175.** 4 **176.** 24 **177.** 24 **178.** 14 **179.** $4,205,000 **180.** $2,129,000 **181.** 53 full boxes with 18 left over **182.** $86

Chapter 1 Test **1.** eighty-two thousand, four hundred twenty-six **2.** 402,550 **3.** 141 **4.** 113 **5.** 14,880 **6.** 766 R 42 **7.** 200 **8.** 10 **9.** 0 **10.** undefined **11.** 33 **12.** 21 **13.** 8 **14.** 36 **15.** 5,698,000 **16.** 11,200,000 **17.** 52,000 **18.** 13,700 **19.** 1600 **20.** 92 **21.** 122 **22.** 1605 **23.** 7 R 2 **24.** $17 **25.** $126 **26.** 360 cal **27.** $7905 **28.** 20 cm; 25 sq cm **29.** 60 yd; 200 sq yd

Chapter 2 Integers and Introduction to Variables

Section 2.1

Vocabulary, Readiness & Video Check **1.** expression **3.** expression; variables **5.** multiplication

Exercise Set 2.1 **1.** 28; 14; 147; 3 **3.** 152; 152; 0; undefined **5.** 57; 55; 56; 56 **7.** 9 **9.** 26 **11.** 6 **13.** 3 **15.** 117 **17.** 94 **19.** 5 **21.** 626 **23.** 20 **25.** 4 **27.** 4 **29.** 0 **31.** 33 **33.** 121 **35.** 121 **37.** 100 **39.** 60 **41.** 4 **43.** 16, 64, 144, 256 **45.** $x + 5$ **47.** $x + 8$ **49.** $20 - x$ **51.** $512x$ **53.** $x \div 2$ or $\frac{x}{2}$ **55.** $5x + (17 + x)$ **57.** $5x$ **59.** $11 - x$ **61.** $x - 5$ **63.** $6 \div x$ or $\frac{6}{x}$ **65.** $50 - 8x$ **67.** hundreds **69.** thousands **71.** incorrect; $2(0) + 3(7) = 0 + 21 = 21$ **73.** correct **75.** 274,657 **77.** 777 **79.** $5x$ **81.** As t gets larger, $16t^2$ gets larger.

Section 2.2

Vocabulary, Readiness & Video Check **1.** integers **3.** inequality symbols **5.** is less than; is greater than **7.** absolute value **9.** number of feet a miner works underground **11.** negative **13.** opposite of

Exercise Set 2.2 **1.** -1445 **3.** $+14,433$ **5.** $+120$ **7.** $-11,810$ **9.** -3140 million **11.** $-160, -147$; Guillermo **13.** -2 **15.** (number line) **17.** (number line) **19.** (number line) **21.** (number line) **23.** > **25.** < **27.** > **29.** < **31.** 5 **33.** 8 **35.** 0 **37.** 5 **39.** -5 **41.** 4 **43.** -23 **45.** 85 **47.** 7 **49.** -20 **51.** -3 **53.** 43 **55.** -15 **57.** 33 **59.** 6 **61.** -2 **63.** 32 **65.** -7 **67.** < **69.** < **71.** = **73.** < **75.** > **77.** < **79.** > **81.** < **83.** 31; -31 **85.** 28; 28 **87.** Caspian Sea **89.** Lake Superior **91.** iodine **93.** oxygen **95.** 13 **97.** 35 **99.** 360 **101.** $-|-8|, -|3|, 2^2, -(-5)$ **103.** $-|-6|, -|1|, -1, -(-6)$ **105.** $-10, -|-9|, -(-2), |-12|, 5^2$ **107.** d **109.** 5 **111.** false **113.** true **115.** false **117.** answers may vary **119.** no; answers may vary

Section 2.3

Calculator Explorations **1.** -159 **3.** 44 **5.** $-894,855$

Vocabulary, Readiness & Video Check **1.** 0 **3.** a **5.** 5 **7.** -35 **9.** 0 **11.** 0 **13.** Negative; the numbers have different signs and the sign of the sum is the same as the sign of the number with the larger absolute value, -6. **15.** The diver's current depth is 231 feet below the surface.

Exercise Set 2.3

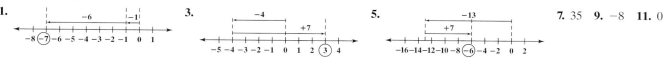

13. 4 **15.** 2 **17.** -2 **19.** -9 **21.** -24 **23.** -57 **25.** -223 **27.** 0 **29.** 7 **31.** -3 **33.** -9 **35.** 30 **37.** 20 **39.** 51 **41.** -33 **43.** -20 **45.** -125 **47.** -7 **49.** -246 **51.** 16 **53.** 13 **55.** -33 **57.** -21 **59.** 21 **61.** -45 **63.** 9 **65.** 0 **67.** 0 **69.** -103 **71.** -70 **73.** 3 **75.** -21 **77.** 17 **79.** -10 **81.** $0 + (-215) + (-16) = -231$; 231 ft below the surface **83.** Dufner: -7; Furyk: -2 **85.** $41,849,000,000 **87.** $62,959,000,000 **89.** 2°C **91.** $-$12,198 **93.** -2°F **95.** -7679 m **97.** 44 **99.** 141 **101.** answers may vary **103.** -3 **105.** -22 **107.** true **109.** false **111.** answers may vary

Section 2.4

Vocabulary, Readiness & Video Check   **1.** $a + (-b)$; b   **3.** $-10 - (-14)$; d   **5.** 0   **7.** 0   **9.** additive inverse   **11.** to follow the order of operations

Exercise Set 2.4   **1.** 0   **3.** 5   **5.** -5 **7.** 14   **9.** 3   **11.** -18 **13.** -14 **15.** -33 **17.** 402   **19.** -14 **21.** -4 **23.** -7 **25.** -42 **27.** -17 **29.** 13   **31.** -38 **33.** -11 **35.** -127 **37.** -11 **39.** 2   **41.** 0   **43.** -1 **45.** -27 **47.** 40   **49.** -22 **51.** -8 **53.** 14   **55.** -11 **57.** 31   **59.** 12   **61.** 20   **63.** 389°F   **65.** 7°F   **67.** 263°F   **69.** $-\$16$ **71.** -12°C   **73.** 154 ft   **75.** 69 ft   **77.** 652 ft   **79.** 144 ft   **81.** $-\$34$ billion   **83.** $-5 + x$ **85.** $-20 - x$ **87.** 5   **89.** 1058   **91.** answers may vary   **93.** 16   **95.** -20 **97.** -4 **99.** 0   **101.** -12 **103.** false   **105.** answers may vary

Integrated Review   **1.** $+29{,}028$ **2.** $-35{,}840$ **3.** -7 **4.** (number line from -4 to 4)   **5.** $>$ **6.** $<$ **7.** $<$ **8.** $>$ **9.** 1   **10.** -4 **11.** 8   **12.** 5   **13.** -6 **14.** 3   **15.** -89 **16.** 0   **17.** 5   **18.** -20 **19.** -10 **20.** -2 **21.** 52   **22.** -3 **23.** 84   **24.** 6   **25.** 1   **26.** -19 **27.** 12   **28.** -4 **29.** -44 **30.** b, c, d   **31.** a, b, c, d   **32.** 10   **33.** -12 **34.** 12   **35.** 10   **36.** 56   **37.** -34

Section 2.5

Vocabulary, Readiness & Video Check   **1.** negative   **3.** positive   **5.** 0   **7.** undefined   **9.** multiplication   **11.** The phrase "lost four yards" in the example translates to the negative number -4.

Exercise Set 2.5   **1.** 12   **3.** -36 **5.** -81 **7.** 0   **9.** 48   **11.** -12 **13.** 80   **15.** 0   **17.** -15 **19.** -9 **21.** 9   **23.** -36 **25.** -64 **27.** -8 **29.** -5 **31.** 7   **33.** 0   **35.** undefined   **37.** -14 **39.** 0   **41.** -15 **43.** -63 **45.** 42   **47.** -24 **49.** 49   **51.** -5 **53.** -9 **55.** -6 **57.** 120   **59.** -1080 **61.** undefined   **63.** -6 **65.** -7 **67.** 3   **69.** -1 **71.** -32 **73.** 180   **75.** 1   **77.** -30 **79.** -1104 **81.** -2870 **83.** -56 **85.** -18 **87.** 35   **89.** -1 **91.** undefined   **93.** 6   **95.** 16; 4   **97.** 0; 0   **99.** $-54 \div 9$; -6 **101.** $-42(-6)$; 252   **103.** $-71 \cdot x$ or $-71x$ **105.** $-16 - x$ **107.** $-29 + x$ **109.** $\dfrac{x}{-33}$ or $x \div (-33)$ **111.** $3 \cdot (-4) = -12$; a loss of 12 yd   **113.** $5 \cdot (-20) = -100$; a depth of 100 feet   **115.** -210°C   **117.** -189°C   **119.** $-\$11$ million per month   **121. a.** $-26{,}932$ movie screens   **b.** -6733 movie screens per year   **123.** 109   **125.** 8   **127.** -19 **129.** -28 **131.** -8 **133.** negative   **135.** $(-5)^{17}$, $(-2)^{17}$, $(-2)^{12}$, $(-5)^{12}$ **137.** answers may vary

Section 2.6

Calculator Explorations   **1.** 48   **3.** -258

Vocabulary, Readiness & Video Check   **1.** division   **3.** average   **5.** subtraction   **7.** base: 3, exponent: 2   **9.** base: 4, exponent: 1; base: 2, exponent: 3   **11.** base: -7, exponent: 5   **13.** base: 5, exponent: 7; base: 10, exponent: 1   **15.** A fraction bar means "divided by" and it is a grouping symbol.   **17.** Finding the average is a good application of both order of operations and adding and dividing integers.

Exercise Set 2.6   **1.** -64 **3.** -64 **5.** 24   **7.** -1 **9.** -7 **11.** -14 **13.** -43 **15.** -8 **17.** -13 **19.** -1 **21.** 4   **23.** -3 **25.** -55 **27.** 8   **29.** 16   **31.** 13   **33.** -77 **35.** 64   **37.** 452   **39.** 129   **41.** 3   **43.** -4 **45.** 4   **47.** 16   **49.** -27 **51.** 34   **53.** 65   **55.** -59 **57.** -7 **59.** -61 **61.** -11 **63.** 36   **65.** -117 **67.** 30   **69.** -3 **71.** -59 **73.** -30 **75.** 1   **77.** -12 **79.** 0   **81.** -20 **83.** 9   **85.** -16 **87.** 1   **89.** -50 **91.** -2 **93.** -19 **95.** 18   **97.** -1 **99.** no; answers may vary   **101.** 4050   **103.** 45   **105.** 32 in.   **107.** 30 ft   **109.** $2 \cdot (7 - 5) \cdot 3$ **111.** $-6 \cdot (10 - 4)$ **113.** no; answers may vary   **115.** answers may vary   **117.** 20,736   **119.** 8900   **121.** 9

Chapter 2 Vocabulary Check   **1.** opposites   **2.** signed   **3.** absolute value   **4.** integers   **5.** variable   **6.** negative   **7.** positive

Chapter 2 Review   **1.** 5   **2.** 17   **3.** undefined   **4.** 0   **5.** 121   **6.** 2   **7.** 4   **8.** 20   **9.** $x - 5$ **10.** $x + 7$ **11.** $10 \div x$ or $\dfrac{10}{x}$ **12.** $5x$ **13.** -1435 **14.** $+11{,}239$ **15.** (number line from -8 to 10)   **16.** (number line from -8 to 8)   **17.** 12   **18.** 0   **19.** -6 **20.** 9   **21.** -9 **22.** 2   **23.** $>$ **24.** $<$ **25.** $>$ **26.** $>$ **27.** 12   **28.** -3 **29.** false   **30.** true   **31.** true   **32.** true   **33.** 2   **34.** 14   **35.** 4   **36.** 17   **37.** -23 **38.** -22 **39.** -21 **40.** -70 **41.** 0   **42.** 0   **43.** -151 **44.** -606 **45.** -20° C   **46.** -150 ft   **47.** -8 **48.** -11 **49.** 99°F   **50.** 110°F   **51.** 8   **52.** -16 **53.** -11 **54.** -27 **55.** 20   **56.** 8   **57.** 0   **58.** -32 **59.** 0   **60.** -7 **61.** -10 **62.** -9 **63.** $-\$25$ **64.** 692 ft   **65.** -3°F   **66.** -4°F   **67.** true   **68.** false   **69.** true   **70.** true   **71.** 21   **72.** -18 **73.** -64 **74.** 60   **75.** 25   **76.** -1 **77.** 0   **78.** 24   **79.** -5 **80.** 3   **81.** 0   **82.** undefined   **83.** -20 **84.** -9 **85.** 38   **86.** -5 **87.** $(-5)(2) = -10$ **88.** $(-50)(4) = -200$ **89.** -18°F   **90.** -3°F   **91.** 28°F   **92.** -26°F   **93.** 49   **94.** -49 **95.** -32 **96.** -32 **97.** 0   **98.** -8 **99.** -16 **100.** 35   **101.** -28 **102.** -44 **103.** 3   **104.** -1 **105.** 7   **106.** -17 **107.** 39   **108.** -26 **109.** 7   **110.** -80 **111.** -2 **112.** -12 **113.** -3 **114.** -35 **115.** -5 **116.** 5   **117.** -1 **118.** -7 **119.** 4   **120.** -4 **121.** 3   **122.** 108   **123.** 16   **124.** -16 **125.** -15 **126.** -19 **127.** 48   **128.** -21 **129.** 21   **130.** -5 **131.** Elevator D   **132.** Elevator B   **133.** 13,118 ft   **134.** -27°C   **135.** 2   **136.** 4   **137.** 3   **138.** 37   **139.** -5 **140.** -25 **141.** -20 **142.** 17

Chapter 2 Test **1.** 3 **2.** −6 **3.** −100 **4.** 4 **5.** −30 **6.** 12 **7.** 65 **8.** 5 **9.** 12 **10.** −6 **11.** 50 **12.** −2 **13.** −11 **14.** −46 **15.** −117 **16.** 3456 **17.** 28 **18.** −213 **19.** −1 **20.** −2 **21.** 2 **22.** −5 **23.** −32 **24.** −12 **25.** −3 **26.** 5 **27.** −1 **28.** −54 **29.** 1 **30.** −17 **31.** 88 ft below sea level **32.** 45 or $45 **33.** 31,642 or 31,642 ft **34.** 3820 ft below sea level **35.** −4 **36. a.** $17x$ **b.** $20 - 2x$

Cumulative Review **1.** hundred-thousands; Sec. 1.2, Ex. 1 **2.** hundreds; Sec. 1.2 **3.** thousands; Sec. 1.2, Ex. 2 **4.** thousands; Sec. 1.2 **5.** ten-millions; Sec. 1.3, Ex. 3 **6.** hundred-thousands; Sec. 1.2 **7. a.** < **b.** > **c.** >; Sec. 2.2, Ex. 3 **8. a.** > **b.** < **c.** >; Sec. 2.2 **9.** 39; Sec. 1.3, Ex. 3 **10.** 39; Sec. 1.3 **11.** 7321; Sec. 1.4, Ex. 2 **12.** 3013; Sec. 1.4 **13.** 36,184 mi; Sec. 1.4, Ex. 5 **14.** $525; Sec. 1.4 **15.** 570; Sec. 1.5, Ex. 1 **16.** 600; Sec. 1.5 **17.** 1800; Sec. 1.5, Ex. 5 **18.** 5000; Sec. 1.5 **19. a.** $3·4 + 3·5$ **b.** $10·6 + 10·8$ **c.** $2·7 + 2·3$; Sec. 1.6, Ex. 2 **20. a.** $5·2 + 5·12$ **b.** $9·3 + 9·6$ **c.** $4·8 + 4·1$; Sec. 1.6 **21.** 78,875; Sec. 1.6, Ex. 5 **22.** 31,096; Sec. 1.6 **23. a.** 6 **b.** 8 **c.** 7; Sec. 1.7, Ex. 1 **24. a.** 7 **b.** 8 **c.** 12; Sec. 1.7 **25.** 741; Sec. 1.7, Ex. 4 **26.** 456; Sec. 1.7 **27.** 12 cards; 10 cards left over; Sec. 1.7, Ex. 11 **28.** $9; Sec. 1.7 **29.** 81; Sec. 1.9, Ex. 5 **30.** 125; Sec. 1.9 **31.** 6; Sec. 1.9, Ex. 6 **32.** 4; Sec. 1.9 **33.** 180; Sec. 1.9, Ex. 8 **34.** 56; Sec. 1.9 **35.** 2; Sec. 1.9, Ex. 16 **36.** 5; Sec. 1.9 **37.** 15; Sec. 2.1, Ex. 1 **38.** 14; Sec. 2.1 **39. a.** 2 **b.** 8 **c.** 0; Sec. 2.2, Ex. 4 **40. a.** 4 **b.** 7; Sec. 2.2 **41.** 21; Sec. 2.3, Ex. 7 **42.** 5; Sec. 2.3 **43.** 22; Sec. 2.4, Ex. 12 **44.** 5; Sec. 2.4 **45.** −21; Sec. 2.5, Ex. 1 **46.** −10; Sec. 2.5 **47.** 0; Sec. 2.5, Ex. 3 **48.** −54; Sec. 2.5 **49.** −16; Sec. 2.6, Ex. 8 **50.** −27; Sec. 2.6

Chapter 3 Fractions and Mixed Numbers
Section 3.1

Vocabulary, Readiness & Video Check **1.** fraction; denominator; numerator **3.** improper; proper; mixed number **5.** equal; improper **7.** whole number; fraction **9.** The fraction is equal to 1.

Exercise Set 3.1 **1.** numerator: 1; denominator: 2; proper **3.** numerator: 10; denominator: 3; improper **5.** numerator: 15; denominator: 15; improper **7.** $\frac{1}{3}$ **9. a.** $\frac{11}{4}$ **b.** $2\frac{3}{4}$ **11. a.** $\frac{23}{6}$ **b.** $3\frac{5}{6}$ **13.** $\frac{7}{12}$ **15.** $\frac{3}{7}$ **17.** $\frac{4}{9}$ **19. a.** $\frac{4}{3}$ **b.** $1\frac{1}{3}$ **21. a.** $\frac{11}{2}$ **b.** $5\frac{1}{2}$ **23.** $\frac{1}{6}$ **25.** $\frac{5}{8}$

27. **29.** **31.** **33.** $\frac{42}{131}$ of the students **35. a.** 89 students **b.** $\frac{89}{131}$ of the students **37.** $\frac{7}{44}$ of the U.S. presidents **39.** $\frac{10}{19}$ of the tropical storms **41.** $\frac{11}{31}$ of the month **43.** $\frac{10}{31}$ of the class **45. a.** $\frac{33}{50}$ of the states **b.** 17 states **c.** $\frac{17}{50}$ of the states **47. a.** $\frac{21}{50}$ of the marbles **b.** 29 marbles **c.** $\frac{29}{50}$ of the marbles

49. **51.** **53.**

55. **57.** 1 **59.** −5 **61.** 0 **63.** 1 **65.** undefined **67.** 3 **69.** 9 **71.** 125 **73.** 7^5 **75.** $2^3 · 3$ **77.** $\frac{-11}{2}; \frac{11}{-2}$ **79.** $\frac{13}{-15}; -\frac{13}{15}$ **81.** answers may vary **83.** **85.** 7 **87.** $\frac{18,000}{135,000}$ **89.** $\frac{6}{36}$ of the countries

Section 3.2

Calculator Explorations **1.** $\frac{4}{7}$ **3.** $\frac{20}{27}$ **5.** $\frac{8}{15}$ **7.** $\frac{2}{9}$

Vocabulary, Readiness & Video Check **1.** prime factorization **3.** prime **5.** equivalent **7.** yes, yes, yes **9.** $3 · 5$ **11.** $2 · 3$ **13.** 2^2 **15.** Check that every factor is a prime number and check that the product of the factors is the original number. **17.** You can simplify the two fractions and then compare them. $\frac{3}{9}$ and $\frac{6}{18}$ both simplify to $\frac{1}{3}$, so the original fractions are equivalent.

Exercise Set 3.2 **1.** $2^2 · 5$ **3.** $2^4 · 3$ **5.** $3^2 · 5$ **7.** $2 · 3^4$ **9.** $2 · 5 · 11$ **11.** $5 · 17$ **13.** $2^4 · 3 · 5$ **15.** $2^2 · 3^2 · 23$ **17.** $\frac{1}{4}$ **19.** $\frac{2}{21}$ **21.** $\frac{7}{8}$ **23.** $\frac{2}{3}$ **25.** $\frac{7}{10}$ **27.** $-\frac{7}{9}$ **29.** $\frac{3}{5}$ **31.** $\frac{27}{64}$ **33.** $\frac{5}{8}$ **35.** $-\frac{5}{8}$ **37.** $\frac{3}{2}$ **39.** $\frac{3}{4}$ **41.** $\frac{5}{14}$ **43.** $\frac{3}{14}$ **45.** $-\frac{11}{17}$ **47.** $\frac{7}{8}$ **49.** 14

51. equivalent **53.** not equivalent **55.** equivalent **57.** equivalent **59.** not equivalent **61.** not equivalent **63.** $\frac{1}{4}$ of a shift **65.** $\frac{1}{2}$ mi **67. a.** $\frac{8}{25}$ **b.** 34 states **c.** $\frac{17}{25}$ **69.** $\frac{5}{12}$ of the wall **71. a.** 8 **b.** $\frac{4}{25}$ **73.** $\frac{13}{232}$ of U.S. astronauts **75.** 364 **77.** 2322 **79.** 2520

81. answers may vary   **83.** $\frac{3}{5}$ **85.** $\frac{9}{25}$ **87.** $\frac{1}{25}$ **89.** $2^2 \cdot 3^5 \cdot 5 \cdot 7$ **91.** answers may vary   **93.** no; answers may vary   **95.** $\frac{3}{50}$

97. answers may vary   **99.** $\frac{7}{100}$ **101.** answers may vary   **103.** 786, 222, 900, 1470   **105.** 6; answers may vary

Section 3.3

Vocabulary, Readiness & Video Check   **1.** $\frac{a \cdot c}{b \cdot d}$ **3.** $\frac{2 \cdot 2 \cdot 2}{7}; \frac{2}{7} \cdot \frac{2}{7} \cdot \frac{2}{7}$ **5.** $\frac{a \cdot d}{b \cdot c}$ **7.** $\frac{2}{15}$ **9.** $\frac{6}{35}$ **11.** $\frac{9}{8}$ **13.** We have a negative fraction times a negative fraction, and a negative number times a negative number is a positive number.   **15.** numerator; denominator

17. radius $= \frac{1}{2} \cdot$ diameter

Exercise Set 3.3   **1.** $\frac{12}{77}$ **3.** $-\frac{9}{50}$ **5.** $\frac{1}{15}$ **7.** $\frac{6}{35}$ **9.** $\frac{5}{28}$ **11.** $\frac{1}{70}$ **13.** 0   **15.** $\frac{18}{55}$ **17.** $\frac{1}{56}$ **19.** $\frac{1}{125}$ **21.** $\frac{4}{9}$ **23.** $-\frac{4}{27}$ **25.** $\frac{4}{5}$ **27.** $-\frac{1}{6}$

29. $-\frac{16}{9}$ **31.** $-\frac{1}{6}$ **33.** $\frac{1}{100}$ **35.** $\frac{35}{36}$ **37.** $\frac{8}{45}$ **39.** undefined   **41.** 0   **43.** $\frac{10}{27}$ **45.** $-\frac{1}{4}$ **47.** $\frac{3}{4}$ **49.** $\frac{9}{16}$ **51.** $\frac{77}{2}$ **53.** $\frac{8}{3}$ **55.** $-\frac{1}{36}$

57. a. $\frac{1}{3}$ **b.** $\frac{12}{25}$ **59. a.** $-\frac{36}{55}$ **b.** $-\frac{44}{45}$ **61.** 50   **63.** 20   **65.** 112   **67.** 69 million   **69.** 868 mi   **71.** $\frac{3}{16}$ in.   **73.** $1838

75. 50 libraries   **77.** $\frac{1}{14}$ sq ft   **79.** 3840 mi   **81.** 2400 mi   **83.** 201   **85.** 196   **87.** answers may vary   **89.** $\frac{2}{5}$ **91.** 5

93. 39,239,250 people   **95.** 132,000,000 adults

Section 3.4

Vocabulary, Readiness & Video Check   **1.** like; unlike   **3.** $-\frac{a}{b}$ **5.** least common denominator (LCD)   **7.** unlike   **9.** like

11. like   **13.** unlike   **15.** numerators; denominator   **17.** $P = \frac{5}{12} + \frac{7}{12} + \frac{5}{12} + \frac{7}{12}$; 2 meters   **19.** Multiplying by 1 does not change the value of the fraction.

Exercise Set 3.4   **1.** $\frac{3}{7}$ **3.** $\frac{1}{5}$ **5.** $\frac{2}{3}$ **7.** $-\frac{1}{4}$ **9.** $-\frac{1}{2}$ **11.** $\frac{13}{11}$ **13.** $\frac{7}{13}$ **15.** $-\frac{1}{9}$ **17.** $\frac{6}{11}$ **19.** $\frac{3}{5}$ **21.** 1   **23.** $\frac{3}{4}$ **25.** $-\frac{10}{3}$ **27.** $\frac{4}{5}$

29. $-\frac{19}{33}$ **31.** $\frac{13}{21}$ **33.** $\frac{9}{10}$ **35.** $-\frac{13}{14}$ **37.** $-\frac{3}{4}$ **39.** $\frac{5}{4}$ **41.** $\frac{2}{5}$ **43.** 1 in.   **45.** 2 m   **47.** $\frac{7}{10}$ mi   **49.** $\frac{13}{50}$ **51.** $\frac{7}{25}$ **53.** $\frac{1}{50}$ **55.** 45

57. 72   **59.** 150   **61.** 168   **63.** 126   **65.** 168   **67.** 14   **69.** 20   **71.** 25   **73.** 56   **75.** 8   **77.** 20

79. $\frac{86}{100}, \frac{80}{100}, \frac{58}{100}, \frac{68}{100}, \frac{12}{100}, \frac{84}{100}, \frac{68}{100}, \frac{81}{100}, \frac{90}{100}, \frac{79}{100}, \frac{74}{100}, \frac{78}{100}$ **81.** drugs, health and beauty aids   **83.** 9   **85.** 81   **87.** 24

89. $\frac{2}{7} + \frac{9}{7} = \frac{11}{7}$ **91.** answers may vary   **93.** 1; answers may vary   **95.** 814   **97.** answers may vary   **99.** a, b, and d

Integrated Review   **1.** $\frac{3}{7}$ **2.** $\frac{5}{4}$ or $1\frac{1}{4}$ **3.** $\frac{73}{85}$ **4.**   **5.** -1 **6.** 17   **7.** 0

8. undefined   **9.** $5 \cdot 13$ **10.** $2 \cdot 5 \cdot 7$ **11.** $3^2 \cdot 5 \cdot 7$ **12.** $3^2 \cdot 7^2$ **13.** $\frac{1}{7}$ **14.** $\frac{6}{5}$ **15.** $-\frac{14}{15}$ **16.** $-\frac{9}{10}$ **17.** $\frac{2}{5}$ **18.** $\frac{3}{8}$ **19.** $\frac{11}{14}$ **20.** $\frac{7}{11}$

21. not equivalent   **22.** equivalent   **23. a.** $\frac{1}{25}$ **b.** 48   **c.** $\frac{24}{25}$ **24. a.** $\frac{29}{108}$ **b.** 395   **c.** $\frac{79}{108}$ **25.** 30   **26.** 14   **27.** 90   **28.** $\frac{28}{36}$

29. $\frac{55}{75}$ **30.** $\frac{40}{48}$ **31.** $\frac{6}{5}$ **32.** $\frac{3}{5}$ **33.** $\frac{3}{5}$ **34.** $\frac{27}{20}$ **35.** $\frac{9}{35}$ **36.** $\frac{12}{35}$ **37.** $\frac{98}{5}$ **38.** $\frac{9}{250}$ **39.** $-\frac{28}{45}$ **40.** $\frac{10}{11}$ **41.** $-\frac{2}{3}$ **42.** $-\frac{2}{45}$

43. 24 lots   **44.** $\frac{3}{4}$ ft

Section 3.5

Calculator Explorations   **1.** $\frac{37}{80}$ **3.** $\frac{95}{72}$ **5.** $\frac{394}{323}$

Vocabulary, Readiness & Video Check   **1.** equivalent; least common denominator   **3.** $\frac{4}{24}; \frac{15}{24}; \frac{19}{24}$ **5.** expression   **7.** 6   **9.** 12

11. 56   **13.** 12   **15.** The fractions are unlike, so the numerators cannot be combined.   **17.** $\frac{3}{2} + \frac{1}{3}$; 6

Exercise Set 3.5   **1.** $\frac{5}{6}$ **3.** $\frac{1}{6}$ **5.** $-\frac{4}{33}$ **7.** $-\frac{3}{14}$ **9.** $\frac{3}{5}$ **11.** $\frac{19}{12}$ **13.** $\frac{11}{36}$ **15.** $\frac{12}{7}$ **17.** $\frac{89}{99}$ **19.** $\frac{1}{2}$ **21.** $\frac{13}{27}$ **23.** $-\frac{8}{33}$ **25.** $\frac{3}{14}$

27. $\frac{1}{35}$ **29.** $-\frac{11}{36}$ **31.** $\frac{1}{20}$ **33.** $\frac{33}{56}$ **35.** $\frac{17}{16}$ **37.** $\frac{8}{9}$ **39.** $-\frac{11}{30}$ **41.** $-\frac{53}{42}$ **43.** $\frac{11}{18}$ **45.** $\frac{98}{143}$ **47.** $-\frac{11}{60}$ **49.** $\frac{19}{20}$ **51.** $\frac{56}{45}$ **53.** $-\frac{5}{24}$

55. $-\frac{9}{1000}$ **57.** $<$ **59.** $>$ **61.** $>$ **63.** $\frac{13}{12}$ **65.** $\frac{1}{4}$ **67.** $\frac{11}{6}$ **69.** $\frac{34}{15}$ cm   **71.** $\frac{17}{10}$ m   **73.** $x + \frac{1}{2}$ **75.** $-\frac{3}{8} - x$ **77.** $\frac{7}{100}$ mph

79. $\frac{5}{8}$ in.   **81.** $\frac{49}{100}$ of students   **83.** $\frac{47}{32}$ in.   **85.** $\frac{19}{25}$ **87.** $\frac{17}{50}$ **89.** $\frac{81}{100}$ **91.** 20   **93.** -6 **95.** $\frac{3}{5} + \frac{4}{5} = \frac{7}{5}$ **97.** $\frac{223}{540}$ **99.** $\frac{49}{44}$

101. answers may vary   **103.** standard mail

Section 3.6

Vocabulary, Readiness & Video Check   **1.** complex   **3.** division   **5.** addition   **7.** distributive property   **9.** Since x is squared and the replacement value is negative, we use parentheses to make sure the whole value of x is squared. Without parentheses, the exponent would not apply to the negative sign.   **11.** division

Exercise Set 3.6   **1.** $\frac{1}{6}$ **3.** $\frac{7}{3}$ **5.** $\frac{1}{6}$ **7.** $\frac{23}{22}$ **9.** $\frac{2}{13}$ **11.** $\frac{17}{60}$ **13.** $\frac{5}{8}$ **15.** $\frac{35}{9}$ **17.** $-\frac{17}{45}$ **19.** $\frac{11}{8}$ **21.** $\frac{29}{10}$ **23.** $\frac{27}{32}$ **25.** $\frac{1}{100}$ **27.** $\frac{9}{64}$

29. $\frac{7}{6}$ **31.** $-\frac{2}{5}$ **33.** $-\frac{2}{9}$ **35.** $\frac{11}{9}$ **37.** $\frac{5}{2}$ **39.** $\frac{7}{2}$ **41.** $\frac{9}{20}$ **43.** $-\frac{13}{2}$ **45.** $\frac{9}{25}$ **47.** $-\frac{5}{32}$ **49.** 1   **51.** $-\frac{2}{5}$ **53.** $-\frac{11}{40}$ **55.** $\frac{7}{3}$ **57.** $\frac{18}{5}$

59. $\frac{53}{8}$ **61.** $\frac{83}{7}$ **63.** $\frac{187}{20}$ **65.** $\frac{500}{3}$ **67.** $3\frac{2}{5}$ **69.** $4\frac{5}{8}$ **71.** $3\frac{2}{15}$ **73.** 15   **75.** $1\frac{7}{175}$ **77.** $6\frac{65}{112}$ **79.** $\frac{7}{2}$ or $3\frac{1}{2}$ **81.** $\frac{49}{6}$ or $8\frac{1}{6}$

83. answers may vary   **85.** no; answers may vary   **87.** $\frac{5}{8}$ **89.** $\frac{11}{56}$ **91.** halfway between a and b **93.** false   **95.** true   **97.** true

99. addition; answers may vary   **101.** subtraction, multiplication, addition, division   **103.** division, multiplication, subtraction, addition

105. $-\frac{77}{16}$ **107.** $-\frac{55}{16}$

Section 3.7

Calculator Explorations   **1.** $\frac{280}{11}$ **3.** $\frac{3776}{35}$ **5.** $26\frac{1}{14}$ **7.** $92\frac{3}{10}$

Vocabulary, Readiness & Video Check   **1.** mixed number   **3.** round   **5.** The denominator of the mixed number we're graphing, $-3\frac{4}{5}$, is 5.   **7.** The fractional part of a mixed number should always be a proper fraction.   **9.** We're adding two mixed numbers with unlike signs, so the answer has the sign of the mixed number with the larger absolute value, which in this case is negative.

Exercise Set 3.7   **1.**   **3.**   **5.** b   **7.** a   **9.** $\frac{8}{21}$ **11.** $4\frac{3}{8}$ **13.** $7\frac{7}{10}$; 8

15. $23\frac{31}{35}$; 24   **17.** $12\frac{1}{2}$ **19.** $5\frac{1}{2}$ **21.** $18\frac{2}{3}$ **23.** a   **25.** b   **27.** $6\frac{2}{3}$; 7   **29.** $13\frac{11}{14}$; 14   **31.** $17\frac{7}{25}$ **33.** $47\frac{53}{84}$ **35.** $25\frac{5}{14}$ **37.** $13\frac{13}{24}$

39. $2\frac{3}{5}$; 3   **41.** $7\frac{5}{14}$; 7   **43.** $\frac{24}{25}$ **45.** $3\frac{5}{9}$ **47.** $15\frac{3}{4}$ **49.** 4   **51.** $5\frac{11}{14}$ **53.** $6\frac{2}{9}$ **55.** $\frac{25}{33}$ **57.** $35\frac{13}{18}$ **59.** $2\frac{1}{2}$ **61.** $73\frac{7}{30}$ **63.** $\frac{11}{14}$

65. $5\frac{4}{7}$ **67.** $13\frac{16}{33}$ **69.** $-5\frac{2}{7} - x$ **71.** $1\frac{9}{10} \cdot x$ **73.** $3\frac{3}{16}$ mi   **75.** $9\frac{2}{5}$ in.   **77.** $7\frac{13}{20}$ in.   **79.** $3\frac{1}{2}$ sq yd   **81.** $\frac{15}{16}$ sq in.   **83.** $21\frac{5}{24}$ m

85. no; it will be $\frac{1}{12}$ ft short   **87.** $4\frac{2}{3}$ m   **89.** $9\frac{3}{4}$ min   **91.** $1\frac{4}{5}$ min   **93.** $-10\frac{3}{25}$ **95.** $-24\frac{7}{8}$ **97.** $-13\frac{59}{60}$ **99.** $4\frac{2}{7}$ **101.** $-1\frac{23}{24}$

103. $\frac{-73}{1000}$ **105.** 4   **107.** 167   **109.** a, b, c   **111.** Incorrect; to divide mixed numbers, first write each mixed number as an improper fraction.   **113.** answers may vary   **115.** answers may vary   **117.** answers may vary

Chapter 3 Vocabulary Check   **1.** reciprocals   **2.** composite number   **3.** equivalent   **4.** improper fraction   **5.** prime number **6.** simplest form   **7.** proper fraction   **8.** mixed number   **9.** numerator; denominator   **10.** prime factorization   **11.** undefined **12.** 0   **13.** like   **14.** least common denominator   **15.** complex fraction   **16.** cross products

Chapter 3 Review   **1.** proper fraction   **2.** improper fraction   **3.** proper fraction   **4.** mixed number   **5.** $\frac{2}{6}$ **6.** $\frac{4}{7}$ **7.** $\frac{7}{3}$ or $2\frac{1}{3}$

8. $\frac{13}{4}$ or $3\frac{1}{4}$ **9.** $\frac{11}{12}$ **10. a.** 108   **b.** $\frac{108}{131}$ **11.** -1 **12.** 1   **13.** 0   **14.** undefined   **15.**

16. (number line with $\frac{4}{7}$ marked) **17.** (number line with $\frac{5}{4}$ marked) **18.** (number line with $\frac{7}{5}$ marked) **19.** $2^2 \cdot 17$ **20.** $2 \cdot 3^2 \cdot 5$

21. $5 \cdot 157$ **22.** $3 \cdot 5 \cdot 17$ **23.** $\frac{3}{7}$ **24.** $\frac{5}{9}$ **25.** $-\frac{1}{3}$ **26.** $-\frac{1}{2}$ **27.** $\frac{29}{32}$ **28.** $\frac{18}{23}$ **29.** 8 **30.** 6 **31.** $\frac{2}{3}$ of a foot **32.** $\frac{3}{5}$ of the cars

33. no **34.** yes **35.** $-\frac{3}{10}$ **36.** $\frac{5}{14}$ **37.** 9 **38.** $\frac{5}{3}$ **39.** $-\frac{1}{27}$ **40.** $\frac{25}{144}$ **41.** $\frac{2}{15}$ **42.** $-\frac{63}{10}$ **43.** $\frac{1}{7}$ **44.** $\frac{23}{14}$ **45.** -2 **46.** $\frac{15}{4}$ **47.** $-\frac{5}{6}$

48. $\frac{27}{2}$ **49.** $\frac{12}{7}$ **50.** $-\frac{15}{2}$ **51.** $\frac{77}{48}$ sq ft **52.** $\frac{4}{9}$ sq m **53.** $\frac{10}{11}$ **54.** $\frac{2}{3}$ **55.** $-\frac{1}{3}$ **56.** $\frac{4}{5}$ **57.** $\frac{1}{21}$ **58.** $-\frac{1}{15}$ **59.** 21 **60.** 24 **61.** 20

62. 35 **63.** 49 **64.** 45 **65.** 40 **66.** 10 **67.** $\frac{3}{4}$ of his homework **68.** $\frac{3}{2}$ mi **69.** $\frac{11}{18}$ **70.** $\frac{7}{26}$ **71.** $-\frac{1}{12}$ **72.** $-\frac{5}{12}$ **73.** $-\frac{15}{14}$ **74.** $\frac{7}{36}$

75. $\frac{91}{150}$ **76.** $\frac{5}{18}$ **77.** $\frac{19}{9}$ m **78.** $\frac{3}{2}$ ft **79.** $\frac{21}{50}$ of the donors **80.** $\frac{1}{4}$ yd **81.** $<$ **82.** $>$ **83.** $>$ **84.** $<$ **85.** $\frac{4}{7}$ **86.** $\frac{3}{11}$ **87.** -2

88. -7 **89.** $\frac{4}{9}$ **90.** $-\frac{3}{10}$ **91.** $3\frac{3}{4}$ **92.** 3 **93.** 1 **94.** $31\frac{1}{4}$ **95.** $\frac{11}{5}$ **96.** $\frac{35}{9}$ **97.** $\frac{8}{13}$ **98.** $-\frac{1}{27}$ **99.** $\frac{29}{110}$ **100.** $-\frac{1}{7}$ **101.** $45\frac{16}{21}$

102. $20\frac{7}{24}$ **103.** $4\frac{19}{35}$ **104.** $2\frac{51}{55}$ **105.** Exact: $5\frac{1}{5}$
Estimate: 6 **106.** Exact: $5\frac{5}{11}$
Estimate: 8 **107.** $5\frac{1}{4}$ **108.** $2\frac{29}{46}$ **109.** 22 mi **110.** $36\frac{2}{3}$ g

111. Each measurement is $4\frac{1}{4}$ in. **112.** $\frac{7}{10}$ yd **113.** $-27\frac{5}{14}$ **114.** $-\frac{33}{40}$ **115.** $1\frac{5}{27}$ **116.** $-3\frac{15}{16}$ **117.** $\frac{7}{12}$ **118.** $\frac{1}{4}$ **119.** 9

120. $\frac{27}{2}$ or $13\frac{1}{2}$ **121.** $\frac{1}{6}$ **122.** $\frac{1}{5}$ **123.** $\frac{11}{12}$ **124.** $\frac{27}{55}$ **125.** Exact: 10
Estimate: 8 **126.** Exact: $12\frac{3}{4}$
Estimate: 12 **127.** $2\frac{1}{3}$ **128.** $6\frac{2}{5}$ **129.** $13\frac{5}{12}$ **130.** $12\frac{3}{8}$

131. $3\frac{16}{35}$ **132.** $8\frac{1}{21}$ **133.** $\frac{11}{25}$ **134.** $\frac{1}{144}$ **135.** $-\frac{1}{12}$ **136.** $6\frac{7}{20}$ lb **137.** $\frac{47}{61}$ in. **138.** $44\frac{1}{2}$ yd **139.** $40\frac{1}{2}$ sq ft

Chapter 3 Test **1.** $\frac{7}{16}$ **2.** $\frac{23}{3}$ **3.** $18\frac{3}{4}$ **4.** $\frac{4}{35}$ **5.** $-\frac{3}{5}$ **6.** not equivalent **7.** equivalent **8.** $2^2 \cdot 3 \cdot 7$ **9.** $3^2 \cdot 5 \cdot 11$ **10.** 72

11. $\frac{8}{9}$ **12.** $-\frac{2}{3}$ **13.** $\frac{4}{3}$ or $1\frac{1}{3}$ **14.** $-\frac{4}{3}$ or $-1\frac{1}{3}$ **15.** $\frac{8}{21}$ **16.** $\frac{13}{24}$ **17.** $\frac{16}{45}$ **18.** 16 **19.** $\frac{1}{7}$ **20.** $\frac{7}{50}$ **21.** $\frac{4}{11}$ **22.** 9 **23.** $\frac{3}{4}$ **24.** $14\frac{1}{40}$

25. $16\frac{8}{11}$ **26.** $\frac{64}{3}$ or $21\frac{1}{3}$ **27.** $22\frac{1}{2}$ **28.** $-\frac{5}{3}$ or $-1\frac{2}{3}$ **29.** $\frac{9}{16}$ **30.** $\frac{3}{8}$ **31.** $\frac{11}{12}$ **32.** $\frac{76}{21}$ or $3\frac{13}{21}$ **33.** $\frac{5}{2}$ or $2\frac{1}{2}$ **34.** $\frac{4}{31}$ **35.** $3\frac{3}{4}$ ft **36.** $\frac{23}{50}$

37. $\frac{13}{50}$ **38.** $2820 **39.** perimeter: $3\frac{1}{3}$ ft; area: $\frac{2}{3}$ sq ft **40.** 24 mi

Cumulative Review **1.** one hundred twenty-six; Sec. 1.2, Ex. 5 **2.** one hundred fifteen; Sec. 1.2 **3.** twenty-seven thousand thirty-four; Sec. 1.2, Ex. 6 **4.** six thousand five hundred seventy-three; Sec. 1.2 **5.** 159; Sec. 1.3, Ex. 1 **6.** 631; Sec. 1.3 **7.** 514; Sec. 1.4, Ex. 3 **8.** 933; Sec. 1.4 **9.** 278,000; Sec. 1.5, Ex. 2 **10.** 1440; Sec. 1.5 **11.** 57,600 megabytes; Sec. 1.6, Ex. 11 **12.** 1305 mi; Sec. 1.6 **13.** 7089 R 5; Sec. 1.7, Ex. 7 **14.** 379 R 10; Sec. 1.7 **15.** 7^3; Sec. 1.9, Ex. 1 **16.** 7^2; Sec. 1.9 **17.** $3^4 \cdot 17^3$; Sec. 1.9, Ex. 4 **18.** $9^4 \cdot 5^2$; Sec. 1.9 **19.** 8; Sec. 2.1, Ex. 2 **20.** 52; Sec. 2.1 **21.** -7188; Sec. 2.2, Ex. 1 **22.** -21; Sec. 2.2 **23.** -4; Sec. 2.3, Ex. 3 **24.** 5; Sec. 2.3 **25.** 3; Sec. 2.4, Ex. 9 **26.** 10; Sec. 2.4 **27.** 25; Sec. 2.5, Ex. 8 **28.** -16; Sec. 2.5 **29.** -2; Sec. 2.6, Ex. 5 **30.** 25; Sec. 2.6

31. $3 \cdot 3 \cdot 5$ or $3^2 \cdot 5$; Sec. 3.2, Ex. 1 **32.** $2 \cdot 2 \cdot 23$ or $2^2 \cdot 23$; Sec. 3.2 **33.** $\frac{10}{33}$; Sec. 3.3, Ex. 1 **34.** $\frac{2}{35}$; Sec. 3.3 **35.** $\frac{1}{8}$; Sec. 3.3, Ex. 2

36. $\frac{3}{25}$; Sec. 3.3 **37.** $\frac{2}{5}$; Sec. 3.1, Ex. 3 **38.** $2^2 \cdot 3 \cdot 13$; Sec. 3.2 **39. a.** $\frac{38}{9}$ **b.** $\frac{19}{11}$; Sec. 3.6, Ex.8 **40.** $\frac{39}{5}$; Sec. 3.6

41. $\frac{7}{11}$; Sec. 3.2, Ex. 5 **42.** $\frac{2}{3}$; Sec. 3.2 **43.** $2\frac{11}{12}$; Sec. 3.7, Ex. 2 **44.** $\frac{8}{3}$ or $2\frac{2}{3}$; Sec. 3.7 **45.** $\frac{5}{12}$; Sec. 3.3, Ex. 10 **46.** $\frac{11}{56}$; Sec. 3.7

Chapter 4 Decimals

Section 4.1

Vocabulary, Readiness & Video Check **1.** words; standard form **3.** decimals **5.** tenths; tens **7.** tens **9.** tenths **11.** as "and" **13.** Reading a decimal correctly gives you the correct place value, which tells you the denominator of your equivalent fraction. **15.** When rounding, we look to the digit to the right of the place value we're rounding to. In this case, we look to the hundredths-place digit, which is 7.

Exercise Set 4.1 **1.** six and fifty-two hundredths **3.** sixteen and twenty-three hundredths **5.** negative two hundred five thousandths **7.** one hundred sixty-seven and nine thousandths **9.** three thousand and four hundredths **11.** one hundred five and six tenths **13.** two and forty-three hundredths

15.

Preprinted Name		Current date
Preprinted Address		DATE

PAY TO THE
ORDER OF R.W. Financial $ | 321.42 |
Three hundred twenty-one and 42/100   DOLLARS

FOR _____   Signature

17.

Preprinted Name		Current date
Preprinted Address		DATE

PAY TO THE
ORDER OF Verizon $ | 91.68 |
Ninety-one and 68/100   DOLLARS

FOR _____   Signature

19. 6.5   **21.** 9.08   **23.** -705.625 **25.** 0.0046   **27.** $\frac{3}{10}$ **29.** $\frac{27}{100}$ **31.** $\frac{2}{5}$ **33.** $5\frac{2}{5}$ **35.** $-\frac{29}{500}$ **37.** $7\frac{1}{125}$ **39.** $15\frac{401}{500}$ **41.** $\frac{601}{2000}$

43. 0.8; $\frac{8}{10}$ or $\frac{4}{5}$ **45.** seventy-seven thousandths; $\frac{77}{1000}$ **47.** $<$ **49.** $<$ **51.** $<$ **53.** $=$ **55.** $<$ **57.** $>$ **59.** $<$ **61.** $>$ **63.** 0.6

65. 98,210   **67.** -0.23 **69.** 0.594   **71.** 3.1   **73.** 3.142   **75.** $27   **77.** $0.20   **79.** 0.7 in.   **81.** 2.07 min   **83.** $68   **85.** 225 days

87. 5766   **89.** 35   **91.** b   **93.** a   **95.** answers may vary   **97.** 7.12   **99.** $\dfrac{26,849,577}{100,000,000,000}$ **101.** answers may vary

103. answers may vary   **105.** 0.26499, 0.25786   **107.** 0.10299, 0.1037, 0.1038, 0.9   **109.** $3600 million

Section 4.2

Calculator Explorations   1. 328.742   **3.** 5.2414   **5.** 865.392

Vocabulary, Readiness & Video Check   1. last   **3.** vertically   **5.** 0.5   **7.** 1.26   **9.** 8.9   **11.** 0.6   **13.** Lining up the decimal points also lines up place values, so we only add or subtract digits in the same place values.   **15.** So that the subtraction can be written vertically with decimal points lined up

Exercise Set 4.2   1. 3.5   **3.** 6.83   **5.** 27.0578   **7.** 91.204   **9.** -8.57 **11.** 11.16   **13.** Exact: 465.56; Estimate: $\begin{array}{r}230\\+\ 230\\\hline 460\end{array}$ **15.** Exact: 115.123;

Estimate: $\begin{array}{r}100\\6\\+\ \ 9\\\hline 115\end{array}$ **17.** 56.432   **19.** 6.5   **21.** 15.3   **23.** 598.23   **25.** Exact: 1.83; Estimate: $6-4=2$ **27.** 861.6   **29.** Exact: 876.6;

Estimate: $\begin{array}{r}1000\\-\ 100\\\hline 900\end{array}$ **31.** 194.4   **33.** -6.32 **35.** -6.4 **37.** 3.1   **39.** 2.9988   **41.** 16.3   **43.** 3.1   **45.** -5.62 **47.** 776.89   **49.** -549.8

51. 861.6   **53.** 512.101   **55.** 0.088   **57.** -180.44 **59.** -1.1 **61.** 3.81   **63.** 3.39   **65.** 1.61   **67.** $7.52   **69.** $-$0.42   **71.** 28.56 m
73. 14.36 in.   **75.** 195.8 mph   **77.** 5.2 billion or 5,200,000,000   **79.** $2042.5 million   **81.** 326.3 in.   **83.** 67.44 ft   **85.** 13.462 mph

87. Switzerland   **89.** 7.94 lb   **91.**

Country	Pounds of Chocolate per Person
Switzerland	26.24
Ireland	21.83
UK	20.94
Austria	19.40
Belgium	18.30

93. 46   **95.** $\frac{4}{9}$ **97.** incorrect; $\begin{array}{r}9.200\\8.630\\+\ 4.005\\\hline 21.835\end{array}$ **99.** 6.08 in.

101. $1.20   **103.** 1 nickel, 1 dime, and 2 pennies; 3 nickels and 2 pennies; 1 dime and 7 pennies; 2 nickels and 7 pennies
105. answers may vary   **107.** answers may vary

Section 4.3

Vocabulary, Readiness & Video Check   1. sum   **3.** circumference   **5.** right; zeros   **7.** 4   **9.** 4   **11.** 8   **13.** We need to learn where to place the decimal point in the product.   **15.** We just need to know how to move the decimal point. 100 has two zeros, so we move the decimal point two places to the right.   **17.** We used an approximation for π. The exact answer is 10π cm.

Exercise Set 4.3   1. 1.36   **3.** 0.6   **5.** -17.595 **7.** 55.008   **9.** Exact: 28.56; Estimate: $7\times 4=28$ **11.** 0.1041   **13.** Exact: 8.23854;
Estimate: $\begin{array}{r}1\\\times\ 8\\\hline 8\end{array}$ **15.** 11.2746   **17.** 65   **19.** 0.83   **21.** -7093 **23.** 70   **25.** 0.0983   **27.** 0.02523   **29.** 0.0492   **31.** 14,790   **33.** 1.29

35. −9.3762   **37.** 0.5623   **39.** 36.024   **41.** 1,500,000,000   **43.** 49,800,000   **45.** −0.6   **47.** 17.3   **49.** 10π cm $\approx$ 31.4 cm
51. 18.2π yd $\approx$ 57.148 yd   **53.** $715.20   **55.** 24.8 g   **57.** 11.201 sq in.   **59.** 250π ft $\approx$ 785 ft   **61.** 135π m $\approx$ 423.9 m
63. 64.9605 in.   **65. a.** 62.8 m and 125.6 m   **b.** yes   **67.** $708   **69.** 786.9 Canadian dollars   **71.** 1024.67 New Zealand
dollars   **73.** 486   **75.** −9   **77.** 3.64   **79.** 3.56   **81.** −0.1105   **83.** 3,831,600 mi   **85.** answers may vary   **87.** answers may vary

Section 4.4

Calculator Explorations   **1.** not reasonable   **3.** reasonable

Vocabulary, Readiness & Video Check   **1.** quotient; divisor; dividend   **3.** left; zeros   **5.** 5.9   **7.** 0   **9.** 1   **11.** undefined
13. a whole number   **15.** We just need to know how to move the decimal point. 1000 has three zeros, so we move the decimal point
in the decimal number three places to the left.   **17.** We want the answer rounded to the nearest tenth, so we go to one extra place
value, to the hundredths place, in order to round.

Exercise Set 4.4   **1.** 4.6   **3.** 0.094   **5.** 300   **7.** 2.6   **9.** Exact: 6.6; Estimate: $6\overline{)36}$ **11.** 0.413   **13.** −600   **15.** 7   **17.** 4.8   **19.** 2100
21. 5.8   **23.** 5.5   **25.** Exact: 9.8; Estimate: $7\overline{)70}$ **27.** 9.6   **29.** 45   **31.** 54.592   **33.** 0.0055   **35.** 179   **37.** 23.87   **39.** 114.0
41. 0.54982   **43.** 2.687   **45.** −0.0129   **47.** 12.6   **49.** 1.31   **51.** 0.045625   **53.** 0.413   **55.** −8   **57.** −7.2   **59.** 1400   **61.** 30
63. −58,000   **65.** −0.69   **67.** 0.024   **69.** 65   **71.** −5.65   **73.** −7.0625   **75.** 11 qt   **77.** 5.1 m   **79.** 11.4 boxes   **81.** 24 tsp   **83.** 8 days
85. 146.6 mi per week   **87.** $3980.77 per hr   **89.** $\dfrac{9}{10}$ **91.** $\dfrac{1}{20}$ **93.** 4.26   **95.** 1.578   **97.** −26.66   **99.** 904.29   **101.** c
103. b   **105.** 85.5   **107.** 8.6 ft   **109.** answers may vary   **111.** 65.2−82.6 knots   **113.** 27.3 m

Integrated Review   **1.** 2.57   **2.** 4.05   **3.** 8.9   **4.** 3.5   **5.** 0.16   **6.** 0.24   **7.** 0.27   **8.** 0.52   **9.** −4.8   **10.** 6.09   **11.** 75.56   **12.** 289.12
13. −24.974   **14.** −43.875   **15.** −8.6   **16.** 5.4   **17.** −280   **18.** 1600   **19.** 224.938   **20.** 145.079   **21.** 0.56   **22.** −0.63   **23.** 27.6092
24. 145.6312   **25.** 5.4   **26.** −17.74   **27.** −414.44   **28.** −1295.03   **29.** −34   **30.** −28   **31.** 116.81   **32.** 18.79   **33.** 156.2   **34.** 1.562
35. 25.62   **36.** 5.62   **37.** 200 mi   **38.** $0.81   **39.** $7.4 billion, or $7,400,000,000

Section 4.5

Vocabulary, Readiness & Video Check   **1.** false   **3.** false   **5.** We place a bar over just the repeating digits, and only 6 repeats in
our decimal answer.   **7.** The fraction bar serves as a grouping symbol.   **9.** $4(0.3) - (-2.4)$

Exercise Set 4.5   **1.** 0.2   **3.** 0.68   **5.** 0.75   **7.** −0.08   **9.** 2.25   **11.** $0.91\overline{6}$ **13.** 0.425   **15.** 0.45   **17.** $-0.\overline{3}$ **19.** 0.4375   **21.** $0.\overline{63}$
23. 5.85   **25.** 0.624   **27.** −0.33   **29.** 0.44   **31.** 0.6   **33.** 0.62   **35.** 0.86   **37.** 0.02   **39.** <   **41.** =   **43.** <   **45.** <   **47.** <   **49.** >
51. <   **53.** <   **55.** 0.32, 0.34, 0.35   **57.** 0.49, 0.491, 0.498   **59.** $5.23, \dfrac{42}{8}, 5.34$ **61.** $0.612, \dfrac{5}{8}, 0.649$ **63.** 0.59   **65.** −3   **67.** 5.29
69. 9.24   **71.** 0.2025   **73.** −1.29   **75.** −15.4   **77.** −3.7   **79.** 25.65 sq in.   **81.** 0.248 sq yd   **83.** 5.76   **85.** 5.7   **87.** 3.6   **89.** 72
91. $\dfrac{5}{2}$ **93.** $= 1$ **95.** > 1 **97.** < 1 **99.** 0.061   **101.** 6300 stations   **103.** answers may vary

Section 4.6

Calculator Explorations   **1.** 32   **3.** 5.568   **5.** 9.849

Vocabulary, Readiness & Video Check   **1.** 10   **3.** squaring   **5.** 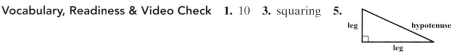   **7.** $\sqrt{49} = 7$ because 7^2 or $7 \cdot 7 = 49$.   **9.** The Pythagorean theorem works only for right triangles.

Exercise Set 4.6   **1.** 2   **3.** 11   **5.** $\dfrac{1}{9}$ **7.** $\dfrac{4}{8} = \dfrac{1}{2}$ **9.** 1.732   **11.** 3.873   **13.** 6.856   **15.** 5.099   **17.** 6, 7   **19.** 10, 11   **21.** 16   **23.** 9.592
25. $\dfrac{7}{12}$ **27.** 8.426   **29.** 13 in.   **31.** 6.633 cm   **33.** 52.802 m   **35.** 117 mm   **37.** 5   **39.** 12   **41.** 17.205   **43.** 44.822   **45.** 42.426
47. 1.732   **49.** 8.5   **51.** 141.42 yd   **53.** 25.0 ft   **55.** 340 ft   **57.** $\dfrac{5}{6}$ **59.** $\dfrac{2}{5}$ **61.** $\dfrac{5}{12}$ **63.** 6   **65.** 10   **67.** answers may vary   **69.** yes
71. $(\sqrt{80} - 6)$ in. $\approx$ 2.94 in.

Chapter 4 Vocabulary Check   **1.** decimal   **2.** numerator; denominator   **3.** vertically   **4.** and   **5.** sum   **6.** square root   **7.** right
triangle; hypotenuse; legs   **8.** standard form

Chapter 4 Review   **1.** tenths   **2.** hundred-thousandths   **3.** negative twenty-three and forty-five hundredths   **4.** three hundred
forty-five hundred-thousandths   **5.** one hundred nine and twenty-three hundredths   **6.** two hundred and thirty-two millionths
7. 2.07   **8.** −503.102   **9.** 16,025.0014   **10.** 14.011   **11.** $\dfrac{4}{25}$ **12.** $\dfrac{11}{20}$ **13.** $-12\dfrac{23}{1000}$ **14.** $25\dfrac{1}{4}$ **15.** >   **16.** =   **17.** <   **18.** >
19. 0.6   **20.** 0.94   **21.** −42.90   **22.** −16.349   **23.** 9.5   **24.** 5.1   **25.** −7.28   **26.** −12.04   **27.** 320.312   **28.** 148.74236   **29.** 1.7
30. 2.49   **31.** −1324.5   **32.** −10.136   **33.** 65.02   **34.** 199.99802   **35.** 52.6 mi   **36.** −5.7   **37.** 22.2 in.   **38.** 38.9 ft   **39.** 72   **40.** 9345
41. −78.246   **42.** 73,246.446   **43.** 887,000,000   **44.** 600,000   **45.** 14π m $\approx$ 43.96 m   **46.** 20π in. $\approx$ 62.8 in.   **47.** 0.088   **48.** 15.825

49. 70 **50.** -0.21 **51.** 8.059 **52.** 30.4 **53.** 0.024 **54.** -9.3 **55.** 7.3 m **56.** 45 months **57.** 0.8 **58.** -0.923 **59.** $2.\overline{3}$ or 2.333

60. $0.21\overline{6}$ or 0.217 **61.** $=$ **62.** $<$ **63.** $<$ **64.** $<$ **65.** 0.832, 0.837, 0.839 **66.** $\frac{5}{8}$, 0.626, 0.685 **67.** 0.42, $\frac{3}{7}$, 0.43 **68.** $\frac{19}{12}$, 1.63, $\frac{18}{11}$

69. -11.94 **70.** 3.89 **71.** 7.26 **72.** 0.81 **73.** 55 **74.** -129 **75.** 6.9 sq ft **76.** 5.46 sq in. **77.** 8 **78.** 12 **79.** $\frac{2}{5}$ **80.** $\frac{1}{10}$ **81.** 13

82. 29 **83.** 10.7 **84.** 93 **85.** 127.3 ft **86.** 88.2 ft **87.** two hundred and thirty-two ten-thousandths **88.** $-16{,}025.014$ **89.** $\frac{231}{100{,}000}$

90. 0.75, $\frac{6}{7}$, $\frac{8}{9}$ **91.** -0.07 **92.** 0.1125 **93.** $>$ **94.** $<$ **95.** 42.90 **96.** 16.349 **97.** \$123.00 **98.** \$3646.00 **99.** -1.7 **100.** 2.49

101. 80.668 **102.** -148.74236 **103.** 8.128 **104.** -7.245 **105.** 4900 **106.** 23.904 **107.** 9600 sq ft **108.** yes **109.** 0.1024 **110.** 3.6

111. 1 **112.** 6 **113.** $\frac{4}{9}$ **114.** $\frac{1}{11}$ **115.** 86.6 **116.** 20.8 **117.** 48.1 **118.** 19.7

Chapter 4 Test **1.** forty-five and ninety-two thousandths **2.** 3000.059 **3.** 17.595 **4.** -51.2 **5.** -20.42 **6.** 40.902 **7.** 0.037

8. 34.9 **9.** 0.862 **10.** $<$ **11.** $<$ **12.** $\frac{69}{200}$ **13.** $-24\frac{73}{100}$ **14.** -0.5 **15.** 0.941 **16.** 1.93 **17.** -6.2 **18.** 11.4 **19.** 7 **20.** 12.530

21. $\frac{4}{5}$ **22.** 4,583,000,000 **23.** 2.31 sq mi **24.** 18π mi ≈ 56.52 mi **25. a.** 9904 sq ft **b.** 198.08 oz **26.** 54 mi

Cumulative Review **1.** eighty-five; Sec. 1.2, Ex. 4 **2.** one hundred seven; Sec. 1.2 **3.** one hundred twenty-six; Sec. 1.2, Ex. 5 **4.** five thousand, twenty-six; Sec. 1.2 **5.** 159; Sec. 1.3, Ex. 1 **6.** 19 in.; Sec. 1.3 **7.** 514; Sec. 1.4, Ex. 3 **8.** 121 R 1; Sec. 1.7 **9.** 278,000; Sec. 1.5, Ex. 2 **10.** $2 \cdot 3 \cdot 5$; Sec. 3.2 **11.** 20,296; Sec. 1.6, Ex. 4 **12.** 0; Sec. 1.6 **13. a.** 7 **b.** 12 **c.** 1 **d.** 1 **e.** 20 **f.** 1; Sec. 1.7, Ex. 2 **14.** 25; Sec. 1.7 **15.** 1038 mi; Sec. 1.8, Ex. 1 **16.** 11; Sec. 1.9 **17.** 81; Sec. 1.9, Ex. 5 **18.** 125; Sec. 1.9 **19.** 81; Sec. 1.9, Ex. 7 **20.** 1000; Sec. 1.9 **21.** 2; Sec. 2.1, Ex. 3 **22.** 6; Sec. 2.1 **23. a.** -11 **b.** 2 **c.** 0; Sec. 2.2, Ex. 5 **24. a.** 7 **b.** -4 **c.** 1; Sec. 2.2 **25.** -23; Sec. 2.3, Ex. 4 **26.** -22; Sec. 2.3 **27.** 180; Sec. 1.9, Ex. 8 **28.** 32; Sec. 1.9

29. -49; Sec. 2.5, Ex. 9 **30.** -32; Sec. 2.5 **31.** 25; Sec. 2.5, Ex. 8 **32.** -9; Sec. 2.5 **33.** $\frac{4}{3}$ or $1\frac{1}{3}$; Sec. 3.1, Ex. 11 **34.** $\frac{11}{4}$ or $2\frac{3}{4}$; Sec. 3.1

35. $\frac{15}{4}$ or $3\frac{3}{4}$; Sec. 3.1, Ex. 12 **36.** $\frac{14}{3}$ or $4\frac{2}{3}$; Sec. 3.1 **37.** $2^2 \cdot 3^2 \cdot 7$; Sec. 3.2, Ex. 3 **38.** 62; Sec. 1.4 **39.** $-\frac{36}{13}$; Sec. 3.2, Ex. 8

40. $\frac{79}{8}$; Sec. 3.6 **41.** equivalent; Sec. 3.2, Ex. 10 **42.** $>$; Sec. 3.5 **43.** $\frac{10}{33}$; Sec. 3.3, Ex. 1 **44.** $1\frac{1}{2}$; Sec. 3.7 **45.** $\frac{1}{8}$; Sec. 3.3, Ex. 2 **46.** 37; Sec. 3.7 **47.** 829.6561; Sec. 4.2, Ex. 2 **48.** 230.8628; Sec. 4.2 **49.** 18.408; Sec. 4.3, Ex. 1 **50.** 28.251; Sec. 4.3

Chapter 5 Ratio, Proportion, and Measurement
Section 5.1

Vocabulary, Readiness & Video Check **1.** unit **3.** division **5.** numerator; denominator **7.** false **9.** false **11.** true
13. We can use "to" as in 1 to 2, a colon as in 1:2, or a fraction as in $\frac{1}{2}$. **15.** We want a unit rate, which is a rate with a denominator of 1. A unit rate tells us how much of the first quantity (\$) will occur in 1 of the second quantity (years).

Exercise Set 5.1 **1.** $\frac{2}{3}$ **3.** $\frac{77}{100}$ **5.** $\frac{463}{821}$ **7.** $\frac{3}{4}$ **9.** $\frac{8}{25}$ **11.** $\frac{12}{7}$ **13.** $\frac{2}{7}$ **15.** $\frac{4}{1}$ **17.** $\frac{10}{29}$ **19.** $\frac{25}{144}$ **21.** $\frac{5}{4}$ **23.** $\frac{17}{40}$ **25.** $\frac{45}{659}$ **27.** $\frac{1}{3}$
29. $\frac{15}{1}$ **31.** $\frac{1 \text{ shrub}}{3 \text{ ft}}$ **33.** $\frac{3 \text{ returns}}{20 \text{ sales}}$ **35.** $\frac{2 \text{ phone lines}}{9 \text{ employees}}$ **37.** $\frac{9 \text{ gal}}{2 \text{ acres}}$ **39.** 75 riders/car **41.** 90 wingbeats/sec **43.** \$50,000/yr
45. 319,368 voters/senator **47.** 300 good/defective **49.** \$60,750/species **51.** \$140,000/house **53. a.** 31.25 computer boards/hr
b. 33.5 computer boards/hr **c.** Suellen **55. a.** ≈ 27.6 miles/gal **b.** ≈ 29.2 miles/gal **c.** the truck **57.** \$11.50 per DVD
59. \$0.17 per banana **61.** 8 oz: \$0.149 per oz; 12 oz: \$0.133 per oz; 12 oz **63.** 16 oz: \$0.106 per oz; 6 oz: \$0.115 per oz; 16 oz
65. 12 oz: \$0.191 per oz; 8 oz: \$0.186 per oz; 8 oz **67.** 100: \$0.006 per napkin; 180: \$0.005 per napkin; 180 napkins **69.** 2.3 **71.** 0.15
73. no; answers may vary **75.** 257; 19.2 **77.** 347; 21.6 **79.** 1.5 steps/ft **81.** answers may vary **83.** no; answers may vary
85. no; $\frac{71}{43}$ **87.** no; $\frac{9}{2}$ **89.** No, the shipment should not be refused. **91. a.** $\frac{6}{25}$ **b.** 38 states **c.** $\frac{6}{19}$

Section 5.2

Vocabulary, Readiness & Video Check **1.** proportion; ratio **3.** true **5.** true **7.** false **9.** true **11.** equals or $=$ **13.** a variable

Exercise Set 5.2 **1.** $\dfrac{10 \text{ diamonds}}{6 \text{ opals}} = \dfrac{5 \text{ diamonds}}{3 \text{ opals}}$ **3.** $\dfrac{3 \text{ printers}}{12 \text{ computers}} = \dfrac{1 \text{ printer} \cdot}{4 \text{ computers}}$ **5.** $\dfrac{6 \text{ eagles}}{58 \text{ sparrows}} = \dfrac{3 \text{ eagles}}{29 \text{ sparrows}}$

7. $\dfrac{2\frac{1}{4} \text{ cups flour}}{24 \text{ cookies}} = \dfrac{6\frac{3}{4} \text{ cups flour}}{72 \text{ cookies}}$ **9.** $\dfrac{22 \text{ vanilla wafers}}{1 \text{ cup cookie crumbs}} = \dfrac{55 \text{ vanilla wafers}}{2.5 \text{ cups cookie crumbs}}$ **11.** true **13.** false **15.** true **17.** true

19. false   **21.** false   **23.** true   **25.** false   **27.** true   **29.** $\frac{8}{12} = \frac{4}{6}$; true   **31.** $\frac{5}{2} = \frac{13}{5}$; false   **33.** $\frac{1.8}{2} = \frac{4.5}{5}$; true   **35.** $\dfrac{\frac{2}{3}}{\frac{1}{5}} = \dfrac{\frac{2}{5}}{\frac{1}{9}}$; false

37. 3   **39.** -9 **41.** 4   **43.** 3.2   **45.** 38.4   **47.** 25   **49.** 0.0025   **51.** 1   **53.** $\frac{9}{20}$ **55.** 12   **57.** $\frac{3}{4}$ **59.** $\frac{35}{18}$ **61.** $<$ **63.** $>$ **65.** $<$
67. possible answers: $\frac{9}{3} = \frac{15}{5}; \frac{5}{15} = \frac{3}{9}; \frac{15}{9} = \frac{5}{3}$ **69.** possible answers: $\frac{6}{1} = \frac{18}{3}; \frac{3}{18} = \frac{1}{6}; \frac{18}{6} = \frac{3}{1}$ **71.** $\frac{d}{b} = \frac{c}{a}; \frac{a}{c} = \frac{b}{d}; \frac{b}{a} = \frac{d}{c}$
73. answers may vary   **75.** 14.9   **77.** 0.07   **79.** 3.163   **81.** 0   **83.** 1400   **85.** 252.5

Section 5.3

Vocabulary, Readiness & Video Check   **1.** There are approximately 102.9 mg of cholesterol in a 5-ounce serving of lobster.

Exercise Set 5.3   **1.** 360 baskets   **3.** 165 min   **5.** 630 applications   **7.** 23 ft   **9.** 270 sq ft   **11.** 25 gal   **13.** 450 km   **15.** 16 bags
17. 15 hits   **19.** 27 people   **21.** 18 applications   **23.** 5 weeks   **25.** $10\frac{2}{3}$ servings   **27.** 37.5 sec   **29. a.** 18 tsp   **b.** 6 tbsp
31. 6 people   **33.** 112 ft; 11-in. difference   **35.** 102.9 mg   **37.** 1248 feet; coincidentally, this is the actual height of the Empire State
Building   **39.** 434 emergency room visits   **41.** 85 million Samsung smartphones   **43.** 2.4 c   **45. a.** 0.1 gal   **b.** 13 fl oz
47. a. 2062.5 mg   **b.** no   **49.** $3 \cdot 5$ **51.** $2^2 \cdot 5$ **53.** $2^3 \cdot 5^2$ **55.** 2^5 **57.** 0.8 ml   **59.** 1.25 ml   **61.** $11 \approx 12$ or 1 dozen;
$1.5 \times 8 = 12$; 12 cups of milk   **63.** $4\frac{2}{3}$ ft   **65.** answers may vary

Integrated Review   **1.** $\frac{9}{10}$ **2.** $\frac{9}{25}$ **3.** $\frac{43}{50}$ **4.** $\frac{8}{23}$ **5.** $\frac{173}{139}$ **6.** $\frac{6}{7}$ **7.** $\frac{7}{26}$ **8.** $\frac{20}{33}$ **9.** $\frac{2}{3}$ **10.** $\frac{1}{8}$ **11.** $\frac{2123}{1324}$ **12.** $\frac{1}{4}$ **13. a.** 4   **b.** $\frac{1}{3}$
14. $\frac{2}{3}$ **15.** $\dfrac{1\ \text{office}}{4\ \text{graduate assistants}}$ **16.** $\dfrac{2\ \text{lights}}{5\ \text{ft}}$ **17.** $\dfrac{16\ \text{computers}}{25\ \text{households}}$ **18.** $\dfrac{9\ \text{students}}{2\ \text{computers}}$ **19.** 55 mi/hr   **20.** 140 ft/sec   **21.** 23 mi/gal
22. 16 teachers/computer   **23.** 3 packs: $0.80 per pack; 8 packs: $0.75 per pack; 8 packs   **24.** 4: $0.92 per battery; 10: $0.99 per battery;
4 batteries   **25.** no   **26.** yes   **27.** 24   **28.** 32.5   **29.** $2.\overline{72}$ or $2\frac{8}{11}$ **30.** 18

Section 5.4

Vocabulary, Readiness & Video Check   **1.** meter   **3.** yard   **5.** feet   **7.** feet   **9.** feet; Feet are the original units and we
want them to divide out.   **11.** The sum of 21 yd 4 ft is correct but is not in a good format since there is a yard in 4 feet. Convert
4 feet $= 1$ yd 1 ft and add again: 21 yd $+ 1$ yd $+ 1$ ft $= 22$ yd 1 ft.   **13.** 1.29 cm and 12.9 mm; These two different-unit lengths are equal.

Exercise Set 5.4   **1.** 5 ft   **3.** 36 ft   **5.** 8 mi   **7.** 102 in.   **9.** $3\frac{1}{3}$ yd   **11.** 33,792 ft   **13.** 4.5 yd   **15.** 0.25 ft   **17.** 13 yd 1 ft   **19.** 7 ft 1 in.
21. 1 mi 4720 ft   **23.** 62 in.   **25.** 26 ft   **27.** 84 in.   **29.** 11 ft 2 in.   **31.** 22 yd 1 ft   **33.** 6 ft 5 in.   **35.** 7 ft 6 in.   **37.** 14 ft 4 in.
39. 83 yd 1 ft   **41.** 6000 cm   **43.** 4 cm   **45.** 0.5 km   **47.** 1.7 m   **49.** 15 m   **51.** 42,000 cm   **53.** 7000 m   **55.** 83 mm   **57.** 0.201 dm
59. 40 mm   **61.** 8.94 m   **63.** 2.94 m or 2940 mm   **65.** 1.29 cm or 12.9 mm   **67.** 12.64 km or 12,640 m   **69.** 54.9 m   **71.** 1.55 km
73. $348\frac{2}{3}$; 12,552   **75.** $11\frac{2}{3}$; 420   **77.** 5000; 0.005; 500   **79.** 0.065; 65; 0.000065   **81.** 342,000; 342,000,000; 34,200,000   **83. a.** $213\frac{2}{3}$ yd
b. 7692 in.   **85.** 10 ft 6 in.   **87.** 5100 ft   **89.** 5.0 times   **91.** 13 ft 11 in.   **93.** 26.7 mm   **95.** 15 ft 9 in.   **97.** 3 ft 1 in.   **99.** 41.25 m
or 4125 cm   **101.** 3.35 m   **103.** 2.13 m   **105.** $121\frac{1}{3}$ yd   **107.** 15 tiles   **109.** $\frac{21}{100}$ **111.** 0.13   **113.** 0.25   **115.** no   **117.** yes   **119.** no
121. Estimate: 13 yd   **123.** answers may vary; for example, $1\frac{1}{3}$ yd or 48 in.   **125.** answers may vary   **127.** 334.89 sq m

Section 5.5

Vocabulary, Readiness & Video Check   **1.** Mass   **3.** gram   **5.** 2000   **7.** 2 lb   **9.** 2 tons   **11.** pounds; Pounds are the units we're
converting to.   **13.** 3 places to the right; 4 g $= 4000$ mg

Exercise Set 5.5   **1.** 32 oz   **3.** 10,000 lb   **5.** 9 tons   **7.** $3\frac{3}{4}$ lb   **9.** $1\frac{3}{4}$ tons   **11.** 204 oz   **13.** 9800 lb   **15.** 76 oz   **17.** 1.5 tons
19. $\frac{1}{20}$ lb   **21.** 92 oz   **23.** 161 oz   **25.** 5 lb 9 oz   **27.** 53 lb 10 oz   **29.** 8 tons 750 lb   **31.** 3 tons 175 lb   **33.** 8 lb 11 oz   **35.** 31 lb 2 oz
37. 1 ton 700 lb   **39.** 0.5 kg   **41.** 4000 mg   **43.** 25,000 g   **45.** 0.048 g   **47.** 0.0063 kg   **49.** 15,140 mg   **51.** 6250 g   **53.** 350,000 cg
55. 13.5 mg   **57.** 5.815 g or 5815 mg   **59.** 1850 mg or 1.85 g   **61.** 1360 g or 1.36 kg   **63.** 13.52 kg   **65.** 2.125 kg
67. 200,000; 3,200,000   **69.** $\frac{269}{400}$ or 0.6725; 21,520   **71.** 0.5; 0.0005; 50   **73.** 21,000; 21,000,000; 2,100,000   **75.** 8.064 kg   **77.** 30 mg
79. 5 lb 8 oz   **81.** 35 lb 14 oz   **83.** 6 lb 15.4 oz   **85.** 144 mg   **87.** 6.12 kg   **89.** 130 lb   **91.** 211 lb   **93.** 0.16   **95.** 0.875   **97.** no
99. yes   **101.** no   **103.** answers may vary; for example, 250 mg or 0.25 g   **105.** true   **107.** answers may vary

Section 5.6

Vocabulary, Readiness & Video Check **1.** capacity **3.** fluid ounces **5.** cups **7.** quarts **9.** 2 pt **11.** 2 gal **13.** 3 qt **15.** 3 c **17.** amount; units **19.** 3 places to the left; 5600 ml = 5.6 L

Exercise Set 5.6 **1.** 4 c **3.** 16 pt **5.** $3\frac{1}{2}$ gal **7.** 5 pt **9.** 8 c **11.** $3\frac{3}{4}$ qt **13.** $10\frac{1}{2}$ qt **15.** 9 c **17.** 23 qt **19.** $\frac{1}{4}$ pt **21.** 14 gal 2 qt **23.** 4 gal 3 qt 1 pt **25.** 22 pt **27.** 13 gal 2 qt **29.** 4 c 4 fl oz **31.** 1 gal 1 qt **33.** 2 gal 3 qt 1 pt **35.** 17 gal **37.** 4 gal 3 qt **39.** 5000 ml **41.** 0.00016 kl **43.** 5.6 L **45.** 320 cl **47.** 0.41 kl **49.** 0.064 L **51.** 160 L **53.** 3600 ml **55.** 19.3 L **57.** 4.5 L or 4500 ml **59.** 8410 ml or 8.41 L **61.** 16,600 ml or 16.6 L **63.** 3840 ml **65.** 162.4 L **67.** 336; 84; 168 **69.** $\frac{1}{4}$; 1; 2 **71.** 1.59 L **73.** 18.954 L **75.** 4.3 fl oz **77.** yes **79.** $0.677 **81.** $\frac{4}{5}$ **83.** $\frac{3}{5}$ **85.** $\frac{9}{10}$ **87.** no **89.** no **91.** less than; answers may vary **93.** answers may vary **95.** 128 fl oz **97.** 1.5 cc **99.** 2.7 cc **101.** 54 u or 0.54 cc **103.** 86 u or 0.86 cc

Section 5.7

Vocabulary, Readiness & Video Check **1.** 1 L ≈ 0.26 gal or 3.79 L ≈ 1 gal

Exercise Set 5.7 **1.** 25.57 fl oz **3.** 218.44 cm **5.** 40 oz **7.** 57.66 mi **9.** 3.77 gal **11.** 13.5 kg **13.** 1.5; $1\frac{2}{3}$; 150; 60 **15.** 55; 5500; 180; 2160 **17.** 3.94 in. **19.** 80.5 kph **21.** 0.008 oz **23.** yes **25.** 2790 mi **27.** 90 mm **29.** 112.5 g **31.** 104 mph **33.** 26.24 ft **35.** 3 mi **37.** 8 fl oz **39.** b **41.** b **43.** c **45.** d **47.** d **49.** 29 **51.** 9 **53.** 5 **55.** 36 **57.** 2.13 sq m **59.** 1.19 sq m **61.** 1.69 sq m **63.** 21.3 mg–25.56 mg **65.** 800 sq m or 8606.72 sq ft

Chapter 5 Vocabulary Check **1.** ratio **2.** proportion **3.** unit rate **4.** unit price **5.** rate **6.** cross products **7.** equal **8.** not equal **9.** Weight **10.** Mass **11.** meter **12.** unit fractions **13.** liter

Chapter 5 Review **1.** $\frac{23}{37}$ **2.** $\frac{14}{51}$ **3.** $\frac{5}{4}$ **4.** $\frac{11}{13}$ **5.** $\frac{7}{15}$ **6.** $\frac{17}{35}$ **7.** $\frac{18}{35}$ **8.** $\frac{35}{27}$ **9. a.** 1 **b.** $\frac{1}{25}$ **10. a.** 15 **b.** $\frac{3}{5}$ **11.** $\frac{5 \text{ pages}}{2 \text{ min}}$ **12.** $\frac{4 \text{ computers}}{3 \text{ hr}}$ **13.** 52 mi/hr **14.** 15 ft/sec **15.** $6.96/CD **16.** $1\frac{1}{3}$ gal/acre **17.** 8 oz: $0.124 per oz; 12 oz: $0.141 per oz; 8-oz size **18.** 18 oz: $0.083; 28 oz: $0.085; 18-oz size **19.** $\frac{16 \text{ sandwiches}}{8 \text{ players}} = \frac{2 \text{ sandwiches}}{1 \text{ player}}$ **20.** $\frac{12 \text{ tires}}{3 \text{ cars}} = \frac{4 \text{ tires}}{1 \text{ car}}$ **21.** no **22.** yes **23.** yes **24.** no **25.** −5 **26.** −15 **27.** $6\frac{3}{4}$ **28.** $7\frac{1}{5}$ **29.** 0.94 **30.** 0.36 **31.** $1\frac{1}{8}$ **32.** $1\frac{3}{7}$ **33.** 14 passes **34.** 35 attempts **35.** 8 bags **36.** 16 bags **37.** $40\frac{1}{2}$ ft **38.** $8\frac{1}{4}$ in. **39.** 9 ft **40.** 18 in. **41.** 17 yd 1 ft **42.** 3 ft 10 in. **43.** 4200 cm **44.** 0.00231 km **45.** 21 yd 1 ft **46.** 7 ft 5 in. **47.** 9.5 cm or 95 mm **48.** 2.45 km **49.** 108.5 km **50.** 0.24 sq m **51.** 4.125 lb **52.** 4600 lb **53.** 3 lb 4 oz **54.** 5 tons 300 lb **55.** 0.027 g **56.** 40,000 g **57.** 3 lb 9 oz **58.** 33 lb 8 oz **59.** 4 lb 4 oz **60.** 9 tons 1075 lb **61.** 8 qt **62.** 5 c **63.** 7 pt **64.** 72 c **65.** 4 qt 1 pt **66.** 3 gal 3 qt **67.** 3800 ml **68.** 0.042 dl **69.** 1 gal 1 qt **70.** 736 ml or 0.736 L **71.** 10.88 L **72.** yes **73.** 22.96 ft **74.** 10.55 m **75.** 4.55 gal **76.** 8.27 qt **77.** 425.25 g **78.** 10.35 kg **79.** 109 yd **80.** 180.4 lb **81.** 3.18 qt **82.** 2.36 in. **83.** $\frac{3}{5}$ **84.** $\frac{1}{8}$ **85.** $\frac{1 \text{ teacher}}{9 \text{ students}}$ **86.** $\frac{1 \text{ nurse}}{4 \text{ patients}}$ **87.** 34 miles/hour **88.** 2 gallons/cow **89.** 4 oz: $1.235; 8 oz: $1.248; 4-oz size **90.** 12 oz: $0.054; 64 oz: $0.047; 64-oz size **91.** $\frac{2 \text{ cups cookie dough}}{30 \text{ cookies}} = \frac{4 \text{ cups cookie dough}}{60 \text{ cookies}}$ **92.** $\frac{5 \text{ nickels}}{3 \text{ dollars}} = \frac{20 \text{ nickels}}{12 \text{ dollars}}$ **93.** 1.6 **94.** 25 **95.** 13,200 ft **96.** 10.75 ft **97.** 4 tons 200 lb **98.** 500 cm **99.** 1.4 g **100.** 0.000286 km **101.** 9117 m or 9.117 km **102.** 8 gal 2 qt

Chapter 5 Test **1.** $\frac{15}{2}$ **2.** $\frac{3 \text{ in.}}{10 \text{ days}}$ **3.** $\frac{43}{50}$ **4.** $\frac{47}{78}$ **5.** $\frac{197}{62}$ **6.** 81.25 km/hr **7.** 28 students/teacher **8.** 9 in./sec **9.** 8-oz size **10.** true **11.** 5 **12.** $4\frac{4}{11}$ **13.** −8 **14.** $\frac{7}{8}$ **15.** $49\frac{1}{2}$ ft **16.** $3\frac{3}{4}$ hr **17.** $53\frac{1}{3}$ g **18.** 23 ft 4 in. **19.** 10 qt **20.** 1.875 lb **21.** 0.04 g **22.** 36 mm **23.** 830 ml **24.** 3 lb 13 oz **25.** 2 gal 3 qt **26.** 2.256 km or 2256 m **27.** 5.6 m **28.** 4 gal 3 qt **29.** 91.4 m **30.** 16 ft 6 in. **31.** 493 ft 6 in. **32.** 150.368 m **33.** 3.1 mi

Cumulative Review **1. a.** 3 **b.** 15 **c.** 0 **d.** 70; Sec. 1.4, Ex. 1 **2. a.** 0 **b.** 20 **c.** 0 **d.** 20; Sec. 1.6 **3.** 249,000; Sec. 1.5, Ex. 3 **4.** 249,000; Sec. 1.5 **5. a.** 200 **b.** 1230; Sec. 1.6, Ex. 3 **6.** 373 R 24; Sec. 1.7 **7.** $6171; Sec. 1.8, Ex. 3 **8.** 16,591 feet; Sec. 1.8 **9.** $2^4 \cdot 5$; Sec. 3.2, Ex. 2 **10.** 8; Sec. 1.9 **11.** $\frac{3}{5}$; Sec. 3.2, Ex. 4 **12.** 243; Sec. 1.9 **13.** $-\frac{1}{8}$; Sec. 3.3, Ex. 5 **14.** $15\frac{3}{8}$; Sec. 3.7 **15.** $\frac{2}{5}$; Sec. 3.3, Ex. 6 **16.** $\frac{5}{54}$; Sec. 3.3 **17.** $\frac{5}{7}$; Sec. 3.4, Ex. 1 **18.** $\frac{19}{30}$; Sec. 3.4 **19.** 2; Sec. 3.4, Ex. 3 **20.** $\frac{4}{5}$; Sec. 3.4 **21.** 14; Sec. 3.4, Ex. 12 **22.** $\frac{49}{50}$; Sec. 3.5 **23.** $\frac{15}{20}$; Sec. 3.4, Ex. 16 **24.** yes; Sec. 3.2 **25.** $-\frac{8}{33}$; Sec. 3.5, Ex. 5 **26.** $7\frac{47}{72}$; Sec. 3.7 **27.** $\frac{1}{6}$ hr; Sec. 3.5, Ex. 11

28. 27; Sec. 1.9   **29.** $7\frac{17}{24}$; Sec. 3.7, Ex. 9   **30.** $\frac{16}{27}$; Sec. 3.6   **31.** $<$; Sec. 3.5, Ex. 7   **32.** 14,000,000; Sec. 1.6   **33.** negative five and eighty-two hundredths; Sec. 4.1, Ex. 1b   **34.** 0.075; Sec. 4.1   **35.** 736.2; Sec. 4.1, Ex. 15   **36.** 736.236; Sec. 4.1   **37.** 25.454; Sec. 4.2, Ex. 1   **38.** 681.24; Sec. 4.2   **39.** 0.8496; Sec. 4.3, Ex. 2   **40.** 0.375; Sec. 4.5   **41.** -0.052; Sec. 4.4, Ex. 3   **42.** $\frac{79}{10}$; Sec. 4.5   **43.** -3.7; Sec. 4.5, Ex. 12   **44.** 3; Sec. 5.2   **45.** $\frac{4}{9}, \frac{9}{20}$, 0.456; Sec. 4.5, Ex. 10   **46.** 140 m/sec; Sec. 5.1   **47.** $\frac{3}{2}$; Sec. 5.1, Ex. 2   **48.** $\frac{1}{3}$; Sec. 5.1   **49.** $\frac{26}{31}$; Sec. 5.1, Ex. 3   **50.** $\frac{1}{10}$; Sec. 5.1

Chapter 6 Percent

Section 6.1

Vocabulary, Readiness & Video Check   **1.** Percent   **3.** percent   **5.** 0.01   **7.** 13%   **9.** 87%   **11.** 1%   **13.** Percent means "per 100."   **15.** In both cases, we multiply the number by 1 in the form of 100%.

Exercise Set 6.1   **1.** 96%   **3. a.** 75%   **b.** 25%   **5.** football; 37%   **7.** 50%   **9.** 0.41   **11.** 0.06   **13.** 1.00 or 1   **15.** 0.613   **17.** 0.028   **19.** 0.006   **21.** 3.00 or 3   **23.** 0.3258   **25.** $\frac{3}{25}$ **27.** $\frac{1}{25}$ **29.** $\frac{9}{200}$ **31.** $\frac{7}{4}$ or $1\frac{3}{4}$ **33.** $\frac{1}{16}$ **35.** $\frac{31}{300}$ **37.** $\frac{179}{800}$ **39.** 0.3%   **41.** 22%   **43.** 530%   **45.** 5.6%   **47.** 33.28%   **49.** 300%   **51.** 70%   **53.** 70%   **55.** 40%   **57.** 34%   **59.** $37\frac{1}{2}$%   **61.** $77\frac{7}{9}$%   **63.** 250%   **65.** 190%   **67.** 63.64%   **69.** 26.67%   **71.** $0.35, \frac{7}{20}$; $20\%, 0.2$; $50\%, \frac{1}{2}$; $0.7, \frac{7}{10}$; $37.5\%, 0.375$ **73.** $0.4, \frac{2}{5}$; $23.5\%, \frac{47}{200}$; $80\%, 0.8$; $0.333\overline{3}, \frac{1}{3}$; $87.5\%, 0.875$; $0.075, \frac{3}{40}$ **75.** $2, 2$; $280\%, 2\frac{4}{5}$; $7.05, 7\frac{1}{20}$; $454\%, 4.54$ **77.** $0.67; \frac{67}{100}$ **79.** $0.554; \frac{277}{500}$ **81.** 18%   **83.** 0.38   **85.** 0.394   **87.** $0.005; \frac{1}{200}$ **89.** $0.142; \frac{71}{500}$ **91.** $0.079; \frac{79}{1000}$ **93.** $n = 15$ **95.** $n = 12$ **97.** 77%   **99.** 107.8%   **101. a.** 52.9%   **b.** 52.86%   **103.** b, d   **105.** 4%   **107.** personal care aides   **109.** 0.617   **111.** 75%   **113.** 80%   **115.** greater   **117.** answers may vary   **119.** 0.266; 26.6%   **121.** 1.155; 115.5%

Section 6.2

Vocabulary, Readiness & Video Check   **1.** is   **3.** amount; base; percent   **5.** greater   **7.** percent: 42%; base: 50; amount: 21   **9.** percent: 125%; base: 86; amount: 107.5   **11.** "of" means multiplication; "is" means equals; "what" (or some equivalent) means the unknown number

Exercise Set 6.2   **1.** $18\% \cdot 81 = n$ **3.** $20\% \cdot n = 105$ **5.** $0.6 = 40\% \cdot n$ **7.** $n \cdot 80 = 3.8$ **9.** $n = 9\% \cdot 43$ **11.** $n \cdot 250 = 150$ **13.** 3.5   **15.** 28.7   **17.** 10   **19.** 600   **21.** 110%   **23.** 34%   **25.** 1   **27.** 645   **29.** 500   **31.** 5.16%   **33.** 25.2   **35.** 35%   **37.** 35   **39.** 0.624   **41.** 0.5%   **43.** 145   **45.** 63%   **47.** 4%   **49.** $n = 30$ **51.** $n = 3\frac{7}{11}$ **53.** $\frac{17}{12} = \frac{n}{20}$ **55.** $\frac{8}{9} = \frac{14}{n}$ **57.** c   **59.** b   **61.** Twenty percent of some number is eighteen and six-tenths.   **63.** b   **65.** c   **67.** c   **69.** a   **71.** a   **73.** answers may vary   **75.** 686.625   **77.** 12,285

Section 6.3

Vocabulary, Readiness & Video Check   **1.** amount; base; percent   **3.** amount   **5.** amount: 12.6; base: 42; percent: 30   **7.** amount: 102; base: 510; percent: 20   **9.** 45 follows the word "of," so it is the base.

Exercise Set 6.3   **1.** $\frac{a}{45} = \frac{98}{100}$ **3.** $\frac{a}{150} = \frac{4}{100}$ **5.** $\frac{14.3}{b} = \frac{26}{100}$ **7.** $\frac{84}{b} = \frac{35}{100}$ **9.** $\frac{70}{400} = \frac{p}{100}$ **11.** $\frac{8.2}{82} = \frac{p}{100}$ **13.** 26   **15.** 18.9   **17.** 600   **19.** 10   **21.** 120%   **23.** 28%   **25.** 37   **27.** 1.68   **29.** 1000   **31.** 210%   **33.** 55.18   **35.** 45%   **37.** 75   **39.** 0.864   **41.** 0.5%   **43.** 140   **45.** 9.6   **47.** 113%   **49.** $\frac{7}{8}$ **51.** $3\frac{2}{15}$ **53.** 0.7   **55.** 2.19   **57.** answers may vary   **59.** no; $a = 16$ **61.** yes   **63.** answers may vary   **65.** 12,011.2   **67.** 7270.6

Integrated Review   **1.** 12%   **2.** 68%   **3.** 12.5%   **4.** 250%   **5.** 520%   **6.** 800%   **7.** 6%   **8.** 44%   **9.** 750%   **10.** 325%   **11.** 3%   **12.** 5%   **13.** 0.65   **14.** 0.31   **15.** 0.08   **16.** 0.07   **17.** 1.42   **18.** 4   **19.** 0.029   **20.** 0.066   **21.** $0.03; \frac{3}{100}$ **22.** $0.05; \frac{1}{20}$ **23.** $0.0525; \frac{21}{400}$ **24.** $0.1275; \frac{51}{400}$ **25.** $0.38; \frac{19}{50}$ **26.** $0.45; \frac{9}{20}$ **27.** $0.123; \frac{37}{300}$ **28.** $0.167; \frac{1}{6}$ **29.** 8.4   **30.** 100   **31.** 250   **32.** 120%   **33.** 28%   **34.** 76   **35.** 11   **36.** 130%   **37.** 86%   **38.** 37.8   **39.** 150   **40.** 62

Section 6.4

Vocabulary, Readiness & Video Check   **1.** The price of the home was $175,000.

Exercise Set 6.4   **1.** 1600 bolts   **3.** 8.8 lb   **5.** 14%   **7.** 91,800 businesses   **9.** 45%   **11.** 496 chairs; 5704 chairs   **13.** 108,680 physician assistants   **15.** 9900.24 thousand or approximately 9900 thousand   **17.** 29%   **19.** 50%   **21.** 12.5%   **23.** 29.2%   **25.** $175.000   **27.** 31.2 hr   **29.** increase: $867.87; new price: $20,153.87   **31.** 40 ft   **33.** increase: $1210; tuition in 2013–2014: $9616   **35.** increase: 128,760 associate degrees; 2020–2021: 1,016,760 associate degrees   **37.** 30; 60%   **39.** 52; 80%   **41.** 2; 25%   **43.** 120; 75%   **45.** 44%   **47.** 1.3%   **49.** 142.0%   **51.** 10.3%   **53.** 139.5%   **55.** 18.8%   **57.** 19.7%   **59.** 55.6%   **61.** 4.56   **63.** 11.18   **65.** 58.54   **67.** The increased number is double the original number.   **69.** percent of increase $= \dfrac{30}{150} = 20\%$ **71.** False; the percents are different.

Section 6.5

Vocabulary, Readiness & Video Check   **1.** sales tax   **3.** commission   **5.** sale price   **7.** We rewrite the percent as an equivalent decimal.   **9.** Replace "amount of discount" in the second equation with "discount rate · original price": sale price = original price − (discount rate · orginal price).

Exercise Set 6.5   **1.** $7.50   **3.** $858.93   **5.** 7%   **7. a.** $120   **b.** $130.20   **9.** $117; $1917   **11.** $485   **13.** 6%   **15.** $16.10; $246.10   **17.** $53,176.04   **19.** 14%   **21.** $4888.50   **23.** $185,500   **25.** $8.90; $80.10   **27.** $98.25; $98.25   **29.** $143.50; $266.50   **31.** $3255; $18,445   **33.** $45; $255   **35.** $27.45; $332.45   **37.** $3.08; $59.08   **39.** $7074   **41.** 8%   **43.** 1200   **45.** 132   **47.** 16   **49.** d   **51.** $4.00; $6.00; $8.00   **53.** $7.20; $10.80; $14.40   **55.** a discount of 60% is better; answers may vary   **57.** $26,838.45

Section 6.6

Calculator Explorations   **1.** 1.56051   **3.** 8.06231   **5.** $634.49

Vocabulary, Readiness & Video Check   **1.** simple   **3.** Compound   **5.** Total amount   **7.** principal   **9.** The denominator is the total number of payments. We are asked to find the monthly payment for a 4-year loan, and since there are 48 months in 4 years, there are 48 total payments.

Exercise Set 6.6   **1.** $32   **3.** $73.60   **5.** $750   **7.** $33.75   **9.** $700   **11.** $101,562.50; $264,062.50   **13.** $5562.50   **15.** $14,280   **17.** $46,815.37   **19.** $2327.14   **21.** $58,163.65   **23.** 2915.75   **25.** $2938.66   **27.** $2971.89   **29.** $260.31   **31.** $637.26   **33.** 32 yd   **35.** 35 m   **37.** answers may vary   **39.** answers may vary

Chapter 6 Vocabulary Check   **1.** of   **2.** is   **3.** Percent   **4.** Compound interest   **5.** $\dfrac{\text{amount}}{\text{base}}$ **6.** 100%   **7.** 0.01   **8.** $\dfrac{1}{100}$ **9.** base; amount   **10.** Percent of decrease   **11.** Percent of increase   **12.** Sales tax   **13.** Total price   **14.** Commission   **15.** Amount of discount   **16.** Sale price

Chapter 6 Review   **1.** 37%   **2.** 77%   **3.** 0.83   **4.** 0.75   **5.** 0.735   **6.** 0.015   **7.** 1.25   **8.** 1.45   **9.** 0.005   **10.** 0.007   **11.** 2.00 or 2   **12.** 4.00 or 4   **13.** 0.2625   **14.** 0.8534   **15.** 260%   **16.** 102%   **17.** 35%   **18.** 5.5%   **19.** 72.5%   **20.** 25.2%   **21.** 7.6%   **22.** 8.5%   **23.** 71%   **24.** 65%   **25.** 400%   **26.** 900%   **27.** $\dfrac{1}{100}$ **28.** $\dfrac{1}{10}$ **29.** $\dfrac{1}{4}$ **30.** $\dfrac{17}{200}$ **31.** $\dfrac{51}{500}$ **32.** $\dfrac{1}{6}$ **33.** $\dfrac{1}{3}$ **34.** $1\dfrac{1}{10}$ **35.** 20%   **36.** 70%   **37.** $83\dfrac{1}{3}\%$ **38.** 60%   **39.** 125%   **40.** $166\dfrac{2}{3}\%$ **41.** 6.25%   **42.** 62.5%   **43.** 100,000   **44.** 8000   **45.** 23%   **46.** 114.5   **47.** 3000   **48.** 150%   **49.** 418   **50.** 300   **51.** 64.8   **52.** 180%   **53.** 110%   **54.** 165   **55.** 66%   **56.** 16%   **57.** 20.9%   **58.** 106.25%   **59.** $206,400   **60.** $13.23   **61.** $263.75   **62.** $1.15   **63.** $5000   **64.** $300.38   **65.** discount: $900; sale price: $2100   **66.** discount: $9; sale price: $81   **67.** $160   **68.** $325   **69.** $30,104.61   **70.** $17,506.54   **71.** $80.61   **72.** $32,830.10   **73.** 0.038   **74.** 0.245   **75.** 0.009   **76.** 54%   **77.** 9520%   **78.** 30%   **79.** $\dfrac{47}{100}$ **80.** $\dfrac{8}{125}$ **81.** $\dfrac{7}{125}$ **82.** $37\dfrac{1}{2}\%$ **83.** $15\dfrac{5}{13}\%$ **84.** 120%   **85.** 268.75   **86.** 110%   **87.** 708.48   **88.** 134%   **89.** 300%   **90.** 38.4   **91.** 560   **92.** 325%   **93.** 26%   **94.** $6786.50   **95.** $617.70   **96.** $3.45   **97.** 12.5%   **98.** $1491   **99.** $17,951.01   **100.** $11,687.50

Chapter 6 Test   **1.** 0.85   **2.** 5   **3.** 0.008   **4.** 5.6%   **5.** 610%   **6.** 39%   **7.** $\dfrac{6}{5}$ or $1\dfrac{1}{5}$ **8.** $\dfrac{77}{200}$ **9.** $\dfrac{1}{500}$ **10.** 55%   **11.** 37.5%   **12.** $155\dfrac{5}{9}\%$ **13.** 33.6   **14.** 1250   **15.** 75%   **16.** 38.4 lb   **17.** $56,750   **18.** $358.43   **19.** 5%   **20.** discount: $18; sale price: $102   **21.** $395   **22.** 1%   **23.** $647.50   **24.** $2005.63   **25.** $427

Cumulative Review   **1.** 206 cases; 12 cans; yes; Sec. 1.8, Ex. 2   **2.** 31,084; Sec. 1.6   **3. a.** $4\dfrac{2}{7}$ **b.** $1\dfrac{1}{15}$ **c.** 14; Sec. 3.6, Ex. 9   **4. a.** $\dfrac{19}{7}$ **b.** $\dfrac{101}{10}$ **c.** $\dfrac{43}{8}$; Sec. 3.6   **5.** $-\dfrac{10}{27}$; Sec. 3.2, Ex. 6   **6.** 44; Sec. 1.7   **7.** $\dfrac{23}{56}$; Sec. 3.3, Ex. 4   **8.** 76,500; Sec. 1.5   **9.** $\dfrac{4}{5}$; Sec. 3.3, Ex. 11   **10.** $\dfrac{15}{4}$; $3\dfrac{3}{4}$; Sec. 3.1   **11.** $\dfrac{4}{5}$ in.; Sec. 3.4, Ex. 10   **12.** 50; Sec. 1.9   **13.** 60; Sec. 3.4, Ex. 13   **14.** $\dfrac{1}{3}$; Sec. 3.4   **15.** $\dfrac{2}{3}$; Sec. 3.5, Ex. 1

16. 340; Sec. 3.3   **17.** $3\frac{5}{14}$; Sec. 3.7, Ex. 13   **18.** 33; Sec. 1.9   **19.** $\frac{3}{20}$; Sec. 3.7, Ex. 6   **20.** $33\frac{27}{40}$; Sec. 3.7   **21.** $\frac{1}{16}$; Sec. 3.3, Ex. 8b

22. $6\frac{3}{8}$; Sec. 3.7   **23.** -0.625; Sec. 4.5, Ex. 2   **24.** 0.09; Sec. 4.5   **25.** 3.14; Sec. 4.5, Ex. 4   **26.** 0.0048; Sec. 4.5   **27.** $41,568; Sec. 4.1, Ex. 18

28. 27.94; Sec. 4.2   **29.** 829.6561; Sec. 4.2, Ex. 2   **30.** 1248.3; Sec. 4.3   **31.** 18.408; Sec. 4.3, Ex. 1   **32.** 76,300; Sec. 4.3   **33.** 0.7861; Sec. 4.4,

Ex. 8   **34.** 1.276; Sec. 4.4   **35.** -0.012; Sec. 4.4, Ex. 9   **36.** 50.65; Sec. 4.4   **37.** 7.236; Sec. 4.5, Ex. 11   **38.** 0.191; Sec. 4.5   **39.** 0.25; Sec. 4.5,

Ex. 1   **40.** $0.\overline{5} \approx 0.556$; Sec. 4.5   **41.** no; Sec. 5.2, Ex. 3   **42.** 0.052 per tortilla; 0.058 per tortilla; 18-tortilla pkg is better buy; Sec. 5.1

43. $17\frac{1}{2}$ mi; Sec. 5.3, Ex. 1   **44. a.** 0.07   **b.** 2   **c.** 0.005; Sec. 6.1   **45.** $n = 25\% \cdot 0.008$; Sec. 6.2, Ex. 3   **46.** 37.5% or $37\frac{1}{2}$ %; Sec. 6.1

Chapter 7 Statistics and Probability

Section 7.1

Vocabulary, Readiness & Video Check   1. bar   **3.** line   **5.** Count the number of symbols and multiply this number by how much each symbol stands for (from the key).   **7.** bar graph

Exercise Set 7.1   1. Kansas   **3.** 5.5 million or 5,500,000 acres   **5.** South Dakota   **7.** 4.0 million or 4,000,000 acres   **9.** 66,000

11. 2006   **13.** 12,000   **15.** 66,000 wildfires/year   **17.** September   **19.** 76   **21.** $\frac{1}{38}$ **23.** Tokyo, Japan; about 34.8 million or

34,800,000   **25.** New York, U.S.; 21.6 million or 21,600,000   **27.** approximately 2 million or 2,000,000

29.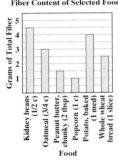

31.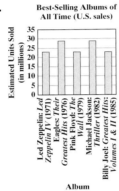

33. 15 adults   **35.** 61 adults   **37.** 24 adults   **39.** 12 adults

41. $\frac{9}{100}$ **43.** 20 to 44   **45.** 109 million or 109,000,000

47. 23 million or 23,000,000   **49.** answers may vary   **51.** |; 1

53. ⵎ⵰ |||; 8   **55.** ⵎ⵰ |; 6   **57.** ⵎ⵰ |; 6   **59.** ||; 2

61.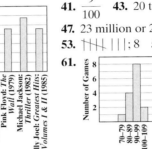

63. 50   **65.** 2011, 2013, 2014   **67.** 2006, 2008
69. 2006, 2008, 2012   **71.** 3.6   **73.** 6.2   **75.** 25%
77. 34%   **79.** 83°F   **81.** Sunday; 68°F
83. Tuesday; 13°F   **85.** answers may vary

Section 7.2

Vocabulary, Readiness & Video Check   1. circle   **3.** 360   **5.** 100%

Exercise Set 7.2   1. parent or guardian's home   **3.** $\frac{9}{35}$ **5.** $\frac{9}{16}$ **7.** Asia   **9.** 37%   **11.** 17,100,000 sq mi   **13.** 2,850,000 sq mi

15. 55%   **17.** nonfiction   **19.** 31,400 books   **21.** 27,632 books   **23.** 25,120 books

25.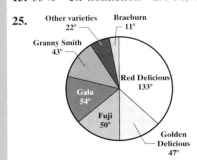

27.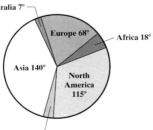

29. $2^2 \times 5$ **31.** $2^3 \times 5$ **33.** 5×17 **35.** Pacific; answers may vary   **37.** 129,600,002 sq km
39. 55,542,858 sq km   **41.** 602 respondents
43. 2276 respondents   **45.** $\frac{301}{837}$
47. no; answers may vary

Integrated Review   1. 700,000   **2.** 500,000   **3.** registered nurses, retail salespersons, home health aides   **4.** food workers
5. Oroville Dam; 755 ft   **6.** New Bullards Bar Dam; 635 ft   **7.** 15 ft   **8.** 4 dams   **9.** Thursday and Saturday; 100°F   **10.** Monday; 82°F
11. Sunday, Monday, and Tuesday   **12.** Wednesday, Thursday, Friday, and Saturday   **13.** 70 qt containers   **14.** 52 qt containers
15. 2 qt containers   **16.** 6 qt containers   **17.** ||; 2   **18.** |; 1   **19.** |||; 3   **20.** ⵎ⵰ |; 6   **21.** ⵎ⵰; 5

22.

Section 7.3

Vocabulary, Readiness & Video Check **1.** average **3.** mean (or average) **5.** grade point average **7.** Place the data numbers in numerical order (or verify that they already are).

Exercise Set 7.3 **1.** mean: 21; median: 23; no mode **3.** mean: 8.1; median: 8.2; mode: 8.2 **5.** mean: 0.5; median: 0.5; mode: 0.2 and 0.5 **7.** mean: 370.9; median: 313.5; no mode **9.** 1911.6 ft **11.** 1601 ft **13.** answers may vary **15.** 2.79 **17.** 3.64 **19.** 6.8 **21.** 6.9 **23.** 88.5 **25.** 73 **27.** 70 and 71 **29.** 9 rates **31.** $\frac{1}{3}$ **33.** $\frac{3}{5}$ **35.** $\frac{11}{15}$ **37.** 35, 35, 37, 43 **39.** yes; answers may vary

Section 7.4

Vocabulary, Readiness & Video Check **1.** outcome **3.** probability **5.** 0 **7.** The number of outcomes equals the ending number of branches drawn.

Exercise Set 7.4 **1.**

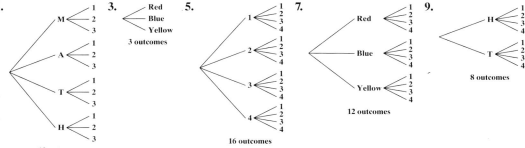

11. $\frac{1}{6}$ **13.** $\frac{1}{3}$ **15.** $\frac{1}{2}$ **17.** $\frac{2}{3}$ **19.** $\frac{1}{3}$ **21.** 1 **23.** $\frac{2}{3}$ **25.** $\frac{1}{7}$ **27.** $\frac{2}{7}$ **29.** $\frac{4}{7}$ **31.** $\frac{19}{100}$ **33.** $\frac{1}{20}$ **35.** $\frac{5}{6}$ **37.** $\frac{1}{6}$ **39.** $\frac{20}{3}$ or $6\frac{2}{3}$
41. $\frac{1}{52}$ **43.** $\frac{1}{13}$ **45.** $\frac{1}{4}$ **47.** $\frac{1}{2}$ **49.** $\frac{5}{36}$ **51.** 0 **53.** answers may vary

Chapter 7 Vocabulary Check **1.** bar **2.** mean **3.** outcomes **4.** pictograph **5.** mode **6.** line **7.** median **8.** tree diagram **9.** experiment **10.** circle **11.** probability **12.** histogram; class interval; class frequency

Chapter 7 Review **1.** 1,250,000 **2.** 1,500,000 **3.** South **4.** Northeast **5.** South **6.** Northeast, Midwest, West **7.** 12% **8.** 2012 **9.** 1992, 2002, 2012 **10.** answers may vary **11.** 962 **12.** 927 **13.** 2000 and 2004 **14.** 1996 and 2000 **15.** 27 **16.** 120 **17.** 1 employee **18.** 4 employees **19.** 18 employees **20.** 9 employees **21.** 𝈂; 5 **22.** |||; 3 **23.** ||||; 4

24. 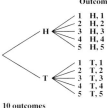 **25.** mortgage payment **26.** utilities **27.** $1225 **28.** $700 **29.** $\frac{39}{160}$ **30.** $\frac{7}{40}$ **31.** 30 **32.** 12 **33.** 21
34. 1 **35.** mean: 17.8; median: 14; no mode **36.** mean: 58.1; median: 60; mode: 45 and 86 **37.** mean: 24,500; median: 20,000; mode: 20,000 **38.** mean: 447.3; median: 420; mode: 400 **39.** 3.25 **40.** 2.57

41.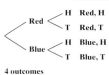

42.

43.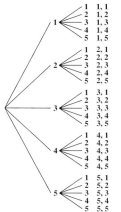

44.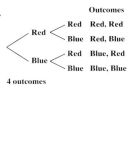

45.

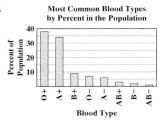

10 outcomes

46. $\frac{1}{6}$ **47.** $\frac{1}{6}$ **48.** $\frac{1}{5}$ **49.** $\frac{1}{5}$ **50.** $\frac{3}{5}$ **51.** $\frac{2}{5}$ **52.** $\frac{1}{2}$ **53.** $\frac{1}{2}$ **54.** mean: 74.4; median: 73; mode: none
55. mean: 48.8; median: 32; mode: none **56.** mean: 454; median: 463.5; mode: 500 **57.** mean: 619.17; median: 647.5; mode: 327 **58.** $\frac{1}{4}$ **59.** $\frac{3}{8}$ **60.** $\frac{1}{4}$ **61.** $\frac{1}{8}$

Chapter 7 Test **1.** $225 **2.** 3rd week; $350 **3.** $1100 **4.** June, August, September **5.** February; 3 cm **6.** March and November
7.

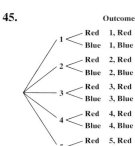

Most Common Blood Types by Percent in the Population

8. 3.0% **9.** 2004, 2005, 2007 **10.** 2003–2004, 2004–2005, 2006–2007, 2008–2009, 2010–2011
11. $\frac{17}{40}$ **12.** $\frac{31}{22}$ **13.** 22,744,700 people **14.** 8,337,500 people **15.** 9 students **16.** 11 students

17.

Class Intervals (Scores)	Tally	Class Frequency (Number of Students)
40–49	|	1
50–59	|||	3
60–69	||||	4
70–79	⊬⊬|	5
80–89	⊬⊬||||	8
90–99	||||	4

18.

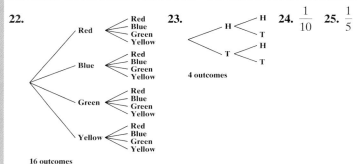

19. mean: 38.4; median: 42; no mode **20.** mean: 12.625; median: 12.5; mode: 12 and 16 **21.** 3.07

22.

Red — Red, Blue, Green, Yellow
Blue — Red, Blue, Green, Yellow
Green — Red, Blue, Green, Yellow
Yellow — Red, Blue, Green, Yellow

16 outcomes

23.

H — H, T
T — H, T

4 outcomes

24. $\frac{1}{10}$ **25.** $\frac{1}{5}$

Cumulative Review **1.** one hundred six million, fifty-two thousand, four hundred forty-seven; Sec. 1.2, Ex. 7 **2.** two hundred seventy-six thousand, four; Sec. 1.2 **3.** 13 in.; Sec. 1.3, Ex. 5 **4.** 18 in.; Sec. 1.3 **5.** 726; Sec. 1.4, Ex. 4 **6.** 9585; Sec. 1.4
7. 249,000; Sec. 1.5, Ex. 3 **8.** 844,000; Sec. 1.5 **9.** 200; Sec. 1.6, Ex. 3a **10.** 29,230; Sec. 1.6 **11.** 208; Sec. 1.7, Ex. 5 **12.** 86; Sec. 1.7
13. 7; Sec. 1.9, Ex. 12 **14.** 35; Sec. 1.9 **15.** 26; Sec. 2.1, Ex. 4 **16.** 10; Sec. 2.1 **17. a.** < **b.** > **c.** >; Sec. 2.2, Ex. 3 **18. a.** <
b. >; Sec. 2.2 **19.** 3; Sec. 2.3, Ex. 1 **20.** −7; Sec. 2.3 **21.** −15; Sec. 2.3, Ex. 5 **22.** −4; Sec. 2.3 **23.** 8; Sec. 2.3, Ex. 6 **24.** 17; Sec. 2.3
25. −14; Sec. 2.4, Ex. 2 **26.** −5; Sec. 2.4 **27.** 11; Sec. 2.4, Ex. 3 **28.** 29; Sec. 2.4 **29.** −4; Sec. 2.4, Ex. 4 **30.** −3; Sec. 2.4
31. −2; Sec. 2.5, Ex. 10 **32.** 6; Sec. 2.5 **33.** 5; Sec. 2.5, Ex. 11 **34.** −13; Sec. 2.5 **35.** −16; Sec. 2.5, Ex. 12 **36.** −10; Sec. 2.5
37. $9\frac{3}{10}$; Sec. 3.7, Ex. 11 **38.** $11\frac{1}{3}$; Sec. 3.7 **39.** $\frac{\$180}{1\ \text{week}}$; Sec. 5.1, Ex. 7 **40.** $\frac{68\ \text{mi}}{1\ \text{hr}}$ or 68 mi/hr; Sec. 5.1 **41.** $2\frac{1}{3}$ yd; Sec. 5.4, Ex. 2
42. 5000 pounds; Sec. 5.5 **43.** 0.0235 g; Sec. 5.5, Ex. 8 **44.** 1060 mm; Sec. 5.4 **45.** $\frac{1}{6}$; Sec. 4.6, Ex. 2 **46.** $\frac{1}{5}$; Sec. 4.6
47. 14 and 77; Sec. 7.3, Ex. 5 **48.** 56; Sec. 7.3 **49.** $\frac{1}{4}$; Sec. 7.4, Ex. 3 **50.** $\frac{3}{5}$; Sec. 7.4

Chapter 8 Introduction to Algebra

Section 8.1

Vocabulary, Readiness & Video Check **1.** expression; term **3.** combine like terms **5.** variable; constant **7.** associative **9.** numerical coefficient **11.** unlike **13.** like **15.** unlike **17.** like **19.** multiplication **21.** the distributive property first and then the associative property of multiplication **23.** addition; multiplication; $P = $ perimeter, $A = $ area

Exercise Set 8.1 **1.** -3 **3.** -2 **5.** 4 **7.** -3 **9.** -10 **11.** 133 **13.** -15 **15.** -4 **17.** $-\frac{4}{3}$ or $-1\frac{1}{3}$ **19.** -12 **21.** $\frac{3}{2}$ or $1\frac{1}{2}$ **23.** -10.6 **25.** $8x$ **27.** $-4n$ **29.** $-2c$ **31.** $-4x$ **33.** $13a - 8$ **35.** $-0.9x + 11.2$ **37.** $2x - 7$ **39.** $-5x + 4y - 5$ **41.** $\frac{1}{2} - \frac{53}{60}x$ **43.** $-4.8m - 4.1$ **45.** $30x$ **47.** $-22y$ **49.** $-4.2a$ **51.** $-4a$ **53.** $2y + 4$ **55.** $15a - 40$ **57.** $-12x - 28$ **59.** $6x - 0.12$ **61.** $-4x - \frac{3}{2}$ **63.** $2x - 9$ **65.** $27n - 20$ **67.** $7w + 15$ **69.** $-11x + 8$ **71.** $(14y + 22)$ m **73.** $(11a + 12)$ ft **75.** $(-25x + 55)$ in. **77.** $36y$ sq in. **79.** $(32x - 64)$ sq km **81.** $(60y + 20)$ sq mi **83.** 4700 sq ft **85.** 64 ft **87.** $(3x + 6)$ ft **89.** \$360 **91.** 78.5 sq ft **93.** $23°F$ **95.** 288 cu in. **97.** $91.2x$ cu in. **99.** -3 **101.** 8 **103.** 0 **105.** incorrect; $15x - 10$ **107.** incorrect; $7x - x - 2$ or $6x - 2$ **109.** distributive **111.** associative **113.** answers may vary **115.** $(20x + 16)$ sq mi **117.** 23,506.2 sq in. **119.** $4824q + 12,274$

Section 8.2

Vocabulary, Readiness & Video Check **1.** equivalent **3.** simplifying **5.** addition **7.** We can add the same number to both sides of an equation or subtract the same number from both sides of an equation and we'll have an equivalent equation.

Exercise Set 8.2 **1.** yes **3.** no **5.** yes **7.** no **9.** 18 **11.** -8 **13.** 9 **15.** -16 **17.** 3 **19.** $\frac{1}{8}$ **21.** 6 **23.** 8 **25.** 5.3 **27.** -1 **29.** -20.1 **31.** 2 **33.** 0 **35.** -28 **37.** -6 **39.** 24 **41.** -30 **43.** 12 **45.** 1 **47.** 1 **49.** 1 **51.** subtract $\frac{2}{3}$ from both sides **53.** add $\frac{4}{5}$ to both sides **55.** answers may vary **57.** 162,964 **59.** 1383 yd **61.** \$19,279,000,000

Section 8.3

Vocabulary, Readiness & Video Check **1.** equivalent **3.** simplifying **5.** multiplication **7.** simplified each side of the equation by combining like terms

Exercise Set 8.3 **1.** 4 **3.** -4 **5.** -30 **7.** -17 **9.** 50 **11.** 25 **13.** -30 **15.** $\frac{1}{3}$ **17.** $-\frac{2}{3}$ **19.** -4 **21.** 8 **23.** 1.3 **25.** 2 **27.** 0 **29.** -0.05 **31.** $-\frac{15}{64}$ **33.** $\frac{2}{3}$ **35.** 6 **37.** 0 **39.** -2 **41.** -8 **43.** 1 **45.** 72 **47.** 35 **49.** -28 **51.** 25 **53.** 1 **55.** -10 **57.** 2006 **59.** 9.3 million acres or 9,300,000 acres **61.** addition **63.** division **65.** answers may vary **67.** 6.5 hr **69.** 58.8 mph **71.** -3648 **73.** 67,896 **75.** -48

Integrated Review **1.** expression **2.** equation **3.** equation **4.** expression **5.** simplify **6.** solve **7.** 4 **8.** -6 **9.** 1 **10.** -4 **11.** $8x$ **12.** $-4y$ **13.** $-2a - 2$ **14.** $-2x + 3y - 7$ **15.** $-8x - 14$ **16.** $-6x + 30$ **17.** $5y - 10$ **18.** $15x - 31$ **19.** $(12x - 6)$ sq m **20.** $20y$ in. **21.** 13 **22.** -9 **23.** $\frac{7}{10}$ **24.** 0 **25.** -4 **26.** 25 **27.** -1 **28.** -3 **29.** 6 **30.** 8 **31.** $-\frac{9}{11}$ **32.** 5

Section 8.4

Calculator Explorations **1.** yes **3.** no **5.** yes

Vocabulary, Readiness & Video Check **1.** $3x - 9 + x - 16$; $5(2x + 6) - 1 = 39$ **3.** addition **5.** distributive **7.** the addition property of equality; to make sure we get an equivalent equation **9.** gives; amounts to

Exercise Set 8.4 **1.** 3 **3.** 1.9 **5.** -4 **7.** 100 **9.** $-3,9$ **11.** -4 **13.** -12 **15.** -3 **17.** -1 **19.** -45 **21.** -9 **23.** 6 **25.** -5 **27.** 5 **29.** 8 **31.** 0 **33.** -22 **35.** 6 **37.** -11 **39.** -7 **41.** -5 **43.** 5 **45.** -3 **47.** 2 **49.** -4 **51.** -1 **53.** 4 **55.** 3 **57.** -1 **59.** 8 **61.** 4 **63.** 3 **65.** 64 **67.** $\frac{1}{9}$ **69.** $\frac{3}{2}$ **71.** -54 **73.** 2 **75.** $\frac{9}{5}$ **77.** 270 **79.** 0 **81.** 4 **83.** 1 **85.** $-42 + 16 = -26$ **87.** $-5(-29) = 145$ **89.** $3(-14 - 2) = -48$ **91.** $\frac{100}{2(50)} = 1$ **93.** 97 or 98 million returns **95.** 9 million returns **97.** b **99.** a **101.** $6x - 10 = 5x - 7$ **103.** 0 **105.** -4 **107.** no; answers may vary **109.** $123.26°F$ **111.** $-9.4°F$

$$6x = 5x + 3$$
$$x = 3$$

Section 8.5

Vocabulary, Readiness & Video Check **1.** decreased by **3.** The phrase is "three times the difference of a number and 5." The "difference of a number and 5" translates to the expression $x - 5$, and in order to multiply 3 times this expression, we insert parentheses around the expression.

Exercise Set 8.5 **1.** $x + 5$ **3.** $x + 8$ **5.** $20 - x$ **7.** $512x$ **9.** $\frac{x}{2}$ **11.** $17 + x + 5x$ **13.** $-5 + x = -7$ **15.** $3x = 27$ **17.** $-20 - x = 104$ **19.** $5x$ **21.** $11 - x$ **23.** $2x = 108$ **25.** $50 - 8x$ **27.** $5(-3 + x) = -20$ **29.** $9 + 3x = 33; 8$ **31.** $3 + 4 + x = 16; 9$ **33.** $x - 3 = \frac{10}{5}; 5$ **35.** $30 - x = 3(x + 6); 3$ **37.** $5x - 40 = x + 8; 12$ **39.** $3(x - 5) = \frac{108}{12}; 8$ **41.** $4x = 30 - 2x; 5$ **43.** California: 55 votes; Florida: 29 votes **45.** falcon: 185 mph; pheasant: 37 mph **47.** India: 3.5 million students; Turkey: 2.0 million students **49.** Xbox: $420; games: $140 **51.** 5470 mi **53.** Michigan Stadium: 109,901; Beaver Stadium: 106,572 **55.** California: 309 thousand; Washington: 103 thousand **57.** 43 points **59.** 2010: 275,215; 2020: 808,416 **61.** Spain: 6609 cars per day; Germany: 13,218 cars per day **63.** $225 **65.** United States: 311 million; China: 195 million **67.** 590 **69.** 1000 **71.** 3000 **73.** yes; answers may vary **75.** $216,200 **77.** $549

Chapter 8 Vocabulary Check **1.** simplified; combined **2.** like **3.** variable **4.** algebraic expression **5.** terms **6.** numerical coefficient **7.** evaluating the expression **8.** constant **9.** equation **10.** solution **11.** distributive **12.** multiplication **13.** addition

Chapter 8 Review **1.** -5 **2.** 17 **3.** undefined **4.** 0 **5.** 129 **6.** -2 **7.** 8 cu ft **8.** 64 cu ft **9.** $1800 **10.** $300 **11.** $-15x$ **12.** $-\frac{7}{30}x$ **13.** $-6y - 10$ **14.** $-6a - 7$ **15.** $-8y + 2$ **16.** $4.6x - 11.9$ **17.** $-8y$ **18.** $15y - 24$ **19.** $11x - 12$ **20.** $4x - 7$ **21.** $-5a + 4$ **22.** $12y - 9$ **23.** $(6x - 3)$ sq yd **24.** $28y$ m **25.** yes **26.** no **27.** -2 **28.** 7 **29.** $-\frac{1}{2}$ **30.** $-\frac{6}{11}$ **31.** -6 **32.** -20 **33.** 1.3 **34.** 2.4 **35.** 7 **36.** -9 **37.** 1 **38.** -5 **39.** $-\frac{4}{5}$ **40.** -24 **41.** -120 **42.** 13 **43.** 4 **44.** 3 **45.** 5 **46.** 12 **47.** 63 **48.** 33 **49.** -2.25 **50.** 1.3 **51.** 2 **52.** 6 **53.** 11 **54.** -5 **55.** 6 **56.** 8 **57.** $20 - (-8) = 28$ **58.** $-2 - 19 = -21$ **59.** $\frac{-75}{5 + 20} = -3$ **60.** $-5(-2 + 6) = -20$ **61.** $\frac{70}{x}$ **62.** $x - 13$ **63.** $85 - x$ **64.** $2x + 11$ **65.** $x + 8 = 40$ **66.** $2x - 12 = 10$ **67.** 5 **68.** -16 **69.** bamboos: 36 in.; kelp: 18 in. **70.** incumbent: 14,752 votes; challenger: 3546 votes **71.** 29 **72.** $-\frac{1}{4}$ **73.** $-11x$ **74.** $-35x$ **75.** $22x - 19$ **76.** $-9x - 32$ **77.** $x - 17$ **78.** $3(x + 5)$ **79.** $x - 3 = \frac{x}{4}$ **80.** $6x = x + 2$ **81.** no **82.** yes **83.** -1 **84.** -25 **85.** 13 **86.** -6 **87.** 17 **88.** 7 **89.** -22 **90.** -6 **91.** $-\frac{3}{4}$ **92.** -4 **93.** 2 **94.** $-\frac{8}{3}$ **95.** 12 **96.** -8 **97.** 0 **98.** 0 **99.** 5 **100.** 1

Chapter 8 Test **1.** -1 **2.** $-5x + 5$ **3.** $-6y - 14$ **4.** $14z - 8$ **5.** $4(3x - 1) = (12x - 4)$ sq m **6.** 7 **7.** $-\frac{1}{2}$ **8.** -12 **9.** 40 **10.** 24 **11.** 3 **12.** -2 **13.** -2 **14.** 4.5 **15.** 0 **16.** -2 **17.** $\frac{22}{3}$ **18.** 4 **19.** 0 **20.** 6000 sq ft **21.** 30 sq ft **22. a.** $17x$ **b.** $20 - 2x$ **23.** -2 **24.** 34 points **25.** 244 women

Cumulative Review **1.** hundred-thousands; Sec. 1.2, Ex. 1 **2.** two thousand thirty-six; Sec. 1.2 **3.** 184,046; Sec. 1.3, Ex. 2 **4.** 39; Sec. 1.7 **5.** 13 in.; Sec. 1.3, Ex. 5 **6.** 17; Sec. 1.4 **7.** 14,440,060; Sec. 1.3, Ex. 7 **8.** 5; Sec. 1.9 **9.** 7321; Sec. 1.4, Ex. 2 **10.** 64; Sec. 1.9 **11. a.** Indonesia **b.** 331; Sec. 1.3, Ex. 8 **12.** $-\frac{1}{8}$; Sec. 3.3 **13.** 4; Sec. 2.2, Ex. 6a **14.** -20; Sec. 2.2 **15.** -5; Sec. 2.2, Ex. 6b **16.** 0; Sec. 2.2 **17.** -23; Sec. 2.3, Ex. 4 **18.** -3.6; Sec. 4.2 **19.** -10; Sec. 2.3, Ex. 14 **20.** $\frac{1}{10}$; Sec. 3.5 **21.** -7; Sec. 2.4, Ex. 6 **22.** 5.8; Sec. 4.2 **23.** 1; Sec. 2.4, Ex. 7 **24.** $-\frac{31}{120}$; Sec. 3.5 **25.** 15; Sec. 2.5, Ex. 2 **26.** -9.6; Sec. 4.3 **27.** 24; Sec. 2.5, Ex. 7 **28.** -4; Sec. 3.7 **29.** 3; Sec. 2.6, Ex. 9 **30.** 9.7; Sec. 8.2 **31.** 0.8496; Sec. 4.3, Ex. 2 **32.** 53.1; Sec. 4.3 **33. a.** $\frac{5}{8}$ **b.** $\frac{4}{13}$; Sec. 5.1, Ex. 6 **34.** $\frac{23}{36}$; Sec. 3.5 **35.** 5; Sec. 6.2, Ex. 9 **36.** $\frac{7}{12}$; Sec. 3.3 **37.** 75%; Sec. 6.2, Ex. 11 **38.** $\frac{18}{23}$; Sec. 3.7 **39.** 48 oz; Sec. 5.5, Ex. 2 **40.** 24,000; Sec. 1.5 **41.** 11.3 L or 11,300 ml; Sec. 5.6, Ex. 8 **42.** 0.024; Sec. 4.1 **43.** yes; Sec. 5.2, Ex. 4 **44.** yes; Sec. 5.2 **45.** 22.4 cc; Sec. 5.3, Ex. 2 **46.** 262.5 mi; Sec. 5.3 **47.** 0.046; Sec. 6.1, Ex. 4 **48.** 4.52; Sec. 6.1 **49.** $\frac{1}{3}$; Sec. 6.1, Ex. 11 **50.** $\frac{27}{100}$; Sec. 6.1 **51.** $5 = n \cdot 20$; Sec. 6.2, Ex. 1 **52.** $\frac{5}{20} = \frac{p}{100}$; Sec. 6.3 **53.** sales tax: $6.41; total price: $91.91; Sec. 6.5, Ex. 1 **54.** $1610; Sec. 6.5 **55. a.** 75 clam species **b.** fishes; Sec. 7.1, Ex. 3 **56.** mean: 8; median: 9; mode: 11; Sec. 7.3

Chapter 9 Geometry

Section 9.1

Vocabulary, Readiness & Video Check 1. plane **3.** Space **5.** ray **7.** straight **9.** acute **11.** Parallel; intersecting
13. degrees **15.** vertical **17.** $\angle WUV$, $\angle VUW$, $\angle U$, $\angle x$ **19.** $180° - 17° = 163°$

Exercise Set 9.1 1. line; line CD or line l or $\overleftrightarrow{CD}$ **3.** line segment; line segment MN or $\overline{MN}$ **5.** angle; $\angle GHI$ or $\angle IHG$ or $\angle H$
7. ray; ray UW or $\overrightarrow{UW}$ **9.** $\angle CPR$, $\angle RPC$ **11.** $\angle TPM$, $\angle MPT$ **13.** straight **15.** right **17.** obtuse **19.** acute **21.** 67° **23.** 163°
25. 32° **27.** 30° **29.** $\angle MNP$ and $\angle RNO$; $\angle PNQ$ and $\angle QNR$ **31.** $\angle SPT$ and $\angle TPQ$; $\angle SPR$ and $\angle RPQ$; $\angle SPT$ and $\angle SPR$; $\angle TPQ$
and $\angle QPR$ **33.** 27° **35.** 132° **37.** $m\angle x = 30°$; $m\angle y = 150°$; $m\angle z = 30°$ **39.** $m\angle x = 77°$; $m\angle y = 103°$; $m\angle z = 77°$
41. $m\angle x = 100°$; $m\angle y = 80°$; $m\angle z = 100°$ **43.** $m\angle x = 134°$; $m\angle y = 46°$; $m\angle z = 134°$ **45.** $\angle ABC$ or $\angle CBA$ **47.** $\angle DBE$ or
$\angle EBD$ **49.** 15° **51.** 50° **53.** 65° **55.** 95° **57.** $\frac{9}{8}$ or $1\frac{1}{8}$ **59.** $\frac{7}{32}$ **61.** $\frac{5}{6}$ **63.** $\frac{4}{3}$ or $1\frac{1}{3}$ **65.** 54.8° **67.** false; answers may vary
69. true **71.** $m\angle a = 60°$; $m\angle b = 50°$; $m\angle c = 110°$; $m\angle d = 70°$; $m\angle e = 120°$ **73.** no; answers may vary **75.** 45°; 45°

Section 9.2

Vocabulary, Readiness & Video Check 1. Because the sum of the measures of the angles of a triangle equals 180°, each angle in
an equilateral triangle must measure 60°.

Exercise Set 9.2 1. pentagon **3.** hexagon **5.** quadrilateral **7.** pentagon **9.** equilateral **11.** scalene; right **13.** isosceles
15. 25° **17.** 13° **19.** 40° **21.** diameter **23.** rectangle **25.** parallelogram **27.** hypotenuse **29.** 14 m **31.** 14.5 cm **33.** 40.6 cm
35. 84 in. **37.** cylinder **39.** rectangular solid **41.** cone **43.** cube **45.** rectangular solid **47.** sphere **49.** pyramid **51.** 14.8 in.
53. 13 mi **55.** 72,368 mi **57.** 108 **59.** 12.56 **61.** true **63.** true **65.** false **67.** yes; answers may vary **69.** answers may vary

Section 9.3

Vocabulary, Readiness & Video Check 1. perimeter **3.** π **5.** $\frac{22}{7}$ (or 3.14); 3.14 $\left(\text{or } \frac{22}{7}\right)$ **7.** Opposite sides of a rectangle
have the same measure, so we can just find the sum of the measures of all four sides.

Exercise Set 9.3 1. 64 ft **3.** 120 cm **5.** 21 in. **7.** 48 ft **9.** 42 in. **11.** 155 cm **13.** 21 ft **15.** 624 ft **17.** 346 yd **19.** 22 ft
21. $55 **23.** 72 in. **25.** 28 in. **27.** $36.12 **29.** 96 m **31.** 66 ft **33.** 74 cm **35.** 17π cm; 53.38 cm **37.** 16π mi; 50.24 mi
39. 26π m; 81.64 m **41.** 15π ft; 47.1 ft **43.** 12,560 ft **45.** 30.7 mi **47.** 14π cm $\approx$ 43.96 cm **49.** 40 mm **51.** 84 ft **53.** 23
55. 1 **57.** 6 **59.** 10 **61. a.** width: 30 yd; length: 40 yd **b.** 140 yd **63.** b **65. a.** 62.8 m; 125.6 m **b.** yes **67.** answers may vary
69. 27.4 m **71.** 75.4 m **73.** 6.5 ft

Section 9.4

Vocabulary, Readiness & Video Check 1. The formula for the area of a rectangle; we split the L-shaped figure into two rectangles,
used the area formula twice to find the area of each, and then added these two areas.

Exercise Set 9.4 1. 7 sq m **3.** $9\frac{3}{4}$ sq yd **5.** 15 sq yd **7.** 2.25π sq in. $\approx$ 7.065 sq in. **9.** 17.64 sq ft **11.** 28 sq m **13.** 22 sq yd
15. $36\frac{3}{4}$ sq ft **17.** $22\frac{1}{2}$ sq in. **19.** 25 sq cm **21.** 86 sq mi **23.** 24 sq cm **25.** 36π sq in. $\approx 113\frac{1}{7}$ sq in. **27.** 168 sq ft
29. 128,775 sq ft **31.** 1π sq cm $\approx$ 3.14 sq cm **33.** 128 sq in.; $\frac{8}{9}$ sq ft **35.** 510 sq in. **37.** 168 sq ft **39.** 9200 sq ft **41. a.** 381 sq ft
b. 4 squares **43.** 14π in. $\approx$ 43.96 in. **45.** 25 ft **47.** $12\frac{3}{4}$ ft **49.** perimeter **51.** area **53.** area **55.** perimeter **57.** 12-in. pizza
59. $1\frac{1}{3}$ sq ft; 192 sq in. **61.** 7.74 sq in. **63.** 7056π sq in. $\approx$ 22,155.84 sq in. **65.** 298.5 sq m **67. a.** width: 40 yd; length: 60 yd
b. 2400 sq yd **69.** no; answers may vary

Section 9.5

Vocabulary, Readiness & Video Check 1. surface area **3.** cubic **5.** perimeter **7.** Exact answers are in terms of π, and approximate answers use an approximation for π.

Exercise Set 9.5 1. $V = 72$ cu in.; $SA = 108$ sq in. **3.** $V = 512$ cu cm; $SA = 384$ sq cm **5.** $V = 4\pi$ cu yd $\approx 12\frac{4}{7}$ cu yd;
$SA = (2\sqrt{13}\pi + 4\pi)$ sq yd $\approx$ 35.23 sq yd **7.** $V = \frac{500}{3}\pi$ cu in. $\approx 523\frac{17}{21}$ cu in.; $SA = 100\pi$ sq in. $\approx 314\frac{2}{7}$ sq in.
9. $V = 9\pi$ cu in. $\approx 28\frac{2}{7}$ cu in. **11.** 75 cu cm **13.** $2\frac{10}{27}$ cu in. **15.** $V = 8.4$ cu ft; $SA = 26$ sq ft **17.** 960 cu cm

19. $V = \dfrac{1372}{3}\pi$ cu in. or $457\dfrac{1}{3}\pi$ cu in.; $SA = 196\pi$ sq in.   **21.** $7\dfrac{1}{2}$ cu ft   **23.** 5.25π cu in.   **25.** 7.96 cu m   **27.** $10\dfrac{5}{6}$ cu in.

29. 288π cu ft   **31.** $12\dfrac{4}{7}$ cu cm   **33.** 8.8 cu in.   **35.** 10.648 cu in.   **37.** 25   **39.** 9   **41.** 5   **43.** 20   **45.** 2093.33 cu m

47. no; answers may vary   **49.** 5.5 cu ft; 5.8 cu ft; (b) is larger   **51.** $6\dfrac{2}{3}\pi$ cu in. $\approx$ 21 cu in.

Integrated Review   **1.** 153°; 63°   **2.** $m\angle x = 75°$; $m\angle y = 105°$; $m\angle z = 75°$ **3.** $m\angle x = 128°$; $m\angle y = 52°$; $m\angle z = 128°$
4. $m\angle x = 52°$ **5.** 4.6 in.   **6.** $4\dfrac{1}{4}$ in.   **7.** 20 m; 25 sq m   **8.** 12 ft; 6 sq ft   **9.** 10π cm $\approx$ 31.4 cm; 25π sq cm $\approx$ 78.5 sq cm
10. 32 mi; 44 sq mi   **11.** 54 cm; 143 sq cm   **12.** 62 ft; 238 sq ft   **13.** $V = 64$ cu in.; $SA = 96$ sq in.   **14.** $V = 30.6$ cu ft; $SA = 63$ sq ft
15. 400 cu cm   **16.** $4\dfrac{1}{2}\pi$ cu mi $\approx 14\dfrac{1}{7}$ cu mi

Section 9.6

Vocabulary, Readiness & Video Check   **1.** false   **3.** true   **5.** false   **7.** $\angle M$ and $\angle Y$, $\angle N$ and $\angle X$, $\angle P$ and $\angle Z$; $\dfrac{p}{z} = \dfrac{m}{y} = \dfrac{n}{x}$
9. The ratios of corresponding sides are the same.

Exercise Set 9.6   **1.** congruent; SSS   **3.** not congruent   **5.** congruent; ASA   **7.** congruent; SAS   **9.** $\dfrac{2}{1}$ **11.** $\dfrac{3}{2}$ **13.** 4.5   **15.** 6
17. 5   **19.** 13.5   **21.** 17.5   **23.** 10   **25.** 28.125   **27.** 10   **29.** 520 ft   **31.** 500 ft   **33.** 60 ft   **35.** 14.4 ft   **37.** 52 neon tetras
39. 381 ft   **41.** 4.01   **43.** 1.23   **45.** $3\dfrac{8}{9}$ in.; no   **47.** 8.4   **49.** answers may vary   **51.** 200 ft, 300 ft, 425 ft

Chapter 9 Vocabulary Check   **1.** right triangle; hypotenuse; legs   **2.** line segment   **3.** complementary   **4.** line   **5.** perimeter
6. angle; vertex   **7.** Congruent   **8.** Area   **9.** ray   **10.** polygon   **11.** transversal   **12.** straight   **13.** volume   **14.** vertical
15. adjacent   **16.** obtuse   **17.** right   **18.** acute   **19.** supplementary   **20.** Similar

Chapter 9 Review   **1.** right   **2.** straight   **3.** acute   **4.** obtuse   **5.** 65°   **6.** 75°   **7.** 58°   **8.** 98°   **9.** 90°   **10.** 25°   **11.** $\angle a$ and $\angle b$;
$\angle b$ and $\angle c$; $\angle c$ and $\angle d$; $\angle d$ and $\angle a$ **12.** $\angle x$ and $\angle w$; $\angle y$ and $\angle z$ **13.** $m\angle x = 100°$; $m\angle y = 80°$; $m\angle z = 80°$ **14.** $m\angle x = 155°$;
$m\angle y = 155°$; $m\angle z = 25°$ **15.** $m\angle x = 53°$; $m\angle y = 53°$; $m\angle z = 127°$ **16.** $m\angle x = 42°$; $m\angle y = 42°$; $m\angle z = 138°$ **17.** 103°
18. 60°   **19.** 60°   **20.** 65°   **21.** $4\dfrac{1}{5}$ m   **22.** 7 ft   **23.** 9.5 m   **24.** $15\dfrac{1}{5}$ cm   **25.** cube   **26.** cylinder   **27.** pyramid   **28.** rectangular
solid   **29.** 18 in.   **30.** 2.35 m   **31.** pentagon   **32.** hexagon   **33.** equilateral   **34.** isosceles, right   **35.** 89 m   **36.** 30.6 cm
37. 36 m   **38.** 90 ft   **39.** 32 ft   **40.** 440 ft   **41.** 5.338 in.   **42.** 31.4 yd   **43.** 240 sq ft   **44.** 140 sq m   **45.** 600 sq cm   **46.** 189 sq yd
47. 49π sq ft $\approx$ 153.86 sq ft   **48.** 82.81 sq m   **49.** 1248 sq cm   **50.** 144 sq m   **51.** 432 sq ft   **52.** 130 sq ft   **53.** $V = 15\dfrac{5}{8}$ cu in.;
$SA = 37\dfrac{1}{2}$ sq in.   **54.** $V = 84$ cu ft; $SA = 136$ sq ft   **55.** $V = 20{,}000\pi$ cu cm $\approx$ 62,800 cu cm   **56.** $V = \dfrac{1}{6}\pi$ cu km $\approx \dfrac{11}{21}$ cu km
57. $2\dfrac{2}{3}$ cu ft   **58.** 307.72 cu in.   **59.** $7\dfrac{1}{2}$ cu ft   **60.** 0.5π cu ft or $\dfrac{1}{2}\pi$ cu ft   **61.** $37\dfrac{1}{2}$ **62.** $13\dfrac{1}{3}$ **63.** 17.4   **64.** 12   **65.** 33 ft
66. $x = \dfrac{5}{6}$ in.; $y = 2\dfrac{1}{6}$ in.   **67.** 108°   **68.** 89°   **69.** 82°   **70.** 78°   **71.** 95°   **72.** 57°   **73.** 13 m   **74.** 12.6 cm   **75.** 22 dm   **76.** 27.3 in.
77. 194 ft   **78.** 1624 sq m   **79.** 9π sq m $\approx$ 28.26 sq m   **80.** $346\dfrac{1}{2}$ cu in.   **81.** 140 cu in.   **82.** 1260 cu ft   **83.** 28.728 cu ft   **84.** $6\dfrac{1}{2}$ **85.** 12

Chapter 9 Test   **1.** 12°   **2.** 56°   **3.** 57°   **4.** $m\angle x = 118°$; $m\angle y = 62°$; $m\angle z = 118°$ **5.** $m\angle x = 73°$; $m\angle y = 73°$; $m\angle z = 73°$
6. 6.2 m   **7.** $10\dfrac{1}{4}$ in.   **8.** 26°   **9.** circumference = 18π in. $\approx$ 56.52 in.; area = 81π sq in. $\approx$ 254.34 sq in.   **10.** perimeter = 24.6 yd;
area = 37.1 sq yd   **11.** perimeter = 68 in.; area = 185 sq in.   **12.** $62\dfrac{6}{7}$ cu in.   **13.** $V = 30$ cu ft; $SA = 62$ sq ft   **14.** 16 in.
15. 18 cu ft   **16.** 62 ft; $115.94   **17.** 198.08 oz   **18.** 7.5   **19.** 69 ft

Cumulative Review   **1.** negative fifty and eighty-two hundredths; Sec. 4.1, Ex. 1b   **2.** $\dfrac{53}{66}$; Sec. 3.5   **3.** 736.2; Sec. 4.1, Ex. 15
4. 700; Sec. 4.1   **5.** 47.06; Sec. 4.2, Ex. 3   **6.** $-1\dfrac{9}{11}$; Sec. 3.7   **7.** 76.8; Sec. 4.3, Ex. 5   **8.** $\dfrac{7}{66}$; Sec. 3.3   **9.** $-76{,}300$; Sec. 4.3, Ex. 7
10. $11\dfrac{1}{2}$; Sec. 3.7   **11.** 38.6; Sec. 4.4, Ex. 1   **12.** 0.567; Sec. 4.4   **13.** -3.7; Sec. 4.5, Ex. 12   **14.** $\dfrac{3}{5}$ or 0.6; Sec. 4.5   **15.** >; Sec. 4.5, Ex. 8
16. <; Sec. 4.5   **17.** $\dfrac{26}{31}$; Sec. 5.1, Ex. 3   **18.** $\dfrac{16}{45}$; Sec. 3.5   **19.** yes; Sec. 5.2, Ex. 2   **20.** $\dfrac{35}{2}$ or $17\dfrac{1}{2}$; Sec. 5.2   **21.** 25%; Sec. 6.1, Ex. 1
22. 10; Sec. 8.3   **23.** $\dfrac{19}{1000}$; Sec. 6.1, Ex. 9   **24.** $\dfrac{13}{50}$; Sec. 6.1   **25.** $\dfrac{5}{4}$ or $1\dfrac{1}{4}$; Sec. 6.1, Ex. 10   **26.** $\dfrac{28}{5}$ or $5\dfrac{3}{5}$; Sec. 6.1   **27.** 255; Sec. 6.2, Ex. 8

28. 15%; Sec. 6.2 or 6.3   **29.** 52; Sec. 6.3, Ex. 9   **30.** 9; Sec. 1.9   **31.** 775 freshmen; Sec. 6.4, Ex. 3   **32.** $2.25/sq ft; Sec. 5.1
33. $3210; Sec. 6.5, Ex. 3   **34.** 35 exercises; Sec. 5.3   **35.** 96 in.; Sec. 5.4, Ex. 1   **36.** 2 yd 2 ft 4 in.; Sec. 5.4   **37.** 3200 g; Sec. 5.5, Ex. 7
38. 0.07 m; Sec. 5.4   **39.** 3 gal 3 qt; Sec. 5.6, Ex. 3   **40.** 70,052; Sec. 1.2   **41.** 50°; Sec. 9.2, Ex. 1   **42.** 33 m; Sec. 9.3   **43.** 28 in.;
Sec. 9.3, Ex. 1   **44.** −4; Sec. 8.4   **45.** $\frac{2}{5}$; Sec. 4.6, Ex. 3   **46.** $\frac{3}{4}$; Sec. 4.6

Appendix A   Tables

A1: One Hundred Addition Facts   **1.** 5   **3.** 5   **5.** 12   **7.** 8   **9.** 14   **11.** 12   **13.** 10   **15.** 9   **17.** 11   **19.** 18   **21.** 10   **23.** 10   **25.** 17
27. 13   **29.** 12   **31.** 16   **33.** 7   **35.** 4   **37.** 8   **39.** 8   **41.** 6   **43.** 16   **45.** 9   **47.** 12   **49.** 7   **51.** 2   **53.** 11   **55.** 8   **57.** 5   **59.** 14
61. 13   **63.** 7   **65.** 7   **67.** 13   **69.** 5   **71.** 4   **73.** 16   **75.** 14   **77.** 1   **79.** 4   **81.** 10   **83.** 8   **85.** 5   **87.** 10   **89.** 11   **91.** 12   **93.** 9
95. 9   **97.** 14   **99.** 9

A2: One Hundred Multiplication Facts   **1.** 1   **3.** 56   **5.** 32   **7.** 28   **9.** 4   **11.** 63   **13.** 6   **15.** 30   **17.** 24   **19.** 18   **21.** 40   **23.** 32
25. 54   **27.** 56   **29.** 54   **31.** 2   **33.** 36   **35.** 12   **37.** 36   **39.** 12   **41.** 48   **43.** 8   **45.** 0   **47.** 27   **49.** 15   **51.** 45   **53.** 0   **55.** 81
57. 0   **59.** 0   **61.** 18   **63.** 3   **65.** 36   **67.** 63   **69.** 35   **71.** 42   **73.** 40   **75.** 0   **77.** 9   **79.** 15   **81.** 5   **83.** 0   **85.** 21   **87.** 0
89. 16   **91.** 0   **93.** 4   **95.** 6   **97.** 18   **99.** 0

Appendix B   Exponents and Polynomials

Exercise Set Appendix B.1   **1.** $-5x - 24$ **3.** $-9z^2 - 2z + 6$ **5.** $9y^2 + 25y - 40$ **7.** $-4.3a^4 - 2a^2 + 9$ **9.** $4a - 8$
11. $-2x^2 + 4x + 1$ **13.** $-20y^3 + 12y^2 - 4$ **15.** $-x + 16$ **17.** $8y^2 - 10y - 8$ **19.** $5x - 12$ **21.** $y - 4$ **23.** $4x^2 + x - 16$
25. $9x - 4.1$ **27.** $15a - 7$ **29.** $-15y + 3.6$ **31.** $19t^2 - 11t + 3$ **33.** $b^3 - b^2 + 7b - 1$ **35.** $-5x^2 - 11x + 13$ **37.** $\frac{9}{7}$
39. 1   **41.** -5 **43.** -8 **45.** 20   **47.** 25   **49.** 50   **51.** 576 ft   **53.** $3200   **55.** 909 ft   **57.** $(8x + 2)$ in.   **59.** $(4x - 15)$ units
61. 20; 6; 2   **63.** 7.2752   **65.** 29 ft; -243 ft; answers may vary

Exercise Set Appendix B.2   **1.** x^{14} **3.** a^7 **5.** $15z^5$ **7.** $-40x^2$ **9.** $25x^6y^4$ **11.** $28a^5b^6$ **13.** $42x^3$ **15.** $12a^{17}$ **17.** x^{15} **19.** z^{20}
21. b^{62} **23.** $81a^4$ **25.** $a^{33}b^{24}$ **27.** $121x^6y^{12}$ **29.** $-24y^{22}$ **31.** $256x^9y^{13}$ **33.** $16x^{12}$ sq in.   **35.** $12a^4b^5$ sq m   **37.** $18{,}003{,}384a^{45}b^{30}$
39. $34{,}867.84401x^{50}$ **41.** $a^{127}b^{86}c^{25}$ **43.** answers may vary

Exercise Set Appendix B.3   **1.** $27x^3 - 9x$ **3.** $-20a^3 + 30a^2 - 5a$ **5.** $42x^4 - 35x^3 + 49x^2$ **7.** $x^2 + 13x + 30$ **9.** $2x^2 + 2x - 24$
11. $36a^2 + 48a + 16$ **13.** $a^3 - 33a + 18$ **15.** $8x^3 + 2x^2 - 55x + 50$ **17.** $x^5 + 4x^4 + 6x^3 + 7x^2 + 2x$ **19.** $-30r^2 + 20r$
21. $-6y^3 - 2y^4 + 12y^2$ **23.** $x^2 + 14x + 24$ **25.** $4a^2 - 9$ **27.** $x^2 + 10x + 25$ **29.** $b^2 + \frac{7}{5}b + \frac{12}{25}$ **31.** $6x^3 + 25x^2 + 10x + 1$
33. $49x^2 + 70x + 25$ **35.** $4x^2 - 4x + 1$ **37.** $8x^5 - 8x^3 - 6x^2 - 6x + 9$ **39.** $x^5 + 2x^4 + 3x^3 + 2x^2 + x$ **41.** $10z^4 - 3z^3 + 3z - 2$
43. $(y^3 - 3y^2 - 16y - 12)$ sq ft   **45.** $(x^4 - 3x^2 + 1)$ sq m   **47.** answers may vary

Appendix C   Inductive and Deductive Reasoning

Exercise Set Appendix C   **1.** inductive   **3.** deductive   **5.** inductive   **7.** 81   **9.** C   **11.** 36   **13.** E   **15.** 99,999   **17.** 9   **19.** 74   **21.** 241
23.   **25.**   **27.**   **29.**   **31.**   **33.**   **35.** answers may vary   **37.**

39.   **41.**   **43.**   **45.** Joey   **47.** 15 people   **49.** Donna   **51.** Julio
53. 5 people

Practice Final Exam   **1.** 113   **2.** 33   **3.** $\frac{64}{3}$ or $21\frac{1}{3}$ **4.** $\frac{3}{4}$ **5.** 8   **6.** 40.902   **7.** -6.2 **8.** $16\frac{8}{11}$ **9.** $\frac{45}{2}$ or $22\frac{1}{2}$ **10.** -20.42
11. $\frac{3}{8}$ **12.** -117 **13.** 12   **14.** 2   **15.** $-\frac{5}{3}$ or $-1\frac{2}{3}$ **16.** 52,000   **17.** 34.9   **18.** 0.941   **19.** 0.85   **20.** 610%   **21.** 37.5%
22. $\frac{1}{500}$ **23.** perimeter: $3\frac{1}{3}$ ft; area: $\frac{2}{3}$ sq ft   **24.** $\frac{15}{2}$ **25.** $\frac{3 \text{ in.}}{10 \text{ days}}$ **26.** 81.25 km/hr   **27.** 8-oz size   **28.** $\frac{48}{11}$ or $4\frac{4}{11}$ **29.** 11.4
30. 24 mi   **31.** $53\frac{1}{3}$ g   **32.** $17   **33.** 1250   **34.** 75%   **35.** 38.4 lb   **36.** discount: $18; sale price: $102   **37.** 0.04 g   **38.** 10 qt
39. 16 ft 6 in.   **40.** mean: 38.4; median: 42; no mode   **41.** 11 students   **42.** -1 **43.** $14z - 8$ **44.** 40   **45.** 24   **46.** -2
47. 244 women   **48.** 12°   **49.** $m\angle x = 118°; m\angle y = 62°; m\angle z = 118°$ **50.** 26°   **51.** 7.5 or $7\frac{1}{2}$

Index

Photo Credits